KU-006-664

CONSERVATION BIOLOGY

SECOND EDITION

FOR THE COMING DECADE

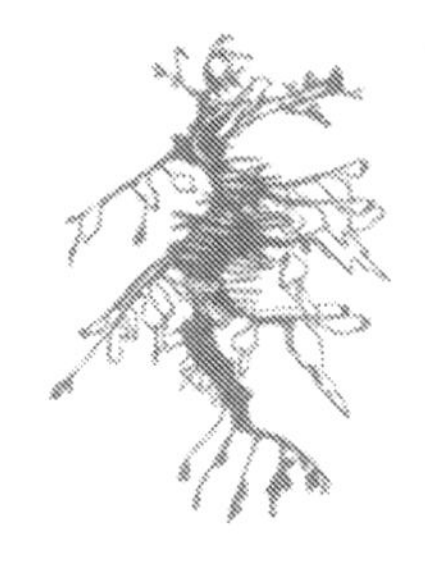

JOIN US ON THE INTERNET

WWW:
http://www.thomson.com
EMAIL:
findit@kiosk.thomson.com

thomson.com is the on-line portal for the products, services, and resources available from International Thomson Publishing (ITP). This Internet kiosk gives users immediate access to more than 34 ITP publishers and over 20,000 products. Through *thomson.com* Internet users can search catalogs, examine subject-specific resource centers, and subscribe to electronic discussion lists. You can purchase ITP products from your local bookseller, or directly through *thomson.com.*

Visit Chapman & Hall's Internet Resource Center for information on our new publications, links to useful sites on the World Wide Web, and the opportunity to join our e-mail mailing list. Point your browser to:
http://www.chaphall.com
or
http://www.thomson.com/chaphall/lifesce.html
for Life Sciences

CONSERVATION BIOLOGY

SECOND EDITION

FOR THE COMING DECADE

EDITED BY PEGGY L. FIEDLER AND
PETER M. KAREIVA

CHAPMAN & HALL

I(T)P® International Thomson Publishing

New York • Albany • Bonn • Boston • Cincinnati • Detroit • London • Madrid • Melbourne
Mexico City • Pacific Grove • Paris • San Francisco • Singapore • Tokyo • Toronto • Washington

Cover design: Diane C. Fiedler

Printed in the United States of America

Chapman & Hall
115 Fifth Avenue
New York, NY 10003

Chapman & Hall
2-6 Boundary Row
London SE1 8HN
England

Thomas Nelson Australia
102 Dodds Street
South Melbourne, 3205
Victoria, Australia

Chapman & Hall GmbH
Postfach 100 263
D-69442 Weinheim
Germany

International Thomson Editores
Campos Eliseos 385, Piso 7
Col. Polanco
11560 Mexico D.F
Mexico

International Thomson Publishing–Japan
Hirakawacho-cho Kyowa Building, 3F
1-2-1 Hirakawacho-cho
Chiyoda-ku, 102 Tokyo
Japan

International Thomson Publishing Asia
221 Henderson Road #05-10
Henderson Building
Singapore 0315

1 2 3 4 5 6 7 8 9 10 XXX 01 00 99 98

Library of Congress Cataloging-in-Publication Data

Conservation biology : for the coming decade / edited by Peggy L.
Fiedler and Peter M. Kareiva -- 2nd ed.
p. cm.
Includes bibliographical references and index.
ISBN 0-412-09651-X (hardcover : alk.paper).--ISBN 0-412-09661-7
(pbk.)
1. Conservation biology. I. Fiedler, Peggy Lee. II. Kareiva,
Peter M.
QH75.C6615 1997
333.95'16--dc21 97-13981
CIP

British Library Cataloguing in Publication Data available

To order this or any other Chapman & Hall book, please contact **International Thomson Publishing, 7625 Empire Drive, Florence, KY 41042.** Phone: (606) 525-6600 or 1-800-842-3636. Fax: (606) 525-7778. e-mail: order@chaphall.com.

For a complete listing of Chapman & Hall titles, send your request to **Chapman & Hall, Dept. BC, 115 Fifth Avenue, New York, NY 10003.**

Everyone, real or invented, deserves the open destiny of life.

Grace Paley. 1994. A conversation with my father.
In *The Collected Stories*. Farrar, Straus, and Giroux, New York.

TO OUR NEXT GENERATION,

GARRETT, JENA LOUISE, CELINA, AND ISAAC .

CONTRIBUTORS

• • •

Douglas T. Bolger
Environmental Studies Program
Dartmouth College
HB 6182
Hanover, New Hampshire 03755

C. Ron Carroll
Institute of Ecology
University of Georgia
Athens, Georgia 30602

Ted J. Case
Department of Biology C–0116
University of California
La Jolla, California 92093

Gary C. Chang
Department of Zoology
University of Washington
Seattle, Washington 98195

Alastair Culham
Botany Department
The University of Reading
Whiteknights, Reading, RG6 6AS
United Kingdom

Anne M. Dix
Institute of Ecology
University of Georgia
Athens, Georgia 30602.

Andrew R. Dyer
Mitrani Center, Blaustein Institute for
Desert Research
Ben-Gurion University
Midreshet Ben-Gurion, Midreshet Ben-Gurion,
Israel 84990

Steven G. Fancy, Ph.D.
U.S. Geological Survey Biological Resources
Division
Pacific Islands Ecosystems Research Center
P.O. Box 44
Hawaii National Park, Hawaii 96718

Peggy L. Fiedler
Department of Biology
San Francisco State University
San Francisco, California 94132

Nancy Fredricks
Gifford Pinchot National Forest
6926 E. Fourth Plain Blvd.
Vancouver, Washington 98668.

Lauri K. Freidenburg
Ecology and Evolutionary Biology
University of Connecticut
Storrs, Connecticut 06269

Martha J. Groom
Department of Zoology
North Carolina State University
Raleigh, North Carolina 27695

Ellen Gryj
Department of Zoology
University of Washington
Seattle, Washington 98195

Edward O. Guerrant, Jr.
The Berry Botanic Garden
11505 S.W. Summerville Avenue
Portland, Oregon 97219

Clare E. Hankamer
Conservation Projects Development Unit
Royal Botanical Gardens
Kew, Richmond, Surrey TW9 3AB
United Kingdom

Cynthia G. Hays
Department of Biological Sciences
Florida State University
Tallahassee, Florida 32306

Stephen D. Hopper
Director, Kings Park and Botanic Garden
West Perth, Western Australia 6005, Australia

Christopher E. Jordan
Evironmental, Population, and Organismic
Biology
University of Colorado
Boulder, Colorado 80304

Peter M. Kareiva
Department of Zoology
University of Washington
Seattle, Washington 98195

James S. Kettler
Graduate School of Environmental Studies
Bard College
Annandale-on-Hudson, NY 12504

B.E. Knapp
348 Iowa St.
Ashland, Oregon 97520

Eric E. Knapp
Department of Agronomy and Range Science
University of California
Davis, California 95616

Robert A. Leidy
Wetlands Section
U.S. Environmental Protection Agency, Region IX
75 Hawthorne Street
San Francisco, California 94105

Joseph L. Lipar
Department of Biology and Center for the Integrative Study of Animal Behavior
Indiana University
Bloomington, Indiana 47405

Michael Maunder
Conservation Projects Development Unit
Royal Botanical Gardens
Kew, Richmond, Surrey TW9 3AB United Kingdom

Eli Meir
Department of Zoology
University of Washington
Seattle, Washington 98195

Eric S. Menges
Archbold Biological Station
P.O. Box 2057
Lake Placid, Florida 33852

Dominic Moran
Center for Social and Economic Research on the Global Environment (CSERGE)
University College London
Gower Street
London WC1E 6BT, United Kingdom

Peter B. Moyle
Department of Wildlife, Fisheries, and Conservation Biology
University of California
Davis, California 95616

José M. Orensanz
School of Fisheries
University of Washington
Seattle, Washington 98195

Ingrid M. Parker
Department of Integrative Biology
3060 Valley Life Sciences Bldg.
University of California
Berkeley, California 941332

Ana Parma
International Pacific Halibut Commission Seattle
P.O. Box 95009
Seattle, Washington 98145

Miguel A. Pascual
Universidad Nacional de la Patagonia S.J.B.
Boulevard Brown s/n
9120 Puerto Madryn, Argentina

Bruce M. Pavlik
Department of Biology
Mills College
5000 MacArthur Blvd.
Oakland, California 94613

David Pearce
Center for Social and Economic Research on the Global Environment (CSERGE)
University College London
Gower Street
London WC1E 6BT, United Kingdom

Sarah H. Reichard
Department of Zoology
University of Washington
Seattle, Washington 98195

Adam D. Richman
Department of Biology C–0116
University of California
La Jolla, California 92093

Mary H. Ruckelshaus
Department of Biological Sciences
Florida State University
Tallahassee, Florida 32306

Sergio L. Saba
Universidad Nacional de la Patagonia S.J.B.
Boulevard Brown s/n
9120 Puerto Madryn, Argentina

Stephan J. Schoech
Department of Biology and Center for the Integrative Study of Animal Behavior
Indiana University
Bloomington, Indiana 47405

Cheryl B. Schultz
Department of Zoology
University of Washington
Seattle, Washington 98195

Eleanor K. Steinberg
Department of Zoology
University of Washington
Seattle, Washington 98195

Thomas B. Smith
Department of Biology
San Francisco State University
San Francisco, California 94132

Joy Zedler
Pacific Estuarine Research Laboratory
San Diego State University
San Diego, California 92182

CONTENTS

• • •

PREFACE

• • •

"Why do we need yet another edited volume on conservation biology?"

There is a changing of the guard in conservation biology—the founders of the field are being replaced by a new generation of scientists who started graduate school knowing they wanted to be conservation biologists (as opposed to population biologists) and who have been steeped simultaneously in theory and application. Early debates about the relative merits of "single large or several small" (SLOSS) reserves happily have been buried. Demographic matrices are routine rather than innovative. This new breed of conservation biologists is more quantitative, more practical, more accustomed to politics, litigation, and economics, and more skeptical than conservation biology's "founding fathers" ("mothers" were indeed lacking in those early circles of the Society for Conservation Biology). Yet there is a tendency for our current textbooks and edited volumes to trot out the same old names and a lot of the same old ideas.

In contrast, we wanted to produce a more vital book that reflects what this new generation of conservation biologists is doing and thinking. To achieve this vitality, we sought a mix of contributions from both the old school and the new breed of conservation biologists. This book is meant to represent a challenge to conventional conservation biology. Authors were encouraged to question prevailing wisdom. The result is a text that not everyone will agree with entirely, but we think this volume uniquely captures where conservation biology is heading in the future, and exposes some weaknesses of widely accepted "principles."

PROLOGUE

• • •

The Next Generation

The number of degree-granting programs in conservation biology is on the rise and enrollments in conservation courses are skyrocketing. Many of our brightest undergraduate biology majors select conservation as their field of choice, and universities around the world are hiring so-called "conservation biologists." These are indications of a vigorous and healthy scientific field. But conservation biology as a science easily can be criticized. First of all, it is a field that is all too easily trivialized—made banal by the canonization of obvious "principles," such as large reserves are better than small reserves. It is also a field marked by subtle tensions between those who place a premium on actually solving problems, however mundane, and those who place a premium on the originality or cleverness of the problems they solve, however impractical. Finally, it is a field laden with alarmist half-truths and blatant exaggerations about crises. For example, the conservation biology literature is replete with "estimates" of extraordinary rates of extinction that read more like tabloid headlines than anything grounded in hard data. Extinction clearly is a major problem; however, we have little data documenting this fact. If conservation biology is to thrive, it must move beyond platitudes to solutions, it must resolve the tension between immediate applied questions and more farsighted basic research, and it must develop and maintain scientific credibility even though most of its practitioners have almost religious beliefs regarding issues such as the supreme importance of biodiversity.

In this book we have attempted to unite new ideas under our vision of what is needed for conservation biology to meet its challenges. First, we have emphasized topics that are part of the daily activities of conservation biologists responsible for resource management. Thus, many chapters address restoration and invasions because a great deal of modern conservation involves recovering lost habitat and attempting to thwart exotic invaders. Similarly, several chapters touch on the mechanics of population viability analyses that are now a routine facet of single-species conservation efforts. Secondly, we have encouraged a "case history" approach because we are suspicious of broad generalizations and think much of practical importance can be learned by the details of real-world examples. Our last bias is the emphasis we place on quantitative analyses. It is our view that models and mathematical analyses can pinpoint limitations in existing data, and guide research towards aspects of biology that are most likely critical to the dynamics of a species or an ecosystem.

The first section of the book concerns "single species conservation." Here the emphasis is on populations and on concepts from mainstream population ecology and evolutionary genetics that have found a home in conservation. Population viability analyses, ecological genetics, and studies of fragmentation provide the mainstays for the application of population biology to conservation. It is interesting to note the extent to which a "single-species approach" dominates academic conservation biology—just scan any volume of the journal *Conservation Biology*. The irony is that whereas basic ecological theory emphasizes species interactions (few contemporary ecology graduate students would consider doctoral theses on single-species dynamics), conservation biology almost never draws on that huge core of ecology. This neglect of species interactions may not be misguided given the practical concerns of conservation. It could simply reflect the fact that the direct effects of habitat loss or environmental degradation overwhelm any effects of species interactions.

The second section of the book escapes the confines of population biology and addresses big issues such as threats to marine biodiversity and the state of faunas and floras on the scale of entire continents. Instead of hand-waving estimates about extinction, we learn what the hard data have to say about endangered species. Causes of endangerment in aquatic systems will come as no surprise to anyone—*we* are the cause of endangerment through habitat destruction and pollution. By abandoning single species studies and looking at large-scale patterns, the contributors identify broad sources of extinction with respect to particular taxa or particular regions of the world. For example, many endangered species of butterflies are at risk because we have decimated populations of their food plants, making it clear that we cannot save single species without saving webs of interaction. Instead of narrowly worrying about particular species of fish in specific bodies of water, we learn that dams and pollutants repeatedly are the villains when it comes to aquatic extinctions. Clearly, an efficient conservation approach would be to target dams and pollution everywhere, and to move quickly toward the removal of selected dams and the mitigation of pollution as opposed to focusing on one threatened aquatic species at a time.

An appreciation of the need for large-scale remedies and for the broad scope of ecosystem degradation leads into the third section of the book—restoration ecology. At this stage of its development, the science of restoration ecology lacks a theoretical foundation. It is defined almost entirely by practice and very little by experimentation. Interestingly, however, the contributions in this section all point out areas in which experiments and theory might assist our struggle to restore damaged or lost ecosystems. Two features stand out when discussing restoration ecology: (1) it requires that we salvage and reinstate ecological processes and not just species as though they exist in a vacuum, and (2) exotic invaders are a major source of degradation in many of the systems we seek to restore. Both restoration and invasion ecology are haunted by a surplus of anecdotes and a deficit of solid evidence. We lack "before and after" studies documenting the impacts of exotic invaders, and we similarly lack critical analyses of restoration projects. Maybe this is understandable in a field where practical success is what most matters; however, we can only improve our rate of success if we understand our failures.

Saving species on the brink of extinction and the restoration of degraded systems represent economic and political challenges as much as they do scientific challenges. The fourth section of this book considers economics, sustainable development, and decision analysis in the context of conservation. These may not be areas that hold much interest for most biologists, but familiarity with economic and political perspectives can add important practical breadth to those interested in actually applying conservation knowledge to conservation practice. After all, conservation is ultimately about choices, and choices are political and always involve cost-benefit analysis. We must choose where to put our reserves, what research to conduct, what development to encourage, and what development to block. If we keep these choices in mind as we conduct traditional hypothesis-driven research, we may be able to ask questions in a way that better serves our ultimate goals. Thus, instead of simply identifying the habitat requirements of an endangered species, we can ask how tolerant that species might be of certain forms of human disturbance, and whether the economic value of that disturbance is so great that different solutions need to be found. Ultimately, successful conservation depends upon public support—and this means communicating value, trade-offs, and options to the public—something at which scientists can be hopelessly inept, given their esoteric culture of jargon and disconnection from mainstream social interactions. Obviously all scientists cannot, and should not, enter the interface of policy and science, though some must. But all students of conservation should be introduced to potential opportunities for interdisciplinary efforts that blend biology and policy.

Admittedly, conservation biology is very trendy. This is not necessarily a bad thing. However, a downside of trendiness is that substantive advances and questions can be neglected in favor of what looks new and exciting. We close the book with a section on what we have identified and embraced as some of the trendiest areas of contemporary conservation: molecular genetics, conservation endocrinology, landscape ecology, and global climate change. Our goal is not to glorify scientific fashion, but to scrutinize it, and all of the contributors to this section conclude with a somewhat skeptical evaluation of each "hot topic" in conservation. One common aspect of these topics is that they all seem to involve new ways of looking at old problems. Molecular genetics may help us better understand the viability of threatened populations; endocrinology may help us anticipate population risks; landscape models provide us with a potential tool to examine threats to species in fragmented landscapes; and global climate change draws our attention to the importance of preserving species interactions and not just species.

Obviously, like any book, there is much of value we leave out, some of it by accident, some of it because invited authors failed to make deadlines or to meet our editorial standards, and some of it by design. We conclude the book in a rather unorthodox way—i.e., by highlighting what is missing. This serves two purposes: (1) it points the reader to a literature she or he may want to visit, and (2) it offers one last chance to offer our personal slant on what is most important to beginning conservation biologists anxious to familiarize themselves with what is going on in the field, as well as what readings they might explore should they want to conduct research and to make a difference in the practice of conservation.

SAVING SPECIES THROUGH POPULATION BIOLOGY AND VIABILITY ANALYSES: A MORASS OF MATH, MYTH, AND MISTAKES?

Population biology will always be the core of conservation biology. The reason for this is simple. Ultimately, we aim to save species, and this task requires that we consider reproduction, survival and dispersal rates, evolutionary vigor, and extinction risks. These factors are the domain of population biology. Indeed, anyone receiving an advanced degree in conservation biology is likely to have been exposed to course work in population biology; many will have used some form of population viability analysis (affectionately known as "PVA"). But these standard subjects have become much more sophisticated than treatments in most textbooks suggest. We asked a wide range of population biologists engaged in everything from academic research to "on-the-ground" conservation to review their area of population-oriented conservation biology. The response overwhelmingly indicates a rejection of the simplicity of our early applications of population theory.

It is not the case that early ideas in conservation population biology were wrong, however. Smaller populations *have* greater extinction risks. Fragmentation and habitat loss *is* bad. Demographic rates summarized in transition matrices *can* be informative. But these truisms do not get us very far in the practice of conservation. The question should not be whether small populations are at greater risk than large populations, but rather, what is the marginal gain (in terms of reducing extinction risk) of each incremental increase in population size? Similarly, while it is easy to appreciate the merits of a diverse seed bank, it is much harder to decide on a practical protocol for gathering seeds to be included in a seed bank (i.e., if there is space for 100 seeds of each species, how should the sample be apportioned among individual plants, families of plants, populations, and regions?). Similarly, when collecting demographic data a plea is always made for long-term studies, but those studies are expensive. Exactly how much insight is gained by going from a two-year to a three-year demographic study? These are the sorts of questions asked by contributors to this section—hard practical questions that bring somewhat vague theory into the realm of real-world practical constraints.

The answers to these hard questions will not be pleasing to those who hope for simple rules to follow. It turns out in all cases that the answers depend on the details, although the influence of details is nicely illuminated by general theory. Guerrant and Pavlik show how strategies for collecting seeds and reintroducing plants depend critically on details regarding a plant's breeding system, life cycle, and habitat requirements. Similarly Menges points out that the extinction risks faced by plants depend on variance-to-mean ratios in demographic rates. Fiedler and her colleagues discover that a clear-cut demographic distinction between rare versus common plants is not apparent—instead the distinction takes form only after careful analyses of projection matrices in all their arcane detail. The management conclusions to emerge from PVA, even when using the same baseline data, is shown by Groom and Pascual to depend on the source of variation and on the mechanics of how variation is incorporated into models. Similarly, the effects of fragmentation can depend on exactly what variables are considered. For example, Freidenburg points out that whereas edge effects with respect to light penetrate only 50 meters into forest fragments, edge effects with respect to airflow commonly penetrate at least 200 meters into forests. When one realizes that many forest patches have diameters of less than a half a kilometer, these specific differences translate into huge contrasts with respect to how much core habitat we conclude is remaining for many of our North American forests.

A second advance evident in this section is a greatly sharpened view of variability. Certainly the importance of variation was not lost on the founders of conservation biology; however, earlier treatments tended to be anecdotal and casual—at best distinguishing between demographic stochasticity and environmental stochasticity. Contemporary viability analyses, investigations of genetic viability, and studies of fragmentation all deal with a much more elaborate classification and portrait of variability. Indeed, most of the contributions to this section include analyses that show precisely how the magnitude and character of variability alter a population's likely fate and dictate different management solutions.

Finally, although the dependence on detail is unarguable, equally clear is the realization that all detail is not crucial, or that in some situations the details will not matter. Thus, Groom and Pascual report that the results of PVA are typically insensitive to exactly how many stage-classes are used in the projection matrices. Menges finds (through exhaustive computer simulation) that environmental stochasticity matters little to extinction risks once population grow by more than 15% per year. And Fiedler and her colleagues demonstrate that the same basic format for a projection matrix aptly captures the demographic processes of nearly a dozen different species, albeit species in one taxonomic group.

The protection and management of single species is the branch of conservation biology most steeped in abstract theory—life history theory, metapopulation theory, minimum viable population theory, and so forth. Although these general models rarely help solve particular problems, this does not mean that population theorists are without value to the practice of conservation. Indeed, conservation practitioners have much to gain from combining forces with theoreticians to determine the optimal level of detail and specificity for studies aimed at protecting species.

CHAPTER 1

The Analysis of Population Persistence: An Outlook on the Practice of Viability Analysis

MARTHA J. GROOM
and MIGUEL A. PASCUAL

One of the critical challenges in conservation biology is to develop quantitative methods for evaluating the fate of populations that are threatened by human activities (Soulé 1987). Predicting population responses to various perturbations, such as habitat destruction, harvest, or supplementation via reintroduction, requires some practical analyses of population viability. These "population viability analyses" (PVAs) have come into increasing usage, and every indication is that their importance will rise in the future. A recent National Research Council panel convened to evaluate the Endangered Species Act vigorously recommended even greater reliance on viability models (NRC 1996), as have many other groups of biologists seeking to improve management of endangered and rare species (e.g., Carroll *et al.* 1996; Mangel *et al.* 1996; Ruggiero, Hayward, and Squires 1995; Schemske *et al.* 1994).

PVA can take many forms, but it includes a projection of populations into the future based on what is known about demographic rates and their determinants, as well as the probabilistic effects of stochastic events or environmental variability. As originally described (Gilpin and Soulé 1986) and subsequently employed, PVA has been limited typically to cases which best fit Graeham Caughley's "small population paradigm" (Caughley 1994). However, because not all major issues in conservation concern small populations, we prefer to use the term PVA more inclusively. We use it here to refer to any quantitative analysis of population viability, regardless of the current size or economic status of the species in question.

At its best, PVA is a rigorous form of risk analysis, and draws as one of its strengths the explicit consideration of uncertainty regarding the response of populations to management strategies. Yet many examples of PVAs fall short of this ideal, in part because of limitations of methodology and in data. In this chapter, we review the elements that are used in a PVA, discuss what is known about the effects of model structure on the viability results, and suggest ways to cope with uncertainty.

CONCERNS ABOUT DATA LIMITATION

It should be acknowledged at the outset that PVAs are universally limited by the quality of the data they use. Most endangered species have very few, if any, data available to use in performing a PVA. In a survey of threatened bird species, Green and Hirons (1991) found only 58 datasets were suitable for population modeling (roughly 2% of all threatened bird species). Obviously, the scarcity of good data severely constrains our ability to construct accurate viability models.

One way to evaluate how much data limitations may affect the accuracy of PVA is to investigate the effects of incomplete data on model outcome. For example, sampling from detailed simulated data sets, one could estimate the viability of the population using the full range of data available, and compare the aptness of the answer to analyses that sample only portions of the data over a short time span. Using just such an approach, Goldwasser, Groom, and Kareiva (1993) found that models built upon sparse data had no bias, consistently yielding neither optimistic nor pessimistic results. However, these researchers, as well as others (e.g., Wennegren, Ruckelshaus, and Kareiva 1995; Ruckelshaus, Hartway, and Kareiva in press) have demonstrated that data limitations indeed lead to great inaccuracy in predictions, although they also identify circumstances in which their models are reasonably accurate (e.g., according to the degree of error in parameter estimates, or the inclusion of particular model features). Clearly, we need many more studies of this type to define the limits to the reliability of PVA in the face of data limitations.

On consideration, one may ask when is it worth attempting to put together a viability model. The major reasons to hesitate are that (1) an inaccurate PVA may confuse the issues, ultimately leading to political inaction, or worse, errant policy; and (2) the PVA may be a waste of scarce financial resources. In general, we would consider it worth creating a PVA if there are enough data that a model could help either guide future data collection or provide a framework that highlights uncertainties and risks. A PVA may in itself indicate whether the existing data are sufficient to distinguish between widely disparate outcomes. Many of the existing PVAs have served to clarify conservation issues (e.g., the various models of northern spotted owls), and have aided in establishing new management practices. In one of the most inspiring cases, the model of Crouse, Crowder, and Caswell (1987) was instrumental in establishing regulations requiring trawl fisheries to use Turtle Exclusion Devices (TEDs) that reduce bycatch mortality in endangered sea turtles. A model does not need to have 95% accuracy to provide meaningful guidelines for conservation or sustainable development.

MODEL STRUCTURE AND THE OUTCOMES OF PVA

The practice of performing a PVA inevitably involves many seemingly arbitrary decisions that are driven by logistics, sparse data, and convention. Developing a PVA requires not only decisions about what model structure should represent the dynamics of the population (e.g., aggregated vs. stage-structured, linear or non-linear, deterministic vs. stochastic, etc.), but also the choice of a meaningful model output to represent population persistence.

Two general approaches have been used to assess population persistence. In the first, structured models are constructed where the population is divided into ages, sizes, or stages, most commonly using a matrix formulation. Given the best current estimates of demographic rates for each category, the model is deterministically projected into the future to evaluate population growth (e.g., Caswell 1989; Manly 1990). This approach serves to identify the trajectory (increasing, decreasing, or stable) of the population given current conditions. An extremely useful tool in such analyses is to perform elasticity or sensitivity analyses, where the effect of small changes in individual demographic rates on population growth is evaluated (Caswell 1989). Elasticity analysis can indicate where changes in demographic rates (which might result from an alteration in management practices) will serve to revert downward trends or enhance population growth (e.g., Crouse, Crowder, and Caswell 1987; Crowder *et al.* 1994).

The second approach consists of constructing stochastic models that incorporate factors that affect population size. Results from this approach are probabilistic predictions about the future state of the population. While very simple models provide analytic results of population persistence (e.g., Goodman 1987a, 1987b; Belovsky 1987), more realistic models require computer simulation of multiple population trajectories to evaluate the distribution of population states in the future (Burgman, Ferson, and Akcakaya 1993).

To help us characterize the methods and models used in PVA, we surveyed four journals (*Biological Conservation, Conservation Biology, Ecological Applications,* and *Journal of Wildlife Management*) for papers that contained PVAs over a ten-year span from 1987–1996. We included in our sample only papers that considered real populations and employed a mathematical model to explore the effect of human perturbations on population fates. From this survey, we obtained 58 papers (Table 1.1). Over this ten-year time frame, PVAs increased in usage dramatically (from ten in the first five years to 48 in the last five years of our sample). Below we discuss in greater detail some of the common practices in PVA, and illustrate how model structure can affect PVA predictions.

OUTPUTS OF VIABILITY ANALYSIS

The aim of PVA is to evaluate when a population will fail or prosper in response to specific circumstances. Therefore, the output of PVA must include a measure of extinction risk over some time horizon for a given set of conditions. In practice, we seek to estimate the probabilities that a population will shrink to levels low enough to entail a substantial risk for its persistence, the quasiextinction threshold (Ginzburg *et al.* 1982), rather than true extinction (when N=0). The difficulty lies in defining an appropriate threshold and time horizon, because obviously the probability of quasiextinction for any population increases with longer times and with higher thresholds. Figure 1.1 shows the quasiextinction probabilities for a hypothetical population as a function of time and the threshold population size. The object of typical viability analyses is to use population models to portray some aspect of this quasiextinction surface. Simplest results are generated by fixing both the time horizon and the threshold, and reporting the corresponding probability of quasiextinction (e.g., the point A in Figure 1.1). Alternatively, the time horizon can be fixed and the probability that the population will drop below different threshold values is evaluated (line B in Figure 1.1), generating a "quasiextinction curve" (Ginzburg *et al.* 1982). Finally, by fixing the threshold, one can generate the probability distribution of times to extinction (line C in Figure 1.1), and find the mean or median of this distribution (point D in Figure 1.1).

For approaches that entail fixing a population threshold and/or time horizon, the choice of an appropriate reference point will depend on the specific population and question at hand, as well as on how risk-prone or averse are those responsible for making

Table 1.1

Summary of model structures used in 58 PVAs published between 1987 and 1996 in Biological Conservation, Conservation Biology, Ecological Applications *or the* Journal of Wildlife Management. *1. Purpose of PVA according to type of risk evaluated. cons mgmt = effect of practices specifically designed to enhance population of interest, habitat = effect of changes in land use, harvest = effect of harvest (intentional or by-catch), viability = extinction risk given particular conditions (but not affected by changes in land use or harvests). 2. G=aggregated model, A= age-structured, St= stage-structured, Z=size-structured. 3. L=linear, usually matrix, +K=carrying capacity imposed as a ceiling, DD = density dependent (usually non-linear model), A=Allee effect imposed. 4. D=demographic stochasticity, E= environmental stochasticity, C=catastrophes. 5. S=spatial model, G=genetic effects modeled, TS= time series data used, CP=commercial software package used. 6. SA= sensitivity analysis. 7. Journal references BC =* Biological Conservation, *CB =* Conservation Biology, *EA =* Ecological Applications, *JWM =* Journal of Wildlife Management. *Surveyed publications are indicated with an asterisk following full citation in reference list. Only when the main purpose of the paper was some form of PVA were the works included in this list.*

Year	Species	Purpose[1]	General Structure[2]	Linear?[3]	Stochastic?[4]	>1 Model Form?	S A[5]	Other Features[6]	Author(s)	Journal[7]
1987	*Ursus maritimus*	harvest	A	L			Y		Taylor et al.	JWM 51:811
1988	*Picoides borealis*	viability	A	L				G	Reed et al.	JWM 52:563
	Odocoileus virginianus	harvest	G	DD		Y			Lancia et al.	JWM 52:589
1989	*Vicugna vicugna*	harvest	St	L+K					Cattan & Glade	BC 49:131
	Strix occidentalis caurina	habitat	St	L, DD	E	Y		S	Doak	CB 3:389
1990	*Strix occidentalis caurina*	habitat	A	L			Y		Noon & Biles	JWM 54:18
	Pedicularis furbishiae	viability	St	L	E	Y	Y	S	Menges	CB 4:52
	Gymnobelideus leadbeateri	habitat	St	L+K	D		Y	CP	Lindenmayer et al.	BC 54:133
1991	*Cercocebus galeritus g.*	viability	G	DD		Y		S, G	Kinnaird & O'Brien	CB 5:203
	Haliaeetus leucocephalis	viability	G	DD					Montopoli & Andersen	JMW 55:290
1992	*Monachus monachus*	viability	A	L	D, S		Y		Durant & Harwood	BC 61:81
	Hippotragus equinus	viability	A	L	D, S, C	Y	Y		Beudels et al.	BC 61:107
	Crocidura russula	viability	G	DD	D, S				Burgman et al.	BC 61:117
	Strix occidentalis caurina	habitat	St	L+K, A	E	Y	Y	S	Lamberson et al.	CB 6:505
	Melanerpes formicivorus	viability	St	L+K, DD	E	Y			Stacey & Taper	EA 2:18
	Aimophila aestivalis	habitat	St	L, DD	D		Y	S	Pulliam et al.	EA 2:165
	Falco peregrinus	viability	St	L		Y	Y	S	Wootton & Bell	EA 2:307
1993	*Felis concolor*	viability	St	DD, A	D				Beier	CB 7:94
	Picoides borealis	cons mgmt	A	L+K	D			G, CP	Haig et al.	CB 7:289
	Chadrius melodus	cons mgmt	St	L	E		Y		Ryan et al.	CB 7:581
	Loxodonta africana	harvest	A	L, DD	E		Y		Armbruster & Lande	CB 7:602
	Haliaeetus leucocephalis	cons mgmt	A	L	E			CP	Wood & Collapy	JMW 57:1
1994	*Lichenostomus melanops c.*	viability	St	L+K	D	Y	Y		McCarthy et al.	BC 67:135
	Loxodonta africana	viability	G	DD			Y	TS	Burton	BC 70:183
	Petauroides volans	viability	A	L+K	D, C			S, CP	Possingham et al.	BC 70:227
	Aepyceros melanpus	harvest	A	L			Y		Ginsberg & Milner-Gulland	CB 8:157

(continued)

Table 1.1 *(continued)*

Year	Species	Purpose[1]	General Structure[2]	Linear?[3]	Stochastic?[4]	>1 Model Form?	S A[5]	Other Features[6]	Author(s)	Journal[7]
	Strix occidentalis caurina	habitat	St	L+K, A	E			S	Lamberson et al.	CB 8:185
	Dipidomys stephensi	viability	A	L			Y		Price & Kelly	CB 8:810
	Martes americana	viability	St	L+K	D, E	Y		S	Schneider & Yodzis	CB 8:1058
	Eumetopia jubatus	viability	A	L	E	Y			Pascual & Adkison	EA 4:393
	Caretta caretta	cons mgmt	St	L			Y		Crowder et al.	EA 4:437
	Gopherus agassizii	viability	St	L	D, E	Y	Y		Doak et al.	EA 4:446
	Stizostedion vitreum v.	harvest	G	DD	D	Y	Y	TS	Carpenter et al.	EA 4:822
	Picoides borealis	cons mgmt	St	L			Y		Heppell et al.	JWM 58:479
1995	*Puffinus griseus*	viability	A	L	D, C		Y	CP	Hamilton & Moller	BC 73:107
	Trichosurus caninus	viability	A	L+K	D		Y	S, CP	Lindenmayer & Lacy	BC 73:119
	Macrotis lagotis	cons mgmt	A	L	D, E, C		Y	S, CP	Southgate & Possingham	BC 73:151
	Petaurus australis	viability	A	L+K	D, E		Y	CP	Goldingay & Possingham	BC 73:161
	Gymnobelideus leadbeateri	habitat	A	L+K	E, C		Y	S, CP	Lindenmayer & Possingham	BC 73:239
	Aimophila aestivalis	habitat	St	L, DD	D, E		Y	S	Liu et al.	CB 9:62
	Monachus schauinslani	cons mgmt	A	L, DD	D, E	Y			Starfield et al.	CB 9:166
	Panthera tigris tigris	harvest	St	DD	D			S	Kenney et al.	CB 9:1127
	Gymnobelideus leadbeateri	habitat	A	L	D, E			S, G, CP	Lindenmayer & Lacy	EA 5:164
	Trichosurus caninus, Petauroides volans	habitat	A	L	D			S, G, CP	Lindenmayer & Lacy	EA 5:183
	Equus hamionus	viability	A	L	D				Saltz & Rubenstein	EA 5:327
	Thrinax radiata, Coccothrinax readii	harvest	St	L			Y		Olmstead & Alvarez-Buylla	EA 5:484
	Rostrhamus sociabilis p.	habitat	St	L	E	Y			Bessinger	EA 5:618
	Strix occidentalis caurina	habitat	G	L	C	Y			Anderson & Mahato	EA 5:639
	Haliaeetus leucocephalis	viability	A	L			Y		Bowman et al.	JWM 59:317
	Picoides borealis	habitat	St	L	E		Y	CP	Maguire et al.	JWM 59:533
1996	*Ursus americanus*	harvest	A	L	D		Y		Powell et al.	CB 10:224
	Gymnobelideus leadbeateri	habitat	St	L+K	D, E, C			S, CP	Lindenmayer & Possingham	CB 10:235
	Parex quinquefolium, Allium tricoccum	harvest	St	L	E		Y		Nantel et al	CB 10:608
	Ursus arctos horribilis	habitat	A	L, DD		Y			Mattson et al.	CB 10:1013
	Diceros bicornis	harvest	A	L		Y	Y	G, CP	Moehlman et al.	CB 10:1107
	Salvelinus fontinalis	viability	Z	L, DD		Y	Y		Marschall & Crowder	EA 6:152
	Strongylocentrotus franciscanus	harvest	Z	L, A	E	Y			Pfister & Bradbury	EA 6:298
	Kinosternon flavescens, Lepidochelys kempi	cons mgmt	A/St	L			Y		Heppell et al.	EA 6:556

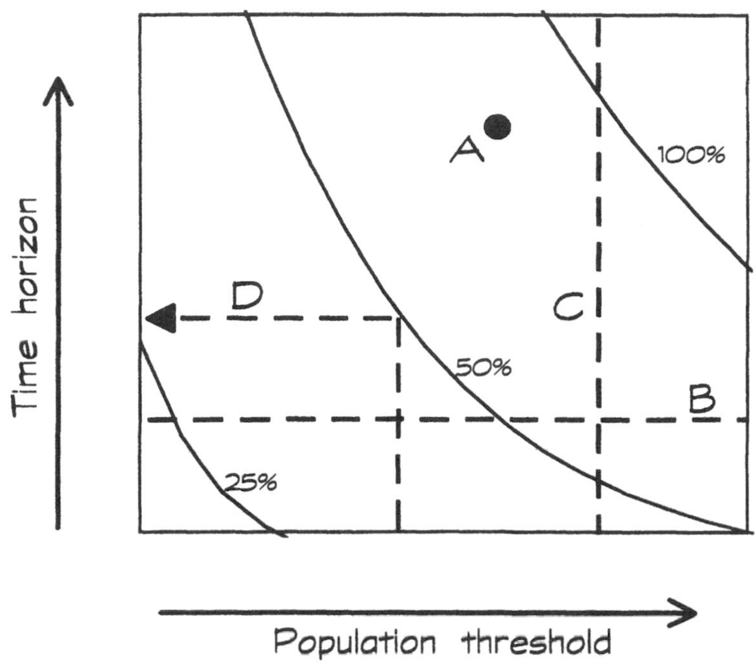

Figure 1.1
Quasiextinction contours for a hypothetical population. The contour lines show the probabilities of the population dropping below a given threshold in a given time horizon. Results of viability analyses typically report some partial attribute of this profile: A. probability of quasiextinction for a given time horizon and threshold, B. quasiextinction curves, C. cumulative distribution of time to extinction, and D. median time to extinction for a given time horizon.

management decisions (Shaffer 1981; Boyce 1992). However, fixing specific reference points can hide important information. For example, the probability of extinction for two different management strategies can change in relative favorability as the time horizon or threshold changes (e.g., Quinn and Hastings 1987). Also, because the distributions of extinction times are typically skewed (Dennis, Munholland, and Scott 1991), reporting mean values can underestimate extinction risk. Further, reporting the mean or median without a measure of variability can hide a good deal of information, especially if there is substantial uncertainty in the results (Groom, Goldwasser, and Kareiva 1993). Because modern computing facilities allow us to do extensive simulations, information provided by PVAs does not need to be artificially constrained by fixing reference points. PVA results may be best reported by quasiextinction contours such as those in Figure 1.1, or by displaying the frequency distribution of predicted times to extinction or population sizes among replicate runs of a projection model.

Population size projections can be useful because they can be compared with new data, whereas predictions about extinction probabilities are impractical for this use. Accompanied by 95% confidence envelopes (for methodology see Heyde and Cohen 1985; Alvarez-Buylla and Slatkin 1991), projected population profiles can illuminate cases where short-term behavior differs markedly from long-term expectations. For example, Doak, Kareiva, and Klepetka (1994) found that short-term increases fell well within the 95% confidence interval in a matrix projection model for desert tortoises, although their model predicted an inexorable decline to extinction. If only the mean prediction had been reported and then real populations were seen to increase, the model's long-term prediction of extinction probably would be doubted. Worse, without an apt representation of uncertainty, the fact that a population happened to increase over a few years could be given too much credence—perhaps even to the point of suggesting there is "nothing to worry about."

AGGREGATED VS. STRUCTURED MODELS

The decision of whether to use an aggregated or a structured model should balance sampling errors with estimation errors (Vandermeer 1979; Ludwig and Walters 1985; Moloney 1986). Environmental variability, demographic stochasticity, and small sample sizes often lead to large errors in the estimation of particular age-specific demographic rates. Thus, a completely aggregated or moderately aggregated stage-based model may make up for its simplification by minimizing parameter estimation errors (Ludwig and Walters 1985). In practice, virtually all PVAs are structured (88% of the models we surveyed were structured), with aggregated models used most often when more detailed data were not available.

Most PVAs partition populations into stages or ages. The choice to use a stage-classified population model usually is dictated by the impracticality of accurately aging members of long-lived species, the difficulty of obtaining age-specific demographic rates for particular age classes, and the intractability of a model that employs a matrix with dimensions as great as the oldest possible age class (e.g., a 72×72 model for loggerhead sea turtles, Crouse *et al.* 1987). However, for many species, stage or size is a better determinant of life-history characteristics than age. Simulation studies of age- versus stage-structured models for *Daphnia* have revealed a remarkable agreement for the two different modeling approaches when only long-run behaviors (e.g., asymptotic rates of population growth) are compared (Law and Edley 1990). Similarly, in an analysis of the effects of model structure on population projections for spotted owls, Goldwasser, Groom, and Kareiva (1993) contrasted a full age-structured model with the conventional three-stage model (juvenile, subadult, and adult), and found that the models produced virtually indistinguishable projections of viability. Still, it is unclear how much the choice of a particular division of the population into ages or stages may affect the model's predictions. It would be most prudent to investigate potential effects by building models that aggregate the population in different ways.

MODELING STOCHASTIC EFFECTS

It is well known that increasing environmental variability promotes extinction (e.g., Goodman 1987a, 1987b; Tuljapurkar 1990), and examinations of this issue in PVA consistently have confirmed these predictions (e.g., Menges 1992; Doak, Kareiva, and Klepetka 1994). For example, using 28 published matrices for perennial plant populations, Menges (1992) found that demographic stochasticity had negligible effects on population persistence, but that environmental variability could profoundly increase extinction probabilities, particularly for species with low finite rates of increase under constant conditions. The incorporation of stochastic effects, particularly environmental and catastrophic variation, should be *de rigeur* in PVA, and perhaps is becoming so. Although many of the earlier PVAs assumed a constant environment, most recent efforts include variability in some fashion (30% of the PVAs were stochastic in the first 5 years of our sample compared with 75% in the next five years; Table 1.1).

Deterministic models have been attractive because they are simple to construct and analyze, and because methods for performing elasticity and sensitivity analyses are well developed (e.g., Caswell 1989). However, methods for performing elasticity analyses on stochastic simulations are being developed (McCarthy, Burgman, and Ferson 1995; Ferson and Burgman 1995; Ferson and Uryasev 1996), which can be expected to be a major advance for PVA. Once these methods become established, deterministic models may be used only to aid in constructing more complex models.

The simplest form of stochasticity included in PVAs is demographic stochasticity, which views the fate of each individual organism as a random event governed by a constant probability process. Viability analyses typically include demographic stochasticity using standard Monte Carlo methods that first assign the fate of each individual according to the probability of survival, growth, or reproduction given in a projection matrix or according to an estimated function, and then sum the outcomes for all members of a population (e.g., Shaffer 1981; Menges 1990). Because population means and variances become increasingly accurate descriptors of the collection of individual events as population size increases, it may be necessary to explicitly model demographic stochasticity only at small population sizes (Leigh 1981; Goodman 1987b), or in spatial models (see below). However, some authors have found that demographic uncertainty can play a key role in their simulations (McCarthy, Franklin, and Burgman 1994; McCarthy, Burgman, and Ferson 1995), and maintain that models that showing a negligible effect of demographic stochasticity only do so because the model structure used underemphasizes this type of variation.

Environmental stochasticity usually affects PVA predictions, regardless of population size (although it may not be reflected in all vital rates; McCarthy, Burgman, and Ferson 1995). Environmental stochasticity was included in 22/39 stochastic PVAs. Environmental variability is typically added to projection models either by drawing demographic parameters from arrays of values representing the data (e.g., Bierzychudek 1982; Stacey and Taper 1992), or by treating demographic parameters as random variables before projecting forward in time (e.g., Bayliss 1989; Burgman, Ferson, and Akcakaya 1993; Burton 1994; Nordheim, Hogg, and Chen 1989; McKelvey, Noon, and Lamberson 1992; Carpenter *et al.* 1994). Catastrophic variation was rarely included in PVAs, appearing only seven times in our survey. In most cases, catastrophes were modeled as a sharp, randomly occurring decreases in vital rates. Harris, Maguire, and Shaffer (1987) exhorted PVA practitioners to run large numbers of replicate simulations (1,000 or more) to ensure accurate results. However, the majority of authors did not heed this advice: only nine of the studies we surveyed performed ≥ 1000 replicates, and an additional 13 used 500 replicates to determine the distribution of outcomes. A few studies (particularly those using spatial models) only used ten to 50 replicates (e.g., Doak 1989; Pulliam, Dunning, and Liu 1992; Lamberson *et al.* 1994), perhaps too few to assure a high degree of confidence.

Environmental or catastrophic stochasticity entails an enormous variety of decisions regarding how that variability can be incorporated into a model. The major issues include: what probability distribution should be used to describe each parameter's variability; should the temporal sequence of variation be purely random, correlated, or cyclic; should demographic parameters (e.g., fecundity or juvenile survival) be allowed to vary independently of one another or forced to vary in parallel; and should variation among individuals in their demographic rates be included?

Theoretical studies have revealed that the exact manner in which environmental variability is modeled dramatically influences the long-run population growth for simulated populations. For instance, temporal autocorrelation in variable demographic rates can dramatically reduce long-run population growth rates compared to purely random sequences of variation (Tuljapurkar 1990). Demographic variation among individuals within the same stage class also dramatically alters population growth in a manner determined largely by the skewness of the probability density describing demographic variance (Slade and Levenson 1984). The particular probability distribution chosen to represent environmental variation also is likely to influence population projections, although no one has systematically explored the consequences of using even the most commonly used distributions (normal, lognormal, beta, and gamma). Finally, recent theoretical work has shown that catastrophes (rare events that occur with a low binomial probability, but with a large effect when they occur) can enhance likelihood of extinction by an order of magnitude (Mangel and Tier 1993).

Clearly, authors of PVAs need to better address the extent to which specific results depend on the details of how environmental variation is incorporated in the analysis. Few of the papers we surveyed included a discussion of the form of stochasticity and its potential effects on the results. Those that did found that the details mattered. For example, Doak, Kareiva, and Klepetka (1994) found that correlation in vital rates of desert tortoises increased the probability of extinction. Beissinger (1995) reported that periodic fluctuation in vital rates exaggerated negative trends in snail kite populations compared with random fluctuations. Similarly, Goldwasser, Groom, and Kareiva (1993) found that population growth rates of northern spotted owls were much more variable when environmental variation was included as a correlated or periodic variable than when it was modeled as a random variable, although there was no effect on the mean prediction.

SPATIAL STRUCTURE

Many populations interact over a complex environment, and would be more accurately represented in a spatial model. Spatial models must keep track of the sizes of separate subpopulations, the rates of migration between the subpopulations, and may also simulate differing effects of environmental conditions on demographic rates in each subpopulation, which may or may not be correlated temporally. Clearly, such models are complex to construct and analyze, and are extremely data-hungry. In rare cases when a great deal is known about a population, spatially explicit models have been constructed (e.g., Pulliam, Dunning, and Liu 1992, or the GIS-based approach of Possingham *et al.* 1994 or Liu, Dunning, and Pulliam 1995). However, there are rarely sufficient data to put together even a well-founded non-spatial model, so development of spatial models has lagged behind other approaches.

Only 29% of the models we surveyed include a spatial component, and the majority of these concern one of a few species; the northern spotted owl in the Pacific Northwest, Bachman's sparrow in the Southeast, or threatened arboreal marsupials in Southeastern Australia. Most often spatial models were used in a very abstract manner, as in the pseudospatial approach of the classic Levins' (1969, 1970) metapopulation construct (e.g., Lande 1987; Hanski and Thomas 1994), or in spatial "island" models that incorporate variation in dispersal rates (e.g., Doak 1989; Lamberson *et al.* 1992; Schneider and Yodzis 1994; Southgate and Possingham 1995). It may not be the case that a more explicit model is more accurate than a simpler one. Anderson and Mahato (1995) found that a relatively simple model of spotted owls yielded qualitatively similar results to the extremely complex models of Lamberson *et al.* (1992, 1994). However, spatial modeling is too young to make any definitive statement in this regard.

The subdivision of the population can have a negative effect on viability projections, due to the greater probability that some or all of the subpopulations become so small that they are vulnerable to demographic stochasticity (e.g., Menges 1990; Kinnaird and O'Brien 1991; Possingham *et al.* 1994). However, more theoretical explorations of the issue predict that subdivision can stabilize population fluctuations, therefore increasing the probability of persistence (e.g., Levins 1970; Richter-Dyn and Goel 1974). This can be true, so long as variation in vital rates or the occurrence of catastrophes are uncorrelated among subpopulations. However, many of these models do not take into account the effects of demographic stochasticity, or the probable correlation in environmental stochasticity among closely situated subpopulations (Harrison and Quinn 1989; Gilpin 1990). Spatial models generally may predict lower population viability than non-spatial models, except where catastrophic effects on populations are uncorrelated spatially.

Using a general model consisting simply of patches of either poor or good habitat, Doak (1995) found that the ability to detect population declines plummets as the amount of suitable habitat shrinks. His result raises a red flag that would not have been predicted in a non-spatial approach. At least heuristically, building spatial models can provide valuable lessons about what factors to worry about, and may provide guidance on the design of monitoring programs to detect population declines.

However, the use of spatial models for specific viability predictions may be more limited. Wennegren, Ruckelshaus, and their colleagues (Wennegren *et al.* 1995; Ruckelshaus, Hartway, and Kareiva in press) have found that spatially explicit models are particularly sensitive to the quality of dispersal data. This group of researchers has found that very low parameter estimation errors (of <10%) in dispersal data are required to provide reasonably accurate results from spatially explicit models. Their results are dependent on the amount of suitable habitat presumed to be available. For landscapes with more than 30% suitable area, larger errors in dispersal data do not corrupt the accuracy of model predictions, but in highly fragmented applications, data limitations may make spatial PVA a fruitless exercise. Many more exercises such as this one are needed to delineate when the use of a spatial model can be expected to provide more accurate viability predictions than non-spatial approaches.

DENSITY-DEPENDENCE

Most PVA analyses (88% in our survey) used models that predict exponential population growth or shrinkage (Table 1.1). Thirteen studies included a carrying capacity, but otherwise retained a linear form to their model, and the assumption of exponential growth away from K. Only nine (16%) studies included some type of feedback on demographic rates as a result of changes in density. Since we teach undergraduates in ecology that simple linear models are patently false, one has to wonder why they are the standard for PVA. The traditional answer has three components: (1) data on density-dependence are scarce. While it is generally possible to estimate survival and fertility and, sometimes, get an estimate of how variable these rates are, it is extremely difficult to estimate how survival and reproduction change in response to population density. Discriminating density-dependent from density-independent variability requires long-time series of population condition, as well as variation in population sizes. Particularly for declining species, such data rarely are available; (2) including density dependence only tends to make the picture more optimistic for species, so we might as well go with the more pessimistic model as a way of guaranteeing that we err on the safe side (Ferson, Ginzburg, and Silvers 1989; Ginzburg, Ferson, and Akcakaya 1990); and (3) at the densities typical of species in the position of "endangerment," density-dependent competition for limiting resources is nearly impossible to quantify, and may be in fact negligible. In addition, some researchers have found evidence that density-dependent effects, though present, are swamped by the effects of environmental fluctuations (e.g., Croze, Hillman, and Lang 1981) or human-caused mortality (e.g., Mattson *et al.* 1996) on population trends.

However, we take issue with neglecting density-dependence. Clearly, ignoring one type of density-dependence, the lowered fecundity for populations at low density or of small size, known as the Allee effect, is not in the least conservative. On the contrary, populations that experience Allee effects are more vulnerable at low population densities, and can experience striking non-linear changes in fecundity as their density drops below some threshold level (Dennis 1989). Reproductive success in an annual plant was found to drop abruptly to zero when small groups were isolated by as little as 50–100m from other populations (Groom 1996). Survival can also be abruptly lower at low densities. Campagna *et al.* (1992) found that southern sea lions (*Otaria byronia*) breeding in isolated pairs had much lower infant survivorship (≈ 40% lower) than individuals breeding in larger groups consisting of one male and several females. To some degree, Allee effects can be addressed through the use of quasiextinction probabilities. By adjusting the critical population threshold above that at which Allee effects are felt by the species, this type of dangerous non-linearity in population dynamics can be taken into account. Allee effects can also be handled more directly by reducing fecundity or survival in model algorithms to zero or near-zero when densities are low (e.g., Price and Kelly 1994; Pfister and Bradbury 1996).

Neglect of density-dependence also has important implications for sensitivity analyses of viability models. PVAs routinely report either numerical experiments or analytical exercises that explore which aspects of a species' demography play the most important role in governing the rate of population growth. Superficially, such a sensitivity analysis might appear to yield information about "limiting factors" for population growth. But in fact, a transition matrix says nothing about what biologically might limit a population's success—it can tell us only what is gained with respect to the finite rate of population multiplication as a consequence of some small percent change in a particular demographic parameter. In other words, we need to be careful to distinguish between an algebraic and a biological examination of limits to population change.

A hypothetical example illustrates this point. Many plants have extraordinarily high mortality rates as seedlings, and can live for long time periods as reproductive adults. A matrix analysis of this hypothetical demography would reveal that the long-run population growth rate for such plants is not much influenced by juvenile survivorship compared to adult survival or reproduction. Indeed, elasticities as defined by Caswell (1989) would probably be comparatively low for seedling survival, which could lead one to conclude seedling survival's importance was negligible. However, the reason seedlings suffer such high mortality might be a shortage of "safe sites" (Harper 1977), and the way of increasing plant density and viability could well be a manipulation of soils to make more safe sites available. In this way, even minor increases in seedling establishment via placement in safe sites might have a significant effect on population growth.

A potential solution to this dilemma is to use sensitivity analyses in conjunction with a density-dependent model. Another is to acknowledge the limitation of using a linear model. Thus, even if density dependence is hard to detect with scarce populations, one should still attempt to identify what critical resources are limiting a population's rate of growth (Schemske *et al.* 1994). We conclude that while it may be impractical to include any density dependence in a viability model, modelers should recognize that such a neglect could constrain the way in which a model can be interpreted.

HOW ROBUST ARE PARTICULAR MODEL STRUCTURES?

In previous sections we reviewed some of the decisions with which we are confronted when modeling the dynamics of populations, as well as a number of results that show how viability analysis can be very sensitive to details in model structure. Most viability analyses rely on the description of the probabilistic behavior of the population, with particular emphasis on extreme behaviors. Small changes in parameter values or model structure that might not produce significant changes in the estimated average population growth can still affect these extreme behaviors in substantial ways. Viability analysis, therefore, is expected to be consistently less robust to the structural complexity of models than analyses that rely more heavily on the average behavior of the population. Perhaps more critical, model mis-specification can affect assessments of different human interventions (Pascual, Kareiva, and Hartway in press). The bottom line is that we should be especially scrupulous with our appraisal of model uncertainty in viability analyses.

When confronted with the task of evaluating the state of a natural system or the possible consequences of a management action, we tend to harbor the comforting feeling that good experimental and sampling protocols, adequate resources, and long-term studies will allow us to accumulate enough data to resolve all major biological uncertainties. Our collective experiences, however, tell us that no matter how much data we gather, we will always find limits in our ability to establish the importance of processes, or value of crucial population parameters. Populations are complex entities and anticipating their response to possible management interventions typically requires us to predict how these populations will respond to events or situations never observed before. A good illustration of the pitfalls of constructing predictive models is provided by the field of fisheries management, where very large population data sets are not uncommon but uncertainties abound (see for example Deriso, Hoag, and McCaughran 1986; Myers and Cadigan 1995).

One of the adages of wise model building is to keep models simple. The ecological and resource management literature offers several examples of simple models efficiently approximating the behavior of complex systems. For example, Barlow and Clout (1983) tried several logistic-type models and an age-structured model to investigate optimal harvest levels of possum in New Zealand. They found that several models yielded similar population growth curves, which in turn resulted in similar estimates of maximum sustainable yield. Based on simulated data, Ludwig and Walters (1985) showed that a simple logistic model can provide perfectly good estimates of optimum harvest rates of more complex populations when data are limited. Perhaps more remarkable, they showed that when data are extremely uninformative the simple approximation can perform better than the true complex model used to generate the data.

Of course, simple models are not always superior, and many management questions are extremely sensitive to the details of model structure. For example, Burgman, Ferson, and Akcakaya (1993) estimated contrasting persistence estimates of a suburban population of shrews in Switzerland with two apparently suitable models, a Ricker and a logistic model, and found that the logistic was much more extinction-prone. McCarthy, Franklin, and Burgman (1994) found that adding detail on mating limitations for the helmeted honeyeater, a bird inhabiting Eucalyptus forests in Southern Australia, resulted in more pessimistic assessments of population persistence. Pascual, Kareiva, and Hilborn (in press) fitted several dynamic models to data for the Serengeti wildebeest. They not only found that different models produced extremely different quasiextinction profiles, but produced contrasting estimates of the added risk (*sensu* Burgman, Ferson, and Akcakaya 1993) due to exploitation. Ludwig (1996) found that changes in the value of parameters in demographic models within ranges well supported by the data, had a large effect on estimated population persistence. Yet, each of these examples are particular results that may not have broad application to all PVAs. Clearly more analyses of this kind are needed, both as a means of evaluating the robustness of particular models, and as an aid to developing more general guidelines.

Reporting the statistical power of our estimates can also be used to judge the reliability of particular analyses (Taylor and Gerrodette 1993; Shrader-Frechette and McCoy 1993). This point is illustrated by analyses of Taylor and Gerrodette (1993), who used data on the Vaquita and spotted owls to show that as sample sizes dropped, the ability to detect changes in population size, and declines in particular, also dropped. Indeed, they found that in these cases it is easier to detect changes in population size directly from censuses, than by any kind of viability projection.

MODEL JUSTIFICATION AND THE CASE FOR USING MULTIPLE MODELS

In fisheries and other harvest applications, a thorough justification of the chosen model structure is expected. In contrast, the norm in PVA appears to be to put together a reasonable representation of one's best understanding of a population's dynamics, with only a handful of authors justifying their choices (e.g., Stacey and Taper 1992; Pascual and Adkison 1994; Carpenter *et al.* 1994; Beissinger 1995; Anderson and Mahato 1995). For example, only rarely do authors examine available time series data to evaluate the fit of the model they chose (e.g., Beier 1992; Carpenter *et al.* 1994). This lack of model justification is somewhat excusable if we recall the fact that in conservation applications model validation is hampered by chronic data limitations (time series data often are not available) and limitations in understanding the behavior of the system outside of its historical states (where extrapolation may be inappropriate since future trends may have no relation to past patterns). Validation of models by contrast with the realized behavior of the population is also difficult because of the intrinsic nature of PVA, in which our predictions are given in probabilistic terms. Yet justification of the choices made in picking a model structure adds rigor to PVA.

Typically, models are selected either by statistical approaches that tell us which is the "best" model, or on the basis of the particulars of the species' biology. The statistical methods we are most familiar with (e.g., ANOVA, likelihood ratio test, Akaike information criteria) stress model selection based on the parsimony principle. This principle says that we should favor the simplest model that is supported by the available data. The parsimony principle arises from the fact that as the number of parameters increases (although the model may better describe the true complexities in the study system) the uncertainty associated with estimating all these parameters increases. This tradeoff, theory suggests, should dictate what is the best model choice. But in management problems, the value of this approach is arguable. First, this technique judges models' exclusively based on descriptive power when, in reality, we are interested in their management predictions. Second, parsimony emphasizes the use of a single model, when there may be several probable models with contrasting management implications.

Figure 1.2 shows the fit of three different models to the time series of observed population size of the Serengeti wildebeest and their predictions in terms of population persistence. The three models fit the data well, but produce very different predictions of both basic quasiextinction probabilities and the relative added risks due to harvest. This example shows that evaluating predictive models based on how well they describe the data can be misleading. In fact, if we were going to select a single "best model" in terms of parsimony, the Ricker model would be selected because it provides the best fit with economy of parameters. But, we could also justify using the age-structured model because it matches more closely our mechanistic ideas of population processes—i.e., how survival and recruitment are affected by the environment. Traditional model selection methods do not provide a formal way to incorporate this type of biological information into the process of model evaluation.

We would like to argue for the further step of building and analyzing multiple models to obtain the best possible portrait of future outcomes of a management strategy or a particular disturbance. Returning to the example in Figure 1.2, we can see that by stressing the selection of a single model we risk discarding reasonable alternatives, producing a distorted assessment of the actual risks associated with management procedures. A logical alternative to using a single model is to admit our uncertainty and evaluate management actions across alternative models. One way of doing this is to purposely search for all conceivable models and search for management alternatives that are robust to the choice of model. A weakness of this approach is that there are no formal limits to the scenarios we can conceive (Ludwig 1996) and our results may end up depending too heavily on our personal biases. However, we are likely to learn much from the exercise of imagining possible scenarios. By evaluating the predictions of alternative representation(s) we learn about what the uncertainties are and can determine which are the critical modeling decisions. Another alternative is to perform a meta-analysis (see for example, Fernandez-Duque and Valeggia 1994; Arnqvist and Wooster 1995), where extinction risk could be evaluated in light of multiple pieces of evidence, which in this case consist of competing viability models. Meta-analysis has only begun to be applied to ecological and conservation issues in the past few years, and can be expected to develop into an increasingly robust analytical method in the future.

One-third of the PVAs we surveyed explored two or more structurally different models in their analyses. Excellent examples include the use different structures to explore different effects (such as density dependence or spatial subdivision) independently from a simpler model, or employing a simple model to determine reasonable limits for unknown parameter values before employing a more sophisticated model (e.g., Heppell, Walters, and Crowder 1994). However, few authors discussed the implications of using alternative model structures for the interpretation of their PVAs. To their credit, a large number of authors employed sensitivity analyses, which may broaden the insight to be gained in PVA. Yet even though sensitivity analyses are extremely useful for examining how changes in particular parameters may affect a population, they do not help us cope with the problem of uncertainty in our data or potential bias caused by adopting a given model the structure. We need to know if different structural models provide different projections of population viability, and how to weigh such different models.

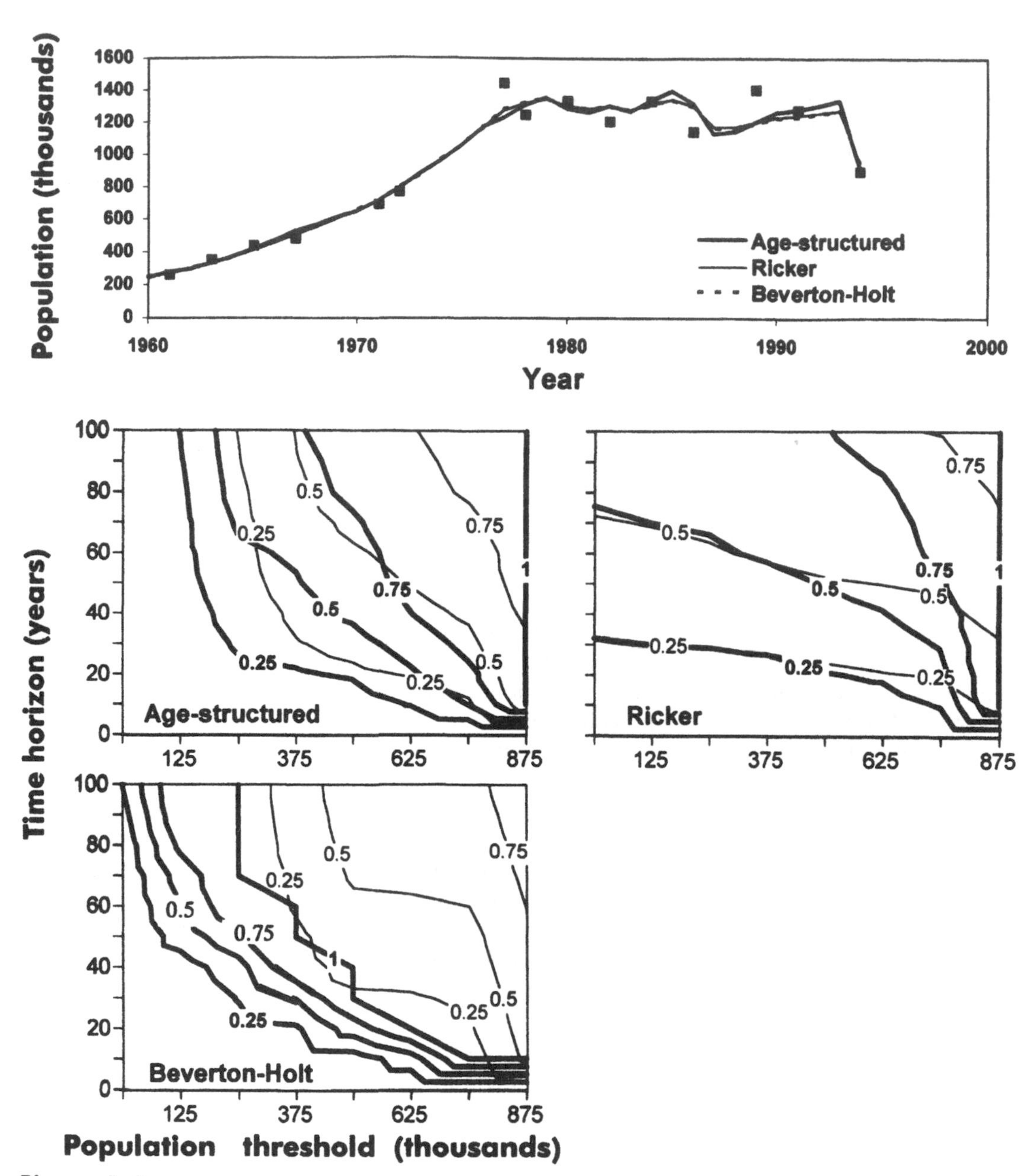

Figure 1.2

Quasiextinction profiles for Serengeti wildebeest under three different models, with and without harvest (model fit following Pascual et al., *in press). Upper panel: fit of three models (Age structured, Ricker, and Beverton and Holt) with time series data on wildebeest population sizes. Lower three panels: quasiextinction contours for all three models. —— indicate quasiextinction probabilities without harvest. ▬▬ indicate quasiextinction probabilities if an annual harvest of 6% is assumed. The three models provide very similar descriptions of the observed data but extremely different portraits of potential risk to wildebeest populations from different harvest levels.*

VIABILITY ANALYSIS AS A PROBLEM IN DECISION ANALYSIS

In typical conservation problems we may have available a variety of management alternatives to maintain or improve the chances of population persistence. Typical options include constructing reserves, culling, reintroducing individuals from captive breeding, or protecting critical habitat (Maguire, Seal, and Brussard 1987). Therefore, we are faced with the task of evaluating management options in the face of significant uncertainty. Regarded in this way, a viability analysis is best stated as a decision analysis problem (Raiffa 1970). If we represent our uncertainty as a collection of hypotheses about the nature and functioning of the system under scrutiny, this problem can be conceptualized as a decision table where the columns represent those hypotheses and the rows represent the management options under consideration. Individual cells contain the object of our analysis: the outcome of applying a given management strategy if a given state of the system were true. The outcomes in the decision table will typically be estimated by stochastic simulations. An alternative representation to this table format is provided by decision trees (e.g., Maguire 1991).

	Possible "States of Nature"		
	State 1	State 2	State 3
Management Option 1	Outcome 1,1	Outcome 1,2	Outcome 1,3
Management Option 2	Outcome 2,1	Outcome 2,2	Outcome 2,3
Management Option 3	Outcome 3,1	Outoome 3,2	Outcome 3,3

In the best-case scenario, the exploration of the decision table may reveal that some management options are clearly superior across all models (or "states"). When this is the case, our management ability is not impaired by the uncertainty about population behavior (Starfield, Roth, and Ralls 1995). However, when alternative options give contrasting results across hypotheses, a judgment is required about the best course of action. The value of a given option can be calculated as the average of outcomes across hypotheses weighted by the probability of each hypotheses (e.g., Maguire 1991). Pleading complete ignorance about the state of the system is represented by giving all hypotheses the same probability. Alternatively, probabilities might be derived from experiments, knowledge from other systems, or empirical observations. The problem with using the average performance to judge the value of a given management option is that it averages over good and bad outcomes and can hide undesirable extreme outcomes (Starfield and Herr 1991). A more "risk-averse" approach is to use indicators that reduce these unpleasant surprises. An extreme risk-averse approach, for example, is to follow the recommendations predicted by assuming the most pessimistic state of nature.

Bayesian statistics can provide a formal framework to make statements about how likely are alternative hypotheses concerning the state of nature (Box and Tiao 1973), thereby providing a formal way to assign probabilities to alternative models. Within this framework, a prior statement of our knowledge —or lack of it— about model parameters is combined with the evidence provided by the data to assign probabilities to the alternative hypotheses. There is significant discussion among statisticians about the validity of Bayesian estimation as compared to traditional or "frequentist" approaches, the main contention involving the definition and formulation of "prior knowledge" (e.g., Edwards, Lindman, and Savage 1963). A discussion of these issues is beyond the objectives of this chapter. It suffices to mention here that this approach is beginning to be applied to ecological problems, and some recent papers propose Bayesian treatments of population viability problems (Pascual and Hilborn 1995; Ludwig 1996).

The simple schematic view of decision analysis provided here concentrates on the case in which we look at a single outcome, such as the probability of early collapse (Ludwig 1996). In most cases we are interested in evaluating multiple objectives, such as the trade-off between extinction probability and cost of management (Maguire, Seal, and Brussard 1987), harvest and extinction probability (Pascual and Hilborn 1995), or several population state indices (Starfield, Roth, and Ralls 1995). Ralls and Starfield (1995) describe a method to integrate multiple objectives in which alternative model predictions are ranked by order of importance for management.

Decision analysis approaches allow us to examine management choices, implementation issues, and biological uncertainty within a unified framework. Not only do they allow us to identify appropriate management actions, major uncertainties, and associated costs, but they also provide the building blocks to evaluate how those uncertainties can be reduced by deliberately "experimenting" with our management options (Walters 1986). This area of resource management, called adaptive management, has a large potential to derive long-term conservation plans. As conservation biologists become more aware of the tools provided by Bayesian estimation and other methods of decision analysis, we should see an increase in their application to specific cases.

FUTURE DIRECTIONS

For PVA to grow in its utility as a conservation tool, we must meet several challenges. First, it is important to recognize that not all data will have the same value in supporting a viability analysis, or in helping define the uncertainty involved in PVA. For example, detailed age-structured data may be less important than data about the form and strength of density-dependence. In addition, time series of population states may be especially valuable. Time series data provide information on the stochastic behavior of population(s), and allow us to evaluate alternative models by contrasting them with the actual behavior of the population. Importantly, PVA itself can help determine where data needs are greatest, and in this way guide future data collection. By analyzing which variables most influence model predictions within a single model, or better still, looking for consistencies across multiple models, it should be possible to pinpoint the greatest data needs.

Further advances may be gained through the use of increasingly sophisticated sensitivity analyses. Importantly, it must be recognized what these analyses will and will not tell us. In general, sensitivity analyses do not tell us which demographic improvements are actually biologically or practically feasible. Sensitivity analyses don't show us what to do, but rather show where change in parameter values of equivalent magnitude will have the biggest effect. For example, in an explicit investigation of the utility of PVA, Hamilton and Moller (1995) found that predator control applied early in the nesting season might have greatest effect in helping threatened populations of sooty shearwaters, because their sensitivity analyses showed that preservation of adults has the greatest effect on population growth rates. However, this finding is no guarantee of success in following such a policy. Further, sensitivity analyses are strongly influenced by the quality of our data: a result showing a particular parameter has no effect on population growth may be accurate, or on the contrary, may indicate only our ignorance about the parameter. Translating the perspective gained through sensitivity analyses into practice involves two crucial steps:

recognizing the limits imposed by data quality, and recognizing that whatever might have the largest effect may not be achievable, and visa versa. For example, while sensitivity analysis may show that improvements in juvenile survival can have large impacts on population growth rates, the population size of breeders may be quite insensitive to juvenile survival if breeding habitat is limited (Green and Hirons 1991). Improvements both in methodology and in linking results to management action will be necessary to make PVA most useful.

To improve our ability to cope with the tremendous uncertainties we face in the analysis of wild populations, we favor candid recognition of our uncertainty and purposely seeking competing representations (i.e., we must abandon the "best-single-model" approach). Only 8% of the PVAs we surveyed considered alternate model structures in their analysis. This tendency has to change and PVA practitioners consistently need to consider management alternatives across different possible models. Ideally, it would be best to adopt a decision theoretical framework when making management decisions, but we realize this goal is a long way from being attained.

Finally, we advocate making much greater use of experimental or adaptive management approaches (intentional manipulations of natural populations, *sensu* Walters and Hilborn 1976; Walters 1986), rather than to define a "minimum viable population" or an "extinction probability." The most appropriate and informative use of PVA is to guide management decisions. We rarely will have sufficient data to support the use of PVA to set some definitive estimate of persistence. In contrast, nearly every management question we face involves considerable uncertainty and often high stakes in terms of species extinction or human prosperity. In the face of such uncertainty, PVA can overcome the gaps between our direct knowledge of a species or system and allow us to assess the array of possible responses to changes in the environment or resource management schemes. What is needed are estimations of the most probable outcomes, with respect to critical appraisals of a model's predictive power.

At present most PVA discussion is limited to conservation concerns, such as with small populations of endangered species, where experimentation may be impractical and dangerous. However, much can be learned about the limit to PVA approaches through comparison with other species or other refinement of our methods via experimentation and adaptive management (e.g., Williams, Johnson, and Wilkins 1996). Because experiments are not an option in most cases, analysis of case studies, simulations, and meta-analysis are particularly valuable. The judicious use of PVA can be a significant aid in discrimination among management options, and guide refinements to our management strategies and data gathering protocols.

ACKNOWLEDGMENTS

This paper grew out of an earlier report to the California Forestry Association (Groom, Goldwasser, and Kareiva 1993), and out of conversations between us and our colleagues over a period of years. We would like to acknowledge particularly the contributions of our discussions with D. Doak, L. Goldwasser, S. Heppell, R. Hilborn, and P. Kareiva. We are grateful to the editors, D. Grünbaum, and S. Heppell for their thoughtful criticisms. Financial support came from the California Forestry Association and the Department of Zoology, NC State University (M. Groom), and NSF DEB 9402314 to P. Kareiva (M. Pascual).

LITERATURE CITED

Alvarez-Buylla, E. and M. Slatkin. 1991. Finding confidence limits on population growth rates. *Trends in Ecology and Evolution* 6:221–224.*

Anderson, M.C. and D. Mahato. 1995. Demographic models and reserve designs for the California spotted owl. *Ecological Applications* 5:639–647.*

Armbruter, P. and R. Lande. 1993. A population viability analysis for African Elephant (*Loxodonta africana*): How big should reserves be? *Conservation Biology* 7:602–610.*

Arnqvist, G. and D. Wooster. 1995. Meta-analysis: Synthesizing research findings in ecology and evolution. *Trends in Ecology and Evolution* 10:236–240.

Barlow, N.D. and M.N. Clout. 1983. A comparison of three-parameter, single-species population models, in relation to the management of bushtail possum in New Zealand. *Oecologia* 60:250–258.

Bayliss, P. 1989. Population dynamics of magpie geese in relation to rainfall and density: Implications for harvest models in fluctuating environments. *Journal of Applied Ecology* 26:913–924.

Beier, P. 1993. Determining minimum habitat area and habitat corridors for cougars. *Conservation Biology* 7:94–108.*

Beissinger, S.R. 1995. Modeling extinction in periodic environments: Everglades water levels and snail kite population viability. *Ecological Applications* 5:618–631.*

Belovsky, G. 1987. Extinction models and mammalian persistence. In *Viable Populations for Conservation*, ed. M.E. Soulé, 35–57. Cambridge: Cambridge University Press.

Beudels, R.C., S.M. Durant, and J. Harwood. 1992. Assessing the risks of extinction for local populations of roan antelope *Hippotragus equinus*. *Biological Conservation* 61:107–116.*

Bierzychudek, P. 1982. The demography of jack-in-the-pulpit—a forest perennial that changes sex. *Ecological Monographs* 52:335–351.

Bowman, T.D., P.F. Schempf, and J.A. Bernatowicz. 1995. Bald eagle survival and population dynamics in Alaska after the *Exxon Valdez* oil spill. *Journal of Wildlife Management* 59:317–324.*

Box, G.E.P. and G.C. Tiao. 1973. *Bayesian inference in statistical analysis*. Reading, MA: Addison-Wesley.

Boyce, M. 1992. Population viability analysis. *Annual Review of Ecology and Systematics* 23:481–506.

Burgman, M., D. Cantoni, and P. Vogel. 1992. Shrews in suburbia: an application of Goodman's extinction model. *Biological Conservation* 61:117–123.*

Burgman, M.A., S. Ferson, and H.R. Akcakaya. 1993. *Risk assessment in conservation biology*. London: Chapman and Hall.

Burton, M.P. 1994. Alternative projections of the decline of the African elephant. *Biological Conservation* 70:183–188.*

Campagna, C., C. Bisioli, F. Quintana, and A. Vila. 1992. Group breeding in sea lions: pups survive better in colonies. *Animal Behavior* 43:541–54.

Carroll, R., C. Augspurger, A. Dobson, J. Franklin, G. Orians, W. Reid, R. Tracy, D. Wilcove, and J. Wilson. 1996. Strengthening the use of science in achieving the goals of the Endangered Species Act: An assessment by the Ecological Society of America. *Ecological Applications* 6:1–11.

Carpenter, S.R., A. Muñoz del Rio, S. Newman, P.W. Rasmusen, and A.M. Johnson. 1994. Interactions of anglers and walleyes in Escanaba Lake, Wisconsin. *Ecological Applications* 4:822–832.*

Caswell, H. 1989. *Matrix population models*. Sunderland, MA: Sinauer Press.

Cattan, P.E. and A.A. Glade. 1989. Management of the vicuña *Vicugna vicugna* in Chile: Use of a matrix model to asses harvest rates. *Biological Conservation* 49:131–140.*

Caughley, G. 1994. Directions in conservation biology. *Journal of Animal Ecology* 63:215–244.

Crouse, D., L. Crowder, and H. Caswell. 1987. A stage-based population model for loggerhead sea-turtles and implications for conservation. *Ecology* 68:1412–1423.

Crowder, L.B., D.T. Crouse, S.S. Heppell, and T.H. Martin. 1994. Predicting the impact of turtle excluder devices in loggerhead sea turtle populations. *Ecological Applications* 4:437–445.*

Croze, H., A.K.K. Hillman, and E.M. Lang. 1981. Elephants and their habitats: How do they tolerate each other. In *Dynamics of Large Mammal Populations,* eds. C.W. Fowler and T.D. Smith, 297–31. New York: Wiley.

Dennis, B. 1989. Allee effects: Population growth, critical density, and the chance of extinction. *Natural Resource Modeling* 3:481–538.

Dennis, B., P.L. Munholland, and J.M. Scott. 1991. Estimation of growth and extinction parameters for endangered species. *Ecological Monographs* 61:115–143.

Deriso, R.B., S.H. Hoag, and D.A. McCaughran. 1986. Two hypotheses about factors controlling production of Pacific halibut. *International North Pacific Fisheries Commission Bulletin* 47:167–173.

Doak, D.F. 1989. Spotted owls and old growth logging in the Pacific Northwest. *Conservation Biology* 3:389–396.*

Doak, D.F. 1995. Source-sink models and the problem of habitat degradation: General models and applications to the Yellowstone grizzly. *Conservation Biology* 9:1370–1379.

Doak, D.F., P. Kareiva, and B. Klepetka. 1994. Modeling population viability for the desert tortoise in the Western Mojave Desert. *Ecological Applications* 4:446–460.*

Durant, S.M. and J. Harwood. 1992. Assessment of monitoring and management strategies for local populations of the Mediterranean monk seal *Monachus monachus. Biological Conservation* 61:81–92.*

Edwards, W., H. Lindman, and L.J. Savage. 1963. Bayesian inference for psychological research. *Psychological Review* 70:193–242.

Fernandez-Duque, E. and C. Valeggia. 1994. Meta-Analysis: A valuable tool in conservation research. *Conservation Biology* 8:555–561.

Ferson, S. and M.A. Burgman. 1995. Correlations, dependency bounds and extinction risk. *Biological Conservation* 73:101–105.

Ferson, S., L. Ginzburg, and A. Silvers. 1989. Extreme event risk analysis for age-structured populations. *Ecological Modeling* 47:175–87.

Ferson, S. and S. Uryasev. 1996. Sensitivity of extinction risk. *Supplement to the Bulletin of the Ecological Society of America* 77:140.

Gilpin, M.E. 1990. Extinction of finite metapopulations in correlated environments. In *Living in a patchy environment*, eds. B. Shorrocks and I.R. Swingland, 177–186. Oxford: Oxford Science Publications.

Gilpin, M.E. and M.E. Soulé. 1986. Minimum viable populations: The processes of population extinction. In *Conservation Biology: Science of scarcity and diversity*, ed. M.E. Soulé, 13–34. Sunderland, MA: Sinauer.

Ginsberg, J.R. and E.J. Milner-Gulland. 1994. Sex-biased harvesting and population dynamics in ungulates: Implications for conservation and sustainable use. *Conservation Biology* 8:157–166.*

Ginzburg, L.R., B. Slobodkin, K. Johnson, and A.G. Bindman. 1982. Quasiextinction probabilities as a measure of impact on population growth. *Risk Analysis* 2:171–181.

Ginzburg, L.R., S. Ferson, and H.R. Akcakaya. 1990. Reconstructability of density dependence and the conservative assessment of extinction risks. *Conservation Biology* 4:63–70.

Goldingay, R. and H. Possingham. 1995. Area requirements for viable populations of the Australian gliding marsupial *Petaurus australis. Biological Conservation* 73:161–167.*

Goldwasser, L., M.J. Groom, and P. Kareiva. 1993. The effects of model structure and annual variability on population viability analyses of the Northern Spotted Owl (*Strix occidentalis caurina*). Technical Report to the California Forestry Association, Sacramento, CA.

Goodman, D. 1987a. Consideration of stochastic demography in the design and management of biological reserves. *Natural Resources Modeling* 1:205–234.

Goodman, D. 1987b. The demography of chance extinction. In *Viable populations for conservation*, ed. M.E. Soulé, 59–68. Cambridge: Cambridge University Press.

Green, R.E. and G.J.M. Hirons. 1991. The relevance of population studies to the conservation of threatened birds. In *Bird Population Studies: Relevance to conservation and management*, eds. C.M. Perrins, J.-D. Lebreton, and G.J.M. Hirons, 594–633. New York: Oxford University Press.

Groom, M.J., L. Goldwasser, and P. Kareiva. 1993. Evaluating the influence of model structure on the results of population viability simulations. Technical Report to the California Forestry Association, Sacramento, CA.

Groom, M.J. 1996. Reproductive failure in an annual plant when critical isolation thresholds are exceeded. In review. *Nature.*

Haig, S.M., J.R. Belthoff, and D.H. Allen. 1993. Population viability analysis for a small population of Red-Cockaded Woodpeckers and an evaluation of enhancement strategies. *Conservation Biology* 7:289–301.*

Hamilton, S. and H. Moller. 1995. Can PVA models using computer packages offer useful conservation advice? Sooty shearwaters *Puffinus griseus* in New Zealand as a case study. *Biological Conservation* 73:107–117.*

Hanski, I. and C.D. Thomas. 1994. Metapopulation dynamics and conservation: a spatially explicit model applied to butterflies. *Biological Conservation* 68:167–180.

Harper, J. 1977. *Population biology of plants.* New York: Academic Press.

Harris, R.B., L.A. Maguire, and M.L. Shaffer. 1987. Sample sizes for minimum viable population estimation. *Conservation Biology* 1:72–75.

Harrison, S. and J. Quinn. 1989. Correlated environments and the persistence of metapopulations. *Oikos* 56:293–298.

Heppell, S.S., L.B. Crowder, and D.T. Crouse. 1996. Models to evaluate headstarting as a management tool for long-lived turtles. *Ecological Applications* 6:556–565.*

Heppell, S.S., J.R. Walters, and L.B. Crowder. 1994. Evaluating management alternatives for red-cockaded woodpeckers: A modeling approach. *Journal of Wildlife Management* 58:479–487.*

Heyde, C.C. and J.E. Cohen. 1985. Confidence intervals for demographic projections based on products of random matrices. *Theoretical Population Biology* 27:120–53.

Kenney, J.S., J.L.D. Smith, A.M. Starfield, and C.W. McDougal. 1995. The long-term effects of tiger poaching on population viability. *Conservation Biology* 9:1127–1133.*

Kinnaird, M.F. and T.G. O'Brien. 1991. Viable populations for an endangered forest primate, the Tana River crested mangabey (*Cercocebus galeritus galeritus*). *Conservation Biology* 5:203–213.*

Lamberson, R.H., K. McKelvey, B.R. Noon, and C. Voss. 1992. A dynamic analysis of northern spotted owl viability in a fragmented forest landscape. *Conservation Biology* 6:505–512.*

Lamberson, R.H., B.R. Noon, C. Voss, and K.S. McKelvey. 1994. Reserve design for territorial species: the effects of patch size and spacing on the viability of the northern spotted owl. *Conservation Biology* 8:185–195.*

Lancia, R.A., K.H. Pollock, J.W. Bishir, and M.C. Conner. 1988. A white-tailed deer harvesting strategy. *Journal of Wildlife Management* 54:589–595.*

Lande, R. 1987. Extinction thresholds in demographic models of territorial populations. *American Naturalist* 130:624–635.

Law, R., and M.T. Edley. 1990. Transient dynamics of populations with age- and size-dependent vital rates. *Ecology* 71:1863–70.

Leigh, E.G., Jr. 1971. The average lifetime of a population in a varying environment. *Journal of Theoretical Biology* 90:213–239.

Levins, R. 1970. Extinction. In *Lectures on mathematics in life sciences*, ed. M. Gerstenhaber, 77–107. Providence: American Mathematical Society.

Lindenmayer, D.B., R.B. Cunningham, M.T. Tanton, and A.P. Smith. 1990. The conservation of arboreal marsupials in the montane ash forests of the Central Highlands of Victoria, southeast Australia. II. The loss of trees with hollows and its implications for the conservation of Leadbeater's Possum *Gymnobelideus leadbeateri* McCoy (Marsupialia: Petauridae). *Biological Conservation* 54:133–145.*

Lindenmayer, D.B. and R.C. Lacy. 1995a. A simulation study of the impacts of population subdivision on the mountain brushtail possum *Trichosurus caninus* in Southeastern Australia. I. Demographic stability and populations persistence. *Biological Conservation* 73:119–129.*

Lindenmayer, D.B. and R.C. Lacy. 1995b. Metapopulation viability of Leadbeater's possum, *Gymnobelideus leadbeateri*, in fragmented old-growth forests. *Ecological Applications* 5:164–182.*

Lindenmayer, D.B. and R.C. Lacy. 1995c. Metapopulation viability of arboreal marsupials in fragmented old-growth forests: comparisons among species. *Ecological Applications* 5:183–199.*

Lindenmayer, D.B. and H.P. Possingham. 1995. The conservation of arboreal marsupials in the montane ash forests of the central highlands of Victoria, south-eastern Australia. VII. Modeling the persistence of Leadbeater's possum in response to modified timber harvesting practices. *Biological Conservation* 73:239–257.*

Lindenmayer, D.B. and H.P. Possingham. 1996. Ranking conservation and timber management options for Leadbeater's possum in Southeast Australia using Population Viability Analysis. *Conservation Biology* 10:235–251.*

Liu, J., J.B. Dunning, and H.R. Pulliam. 1995. Potential effects of a forest management plan on Bachman's sparrows (*Aimophila aestivalis*): linking a spatially explicit model with GIS. *Conservation Biology* 9:62–75.*

Ludwig, D. 1996. Uncertainty and the assessment of extinction probabilities. *Ecological Applications*, in press.

Ludwig, D. and C.J. Walters. 1985. Are age-structured models appropriate for catch-effort data? *Canadian Journal of Fisheries and Aquatic Sciences* 42:1066–1072.

Maguire, L.A. 1991. Risk analysis for conservation biologists. *Conservation Biology* 5:123–125.

Maguire, L.A., U.S. Seal, and P.F. Brussard. 1987. Managing critically endangered species: the Sumatran rhino as a case study. In *Viable Populations for Conservation*, ed. M.E. Soulé, 141–158. Cambridge: Cambridge University Press.

Maguire, L.A., G.F. Wilhere, and Q. Dong. 1995. Population viability analysis for red-cockaded woodpeckers in the Georgia piedmont. *Journal of Wildlife Management* 59:533–542.*

Mangel, M., L.M. Talbot, G.K. Meffe, M.T. Agardy, D.L. Alverson, J. Barlow, D.B. Botkin, G. Budowski, T. Clark, J. Cooke, R.H. Crozier, P.K. Dayton, D.L. Elder, C.W. Fowler, S. Funtowicz, J. Giske, R.J. Hofman, S.J. Holt, S.R. Kellert, L.A. Kimball, D. Ludwig, K. Manusson, B.S. Malayang, III, C. Mann, E.A. Norse, S.P. Northridge, W.F. Perrin, C. Perrings, R.M. Peterman, M.P. Sissenwine, T.D. Smith, A. Starfield, R.J. Taylor, M.F. Tillman, C. Toft, J.R. Twiss, Jr., J. Wilen, and T.P. Young. 1996. Principles for the conservation of wild living resources. *Ecological Applications* 6:338–362.

Mangel, M. and C. Tier. 1993. A simple direct method for finding persistence times of populations and application to conservation problems. *Proceedings of the National Academy of Sciences* 90:1083–1086.

Manly, B.F.J. 1990. *Stage-structured populations: Sampling, analysis and simulation*. Chapman and Hall, London.

Marschall, E.A., and L.B. Crowder. 1996. Assessing population responses to multiple anthropogenic effects: A case study with brook trout. *Ecological Applications* 6:152–167.*

Mattson, D.J., S. Herrero, K.G. Wright, and C.M. Pease. 1996. Science and the management of Rocky Mountain grizzly bears. *Conservation Biology* 10:1013–1025.*

McCarthy, M.A., D.C. Franklin, and M.A. Burgman. 1994. The importance of demographic uncertainty: An example from the helmeted honeyeater *Lichenostomus melanops cassidix*. *Biological Conservation* 67:135–142.*

McCarthy, M.A., M.A. Burgman, and S. Ferson. 1995. Sensitivity analysis for models of population viability. *Biological Conservation* 73:93–100.

McKelvey, K., B.R. Noon, and R.H. Lamberson. 1992. Conservation planning for species occupying fragmented landscapes: The case of the Northern Spotted Owl. In *Biotic interactions and global change.*, eds. P. Kareiva, J.G. Kingsolver, and R.B. Huey, 424–450. Sunderland, MA: Sinauer.

Menges, E.S. 1990. Population viability analysis for an endangered plant. *Conservation Biology* 4:52–62.*

Menges, E.S. 1992. Stochastic modeling of extinction in plant populations. In *Conservation Biology: The theory and practice of nature conservation, preservation and management.*, ed. P.L. Fiedler and S.K. Jain, 253–276. New York: Chapman & Hall.

Moehlman, P.D., G. Amato, and V. Runyoro. 1996. Genetic and demographic threats to the Black Rhino population in the Ngorongoro Crater. *Conservation Biology* 10:1107–1114.*

Moloney, K.A. 1986. A generalized algorithm for determining category size. *Oecologia* 69:176–180.

Montopoli, G.J. and D.A. Anderson. 1991. A logistic model for the cumulative effects of human intervention on bald eagle habitat. *Journal of Wildlife Management* 55:290–293.*

Myers, R.A. and N.G. Cadigan. 1995. Was an increase in natural mortality responsible for the collapse of northern cod? *Canadian Journal of Fisheries and Aquatic Sciences* 52:1274–1285.

Nantel, P., D. Gagnon, and A. Nault. 1996. Population viability analysis of American Ginseng and Wild Leek harvested in stochastic environments. *Conservation Biology* 10:608–621.*

National Research Council. 1996. *Science and the Endangered Species Act*. Washington, D.C.: US Government Printing Office.

Noon, B.R. and C.M. Biles. 1990. Mathematical demography of spotted owls in the Pacific Northwest. *Journal of Wildlife Management* 54:18–26.*

Nordheim, E.V., D.B. Hogg, and S. Chen. 1989. Leslie matrices for insect populations with overlapping generations. In *Estimation and analysis of insect populations. Lecture Notes in Statistics 55*, eds. L.L. McDonald, B.F.J. Manly, J.A. Lockwood, and J.A. Logan, 289–98. Berlin: Springer Verlag.

Olmstead, I. and E.R. Alvarez-Buylla. 1995. Sustainable harvesting of tropical trees: Demography and matrix models of two palm species in Mexico. *Ecological Applications* 5:484–500.*

Pascual, M.A. and M.D. Adkison. 1994. The decline of the Steller sea lion in the northeast Pacific: Demography, harvest or environment. *Ecological Applications* 4:393–403.*

Pascual, M.A. and R. Hilborn. 1995. Conservation of harvested populations in fluctuating environments: the case of the Serengeti wildebeest. *Journal of Applied Ecology* 32:468- 480.

Pascual, M.A., P.Kareiva, and R. Hilborn. In press. The influence of model structure on conclusions about the viability and harvesting of Serengeti wildebeest. *Conservation Biology*.

Pfister, C.A. and A. Bradbury. 1996. Harvesting red sea urchins: Recent efforts and future predictions. *Ecological Applications* 6:298–310.*

Possingham, H.P., D.B. Lindenmayer, T.W. Norton, and I. Davies. 1994. Metapopulation analysis of the greater glider *Petauroides volans* in a wood production area. *Biological Conservation* 70:227–236.*

Powell, R.A., J.W. Zimmerman, D.E. Seaman, and J.F. Gilliam. 1996. Demographic analysis of a hunted black bear population with access to a refuge. *Conservation Biology* 10:224–234.*

Price, M.V. and P.A. Kelly. 1994. An age-structured demographic model for the endangered Stephens' kangaroo rat. *Conservation Biology* 8:810–821.*

Pulliam, H.R., J.B. Dunning, Jr., and J. Liu. 1992. Population dynamics in complex landscapes: A case study. *Ecological Applications* 2:165–177.*

Quinn, J.F. and A. Hastings. 1987. Extinction in sub-divided habitat. *Conservation Biology* 1:198–209.

Raiffa, H. 1970. *Decision analysis*. Reading, MA: Addison-Wesley.

Ralls, K. and A.M. Starfield. 1995. Choosing a management strategy: Two structured decision-making methods for evaluating the predictions of stochastic simulation models. *Conservation Biology* 9:175–181.

Reed, J.M., P.D. Doerr, and J.R. Walters. 1988. Minimum viable population size of the red-cockaded woodpecker. *Journal of Wildlife Management* 52:385–391.*

Richter-Dyn, N. and N.S. Goel. 1972. On the extinction of a colonizing species. *Theoretical Population Biology* 3:406–433.

Ruckelshaus, M., C. Hartway, and P. Kareiva. In press. Assessing the data requirements of spatially explicit dispersal models. *Conservation Biology*.

Ruggiero, L.F., G.D. Hayward, and J.R. Squires. 1994. Viability analysis in biological evaluations: Concepts of population viability analysis, biological population and ecological scale. *Conservation Biology* 8:364–372.

Ryan, M.R., B.G. Root, and P.M. Mayer. 1993. Status of piping plovers in the Great Plains of North America: A demographic simulation model. *Conservation Biology* 7:581–585.*

Saltz, D. and D. Rubenstein. 1995. Population dynamics of a reintroduced asiatic wild ass (*Equus hemionus*) herd. *Ecological Applications* 5:327–335.*

Schemske, D.W., B.C. Husband, M.H. Ruckelshaus, C. Goodwillie, I.M. Parker, and J.G. Bishop. 1994. Evaluating approaches to the conservation of rare and endangered plants. *Ecology* 75:584–606.

Schneider, R.R. and P. Yodzis. 1994. Extinction dynamics in the American marten (*Martes americana*). *Conservation Biology* 7:1058–1068.*

Shaffer, M.L. 1981. Minimum population size for species conservation. *BioScience* 31:131–34.

Shrader-Frechette, K.S. and E.D. McCoy. 1993. *Method in ecology: Strategies for conservation.* New York: Cambridge University Press.

Slade, N. and H. Levenson. 1984. The effect of skewed distributions of vital statistics on growth of age-structured populations. *Theoretical Population Biology* 26: 361–366.

Soulé, M.E., ed. 1987. *Viable populations for conservation.* New York: Cambridge University Press.

Southgate, R. and H.P. Possingham. 1995. Modeling the reintroduction of the greater bilby *Macrotis lagotis* using the metapopulations model analysis of the likelihood of extinction. *Biological Conservation* 73:151–160.*

Stacey, P.B. and M. Taper. 1992. Environmental variation and the persistence of small populations. *Ecological Applications* 2:18–29.*

Starfield, A.M. and A.M. Herr. 1991. A response to Maguire. *Conservation Biology* 5:435.

Starfield, A.M., J.D. Roth, and K. Ralls. 1995. "Mobbing" in Hawaiian monk seals (*Monachus schauinslani*): The value of simulation modeling in the absence of apparently crucial data. *Conservation Biology* 9:166–174.*

Taylor, B.L. and T. Gerrodette. 1993. The uses of statistical power in conservation biology: The Vaquita and Northern Spotted Owl. *Conservation Biology* 7:489–500.

Taylor, M.K., D.P. DeMaster, F.L. Bunnell, and R.E. Schweinsburg. 1987. Modeling the sustainable harvest of female polar bears. *Journal of Wildlife Management* 51:811–820.*

Tuljapurkar, S.D. 1990. *Population dynamics in variable environments. Lecture notes in biomathematics no. 85.* New York: Springer-Verlag.

Vandermeer, J. 1979. Choosing category size in a stage projection matrix. *Oecologia* 32:199–225.

Walters, C.J. 1986. *Adaptive management of renewable resources.* New York: Macmillan.

Walters, C.J. and R. Hilborn. 1976. Adaptive control of fishing. *Journal of the Fisheries Research Board of Canada* 33:145–159.

Wennergren, U., M. Ruckelshaus, and P. Kareiva. 1995. The promise and limitations of spatial models in conservation biology. *Oikos* 74:349–356.

Williams, B.K, F.A. Johnson, and K. Wilkins. 1996. Uncertainty and the adaptive management of waterfowl harvests. *Journal of Wildlife Management* 60:223–232.

Wood, P.B. and M.W. Collopy. 1993. Effects of egg removal on Bald Eagle productivity in Northern Florida. *Journal of Wildlife Management* 57:1–9.*

Wootton, J.T. and D.A. Bell. 1992. A metapopulation model of the Peregrine Falcon in California: viability and management strategies. *Ecological Applications* 2:307–321.*

CHAPTER

2

Rare Plant Demography: Lessons from the Mariposa Lilies (*Calochortus*: Liliaceae)

PEGGY L. FIEDLER
B.E. KNAPP
and NANCY FREDRICKS

Rare plants have been intensively studied and described in the past two decades (e.g., Falk and Holsinger 1991; Given 1994; Frankel, Brown, and Burdon 1995; Falk, Millar, and Olwell 1996), in part due to the recovery mandates under the U.S. Endangered Species Act of 1973. The great majority of past work on rare plants has been genetic in nature (Fiedler unpublished), with demographic studies only recently becoming central to rare plant conservation. Indeed, an understanding of population dynamics along with general life history characteristics is accepted as fundamental to rare plant protection efforts, including restoration (Schemske *et al.* 1994; Falk, Millar, and Olwell 1996). For example, Pavlik (1994, 1995, 1996; chapter 5) has argued repeatedly that rare plant reintroduction efforts must be demographically-based, so that consequent variation in births, deaths, and fecundity can be understood in the context of individual life histories.

The conventional methodology for summarizing population data is the transition matrix (Leslie 1945), although these data can be summarized in other informative ways, such as the life table or life history graph. Use of transition matrices in rare plant protection is commonplace, more than a decade after Bierzychudek's (1982) pioneering work on the herbaceous perennial jack-in-the-pulpit (*Arisaema triphyllum*). Recent plant projection matrix studies have become increasingly insightful with respect to conservation issues. For example, Horvitz and Schemske (1995) and Oostermeijer *et al.* (1996) have revealed varying spatiotemporal aspects of multiple populations of a single species; Cipollini, Wallace-Senft, and Whigham (1994) elegantly integrated patch dynamics, seed dispersal and sex ratios; and Silvertown and his colleagues (Silvertown, Franco, and McConway 1992; Silvertown *et al.* 1993; Silvertown, Franco, and Menges 1996) have explored various matrix statistics to uncover broad patterns of life history characteristics. However, data collection for transition matrix modeling remains a labor-intensive, very long-term, and "low-tech" commitment; such population data are generally sparse, particularly given the large number of plant taxa that need protection. Thus, while transition matrices are now the *modus operandi* for summarizing population data, there are still fewer than 100 published studies for all vascular plants (see Silvertown *et al.* 1993; Silvertown, Franco, and Menges 1996); many fewer exist for rare species.

A transition matrix combines age-specific, size-specific, or both (e.g., Law 1983; Young 1984, 1985) rates of reproduction, survival, and growth into a series of linear equations describing individual probabilities of transition to other ages or sizes, as well as probabilities of death and reproduction during a defined projection interval or time step (Leslie 1945; Caswell 1989). While age determines the vital rates for many animals, the size of plants is usually the primary determinant of their individual mortality, reproductive effort, and growth (Harper 1977; Werner and Caswell 1977; Hanzawa and Kalisz 1993). Importantly for plants, environmental factors may produce different responses in plants of different sizes, but individuals within a specific size range are likely to exhibit similar vital rates if they experience the same conditions. Vital rates for any population can therefore be expressed as rates for stages or size categories of plants (Lefkovitch 1965).

The most commonly calculated statistic derived from the transition matrix is the dominant latent root of the matrix, λ, which is the finite rate of population increase after a stable size distribution has been achieved. Thus lambda describes an increment of change for each size class during each transition. A $\lambda > 1.0$ is interpreted as an increasing population; $\lambda < 1.0$ as a decreasing population; and $\lambda = 1.0$ as a stable population neither decreasing or increasing in numbers. Some biologists have interpreted λ more broadly as a general measure of fitness for a particular set of population characteristics from which λ is derived (e.g., Silvertown *et al.* 1993).

Several other properties of the transition matrix provide additional measures of demographic behavior useful in conservation efforts. For example, when a transition matrix is multiplied by a vector describing the proportion of plants in a population in each stage (i.e., the stage distribution), the result is another vector describing the expected population size and structure at the start of the next time interval. Continued multiplication of the matrix by the stage distribution vectors will eventually result in a geometrically increasing (or decreasing) population with constant proportions in each stage. Thus, the right eigenvector associated with λ can be interpreted as the stable stage distribution, ω, that describes the proportion of individuals in each size-class of a stable population. Any observed size distribution can thus be compared with that predicted from the transition model. The sum of positive differences between the two distributions is known as the Index of Dissimilarity (Keyfitz 1977). In employing the transition matrix model, it is important to remember that the "stability" of the stable age distribution does not necessarily imply that a population is maintaining itself, only that the rate of change in a population is constant. It is possible to have a stable distribution and associated eigenvalue for a declining population.

The left eigenvector associated with λ, υ, provides stage-specific reproductive values (see Caswell 1989). These are particularly important in the interpretation of the transition matrix model because they determine in part the sensitivity of the population multiplication rate to changes in individual transitions. Sensitivity values are used to the estimate relative importance to λ of vital rates, selection pressures on the population, and effects of errors in estimates of transition frequencies. Thus, the sensitivity of the population growth rate (λ) to any transition probability in the matrix is defined as the change in λ that would result from a change in the transition probability (see Caswell 1989). Sensitivities are a function of the reproductive value of the stages and stable stage distribution.

De Kroon *et al.* (1986) have pointed out two difficulties with interpretation of the sensitivities. Fecundity values usually are measured on a different scale than probabilities of transition, so a change in the matrix entry representing fecundity will have a smaller effect on the dominant eigenvalue than a comparable change in any transition probability. In addition, transitions that are extremely rare (or impossible) can have high sensitivity values. Sensitivities also depend upon population growth rates and therefore are not directly comparable between matrices with different growth rates. Elasticity values were developed to resolve this difficulty, and they can be thought of as proportional sensitivities. Thus, elasticity is a measure of the sensitivity of λ to changes in each matrix element, standardized so that elements representing survival probabilities range from 0 to 1 (de Kroon *et al.* 1986). Elasticities must sum to 1.0, providing a statistic of the relative importance of transitions representing growth, survival, and reproduction.

An important distinction should be made between sensitivities, which indicate the degree to which a change in a particular transition will affect the eigenvalue, and elasticities, which are related but are interpreted primarily as the contribution of any particular matrix entry to the eigenvalue. Sensitivities could be high for a transition that was measured as very improbable (or as not occurring), and that would therefore have a low elasticity value. One must decide in such a case whether the measurement of the transition probability was accurate in the long term and the transition unimportant, or if the transition is, in fact, limiting to population growth, even though it did not contribute substantially to the eigenvalue as measured.

While it is useful to consider constant rates of population change as summarized by λ for comparative purposes and to generate hypotheses, the underlying simple model should not be used to forecast populations into the future. The main reason for this is that demographic rates and thus projection matrices do not stay constant through time. For example, density dependence of vital rates and the greater impact of environmental stochasticity in small populations would change the rate of population decline as the population size decreased (see Menges 1992; chapter 3). The same can be said for most real populations that are expanding. Rates of mortality and reproduction may fluctuate as the populations of predators, pollinators, and competitors change over time and space. Different probabilities of transition for different years also may result from different abiotic or historical events, such as extreme weather or forest fires. Data collected in different years are likely to produce different stable distributions and multiplication rates for the same population. Geometric increase and the stable stage distribution are asymptotic results that are probably not found in many real populations. This may be particularly true of rare species because of their small population sizes and vulnerability to environmental stochasticity (Vandermeer 1982; Menges 1992; chapter 3).

For these reasons, among others, transition matrices are not necessarily predictors of the future of a rare plant population. It is perhaps more meaningful and more useful in the context of conservation to ask why the matrix produces the results that it does. The response of populations to conditions and events that occurred during any particular study depends upon the life history characteristics of that species, which will limit its response to, or tolerance of, varying conditions. One way to use transition matrix models is to compare the effects of different conditions on the model, whether through comparison of different treatments or different populations, through following a single population over a number of years, or through simulation by manipulating the matrix itself. In this chapter, we examine 28 matrices and their mathematical properties calculated for 11 populations of eight species of *Calochortus* studied over nearly two decades to learn what the modeling technique tells us about its utility in describing the demography of rare plants.

BACKGROUND

Calochortus comprises approximately 70 taxa, distributed from British Columbia (Canada) to Guatemala, and from California to Nebraska (Fiedler and Zebell 1997). It is a genus characterized primarily by xerophytic bulbous perennials. Most taxa inhabit dry, rocky habitats; of the 70 taxa, only 11 species are described as preferring wet sites. As currently recognized, the genus is divided into three sections, with unique combinations of life history characteristics, geographic distributions, centers of origins, and evolutionary histories (Ownbey 1940).

The genus has proven to be an exemplary group to explore plant life histories using projection matrices. The genus is well suited to demographic studies because individual plants are easy to identify and follow over time, the basal leaf provides a reliable and convenient measure of plant size, dispersal distances are short, little or no bulb dormancy has been seen for the species studied, nor have dormant seed banks been documented. To date, eight species of mariposa lilies have been studied using transition matrices, and of these all but one, *Calochortus albus*, are rare. Six of the seven rare species (*C. coxii*, *C. howellii*, *C. obispoensis*, *C. pulchellus*, *C. tiburonensis*, *C. umpquaensis*) are endemic to serpentinite substrates, and therefore are highly restricted edaphic endemics. In addition, the geographic distribution of most of these rare species is extremely limited within the serpentinite environments (Figure 2.1). For example, known individual populations of *C. coxii* span no more 16 km (Fredricks 1992); *C. tiburonensis* is limited to less than 50 ha of boulder-strewn bunchgrassland within a 140 ha nature preserve.

The *Calochortus* species studied also are closely related. All but one of the eight species belong to the same section of the genus (*Calochortus*), three belong to subsection *Nitidi* within this section, two belong to subsection *Eleganti*, and two to subsection *Pulchelli* (see Table 2.1). Thus the demographic comparisons made in this chapter are among closely related taxa largely within a single clade. These population data are unique in the study of a plant genus, particularly one with a relatively large number of rare, localized, and endemic taxa.

SPECIES DESCRIPTIONS

Calochortus albus (Figure 2.2a) is a widespread species common to annual grasslands, woodlands, chaparral, and their ecotones throughout the central and southern Coast Ranges and the central Sierra Nevada foothills of California. Population sizes are often extremely large, with tens of thousands of individuals found within a few square meters (Fiedler personal observation). *Calochortus pulchellus* (Figure 2.2b) is a perennial herb of open gray pine (*Pinus sabiniana*)/live oak (*Quercus agrifolia*) woodlands on the northern slopes of Mt. Diablo, the northern terminus of the Diablo Range of central California. Populations are disjunct and often large within this small geographic area.

Calochortus umpquaensis (Figure 2.2c) is restricted to several different habitats within an adjacent 80 km span of serpentinite soils, extending in a southwesterly direction from Ace Williams Mountain to Sexton Mountain in southwestern Oregon (Figure 2.1). This mariposa lily is found within a rather broad continuum of habitats, from closed canopy coniferous forests (e.g., *Pinus jeffreyi*, *Pseudotsuga menziesii*, *Calocedrus decurrens*, and *Arbutus menziesii*) to rather open, species-rich, grass-forb meadows (Fredricks 1992).

Figure 2.1
Generalized geographic distribution of the eight Calochortus *species studied.*

Populations of *C. coxii* (Figure 2.2d) are found within a narrow 50 km long band of serpentinite running northeast to southwest in southwestern Oregon (Fredricks 1992). *Calochortus coxii* occurs in ecotonal as well as grassland habitats similar to those occupied by *C. umpquaensis*. The third southwestern Oregon endemic, *Calochortus howellii* (Figure 2.2e), is restricted to the Illinois drainage basin within the Siskiyou Mountains. This serpentine endemic can be found in several community types, with the higher elevation populations found in mixed conifer/hardwood forests dominated by *Pinus jeffreyi*, *P. lambertiana*, *Pseudotsuga menziesii*, *Lithocarpus densiflora*, *Quercus vaccinifolia*, and *Vaccinium occidentale* (Fredricks 1992). *Calochortus howellii* also can be found in shrub-dominated sites, but is most common in lower elevation grass/forb savannahs with an open understory.

Calochortus obispoensis (Figure 2.2f) is an unusual species whose taxonomic placement within section *Cyclobothra* is now in question (Patterson, Givnish, and Sytsma 1996). This species consists of several isolated populations in the Santa Lucia Mountains

Table 2.1
Selected characteristics of the eight species of Calochortus[1] *studied.*

Species[2]	Taxonomy[3] (Section/Subsection, *sensu* Ownbey 1940)	Geographic Range[4]	Habitat/Elevation
C. albus	*Calochortus//Pulchelli*	Coast Range and Sierra Nevada foothills	annual grasslands, woodlands, chaparral, ecotones/0—2000 m
C. coxii	*Calochortus//Nitidi/Eleganti*(?)	Southwest Oregon (Douglas County)	serpentinite mixed conifer forests, grasslands, ecotones/244–847m
C. howellii	*Calochortus//Nitidi*	Illinois River drainage, Southwest Oregon (Douglas & Josephine cos.)	serpentinite open mixed conifer forests, chaparral, grass-forb savannah/ 390–730 m
C. obispoensis	*Cyclobothra//Weediani*	Santa Lucia Mountains (San Luis Obispo County)	serpentinite grassland and chaparral/100–500 m
C. pulchellus	*Calochortus//Pulchelli*	Mt. Diablo (Contra Costa County)	wooded slopes and chaparral/200–800 m
C. tiburonensis	*Calochortus//Nitidi*?	Ring Mountain (Marin County)	serpentinite grassland/ 50 –150 m
C. umpquaensis	*Calochortus//Nitidi*	Umpqua River drainage, Southwest Oregon (Douglas County)	open forests and forest-meadow ecotones on serpentinite/244–830 m
C. westonii	*Calochortus//Eleganti*	Southern Sierra Nevada (Kern & Tulare cos.)	meadows, open woodlands/ 1500–2000 m

of central coastal California. *Calochortus obispoensis* is restricted to open, dry serpentinite slopes dominated by native bunchgrasses and other perennial herbs. *Calochortus tiburonensis* (Figure 2.2g) is another enigmatic member of the genus, having characteristics that align it with two of the three generic sections. It is restricted to the rocky and boulder-strewn bunchgrasslands along the upper slopes of Ring Mountain, a serpentinite-capped mountain within the northern San Francisco Bay region.

Calochortus westonii (Figure 2.2h) is restricted to the Greenhorn Mountain region in the southern Sierra Nevada. It is found primarily in the margins of dry to wet meadows in mixed conifer (*Pinus ponderosa/Abies magnifica/Pseudotsuga menziesii*)—black oak (*Quercus kelloggii*) forests, at elevations ranging from 1,300 to 2,400 m. *Calochortus westonii* also occurs, however, in post-fire montane chaparral and forested ridgetops and first order draws. Populations of *C. westonii*, like those of many rare species of *Calochortus*, are disjunct within the range. A comparison of the *Calochortus* species discussed in this chapter is found in Table 2.1.

TRANSITION MATRIX MODELING IN *CALOCHORTUS*

Methods of the transition matrix calculations for *Calochortus albus*, *C. obispoensis*, *C. pulchellus*, and *C. tiburonensis* can be found in Fiedler (1987); for *C. coxii*, *C. howellii*, and *C. umpquaensis*, methods can be found in Fredricks (1992); and for *C. westonii*, in Knapp (1996). However, transition matrices for *C. albus*, *C. obispoensis*, *C. pulchellus*, and *C. tiburonensis* were modified slightly from Fiedler (1987) and Fredricks (1992) to

Figure 2.2
Flowers of the eight Calochortus *species studied: (a)* C. albus; *(b)* C. pulchellus; *(c)* C. coxii; *(d)* C. howellii; *(e)*; C. umpquaensis; *(f)* C. obispoensis; *(g)* C. tiburonensis; *(h)* C. westonii. *Photos (c), (d), and (e) taken by Jeanette Sainz.*

facilitate comparisons. For complete consistency, all lambda values were recalculated using Maple V, Release 3 (Waterloo Maple 1995). Reproductive values for *Calochortus albus*, *C. obispoensis*, *C. pulchellus*, *C. tiburonensis*, and *C. westonii* were computed using RAMAS/stage (Ferson 1991). Sensitivities and elasticities for *C. albus*, *C. coxii*, *C. obispoensis*, *C. pulchellus*, *C. tiburonensis*, and *C. umpquaensis* also were computed using RAMAS/stage (Ferson 1991). These data are compared with those for *C. howellii* (Fredricks 1992) and *C. westonii* (Knapp 1996). All subsequent statistical analyses were calculated using SYSTAT.

Transition Matrices and Population Growth Rates

For the comparative analyses reported here, entries in Fiedler's and Fredrick's original matrices were modified slightly, by adding or subtracting 0.01 to any observed transitions that were recorded as happening with certainty (i.e., with a probability 1.0 or 0.0). We made this correction so that all life stages could be connected and low sample sizes would not preclude events we knew were biologically possible, albeit rare. In addition,

Figure 2.2 ***(continued)***

reproduction in Fiedler's (1987) matrices was recalculated to conform with Knapp (1996) and Fredricks (1992) using an "establishment rate" factor based on loss of seeds and seedlings between dispersal from the capsule and seedling establishment.

An important question in a transition matrix analysis is: "What is the appropriate temporal scale on which population dynamics should be measured?" For example, the six transitions for *C. howellii* and the multiple population comparisons indicate that considerable spatial and temporal variation in vital rates for a given species is commonplace (see also Horvitz and Schemske 1995; chapter 3). Not all transition rates will vary to the same degree, however, or with the same degree of predictability, or with the same importance to the population model. In most of these *Calochortus* matrices, reproduction is both more limiting to population growth, more variable in time and space, and perhaps less predictable than is mortality. Therefore, factors that affect reproductive success may require investigation on a different spatio-temporal scale than those affecting mortality in order to create a generalizable model.

Another consideration is the accuracy of matrices in describing plant growth rates. A stage that consists of a single cohort might make the transition all at once, creating the false impression of rapid growth (or of very low growth for transition years that fall between cohorts), if only two transitions are measured, as is typically the case in these three-year studies of *Calochortus*. The model's requisite imposition of categories (discrete classes) on a continuous variable (size/age) should be carely weighed in the biological intepretation of the model's results (see Moloney 1986). This point is well illustrated by Fredricks (1992), who observed that in the course of a nine-year study, no seedling matured to reproductive status, even though the corresponding projection matrix predicted that a few seedlings could mature to adulthood in as few as six years. This discrepancy does not affect the eigenvalues in this case, largely because of the relative insensitivity of the *C. howellii* matrices to transitions affecting seedlings. Category size determination could be a source of error, however, in projection matrix models for species with higher reproductive rates (e.g., *C. albus*, *C. westonii*, and *C. umpquaensis*).

Twenty-eight matrices for eight species of *Calochortus* suggest that the values of each stage transition vary by species, habitat, and year (Table 2.2). *Calochortus albus*, the common species, exhibits high rates of transition to reproductive stages and high reproductive rates. Four rare species, *Calochortus pulchellus*, *C. obispoensis*, *C. tiburonensis*, and *C howellii*, exhibit low rates of transition, low reproductive rates, and low mortality. The remaining four rare *Calochortus* species, *C. coxii*, *C. howellii*, *C. umpquaensis*, and *C. westonii*, all exhibit high transition rates overall, but particularly from stage 1 (seedling) to stage 2 (juvenile). *Calochortus coxii* shows very low reproduction but high survival rates, while conversely, *C. westonii* shows very high reproduction coupled with low survival in all stages. For *C. umpquaensis* and *C. westoni*, reproduction and mortality vary with habitat. *Calochortus coxii* demonstrates variation in transitions between populations. Finally, while there is considerable variation between years for most species, especially in reproduction, the matrices for *C. howellii* are remarkably consistent over time.

Population growth rates vary among the rare species, from 0.54 for *C. westonii* to 1.05 for *C. umpquaensis* (year 1 ecotone) and *C. howellii* (year 2). Inspection of the matrices suggests that the larger lambda value for a given species is usually due to increased reproduction. *Calochortus* researchers have documented that the fate of the majority of the flowers and fruits produced for all species, including the common *C. albus*, is to be either completely grazed, or to undergo ovule abortion due primarily to sawfly larvae damage (Fiedler 1987; Fredricks 1992; Knapp 1996). This is particularly true for those *Calochortus* species that prefer open, rocky substrates (e.g., *C. obispoensis*, *C. persistens*, *C. greenei* [Knapp unpublished]). Thus, when herbivore pressure on flowers and fruits is relatively low, reproductive values rise accordingly. Once seeds germinate and seedlings establish, however, the seedling size class typically demonstrates moderate to high rates of survival (e.g., *C. tiburonensis* year 2, *C. howellii*, *C. umpquaensis*) and low to moderate rates of transition (e.g., *C. coxii* Smith/year 2, *C. howellii* years 3 and 6, *C. tiburonensis* year 2, *C. pulchellus*). That the seedling size class is limiting to population growth for *Calochortus* parallels the empirical results for several other rare taxa (e.g. *Tetramolopium arenarium*; Aplet, Laven, and Shaw 1994), but not all (e.g., *Panax quinqefolium*; Charron and Gagnon 1991).

The larger lambda values of rare species in "good" reproductive years are comparable to the first year's λ of the common species, *C. albus*. This suggests that in certain

Table 2.2
Transition matrices and associated λ for seven rare and one common species of Calochortus.

	Year 1	Year 2	Source
C. albus			Fiedler 1987
(Common)	0.67 0.00 0.65 4.30	0.32 0.03 0.22 1.20	
	0.33 0.71 0.20 0.00	0.32 0.69 0.12 0.00	
	0.00 0.13 0.70 0.40	0.00 0.15 0.62 0.00	
	0.00 0.00 0.02 0.04	0.00 0.00 0.02 0.99	
	λ = **1.04**	λ = **1.01**	
C. coxii			Fredricks 1992
Smith	0.36 0.05 0.06	0.46 0.09 0.02	
	0.64 0.71 0.12	0.27 0.75 0.25	
	0.00 0.19 0.80	0.09 0.10 0.68	
	λ = **0.97**	λ = **0.93**	
Bilger	0.50 0.00 0.04	0.27 0.02 0.04	
	0.40 0.70 0.02	0.46 0.50 0.22	
	0.00 0.27 0.84	0.18 0.30 0.64	
	λ = **0.92**	λ = **0.82**	
Both	0.43 0.02 0.05	0.36 0.06 0.03	
	0.52 0.71 0.09	0.36 0.65 0.24	
	0.00 0.22 0.82	0.14 0.18 0.66	
	λ = **0.96**	λ = **0.90**	
C. obispoensis			Fiedler 1987
	0.00 0.00 0.28	0.00 0.00 0.07	
	0.00 0.94 0.36	0.99 0.96 0.84	
	0.06 0.05 0.63	0.00 0.01 0.13	
	λ = **1.00**	λ = **0.97**	
C. pulchellus			Fiedler 1987
	0.49 0.01 0.91 0.70	0.50 0.01 0.13 0.49	
	0.01 0.84 0.14 0.00	0.07 0.85 0.02 0.12	
	0.00 0.15 0.63 0.29	0.00 0.11 0.89 0.04	
	0.00 0.00 0.22 0.70	0.00 0.00 0.07 0.37	
	λ = **1.00**	λ = **0.96**	
C. tiburonensis			Fiedler 1987
	0.55 0.00 0.02	0.93 0.00 0.00	
	0.42 0.87 0.03	0.05 0.86 0.04	
	0.00 0.12 0.94	0.00 0.04 0.95	
	λ = **0.99**	λ = **0.97**	
C. umpquaensis			Fredricks 1992
Forest	0.61 0.16 0.06	0.71 0.03 0.01	
	0.22 0.61 0.06	0.28 0.74 0.09	
	0.00 0.23 0.92	0.00 0.16 0.85	
	λ = **0.99**	λ = **0.98**	
Meadow	0.39 0.02 0.66	0.66 0.06 0.36	
	0.62 0.57 0.24	0.24 0.49 0.12	
	0.00 0.19 0.57	0.00 0.20 0.57	
	λ = **0.99**	λ = **0.95**	

(continued)

Table 2.2 *(continued)*

	Year 1	Year 2		Source
Ecotone	0.44 0.04 0.05	0.21 0.05 0.02		
	0.44 0.64 0.13	0.64 0.54 0.14		
	0.00 0.29 0.84	0.00 0.29 0.77		
	λ = **1.05**	λ = **0.92**		
C. westonii				Knapp 1996
	0.13 1.29 2.07	0.36 0.19 0.71		
	0.19 0.23 0.03	0.11 0.16 0.01		
	0.06 0.28 0.32	0.01 0.13 0.30		
	λ = **0.93**	λ = **0.54**		
C. howellii				Fredricks 1992
	Year 1	**Year 2**	**Year 3**	
	0.50 0.00 0.01	0.75 0.00 0.09	0.79 0.01 0.10	
	0.50 0.44 0.05	0.25 0.58 0.06	0.06 0.43 0.01	
	0.00 0.54 0.93	0.00 0.43 0.93	0.00 0.56 0.98	
	λ = **0.99**	λ = **1.05**	λ = **1.02**	
	Year 4	**Year 5**	**Year 6**	
	0.77 0.02 0.10	0.67 0.00 0.04	0.84 0.11 0.05	
	0.13 0.05 0.05	0.11 0.48 0.04	0.07 0.65 0.10	
	0.00 0.44 0.92	0.02 0.43 0.92	0.00 0.21 0.86	
	λ = **0.97**	λ = **0.97**	λ = **0.96**	
	mean λ = 0.99; SD = 0.34			

years, rare species can and do exhibit episodic spurts of reproductive "boom." Further generalizations about intrinsic population growth rates require autecological data, such as intrinsic fruit and flower production rates, minimum size to reproduction, and so forth.

Our 28 matrices also indicate that reproductive failure is slow and difficult to detect in the field. A population that is not reproducing may appear stable for many years because reproduction and establishment rates are normally low and variable, and the number of seedlings present are small, despite a high rate survival of mature plants. Such a population would be highly vulnerable to stochastic events, especially catastrophes affecting normally surviving adults. These represent events outside of our studies, and ones hardest to anticipate but possibly most important. Thus, while our *Calochortus* matrices are accurately reflecting the importance of adult survival to population persistence, it will be necessary in the future to distinguish between normal, sustainable variability in reproduction, and continuing reproductive failure. More generally, high survival in perennials translates to acceptable reproductive failure in some years. Continued lack of recruitment threatens long-term population persistence, yet it is hard to distinguish from a chance run of failures between rare bouts of success.

A common complaint is that adequate long-term demographic data are lacking, for it is self-evident that for a particular population, one gains accuracy by adding years to any estimation of projection matrices. A more subtle question concerns comparisons between species, or among populations of the sample species. To approach this question, we multiplied together the matrices in each of the 11 populations of the eight *Calochortus* species to obtain a rate of increase (i.e., λ) using the "complete" data set for each

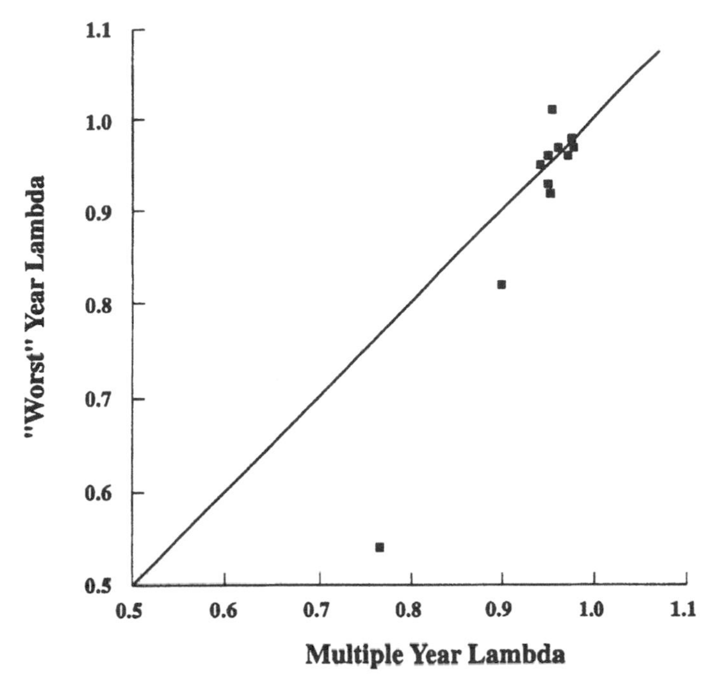

Figure 2.3
Relationship between the square root of the lambda derived from the square root of the average matrix and the lambda from the "worst" year's transition (n = 12).

population. Lambda values subsequently were calculated as the square root of lambda obtained from the product matrix (the 6th root in the case of *C. howellii*), and plotted against the single "worst" year lambda for each population. From here, we then asked: "How far off from the prediction made from the complete data set is the prediction obtained from a one-year time step matrix?" Alternatively, "Does the population growing at the fastest (or slowest) rate according to all the data still turn out to be growing at the fastest (or slowest) rate when only a one-year time step matrix is used?"

Points plotted for most of the species more or less fall on the 45° diagonal, suggesting that the single and multiyear matrices provide similar, if not biologically indistinguishable, projections (Figure 2.3). This is true even for *C. howellii*, the species with largest data set (i.e., six matrices). Points that deviated significantly from the diagonal are those for *C. albus*, *C. westonii*, and the Bilger population of *C. coxii*. Lambda values predicted by a single matrix of *Calochortus albus* apparently give a "better" (i.e., higher) λ than the multiple year matrix, while the reverse is true for *C. westonii* and the Bilger population of *C. coxii*. We caution, however, that while for most *Calochortus* populations the one-year time step matrix provides the same prediction as the multiple-year time step, it would be foolish to conclude that a two-year/one matrix study is sufficient to understand a population's long-term behavior.

Stable Stage Distribution. Keyfitz's (1977) Index of Dissimilarity can be a useful calculation for interpreting stable stage distributions. This sum of the positive differences between the observed and the stable size distributions provides insight into how the distribution of individuals within a size class differs from what would be expected, given the "stable" stage distribution. For example, a large difference between the stable and actual distributions suggests that vital rates (as incorporated within the transition probabilities) of the matrix are not those that have operated overtime to produce the current distribution. Results for the eight *Calochortus* studied indicate considerable variation in the predicted stable distribution among species, between populations in different habitats, and between years within a population. The greatest positive contribution to the Index is made by seedlings and juveniles in every species except *C. tiburonensis* and *C. pulchellus*, indicating that for most species studied, there was a lower proportion of immature plants than the model predicts. This likely reflects transient changes in reproductive rates and seedling survival, which would be expected to create fluctuating stage distributions from year to year. Such differences are probably not significant over the long term, but suggest episodic seedling recruitment patterns are important.

We examined the relationship between the "damping ratio" (λ_1/λ_2; see Caswell 1989) and the Index of Dissimilarity (Keyfitz 1977) in a correlation analysis for each matrix (n = 28) and for each population (n = 11; Figure 2.4). Our results reveal no correlation ($r^2 = 0.47$; $r^2 = 0.40$, respectively). Some populations exhibited high damping ratios, but also high departures from the stable size distribution, even in many cases for the same population between years. Thus, for *Calochortus*, no inference can be drawn from the damping ratio regarding the likelihood of being close to a stable size distribution.

We also found no additional significant correlation between either the Keyfitz Index or the damping ratio and measures of reproduction, mortality, or growth derived from matrices. Vital rates for these eight species and 11 populations of *Calochortus* appear to act in various combinations to determine the rate of convergence, with no one factor contributing substantially more than the others in a way generalizable across species, populations in different habitats, or years within a population. While many of the matrices produced similar damping ratios, they all apparently did so for different reasons.

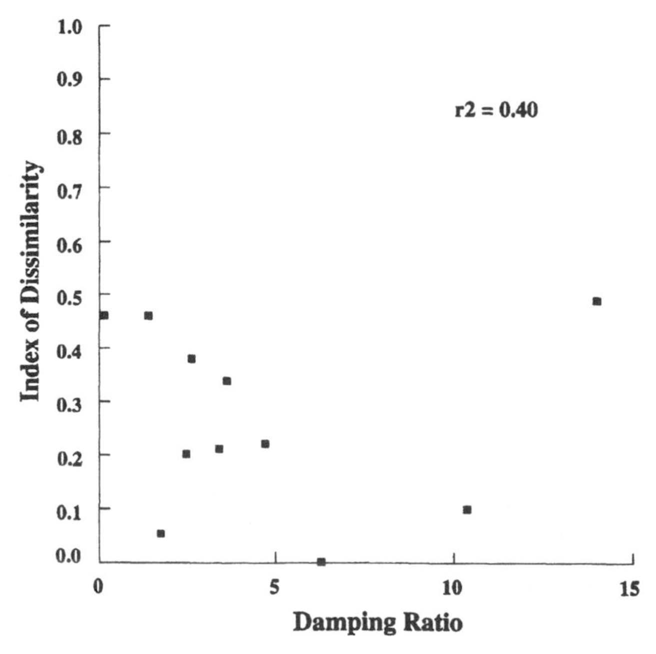

Figure 2.4

Regression analysis between Keyfitz's Index of Dissimilarity and the damping ratio (dominant eigenvalue/subdominant eigenvalue, λ_1/λ_2) (n = 11).

Reproductive Values. The reproductive value of an individual is defined as its probable contribution through current and expected reproduction to future population growth (Fisher 1930). It depends upon the risk of mortality, probability of reproduction, generation time, and population growth rate. The reproductive value is an important component determining the sensitivities and elasticities of matrix entries. Specifically, as the reproductive value of a stage increases, the matrix becomes more sensitive to transitions that affect the number of individuals entering and remaining in that stage. Reproductive values are particularly useful in interpreting sensitivities and elasticities.

For the majority of *Calochortus* species studied, adult reproductive values are low and are nearly equivalent to the reproductive values of seedlings. This result is primarily because the *Calochortus* populations studied are in apparent decline. More specifically, the low reproductive values of adults are due to the coupling of low reproductive potential with low rates of transition to reproductive size. However, those matrices that do result in higher reproductive values for adults relative to seedlings do so through several different combinations of life history variables, i.e, either increased adult reproductive value by increased population growth rate, or decreased seedling reproductive value even in a declining population. Thus, relatively high reproductive values for mature plants can result from (a) a relatively high reproductive potential, moderate seedling growth and survival, and high adult survival (representing characteristics of an expanding population, e.g., *C. albus*); (b) exceedingly low seedling survival, moderate to low reproductive output, and high adult survival (e.g., *C. obispoensis*); (c) exceedingly low rates of transition from stage 2 (non-reproductive) to stage 3 (reproductive), and moderate to low reproductive output (e.g., *C. pulchellus*); or (d) high seedling survival and fast rates of transition to reproductive stages, and moderate to low reproductive potential (e.g., *C. umpquaensis*). Such a finding supports Fiedler's (1987) early contention that rare species, even closely related congeners, may behave a bit differently.

Sensitivity and Elasticity Matrices

Elasticities and sensitivities vary considerably in space and time (Oostermeijer *et al.* 1966; Silvertown, Franco, and Menges 1966). Nearly all the *Calochortus* matrices examined here are relatively insensitive to changes affecting growth and survival of stage 1 (seedlings). The exceptions are found in *C. albus*, *C. umpquaensis*/meadow, and *C. westonii*, which are species with high reproductive rates, combined, in the case of the two rare species, with high adult mortality. *Calochortus howellii* is also somewhat sensitive to changes in stage 1 in years 3, 4, and 6, which are years with demonstrably higher reproduction. In contrast, most of the *Calochortus* matrices are particularly sensitive to transitions representing survival of reproductive individuals, and somewhat sensitive to changes in reproduction.

Elasticities (Figure 2.5) demonstrate that the later (older/larger) stages are making the greatest contribution to population growth rates, and that their contribution depends largely upon survival and not on sexual reproduction. For *C. albus* and *C. westonii*, earlier stages also make a notable contribution to population growth. For almost all species, the matrix diagonal representing stasis overwhelmingly contributes to the eigenvalue. *Calochortus westonii* is the exception because of its relatively faster growth rates and higher reproduction.

The large component attributed to stasis does not mean that reproduction and plant growth are not important. Indeed, the fact that stasis is contributing most to the value of

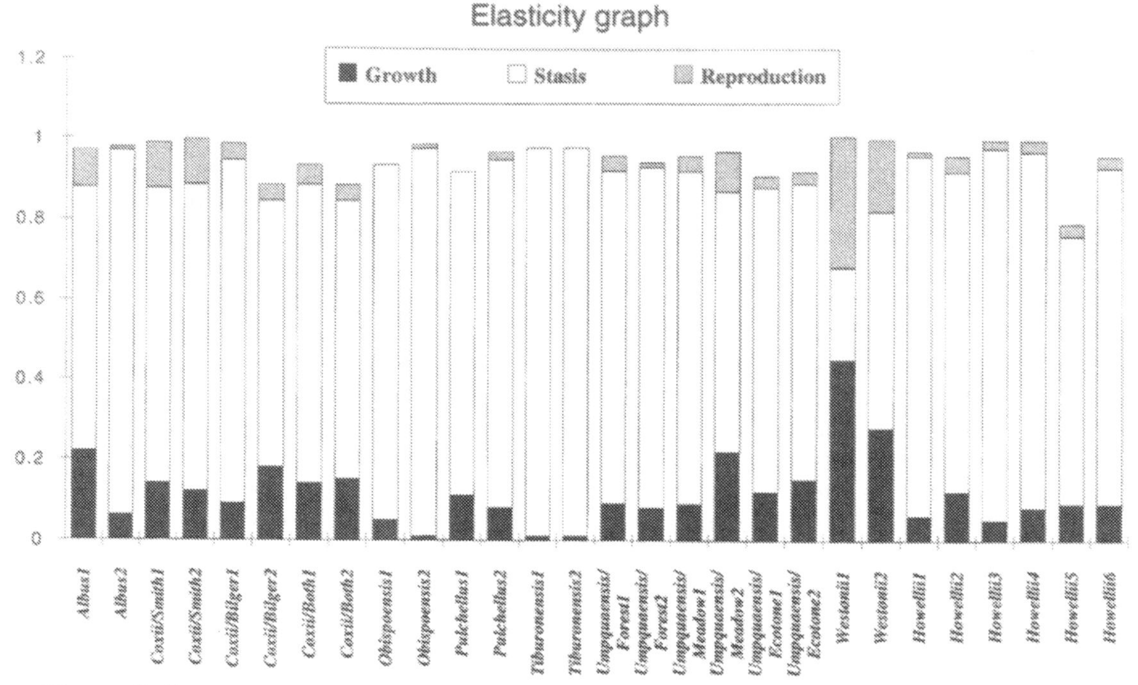

Figure 2.5
Proportional contribution of growth, stasis, and reproduction elasticities to the dominant eigenvalue. (Retrogressions were ignored in the calculations.)

lambda may be "danger sign" when considering the meaning of the eigenvalue. In other words, a population cannot persist in the long term if its individuals neither grow nor reproduce. For the eight *Calochortus* species studied to date, either plant growth or reproduction is limiting population growth for every matrix.

To explore further the relationship among the elasticities, we questioned whether a significant positive correlation exists between the composite elasticities for growth and reproduction. This translates biologically to asking whether increased reproduction is dependent upon plant growth rates. Results from a correlation analysis suggest a highly significant relationship ($r^2 = 0.93$; Figure 2.6). Reproduction will be limiting to population growth when plant growth rates are high; increases in reproduction will have little effect on population growth if plant growth rates are very low. For example, for *C. tiburonensis* year 2, increasing the reproductive rate increases the sensitivity of the matrices to seedling transitions and decreases sensitivity to reproduction dramatically until reproduction is approximately 0.10, after which the change in sensitivity to reproduction is small, while the sensitivity to seedling survival and growth continues to increase. For *C. pulchellus*, growth is so limiting that no change in reproduction has an appreciable effect on the matrix. The sensitivity to reproduction for *C. obispoensis* also is limited strongly by rates of transition to reproductive size. This result indicates that for these three rare species, a point exists at which the matrices are limited by plant growth rates, and this happens at a very low reproductive rate. A similar analysis examining the composite elasticities

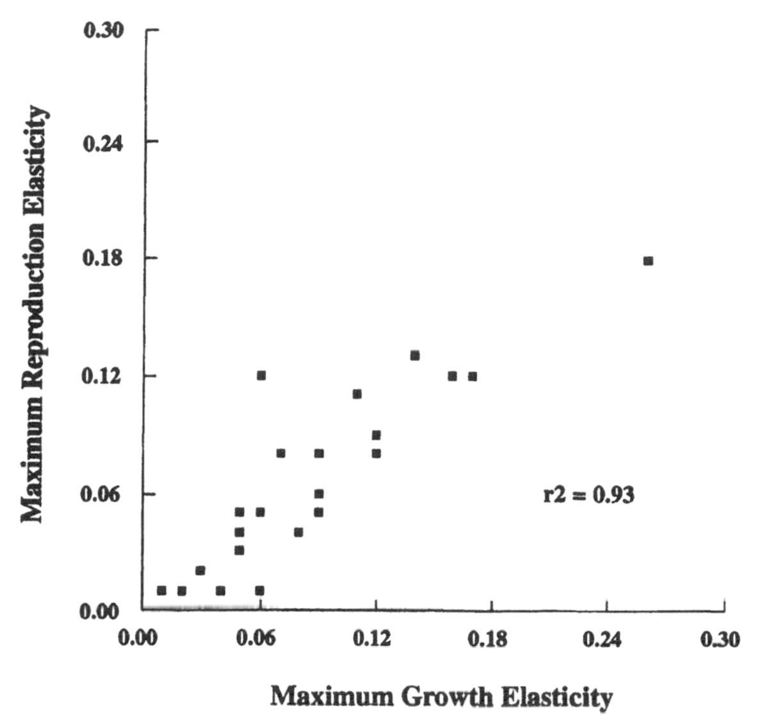

Figure 2.6
Regression analysis between the sum of reproduction elasticities and the lambda from the sum of reproduction elasticities (n = 22).

for stasis and reproduction did not reveal any pattern. Stasis and reproduction are not correlated ($r^2 = 0.17$), either positively or negatively.

One of the most creative and potentially useful approaches to understanding elasticities is the ***G/L/F*** triangle developed from composite elasticity values of growth (***G***), stasis (or retrogression, ***L***), and reproduction (***F***) (Silvertown, Franco, and McConway 1992; Silvertown *et al.* 1993; Silvertown, Franco, and Menges 1996). Seeking to find broad patterns in the relative importance to lambda in these three statistics, Silvertown and his colleagues examined elasticity values for virtually all of the plant project matrices published in the relevant botanical or ecological English-language journals. Their extensive analyses and subsequent results suggest that the relative importance of ***G***, ***L***, and ***F*** varies systematically between the "functional groups" of semelparous herbs, iteroparous herbs of open habitats, forest herbs, shrubs, and trees.

Interestingly, a comparison of elasticity components of projection matrices for multiple populations of two perennial species, *Cirsium vulgare* and *Pedicularis furbishae*, revealed a gradient in lambda values toward larger lambdas for populations with higher growth and reproductive rates (Silvertown *et al.* 1996). The authors suggest that this gradient mirrors that found among the functional groups, providing a basis for managing populations by optimizing the appropriate successional state of the habitat for the life history of any particular species. However, plotting the eight *Calochortus* species in the ***G/L/F*** triangle does not reveal any such pattern among the rare or common taxa (Figure 2.7). All the matrices, except for *C. westonii* year 1, fall squarely into the portion of the

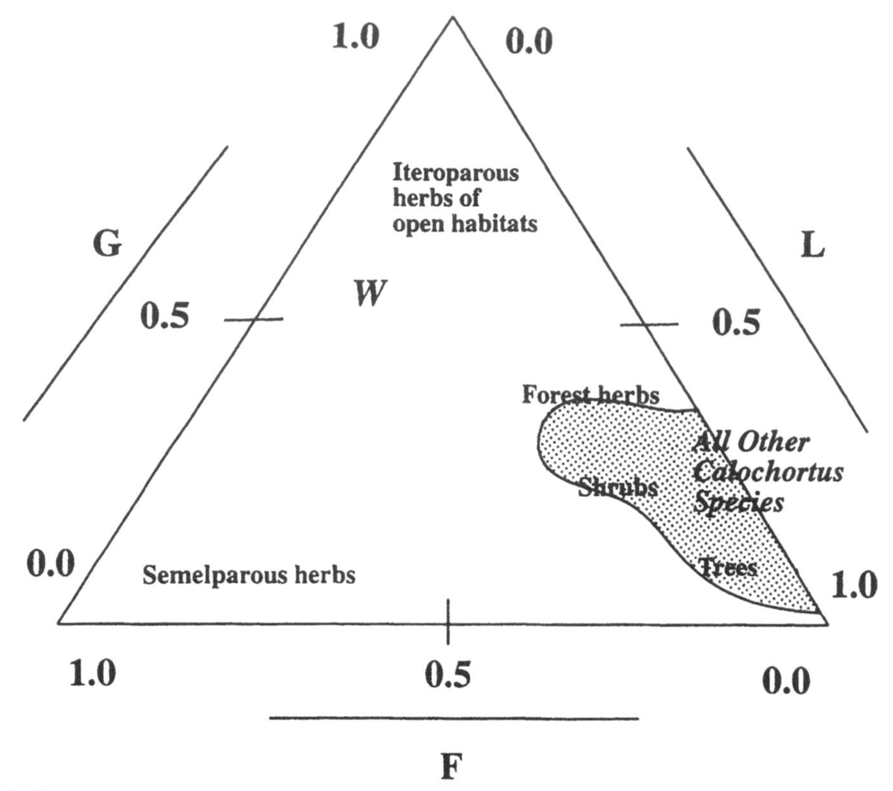

Figure 2.7
G/L/F triangle for the eight species of Calochortus *in this study (after Silvertown* et al. *1993). **W** =* C. westonii.

triangle in which stasis is most important, along with long-lived trees, shrubs, and forest herbs. We suspect that while the ***G/L/F*** triangle may be broadly useful, primarily heuristically, generalizations supporting ***G/L/F*** triangle apparently do not extend to *Calochortus.* Such a conclusion further suggests that utility of the ***G/L/F*** triangle offered by Silvertown, Franco, and Menges (1996) is limited for managing *Calochortus* populations.

DISCUSSION AND SUMMARY

Calochortus albus, the only mariposa lily species studied with a relatively large geographic distribution and commonly large population sizes, is characterized by high reproduction, high transition rates to reproductive stages, high survival rates overall, and both matrices resulting in $\lambda > 1.0$. These results are not surprising, but as a whole remain in strong contrast with all other *Calochortus* species studied. *Calochortus obispoensis* and *C. tiburonensis,* for example, are characterized by low reproduction and very low transition rates, but high survival rates. Low reproduction and very low transitions are also true of *C. pulchellus,* but this mariposa lily demonstrates high adult survival coupled with low seedling survival. Thus, while there appear to be real differences between the rare and common species, we remain cautious in this interpretation until additional common *Calochortus* taxa are studied. *Calochortus albus* may well be more common because it has relatively high reproductive rates, fairly high growth rates, and high survival, while all rare species of *Calochortus* studied have one or more of these factors low.

Lessons learned from the similarities among *Calochortus* taxa include the conclusions that (1) reproduction, because it is so uncertain, is the limiting factor for most of the rare mariposa species, combined in some cases with exceptionally low plant growth rates; and (2) because of the limitations on reproduction, all *Calochortus* species rely on high survival rates of reproductive plants for population persistence. All the life histories of the *Calochortus* studied to date are limited by environmental factors, primarily vertebrate predation on fruits and limited pollination, both events resulting in low seed set curtailing sexual reproduction. In addition, a general pattern across the species studied in more than one habitat (i.e., *C. coxii*, *C. umpquaensis*, and *C. westonii*) is that reproductive output varies considerably across habitats. Individual survival makes the greatest contribution to the growth rate in most cases; reproduction and mortality appear episodic; and reproductive success and seedling establishment appear to limit populations to certain habitats. This life history is typical of long-lived perennials.

In effect, we are looking for "optimal" life histories in our set of matrices that provide few examples of populations that are stable, much less ones that are expanding. This is in itself an interesting result, provoking important questions. Is the observed pattern of general instability and decline in *Calochortus* populations, punctuated by an occasional good year, generally characteristic of rare plant species? Is it characteristic of the mariposa lilies? Or could these results reflect recent phenomena in the history of the taxa, related to habitat destruction, increased herbivore populations, introduced species, fire suppression, cattle grazing, or climate change? Despite two decades of observations and analyses, we still cannot answer unequivocally these dramatically important questions.

Within this set of closely related species, there is considerable variation in the relative importance of growth, reproduction, and survival, and in how these factors combine to produce the dynamics reflected in the matrices. While some generalizations are possible, the projection matrices illustrate the importance of understanding the natural history of the species under study. For example, *C. tiburonensis* will respond to disturbance very differently than *C. westonii* or *C. coxii*, and even these latter two species, although found in similar habitats, exhibit some important differences in their population behavior.

The variability found between sampling years and populations for many of these species suggests that matrix models require a great deal of caution in their interpretation. It is arguable that a transition matrix model is, to a large extent, a useful fiction, particularly for rare species that may never actually reach a stable distribution. The extent to which such results are useful will depend upon how systematically they are arrived. The variation in time and space that is quantified by annual projection matrices can provide important information for conservation if its sources and effects are understood (Horvitz and Schemske 1995; Silvertown, Oostermeijer *et al.* 1966; Silvertown, Franco, and Menges 1996). The degree of variability in vital rates may be an important aspect of life history and is in itself, worthy of investigation when comparing species or populations (see chapter 3). To this end, design of the field study, monitoring protocol, or monitoring program is of greatest importance in elucidating demographic patterns and their relationship to habitat and environmental variability. Of equal importance is how the matrices and size categories are constructed, as first demonstrated for *Calochortus* by Fredricks (1992).

In the end, translating the population biology of rare species to management remains a challenge. For example, a management plan that focuses on increasing reproductive success (i.e., the major limitation to population growth for almost all *Calochortus* taxa studied) could be difficult to implement over time and would be of little use for populations that are also limited by plant growth rates. In such cases, ensuring sufficient survival through protection and management of habitat may be the only real option to ensure long-term population persistence. The immediate implication of this management approach is that populations must be administratively protected *in situ* first, and protected against grazing by generalist herbivores (e.g., deer, rabbits, gophers, livestock) second.

Silvertown *et al.* (1993, pp. 465–66) suggest that the *modus operandi* in the contemporary school of plant ecology is the comparative method. Such an approach "is founded on the principle that similar environments exert similar selective forces on different species, leading to convergent evolution and adaptive patterns that transcend taxonomic limits." However, despite very similar life history—geographic restriction to serpentinite environments and narrow taxonomic circumscription—*Calochortus* rarities defy the generalizations sought by conservationists.

We believe that our results can be interpreted more broadly. Transition matrices, stable stage distributions, reproductive values, and elasticities can mean a number of different things, even for very closely related taxa. The rare *Calochortus* species studied here will require differing, species-specific, and for some mariposa lilies, population-specific management strategies to ensure long-term persistence. Indeed, we suggest that life histories are idiosyncratic, and that math is no substitute for natural history. Documentation of spatial and temporal variability of population dynamics can be a real asset to rare plant protection, but only if one uses the most appropriate analytical technique, and uses it correctly.

ACKNOWLEDGMENTS

We wish to thank Peter Kareiva for his great insight and editorial skills, and for his willingness to light the way. We also extend our gratitude to Paulette Bierzychudek, Ed Guerrant, and Eric Menges for their cheerful assistance in dramatically improving the manuscript.

LITERATURE CITED

Aplet, G.H., R.D. Laven, and R.B. Shaw. 1994. Application of transition matrix models to the recovery of the rare shrub, *Tetramolopium arenarium* (Asteraceae). *Natural Areas Journal* 14:99–106.

Bierzychudek, P. 1982. The demography of jack-in-the-pulpit, a forest perennial that changes sex. *Ecological Monographs* 52:335–351.

Caswell, H. 1989. *Matrix population models*. Sunderland, MA: Sinauer Associates, Inc.

Charron, D. and D. Gagnon. 1991. The demography of northern populations of *Panax quinquefolium*. *Journal of Ecology* 79:431–445.

Cipollini, M.L., D.A. Wallace-Senft, and D.F. Whigham. 1994. A model of patch dynamics, seed dispersal, and sex ratio in the dioecious shrub *Lindera benzoin* (Lauraceae). *Journal of Ecology* 82:621–633.

deKroon, H., A. Plaisier, J. van Groenendael, and H. Caswell. 1986. Elasticity: The relative contribution of demographic parameters to population growth rate. *Ecology* 67:1427–1431.

Falk, D.A. and K.E. Holsinger, eds. 1991. *Genetics and conservation of rare plants*. Oxford: Oxford University Press.

Falk, D.A., C.I. Millar, and M. Olwell, eds. 1996. *Restoring diversity. Strategies for reintroduction of endangered plants*. Washington, D.C.: Island Press.

Ferson, S. 1991. *RAMAS/stage: Generalized stage-based modeling for population dynamics*. Seatuket, NY: Applied Biomathematics.

Fiedler, P.L. 1987. Life history and population dynamics of rare and common mariposa lilies. *Journal of Ecology* 75:977–995.

Fiedler, P.L. and R.K. Zebell. 1997. *Calochortus*. Treatment for Volume 11, *Flora of North America*. N. Morin, convening editor. Cambridge: Cambridge University Press.

Fisher, R.A. 1930. *The genetical nature of natural selection*. Oxford: Clarendon Press.

Frankel, O.H., A.H.D. Brown, and J.J. Burdon. 1995. *The conservation of plant diversity*. Cambridge: Cambridge University Press.

Fredricks, N.A. 1992. Population biology of rare mariposa lilies (*Calochortus*: Liliaceae) endemic to serpentine soils in southwestern Oregon. Ph.D. dissertation, Oregon State University, Corvallis.

Given, D.R. 1994. *Principles and practice of plant conservation*. Portland: Timber Press.

Hanzawa, F.M. and S. Kalisz. 1993. The relationship between age, size and reproduction in *Trillium grandiflorum* (Liliaceae). *American Journal of Botany* 80:405–410.

Harper, J.L. 1977. *Population biology of plants*. London: Academic Press.

Horvitz, C.C. and D.W. Schemske. 1995. Spatiotemporal variation in demographic transitions of a tropical understory herb: Projection matrix analysis. *Ecological Monographs* 65:155–192.

Keyfitz, N. 1977. *Index to the mathematics of populations*. Reading, MA: Addison-Wesley.

Knapp, B.E. 1996. Natural history and population dynamics of *Calochortus westonii*. M.A. thesis, San Francisco State University, San Francisco.

Law, R. 1983. A model for the dynamics of a plant population containing individuals classified by age and size. *Ecology* 64: 224–230.

Lefkovitch. L.P. 1965. The study of population growth in organisms grouped by stages. *Biometrics* 1:1–18.

Leslie, P.H. 1945. On the use of matrices in certain population mathematics. *Biometrika* 32: 183–212.

Menges, E.S. 1992. Stochastic modeling of extinction in plant populations. In *Conservation biology: The theory and practice of nature conservation, preservation, and management*, eds. P.L. Fiedler and S.K. Jain, 253–275. New York: Chapman and Hall.

Moloney, K.A. 1986. A generalized algorithm for determining category size. *Oecologia* 69:176–180.

Ownbey, M. 1940. A monograph of the genus *Calochortus*. *Annuals of the Missouri Botanic Garden* 27:371–561.

Oostermeijer, J.G., M.L. Brugman, E.R. De Boer, and H.C.M. Den Nijs. 1996. Temporal and spatial variation in the demography of *Gentiana pneumonanthe*, a rare perennial herb. *Journal of Ecology* 84:153–166.

Pavlik, B.M. 1994. Demographic monitoring and the recovery of endangered plants. In *Recovery and restoration of endangered species*, eds. M. Bowles and C.J. Whelan, 322–350. Cambridge: Cambridge University Press.

Pavlik, B.M. 1995. The recovery of an endangered plant II. A three-phased approach to restoring populations. In *Restoration ecology in Europe*, eds. K.M. Urbanska and K. Grodzinska, 49–69. Zurich, Switzerland: Geobotanical Institute SFIT.

Pavlik, B.M. 1996. Conserving plant species diversity: The challenge of recovery. In *Biodiversity in managed landscapes—theory and practice*, eds. R.C. Szaro and D.W. Szaro, 359–376. New York: Oxford University Press.

Patterson, T.B., T.J. Givnish, and K.J. Sytsma. 1996. Preliminary molecular phylogeny for *Calochortus* (Liliaceae s.l.) based on cpDNA spacer sequences. *American Journal of Botany Suppl.* 83:185.

Schemske, D.W., B.C. Husband, M.H. Ruckelshaus, C. Goodwillie, I.M. Parker, and J.G. Bishop. 1994. Evaluating approaches to the conservation of rare and endangered plants. *Ecology* 75:584–606.

Silvertown, J., M. Franco, and K. McConway. 1992. A demographic interpretation of Grime's triangle. *Functional Ecology* 6:130–136.

Silvertown, J., M. Franco, and E. Menges. 1996. Interpretation of elasticity matrices as an aid to the management of plant populations for conservation. *Conservation Biology* 10:591–597.

Silvertown, J., M. Franco, I. Pisanty, and A. Mendoza. 1993. Comparative plant demography—relative importance of life-cycle components to the finite rate of increase in woody and herbaceous perennials. *Journal of Ecology* 81:465–476.

Vandermeer, J. 1982. To be rare is to be chaotic. *Ecology* 63:1167–1168.

Waterloo Maple Software. 1995. *Maple V, Release 3, Student Edition*. Pacific Grove: Brooks/Cole Publishing Company.

Werner, P.A. and H. Caswell. 1977. Population growth rates and age versus stage-distribution models for teasel (*Dipsacus sylvestris* Huds.). *Ecology* 58:1103–1111.

Young, T.P. 1984. Comparative demography of semelparous *Lobelia telekii* and iteroparous *Lobelia keniensis* on Mount Kenya. *Journal of Ecology* 72:637–650.

Young, T.P. 1985. *Lobelia telekii* herbivory, mortality, and size a reproduction: Variation with growth rate. *Ecology* 66:1879–1883.

CHAPTER

3

Evaluating Extinction Risks in Plant Populations

ERIC S. MENGES

The ability to predict extinction risks is crucial to answering several recurring questions involving the conservation of plant species. Is a given reserve large enough to support a viable population? Is active intervention necessary to rescue a declining population (e.g., Burgman and Lamont 1992)? Should a small, unprotected population be ignored because it is doomed to extinction? Unfortunately, there are usually insufficient empirical data to answer these questions. One approach in the face of scarce data is the application of viability assessments based on stochastic population modeling. In this chapter, I explore stochastic viability models as a tool for gaining insight into which management regimes might best enhance population viability.

THE CONCEPTUAL BASIS: EXTINCTION PROBABILITY AND MINIMUM VIABLE POPULATION

Although systematic trends threaten many species with extinction, the concept of a minimum viable population applies mainly to stochastic sources of extinction. Among these stochastic sources are environmental stochasticity, demographic stochasticity, natural catastrophes, and genetic stochasticity (Shaffer 1981). All stochastic factors increase in importance as population size drops, with environmental stochasticity and natural catastrophes being more important than demographic stochasticity for most population sizes (Shaffer 1987). A minimum viable population (MVP) is sufficiently large to produce some specified, suitably low probability of extinction over a long time period.

The conceptual basis of MVP was first developed largely within a population genetics framework (Frankel and Soulé 1981). Uncommon species, endemic species, and small populations tend to have lower genetic variation (Karron 1987; Hamrick *et al.* 1990; Billington 1991; van Treuren *et al.* 1991; Prober and Brown 1994; Raijmann *et al.* 1994). Genetic bottlenecks may lead to loss of within-population genetic variation, causing inbreeding depression or loss of evolutionary flexibility (Huenneke 1991; Oostermeijer, van Eijck, and den Nijs 1994; Oostermeijer *et al.* 1995). General estimates of genetic MVP range from about 50 to 5,000 (Franklin 1980; Lande 1995; Lynch, Conery, and

Bürger 1995). However, these general guidelines may not apply well to plant populations because of small neighborhood sizes (Meagher 1986) and local genetic differentiation (e.g., Silander 1985). Also, species capable of inbreeding, apomixis, and vegetative reproduction (e.g., Jain 1976) may purge deleterious recessives. Unfortunately, although coordinated ecological and genetic studies are increasingly common for plants (Burgman and Lamont 1992; Widen 1993; Oostermeijer, van't Veer, and de Nijs 1994; Oostermeijer *et al.* 1995; Ouborg and van Treuren 1994), little evidence directly links genetic composition of plant populations to growth rate or survival in the wild (Schemske *et al.* 1994).

An MVP defined by genetic constraints often may be secondary to the demographic difficulties of small population size in varying environments (Lande 1988; Menges 1991; Heschel and Paige 1995). A good example can be found in *Banksia cuneata*, where fecundity reductions due to inbreeding in small populations have been found to increase extinction risk only in populations smaller than ten individuals. In contrast, rainfall patterns following periodic fire (environmental stochasticity) have far more important effects on *Banksia cuneata* extinction over a wider range of population sizes (Burgman, Ferson, and Akcakaya 1993). In general, extinctions of small populations may occur long before deleterious genetic changes become evident.

PREDICTING POPULATION DYNAMICS

Assessing extinction risks due to nongenetic factors requires an understanding of population dynamics and the factors that drive changes in population size. Matrix-projection techniques are often used to project population dynamics, with demographic parameters defined by an invariant age- or stage-projection matrix (Leslie 1945; Lefkovitch 1965). Using matrix projection, one can easily obtain equilibrium demographic properties of a population, including finite rate of increase (λ), age structure, reproductive values, and sensitivity of population growth and extinction to various vital rates (van Groenendael and Slim 1988; Caswell 1989; Cochrane and Ellner 1992; Burgman, Ferson,and Akcakaya 1993; McCarthy, Burgman, and Ferson 1995). Sensitivity analyses (e.g., elasticity; deKroon *et al.* 1986; Caswell 1989) can identify which demographic processes (e.g., seed germination, stage-specific growth, stasis, size-specific reproductive output) have large effects on population growth, which in turn may be key in recovering endangered species (Schemske *et al.* 1994). Uncertainty in population growth rates can also be assessed (Caswell 1989; Alvarez-Buyulla 1994; Kalisz and McPeek 1992; Alvarez-Buylla and Slatkin 1991, 1994; Doak, Kareiva, and Klepetka 1994).

The major drawback to deterministic matrix projection is the assumption of unchanging conditions, which leads to unrealistic long-term predictions of exponential growth. Of course, demographic parameters such as mortality, growth, reproductive status, and reproductive output vary over time in all populations (e.g., Carlsson and Callaghan 1991; Bengtsson 1993; Nault and Gagnon 1993; Horvitz and Schemske 1995). Various types of stochastic solutions provide better insight into population dynamics and extinction risk given changing environments and consequent demography.

Quantification of risk is best accomplished by analyzing extinction probability, rather than expected time to extinction. This is because our practical concern is with risks on a finite time scale and risks measured by the mean time to extinction can be severely at odds with the probability of extinction over 50 years or 100 years. For a given life history, defined variation in demographic parameters, and a time frame of interest, the probability of extinction is readily calculated from replicate stochastic simulations. From a judgment of "acceptable" extinction probabilities, a demographic minimum viable population (MVP) can be calculated. These are the basic approaches of this paper.

The use of stage-structured, stochastic models to analyze population dynamics is now commonplace in ecology. A lingering problem, however, is that long-term studies are required for accurate forecasting (Holsinger 1995). Fortunately even short-term studies can provide insights—especially when compared to no analyses at all. For example, stochastic simulations have been used to compare population viability across various habitats (van Groenendael and Slim 1988; Menges 1990). Management regimes can also be contrasted for their effects on demography, as when Silva *et al.* (1991) evaluated effects of individual fires and fire frequency on population growth in a savanna grass using matrix models. Similarly, Burgman and Lamont (1992) combined environmental, demographic, and genetic factors in a stochastic demographic simulation to evaluate effects of various fire management tactics on population size and extinction risk in *Banksia cuneata*. Extinction thresholds have been modeled for various harvesting levels in American ginseng and wild leek (Nantel, Gagnon, and Nault 1996). The database is now large enough that in this paper I am able to explore the consequences of environmental and demographic stochasticity, and of different levels of environmental variation, on persistence of 28 populations of perennial herbs and trees.

METHODS

Stochastic Modeling of Population Growth: General Protocol and Assumptions

I used a simulation model (POPPROJ) to project population growth of various species, generate variation in demographic parameters, and calculate various statistics, particularly extinction probability. The program used standard matrix algebra to project the stage distribution at each time-step (a year) by multiplying the previous stage distribution by a projection matrix.

Stochastic runs used projection matrices that varied with time. For this study, two types of variation were simulated. **Environmental stochasticity** represents events that affect the entire population but impact different stages uniquely (Lewontin and 1969; May 1973; Cohen 1979a). Once the year's matrix was set, it was applied to the entire population. **Demographic stochasticity** represents chance events affecting individuals in a very small population, even in a constant environment (May 1973; Kieding 1975). For example, the fact that individuals either survive or die according to a binomial process means that there is always some chance that all the individuals in a population may die, far in excess of the expected mean rate. Such excess death by chance alone is what causes small populations to be at risk due to demographic stochasticity. As populations get larger, the "law of averages" increasingly wins out over chance runs of bad luck.

To simulate environmental stochasticity, values for each non-zero matrix element were drawn from a normal distribution with specified means and variances for each element. Unreasonable values (e.g., reproductive output less than zero, survivorship less than zero or greater than one) were truncated to zero or one. I did not examine different statistical distributions for the variation in demographic parameters, but it would be valuable to conduct such an analysis in future studies.

Environmental stochasticity was modeled without covariance, autocorrelation, or density-dependence. First, each matrix element was assumed to vary independently of all other elements. Second, zero autocorrelation among conditions was assumed, i.e., the effects of severe environments were not carried over from one year to another. Finally, no density-dependent regulation was assumed, because it often has only weak effects on wild plant populations (e.g., Fowler 1995) and requires long data sets for estimation (Ginzburg, Ferson, and Akcakaya 1990, see chapter 1). Inclusion of these features might affect extinction probability.

Projection Matrices and Simulation Procedure

In this chapter, explorations of stochastic effects on population growth are based on 28 published projection matrices corresponding to eight different species representing "biennials," perennial iteroparous herbs, and trees (Table 3.1). One particular feature of life histories was emphasized for comparisons: λ, the finite rate of increase under equilibrium, deterministic conditions. In most cases, the projection matrix represents a particular population in a given year. Because variation in demographic parameters has rarely been fully characterized, the type and quantity of environmental stochasticity is explored as a sensitivity analysis.

Although the POPPROJ program is flexible in many ways, certain conventions were used to permit comparisons among species and types of variation. Changes in this protocol are not thought to be important enough to qualitatively affect results. Most simulations were begun with 200 individuals arranged in a stable stage distribution. When population size dipped to ten individuals or fewer, the population was termed extinct (quasiextinct; Ferson and Burgman 1995) and the run terminated. All simulations were terminated after 100 years. Because demographic stochasticity was slow to run at large population sizes, runs were terminated when population size reached 5,000. Each stochastic simulation produced different results. I generally performed 50 simulations using each set of input parameters. (Thus, statements of "zero extinction probability" refer specifically to 0/50 extinctions in 100 years; the true extinction probability may be non-zero but is certainly not large). Additional simulations (up to 200) suggested that 50 simulations provided sufficient precision for extinction probabilities. Statistics for final population size and

Table 3.1
Plant projection matrices used in this study.

Reference	Life History[1]	Matrix Type	Species	No. of Matrices[2]	No. of Classes	Range of λ^3
Enright and Ogden 1979[4]	I,T	Stage	*Araucaria cunninghamii*	3	4	0.39–1.02
Enright 1982	I,T	State	*Araucaria hunsteinii*	4	8–14	0.99–1.05
Bierzychudek 1982	I,P	Stage	*Arisaema triphyllum*	4	7	0.85–1.32
Piñero et al. 1984	I,T	Stage	*Astrocaryum mexicanum*	4	7	0.99–1.03
Burns and Ogden 1985	I,T	Stage	*Avicennia marina*	1	7	1.22
Fiedler 1987	I,P	Stage	*Calochortus* spp.	8	4–5	0.96–1.39
Werner and Caswell 1977	S	Stage & age	*Dipsacus sylvestris*	16	7–8	0.0–2.60
Hartshorn 1975	I,T	Stage	*Pentaclethra macroloba*	1	15	1.002

[1]I = iteroparous; S = semelparous herb; P = perennial herb; T = tree.
[2]Representing different species, populations, and/or years.
[3]λ = finite rate of increase, assuming deterministic population growth.
[4]Another species reported overlaps with Enright (1982).

years to extinction apply only to nonextinct populations and extinct populations, respectively. These values are reported only if based on sample sizes of five or more.

The degree of environmental stochasticity was varied by changing an arbitrary base level of variation. This base variation index is a variance/mean ratio of 0.01 for seed production and 0.0001 for mortality, growth, and clonal reproduction. This base environmental stochasticity produced little change relative to the deterministic case. To change levels of environmental stochasticity, variances were systematically increased from base levels by a multiplier (variation index). I also collected projection matrices from literature studies that included at least three annual matrices from one or more populations, and calculated variance/mean ratios for each matrix element and each population, to provide a basis for evaluating modeled levels of environmental stochasticity.

RESULTS

As expected, introduction of stochasticity resulted in year-to-year variation in population sizes (Figure 3.1) and stage structure. Demographic stochasticity resulted in minor fluctuations around deterministic, exponential growth, whereas environmental stochasticity resulted in more marked fluctuations and a lowering of the average growth rate. As environmental stochasticity increased, populations were increasingly likely to become extinct within 100 years (Table 3.2). This relationship generally exhibited thresholds; a minimum level of variation was necessary to generate a nonzero extinction probability (Figure 3.2). Likewise, populations suffering > 80 percent extinction probability were not particularly sensitive to further increases in environmental stochasticity. Between upper and lower

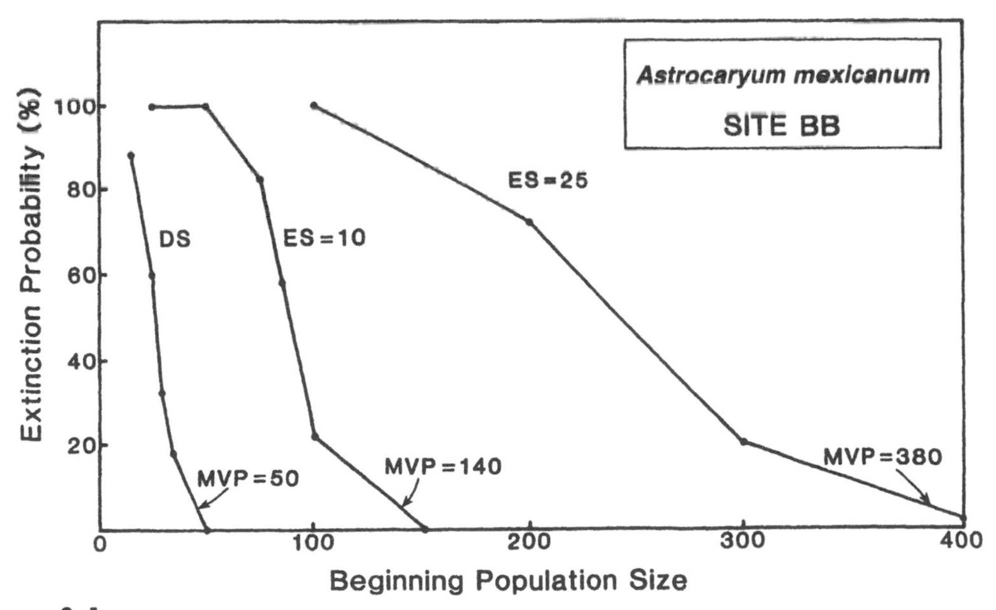

Figure 3.1
Environmental stochasticity (ES) causes variation in population size with time and also slows average population growth. ES-10 refers to V/M ratio of 0.0010 for non-fecundity elements, 0.1000 for fecundity. ES-5 is half of this. Demographic stochasticity (DS) has minor effects relative to the deterministic case, and these effects are dampened as population size increases. This graph shows individual simulations with Astrocaryum mexicanum.

Table 3.2

Overall effects of environmental (ES) versus demographic (DS) stochasticity on extinction probability (EP) for the species analyzed in this paper.

Type of Stochasticity and Relative Level	V/M[1]	Number of Populations 0–9% EP	10%–49% EP	50%–90% EP	>90% EP
ES, weak	0.0010	21	0	0	3
ES, moderately weak	0.0025	17	3	1	3
ES, moderate	0.0100	13	0	0	11
ES, strong	0.0500	11	1	0	12
ES, very strong	0.1000	9	2	1	12
ES, extreme	0.3000	2	2	4	16
DS, $\lambda > 1$	λ	17	3	1	0
DS, $1 <<< 1$	λ	2	0	0	3

[1]Variance/mean ratio for non-fecundity elements used in simulations. For fecundity elements, V/M used is 100 times higher.

thresholds, extinction probability increased linearly, with more sensitive species showing faster increases in extinction probability with increases in environmental stochasticity (Figure 3.2). Increasing environmental stochasticity also decreased the average time to extinction and decreased final population size.

Sensitivity to environmental variation was a function of life history. Species with

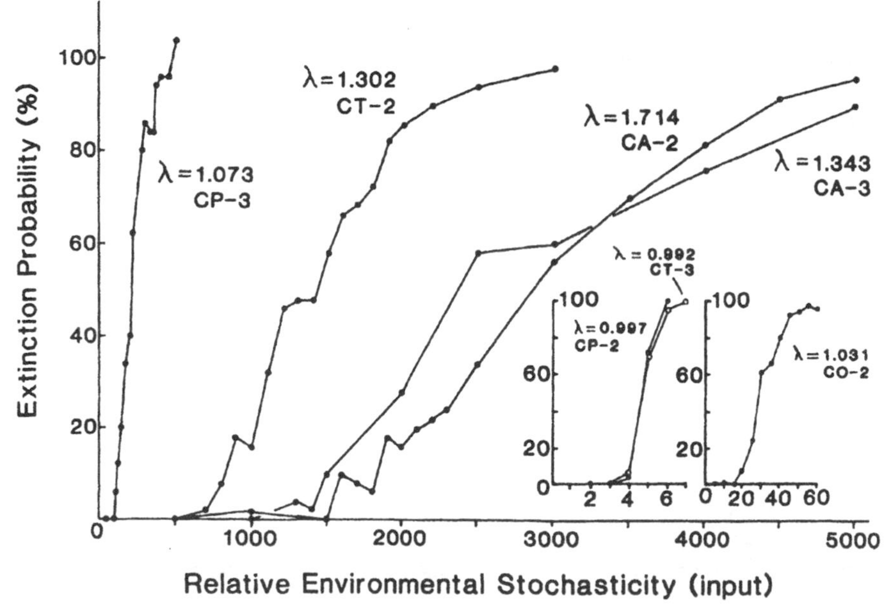

Figure 3.2

Extinction probabilities increase with the level of environmental stochasticity for four species of Calochortus *studied by Fiedler (1987). CP =* C. pulchellus, *CT =* C. tiburonensis, *CA =* C. albus, *CO =* C. obispoensis; *2 = 1982–1983, 3 = 1983–1984. Species with the lowest λ are most sensitive to environmental stochasticity. Note that the two insets at lower right have different X axes. NES values are 1,000 * V/M.*

$\lambda \leq 1$ were sensitive to variation two orders of magnitude less than were populations with $\lambda > 1.2$ (Figure 3.2). The variation level necessary to produce a given level of extinction probability increased sharply with increasing λ for most life histories. For example, populations with $\lambda < 1$ (declining populations) were sensitive to environmental stochasticity with variance to mean ratios < 0.0025. In contrast, fairly stable populations ($1.0 > \lambda > 1.1$ with one exception) required variance/mean ratios in excess of 0.005 before environmental stochasticity had much of an effect. Finally, rapidly expanding populations ($\lambda > 1.15$) were relatively insensitive to environmental stochasticity; only environmental stochasticity with a variance/mean ratio in excess of 0.14 was was sufficient to induce a 50% extinction probability for these species. (See Table 3.3 for estimates of variance/mean ratios in natural populations.)

Other factors besides λ affected population sensitivity to environmental stochasticity. I selected eleven populations that were unusually sensitive or insensitive for their λ (as judged by scatter of extinction probability vs. environmental stochasticity) and analyzed their life histories relative to those with similar λ but more typical sensitivities to environmental stochasticity. Less sensitive populations had the same or a greater number of life history stages (four of six populations). Because environmental stochasticity was applied independently to matrix elements, multiple stages buffered populations against catastrophic mortality or reproduction failure. Real-world populations probably have several matrix elements affected by environmental stochasticity in a correlated manner, and may not be buffered by multiple stages. Populations sensitive to environmental stochasticity for their λ also reproduced at earlier stages (5/5), and had a greater proportion of nonzero elements (3/5). These characteristics are found in short-lived species. However, two populations of *Arisaema triphyllum* with $\lambda > 1.2$ were insensitive to environmental stochasticity, showing no extinction until extremely high levels (environmental stochasticity characterized by variance to mean ratios greater than ten). This species reproduces both clonally and by seed, and the multiple reproduction modes buffer the species against reproductive failure under environmental stochasticity. Positive covariances among matrix elements, not modeled here, would reduce this buffering effect.

The threat posed to population survival by environmental variation appeared almost entirely due to variation in mortality, growth, and reproduction status and not to variation in reproductive output. In fact, the addition of reproductive variation to environmental stochasticity in mortality and growth did not consistently increase or decrease extinction probability. Generally, environmental stochasticity in reproductive output by itself was not sufficient to generate extinctions.

Most published studies do not provide sufficient detail (matrices for each year of the study) to calculate variance/mean ratios (V/M) for matrix elements, as measures of observed environmental stochasticity. I assembled information from studies of eleven species with recently published studies (Table 3.3). Median V/M for all matrix elements was consistently in the 0.02–0.19 range, corresponding to moderate to high environmental stochasticity (multipliers 200–1900). However, V/M for certain non-fecundity matrix elements included high environmental stochasticity (V/M values > 0.3 for all species, multiplier > 3000). In contrast, based on the available data, variation in fecundity is orders of magnitude higher than variation in non-fecundity elements (Table 3.3). Demographic stochasticity had significant effects on extinction probability (for an initial population size of 200) for only a few populations. Extinction due to demographic stochasticity was more likely for species with $\lambda \sim 1$ (Figure 3.3). In these species, chance variation in individual fates can be crucial to the survival of small populations. Populations sensitive to

Table 3.3

Variance/mean ratios for demographic parameters, derived from published studies on plants (ordered by date). l.f: S: # sites I: # intervals St: # stages v/m: median variance to mean ratio across elements CD: Combined fecundity and other data; can't separate NF: No fecundity data for every year.

Citation	Species	S	I	St	v/m	Range: All Parameters	Range: All but Fecundity	Range: Fecundity
O'Connor 1993	*Aristida bipartita*	2	4	4	0.07–0.08	0.01–0.35	CD	CD
O'Connor 1993	*Bothriochloa insculpta*	2	4	5	0.07–0.13	0.001–0.97	CD	CD
O'Connor 1993	*Heteropogon contortus*	2	4	5	0.11–0.12	0.004–0.85	CD	CD
O'Connor 1993	*Setaria incrassata*	2	4	4	0.02–0.11	0.004–1.50	CD	CD
O'Connor 1993	*Themeda triandra*	2	4	6	0.09–0.19	0.003–2.23	CD	CD
O'Connor 1993	*Digitaria eriantha*	2	4	3	0.08–0.11	0.003–1.19	CD	CD
Nault and Gagnon 1993	*Allium tricoccum*	1	4	15	0.016	0.0007–16.6	0.0007–0.18	8.61–16.6
Alvarez-Buylla 1994	*Cecropia obtusifolia*	1	3 (7)	8	0.06	0.000006–26992	0.000006–0.10	1468–26992
Horvitz and Schemske 1995	*Calathea ovandensis*	4	5	8	0.06–0.14	0.0003–28.02	0.0003–1.0	0.79–28.02
Guerrant 1995	*Erythronium elegans*	4	4	11	0.03–0.07	NF	0.0001–1.141	NF
Menges and Dolan in review	*Silene regia*	13	2–6	5	0.02–0.11	NF	0.0004–0.5	NF

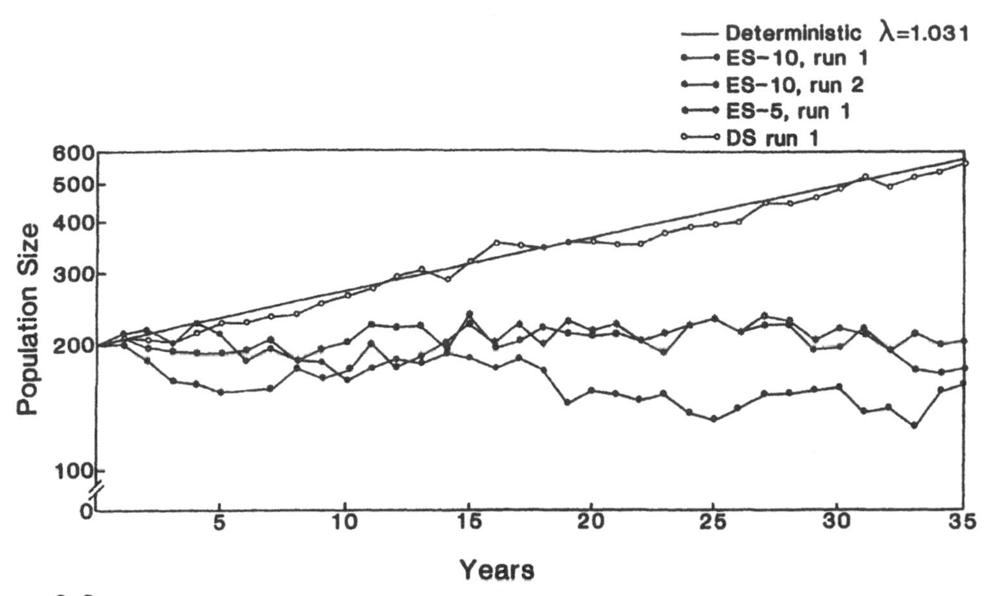

Figure 3.3
Demographic stochasticity increases extinction probability to modest levels only in populations with delayed reproduction and modest λ.

demographic stochasticity were also particularly sensitive to environmental stochasticity, which has its greatest effects on very small populations. The following section explores the effect of initial population size on extinction probability, under both types of stochasticity.

Stochasticity puts small populations at risk, while larger populations may be able to buffer stochasticity, at least for the short term. For the purposes of these analyses, I defined minimum viable population (MVP) as the minimum initial population size necessary to reduce extinction probability below 5% for 100 years. Initial size was then varied iteratively, with other simulation conventions unchanged.

Patterns were similar for all species and projection matrices. For example, for *Astrocaryum mexicanum* at Site BB, demographic stochasticity produces an MVP equal to 50, whereas modest environmental stochasticity produces an MVP greater than 100 (Figure 3.4). As expected, a greater MVP is required to buffer larger environmental stochasticity.

DISCUSSION

Stochasticity and Population Growth

Environmental stochasticity clearly produces greater fluctuations and poses a greater threat of extinction than demographic stochasticity unless populations become very small (Shaffer 1987; Lande 1993). Therefore, a minimum viable population (MVP) necessary to counter environmental stochasticity will be larger than that required to counter demographic stochasticity only. Increasing environmental stochasticity increases variation in population trends and increases extinction probability (see also Tuljapurkar and Orzack 1980; Caswell 1989; Nunney and Campbell 1993; Doak, Kareiva, and Klepetka 1994; Wissel and Zaschke 1994). Most models suggest environmental stochasticity slows population growth (but see Cohen 1979b). Other effects of increasing environmental stochasticity

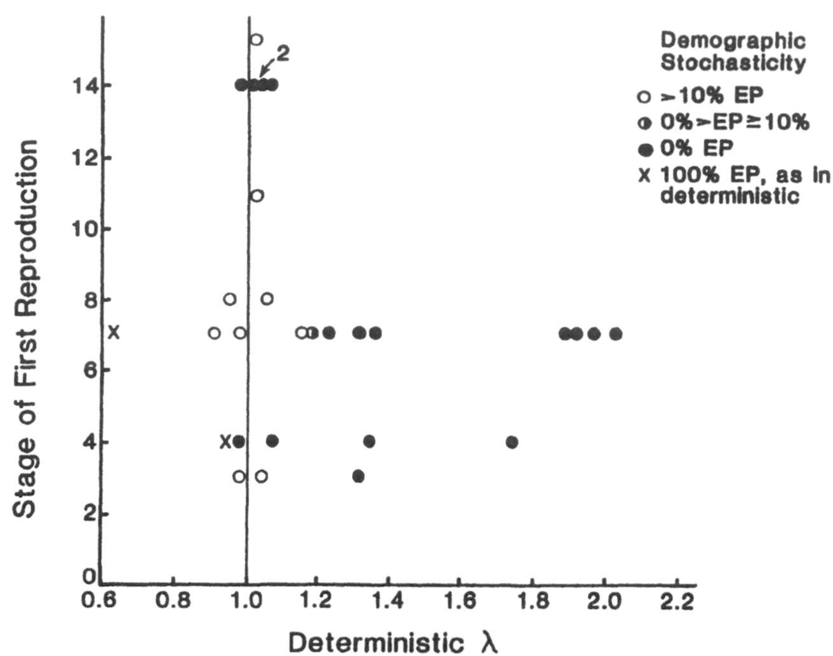

Figure 3.4
Minimum viable population (MVP) sizes necessary to limit extinction probabilities to less than 5 percent for Astrocaryum mexicanum *at site BB (Piñero, Martinez-Ramos, and Sarukhan 1984), for demographic stochasticity and two levels of environmental stochasticity. (ES 25 is a V/M ratio of 0.0025 for non-fecundity elements, 0.25 for fecundity; ES 10 ratios are 0.0010 and 0.10 respectively).*

found in this study include decreased time to extinction (see Goodman 1987), decreased population size, and a greater proportion of years with declining populations. My results support the contention that populations with lowest λ are most sensitive to both demographic stochasticity and to low levels of environmental stochasticity (Nunney and Campbell 1993). Populations with high λ can persist for a long time in the face of moderate environmental stochasticity (Lande 1993). As environmental stochasticity increases, the importance of mean population growth rate as a description of population dynamics and extinction is lessened (e.g., Ginzburg *et al.* 1982; Caswell 1989). In this study, very high levels of environmental stochasticity produce extinction over a large range of life histories.

A comparison of levels of environmental stochasticity used in my simulation models with observed levels in 11 perennial species suggests that real-world levels of environmental stochasticity can produce significant extinction in populations. Moderate levels of environmental stochasticity, which kept modeled populations from growing large and produced extinctions in slowly growing populations ($\lambda < 1.1$), occurred for most matrix elements in most species. High environmental stochasticity levels, which threatened even rapidly growing populations, occurred in many matrix elements in many species. Fecundity variation was huge, although even these levels of fecundity variation rarely produced extinctions.

Populations with low λ may have evolved in "stable" environments having few environmental fluctuations. Are these populations really more sensitive to environmental stochasticity as suggested by this analysis? The answer to this question lies in the definition of environmental stochasticity used in this research: i.e., variation in demographic parameters. An external environmental event (e.g., a drought) can be buffered demographically (Tuljapurkar and Orzak 1980), although its homeostatic reactions may require greater expenditures, such as increased reproductive effort to achieve the same reproductive output (Caswell 1983). Without demographic buffering, populations with low λ are particularly sensitive to environmental stochasticity.

Population models with more life history stages appeared buffered from extinction relative to populations modeled with fewer stages, a likely consequence of the lack of covariance in the model. Positive covariances among matrix elements tend to reduce population growth (Caswell 1989, p. 225; Ferson and Burgman 1995; but see Doak, Kareiva, and Klepetka 1994). Recently, Ferson and Burgman (1995) have proposed a method for estimating bounds on extinction risks when covariances are unknown. Algorithms are also available to choose the number of stages and their cutoffs based on balancing errors of estimation and errors of distribution (Vandermeer 1978; Moloney 1986).

Population behavior is, to a great extent, controlled by "weakest links" in time. Critically unfavorable years may play a major role in causing population changes and extinctions (Bengtsson 1993; McCarthy, Burgman, and Ferson 1995) because long-term population growth depends on geometric means of short-term growth rates (Lewontin and Cohen 1969). Increasing environmental stochasticity increases the proportion of such disastrous years. Populations with reduced variation in demographic parameters may have lower extinction probabilities even if λ is relatively low (cf.Tuljapurkar and Orzack 1980). Reduction of variation in demographic parameters may be a necessary part of species management.

Importance and Implications of Demographic Approaches

Extinction risks due to environmental variations are usually considered to be of primary short-term importance (Lande 1988; Menges 1991; Schemske *et al.* 1994), although demographic approaches also can include the effects of projected deterministic processes (e.g., loss of habitat) that are the main reasons most species are endangered (Holsinger 1995). The flexibility and inclusiveness of demographic approaches in evaluating extinction risk have been proven in a number of studies (see introduction).

In plant species conservation, stochastic demographic modeling can be important in several practical ways. First, by focusing data collection on the dynamics of individuals (vs. measuring changes in population size; see Menges and Gordon 1996) and on quantifying the entire life cycle, it develops a holistic view of the life cycle and maximizes the probability of discovering biologically relevant information. The importance of environmental stochasticity, especially in mortality, also underscores the need for long-term monitoring of rare plants, and for a quantification of variation in demographic parameters, possible within the context of many monitoring programs. Because variation is so important, I encourage publication of individual year's matrices and demographic parameters.

Demographic monitoring and modeling also can be used to guide management. In particular, comparisons of experimental treatments for effects on extinction probability can help guide management (e.g., Bullock, Hill, and Silvertown 1994; Olmsted and Alvarez-Buylla 1995; Nantel, Gagnon, and Nault 1996). For example, simulations by Burgman and Lamont (1992) suggested moderate fire frequencies would increase population size but, at the same time, increase extinction risk because post-fire seedlings are susceptible to drought. However, low fire frequencies promote substantial declines in population size. Such a tradeoff is unlikely to be found by intuition or by deterministic approaches. Analyses that identify limiting stages or transitions in plant life histories (e.g., Nault and Gagnon 1993; Bullock, Hill, and Silvertown 1994) can also suggest mitigation strategies (Aplet, Laven, and Shaw 1994).

Alternatives to Full-blown Viability Assessments

The type of demographic modeling described in this paper requires a large investment in both data collection and modeling, and will not be possible for many species. In some cases, careful consideration of age or stage structures across multiple populations can provide insight into likely population trends (e.g., Oostermeijer, van't Veer, and den Nijs 1994), although precise forecasts are not possible. Some intermediate demographic parameters may provide additional information (non-integrated trend analysis; Pavlik 1994) and allow comparisons among populations or management treatments.

Analyses of general patterns among projection matrices could add generality and suggest how short-cut analyses could proceed. Elasticity analysis, which examines the proportional effects of small changes in matrix coefficients on λ (deKroon *et al.* 1986; Caswell 1989) can be compared among species. Single elasticity terms can be summed to calculate the elasticities of "loops" or sequences of life history traits (van Groenendael *et al.* 1994) or summed across broad classes of growth, fecundity, and survival (Enright and Watson 1991; Silvertown *et al.* 1993; Bullock, Hill, and Silvertown 1994; Silvertown, Franco, and Menges 1996). Silvertown *et al.* (1993, 1996) use the latter approach to generalize which parts of species' life cycles seem to be important to finite rates of increase (λ). For example, across species, λ for forest herbs were relatively sensitive to survival terms (Figure 2.2 in Silvertown, Franco, and Menges 1996), as compared for herbs of open habitats, which were more sensitive to growth terms.

Can such broad patterns of elasticities be used to focus efforts on critical aspects of life histories? There are at least three reasons to be cautious. First, certain species may be exceptions to general trends. For example, some forest herbs are shade-tolerant and some are gap specialists, and growth may be more important to the latter. Second, elasticities vary over time and among populations of a given species (e.g., Bullock, Hill, and Silvertown 1994; Horvitz and Schemske 1995; Lesica and Shelley 1995; Silvertown, Franco, and Menges 1996). Finally, management interventions aimed at augmenting processes with high elasticities may not be effective (Silvertown, Franco, and Menges 1996; Oostermeijer *et al.* 1996) because life history traits are not free to vary independently (van Groenendael and Slim 1988) as implied by elasticity analysis. For example, in declining populations of *Pedicularis furbishiae*, although elasticity terms for survival were highest, controlling mortality might be a poor management choice. Expanding populations had rapid plant growth, and controlling mortality would involve curtailing natural disturbances which provide ideal conditions for growth. This species appears to thrive in landscape with an intermediate disturbance frequency, and limiting disturbances to reduce mortality would be unwise (Menges 1990).

Larger-Scale Approaches

Evaluating extinction risks for individual populations is increasingly being integrated with community, ecosystem, and landscape-level approaches. Successional dynamics, changes in disturbance regimes, and habitat destruction can be used as drivers for demographic projections. For example, several studies have linked gap dynamics or successional models to demographic models to evaluate disturbance regime effects on demography (Stohlgren and Rundel 1986, Chapman, Reese, and Clarke 1989; Cipollini, Wallace-Senft, and Whigham 1994; Beissinger 1995). For many species, metapopulation dynamics are being considered in evaluating extinction risk (Hastings and Harrison 1994; Schemske *et al.* 1994; Possingham and Davies 1995), especially in species affected by habitat fragmentation (e.g., Lindenmayer and Lacy 1995). Conservation and management are conducted mainly at larger scales involving landscapes, communities, and metapopulations, and explicit links between alternative management regimes and the population viability of key species will help management be effective.

ACKNOWLEDGMENTS

I have benefited from discussions with and comments from a number of colleagues, including Marlin Bowles, Becky Dolan, Peggy Fiedler, Kent Holsinger, Susan Kalisz, Peter Kareiva, Jeff Karron, Mark McPeek, Gerard Oostermeijer, Mary Palmer, Pedro Quintana-Ascencio, Mark Shaffer, Jonathan Silvertown, and Bruce Wilcox. Stacia Yoon assisted with computer runs, and Pedro Quintana-Ascencio extracted literature values for variance/mean ratios of matrix elements. Marcia Hestand helped prepare the manuscript. Thanks also to the authors of the empirical studies on which this research is based.

LITERATURE CITED

Alvarez-Buylla, E.R. 1994. Density dependence and patch dynamics in tropical rain forests: Matrix models and applications to a tree species. *American Naturalist* 143:155–191.

Alvarez-Buylla, E.R. and M. Slatkin. 1991. Finding confidence limits on population growth rates. *Trends in Ecology and Evolution* 6:221–224.

Alvarez-Buyulla, E.R. and M. Slatkin. 1994. Finding confidence limits on population growth rates: three real examples revised. *Ecology* 75:255–260.

Aplet, G.H., R.D. Laven, and R.B. Shaw. 1994. Application of transition matrix models to the recovery of the rare Hawaiian shrub, *Tetramolopium arenarium* (Asteraceae). *Natural Areas Journal* 114:99–106.

Beissinger, S.R. 1995. Modeling extinction in periodic environments: Everglades water levels and snail kite population viability. *Ecological Applications* 5:618–631.

Bengtsson, K. 1993. *Fumaria procumbens* on Öland—population dynamics of a disjunct species at the northern limit of its range. *Journal of Ecology* 81:745–758.

Bierzychudek, P. 1982. The demography of jack-in-the-pulpit, a forest perennial that changes sex. *Ecological Monographs* 52:335–51.

Billington, H.I. 1991. Effect of population size on genetic variation in a dioecious conifer. *Conservation Biology* 5:115–119.

Bullock, J.M., B.C. Hill, and J. Silvertown. 1994. Demography of *Cirsium vulgare* in a grazing experiment. *Journal of Ecology* 82:101–111.

Burgman, M.A., S. Ferson, and H.R. Akcakaya. 1993. *Risk assessment in conservation biology.* London: Chapman and Hall.

Burgman, M.A. and B.B. Lamont. 1992. A stochastic model for the viability of *Banksia cuneata* populations: Environmental, demographic, and genetic effects. *Journal of Applied Ecology* 29:719–727.

Burns, B.R. and J. Ogden. 1985. The demography of the temperate mangrove [*Aricennia marina* (Forsk.) Vierh.] at its southern limit in New Zealand. *Australian Journal of Ecology* 10:125–133.

Carlsson, B.A. and T.V. Callaghan. 1991. Simulation of fluctuating populations of *Carex bigelowii* tillers classified by type, age, and size. *Oikos* 60:231–240.

Caswell, H. 1983. Phenotypic plasticity in life history traits: Demographic effects and evolutionary consequences. *American Zoologist* 23:35–46.

Caswell, H. 1989. *Matrix population models*. Sunderland, MA: Sinauer Associates.

Caswell, H. and P. Werner. 1978. Transient behavior and life history analysis of teasel (*Dipsacus sylvestris* Huds.). *Ecology* 59:53–66.

Chapman, S.B., R.J. Reese, and R.T. Clarke. 1989. The behavior of populations of the marsh gentian (*Gentiana pneumonanthe*): a modeling approach. *Journal of Applied Ecology* 26:1059–1072.

Cipollini, M.L., D.A. Wallace-Senft, and D.F. Whigmam. 1994. A model of patch dynamics, seed dispersal, and sex ratio in the dioecious shrub *Lindera benzoin* (Lauraceae). *Journal of Ecology* 82:621–633.

Cochrane, M.E. and S. Ellner. 1992. Simple methods for calculating age-based life history parameters for stage-structured populations. *Ecological Monographs* 62:345–364.

Cohen, J.E. 1979a. Comparative statistics and stochastic dynamics of age-structured populations. *Theoretical Population Biology* 16:159–171.

Cohen, J.E. 1979b. Long-run growth rates of discrete multiplicative processes in Markovian environments. *Journal of Mathematical Analysis and Applications* 69:243–51.

DeKroon, H., A. Plaiser, J.M. Groenendael, and H. Caswell. 1986. Elasticity as a measure of the relative contribution of demographic parameters to population growth rate. *Ecology* 67:1427–1431.

Doak, D., P. Kareiva, and B. Klepetka. 1994. Modeling population viability for the desert tortoise in the western Mojave dessert. *Ecological Applications* 4:446–460.

Enright, N.J. 1982. The ecology of *Araucaria* species in New Guinea. III. Population dynamics of sample stands. *Australian Journal of Ecology* 7:227–37.

Enright, N.J. and J. Ogden. 1979. Applications of transition matrix models in forest dynamics: *Araucaria* in Papua New Guinea and *Nothofagus* in New Zealand. *Australian Journal of Ecology* 4:3–23.

Enright, N.J. and A.D. Watson. 1991. A matrix population model analysis for the tropical tree *Araucaria cunninghamii*. *Australian Journal of Ecology* 16:507–520.

Ferson, S. and M.A. Burgman. 1995. Correlations, dependency bounds, and extinction risks. *Biological Conservation* 73:101–105.

Fiedler, P.L. 1987. Life history and population dynamics of rare and common mariposa lilies (*Calochortus* Pursh: Liliaceae). *Journal of Ecology* 75:977–95.

Fowler, N.L. 1995. Density-dependent demography in two grasses: A five-year study. *Ecology* 76:2145–2164.

Frankel, O.H. and M.E. Soulé. 1981. *Conservation and evolution*. Cambridge: Cambridge University Press.

Franklin, I.R. 1980. Evolutionary change in small populations. In *Conservation Biology: an evolutionary-ecological perspective*, eds. M.E. Soulé and B.A. Wilcox, 134–50. Sunderland, MA: Sinauer Associates.

Ginzburg, L.R., L.B. Slobodkin, K. Johnson, and A.G. Bindman. 1982. Quasiextinction probabilities as a measure of impact on growth. *Risk Analysis* 2:171–81.

Ginzburg, L.R., S. Ferson, and H.R. Akcakaya. 1990. Reconstructability of density dependence and the conservative assessment of extinction risks. *Conservation Biology* 4:63–73.

Goodman, D. 1987. The demography of chance extinction. In *Viable populations for conservation*, ed. M.E. Soulé, 11–34. Cambridge: Cambridge University Press.

Guerrant, E.O. 1995. Comparative demography of *Erythronium elegans* in two populations: One thought to be in decline (Lost Prairie) and one presumably healthy (Mt. Hebo): Interim report on three transitions, or ten years of data. Berry Botanical Garden, Portland, Oregon. 73 pp.

Hamrick, J.L., M.J.W. Godt, D.A. Murawski, and M.D. Loveless. 1990. Correlations between species traits and allozyme diversity: Implications for conservation biology. In *Genetics and conservation of rare plants*. eds. D.A. Falk and K.E. Holsinger, 75–86. New York: Oxford University Press.

Hartshorn, G.L. 1975. A matrix model of tree population dynamics. In *Tropical ecological systems: Trends in terrestrial and aquatic research*, eds. F.B. Golley and E. Medina, 454–51. New York: Springer-Verlag.

Hastings, A. and S. Harrison. 1994. Metapopulation dynamics and genetics. *Annual Review of Ecology and Systematics* 25:167–188.

Heschel, M.S. and K.N. Paige. 1995. Inbreeding depression, environmental stress, and population size variation in Scarlet Gilia (*Ipomopsis aggregata*). *Conservation Biology* 9:126–133.

Holsinger, K.E. 1995. Population biology for policy makers. *BioScience* (Supplement):S10-S20.

Horvitz, C.C. and D.W. Schemske. 1995. Spatiotemporal variation in demographic transitions of a tropical understory herb: Projection matrix analysis. *Ecological Monographs* 65:155–192.

Huenneke, L.F. 1991. Ecological implications of genetic variation in plant populations. In *Genetics and conservation of rare plants*, eds. D.A. Falk and K.E. Holsinger, 31–44. New York: Oxford University Press.

Jain, S.K. 1976. The evolution of inbreeding in plants. *Annual Review of Ecology and Systematics* 7:469–495.

Kalisz, S. and M.A. McPeek. 1992. Demography of an age-structured annual: Resampled projection matrices, elasticity analyses, and seed bank effects. *Ecology* 73:1082–1093.

Karron, J.D. 1987. A comparison of levels of genetic polymorphism and self-compatibility in geographically restricted and widespread plant congeners. *Evolutionary Ecology* 1:47–58.

Lande, R. 1988. Genetics and demography in biological conservation. *Science* 241:1455–1460.

Lande, R. 1993. Risks of population extinction from demographic and environmental stochasticity and random catastrophes. *American Naturalist* 142:911–927.

Lande, R. 1995. Mutation and conservation. *Conservation Biology* 9:782–791.

Lefkovitch, L.P. 1965. The study of population growth in organisms grouped by stages. *Biometrics* 21:1–18.

Leigh, E.G., Jr. 1981. The average lifetime of a population in a varving environment. *Journal of Theoretical Biology* 90:213–39.

Lesica, P. and J.S. Shelly. 1995. Effects of reproductive mode on demography and life history in *Arabis fecunda* (Brassicaceae). *American Journal of Botany* 82:752–762.

Leslie, P.H. 1945. On the use of matrices in certain population mathematics. *Biometrika* 33:183–212.

Lewontin, R.L. and D. Cohen.1969. On population growth in a randomly varying environment. *Proceedings of the National Academy of Science* 62:1056–60.

Lindenmayer, D.B. and R.C. Lacy. 1995. A simulation study of the impacts of population subdivision on the mountain brushtail possum *Trichosurus caninus* Ogilby (Phalangeridae: Marsupialia) in south-eastern Australia. 1. Demographic stability and population persistence. *Biological Conservation* 73:119–129.

Lynch, M., J. Conery, and R. Bürger. 1995. Mutation accumulation and the extinction of small populations. *American Naturalist* 146:489–518.

May, R.M. 1973. *Stability and complexity of model ecosystems*. Princeton: Princeton University Press.

McCarthy, M.A., M.A. Burgman, and S. Ferson. 1995. Sensitivity analysis for models of population viability. *Biological Conservation* 73:93–100.

Meagher, T.R. 1986. Analysis of paternity within a natural population of *Chamaelirium luteum*. I. Identification of most likely male parents. *American Naturalist* 128:199–212.

Menges, E. 1990. Population viability analysis for an endangered plant. *Conservation Biology* 4:41–62.

Menges, E.S. 1991. The application of minimum viable population theory to plants. In *Genetics and conservation of rare plants*, eds. D.A. Falk and K.E. Holsinger, 45–61. New York: Oxford University Press.

Menges, E.S. and D.R. Gordon. 1996. Three levels of monitoring intensity for rare plant species. *Natural Areas Journal* 16:227–237.

Menges, E.S. and R.W. Dolan. In review. Demographic viability of populations of *Silene regia* in midwestern prairies: relationships with fire management, genetics, geography, population size, and isolation. Submitted to *Journal of Ecology*.

Moloney, K.A. 1986. A generalized algorithm for determining category size. *Oecologia* 69:176–180.

Nantel, P., D. Gagnon, and A. Nault. 1996. Population viability analysis of American ginseng and wild leek harvested in stochastic environments. *Conservation Biology* 10:608–621.

Nault, A. and D. Gagnon. 1993. Ramet demography of *Allium tricoccum*, a spring ephemeral, perennial forest herb. *Journal of Ecology* 81:101–119.

Nunney, L. and K.A. Campbell. 1993. Assessing minimum viable population size: Demography meets population genetics. *Trends in Ecology and Evology* 8:234–240.

O'Conner, T.G. 1993. The influence of rainfall and grazing on the demography of some African savanna grasses: A matrix modeling approach. *Journal of Applied Ecology* 30:119–132.

Olmsted, I. and E.R. Alvarez-Buylla. 1995. Sustainable harvesting of tropical trees: Demography and matrix models of two palm species in Mexico. *Ecological Applications* 5:484–500.

Oostermeijer, J.G.B., M.W. van Eijck, and J.C.M. den Nijs. 1994. Offspring fitness in relation to population size and genetic variation in the rare perennial plant species *Gentiana pneumonanthe* (Gentianaceae). *Oecologia* 97:289–296.

Oostermeijer, J.G.B, R. van't Veer, and J.C.M. den Nijs. 1994. Population structure of the rare, long-lived perennial *Gentiana pneumonanthe* in relation to vegetation and management in the Netherlands. *Journal of Applied Ecology* 31:428–438.

Oostermeijer, J.G.B., M.W. van Eijck, N.C. van Leeuwen, and J.C.M. den Nijs. 1995. Analysis of the relationship between allozyme heterozygosity and fitness in the rare *Gentiana pneumonanthe* L. *Journal of Evolutionary Biology* 8:739–759.

Oostermeijer, J.G.B., M.L. Brugman, E.R. de Boer, and J.C. M. den Nijs. 1996. Temporal and spatial demographic variation in the rare perennial herb *Gentiana pneumonanthe. Journal of Ecology* 84:153–166.

Ouborg, N.J. and R. van Treuren. 1994. The significance of genetic erosion in the process of extinction 4. Inbreeding load and heterosis in relation to population size in the mint *Salvia pratensis. Evolution* 48:996–1008.

Pavlik, B.M. 1994. Demographic monitoring and the recovery of endangered plants. In *Restoration of endangered species,* eds. M.L. Bowles and C.J. Whelan, 322–350. Cambridge: Cambridge University Press.

Piñero, D., M. Martinez-Ramos, and J. Sarukhan. 1984. A population model of *Astrocaryum mexicanum* and a sensitivity analysis of its finite rate of increase. *Journal of Ecology* 72:977–91.

Possingham, H.P. and I. Davies. 1995. ALEX: A model for the viability analysis of spatially structured populations. *Biological Conservation* 73:143–150.

Prober, S.M. and A.H.D. Brown. 1994. Conservation of the grass box woodlands: Population genetics and fragmentation of *Eucalyptus albens. Conservation Biology* 8:1003–1013.

Raijman, L.E.L., N.C. van Leeuwen, R. Kersten, J.G.B. Oostermeijer, J.C.M. den Nijs, and S.B.J. Menken. 1994. Genetic variation and outcrossing rate in relation to population size in *Gentiana pneumonanthe* L. *Conservation Biology* 8:1014–1026.

Schemske, D.W., B.C. Husband, M.H. Ruckelshaus, C. Goodwillie, I.M. Parker, and J.G. Bishop. 1994. Evaluating approaches to the conservation of rare and endangered plants. *Ecology* 75:584–606.

Shaffer, M.L. 1981. Minimum population sizes for species conservation. *BioScience* 31:131–34.

Shaffer, M.L.1987. Minimum viable populations: Coping with uncertainty. In *Viable populations for conservation*, ed. M.E. Soulé, 69–86. Cambridge: Cambridge University Press.

Silander, J.A. 1985. The genetic basis of the ecological amplitude of *Spartina patens*. II. Variance and correlation analysis. *Evolution* 39:1034–52.

Silva, J.F., J. Raventos, H. Caswell, and M.C. Trevisan. 1991. Population responses to fire in a tropical savanna grass, *Andropogon semiberbis*: a matrix model approach. *Journal of Ecology* 79:345–356.

Silvertown, J., M. Franco, I. Pisanty, and A. Mendoza. 1993. Comparative plant demography: Relative importance of life cycle components to the finite rate of increase in woody and herbaceous perennials. *Journal of Ecology* 81:465–476.

Silvertown, J., M. Franco, and E. Menges. 1996. Interpretation of elasticity matrices as an aid to the managment of plant populations for conservation. *Conservation Biology* 10:591–597.

Stohlgren, T.J. and P.W. Rundel. 1986. A population model for a long-lived resprouting chaparral shrub: *Adenostoma fasciculatum. Ecological Modeling* 34:245–257.

Tuljapurkar, S.D., and S.H. Orzack. 1980. Population dynamics in variable environments. 1. Long-run growth rates and extinction. *Theoretical Population Biology* 18:314–42.

Vandermeer, J. 1978. Choosing category size in a stage projection matrix. *Oecologia* 32:79–84.

van Groenendael, J.M. and P. Slim. 1988. The contrasting dynamics of two populations of *Plantago lanceolata* classified by age and size. *Journal of Ecology* 76:585–599.

van Groenendael, J.M., H.D. de Kroon, S. Kalisz, and S. Tuljapurkar. 1994. Loop analysis: Evaluating life history pathways in population projection matrices. *Ecology* 75:2410–2415.

van Treuren, R., B. Bijlsma, W. van Delden, and N.J. Ouborg. 1991. The significance of genetic erosion in the process of extinction. I. Genetic differentiation in *Salvia prutensis* and *Scabiosa columbaria* in relation to population size. *Heredity* 66:181–189.

Werner, P.A. and H. Caswell. 1977. Population growth rates and age vs. size distribution models for teasel (*Dipsacus silvestris* Huds.). *Ecology* 58:1103.

Widen, B. 1993. Demographic and genetic effects on reproduction as related to population size in a rare, perennial herb, *Senecio integrifolius*. *Biological Journal of the Linnean Society* 50:179–195.

Wissel, C. And S.H. Zaschke. 1994. Stochastic birth and death processes describing minimum viable populations. *Ecological Modeling* 75/76:193–201.

CHAPTER

4

Physical Effects of Habitat Fragmentation

LAURI K. FREIDENBURG

The consequences of habitat fragmentation are widely discussed in the conservation literature, usually with an emphasis on metapopulation dynamics and landscape ecology. Spatial aspects of fragmentation typically receive the greatest attention; in particular, how the size and shape of habitat patches and the distance between patches affect resident species. Relatively little notice has been given to the most obvious and direct effect of fragmentation: the physical changes that accompany the division of continuous habitat into fragments. In terrestrial systems, physical change is first expressed in the increase of perimeter to area ratio for each habitat patch. This change initiates a sequence of related physical changes that can profoundly influence resident flora and fauna. A comparable sequence of events follows fragmentation in aquatic systems. Field research is just beginning to provide links between patterns of fragmentation and the mechanisms that may lead to biotic changes. My aim in this chapter is to highlight the non-spatial effects of fragmentation. I limit my discussion to two types of ecosystems: forests and rivers. The first section considers the direct abiotic impacts of fragmentation on remnant habitat patches, often referred to as "edge effects" in the literature, and illustrates how these physical factors affect various organisms. In the second section I address the indirect effects of fragmentation, the biotic effects we can expect as a secondary consequence of the direct effects of fragmentation. I center many of my examples on recent research in an attempt to keep pace with the burgeoning conservation literature.

DIRECT EFFECTS OF FRAGMENTATION

Habitat fragmentation causes a profound change in the character of the landscape through the creation of habitat patches. The simple act of creating habitat patches can lead to a multitude of physical changes both within the patch itself as well as in the surrounding matrix; these changes directly affect resident species. Here, a physical change includes any change in an abiotic factor (e.g., temperature, sunlight) that directly results from habitat fragmentation. The spatial scale at which a physical change takes place can range from a few meters to thousands of kilometers and from a temporal scale of hours to millennia. This extreme range of potential effects makes it imperative that we understand how habitat modification propagates through the environment.

Terrestrial Systems

The most familiar form of anthropogenic fragmentation is the dissection of forest into isolated patches interspersed with clear cuts, agricultural fields, or human development. This type of fragmentation creates new edges between forest and "open" habitat. Biologists have long recognized the unique properties of forest edge habitat and the increased species diversity associated with it (Leopold 1933; Ghiselin 1977; Yahner 1988). Indeed, the creation and maintenance of such edges has been advocated by wildlife managers hoping to increase abundances of game species (see Yoakum and Dasmann 1971). More recently, however, this benevolent view of edge habitat has virtually disappeared. Edges now are often perceived as hostile environments, having been viewed as reducing the effective area of remnant patches (Kapos 1989; Saunders *et al.* 1991; Meffe and Carroll 1994). In fact, the distribution and abundance of many species differs between edge and interior habitat (e.g., Fraver 1994; Matlack 1994; Sekgororoane and Dilworth 1995), implying that intrinsic differences exist between the edge and interior of habitat patches. Clearly, the creation of edges engenders some fundamental change in the character of a habitat patch.

Current research has documented five primary physical variables that are affected by forest fragmentation: air flow, light (usually measured as photosynthetically active radiation, PAR), temperature, humidity, and soil moisture (Table 4.1). Undoubtedly, there are others. It is evident that these factors do not change independently. As an example, alterations in light and air flow can be expected to influence air temperature or soil moisture significantly within forest patches (Matlack 1993; Chen *et al.* 1995). Physical effects have been quantified by comparing the magnitude of any particular variable at a reference interior station (presumably within the core region) to sampling stations closer to the patch edge. This research has demonstrated that the magnitude of physical effects and their penetration into forest patches change as a function of the variable considered (Chen *et al.* 1995), forest type (Williams-Linera 1990; Chen *et al.* 1992), surrounding matrix (Young and Mitchell 1994; Camargo and Kapos 1995), and aspect (Brothers 1993). Changes in air flow, for instance, can penetrate over 240 m into patches (Chen *et al.* 1995). And although measurable gradients of humidity, temperature, and light often penetrate less than 50 m into patches, the effects can be substantial. Compared with interior sites, forests near edges can be 37% drier (air), 14% warmer (daytime air temperature), and 1,221% brighter (Young and Mitchell 1994).

While it is probably impossible to agree on a single "edge width" (since different

Table 4.1
*Two modes of fragmentation and their reported physical consequences to remnant habitat. *Physical effects are typically heterogeneously distributed in fragments, being greatest at the boundary and diminishing as the distance from edge increases.*

System	Mode of Fragmentation	Effects within Remnant Habitat*	Sources
Forest	Logging	Changed insolation, air temperature, soil temperature, air flow Decreased humidity Changes in hydrology, soil moisture	Geiger (1965), Kapos (1989)
River	Impoundment	Decrease in large woody debris Changes in flow, sediment transport, temperature, substrate, channel morphology, water chemistry, bank erosion	Petts (1984)

physical effects exert their influences over a variety of distances), it appears that physical effects often propagate over 100 m into forest patches. The result is that typical fragmented landscapes may often be predominantly "edge"; where edge is defined by the extent of influence of some physical factor. For example, a representative survey of the literature shows that the physical effects of fragmentation may touch a majority of forest area in many regions (Table 4.2). In fact, for some of these landscapes, most forest patches fall within the size range where they may be all edge. This has potentially important implications for all fragmented landscapes, for if habitat patches consist primarily of edge habitat, then what are the consequences for resident species? A number of studies have shown that forest fragments differ in population abundance or species composition compared with unfragmented forest (e.g., Pahl *et al.* 1988; Matthysen *et al.* 1995). In recent years, explanations for this pattern have tended to emphasize the influence of isolation on rates of invasion and extinction (metapopulation dynamics) and the effect of fragment size on resident populations (island biogeography) (see reviews by Andren 1994; Thomas 1994). It has been proposed, however, that many species may be predominantly affected by changes in their physical environment due to overwhelming edge effects (Saunders *et al.* 1991; Murcia 1995).

Effects on Vegetation. Forest vegetation responds to habitat fragmentation in a variety of ways. In the years following fragmentation, researchers have documented increases in tree stem density and basal area at or near a forest patch edge (Brothers 1993; Young and Mitchell 1994; but see Chen *et al.* 1992 for an exception) and species richness can increase at patch edges (Brothers 1993; Fraver 1994). This increase in species diversity is generally attributed to the establishment of disturbance-adapted plants as well as increased seedling and sapling recruitment (Williams-Linera 1990; Laurance 1991). In addition, shade-intolerant species tend to proliferate near and at edges and decrease towards the patch interior while shade-tolerant species exhibit the reverse trend (Chen *et al.* 1992; Brothers 1993; Fraver 1994; Young and Mitchell 1994). The primary variables driving these patterns, through their influence on edge microclimate, appear to be light and aspect (Matlack 1993; Fraver 1994). Forest type and the surrounding habitat also determine the impact fragmentation will have on habitat patches. Regardless of forest type or surrounding matrix, however, an almost universal response to forest fragmentation is significant tree mortality at patch edges (Lovejoy *et al.* 1986; Williams-Linera 1990; Chen *et al.* 1992). Several factors have been linked to this phenomenon; most common among them are windthrow and associated stresses caused by logging. At a subtler level there are also physiological responses of vegetation to forest fragmentation and edge proximity. Shure and Wilson (1993) examined differences in leaf toughness, phenolic content, and response to herbivores in conspecific trees growing in two different sizes of forest gaps. Forest gaps mimic forest patch edges in their increased exposure to solar radiation, although the size of the gap limits the amount of increased radiation. Trees found in large gaps developed tougher leaves, contained higher levels of phenolics, and better resisted herbivore damage than conspecifics in small gaps (Shure and Wilson 1993).

Studies of vegetation responses to fragmentation provide clear support for the importance of physical changes within remnant patches. While not uniformly understood, in many instances vegetative responses have been solidly linked to physically-mediated mechanisms.

Effects on Animals. Just as plant communities exhibit change associated with habitat fragmentation, so do animal communities. In contrast to the direct ties between

Table 4.2

Estimated prevalence of physical (edge) effects in fragmented landscapes. Calculation of % edge is a conservative estimate based on the assumptions that patches are round and edge effects are limited to 100 m or less. For studies presenting data for more than one time period the most recent survey was used.

Location	Cover Type	% Cover	Study Area	# Fragments	Mean Fragment	% Edge	Reference
Michigan and Wisconsin, USA	hardwood, conifer forest	75	16872 ha	1801	7.0 ha	89	Mladenoff et al. (1993)
Wisconsin, USA	oak forest	10	10000	111	9.0	83	Sharpe et al. (1987)
Oregon, USA	conifer forest	18	26250	298	16.2	69	Ripple et al. (1991)
New York, USA	mixed hardwood forest	25	203280	3088	16.4	68	Hill (1985)
Georgia, USA	pine, oak-pine forest	60	114264	4264	16.4	68	Turner and Ruscher (1988)
China	boreal forest	78	51984	1915	22.8	60	Tian et al. (1995)
New York, USA	mixed hardwood forest	35	11900	161	26.1	57	Smith et al. (1993)
Western Australia	*Eucalyptus* woodland	7	168000	13	134.9	28	Norton et al. (1995)

physical factors and vegetation response, however, evidence connecting changes in animal populations to physical effects of fragmentation is relatively sparse. Some of the best examples of physical effects on animals come from studies of invertebrates. Pajunen *et al.* (1995) compared spider assemblages in different sized patches of old forest surrounded by managed forest. Although patch size had no effect on species composition among patches, there was a distinct forest-to-edge gradient within patches. Small, web-building spiders were found most often in forest interiors while larger, diurnal and nocturnal hunters predominated in forest edges. The authors speculated that both light and vegetation type influenced spider assemblages (see also Huhta 1971).

In one of the few experimental demonstrations of fragmentation effects, Margules *et al.* (1994) measured abundances of two invertebrate species, a scorpion (*Cercophonius squama*) and a terrestrial amphipod (Family Tallitridae), before and after planned fragmentation of formerly contiguous forests. The authors predicted that smaller fragments should have lower abundances of each species. While this prediction was supported for the amphipod, scorpion abundances were unaffected. Margules *et al.* (1994) suggested that amphipods may have succumbed to changes in the physical conditions in remnant patches (changes in soil moisture in particular). Scorpions, on the other hand, may have been more resistant owing to their burrowing habits. In this case, physical effects were influenced by patch size and these effects differentially impacted species.

Even within a taxon, effects of fragmentation can vary. Klein (1989) determined that for carabid beetles, species richness and beetle size declined in forest fragments compared to contiguous forest. These results could be due to a change in soil microclimate at the forest edge, and Klein (1989) cites laboratory and field studies for various insects that support this explanation. In contrast, Halme and Niemela (1993) found that carabid beetle abundance and number of species increased in small forest patches and surrounding open areas when compared to large patches and contiguous forest. These authors concluded that local carabid species tended to favor vegetation found at edges and in open areas.

There at least two reasons why there is less than overwhelming evidence that fragmentation has physically mediated consequences for animal species. First, far fewer studies have examined the possibility that physical factors are important. In addition, animals may be better buffered (via motility, endothermy, etc.) than plants against the scale of physical alterations caused by fragmentation. For many animals the physical effects of fragmentation may be more frequently mediated indirectly (see *Indirect Effects of Fragmentation* below).

Aquatic Systems

While fragmentation is generally considered within the context of terrestrial environments, comparable processes occur when humans modify aquatic habitats. Indeed, the physical effects of fragmenting aquatic systems are widespread and relatively well understood (Baxter 1977; Ward and Stanford 1983; Petts 1984; see chapters 6 and 8). Although aquatic fragmentation occurs in a variety of ways, I use the example of impounded rivers to demonstrate the similarities between aquatic and terrestrial fragmentation and to highlight the importance of the physical effects of this brand of fragmentation.

Damming of a river creates three distinct habitats (Figure 4.1). Immediately upstream of the dam will be a newly created reservoir, upstream of the reservoir will be an unmodified river, and immediately below the dam will be river habitat altered by regulation and other effects of damming. As with terrestrial fragmentation, isolated habitat patches remain (upstream of the reservoir and below the dam) and an intervening habitat matrix

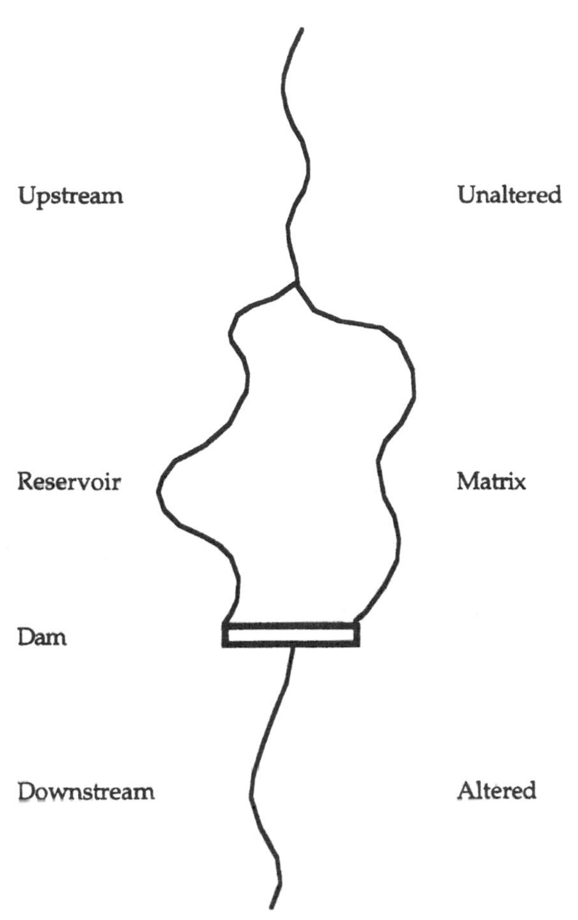

Figure 4.1
A graphical representation of a river fragmented by a dam. On the left-hand side are terms describing the habitat. On the right-hand side are terms describing the physical status of each habitat after fragmentation.

is created (i.e., the reservoir). And, too, a single dam can block fish migrations, flood entire valleys, reverse flow regimes, and reshape river channels (Petts 1984). It is difficult, therefore, to overestimate the potential physical impacts of dams on a global scale.

Historically, dams have played a large role in the development of human societies. Early Egyptians constructed dams to hold water for irrigation, and by Roman times dams were commonly used in the Mediterranean region for flood control and municipal water supply in addition to their agricultural applications (Petts 1984; Allan 1995). Overall, dams have been a feature of the landscape for over 5,000 years (Petts 1984), yet it has been only within the 20th century that technological advances have fostered widespread damming on a wide range of river types. During the peak of dam-building activity (1950–1980), dams were created at a rate of 700 per year worldwide (Petts 1989; Allan 1995). Within North America above Mexico, Europe, and the former Soviet Union, at least 85 of the 139 major river systems were severely fragmented by damming and water diversions (Dynesius and Nilsson 1994). Across this northern temperate region, just one major river system below the 47th parallel, the Pascagoula River, USA, remains unaffected by fragmentation (Dynesius and Nilsson 1994). Within much of the developed world, fragmented rivers are the norm.

Physical Effects of Dams. A dam's size, function, position within a watershed, and geographic location all influence its impact. Physical changes wrought by dams affect the character of an entire watershed. These changes can be loosely categorized into upstream and downstream impacts (Table 4.1). Upstream of a dam are two of the three major habitat types introduced previously, the newly created reservoir and the unaltered river (Figure 4.1). The river upstream of a dam is now isolated by the reservoir, but is otherwise virtually unchanged. In contrast, the reservoir itself undergoes a fundamental physical transformation in response to impoundment. Within a newly created reservoir, depth, volume ("size"), water chemistry, temperature, flow dynamics, and sediment transport are all dramatically modified from the original character of the river. Immediately upon filling, reservoirs often experience oxygen depletion and spikes in phosphorous concentration as submerged terrestrial vegetation decomposes. Flooded soil may yield soluble materials such as gypsum and rock salt; for example, salinity levels in Lake Mead were elevated for 15 years after damming (Baxter 1977; Petts 1984). The increased volume of water in a reservoir can store more heat than a flowing river, and seasonal thermal stratification often occurs. Sediment transported into a low velocity reservoir from a high velocity river differentially drops out of the water column. Large particulate matter can form deltas in the upstream portion of a reservoir while fine sediments can build up behind the dam. Importantly, these changes also have immediate consequences for the physical environment of the river downstream of the dam. Water released from a thermally stratified reservoir can modify considerably downstream temperatures, and the clarified reservoir water can quickly pick up sediment, scouring the riverbanks and armoring the substrate. Changes in water chemistry are also transmitted downstream of dams. Perhaps one of the most striking results of dam construction and operation is the modulation of flow regimes. In general, damming tends to depress seasonal variation in flow, while often increasing flow variation on shorter time scales (Petts 1984; Allan 1995; e.g., five-fold daily variation in rate of flow below a hydroelectric dam in Maine; Trotzky and Gregory 1974). Fragmented riverine systems communicate "edge effects" entirely downstream. As in terrestrial systems, many of these changes gradually decrease farther away from the "edge" created by the dam (Figure 4.2). Thus, river sections far enough downstream from a dam may, in some respects, resemble an unmodified river. Clearly, a single dam has an immense impact on the surrounding river and floodplain and the edge effects generated propagate for many kilometers. Rivers frequently support more than one dam, and so, like fragmented forests, the amount of unaffected river may be substantially less than we imagine.

Biotic Effects of Dams. There is no doubt that perturbations to aquatic systems significantly affect aquatic and semi-aquatic species. In a recent report released by The Nature Conservancy (1996), 13 major plant and animal groups in the U.S. were rated for their conservation status. Of particular note were the large proportion of aquatic species designated vulnerable, imperiled, critically imperiled, or presumed/possibly extinct. Aquatic organisms led all other groups in terms of conservation risk; 67% of all freshwater mussels, 65% of all crayfish, 38% of all amphibians, and 27% of all freshwater fishes were classified as at risk while only 16% of mammals and 14% of birds were designated as at risk.

Within riverine systems, dams are one of the primary causes of endangerment. It is

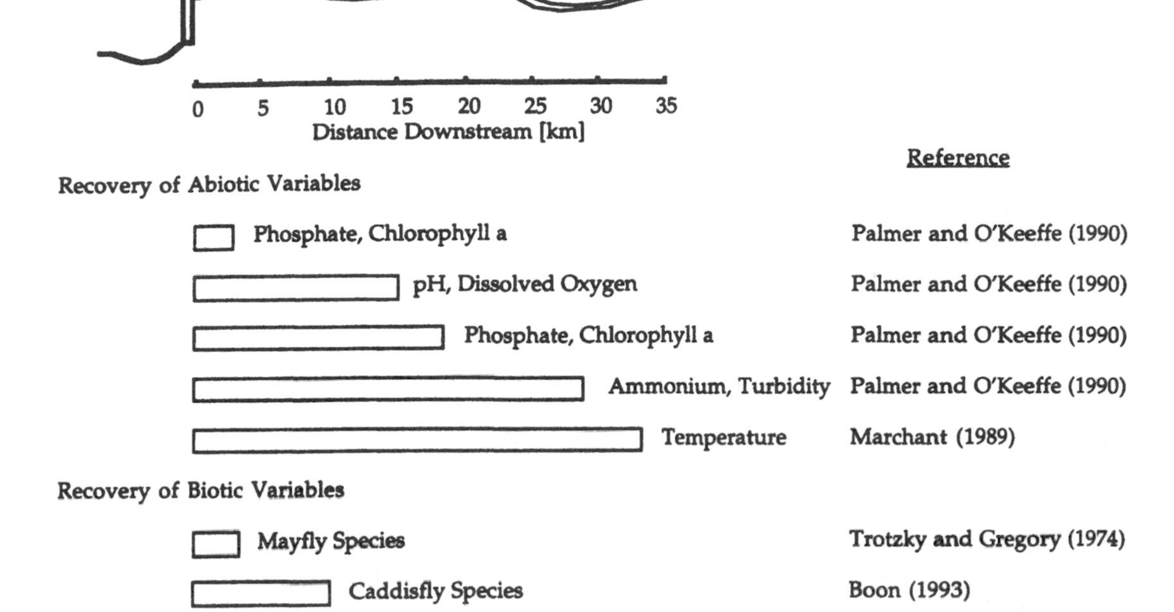

Figure 4.2
Edge widths in dammed rivers. The bars represent the distance downstream it takes for a variable to either attain upstream or pre-dam values or to stabilize. Phosphorus and chlorophyll are presented twice as they were measured under two different flow regimes.

no surprise that the radical changes damming elicits on the form and function of a river often lead to equally dramatic changes in the biota. Densities and biomass of macroinvertebrates can be decreased by orders of magnitude immediately below dams (Trotzky and Gregory 1974; Munn and Brusven 1987; Boon 1993). This response has been attributed to the increased flow fluctuation and associated scour present in dam tailwaters. Temperature can also play a role in macroinvertebrate distribution patterns, because many species have narrow temperature tolerances, and successful completion of many life history stages are often temperature-dependent. Hence, changes in a river's temperature regime may not only define suitable habitat, but also may alter the growth and development of a species (Lehmkuhl 1972; Munn and Brusven 1987). An example of this pattern was seen in Australia. Marchant (1989) documented decreased macroinvertebrate species richness immediately below a dam releasing water from the cool hypolimnion. When dam operation shifted to the release of warmer surface water, species richness began to increase (Marchant 1989).

Plant communities are similarly affected by modifications in downstream physical factors. In some rivers, a highly modified flow regime can stimulate algal growth in downstream riffles and create favorable conditions throughout the river channel for a variety of aquatic angiosperms, while in other rivers it may decimate the resident instream flora (Petts 1984). Riparian plant communities also respond to flow fluctuation. For example, native cottonwood populations in the Colorado River floodplain do not appear to be physiologically suited to the artificially imposed flow regime (Busch and Smith 1995). Within the reservoir itself, the shift from a lotic to lentic environment results in a commensurate shift in biota. The slow-moving waters of an impoundment are conducive to the maintenance of phytoplankton and zooplankton communities (Baxter 1977; Hart 1990) and promote the development of lentic fish assemblages (Petts 1984; Moyle and Leidy 1992; Martinez *et al.* 1994; Travnichek *et al.* 1995).

The Colorado River basin offers a prime example of how dams and their concomitant physical effects influence native biota. Prior to dam construction, the upper Colorado River basin supported 13 native fish species (Table 4.3). Several dams and several decades later, 52 fish species inhabited the upper basin. Changes in fish community composition were facilitated by the creation of lentic habitat in the form of reservoirs and regulated rivers. Martinez *et al.* (1994) sampled fish before and after construction of a dam on a tributary of the Colorado River and documented a significant shift in fish assemblage. Before closure of the dam, native fishes predominated throughout the reservoir basin. In the years following dam closure, 90% of the fishes in the reservoir and 80% of the fishes below the dam were non-native. Of the 13 native fish, six are currently listed by the U.S. Fish and Wildlife Service as endangered or threatened, and only three species are categorized as "abundant" (U.S. Fish and Wildlife Service 1987a, 1987b, 1994; Table 4.3).

Decline of four of the six listed species has been at least partially attributed to dams. Colorado River squawfish (*Ptychocheilus lucius*) and razorback sucker (*Xyrauchen texanus*) depend on an assortment of habitat types to complete their life cycles, and a key component to young of year and juvenile fish survival is the availability of shallow, backwater areas with sand or silt substrates and flooded bottomlands (U.S. Fish and Wildlife Service 1987a, 1987b; Modde *et al.* 1996). The swift, whitewater canyons of the pre-dam Colorado River and its tributaries defined the historic distributions of humpback chub (*Gila cypha*), a species first proposed for listing in 1964 (U.S. Fish and Wildlife Service 1987b; National Biological Service 1995) and subsequently added to the endangered species list in 1974 with the passage of the Endangered Species Act. Current conditions in the basin limit this fish to reaches of remote canyons (U.S. Fish and Wildlife Service 1987b; National Biological Service 1995). Not much is known of the natural history of bonytail chub (*Gila elegans*) except that this fish appears to favor

Table 4.3
Scientific and common names of the thirteen fish species native to the upper Colorado River basin. The status of each fish species is given. Status categories taken from U.S. Fish and Wildlife Service 1987b, 1994.

Scientific Name	Common Name	Status
Ptychocheilus lucius	Colorado squawfish	Listed
Rhynichthys osculus thermalis	Kendall Warm Springs dace	Listed
Xyrauchen texanus	razorback sucker	Listed
Gila cypha	humpback chub	Listed
Gila elegans	bonytail chub	Listed
Oncorhynchus clarki stomias	Greenback cutthroat trout	Listed
Prosopium williamsoni	mountain whitefish	Common
Gila robusta	roundtail chub	Common
Catostomus platyrhynchus	mountain sucker	Common
Cottus bairdi	mottled sculpin	Common
Rhinichthys osculus	speckled dace	Abundant
Catostomus discobolus	bluehead sucker	Abundant
Catostomus latipinnis	flannelmouth sucker	Abundant

eddies and pools (U.S. Fish and Wildlife Service 1987b). What is quite clear, however, is that the preferred habitats for all these fish have virtually disappeared in the regulated rivers of the upper Colorado River basin and this decline can be directly related to dams and their associated impacts on the environment.

INDIRECT EFFECTS OF FRAGMENTATION

The entire field of community ecology is predicated on the notion that a numerical change in one species will have ramifications for other species in an assemblage. Having established that the direct physical effects of fragmentation are common among plants and at least present among animals, ecologists will need to answer the next question: How will the indirectly mediated physical effects of fragmentation reverberate through the communities inhabiting remnant habitat patches? Ecological theory and hard-won experimental data suggest that these indirect effects will be common and often unpredictable (Kareiva 1994). Theory also suggests that even small direct effects of fragmentation may have potentially large, indirect consequences mediated through species interactions.

The case of the brown-headed cowbird (*Molothrus ater*) provides perhaps the most well-known example of how fragmentation can lead to dramatic indirect effects. The cowbird is an obligate nest parasite, generally laying its eggs in the nests of other birds and leaving these "host" parents to raise cowbird young instead of their own (Harris 1988). In pre-settlement North America, the brown-headed cowbird was relegated to the midcontinental grasslands (Wilcove *et al.* 1986). As the forests were logged and the land converted to agricultural crops and forest patches, the amount of suitable brown-headed cowbird habitat increased (Gates and Gysel 1978; Wilcove *et al.* 1986; Harris 1988). These increases in favorable habitat and food availability allowed the cowbird to expand its range and to increase in number. Forest birds once isolated from cowbirds now experience increased levels of nest parasitism for which they have no evolved defenses (Wilcove *et al.* 1986). Firm links have been drawn between the creation of edge habitat surrounding forest fragments, nest parasitism of brown-headed cowbirds, and the population decline of forest songbirds. Robinson *et al.* (1995) examined nesting success and cowbird parasitism for nine songbird species and determined that decreased forest cover was significantly related to increased nest parasitism.

While cowbird nest parasitism provides a signature case of the indirect effects of fragmentation, other compelling examples exist. In a study of mammal assemblages in remnant Australian rainforest, Laurance (1994) found that species composition differed between small (< 20 ha) and large (> 20 ha) patches. Although explanations for responses to fragmentation varied among the 15 species studied, changes in edge-associated vegetation were clearly implicated. In particular, two rodent species (*Rattus leucopus* and *R. fuscipes*) appeared to segregate in response to increased interspecific competition induced by the change in vegetation. The two species were negatively correlated with one another and differentially distributed among patches. *Rattus leucopus* abundance increased in small patches while *R. fuscipes* was found primarily in large patches and in the absence of *R. leucopus*. Further, vegetation patches exhibited gradients in plant communities. Small patches contained more disturbance-adapted plants and had more treefalls than larger patches, and edge-interior gradients in vegetation existed among all patches. Laurance (1994) concluded that a vegetation dependent switch in competitive ability was responsible for the pattern.

Plant-pollinator systems can be particularly vulnerable to habitat fragmentation for two main reasons: (1) plants found in small and/or widely separated patches may not be able to support pollinator populations, and (2) fragmentation may lead to the loss of a specialized pollinator and thus affect the survival of the plant population (Buchmann and Nabhan 1996). In the first case, the spatial aspect of fragmentation results in the decline of a plant species; habitat patches may be too far apart to support a viable pollinator population. The second case, however, describes an indirect, physical effect of fragmentation. Some physical aspect of the fragmented habitat will negatively impact a pollinator species, and in turn, a plant species (or several) may decline, particularly when a specialized plant-pollinator relationship exists. This particular situation may be occurring in Australia, where non-flying mammals are a major pollinator of the Gondwana plant family Proteaceae (Goldingay, Carthew, and Whelan 1991; LaMont, Klinkhamer, and Witkowski 1993). Increased habitat fragmentation has been linked to declines in some of these important mammals (Goldingay *et al.* 1991; Laurance 1994), and these declines may well be reflected in declining populations of certain species within the Proteaceae (LaMont, Klinkhamer, and Witkowski 1993).

Parasite-host relationships also can be affected by habitat fragmentation. A clear case is found in the freshwater mussels of North America. In the family Unionidae most species have parasitic larval stages that depend on a host fish species (Watters 1996), and many mussels require specific fish species to act as the larval host. Additionally, these mussels are sensitive to siltation and changes in temperature and flow (Fisher and LaVoy 1972; Layzer *et al.* 1993). In the southeastern United States, where these mussels predominate, dams are an integral facet of the landscape, and the combination of habitat sensitivity and host specificity leave unionoid mussels especially vulnerable to this type of habitat alteration. Dams often block the upstream migration of host fish species, thereby relegating mussels to a less-than-optimal habitat downstream of the dam (Watters 1996). The result is that more than 60% of freshwater mussels have been extirpated from United States rivers (The Nature Conservancy 1996).

CONCLUSIONS

The theme of this chapter has been to elucidate the connections between the physical effects of habitat fragmentation and their direct and indirect biotic consequences. In recent years, discussion of the physical effects of fragmentation has often been couched in terms of understanding how much of the remnant forest is "edge" and how much is "core." The implicit message has been that physical effects can help us better understand the spatial effects of fragmentation. Issues of patch size and connectivity obviously will be important for many species and will be the major reason for many biotic changes. However, I suggest that the most productive approach to understanding the entire impact of fragmentation is to consider spatial and nonspatial mechanisms of biotic change as equally valid hypotheses for observed or predicted consequences of fragmentation. Where large fractions of remaining habitat are altered substantially, the spatial arrangement of the remaining habitat may have little to do with a species' fate. Because prescriptions for recovery are likely to differ depending on the relative importance of spatial and nonspatial impacts of fragmentation, techniques to elucidate their relative impacts will be important tasks for conservation biologists.

ACKNOWLEDGMENTS

I thank Peter Kareiva, David Skelly, and Sara Windsor for helpful comments on earlier drafts of the manuscript.

LITERATURE CITED

Allan, J.D. 1995. *Stream ecology: structure and function of running waters.* London: Chapman & Hall.

Andren, H. 1994. Effects of habitat fragmentation on birds and mammals in landscapes with different proportions of suitable habitat: A review. *Oikos* 71:355–366.

Baxter, R.M. 1977. Environmental effects of dams and impoundments. *Annual Review of Ecology and Systematics* 8:255–83.

Boon, P.J. 1993. Distribution, abundance and development of Trichoptera larvae in the River North Tyne following the commencement of hydroelectric power generation. *Regulated Rivers: Research and Management* 8:211–224.

Brothers, T.S. 1993. Fragmentation and edge effects in central Indiana old-growth forests. *Natural Areas Journal* 13:268–274.

Buchmann, S.L. and G.P. Nabhan. 1996. *The forgotten pollinators.* Washington, D.C.: Island Press.

Camargo, J.L.C. and V. Kapos. 1995. Complex edge effects on soil moisture and microclimate in central Amazonia forest. *Journal of Tropical Ecology* 11:205–221.

Chen, J., J.F. Franklin, and T.A. Spies. 1992. Vegetation responses to edge environments in old-growth Douglas-fir forests. *Ecological Applications* 2:387–396.

Chen, J., J.F. Franklin, and T.A. Spies. 1995. Growing-season microclimatic gradients from clearcut edges into old-growth Douglas-fir forests. *Ecological Applications* 5:74–86.

Dynesius, M. and C. Nilsson. 1994. Fragmentation and flow regulation of river systems in the northern third of the world. *Science* 266:753–762.

Fisher, S.G. and A. LaVoy. 1972. Differences in littoral fauna due to fluctuating water levels below a hydroelectric dam. *Journal of the Fisheries Research Board of Canada* 29:1472–1476.

Fraver, S. 1994. Vegetation responses along edge-to-interior gradients in the mixed hardwood forests of the Roanoke River Basin, North Carolina. *Conservation Biology* 8:822–832.

Gates, J.E. and L.W. Gysel. 1978. Avian nest dispersion and fledgling success in field-forest ecotones. *Ecology* 59:871–883.

Geiger, R. 1965. *The climate near the ground.* Cambridge, Mass.: Harvard University Press.

Ghiselin, J. 1977. Analyzing ecotones to predict biotic productivity. *Environmental Management* 1:235–238.

Goldingay, R.L., S.M. Carthew, and R.J. Whelan. 1991. The importance of non-flying mammals in pollination. *Oikos* 61:79–87.

Halme, E. and J. Niemela. 1993. Carabid beetles in fragments of coniferous forest. *Annales Zoologici Fennici* 30:17–30.

Harris, L.D. 1988. Edge effects and conservation of biotic diversity. *Conservation Biology* 2:330–332.

Hart, R.C. 1990. Zooplankton distribution in relation to turbidity and related environmental gradients in a large subtropical reservoir: Patterns and implications. *Freshwater Biology* 24:241–263.

Hill, D.B. 1985. Forest fragmentation and its implications in central New York. *Forest Ecology and Management* 12:113–128.

Hoover, J.P., M.C. Brittingham, and L.J. Goodrich. 1995. Effects of forest patch size on nesting success of wood thrushes. *Auk* 112:146–155.

Huhta, V. 1971. Succession in the spider communities of the forest floor after clear-cutting and prescribed burning. *Annales Zoologici Fennici* 8:483–542.

Kareiva, P. 1994. Higher order interactions as a foil to reductionist ecology. *Ecology* 75:1527–1528.

LaMont, B.B., P.G.L. Klinkhamer, and E.T.F. Witkowski. 1993. Population fragmentation may reduce fertility to zero in *Banksia goodii*: A demonstration of the Allee effect. *Oecologia* 94:446–450.

Laurance, W.F. 1991. Edge effects in tropical forest fragments: Application of a model for the design of nature reserves. *Biological Conservation* 57:205–219.

Laurance, W.F. 1994. Rainforest fragmentation and the structure of small mammal communities in tropical Queensland. *Biological Conservation* 69:23–32.

Layzer, J.B., M.E. Gordon, and R.M. Anderson. 1993. Mussels: The forgotten fauna of regulated rivers. A case study of the Caney Fork River. *Regulated Rivers: Research and Management* 8:63–71.

Lehmkuhl, D.M. 1972. Change in thermal regime as a cause of reduction of benthic fauna downstream of a reservoir. *Journal of the Fisheries Research Board of Canada* 29:1329–32.

Leopold, A. 1933. *Game management*. New York: Charles Scribner and Sons.

Marchant, R. 1989. Changes in the benthic invertebrate communities of the Thomson River, southeastern Australia, after dam construction. *Regulated Rivers: Research and Management* 4:71–89.

Martinez, P.J., T.E. Chart, M.A. Trammell, J.G. Wullschleger, and E.P. Bergersen. 1994. Fish species composition before and after construction of a main stem reservoir on the White River, Colorado. *Environmental Biology of Fishes* 40:227–239.

Matlack, G.R. 1993. Microenvironment variation within and among forest edge sites in the eastern United States. *Biological Conservation* 66:185–194.

Matlack, G.R. 1994. Vegetation dynamics of the forest edge-trends in space and successional time. *Journal of Ecology* 82:113–123.

Matthysen, E., L. Lens, S. Van Dongen, G.R. Verheyen, L.A. Wauters, F. Adriaensen, and A. A. Dhondt. 1995. Diverse effects of forest fragmentation on a number of animal species. *Belgian Journal of Zoology* 125:175–183.

Meffe, G.K. and C.R. Carroll. 1994. *Principles of conservation biology*. Sunderland, MA: Sinauer Associates.

Mladenoff, D.J., M.A. White, J. Pastor, and T.R. Crow. 1994. Comparing spatial pattern in unaltered old-growth and disturbed forest landscapes. *Ecological Applications* 3:294–306.

Modde, T., K.P. Burnham, and E.J. Wick. 1996. Population status of the razorback sucker in the Middle Green River (USA). *Conservation Biology* 10:110–119.

Moyle, P.B. and R.A. Leidy. 1992. Loss of biodiversity in aquatic ecosystems: Evidence from fish faunas. In *Conservation Biology: the theory and practice of nature conservation, preservation, and management*, eds. P.L. Fiedler and S.K. Jain, 127–169. New York: Chapman & Hall.

Munn, M.D. and M.A. Brusven. 1987. Discontinuity of Trichopteran (caddisfly) communities in regulated waters of the Clearwater River, Idaho, USA. *Regulated Rivers: Research and Management* 1:61–69.

Murcia, C. 1995. Edge effects in fragmented forests: Implications for conservation. *Trends in Ecology and Evolution* 10:58–62.

Norton, D.A., R.J. Hobbs, and L. Atkins. 1995. Fragmentation, disturbance, and plant distribution: Mistletoes in woodland remnants in the Western Australian wheatbelt. *Conservation Biology* 9:429–438.

Nummelin, M. and I. Hanski. 1989. Dung beetles of the Kibale Forest, Uganda: Comparison between virgin and managed forests. *Journal of Tropical Ecology* 5:349–352.

Pahl, L.I., J.W. Winter, and G. Heinsohn. 1988. Variation in responses of arboreal marsupials to fragmentation of tropical rainforest in northeastern Australia. *Biological Conservation* 46:71–82.

Pajunen, T., Y. Haila, E. Halme, J. Niemela, and P. Punttila. 1995. Ground-dwelling spiders (Arachnida, Araneae) in fragmented old forests and surrounding managed forests in southern Finland. *Ecography* 18:62–72.

Palmer, R.W. and J.H. O'Keeffe. 1990. Downstream effects of impoundments on the water chemistry of the Buffalo River (Eastern Cape), South Africa. *Hydrobiologia* 202:71–83.

Petts, G.E. 1984. *Impounded rivers: Perspectives for ecological management*. Chichester: John Wiley and Sons.

Ripple, W.J., G.A. Bradshaw, and T.A. Spies. 1991. Measuring forest landscape patterns in the Cascade Range of Oregon, USA. *Biological Conservation* 57:73–88.

Robinson, S.K., F.R. Thompson, T.M. Donovan, D.R. Whitehead, and J. Faaborg. 1995. Regional forest fragmentation and the nesting success of migratory birds. *Science* 267:1987–1990.

Saunders, D.A, R.J. Hobbs, and C.R. Margules. 1991. Biological consequences of ecosystem fragmentation: A review. *Conservation Biology* 5:18–29.

Sekgororoane, G.B. and T.G. Dilworth. 1995. Relative abundance, richness, and diversity of small mammals at reduced forest edges. *Canadian Journal of Zoology* 73:1432–1437.

Sharpe, D.M., G.R. Guntenspergen, C.P. Dunn, L.A. Leitner, and F. Stearns. 1987. Vegetation dynamics in a southern Wisconsin agricultural landscape. In *Landscape heterogeneity and disturbance*, ed. M.G. Turner, 137–155. New York: Springer-Verlag.

Shure, D.J. and L.A. Wilson. 1993. Patch size effects on plant phenolics in successional openings of the southern Appalachians. *Ecology* 74:55–67.

Smith, B.E., P.L. Marks, and S. Gardescu. 1993. Two hundred years of forest cover changes in Tompkins County, New York. *Bulletin of the Torrey Botanical Club* 120:229–247.

Stevens, L.E., J.C. Schmidt, T.J. Ayers, and B.T. Brown. 1995. Flow regulation, geomorphology, and Colorado River marsh development in the Grand Canyon, Arizona. *Ecological Applications* 5:1025–1039.

The Nature Conservancy. 1996. *Priorities for conservation: 1996 annual report card for U.S. plant and animal species*. Arlington, VA: The Nature Conservancy.

Thomas, C.D. 1994. Extinction, colonization, and metapopulations: Environmental tracking by rare species. *Conservation Biology* 8:373–378.

Tian, H., H. Xu, and C.A.S. Hall. 1995. Pattern and change of a boreal forest landscape in northeastern China. *Water Air and Soil Pollution* 82:465–476.

Travnichek, V.H., M.B. Bain, and M.J. Maceina. 1995. Recovery of a warmwater fish assemblage after the initiation of a minimum-flow release downstream from a hydroelectric dam. *Transactions of the American Fisheries Society* 124:836–844.

Trotzky, H.M. and R.W. Gregory. 1974. The effects of water flow manipulation below a hydroelectric power dam on the bottom fauna of the upper Kennebec River, Maine. *Transactions of the American Fisheries Society* 103:318–324.

Turner, M.G. and C.L. Ruscher. 1988. Changes in landscape patterns in Georgia, USA. *Landscape Ecology* 1:241–251.

U.S. Fish and Wildlife Service. 1987a. *Recovery implementation program for endangered fish species in the upper Colorado River basin*. Denver: U.S. Dept. of the Interior.

U.S. Fish and Wildlife Service. 1987b. *Final environmental assessment of the recovery implementation program for endangered fish species in the upper Colorado River basin*. Denver: U.S. Department of the Interior.

U.S. Fish and Wildlife Service. 1994. *Recovery implementation program for endangered fish species in the upper Colorado River basin*. Denver: U.S. Department of the Interior.

Ward, J.V. and J.A. Stanford. 1983. The serial discontinuity concept of lotic ecosystems. In *Dynamics of lotic ecosystems*, eds. T.D. Fontaine and S.M. Bartell, 29–42. Ann Arbor, MI: Ann Arbor Science Publishers.

Watters, G.T. 1995. Small dams as barriers to freshwater mussels (*Bivalvia*, Unionoida) and their hosts. *Biological Conservation* 75:79–86.

Wilcove, D., C. McClellan, and A. Dobson. 1986. Habitat fragmentation in the temperate zone. In *Conservation biology: The science of scarcity and diversity*, ed. M. Soulé, 237–56. Sunderland, MA: Sinauer Associates.

Williams-Linera, G. 1990. Vegetation structure and environmental conditions of forest edges in Panama. *Journal of Ecology* 78:356–373.

Yahner, R.H. 1988. Changes in wildlife communities near edges. *Conservation Biology* 2:333–339.

Yoakum, J. and W. Dasmann. 1971. Habitat manipulation practices. In *Wildlife management techniques*, ed. R. Giles, 173–231. Washington, D.C.: The Wildlife Society.

Young, A. and N. Mitchell. 1994. Microclimate and vegetation edge effects in a fragmented podocarp-broadleaf forest in New Zealand. *Biological Conservation* 67:63–72.

CHAPTER

5

Reintroduction of Rare Plants: Genetics, Demography, and the Role of *Ex Situ* Conservation Methods

EDWARD O. GUERRANT, JR.
and BRUCE M. PAVLIK

Many endangered plant species have been reduced to so few populations and such low numbers that timely collection and storage of seed has become imperative. If donor populations become extinct or seriously depleted, then off-site, or *ex situ*, samples can be used to reintroduce or augment populations in the wild. The strategic value of an *ex situ* component for plant conservation has been articulated by Falk (1987, 1990, 1992), who also helped establish the Center for Plant Conservation (CPC). The CPC is a national network of botanic gardens and arboreta that is attempting to assemble a genetically representative collection of our nation's most rare and endangered plants while it is still possible to do so.

If stored seeds are to constitute a meaningful conservation resource, it is necessary to collect genetically representative samples of populations to keep the seeds alive for long periods of time (Wieland 1995), and to develop strategies for using such stocks to establish self-sustaining populations under natural conditions. Not all species have seeds that can be stored, and so for some species it may be necessary to maintain garden populations of growing plants. In principle, sampling considerations are the same as for seed, but maintaining growing collections is a much more formidable challenge. For the sake of brevity, we will hereafter refer only to the considerations surrounding seed storage. The CPC has sponsored two symposia that have led to published guidelines on (1) defining and obtaining a genetically representative sample (CPC 1991), and (2) planning, designing, and evaluating a species reintroduction project of conservation value (Falk, Millar, and Olwell 1996).

In this chapter, we begin by briefly discussing the rationale and need for off-site seed stocks of rare plants and the use of such samples in a reintroduction project. Then we examine the sampling and reintroduction guidelines themselves, emphasizing the underlying relationship to population theory. Finally, we describe case studies of six ongoing plant reintroduction projects, highlighting their experimental and demographic structure as well as real-world difficulties that are frequently encountered. We hope to demonstrate that although genetic sampling and reintroduction are technically challenging, there is much to be gained by careful, controlled, and well documented attempts.

WHY DO WE NEED *EX-SITU* COLLECTIONS AND REINTRODUCTION?

Superficially, it might seem that *ex situ* methods for plant conservation are defeatist, and undermine more important efforts to conserve plants in their native habitats (e.g., Hamilton 1994). This criticism is short-sighted and misleading. *Ex situ* collections and reintroduction represent only one tool in a comprehensive, integrated conservation strategy that seeks to conserve the maximum possible amount of biodiversity (Falk 1987, 1990, 1992). It would be foolhardy to abandon any useful weapon in this all-important battle, given the overwhelming forces of habitat destruction and subsequent rates of extinction. New populations can rise phoenix-like out of the ashes of extinction only if representative germplasm collections exist off-site (Parenti and Guerrant 1990).

Many species and populations are on the brink of extinction. In a review of all U.S. Fish and Wildlife Service (USFWS) recovery plans available in 1992 for plants, Schemske *et al.* (1994) found 91 species for which data on number of populations were available. Over half of these species had five or fewer extant populations, and the modal number of populations per species was one (Figure 5.1). Their finding that many endangered species survive in only one to five populations are similar to the results of other surveys

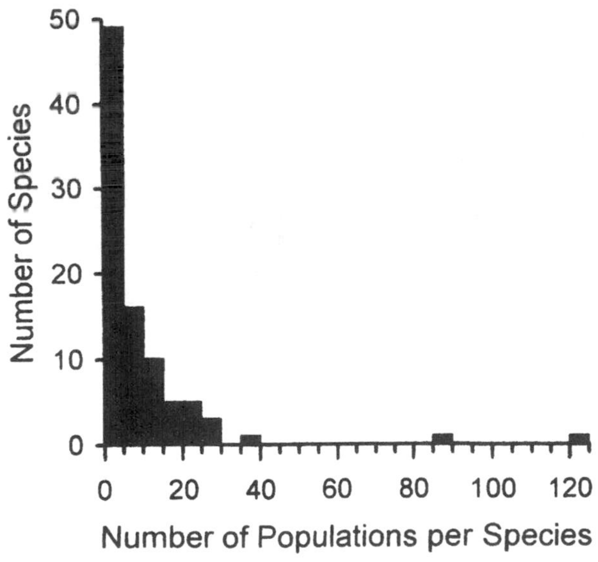

Figure 5.1
Frequency distribution of the number of extant populations for 91 plant species listed by the U.S. Fish and Wildlife Service as threatened or endangered, for which information was available in May 1992. The number of populations per species are presented in groups of five, with 1–5, 6–10, and so on, populations per species. Figure re-drawn from Schemske et al. *(1994).*

in the United States and Australia (CPC 1991; Brown and Briggs 1991; Wilcove, McMillan, and Winston 1993).

Even for species with larger numbers of populations, many populations may consist of very few individuals. The median population size of plants at the time of listing by the USFWS was less than 120 individuals, and the modal category was for population sizes between 11 and 100 individuals (Wilcove, McMillan, and Winston 1993). The listed endangered western lily (*Lilium occidentale*), for example, is known historically from about 58 populations (Guerrant, Schultz, and Imper 1996), between a third and half of which have been extirpated (Figure 5.2). The median population size for extant populations is between 26 and 35 individuals.

It is not reasonable to expect that even the most enlightened, dedicated, and well-funded *in situ* efforts will always be successful in maintaining populations, especially when attemping to rehabilitate small, remnant fragments of endangered species in marginal habitats. *Ex situ* samples can provide a relatively low-cost insurance policy—a hedge against population extirpation and catastrophic loss of habitat—and create management options for the future that would otherwise be unavailable. *Ex situ* collections are not an end in themselves, however. Their ultimate conservation value will come from how they are used in reintroduction and their effect, if any, on the long-term survival prospects of rare species in the wild (Figure 5.3). Reintroduction is used here in the broad sense of the term to refer to any of a number of ways in which stored seed is returned to the wild (Falk, Millar and Olwell 1996). Perhaps the most straightforward form of reintroduction is the release of genetic material to the same site from which it was collected, following

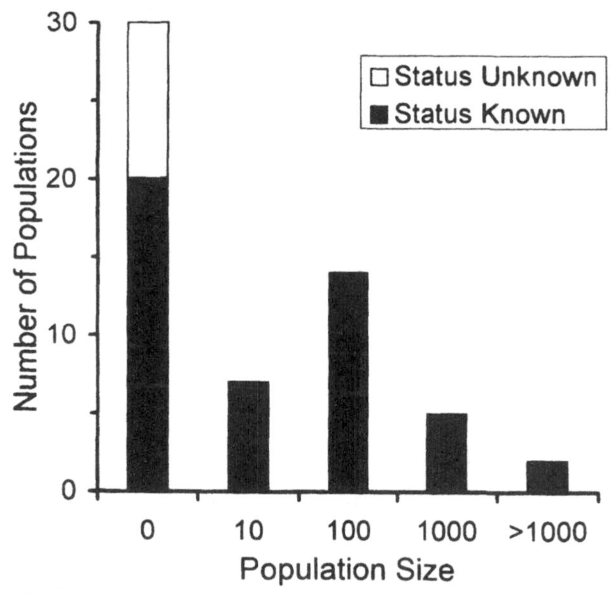

Figure 5.2
Frequency distribution of population sizes for all reported populations of the endangered Lilium occidentale*: extant, extirpated, and for which the current status is unknown (data from Guerrant, Schultz, and Imper 1996). Population size classes are indicated by the upper bound of each group. Groups are 0, 1–10, 11–100, 101–1000, and >1000 individuals per population. The 10 populations for which current status is unknown are indicated above the 20 populations known or presumed to be extirpated. Many if not all of the populations for which current status is unknown have probably been extirpated.*

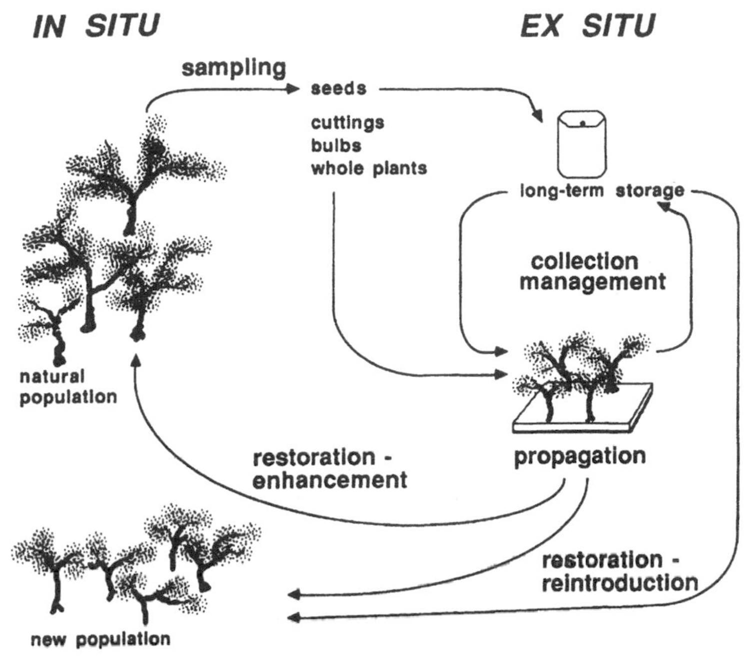

Figure 5.3
Flow diagram illustrating the relationships among natural populations, seeds in ex situ *storage and their use in reintroduction (modified from Brown and Briggs 1991).*

extirpation of the natural population. This was the case in the reintroduction of *Stephanomeria malheurensis*, which will be discussed in detail later in this article. Augmentation, in which stored material is used to increase the population size of an extant natural population, is another form of reintroduction. For *ex situ* samples to be of maximum conservation value, however, they must be genetically representative, thus containing the allelic diversity and evolutionary potential of the original natural population from which they were collected.

OBTAINING GENETICALLY REPRESENTATIVE COLLECTIONS

Given the potential value of seed collections as a conservation tool, biologists must identify priorities for which seeds to collect (Table 5.1). This is essentially a sampling problem and the key issues are: (1) from which species should samples be taken; (2) how many (and which) populations should be sampled; (3) how many individuals per population should be sampled; and (4) how many seeds (or other propagules) should be taken from each female parent? (Falk and Holsinger 1991). Finally, after these questions have been considered and a sampling plan has been developed, a decision has to be made for each population about whether it can withstand, without serious harm, the desired level of collection. If not, collection must be spread over two or more years. The CPC recommendations concerning each of these decisions, along with genetic and demographic considerations, are discussed briefly in turn.

1. Which Species Should be Collected/Sampled?

The degree of endangerment of a species is the single best indicator of priority, but

Table 5.1
Summary of recommended guidelines for genetic sampling for conservation collections of endangered plants (Center for Plant Conservation 1991. Reprinted with permission).

	Decisions			
	Which species to collect?	**How many populations sampled per species?**	**How many individuals sampled per population?**	**How many propagules taken per individual?**
Recommended range	–	1-5	10–50	1–20
Target level of biological organization	Species	Ecotype and population	Individual	Allele
Key consideration	Probability of loss of unique gene pool	Degree of genetic difference among populations	Log of population size	Survivability of propagules
	Potential for restoration or recovery	Population history	Genetic mobility within population	Long-term use of collection
	Factors affecting sampling decisions			
	High degree of endangerment	High diversity/limited gene flow among populations	High diversity among individuals within each population	Low survivability of propagules
Collect more	Experiencing rapid decline	Imminent destruction of populations	Observed microsite variation	Planned use for reintroduction or restoration program
	Few protected sites	High observed ecotypic or site variation	Mixed-mating or outcrossing	
	Biological management required	Isolated populations	Fragmented historical populations	
		Potential for biological management and recovery	Small breeding neighborhood size	
	Recently or anthropogenically reduced	Recent or anthropogenic rarity	Low survivability of propagules	
	Feasibility of successful maintenance in cultivation or storage	Self-fertilization	Extremely large populations	
	Possibility of reintroduction or restoration			

	Economic potential	Herbaceous annual or short-lived perennial	Boreal distribution	
		Early- to mid-successional stage	Gymnosperm or monocot	
		Gravity-, explosively, or animal-dispersed seed	Woody perennial	
		Dicot or monocot	Late-successional stage	
		Temperate–tropical distribution	Animal- or wind-dispersed seed	
		Wind-dispersed seed	Temperate–tropical distribution	
		Outcrossing wind-pollinated	Early- to mid-successional stage	
		Late-successional stage	Dicot	
		Observed similarity among populations	Herbaceous annual or short-lived	
		Long-lived woody perennial	Self-fertilizing	
	High integrity of communities	Gymnosperm	Explosively or gravity-dispersed seed	
	Many protected sites	Boreal–temperate distribution	High survivability of propagules	
	Natural rarity	Protected populations or naturally rare	Large breeding neighborhood size	Low annual reproductive output (indicates multiyear collecting strategy)
	Stable condition	Closely clustered populations		
Collect less/fewer	Low degree of endangerment	Low diversity/extensive gene flow among populations	Low diversity among individuals within each population	High survivability of propagules

Note: Four basic practical decisions are addressed: which species to conserve, and the number of populations, individuals, and propagules to be sampled for each species. For each decision, a recommended range is shown, along with an indication of the relevant level of biological organization and a summary of key considerations. Factors that influence the sampling decision are listed in detail, with the most significant factors shown in italics. Factors suggesting larger or more extensive collections are shown at the top of each column, while those suggesting smaller or less extensive collections are given at the bottom.

determining relative risk can, of course, be highly subjective. Clearly, if only a few sites are known and not protected, or evidence exists that population numbers are declining rapidly, the species can be considered highly endangered. More subtle is whether a species was always rare, or if its rarity is a result of human-mediated habitat destruction.

Species that have always been rare, at least those with historically small or sparse populations, may have ecological or genetic adaptations that facilitate persistence, which remnant fragments of previously widespread species lack (Barrett and Kohn 1991; Menges 1991a, 1991b, 1992). For example, in a long-term study of prairie grasses, Rabinowitz *et al.* (1989) demonstrated that the reproductive output of common species closely tracked environmental fluctuations in rainfall over nine years, whereas reproductive output of sparse species was less variable and thus less likely to suffer complete reproductive failure. Although all endangered plants are in some sense rare, not all rare plants are necessarily endangered (Rabinowitz 1981; Holsinger and Gottlieb 1991).

2. How Many Populations Should be Sampled per Species?

The CPC recommends sampling from one to five populations per species. In practical terms, the decision is often moot, because a large proportion of endangered plants are found in five or fewer populations.

The most important considerations in deciding how many populations to sample are the degree of genetic difference among populations and the history of site disturbance. In general, more populations should be sampled from species with high diversity, or when gene flow among populations is low, than from more genetically uniform species or those with high rates of gene flow among populations. Factors that suggest high among-population diversity are noticeable ecotypic or habitat differentiation, or whether populations are isolated from one another either naturally or from recent habitat fragmentation. Breeding system is another factor to consider. For example, inbreeders typically have less total genetic variation than do outbreeders, but more of what little variation they do have is distributed among populations (Hamrick 1989). Thus, to acquire a given proportion of a species' total variation, a higher proportion of an inbreeder's populations will probably have to be sampled.

The suggested range of one to five populations to be sampled per species may seem surprisingly low. It is based in part on the extensive survey of electrophoretically detectable genetic variation conducted by Hamrick and Godt (1989; Hamrick *et al.* 1991), and on the analysis of Brown and Briggs (1991). The former researchers surveyed 653 published studies of 449 species in 165 genera, finding that on average, approximately 50% of the loci were polymorphic within species. Of those polymorphic loci, approximately 78% of the allozyme diversity occurs within populations. Therefore, the majority of the genetic information in a species can be found within a "typical" population. The remaining 22% of variation that is distributed among populations, however, can be very important with respect to local adaptation and the ability of species or populations to respond to selection. The conservation value of peripheral populations in particular has been discussed by Lesica and Allendorf (1995).

Though in an intuitive sense, the more populations represented the better, CPC recommendations reflect a practical tradeoff between the benefits and probability of acquiring significant new genetic variants with increased sampling, and the costs associated with collection of one species in relation to the needs of other species. In our world of great need and limited resources, it seems reasonable to minimize the number of sampled populations per species, at least until we have basic coverage of all threatened and endangered species.

3. How Many Individuals Should be Sampled per Population?
In order to capture the significant fraction of a species' total genetic information that is found within populations, the CPC recommends sampling ten to 50 individuals per population (CPC 1991). At first glance this might seem like too small a sample, but the allelic content of a sample is proportional to the logarithm of both the population size and the sample size. Consequently, a strong law of diminishing returns is in operation. Monomorphic alleles and those in relatively high frequency are captured in early samples, and the chance of obtaining additional rare alleles declines rapidly with increasing sample size. In a statistical sense, the first ten individuals sampled are as important as the next 90 (Brown and Briggs 1991).

Within the range of ten to 50 individuals recommended, key factors to consider in the design of a sampling scheme are population size and degree of genetic communication among individuals within the population. Conservation biologists should collect seeds from more individuals if there is high diversity among individuals within the population. High diversity among individuals may be indicated by high microsite variation within populations, self-incompatibility, or restricted genetic neighborhoods.

4. How Many Propagules Should be Collected from Each Individual?
There may be as much common sense as science involved in deciding how many seeds to collect from each individual. The CPC recommends one to 20 propagules, with the decision driven largely by expected survivorship of the propagules both in cultivation and in the wild.

Genetic variation is best maintained when each founder individual contributes equally to a new population (Haig, Ballou, and Derrickson 1990; Templeton 1990, 1991; Loebel *et al.* 1992; Hamilton 1994; Guerrant 1996a). In practice, we can equalize the contribution of different female parents of the seeds we collect, simply by keeping seeds from each plant separate. It is next to impossible to equalize the contribution of different males in the population, because paternal fitness can vary considerably among males (Schoen and Stewart 1986). Tonkyn (1993) has recently offered a mathematical optimization technique that can, among other things, assist in choosing mates in a captive population that will best equalize founder representation. Relative to bulk collections from a population, the potential benefits of maintaining separate maternal lines during collection and subsequent handling may be worth the extra effort, especially in extreme cases of collections from relatively few potential founders. This may be especially important for populations in which relatively few individuals contribute disproportionately to reproduction.

5. Under What Circumstances is a Multiyear Collection Plan Indicated?
The preceding four decision criteria appear well founded, but the question remains as to what effect, if any, the sampling itself has on the short-term survival prospects of the sampled populations. If collecting seeds for conservation purposes significantly endangers the population sampled, then indeed the cure may well be worse than the disease, particularly if sampling a rare plant population leads to a false sense of security.

We do not have a good quantitative handle on the magnitude of seed collection that any species can tolerate, even though there are intuitive guides based on life history. More theoretical work clearly is still needed in this arena if conservation biologists are actually to be a force that benefits rather than destroys the plant species we seek to conserve. In the absence of specific information on this issue, it is comforting to know that Menges (1992, p. 266) found that "The threat posed to population survival by environmental variation appeared almost entirely due to variation in mortality, growth, and reproductive status, and not to variation in reproductive output." This conclusion was based on simulation studies done on empirical demographic data in published studies, mostly of iteroparous perennial species.

RESPONSE TO THE GUIDELINES

The CPC Guidelines were primarily intended to assist the participating institutions of the CPC in their joint mission to bring into *ex situ* collections, representative samples of our nation's most imperiled taxa. The guidelines have also found a wider audience. Both the Botanic Gardens Conservation International, which has over 400 member institutions in 75 countries, and the Australian Network for Plant Conservation, have used the CPC genetic sampling guidelines as a basis for their own recommended guidelines (D.A. Falk, personal communication). The guidelines appear to have struck a resonant chord in many of those actively involved with conserving species most in danger. In contrast, Hamilton (1994, p. 47) concludes that "It is not clear, however, that *ex situ* methods will result in significant conservation of genetic variation and evolutionary potential without a great expansion of the scope of biological information used to establish and monitor such programs." Apart from the scientific debate surrounding Hamilton's (1994) critique of the CPC guidelines, there is the practical issue of how to mount effective actions in a less than ideal world. There is always a tradeoff to be made between the potential benefits resulting from better knowledge and the losses resulting from inaction in the absence of perfect knowledge.

The guidelines were designed to replace an expensive, information-greedy, genetic inventory of each taxon (e.g., using allozyme or DNA sequencing data) with a set of standardized, expedient but scientifically based sampling procedures. Although we recognize the potential advantages of having more information to accompany *ex situ* samples, we question the marginal value of much of this information in relation to the resources required to obtain it. We also have reservations about the potential for damage to rare plant populations caused by attempts to obtain the quantitative genetic and ecological information that Hamilton strongly advocates. Clearly, both points of view must be reconciled to the advantage of endangered species.

Given the rapid rate of habitat destruction and the progressive degradation of habitats by exotic species, we maintain it is prudent to collect what we can, while we can, in a manner that is as efficient and responsible as possible. This is not to say that the guidelines cannot be improved; surely they can. In this regard, we see at least two questions that especially merit attention: (1) how large a sample is necessary to support even one reintroduction attempt?, and (2) how many reintroduction attempts is an *ex situ* collection intended to support? Collections must also be large enough to provide seeds for baseline studies (e.g., germination and growth requirements) and for periodic measurements of viability in storage. Plants from these tests may be grown to provide additional seeds, but they could often be genetically inferior to those collected from the wild (Pavlik, Nickrent, and Howald 1993). The growing body of literature on reintroduction attempts, some of which will be described later in this chapter, provides one way to begin to address these questions.

Overall, it seems that the number of seeds recommended for collection may often be much too low to support even a single reintroduction attempt. For example, Wallace (1990) discusses an example from Florida, in which 2,000 seeds of the annual plant *Warea amplexifolia* were sown in an appropriate habitat. Six hundred seedlings emerged, of which only 16 survived to reproduce—a net yield in plants per seed of less than one percent. Dawson (personal communication) reports even lower field germination in two reintroduction attempts with *Astragalus osterhoutii* in Colorado, with 3.3% and 0.3% of the seeds giving rise to seedlings in fall and winter plantings, respectively. Even some highly successful reintroduction projects have suffered major setbacks, with early outplantings experiencing extremely high mortality (e.g., DeMauro 1994). The few recent experiences we have with reintroduction suggest that the absolute ranges of sample sizes recommended may need to be revised upward, and perhaps dramatically so. One strength of the guidelines is, however, that they are couched in relative terms, allowing the user to determine if a sample should be larger or smaller, given the particular circumstances.

REESTABLISHING SELF-SUSTAINING POPULATIONS FROM COLLECTIONS

It is one thing to obtain a genetically representative seed sample from a natural population, and quite another to be able to successfully reestablish a comparable population from such a collection. We may be approaching an adequate theoretical understanding of the former, but the latter is a much more formidable challenge. The transition from population genetic and demographic theory to conservation practice is not an easy one. Theory informs us of the relevant variables and their relationships, yet if we know anything for certain, it is that the phenomena with which we are dealing are highly complex and often unpredictable.

In contrast to the genetic sampling guidelines, which are relatively specific, the reintroduction guidelines (Falk, Millar, and Olwell 1996) are necessarily more generalized. This is due largely to the fact that the choices to be made in reintroduction are contingent on a much more complex set of circumstances. In an effort to clarify circumstances in which reintroduction is an appropriate conservation activity and to improve the overall quality of reintroductions, the guidelines are built around a set of 11 questions pertaining to three project phases: planning, implementation, and evaluation. The planning phase includes the consideration of legal issues, project context (e.g., is it for recovery, mitigation, or primarily to satisfy scientific curiosity?), resource management policies, and the existing scientific literature. Implementation is more technical, requiring decisions about where the reintroduction will take place, the genetic nature of the material to be released, initial population size and structure, and the status of ecosystem-level processes at the reintroduction site. The evaluation phase consists of developing a framework of goals and objectives for the reintroduction and the design of monitoring programs that ultimately supply data for determining success or failure. The guidelines do not attempt to provide recipes, but they do reaffirm our view that reintroductions are necessarily long-term experimental efforts that incorporate both genetic and demographic factors.

Theory: Population Viability and Vulnerability

There is no point in reintroducing plants unless they have a reasonable chance of persistence at population sizes large enough to avoid "mutational meltdown" (Lynch, Conery, and Burger 1995), and to maintain sufficient genetic diversity to respond evolutionarily to changing conditions. In the short term, this means that populations have to be reestablished in a manner that minimizes the risk of rapid extinction due to demographic and environmental stochasticity. In the long term, however, biological success depends on rapidly attaining population sizes large enough to avoid the loss of alleles due to random genetic drift, and increased inbreeding associated with even a short bottleneck in population size (see also Pavlik 1996). Population sizes at which genetic problems become critical appear to be much larger than previously estimated (Lande 1995; Lynch, Conery, and Burger 1995).

Relatively small population sizes, on the order of 50–100 individuals (Lande 1988; Menges 1991b, 1992), may be sufficient to minimize the effects of demographic stochasticity. However, minimizing the deleterious effects of environmental stochasticity may require substantially larger populations and often ensembles of separate populations to increase the chance that different populations experience "bad years" at different times. Worst of all is the possibility of catastrophes, which can reduce a population's "time to extinction" by an order of magnitude, even if the catastrophes occur at rates as low as once in 20 years (Mangel and Tier 1994).

Attempts to re-establish self-sustaining populations must necessarily consider short-term factors that influence the extinction probability of populations and their long-term "genetic health." Early attempts to integrate genetic and demographic parameters have been organized around two related concepts: i.e., minimum viable population (MVP) (Shaffer 1981) and its offshoot population viability (or vulnerability) analysis (PVA) (Gilpin and Soulé 1986). Shaffer (1981, p. 132) states: "A minimum viable population for any given species in any given habitat is the smallest isolated population having a 99% chance of remaining extant for 1,000 years despite the foreseeable effects of demographic, environmental, and genetic stochasticity, and natural catastrophes." The actual values are less important than the idea that discussions of population viability be couched in terms of the probability of survival over a given time period in the context of foreseeable factors that might lead to extinction.

While we know in theory how to predict a population's likelihood of persistence given information on variability in critical demographic parameters, we almost never have reliable estimates of those parameters or their variance. Moreover, short-term positive trends in demographic parameters can be used to support unduly optimistic conclusions. For example, in Menges' (1990) now classic studies of Furbish's lousewort (*Pedicularis furbishiae*), local, rapidly growing populations were still prone to extirpation in the not too distant future for at least two reasons. One is that Furbish's lousewort occupies an early successional habitat along the banks of the St. John's River, and is expected to be outcompeted by late-successional woody vegetation. The other reason is that frequent episodes of bank slumping can simply sweep away existing populations. Even though it can destroy some populations, bank slumping is an important mechanism whereby habitat suitable for colonization by new populations is created. Reintroduction (and efforts to manage existing natural populations) must therefore consider both population size trends, and also the factors underlying expected population growth and persistence (Pavlik 1994).

Although the population is usually considered the basic unit of conservation interest (Fiedler, White, and Leidy 1996), it is perilous to neglect the more inclusive if less well understood unit, the metapopulation (Gilpin 1987). The work of Menges and his co-workers on Furbish's lousewort (see Menges 1990) is both a splendid example of a PVA and a sobering reminder of the central role that metapopulation dynamics plays in the persistence of many plant species. Empirical demographic studies of *P. furbishiae* have suggested that individual populations can grow rather rapidly and thus apparently have a reasonably secure future. However, a conservation strategy that statically protected these populations in some sort of arrested state of distribution and abundance would not be wise (Pickett, Parker, and Fiedler 1992; Fiedler, White, and Leidy 1996). Persistence of the Furbish's lousewort along the St. John's River is dependent on a positive balance between local colonization events that create new populations and local extinctions.

Natural catastrophes are at one extreme in the range of environmental stochasticity that can lead to extinction. A less extreme but still dramatic example of environmental stochasticity that may have contributed to the apparent extinction of *Stephanomeria malheurensis* is annual variation in precipitation (Parenti and Guerrant 1990). *Stephanomeria malheurensis* is known from a single population estimated to comprise a maximum of 750 plants in any one year. However, *S. malheurensis* has experienced nearly 100-fold declines in population sizes from one year to the next, closely correlating with threefold reductions in precipitation (Gottlieb 1979, 1991). This annual plant does not have the apparently adaptive seed dormancy characteristics or germination requirements of its sympatric ancestral species, *S. exigua* ssp. *coronaria* (Gottlieb 1973; Brauner 1988). Thus, with a small, fluctuating soil seed bank and high mortality during fall germination (Gottlieb 1979; Parenti personal communication), the species was apparently extinct by 1985 after years of extremely irregular patterns and amounts of precipitation.

Small populations are also vulnerable to genetic problems. In addition to the well-known loss of alleles due to random genetic drift and inbreeding depression, the accumulation of even mildly deleterious mutations appears to be a much greater threat to small populations than previously thought (Lande 1995; Lynch, Conery, and Burger 1995). Lande (1995) showed the risk of extinction due to the fixation of mildly deleterious mutations may be on a par with that posed by environmental stochasticity. The long-term viability of populations with effective sizes as large as a few thousand may be significantly decreased by the accumulation of such mutations. Lynch, Conery, and Burger (1995) developed a general theoretical model for evaluating such a risk to populations. Those with effective sizes smaller than 100 individuals (and actual sizes smaller than 1,000) are seriously threatened by what they term "mutational meltdown" on timescales of approximately 100 generations. Even the familiar threat to small populations of increased inbreeding depression has become more ominous. In a series of experiments with captive populations of several species of flies, Frankham (1995) empirically investigated the relationship between extinction risk and the inbreeding coefficient. As expected, he found that increased inbreeding led to increased extinction risk. Less obvious and more threatening was the nature of the relationship. Extinction risk gradually increased with increasing inbreeding until a threshold was met at some intermediate level of inbreeding, where the extinction risk rapidly increased. The message from all three of these studies is that small populations may be much more vulnerable to genetic problems than previously thought. The degree to which such models and experiments apply to rare plants in the real world is not clear, but with carefully controlled field studies we could find out.

The implication of these genetic studies is clear: to avoid serious genetic problems, reintroduced populations will need to be quite large. Without having very large and genetically diverse *ex situ* samples and other resources necessary to introduce large populations from the beginning, we may often have to rely on establishing populations that grow rapidly. This may prove to be more difficult than it might seem. Guerrant (1996a) used computer simulation to examine the effects on extinction risk and population growth of differences in the size or stage distribution of the individuals comprising the founding population (Figure 5.4). Relative to using individuals of the smallest size classes in the founding population in a reintroduction, significant reductions in extinction risk appear to result from using the next-to-smallest size class. However, comparable gains in realized population growth rate do not seem to follow until the largest size class of

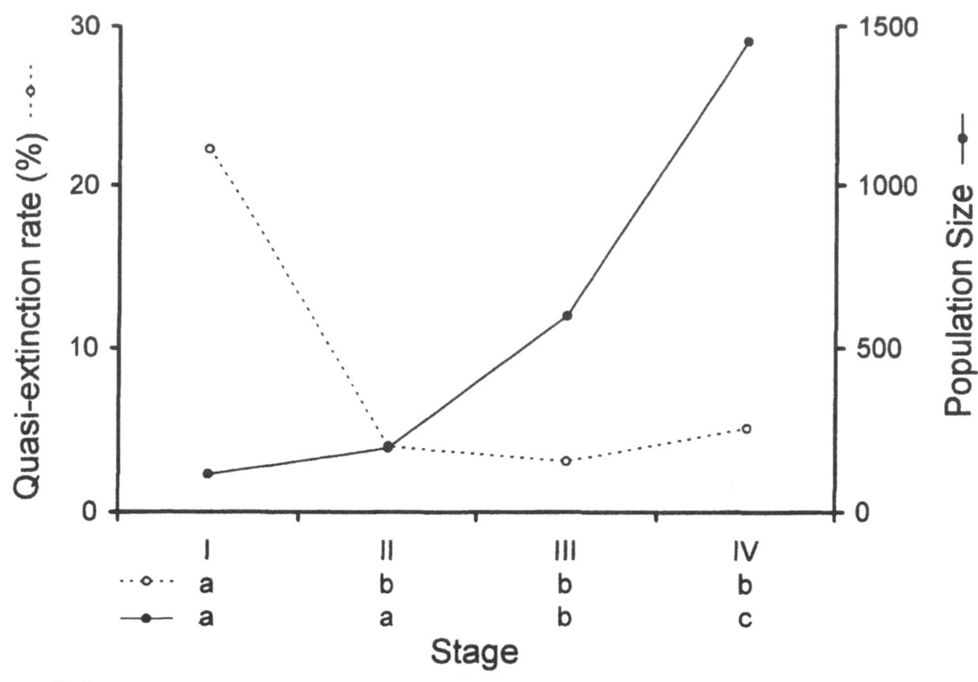

Figure 5.4
Expected quasiextinction rates expressed as an average percentage of 10 simulations of 1,000 iterations each, and average realized population sizes at ten years as a function of different size/stage classes of individuals used to comprise the founding population. Empirical size/stage based transition matrix demographic studies of the rare herbaceous iteroparous perennial Calochortus pulchellus *(Fiedler 1987) were used as the basis for stochastic modeling of the effectsof founding population stage/size distribution on extinction risk and population growth. Each simulation was begun with a founding population of 200 individuals of just the one size/stage class indicated. Quasiextinction occurred when a population dropped to ten individuals at any time over the 100-year time interval modeled. Four size/stage classes were specified, with only the smallest not producing any flowers or fruits. The two rows of letters under each stage refer to the results of Scheffe tests: stages with different letters differ at $p<0.001$ (Figure re-drawn from Guerrant 1996a).*

individuals are used in the founding population. To the degree that these simulations are indicative of what we will face in real reintroductions, it may be much more difficult to establish a population that will grow rapidly than one that merely persists at small sizes.

Practice: Reintroductions as Carefully Monitored Experiments

Efforts to reintroduce rare plants into appropriate habitat have a long history (almost 100 years) in Britain (Birkenshaw 1991), Hawaii (Mehrhoff 1996) and California (Ledig 1996). Unfortunately, the initial absence of conservation theory (e.g., genetic and ecological concerns), detailed records (e.g., monitoring), and a scientific approach (e.g., experimental design), detract significantly from the heuristic value of most early introductions. Although a few of these reintroduced populations persist today (e.g., *Arabis stricta* and *Trinia glauca* in Britain, *Pinus torreyana* in California) and are of conservation value, the majority have disappeared. In some cases, founders were released but a breeding population did not become established. In other cases, the founders were simply overrun by aggressive weeds. For the most part, we do not know why most populations failed, nor do we know why others have apparently succeeded. Consequently, we have learned very little from this long history of reintroductions for lack of theoretical context, careful monitoring, and thoughtful experimental design.

More recent efforts to reintroduce rare plants have benefited greatly from modern syntheses in population genetics (Crow and Kimura 1970; Frankel and Bennett 1970) and population ecology (Harper 1977), including the emergence of conservation biology as a distinct discipline with its own mission and theoretical base (Ehrenfeld 1970; Soulé and Wilcox 1980). A growing emphasis has been placed on demographic processes in natural populations of rare plants (e.g., Meagher Antonovics, and Primack 1978), which suggests the need for demographic monitoring as a conservation tool (Bradshaw 1981; Whitson and Massey 1981; Davy and Jefferies 1981). For example, experimentation with rare plants, especially with regards to founder phenomena *in situ*, has only been recently applied to reintroduction (Maunder 1992). Herein we rely on recent distillations of relevant genetic and ecological theory (Falk and Holsinger 1991; Bowles and Whelan 1994; Falk, Millar, and Olwell 1996), demographic monitoring (Pavlik 1994; Sutter 1996), and experimental design (Schemske *et al.* 1994; Pavlik 1996) in order to review six ongoing reintroduction projects, some apparently successful, others less so. The six projects were selected to illustrate a spectrum of plant life forms, habitat factors, and experimental designs and techniques.

Woody Perennials

Pinus torreyana (Ledig 1996). Torrey pine (Pinaceae) is one of the world's rarest species of pine, known only from a mainland and an insular population along the coast of southern California. Both populations are preserved in relatively small refuges, with the mainland population lying within one of the most rapidly developing regions in the United States. Historic descriptions derived from an 1888 survey indicate that the total number of trees had been low at that time (400 to 500 mature individuals). Currently, the mainland population consists of nearly 6,000 trees, in large part because of "reintroduction efforts" inspired by civic pride early in this century. However, a 1989 outbreak of the ips beetle (*Ips paraconfusus*) has led to widespread and rapid decline in the population, killing more than 840 trees by 1991.

Scientific efforts to conserve and reintroduce Torrey pine began with genetic studies in the early 1980s. Electrophoretic examination of 59 different isozyme loci revealed that every tree in the mainland population was genetically identical, differing from the insular population at only two loci. Recognizing the threat that genetic uniformity posed to this species, the U.S. Forest Service collected nearly 30,000 seeds from 149 trees in 1986 for *ex situ* storage and *ex situ* propagation. Some of these trees were located in stands that were subsequently destroyed by the ips infestation.

Before reintroduction it was necessary to control the advance of the beetle predators. During the spring of 1991 an array of funnel traps containing ips aggregating pheromones was set among stands of dead trees. Over the next 6 months more than 280,000 individual ips were removed from the traps, with progressively fewer found later in the season. At the same time traps containing ips anti-aggregating pheromones were used to repel beetles away from healthy stands. Traps were put out again during the next two years and tree infestation and mortality due to ips were eliminated by the end of 1992. This innovative habitat management on behalf of an endangered tree species owes much to research routinely performed by the Forest Service on common, economically important species.

Outplanting of stored seed began in 1992, and container-grown two-year olds were moved to the field in 1994. Attempts were made to return progeny to their stands of origin, but given the lack of detectable genetic variation this may have been more of a formality than a requirement. Poor germination (2%) due to deterioration of seeds during storage was a problem, but survivorship of container-grown seedlings has been high (98%). In order to learn more about the factors that influence Torrey pine establishment, an additional 4,000 seeds have been sown using an experimental design. The effects of soil type, canopy openings, and slope aspect on growth and survival are being investigated, using the results from demographic monitoring as the metric for evaluating the performance of artificial subpopulations.

Conradina glabra (Gordon 1996). The Apalachicola rosemary (Lamiaceae) was first described in 1962 from the dry sandhills of northern Florida. This small (< 0.8 m tall), ramet-producing shrub probably grew beneath an overstory of longleaf pine (*Pinus elliottii*) and turkey oak (*Quercus laevis*) kept open by occasional wildfire. Most of the rosemary habitat was converted into commercial pine plantations by the late 1950s and six of eight extant sites are now found on land owned by a timber company. Populations of Apalachicola rosemary do not survive in pine plantations and fewer than 600 individuals are now known in the wild. Electrophoretic examination of 11 isozyme loci revealed very high levels of genetic variation, even when compared to mean values for widespread flowering plants (Martin 1992).

Ex situ propagation began in 1987 at Bok Tower Gardens, a participating institution of the Center for Plant Conservation (Wallace 1992). Shoot tip cuttings were taken from as many individuals as could be reached on public right-of-ways (48 donors from each of two populations). More than 75% of the cuttings produced roots in a fast-draining potting soil under greenhouse conditions. Intermittent misting and warm temperatures allowed the plants to grow rapidly and survive transplantation to outdoor beds. Flowering occurred but seedlings were destroyed to eliminate the possibility of gene flow between this and other *Conradina* species in the garden. Within a short time the original cuttings became donors for propagation and records were kept with respect to each clone and its descendants. Logistically, about 1,200 plants could be grown in standard 4-inch nursery containers, transported, and outplanted at any one time.

The chosen reintroduction site was within the historic range for the species, on land owned and managed by The Nature Conservancy (TNC). Although severely impacted by silvaculture in the 1950s, remnants of appropriate habitat could be found. Remnants were evaluated on the basis of soil, topography, vegetation composition, and management history. Three specific localities (microsites) within the most appropriate remnants were then selected for reintroducing six-month-old plants during a few wet days in February of 1991. No water or fertilizer were added. At each microsite, 48 clusters of nine plants were randomly located within a 45 × 80 m grid. Clustering was presumed to promote pollen flow and seed set in this obligate outcrosser, and it would also facilitate monitoring. Management-oriented experiments also were built into the project. For example, to examine the potential impact of competition on survival, growth, and reproduction, all above-ground plant material of other species was clipped at monthly intervals from around half of the clusters during the first two growing seasons. In addition, one of the three reintroduction microsites was subjected to a prescribed burn in its second year and another microsite was burned in its third year.

Demographic monitoring by TNC staff and trained volunteers has been employed to follow the performance of every outplanted individual, to evaluate the management experiments and to record seedling establishment. Survival of plants in the unburned microsites was 95% after two years, indicating that large subpopulations of conservation value could be created. Clipping had no effect on survival but it significantly increased branch growth and flowering. Seedling establishment has been rare, but confined to the clipped plots, primarily in the burned sites. Although the prescribed fire caused substantial mortality, individuals vigorously resprouted and a two to five-year burn cycle will be incorporated into the long-term management regime. That regime will include restoration of common native taxa (e.g., longleaf pine and bunchgrasses) and monitoring of rosemary population dynamics.

Herbaceous Perennials

Hymenoxys acaulis var. *glabra* (DeMauro 1994). Although Lakeside daisy (Asteraceae) originally ranged from central Illinois to Ohio and western Ontario, it was restricted to dolomite or gravel prairie habitats. Such habitats were formed from post-glacial rock outcrops or stony erosion terraces; they tend to be flat, open and relatively dry during the summer. Industrial development and limestone quarrying have destroyed most of these prairies, eliminating or severely reducing populations of Lakeside daisy. After quarrying stopped in portions of a remnant Ohio site, Lakeside daisy was able to re-colonize 900 acres with a population of over a million individuals. An inland Illinois population, however, continued to decline in the early 1970s to fewer than 30 plants. Despite the presence of flowers and insect pollinators, no viable seeds were produced before this small population was finally extirpated by development in 1981. Fortunately, three plants had been collected in 1979 and vegetatively propagated (from rhizomes and caudices) in a private garden.

Efforts to reintroduce Lakeside daisy to Illinois first needed to address the question of reproductive failure during the decade prior to extirpation. Six years worth of hand pollinations had produced no seed from Illinois plants, whereas sexual reproduction was occurring naturally in the large populations that recolonized the Ohio quarry. Detailed genetic studies of plants derived from both populations established that Lakeside daisy had a strong, sporophytic self-incompatibility system that prevented zygote formation between individuals that possessed the same self-incompatibility alleles (or S alleles). Of the 20 plants examined, 15 different forms of the S allele were documented, resulting in fifteen distinct mating groups. Importantly, all of the Illinois plants were found to be in the same mating group and could produce seed only when outcrossed to Ohio plants. It was then clear that reproductive failure in the Illinois population was due to elimination of all but one mating group during the extirpation process (DeMauro 1993).

Such genetic insights were incorporated into the design of the reintroduction. A sampling program was devised to maximize mating group diversity among the founders by obtaining seeds from natural populations in Ohio and Canada and from artificial populations that included hybrids between garden-grown Illinois plants and those from natural populations. The advantage of restoring fertility by incorporating the Illinois genomes into the founders was considered greater than the potential disadvantage of outbreeding depression. Approximately 5,000 seeds were germinated and grown under greenhouse conditions and labeled with respect to their genetic identity. After careful screening of potential sites using geological, historical, land-use, and habitat criteria, two Illinois nature preserves were selected for reintroduction. Transplanting began in the spring of 1988 using 600 rosettes at each site. Many of the important "small decisions" about how to develop the new populations at specific microsites (e.g., those concerning plant density, microtopographic placement, and neighborhood composition) were guided by comparisons to the microsite characteristics of natural populations.

Unfortunately, environmental stochasticity in the form of a severe drought and record high temperatures, killed 95% of the founders during their first two months. An additional 500 plants were reintroduced at each site in the fall and mortality after two years ranged between 20–30%. Herbivores, including insects, deer, and rabbits, consumed two-thirds of all inflorescences during a single year. Despite such setbacks, flowering and seed production occurred at levels comparable to plants in natural populations and more than 500 recruits became established by 1991. Some of those recruits have gone on to flower and produce the third generation of Lakeside daisy in Illinois. A demographic monitoring program that includes information on genetic history of the founders will be used to evaluate these populations over the long-term and to compare their performance to apparently viable natural populations.

Sidalcea nelsoniana (CH2M Hill 1986, 1995). Nelson's checkermallow (Malvaceae) is a rhizomatous herb with tall (50–150 cm), erect stems bearing large numbers of showy pistillate and perfect flowers. In 1984, this checkermallow was known from only six localities, with approximately 110 plants scattered in the Willamette Valley and coastal ranges of western Oregon. A proposed dam and reservoir on public land at Walker Flat would have inundated a significant proportion of the known individuals. A series of extensive field, greenhouse and laboratory studies was begun, involving more than 6,000 hours of professional labor over a ten-year period. Some studies were conducted to determine whether the Walker Flat population was genetically or morphologically distinct from other *S. nelsoniana* populations, and thus deserving of special protection under the U.S. Endangered Species Act (1973, as amended). Comparisons of leaf and seed proteins, pollen grain ultrastructure, and gross morphology failed to demonstrate any significant differences between plants from Walker Flat and plants from other areas (Halse, Rottink, and Mishaga 1989). Consequently, the reservoir project was still considered feasible, but other studies were conducted to determine (1) whether more populations existed within historic range; and (2) if a new, self-perpetuating population could be established in an appropriate, protected site using propagules from Walker Flat.

After two years, more than 33 populations containing nearly 9,000 plants had been documented by thorough field surveys. Wetland, grassy meadow, and even forested upland habitats were found to support populations, indicating that this rare species had a fairly wide ecological amplitude (see also Glad, Halse, and Mishaga 1994). Observations during the field surveys revealed that Nelson's checkermallow was tolerant of mowing, burning, and even some impact from motor vehicles. Vigorous plants could be found in hayfields, pastures, roadsides and abandoned gravel pits, and individuals were commonly associated with the weedy and competitive Canada thistle (*Circium arvense*). The wide ecological amplitude and high tolerance of disturbance suggested that this species might readily be transplanted into a variety of sites for mitigating the losses at Walker Flat.

Reintroduction of Nelson's checkermallow began with site selection and the development of founder populations from seeds and rhizomes. Site selection utilized aerial photographs, descriptions of soil characteristics and associated species, land-use status and field checking. Twelve criteria were applied to an initial list of 19 candidate sites within the known historic range. Some criteria were biological (e.g., community composition, absence of congeners to prevent hybridization, similarity of physical environment), some were logistic (e.g., size of the area, accessibility), and others were legal (e.g., public vs. private ownership, prospects for long-term management). Six sites were chosen for the project, including some that supported small natural populations of Nelson's checker-mallow.

Seeds were collected from six natural populations in 1985, germinated in 4-inch peat pots, and grown for several months under greenhouse conditions. A total of 1,416 seedlings were available by May 1986, with 20 to 826 plants destined for each site (depending on available habitat). Grids were used to position the seedlings at 1m intervals and to facilitate demographic monitoring. Monitored parameters included vegetative growth, survivorship to reproduction, reproductive status, and potential mortality factors. A sub-sample of four plants in each 50 plant grid was used to monitor the number of basal leaves and flowering stems, plant height and width. It was also possible to quantify the extent to which seedling performance depended on source population.

Founders also were grown from 200 rhizome segments collected at Walker Flat in the fall of 1986. Sampling occurred across five microhabitats and larger subpopulations contributed proportionately more to the segment pool. Each donor plant and rhizome segment were marked and labeled to keep records of origin (microhabitat and locality) and inferred degree of relatedness (clones, family groups). Immediate transplantation of the rhizomes into a suitable, protected site was facilitated by first establishing five grids in a large, grassy meadow. The vegetation within those grids varied considerably and included native and exotic grasses, rushes, sedges, and forbs. Each grid received either 25 or 50 segments, which were loosely covered by 2.5 cm of soil and then watered.

Even though the first summer was relatively hot and dry, founders established at all reintroduction sites. Regardless of their origin (seedling or rhizome), between 68–100% of the plants survived their first year (mean of 83%) with pocket gophers and soil drought as the most probable mortality factors. Competition from the surrounding, established plants may have accentuated the drought stress, but those same plants also provided shade and reduced leaf temperatures. Many individuals reproduced during their first year in the field; 15–27% of all plants produced flowers, with small but significant variations between sites. Also, seedlings that were larger when transplanted produced more flowers than smaller seedlings. Variations due to seed source appeared to be weak, but a lack of replication prevented a conclusive statistical analysis.

Although the reintroduction was apparently successful in its first year, the question of whether these new populations are self-perpetuating remains to be answered. Nine successive years of demographic monitoring have demonstrated high survivorship (45–100% of all plants, with survival of transplants from rhizomes slightly higher than survival of seedlings), vigorous vegetative growth, widespread flowering (54–100% of all plants), and fruit production (58% of all plants). However, it has been difficult to determine whether any second generation seedlings have been established from seed produced by the reintroduced populations. This problem arises because the observed seedlings may have originated from nearby natural populations. Further analyses of seed dispersal, seed bank density, seedling distribution patterns, and age structure will be needed to establish the resilience and potential for persistence of these reintroduced populations.

Annuals

Stephanomeria malheurensis (Parenti and Guerrant 1990; Guerrant 1996b; Marshall 1995). During the early 1970s only 750 individuals of the Malheur wirelettuce (Asteraceae) could be found within the sagebrush steppe of Harney County, Oregon.

After a fire in 1972, the exotic annual cheatgrass *(Bromus tectorum)* rapidly invaded the open areas among shrubs and began to displace the rare wirelettuce. Short-term climatic variation also may have contributed to population decline. Relatively few plants were seen during several consecutive years of drought in the late 1970s. This was followed by extremely large amounts of rainfall in 1981–1985, which caused the levels of nearby Malheur and Harney lakes to rise dramatically. These unusually wet conditions coincided with local extirpations of the widespread, sympatric ancestor *S. exigua* ssp. *coronaria.* By 1985, the Malheur wirelettuce apparently was extinct in the wild (Figure 5.5, and Gottlieb 1991).

Stephanomeria malheurensis had been surprisingly well studied in spite of its rarity, recent discovery, and rapid extinction. Dr. Les Gottlieb systematically collected seeds from all portions of the population during the 1970s, carefully maintaining the stock for his evolutionary investigations (Gottlieb 1977, 1978, 1983). Electrophoresis of allozymes from thirteen loci revealed that this highly self-compatible species had, with one possible exception, a subset of the alleles found in its outcrossing ancestor *S. exigua* ssp. *coronaria.* The rare wirelettuce is monomorphic at eleven loci and bears fewer alleles than its common relative at two polymorphic loci. Although artificial selection during off-site propagation could have reduced allelic variation, the largely self-pollinating breeding

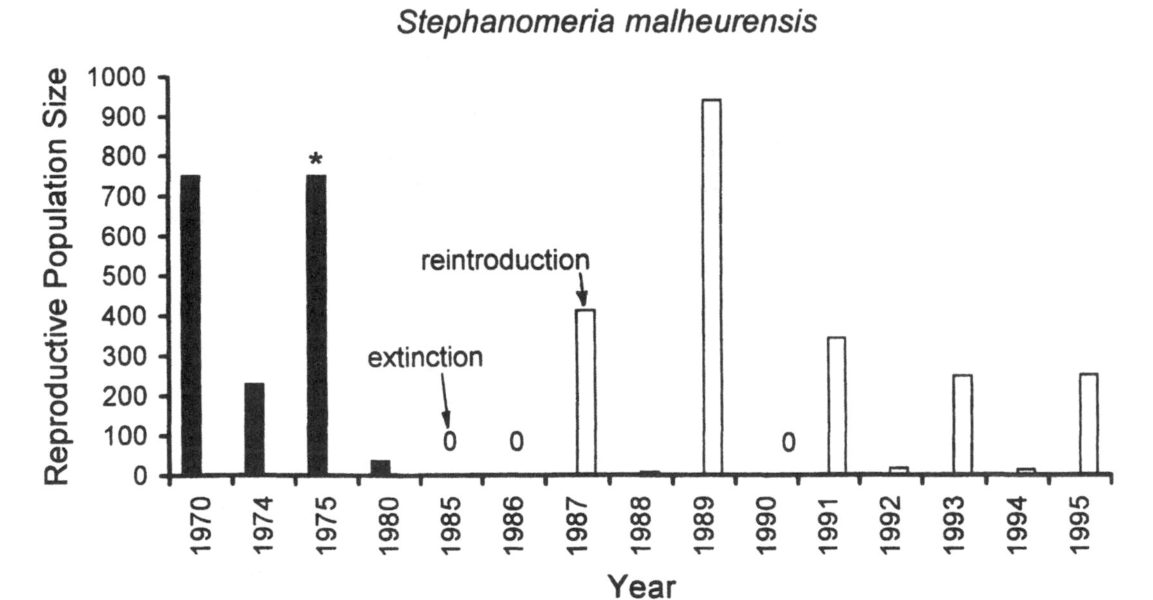

Figure 5.5
Census data from natural (dark bars) and reintroduced (open bars) populations of the rare and endangered annual plant, Stephanomeria malheurensis. *The natural population went extinct (with no detectable soil seed bank) in 1985. Reintroduction at the same site began in 1987 with large subsequent yearly variations in population size. The asterisk (*) above the 1975 bar indicates a conflict among data sources. Gottlieb (1979) estimated the total number of* S. malheurensis *plants was not more than 750 in any one year (with less than half this number actually counted during diligent searches), whereas Marshall (1995) indicated a population size of 1050 in 1975. (Data from Gottlieb 1979; Guerrant 1996b; Marshall 1995; and N. Taylor personal communication).*

system of *S. malheurensis* would have minimized erosion caused by genetic drift and inbreeding depression.

Reintroduction of Malheur wirelettuce at its type locality was made possible by a donation of seed stock from Dr. Gottlieb. One thousand seedlings produced at The Berry Botanic Garden were made available for transplantation in 1987. Four experimental plots were established, each dominated by a single species; big sagebrush (*Artemisia tridentata*), rabbitbrush (*Chrysothamnus viscidiflorus*), Great Basin wildrye (*Elymus cinereus*), and the exotic cheatgrass. Rodent-proof wire mesh extending 10 cm below and 80 cm above the soil surface delineated the 5 × 5 m plots. The cheatgrass plot was twice this size so that half could be hand-weeded to 50% cheatgrass cover, and the other half left as an unweeded control with 50–100% cover. The three native plots were kept free of cheatgrass by hand-weeding. Transplants were watered initially and as needed for another four weeks.

Fates of the founders were followed closely until the end of the first growing season. Overall, survivorship was high and approximately 40,000 seeds were produced. No significant effect of community composition was detected, but plants in the cheatgrass-infested plots were significantly smaller, slower to bolt and flower, and less fecund. Thus, it was determined that site management must address the deleterious effects of weed competition, but controlled burns would only lead to more cheatgrass. Brauner (1988) suggested that small-scale applications of herbicide could be effective and inexpensive if there were not a court-ordered injunction against their use on federal rangelands.

Subsequently, the reintroduced population has fluctuated greatly in size for a variety of reasons (Figure 5.5). In 1988 the sagebrush plot was heavily browsed by small mammals, although the 1989 population was comprised of more than 900 plants (Parenti and Guerrant 1990; Marshall 1995). Cold, dry conditions during the spring of 1990 completely inhibited germination and the seed population remained dormant. However, 385 plants reappeared in 1991, of which 90% survived to set seed. Although cheatgrass competition remains a problem in all plots, other factors have been linked to oscillations in population size and survivorship to reproduction (Marshall 1995). The small amount of available data demonstrated a positive correlation between fall-winter precipitation (September to May) and emergence of Malheur wirelettuce seedlings, suggesting a 17 cm precipitation threshold for simultaneous germination. Survivorship was positively correlated with spring-summer precipitation (May to August), although herbivore activity could have the most significant impact at any one microsite or in any one year. Such extreme, environmentally induced fluctuations in population size will necessitate long-term monitoring and evaluation of this reintroduced population.

Amsinckia grandiflora (Pavlik, Nickrent, and Howald 1993; Pavlik 1995). The large-flowered fiddleneck (Boraginaceae) is a narrow endemic in the Mt. Diablo area of northern California. It has disappeared from most of its original 650 km^2 range, presumably because livestock grazing, fire suppression, and non-native plants have altered the original, perennial grassland habitat. Dense, contiguous swards of European annual grasses, especially wild oats (*Avena fatua*), ripgut (*Bromus diandrus*), and barley (*Hordeum* spp.) have replaced patchy stands of native bunchgrasses, thereby decreasing habitat quantity and quality. During the early 1980s, only one population ("Site 300") was known to exist, fluctuating in size between 23 and 355 individuals (Figure 5.6). Twenty years earlier, however, this same population included thousands of plants (Ornduff 1976). Although two additional natural populations (one large (perhaps 4,000 plants) and one small (<30)) were subsequently discovered, the U.S. Fish and Wildlife Service drafted a recovery plan calling for the creation of new, self-sustaining populations within the historic range, and for the enhancement of natural populations to decrease the probability of near-term extinction.

Very little biological information was available on the large-flowered fiddleneck at the start of the project in 1987. It was thought to have a moderate degree of microhabitat specificity (e.g., north-facing slopes on loamy soils), a grazing-susceptible (caulescent) growth form, and low fecundity (nutlet output per plant) compared to its common congeners. Observations of plants in the field and in experimental gardens suggested that low fecundity and plant size were strongly correlated and that variations in plant size could result from competition with annual grasses. A heterostylous outbreeding system and possible inbreeding depression also were suspected causes of low fecundity (Ornduff 1976; Pantone, Pavlik, and Kelly 1995). Electrophoretic examination of nutlets collected from Site 300 when the population was large revealed low genetic variation, but nutlets propagated off-site at the University of California at Davis had even less variation. Therefore, competition, the relative abundance of "pin" or "thrum" floral morphs, and gene pool composition were identified as the primary variables that might limit the establishment and persistence of *A. grandiflora* populations in the wild.

The reintroduction and enhancement of populations was designed as a three-phased process (Pavlik 1995). The first, experimental phase was used to release a founding population within historic range and to answer the following questions: 1) Does annual grass cover affect the demographic performance of *A. grandiflora*? 2) Can the demographic performance of *A. grandiflora* be affected by manipulating annual grass cover with controlled burns, hand clipping, or a grass-specific herbicide? and 3) Will nutlets from the Site 300 source demonstrate better demographic performance than nutlets from the cultivated source (U.C. Davis)? The second, enhancement phase would use the results of the experimental phase to manipulate the new population and the natural population at Site 300. The objective would be to manage the habitat to increase plant fecundity and consequently, population size. Finally, the third phase, exposure to natural environmental variation in the absence of management, would allow evaluation of population resiliency and the potential for self-sustainability.

The experimental phase of the reintroduction began in the fall of 1989 when 3,460 nutlets from the Site 300 and Davis sources were sown into 20 plots using removable wooden frames. The frame allowed every nutlet to be precisely placed and monitored throughout the year. The frame also allowed identification of the seed source for each seedling. Plots were then burned, clipped, or treated with herbicide to reduce or eliminate annual grasses.

As in the case of *Stephanomeria*, competition from exotic grasses had a significant negative effect on survivorship to reproduction, plant size, and nutlet output. Competition was most effectively reduced by the grass-specific herbicide, which increased plant size and fecundity by a factor of 3.5 relative to control plants. Burning also reduced competition, but was less effective because of annual grass recruitment after the burn. Despite higher levels of genetic variation, plants from the Site 300 source exhibited the same germination, mortality patterns, plant sizes and fecundity as plants from the Davis source. Overall, 1,744 nutlets (50%) germinated during the growing season, of which 1,101 survived to produce an estimated 35,000 new nutlets.

Results from the phase I reintroduction experiments were used to develop site management techniques for enhancing the new population and the declining natural population at Site 300 (Figure 5.6). These techniques apparently have been successful in the short term, allowing both populations to exceed 1,600 plants in 1992 and 1994, respectively.

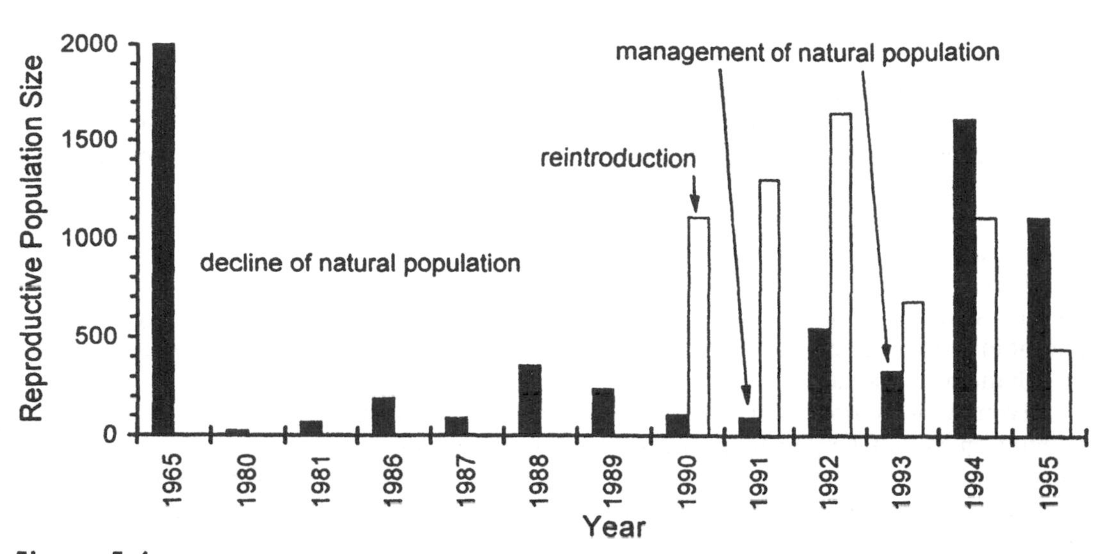

Figure 5.6
Census data from natural (dark bars) and reintroduced (open bars) populations of the rare and endangered annual plant, Amsinckia grandiflora. *The natural population declined from roughly several thousand to as few as 23 individuals. The reintroduction occurred at a separate site and provided the management prescription for the natural population (control of exotic competitors in 1991 and 1993) (Data from Pavlik, Nickrent, and Howard 1993; Pavlik 1995).*

The effects of natural variation in precipitation and temperature have been observed during phase III of the reintroduction. Both populations have either increased or decreased, depending on the timing and the amount of rain. Grass competition appears to be stronger when abundant rains fall during the late autumn and early winter. Weak competition occurs when initial rains are delayed into late winter, followed by peak rainfall in early spring. Once again, variation in environmental parameters appears to play a significant role in the fate of introduced and natural populations.

COMMON THREADS AND DIVERGENT PATHS

First, these projects illustrate a range of experimental approaches to rare plant reintroduction. Rare plants seldom are understood well enough at the beginning of such projects to allow fruitful, *a priori* decisions to be made on reintroduction site characteristics, number of founders, microhabitat preferences, genetic composition, and population structure. Many "educated guesses" are required, but some will have to be tested as hypotheses to assess their relative importance to the new population and the conservation effort as a whole. Pavlik (1994) emphasized the testing of management-oriented hypotheses, i.e., those that identify factors that limit population growth or persistence and that suggest concrete remedies that can be practically implemented in the field. As with any experiment, appropriate controls, adequate replication, and statistical analysis are essential for reaching conclusions (Sutter 1996). In three case studies (*Conradina*, *Stephanomeria*, *Amsinckia*), the effects of weedy competitors and efficacy of various control methods (e.g., fire, hand-clipping, herbicides, restoration of community dominants) were evaluated and incorporated into long-term management plans with a high degree of confidence.

Secondly, demographic monitoring of the outplanted material was the basis for evaluating the reintroduced population and the effects of genetic management or habitat factors. Individual seeds or cuttings from different sources should be marked and monitored to determine whether localized genotypes perform differently at a given site (*Sidalcea* and *Amsinckia*), or to insure that critical alleles controlling the mating system are built into the assemblage of founders (*Hymenoxys*). Regardless of where the outplanted material originated, its abilities to germinate, grow, survive, and reproduce are the most relevant measures of performance within a given habitat. Cruder forms of monitoring, such as photoplots or cover estimates, are inadequate and inappropriate for purposes of creating or enhancing populations.

Even with demographic monitoring, however, the conclusion that a reintroduced population is self-perpetuating and of conservation value is not easily reached. What constitutes a "successful" population? Pavlik (1996) defined success as an argument to be made, based on the fulfillment of taxon-specific objectives, that addressed four major goals: abundance, geographic extent, resilience, and persistence. He suggested that short- and long-term objectives be formulated at the beginning of a reintroduction effort using life history information and studies of similar taxa. For example, a short-term abundance objective might be the completion of the life cycle *in situ*, with more seeds produced by the new population than planted during the reintroduction. A long-term abundance objective would be the attainment of a predetermined MVP size by natural recruitment from second generation seed cohorts. Although not all plant taxa can be modeled to produce a specific MVP, target values could be chosen from a range appropriate for a given set of life history parameters. The careful delineation of objectives also helps define the monitoring program so that crucial parameters for determining success are not missed or ignored.

Ultimately, a minimum of ten years of monitoring and evaluation are required to allow most physical and biological factors at a given reintroduction site to be expressed (Sutter 1996). Extreme variations between years in population size, seed predation, germination, transplant establishment, and fecundity, apparently due to environmental variation, have been demonstrated for both perennial (*Hymenoxys*) and annual (*Stephanomeria*, *Amsinckia*) taxa. Thus, demographic performance in reintroduced populations can span the range from almost complete failure to an order of magnitude increase in propagule production. Such variation may be the rule rather than the exception, and attempts should be made to minimize its impact. In the absence of long-term exposure to variations in rainfall, temperature, competitors, herbivores, seed predators, and disturbance, the six projects reviewed above must be regarded as ongoing and of uncertain conservation value.

The impact of environmental stochasticity on an endangered plant can be ameliorated by adopting a reintroduction strategy that mimics metapopulation structure (Bowles *et al.* 1993; Bowles and McBride 1996) or a patchy spatial distribution. Establishing multiple populations, even if the original range of the species was limited to a single population, has at least two major advantages. First, the probability of catastrophic failure due to unforeseen environmental factors (e.g., herbivory) can be spread over discrete populations in separate localities, thus enhancing long-term persistence overall. Second, the intentional subdivision of propagation material among reintroduction sites can offset the effects of genetic drift in a single, closed population (Templeton 1990, 1991). Although each reintroduced population might be smaller than one large one (i.e., if the seed supply is limited), a greater range of selective forces among sites and the improbability of drift having the same effect in all populations are likely to allow more of the total genetic variation to be preserved. There is, of course, a complex tradeoff with respect to the rate of loss of genetic variation due to different reintroduction strategies. The tradeoff depends in part on the total number of seeds available for the project and the probability of environmental variation causing extreme shifts in population size.

It is important to note that all of these reintroduction efforts took place with seed or vegetative material that had been stored off-site. *Ex situ* propagules were the crucial link between new, created populations and extirpated subpopulations (*Pinus*), populations (*Hymenoxys*) and even whole taxa (*Stephanomeria*). In the case of *Amsinckia*, the 25-year-old collections had preserved genetic variation that was probably lost as the natural population precipitously declined. Some of that variation was subsequently built into the reintroduction with the hope that it might improve population resilience and persistence in the future (Clegg and Brown 1983). The work of the Center for Plant Conservation in developing genetically representative seed collections and storing them off-site is particularly timely given the current rates of species and population loss. Such collections can provide the basis for the reintroduction of a rare plant, thus bridging natural and created populations with an essential genetic "echo" across a wide chasm of time.

The genetic variation contained within samples stored off-site can, however, erode rapidly or be distorted, especially if propagated off-site to counteract declining seed survivorship. It is sobering to consider that what little variation existing in the original *Amsinckia* collections was lowered even further within a few generations of off-site propagation. Four polymorphic loci had been reduced to two or one polymorphic loci, depending on the number of generations and in which experimental garden propagation had taken place. To what degree this was due to selection versus random genetic drift cannot be assessed rigorously. However, different selective pressures on-and off-site cannot be discounted *a priori*, and are almost certainly the norm. The problem of genetic change in captivity must be considered and surmounted by careful collection management (Figure 5.3) if we are to return viable populations to the wild.

One simple way of avoiding selection in captivity is to release propagules directly into protected prospective sites within historic range. Although this offers little direct control over the direction and intensity of selection, the founding gene pool will be molded by conditions *in situ* instead of in the garden. If the fates of the propagules are followed demographically (see Pavlik 1994), we can also obtain information on a major hurdle in the life cycle—the transition from seed bank to established juvenile. Such information often reveals sources of mortality that cannot be seen when reintroduction sites are chosen or cannot be assessed when juvenile plants are transplanted. Microgranivores and herbaceous competitors, for example, can devastate a new seed bank with little or no effect on established individuals. In the case of perennials, a reintroduction of cuttings or seedlings might exhibit indicators of success (e.g., high survivorship, vigorous growth) for years until the first generation of seeds was produced and dispersed onto the site. However, modeling suggests that beginning a reintroduction with cuttings or seedlings may help overcome the problems associated with the seed to juvenile transition, thus leading to higher survivorship and reproductive output of the founders, and higher potential rates of population growth, at least initially (Guerrant 1996). Garden-grown material can be used to create a desirable age- or stage-structure in a reintroduced population that may lower the probability of near-term extirpation. The short-term goal is then to maximize the number of reproducing adults from the available plant material (e.g., *Conradina*,

Stephanomeria) and build the seed bank as rapidly as possible. At least in sexual species, long-term success necessitates that new generations of plants be derived from seed produced and stored *in situ*. Perhaps a dual strategy, with experimental reintroductions of seeds and whole plants at a given site, would minimize off-site selection, assess cryptic mortality factors affecting the seed bank, and maximize on-site reproductive yield from the available material.

The experimental reintroduction or enhancement of rare plant populations also will contribute to the developing science of ecosystem management. Ecosystem prescriptions based solely on the responses of common species are likely to degrade populations of specialists, especially those whose tolerance limits and specific requirements are poorly understood. Documenting the effects of controlled burns, herbicides, insect control, and other habitat manipulations on rare species can provide guidelines for maintaining biological diversity on larger spatial scales.

LOOKING TO THE FUTURE

Ex situ conservation methods can be used to forestall the extinction of plant species but are not developed to the point of being generally reliable. On one hand, the task of obtaining and storing genetically representative seed samples from our nation's most rare and endangered plants is scientifically based and well underway. Given the CPC genetic sampling guidelines and a network of active participants, satisfactory completion of this task appears largely to be a matter of time and resources. However, we have only begun the more formidable task of understanding how to reestablish self-sustaining populations from these samples.

The CPC guidelines for releasing rare plant material back into the wild refer to reintroduction as a complex process that is "exacerbated by a shortage of sound policies, effective models, and strong scientific underpinnings" (Falk, Millar, and Olwell 1996). In an effort to improve the overall quality of reintroductions, the guidelines are built around a set of eleven questions pertaining to three project phases; planning, implementation, and evaluation. The guidelines do not attempt to provide recipes, but they do reaffirm our view that reintroductions are necessarily experimental, long-term efforts.

Learning how to reliably reintroduce populations of endangered plants from *ex situ* samples is a task of immediate importance to applied conservation. But this effort also holds great promise as an avenue for developing ecological and evolutionary theory. Because carefully designed reintroduction attempts involve a wider variety of plant life histories and habitats, we will advance the cause of conserving particular rare species while adding to our understanding of population, community, and ecosystem dynamics.

Ex situ methods are not the ultimate answer to the loss of biodiversity. They are, however, the best option currently available for conserving some of our most endangered plant species in the near future. By learning how to sample, store, and create self-sustaining populations, we can begin to heal damaged communities and ecosystems and restore the complex web of life from which we emerged and upon which we depend.

ACKNOWLEDGMENTS

We thank C. Dawson, M. DeMauro, J. Glad, D. Gordon, R. Halse, and T. Ledig for providing information about their reintroduction projects. We would also like to thank

P. Kareiva, P.L. Fiedler, and an anonymous reviewer for valuable critical commentary on the manuscript.

LITERATURE CITED

Barrett, S.C.H. and J.R. Kohn. 1991. Genetic and evolutionary consequences of small population size in plants: Implications for conservation. In *Genetics and conservation of rare plants*, eds. D.A. Falk and K.E. Holsinger, 3–30. New York: Oxford University Press.

Birkinshaw, C.R. 1991. *Guidance notes for translocating plants as part of recovery plans*. Nature Conservancy Council, CSD Report 1225. Peterborough, England.

Bowles, M. and J. McBride. 1996. Pitcher's thistle (*Circium pitcheri*) reintroduction. In *Restoring diversity: Strategies for reintroduction of endangered plants*, eds. D. Falk, C. Millar, and M. Olwell, 423–431. New York: Island Press.

Bowles, M., R. Flakne, K. McEachern, and N. Pavlovic. 1993. Status and restoration planning for the federally threatened pitcher's thistle (*Circium pitcheri*) in Illinois. *Natural Areas Journal* 13:164–176.

Bowles, M. and C. Whelan, eds. 1994. *Recovery and restoration of endangered species*. Cambridge: Cambridge University Press.

Bradshaw, M.E. 1981. Monitoring grassland plants in Upper Teesdale, England. In *The biological aspects of rare plant conservation*, ed. H. Synge, 241–251. London: John Wiley and Sons.

Brauner, S. 1988. *Malheur wirelettuce (*Stephanomeria malheurensis*) biology and interactions with cheatgrass: 1987 study results and recommendations for a recovery plan*. Unpublished report to Burns District, OR, Bureau of Land Management. Submitted 20 June 1988.

Brown, A.H.D. and J.D. Briggs. 1991. Sampling strategies for genetic variation in *ex situ* collections of endangered plant species. In *Genetics and conservation of rare plants*, eds. D.A. Falk and K.E. Holsinger, 99–119. New York: Oxford University Press.

Center for Plant Conservation. 1991. Appendix. Genetic sampling guidelines for conservation collections of endangered plants. In *Genetics and conservation of rare plants*, eds. D.A. Falk and K.E. Holsinger, 225–238. New York: Oxford University Press.

CH2M Hill. 1986. *Studies of* Sidalcea nelsoniana. Unpublished progress report, City of McMinnville Water and Light Department, McMinnville, Oregon.

CH2M Hill. 1995. *Technical Memorandum:* Sidalcea nelsoniana *monitoring –1995*. Unpublished progress report, City of McMinnville Water and Light Department, McMinnville, Oregon.

Clegg, M.T. and A.H.D. Brown. 1983. The founding of plant populations. In *Genetics and conservation: A reference for managing wild animal and plant populations*, eds. C.M. Schonewald-Cox, S.M. Chambers, B. MacBryde, and L. Thomas, 216–228. London: Benjamin Cummings.

Crow, J.F. and M. Kimura. 1970. *An introduction to population genetics theory*. New York: Harper and Row.

Davy, A.J. and R.L. Jefferies. 1981. Approaches to the monitoring of rare plant populations. In *The biological aspects of rare plant conservation*, ed. H. Synge, 219–232. London: John Wiley and Sons.

DeMauro, M.M. 1993. Relationship of breeding system to rarity in the lakeside daisy (*Hymenoxys acaulis* var. *glabra*). *Conservation Biology* 7:542–550.

DeMauro, M.M. 1994. Development and implementation of a recovery program for the lakeside daisy (*Hymenoxys acaulis* var. *glabra*). In *Recovery and restoration of endangered species*, eds. M. Bowles and C. Whelan, 298–321. Cambridge: Cambridge University Press.

Ehrenfeld, D.W. 1970. *Biological conservation*. New York: Holt, Rinehart and Winston.

Falk, D.A. 1987. Integrated conservation strategies for endangered plants. *Natural Areas Journal* 7:118–23.

Falk, D.A. 1990. Integrated strategies for conserving plant genetic diversity. *Annals of the Missouri Botanical Garden* 77:38–47.

Falk, D.A. 1992. From conservation biology to conservation practice: Strategies for protecting plant diversity. In *Conservation Biology: the theory and practice of nature conservation, preservation, and management*, eds. P.L. Fiedler and S.K. Jain, 397–431. New York: Chapman & Hall.

Falk, D.A. and K.E. Holsinger, eds. 1991. *Genetics and conservation of rare plants*. New York: Oxford University Press.

Falk, D.A., C.I. Millar, and M. Olwell, eds. 1996. *Restoring diversity: Strategies for reintroduction of endangered plants*. New York: Island Press.

Fiedler, P.L. 1987. Life history and population dynamics of rare and common mariposa lilies (*Calochortus* Pursh: Liliaceae). *Journal of Ecology* 75:977–995.

Fiedler, P.L, P.S. White, and R. A. Leidy. 1996. The paradigm shift in ecology and its implications for conservation. In *Enhancing the ecological basis of conservation: Heterogenity, ecosystem function, and biodiversity*, eds. S.T.A. Pickett, R.S. Ostfeld, H. Schchak, and G.E. Likens, 145–160. New York: Chapman & Hall.

Frankel, O.H. and E. Bennett, eds. 1970. *Genetic resources in plants: Their exploration and conservation*. Oxford: Blackwell Scientific.

Frankham, R. 1995. Inbreeding and extinction: A threshold effect. *Conservation Biology* 9:792–799.

Gilpin, M.E. 1987. Spatial structure and population vulnerability. In *Viable populations for conservation*, ed. M.E. Soulé, 125–139. Cambridge: Cambridge University Press.

Gilpin, M.E. and M.E. Soulé. 1986. Minimum viable populations: Processes of species extinction. In *Conservation biology: The science of scarcity and diversity*, ed. M.E. Soulé, 19–34. Sunderland, MA: Sinauer Associates.

Glad, J.B., R.R. Halse, and R. Mishaga. 1994. Observations on distribution, abundance, and habitats of *Sidalcea nelsoniana* Piper (Malvaceae) in Oregon. *Phytologia* 76:307–323.

Gordon, D.R. 1996. Apalachicola rosemary (*Conradina glabra*) reintroduction. In *Restoring diversity: Strategies for reintroduction of endangered plants*, eds. D.A. Falk, C.I. Millar, and M. Olwell, 417–422. New York: Island Press.

Gottlieb, L.D. 1973. Genetic differentiation, sympatric speciation and the origin of a diploid species of *Stephanomeria*. *American Journal of Botany*. 60:545–553.

Gottlieb, L.D. 1977. Phenotypic variation in *Stephanomeria exigua* ssp. *coronaria* (Compositae) and its recent derivative species "*malheurensis*." *American Journal of Botany* 64:873–880.

Gottlieb, L.D. 1978. Allocation, growth rates and gas exchange in seedlings of *Stephanomeria exigua* ssp. *coronaria* and its recent derivative *S. malheurensis*. *American Journal of Botany* 65:970–977.

Gottlieb, L.D. 1979. The origin of phenotype in a recently evolved species. In *Topics in plant population biology*, eds. O.T Solbrig, S. Jain, G.B. Johnson, and P.H. Raven, 264–286. New York: Columbia University Press.

Gottlieb, L.D. 1983. Interference between individuals in pure and mixed cultures of *Stephanomeria malheurensis* and its progenitor. *American Journal of Botany* 70:276–284.

Gottlieb, L.D. 1991. The Malheur wire-lettuce: a rare, recently evolved Oregon species. *Kalmiopsis (Journal of the Native Plant Society of Oregon)* 1:9–13.

Guerrant Jr., E.O. 1996a. Designing populations for reintroduction: Demographic opportunities, horticultural options, and the maintenance of genetic diversity. In *Restoring diversity: Strategies for reintroduction of endangered plants*, eds. D.A. Falk, C.I. Millar, and M. Olwell, 171–207. New York: Island Press.

Guerrant Jr., E.O. 1996b. Experimental reintroduction of *Stephanomeria malheurensis*. In *Restoring diversity: Strategies for reintroduction of endangered plants*, eds. D.A. Falk, C.I. Millar, and M. Olwell, 399–402. New York: Island Press.

Guerrant Jr., E.O., S.T. Schultz, and D. Imper. 1996. *Draft recovery plan for* Lilium occidentale, *the western lily*. Unpublished report prepared for the U.S. Fish and Wildlife Service, Portland field office, Portland, OR.

Haig, S.M., J.D. Ballou, and S.R. Derrickson. 1990. Management options for preserving genetic diversity: Reintroduction of Guam rails to the wild. *Conservation Biology* 4:290–300.

Halse, R.R., B.A. Rottink, and R. Mishaga. 1989. Studies in *Sidalcea* taxonomy. *Northwest Science* 63:154–161.

Hamilton, M.B. 1994. *Ex-situ* conservation of wild plant species: Time to reassess the genetic assumptions and implications of seed banks. *Conservation Biology* 8:39–49.

Hamrick, J.L. 1989. Isozymes and the analysis of genetic structure in plant populations. In *Isozymes in plant biology*, eds. D.E. Soltis and P.S. Soltis, 87–105. Portland: Dioscorides Press.

Hamrick, J.L. and M.J.W. Godt. 1989. Allozyme diversity in plant species. In *Plant population genetics, breeding, and genetic resources*, eds. A.D.H. Brown, M.T. Clegg, A.L. Kahler, and B.S. Weir, 43–63. Sunderland, MA: Sinauer Associates.

Hamrick, J.L., M.J.W. Godt, D.A. Murawski, and M.D. Loveless. 1991. Correlations between species traits and allozyme diversity: Implications for conservation biology. In *Genetics and conservation of rare plants*, eds. D.A. Falk and K.E Holsinger, 75–86. New York: Oxford University Press.

Harper, J.L. 1977. *The population biology of plants*. London: Academic Press.

Holsinger, K.E. and L.D. Gottlieb. 1991. Conservation of rare and endangered plants: Principles and prospects. In *Genetics and conservation of rare plants*, eds. D.A. Falk and K.E. Holsinger, 195–208. New York: Oxford University Press.

Lande, R. 1988. Genetics and demography in biological conservation. *Science* 241:1455–1460.

Lande, R. 1995. Mutation and conservation. *Conservation Biology* 9:782–791.

Ledig, F.T. 1996. *Pinus torreyana* at the Torrey Pine State Reserve, California. In *Restoring diversity: strategies for reintroduction of endangered plants*, eds. D.A. Falk, C.I. Millar, and M. Olwell, 265–271. New York: Island Press.

Lesica, P. and F.W. Allendorf. 1995. When are peripheral populations valuable for conservation? *Conservation Biology* 9:753–760.

Loebel, D.A., R.K. Nurthen, R. Frankham, D.A. Briscoe, and D. Craven. 1992. Modeling problems in conservation genetics using captive *Drosophila* populations: Consequences of equalizing founder representation. *Zoo Biology* 11:319–332.

Lynch, M., J. Conery, and R. Burger. 1995. Mutation accumulation and the extinction of small populations. *American Naturalist* 146:489–518.

Mangel, M. and C. Tier. 1994. Four facts every conservation biologist should know about persistence. *Ecology* 75:607–614.

Marshall, K.A. 1995. Stephanomeria malheurensis *Progress Report 1993–1994*. United States Department of the Interior, Bureau of Land Management, Burns District Office, Hines Oregon.

Martin, E.L. 1992. Patterns of genetic diversity in the rare mint *Conradina glabra* and its nearest relative *Conradina canescens*. M.S. thesis, University of West Florida.

Maunder. M. 1992. Plant reintroduction: An overview. *Biodiversity and Conservation* 1:21–62.

Meagher, T.R., J. Antonovics, and R. Primack. 1978. Experimental ecological genetics in *Plantago*. III. Genetic variation and demography in relation to survival of *Plantago cordata*, a rare species. *Biological Conservation* 14:243–257.

Menges, E.S. 1990. Population viability analysis for an endangered plant. *Conservation Biology* 4:52–62.

Menges, E.S. 1991a. Seed germination percentage increases with population size in a fragmented prairie species. *Conservation Biology* 5:158–64.

Menges, E.S. 1991b. The application of minimum viable population theory to plants. In *Genetics and conservation of rare plants*, eds. D.A. Falk and K.E. Holsinger, 45–61. New York: Oxford University Press.

Menges, E.S. 1992. Stochastic modeling of extinction in plant populations. In *Conservation biology: the theory and practice of nature conservation, preservation, and management*, eds. P.L. Fiedler and S.K. Jain, 253–275. New York: Chapman & Hall.

Mehrhoff, L.A. 1996. The reintroduction of endangered Hawaiian plants. In *Restoring diversity: Strategies for reintroduction of endangered plants*, eds. D.A. Falk, C.I. Millar, and M. Olwell, 101–120. New York: Island Press.

Ornduff, R. 1976. The reproductive system of *Amsinckia grandiflora*, a distylous species. *Systematic Botany* 1:57–66.

Pantone, D.J., B.M. Pavlik, and R.B. Kelley. 1995. The reproductive attributes of an endangered plant as compared to a weedy congener. *Biological Conservation* 71:305–311.

Parenti, R.L. and E.O. Guerrant Jr. 1990. Down but not out: Reintroduction of the extirpated Malheur wirelettuce, *Stephanomeria malheurensis*. *Endangered Species Update* 8:62–63.

Pavlik, B.M. 1994. Demographic monitoring and the recovery of endangered plants. In *Recovery and restoration of endangered species*, eds. M. Bowles and C. Whelan, 322–350. Cambridge: Cambridge University Press.

Pavlik, B.M. 1995. The recovery of an endangered plant. II. A three-phased approach to restoring populations. In *Restoration ecology in Europe*, eds. K.M. Urbanska and K. Grodzinska, 49–69. Zurich: Geobotanical Institute SFIT.

Pavlik, B.M. 1996. Defining and measuring success in rare plant reintroductions. In *Restoring diversity: strategies for reintroduction of endangered plants*, eds. D.A. Falk, C.I. Millar, and M. Olwell, 127–155. New York: Island Press.

Pavlik, B.M., D. Nickrent, and A.M. Howald. 1993. The recovery of an endangered plant. I. Creating a new population of *Amsinckia grandiflora*. *Conservation Biology* 7:510–526.

Pickett, S.T.A., V.T. Parker, and P.L. Fiedler. 1992. The new paradigm in ecology: Implications for conservation biology above the species level. In *Conservation biology: the theory and practice of nature conservation, preservation, and management*, eds. P.L. Fiedler and S.K. Jain, 65–88. New York: Chapman & Hall.

Rabinowitz, D. 1981. Seven forms of rarity. In *The biological aspects of rare plant conservation*, ed. H. Synge, 205–217. Chichester: John Wiley and Sons.

Rabinowitz, D., J.K. Rapp, S. Cairns, and M. Mayer. 1989. The persistence of rare prairie grasses in Missouri: Environmental variation buffered by reproductive output of sparse species. *American Naturalist* 134:525–544.

Schemske, D.W., B.C. Husband, M.H. Ruckelshaus, C. Goodwillie, I.M. Parker, and J.G. Bishop. 1994. Evaluating approaches to the conservation of rare and endangered plants. *Ecology* 75:584–606.

Schoen, D.J. and S.C. Stewart. 1986. Variation in male reproductive investment and male reproductive success in white spruce. *Evolution* 40:1109–1120.

Shaffer, M.L. 1981. Minimum population size for species conservation. *BioScience* 3:131–134.

Soulé, M.E. and B.A. Wilcox. 1980. *Conservation biology: an evolutionary-ecological approach*. Sunderland, MA: Sinauer Associates.

Sutter, R.D. 1996. Monitoring. In *Restoring diversity: strategies for reintroduction of endangered plants*, eds. D.A. Falk, C.I. Millar, and P. Olwell, 235–264. New York: Island Press.

Templeton, A.R. 1990. The role of genetics in captive breeding and reintroduction for species conservation. *Endangered Species Update* 8:14–17.

Templeton, A.R. 1991. Off-site breeding of animals and implications for plant conservation strategies. In *Genetics and conservation of rare plants*, eds. D.A. Falk and K.E. Holsinger, 182–194. New York: Oxford University Press.

Tonkyn, D.W. 1993. Optimization techniques for the genetic management of endangered species. *Endangered Species Update* 10:1–4(+9).

Wallace, S.R. 1990. Central Florida Scrub: Trying to save the pieces. *Endangered Species Update* 8:59–61.

Wallace, S. 1992. Introduction of *Conradina glabra*: a pilot project for the conservation of an endangered Florida endemic. *Botanic Garden Conservation News* 1:34–39.

Whitson, P.D. and J.R. Massey. 1981. Information systems for use in studying population status of threatened and endangered plants. In *The biological aspects of rare plant conservation*, ed. H. Synge, 217–236. London: John Wiley and Sons.

Wieland, G.D. 1995. *Guidelines for the management of orthodox seeds*. St. Louis: Center for Plant Conservation.

Wilcove, D.S., M. McMillan, and K.C. Winston. 1993. What exactly is an endangered species? An analysis of the U.S. endangered species list: 1985–1991. *Conservation Biology* 7:87–93.

SECTION II

BROAD BRUSHES AND TAXONOMIC TOURS: SUMMARIES OF THE STATE OF THE NATURAL WORLD

One of the most maddening aspects of the science of conservation during the last two decades has been the inaccurate, misleading, and unreliable statistics about extinctions and habitat loss promulgated to those writing about conservation. Ecologists included. Some conservationists have both freely offered dire "facts" about the loss of rainforests or coral reefs, and bemoaned how little we actually knew. In the last decade, however, a number of conservation biologists have gone to work and produced reliable, up-to-date, and referenced statistics on the state of the natural world. Thus one piece of good news proffered in this section is that we now have solid data on the past and present fate of many plants, most vertebrates, and some invertebrates on continental scales.

In this section we collected five works from ecologists with nearly two centuries of field experience collectively. Two chapters provide paradigmatic examples of aquatic habitats and their denizens; one chapter speaks to the extraordinary flora of Australia and how a 7,682,300 km^2 chunk of real estate can be unique in a number of ways; another chapter provides a fascinating look at recent reptilian extinctions and why habitat patch size matters; and a final chapter illustrates the still deplorable state of our knowledge of insect biodiversity. There are generalizations to be made from this survey, and they illuminate.

The most common thread woven by our contributors throughout this section is that threats to biodiversity include the same rogues gallery just about everywhere. Some taxonomic groups and some habitats are indeed in reasonable shape, being still abundant in numbers or area. But many regions have endured great declines in species richness, losses of endemic taxa, and concomitant physical degradation.

Most specific to aquatic environments, pollution can be egregious and acute, as the *Exxon Valdez* incident clearly illustrates. Ruckelshaus and Hays explain that even the most closely watched environmental disaster on the west coast of North America did not provide unambiguous data on the impacts of the oil spill to the large marine mammal populations in Prince William Sound. Pollution in aquatic environments is also insidious and chronic, and may be a bigger problem to the maintenance of biodiversity than large, acute oil spills. Sublethal pollutants cause fishes to disappear gradually for a host of reasons, including stress-related diseases. While providing a less tractable statistic than dead sea otters washed ashore, depressed reproductive outputs coupled with crippling heavy metal and pesticides loads in aquatic organisms worldwide may be far more harmful than a single dramatic event like an oil spill.

Another recurrent theme throughout this section is the devastating effects humans have had on a great many taxonomic groups and ecosystems. According to Case and his colleagues, the background rate of reptilian extinctions in recent times has been exaggerated in the presence of humans. For small islands, evidence suggests that humans have increased local extinction rates by an order of magnitude. In aquatic systems, marine or freshwater, habitat loss coupled with overfishing has clearly lead to a number of extinctions of fish species. The problem is exacerbated by the inherent difficulties of obtaining reliable population estimates, statistics that traditionally provide an early warning of species in decline. Ruckelshaus and Hays cogently explain that human effects are particularly pernicious for species with complex life cycles that encompass both marine and freshwater habitats, such as Pacific anadromous salmonids (*Onchorynchus* spp.). Leidy and Moyle provide additional examples for many threatened fishes, with some of the most chilling "horror stories" about biodiversity loss stemming from even "helpful" human meddling. The loss of perhaps 200 species of cichlids due to the introduction of the predatory Nile perch (*Lates niloticus*) into Lake Victoria to "improve" the fisheries needs only to be told once to make the point that humans have irreversably altered aquatic ecosystems on a landscape, indeed continental, scale.

Hopper's contribution on the flora of Australia, however, resonates with a unique perspective. Australia is the "oldest, flattest, driest, and most isolated continent." While it shares the same threats to its flora and fauna as other islands (and other continents), loss of biodiversity is particularly disheartening here because, for example, the continent contains about 10% of the world's flora, with endemism hovering at 85%. Hopper suggests that the state of Western Australia alone holds more species of threatened plants than most countries of the world, due primarily to habitat loss and the invasion of exotic species.

An accounting of insects also provides a sobering look at large-scale phenomena. However, data are slim, as Schulz and Chang reveal that only 6% of our conservation literature is aimed at 80% of our planet's biodiversity. While the absence of natural history information and habitat fragmentation are problems endured by all the taxa in this essay collection, the crux of Schulz and Chang's message is that some problems are particularly vexing to successful protection efforts of insect biodiversity. Such problems include fluctuating population sizes over orders of magnitude not seen in most other taxonomic groupings, the plaguing issue of management scale, and the uncertainty that seminatural landscapes can provide suitable habitat.

Some satisfaction should be derived from the knowledge that, to some degree and for some species and ecosystems, we now know what we have as well as what we don't have. But to paraphrase Daniel Boorstein, former librarian of Congress, the greatest obstacle to our progress in conservation is not our glaring ignorance, but the illusion of knowledge about the natural world. The contributors herein hold no illusions about the scope of their surveys and the generalizations that result, but offer them instead, as simply a place to start.

CHAPTER

6

Conservation and Management of Species in the Sea

MARY H. RUCKELSHAUS
and CYNTHIA G. HAYS

Until recently, conservation biology as a science has largely been a land-locked endeavor, in spite of decades of managed harvest in the sea. Fortunately, scientists and policy-makers are beginning to heed warning signs, manifest most dramatically as extinctions of marine organisms in recent historical time (including the Caribbean monk seal, Stellar's sea cow, a number of marine bird species, and several gastropods; Carlton 1993b; Vermeij 1993). Although global extinction rates in the ocean may be lower than those in terrestrial communities, there is an alarmingly long list of marine species for whom precipitous declines have been well documented (Estes, Duggins, and Rathbun 1989; Upton 1992; Norse 1993; Clark 1994). As biologists examine threats to marine biodiversity, it is becoming clear that conservation in the marine environment presents special challenges distinct from our experience in more traditional terrestrial conservation. Increased communication among terrestrial and marine conservation scientists will improve the prospects of all species through a better understanding of the successes and failures of conservation approaches in both environments.

In this chapter we use five case studies to illustrate a range of threats facing marine habitats and to evaluate approaches for redressing species declines. We conclude with a list of remarkably consistent questions that emerge from these diverse circumstances. Before we turn to the case studies, we briefly discuss differences in the biologies of marine and terrestrial species and outline the myriad environmental threats facing marine organisms.

NATURAL HISTORIES OF ORGANISMS IN THE SEA

The physical characteristics of the oceans clearly distinguish them from terrestrial environments, so not surprisingly, marine organisms exhibit a number of important differences in life history from the terrestrial biota. Consequently, it may not be safe to assume that knowledge we have gleaned from terrestrial conservation efforts will be applicable to marine systems. Perhaps the most striking difference is the presence of currents and tides in the ocean, leading to larger species-dispersal distances (Hedgecock 1986; Palumbi 1992, 1994) and a looser coupling of the dynamics of population and recruitment phases than on land (Fairweather 1991). High dispersal capabilities in marine taxa may offer "insurance" against global extinction, and a corollary observation has been made that endemism is relatively rarer in marine than in terrestrial communities (Vermeij 1993). It has also been shown that the distributions of marine species often shift more rapidly than do those of terrestrial species in response to environmental change (Steele 1991; but see Grosholz 1996 regarding non-indigenous marine taxa). This propensity for rapid distributional shifts may offer marine taxa an alternative to extinction through range alterations. Furthermore, marine predators generally have a greater reproductive output than do terrestrial predators, which in addition to their planktonic phase may buffer them from overexploitation.

These differences in life-history characteristics between marine and terrestrial taxa imply to some that marine species are less vulnerable to extinction than terrestrial species, but other differences between marine and terrestrial systems may suggest the opposite. A number of marine species use more than one discrete habitat during their life cycles, so management efforts directed at just one site may be ineffectual in encouraging population recovery. A dramatic example of such a species is sockeye salmon, *Oncorhynchus nerka*, which spawn in lakes up to 900 miles from the ocean, migrate downriver during a juvenile phase and upriver during a reproductively mature phase, and spend most of their adult lives in a migratory gyre that extends across the Pacific Ocean to Japan.

In terrestrial ecosystems the longest-lived organisms tend to be primary producers (e.g., trees), whereas in the oceans the longest-lived organisms tend to be at the tops of food chains (e.g., whales) (Steele 1985). This difference has important ramifications for population dynamics: because the reproductive values of a long-lived tree and fish species will both be spread out over many years, harvesting one from the bottom of the food web and the other from the top will have vastly different community consequences (Beddington and Cooke 1982; May 1984; Levitan 1992; Kirkwood, Beddington, and Rossouw 1994). Moreover, because of the high variance in reproductive success in many marine species, effective population sizes are expected to be small (Hedgecock and Sly 1990; Hedgecock 1991; Hedrick, Hedgecock, and Hamelberg 1995; but see Palumbi 1994; Ruckelshaus 1996), so the effects of stochastic genetic processes on the genetic diversity of marine species could be especially important.

Thus, in spite of the historical impression that the biology of marine species makes them impervious to human impacts, there is some evidence that this may not be the case. In fact, as we illustrate below, a variety of ocean-dwelling species are in trouble, and because of the distinct biology of marine organisms, we advocate a healthy skepticism in considering strategies whose promise (or lack thereof) is based solely on terrestrial applications.

MAJOR THREATS TO MARINE SPECIES

At the core of most threats to marine biodiversity is the fundamental problem facing all of the earth's ecosystems: increasing growth rates of worldwide human populations (Brown *et al.* 1995). Human activities, including direct and indirect harvest of marine species and modification of nearshore and open-ocean environments as a result of industry, coastal development, and agriculture combine to result in far-reaching impacts on marine environments. Effects on species can occur through both direct (e.g., declines in abundance or biomass) and indirect (e.g., via changes in a predator or competitor) mechanisms. In this section we briefly outline the major sources of population declines in marine taxa (see also Norse 1993; NRC 1995).

Overfishing and its Associated Effects

The National Marine Fisheries Service (NMFS) estimates that as many as 45% of all fish species in the U.S. whose status is known are being overharvested (NMFS 1993, 1994). Many economically important fishes, including the Atlantic bluefin tuna and haddock, have experienced drastic declines due to fishing activity (Safina 1995). Overfishing of the northern cod, *Gadus morhua*, caused the collapse of the Newfoundland fishery, which went from a yearly catch of approximately 250,000 tons in the early 1900s to its commercial extinction in 1992 (Martin 1995). The global harvest of wild fishes, after decades of continued growth, peaked at 82 million metric tons in 1989 (Safina 1995); yearly catch in recent years has stagnated around 78 million metric tons (Parfit 1995).

One reason why total harvested biomass figures have not exhibited greater declines is the opportunistic nature of the fishing industry. As target species become overexploited and decrease in availability, harvesters make up the difference by switching to new species; those previously considered trash fish, such as the Gulf butterfish (*Peprilus burti*) and orange roughy (*Hoplostesthus atlanticus*), become prime targets (Parfit 1995). Unfortunately, even less is known about the life histories of these new target fishes, which makes appropriate management regimes nearly impossible to determine.

Beyond direct effects on target fish populations, fishing can also have powerful indirect effects. Bycatch, habitat destruction, harvesting of predator, prey, or competitor species, and "ghost fishing," in which a number of species continue to be entrapped and killed by lost or abandoned equipment all reduce nontarget populations. Most bycatch goes unreported, but it is clear from available data that it can be a huge source of mortality for nontarget species. Discard/bycatch estimates are as high as 94% in some fisheries (Gulland and Garcia 1984; Jefferson and Curry 1994; Dayton *et al.* 1995). For example, the swordfish fishery in the North Atlantic routinely catches over twice as many sharks as swordfish (Dayton *et al.* 1995). In addition, recreational and commercial harvest of predator species has resulted in dramatic changes in nontarget species (Gulland and Garcia 1984; Moreno, Sutherland, and Lopez 1984, 1986; Castilla and Duran 1985; Carr 1989; Duran and Castilla 1989; Russ and Alcala 1989; Levitan 1992; Yodzis 1994; Butman, Carlton, and Palumbi 1995; Camhi 1995; Dayton *et al.* 1995). Perhaps the best known example of the consequences of overfishing a predator-prey system is the harvest of krill and its group of competing predator species, including whales and seals in the Antarctic (Laws 1977; May *et al.* 1979).

Certain fishing methods can also threaten habitat integrity. For example, trawling (Jones 1992) and hydraulic dredges designed to harvest infaunal bivalves can cause death, injury, and increased risk of predation to nontarget invertebrates left in the wake of trawls (Meyer, Cooper, and Pecci 1981). The physical action of the trawl or dredge can change the sedimentology (Messieh *et al.* 1991) and inhibit the recruitment or survival of epi- and infaunal organisms. In addition, subsistence fishermen often use poisons or dynamite to kill or stun targets. Clearly, these are highly nonspecific fishing methods and can have particularly devastating effects on coral-reef communities (Zann 1982; Munro, Parrish, and Talbot 1987; Norse 1993).

Pollution

There is ample evidence that pollutants result in dramatic impacts on marine species. Human-mediated introductions of toxic chemicals, excess nutrients, and high concentrations of suspended solids into marine waters can have both short- and long-term effects on individuals and communities (Hinrichsen 1990; Davis 1993; GESAMP 1993; Suchanek 1993, 1994). For example, although the dramatic effects of DDT on piscivorous marine birds has declined since its ban in the U.S. (EPA 1975; WHO 1989), biologically significant concentrations of DDT still exist throughout portions of the oceans (Davis 1993). Documented effects of oil spills are relatively straightforward because of the often discrete nature of the pollutant introduction. In 1986, a large oil spill in Panama resulted in direct mortality to seagrasses, coral, and mangrove communities. The chronic exposure of the biota due to hydrocarbon-laden sediments has resulted in continued deleterious effects on growth rates and reproduction of a number of coral species (Jackson *et al.* 1989; Guzman, Jackson, and Weil 1991; Guzman, Burns, and Jackson 1994; Guzman and Holst 1993).

In addition, anthropogenously elevated concentrations of hydrocarbons, PCBs, and heavy metals that have developed over 30 years near McMurdo Station in Antarctica have resulted in clear changes in benthic community composition (Lenihan and Oliver 1995). At larger time scales, diatom community diversity in Chesapeake Bay has dropped over the past 2,000 years, correspondent with increased eutrophication and changes in salinity and turbidity (Cooper 1995).

Habitat Destruction and Fragmentation

Outright destruction of marine habitats occurs primarily in nearshore areas, where human activities such as dredging, commercial trawling for fish, and changes in light and nutrient levels due to agricultural and urban runoff can kill seagrasses, coral, or kelps (Hallock, Muller-Karger, and Halas 1993; Schroeter *et al.* 1993; Fonseca, Kenworthy, and Thayer. 1996). These direct and indirect sources of mortality to species that create habitat for a host of other species can result in loss or modification of entire communities. For example, the decline of coral reefs in the Caribbean and Pacific has resulted in local disappearances of numerous invertebrate species that depend on live reefs for their habitat (Bell 1992; Hallock, Muller-Karger, and Halas 1993; Watson and Ormond 1994).

Habitat destruction on a local scale results in increased fragmentation of remaining healthy habitat. Human activities causing marine habitat fragmentation include recreational SCUBA divers (Hawkins and Roberts 1992) and oil-induced mortality of coral in the Persian Gulf during the Gulf War (Downing and Roberts 1993). Not all marine taxa have large-scale dispersal capabilities; some are just as vulnerable to habitat fragmentation as are many terrestrial species. For example, increasing fragmentation of temperate and tropical coral reefs, kelp beds, and seagrass meadows can make it nearly impossible for dispersing larvae of numerous fish and invertebrate species to locate high-quality settlement sites (Gaines and Roughgarden 1985; Schroeder 1987; Eckman, Duggins, and Sewell 1989; Ilan 1990; Breitburg 1992; Schroeter *et al.* 1993; Chiappone and Sullivan 1994; Sale, Forrester, and Levin 1994; Tupper 1995). If the period over which larvae can persist in the plankton is surpassed before a patch of suitable habitat is encountered, recruitment of the species can be critically limited.

Introduced Species

There is growing concern among biologists about the vast number and increasing rate of nonindigenous species introductions into marine communities. Anthropogenous introductions homogenize distinctly separate biotas with very different evolutionary histories, often with disastrous results for the native communities. There are many routes to invasion of marine systems, including deliberate introductions to enhance fisheries (Randall 1987; Carlton 1992b), such as the global transport of the oysters *Crassostrea virginica* and *C. gigas*. Along with the oysters themselves, the industry has inadvertently transported a plethora of associated nonnative epifauna, parasites, and plants (Druehl 1972; Carlton 1992a). Other sources of nonindigenous species in the ocean include unintentional introductions via canals, escapes or introductions from the aquarium industry, and transport on the hulls and in the ballast water of commercial and recreational ships (Carlton 1985, 1987; Baltz 1991; Meinesz and Hesse 1991; Meinesz *et al.* 1993).

One of the most prevalent mechanisms of introduction is the exchange of ships' ballast water in ports of call. Carlton and Geller (1993) examined ballast water samples from 159 ships coming from Japan into an Oregon harbor and found over 367 different species of living plants and animals. Because of the current high level of global shipping activity, the potential for introductions is staggering: in the United States alone, it is estimated that well over two million gallons of ballast water are released into nonnative ports every hour (Carlton, Reid, and van Leeuwen 1994). Moreover, it has been suggested that recent increases in the number of red tides (toxic phytoplankton blooms), particularly in previously unaffected areas, may be linked to ballast transport of nonnative dinoflagellate cysts (Hallegraeff and Bolch 1992; Hutchings 1992; Hallegraeff 1993).

Global Climate Changes

Documented changes in the global climate include declines in ozone levels and concomitant increases in CO_2, temperature, and solar UV-B radiation (Francis 1990; Kareiva, Kingsolver, and Huey 1993; Peterson, Barber, and Skilleter 1993). Possible ramifications of these changes are sea level elevation, disruption of ocean circulation and upwelling patterns, changes in rainfall patterns, dramatic shifts in species ranges or invasions into northern latitudes, and direct mortality of marine organisms due to elevated temperature or UV-B radiation (Chapman 1988; Ray *et al.* 1992; Norse 1993; Raven 1994; Stanley and Warne 1994; Westbroeck *et al.* 1994; NRC 1995; Gleason and Wellington 1995). For example, in a rocky intertidal community along the central California coast, a northward shift in species ranges was correlated with increases in temperature over a 60-year period (Barry *et al.* 1995). Sea-surface temperatures have been shown to be correlated with intertidal community structure and may have contributed to bleaching of tropical hard and soft corals (Brown and Ogden 1993; Barry *et al.* 1995). In other areas, elevations in temperatures have had relatively minor effects on marine communities (Paine 1993).

CASE STUDIES: APPROACHES TO CONSERVATION OF MARINE SPECIES

The case studies we have chosen illustrate the direct or indirect effects of threats to marine species, the solutions that have been implemented to mitigate these effects, and the degree of success achieved by the solutions. The case studies we discuss are rich in colorful biological and political details; in the following section we distill only pieces of each story, and we encourage the reader independently to dive more deeply into the issues.

Case Study 1: Snake River Salmon

Pacific anadromous salmonids (*Onchorynchus* spp.) exhibit complex life histories that encompass both marine and freshwater stages. Because adults return to their natal streams to reproduce, each species is subdivided into populations, or stocks, with a high degree of reproductive isolation (Ricker 1972). Different stocks may exhibit a great deal of variation in reproductive schedules and other life-history characteristics, and often they exhibit local adaptation to environmental conditions in streams (Taylor 1991; Gharrett and Smoker 1993). For these reasons, conservation efforts have been focused at the level of stocks (Nehlsen, Williams, and Lichatowich 1991; Nehlson 1994). Pacific salmon have experienced general declines in the last decade, primarily due to habitat loss and alteration and secondarily to overfishing. A recent survey by Nehlsen, Williams, and Lichatowich (1991) identified 59 native salmon stocks in the Columbia River Basin in the northwestern U.S. that are now extinct and another 50 that are at risk of extinction. Three of these stocks in the Snake River drainage of the Columbia River Basin are now listed under the federal Endangered Species Act: Snake River spring/summer chinook, fall chinook (*O. tshawytscha*), and sockeye (*O. nerka*) (NMFS 1995). We use this case study to illustrate the challenges associated with managing long-lived single species experiencing threats to their existence in diverse habitats extending from the open ocean to freshwater streams.

Threats to Snake River Salmon. The Snake River is a major tributary of the Columbia River, and has a basin encompassing over 100,000 square miles across six states in the northwestern U.S. (Figure 6.1). At one time, the Columbia River chinook population was considered to be the largest of that species in the world, and the Snake River the most productive salmonid drainage in the system (NMFS 1995). Estimates of the historical abundances of the three stocks listed by the Endangered Species Act (ESA) are staggering: in the late 1800s and early 1900s, approximately 1.5 million spring/summer chinook, 70,000 fall chinook, and 30,000 sockeye returned annually to the Snake River. In 1994, only 1,800 spring/summer and 400 fall chinook adults were counted. The same year, just one sockeye returned to spawn (NMFS 1995). The Snake River sockeye salmon was listed as an endangered species on November 20, 1991; Snake River spring/summer and fall chinook were listed as threatened on April 22, 1992. Because the stocks have continued to decline since listing, the chinook are now in the process of being reclassified as endangered (NMFS 1995).

The decline of the greater Columbia River salmon stocks has been attributed mainly to loss of spawning and rearing habitat and degradation of the migratory routes. Sources of habitat-quality loss include urban development and use of the area's natural resources by mining, agriculture, and logging (for a summary of Snake River Basin habitat quality, see Chapman and Witty 1993). Twelve dams were constructed on the Snake River between 1901 and 1968; six were constructed on the Columbia. Dams resulted in a significant decrease in salmon abundance and distribution, particularly for sockeye salmon (NMFS 1995). After the Hell's Canyon dam complex was built on the Snake River in 1967 without facilities allowing upstream or downstream fish passage, huge spawning areas were eliminated. Harvest of adult Snake River salmon historically has contributed to their decline (NMFS 1995). Today, only fall chinook are harvested at a significant rate as bycatch in mixed-stock fisheries in the Snake and Columbia Rivers and in the open ocean (NMFS 1995). Bycatch of Snake River chinook illustrates a particular problem of mixed

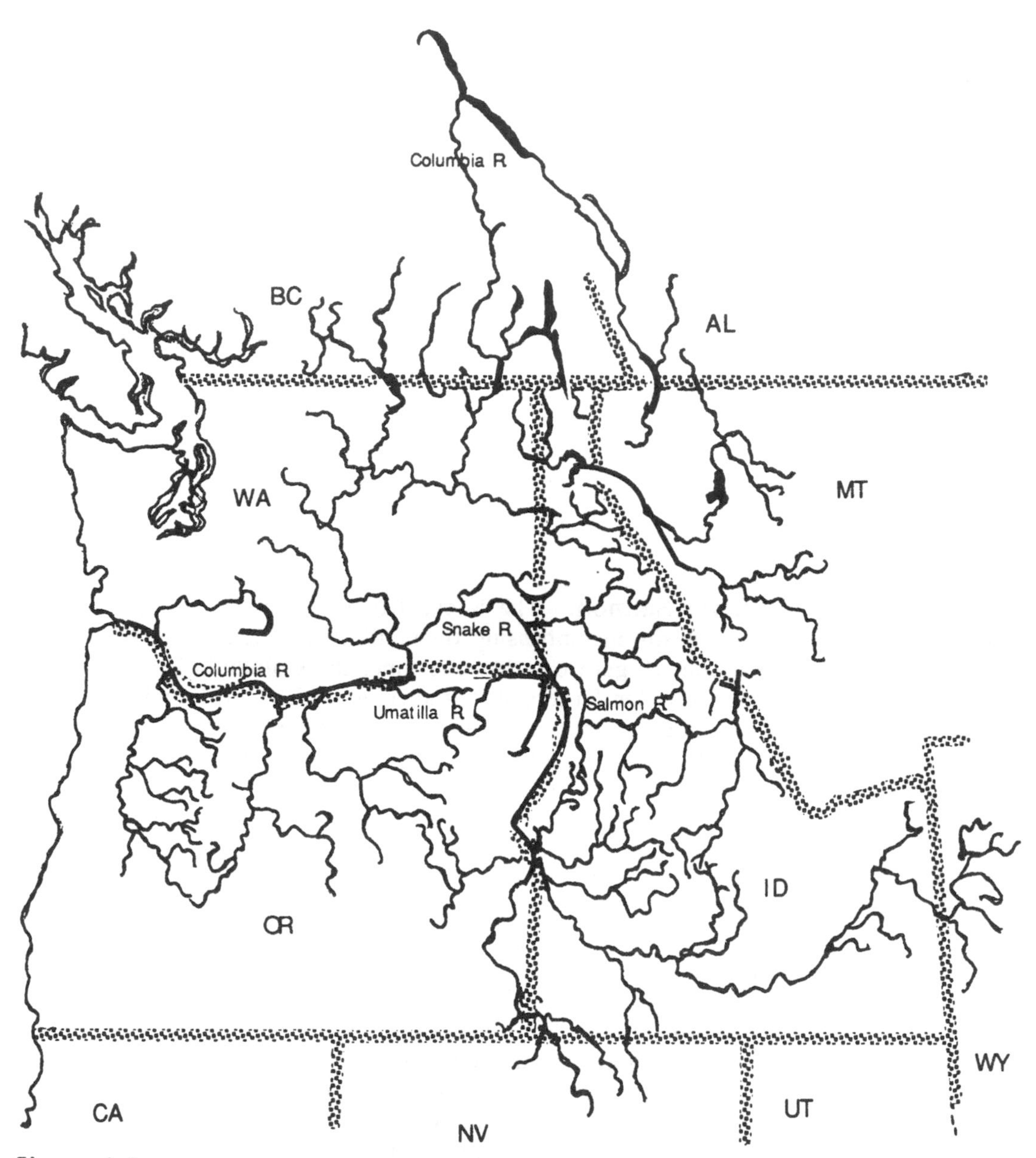

Figure 6.1
The Columbia and Snake River drainages in the northwestern United States and southwestern Canada. Spawning and rearing habitats for the three federally listed Snake River salmon are distributed throughout the drainage, covering an area of greater than 107,000 m^2. Major tributaries in the river basin are depicted, as are selected minor drainages mentioned in the text.

stock fisheries, for which it is difficult to match harvest intensity to the population status of the individual stock.

Another potential factor contributing to the decline of Pacific salmonids is the negative impact of hatchery-reared fish (reviewed by Hindar, Ryman, and Utter 1991; Waples 1991b). Hybridization between native and hatchery fish has been shown to have many deleterious effects; in some cases, it can lead to reduced local adaptation and outbreeding depression (Nickelson *et al.* 1986; Emlen 1991; Hilborn 1992b; Waples 1991b; Fleming 1994). Artificial propagation of salmon in the Columbia River system began over 110 years ago. Two artificial propagation programs in the Columbia River drainage (fall and spring chinook at the Umatilla and Rapid Rivers, respectively; see Figure 6.1) have been shown to affect the listed salmon by allowing genetic introgression between hatchery and native stocks (NMFS 1995). In 1990, Umatilla River hatchery strays constituted 29% of the fall chinook passing through the Lower Granite Dam, above the confluence of the Umatilla River with the Snake (NMFS 1995).

Attempted Solutions: The NMFS Proposed Recovery Plan For Snake River Salmon. To qualify for protection under the Endangered Species Act, a salmon stock must be considered an Evolutionarily Significant Unit (ESU): it must represent a key component of the species' ecological and genetic diversity, and it must be substantially reproductively isolated from other such units (Waples 1991a). Snake River fall and spring/summer chinook and sockeye salmon all qualify as ESUs, because of their genetic distinctness from other stocks in the region and their unique natural-history traits. For example, the Snake River sockeye exist at a higher elevation and travel greater distances to the sea than do sockeye anywhere else in the world. The Snake River spring/summer chinook are composed of more than 30 subpopulations in 12 major subbasins, all of which are considered important in maintaining the integrity of the ESU (SRSRT 1994). The recovery plan for spring/summer chinook therefore includes a metapopulation approach, described below.

The National Marine Fisheries Service (NMFS) appointed the Snake River Salmon Recovery Team to recommend a plan of action for delisting the three salmon stocks. Their recommendations form the framework for the Proposed Recovery Plan (NMFS 1995), as required under the ESA. The general strategy for recovery of the Snake River salmon is to coordinate management efforts among the different agencies and affected interests (including the Fish and Wildlife Service, as well as other federal, state, and tribal groups), to take immediate action to reduce human-induced mortality at all life stages of the listed stocks, and to initiate research programs designed to address the lack of stage-specific information on survivorship of the different stocks (NMFS 1995).

Actions designed to mitigate human-induced mortality to the endangered salmon stocks include reducing harvest of fall chinook, modifying dams to increase both downstream and upstream survival of juveniles and adults, and improving hatchery practices to minimize impacts on natural populations (NMFS 1995). The recovery plan proposes the use of hatcheries to maintain captive broodstock and gene banks for the listed stocks and to supplement dwindling populations. Although these methods may prove to be a valuable tool in the conservation of salmonids, the authors of the recovery plan recognize that artificial propagation is not a substitute for reestablishing viable wild populations. Long-term goals of the recovery plan include salmon habitat protection and restoration through an ecosystem approach that considers conservation at the level of the watershed rather than individual tributaries. The plan requires integrated management of federal and nonfederal land (35% of designated Snake River critical habitat is nonfederal) to provide connectivity between high-quality spawning habitats.

The delisting criteria for Snake River salmon are based on a demographic indicator, the cohort replacement rate, as well as spawner abundance, referred to as escapement (NMFS 1995). The natural cohort replacement rate is defined as the rate at which each subsequent generation replaces the previous one; it must exceed one (indicating a stable or increasing population) for eight consecutive years for delisting of the stock to be considered. Because there are no universally accepted approaches for determining appropriate recovery population sizes (Thomas 1990; Thompson 1991), the numerical escapement goals for each stock were assessed by several different methods. Table 6-1 illustrates how few data exist to help those orchestrating recovery of these endangered species. For example, the ranges of escapement goals for all three salmon species were estimated in part from published "minimum viable population" estimates for bird and vertebrate species (Thompson 1991), estimates that are themselves subject to considerable controversy (Ralls and Taylor 1996). Ultimate criteria for recovery of the Snake River salmon are necessarily based on little more than educated guesses.

Overview: Prospects For Snake River Salmon Recovery. The classic approach to management of marine fisheries is one of a controlled harvest, striving for maximum sustainable yields. There is considerable dissatisfaction with these classic approaches: a committee on fisheries established under the auspices of the Ocean Studies Board of the National Research Council concluded that the definitions of "optimum and maximum sustainable yields" provided no usable guidelines for making decisions (NRC 1994), and many fisheries managers acknowledge problems with the application of such terms as "optimal yield," "minimum viable population," and "maximum sustained yield" (Frank and Leggett 1994; Hilborn, Walters, and Ludwig 1995; NMFS 1995). For example, the Marine Mammal Protection Act amendments in 1984 included authorized taking of certain dolphin populations, in part because it was too complicated to determine the legally-required "optimum sustained population" for eastern tropical Pacific dolphins (Gerrodette and DeMaster 1990).

One of the most extensive efforts at increasing the population sizes of marine species in decline has been the salmonid enhancement program (SEP), an attempt at artificial enhancement of Pacific salmon populations in Canada using hatcheries, spawning channels, and a lake enrichment program. Established in 1977 in British Columbia, the SEP has spent $450 million in 15 years and has had mixed success at best: the total cost of the SEP has not been covered by the landed value of fish resulting from the program (Hilborn and Winton 1993). In spite of some successes, the SEP spent most of its budget on chinook and coho salmon, the two species that have shown continual declines in abundance. A careful evaluation of the program resulted in the conclusion that 15 years is insufficient time to evaluate which technologies are most promising (Hilborn and Winton 1993)!

The federally listed Snake River salmon stocks are characterized by generations approximately four years in length, so it will be at least 12 to 24 years before evidence of recovery can be detected (NMFS 1995). Because of the long-term nature of salmon recovery, the architects of the recovery plan adopted an adaptive management approach. The plan explicitly factors in a critical decision point in 1999. At this time, the effects of improvements made to in-river migration conditions and structural changes at dams designed to increase safe passage will be evaluated. Implementation of subsequent actions are contingent upon the relative successes of the initial improvements. The authors of

Table 6.1

Delisting criteria for Snake River salmon species. All three stocks share the criterion of a cohort replacement rate of two over an 8-year period (see text). Numerical escapement indicates the number of adult salmon returning to natal streams to spawn. The rationale underlying the federally required delisting criteria is documented in the Recovery Plan prepared by the National Marine Fisheries Service (see text).

Salmon stock	Numerical escapement goal	Basis for the goal
Sockeye	8-year geometric mean of > 1,000 natural spawners in Redfish Lake, and 500 natural spawners in 2 additional Snake River basin lakes	(1) threshold population sizes for other federally listed Snake River salmon (2) historical survival rates (3) assumption of high population variability
fall Chinook	8-year geometric mean of > 2,500 natural spawners	(1) estimates of maximum sustainable production (MSP), based on stock recruitment estimates. The numerical escapement goal represents 35% of the most recent estimate of MSP. (2) estimates of the carrying capacity of the habitat. (3) assumption of high population variability
spring/summer chinook	(1) 8-year geometric mean of > 60% of pre-1971 redd (nest) counts[1] in at least 80% of the index areas AND (2) 8-year geometric mean of > 60% of the number of natural spawners past Harbor Dam in the period 1962–1967[3]	(1) subdivision of spring/summer chinook into 12 stocks in 40 subpopulations. This involves a metapopulation approach where a "majority" of subpopulations must meet delisting criteria[2]. (2) estimates of historical abundance, using 1960s data as representative of healthy populations (3) assumption of high population variability was used to set the upper bound on numerical goals; assumption of low variability was used to set the lower bound[4].

[1]Redd counts before 1971 vary from 12–1,184 redds/index area

[2]"Majority" is loosely considered to be attainment of recovery goals in 50% of subpopulations by the NMFS Recovery Plan. Therefore, if 63% of census areas have historical data and 80% of those census areas meet recovery goals, then 50% of the census areas will have met delisting criteria.

[3]60% of the mean number of spawners passing Ice Harbor Dam from 1962–1967 is 31,440 fish.

[4]low variability assumption yields goals of 7,800–9,800; high variability assumption: 43,000–54,000

the plan are cautious in acknowledging that using controlled artificial propagation to preserve listed stocks is truly an experimental conservation approach and that careful monitoring will be necessary to assess the value of this program (Hard *et al.* 1992). Therefore, rather than using hatcheries in an attempt to circumvent the problem of unsuitable spawning habitat, implementation of the recovery plan involves both habitat restoration and supplementation of stocks as approaches to salmon recovery. The biologies of long-lived marine species occupying diverse habitats, such as salmon, increasingly are forcing managers to adjust recovery activities according to what works, an approach from which all species will benefit.

Case Study 2: Red Urchins In California

The red sea urchin (*Strongylocentrotus franciscanus*) is a marine echinoderm found on the west coast of North America from Baja California to Kodiac, Alaska, and south along the Asiatic coast to the tip of Hokkaido Island, Japan (McCauley and Carey 1967). Red urchins are harvested by humans for their gonads (or roe, called "uni" in the Japanese market). Sea urchins are free-spawning, releasing their gametes into the water column, where fertilization occurs. Zygotes develop through several stages of free-swimming larvae before settling into the benthos as juveniles after six to eight weeks (Johnson 1930). The high variability in recruitment and individual growth rates displayed by red sea urchins (Ebert 1993) may make them especially vulnerable to overharvest. In this case study, we illustrate how fishing pressure can be exacerbated by the biology of the harvested species and discuss the indirect community-level effects that can result from removal of an ecologically important species. Perhaps the central message in this case study is the increasing difficulty of identifying the ecological effects attributable solely to direct human harvest of a single species; many taxa in the red urchin community have been perturbed by human activity.

Threats to Urchin Populations. Red sea urchins have been commercially fished in California since 1970. Increases in fishing pressure in the last decade have threatened the stability of urchin populations and thus the sustainability of the resource. Although red sea urchins represent the largest fishery in California by both weight and revenue (Quinn, Wing, and Botsford 1993), there are significant signs of decline. Catch per unit effort has decreased by over 50% from its peak, and post-harvest urchin density has declined in many areas (P. Kalvass personal communication). Landings of red urchins in the southern California fishery began to decline in the early 1980s, and only the spread of the fishery into the northern portion of the state in 1985 allowed landings to continue to increase. Recent landing estimates have declined from 52 million pounds in 1988 to fewer than 20 million in 1994 (P. Haaker, personal communication). Reflecting the intensity of the fishing efforts, scientists estimate that there may be no red urchin stocks in the state that have escaped harvest (Tegner 1989).

Because there are very few data on the ecological consequences of overfishing red urchins, we discuss potential impacts based on what is known of their biology. An important issue in evaluating the impact of the fishery on red urchins is the selectivity of the harvest. Red urchins become sexually mature at approximately 45 mm (Tegner 1989), and the minimum legal size for harvest in California is 83–89 mm (P. Kalvass, personal communication). The selectivity of the fishery may affect reproductive potential by decreasing the number of large individuals in the population. Census data collected by the National Park Service show that the size-frequency distribution of urchins in fished areas shows fewer large individuals than those in preserve areas in the Channel Islands (M. Pentony personal communication). The relationship between size and fecundity is well documented for many marine fishes and invertebrates: large individuals produce more gametes. In red urchins, this relationship is exponential; larger individuals contribute a disproportionately large fraction of gametes to the next generation (Tegner and Levin 1983). The fishery can also compromise reproductive output of an urchin population through the location and timing of harvest, because the intensity of spawning in red urchins is highly spatially and temporally variable (Kato and Schroeter 1985; Tegner 1989; Quinn *et al.* 1993).

A second issue that should be considered is the effect of reduced population sizes on urchin reproduction. The Allee effect, defined as a decrease in per capita population growth rate with small population size (Allee 1931), is especially pronounced in broadcast spawners. In sea urchins like *S. franciscanus*, this effect is manifest at two different stages, before and after dispersal. Pre-dispersal effects involve decreased fertilization success due to small, highly dispersed populations (Pennington 1985; Denny and Shibata 1989; Levitan 1991; Levitan, Sewell, and Chia 1992). Post-dispersal Allee effects are seen as decreased recruitment success at low population densities. Individual red urchin juveniles have been shown to recruit preferentially under the spines of larger urchins (Breen, Carolsfeld, and Yamanaka 1985), apparently seeking protection against predation. Tegner and Dayton (1977) showed that mortality of recruits and small juveniles increased significantly when urchins over 95 mm were experimentally removed from the reef. The results of these studies suggest that the urchin fishery can reduce population-level performance well beyond what might be expected just from the fraction of urchins that is harvested directly.

Attempted Solutions: Regulation of the Red Urchin Fishery. Commercial harvest of *Strongylocentrotus franciscanus* in California is regulated by the state Department of Fish and Game, which has employed several different management approaches. Fishing intensity is limited by the number of licenses issued and by the number of days per week when harvest is permitted (P. Kalvass personal communication). The state has also established a minimum legal size for harvested red urchins at 89 mm in northern California and 83 mm in the southern part of the state (P. Kalvass personal communication). The state of Washington uses a slightly different approach to size regulation: a maximum as well as a minimum legal size for harvest (Sloan 1986). The size maximum was implemented to maintain the reproductive potential of the populations; it has been suggested that this may be a valuable practice for California fisheries as well (Kato and Schroeter 1985; Tegner 1989). The Japanese use minimum size limits in conjunction with two other protective measures: harvest is prohibited during the spawning season and in many small sites that are designated as protected. The use of harvest refugia for California urchins is being considered (Kato and Schroeter 1985; Quinn, Wing, and Botsford 1993), but because of the high variability in the spawning season of red urchins, legislating restrictions on harvest during spawning periods is very difficult (Tegner 1989).

One potential solution to the problem of overharvesting red urchins is stock enhancement. Tegner (1989) provides a good review of possible enhancement methods and their applicability to the California red urchin. Enhancement practices, used extensively in Japan, include habitat improvement (using rocks and other structures to improve flat or soft substrata), collection of wild recruits onto artificial settlement substrata, supplementing natural food levels, and hatchery propagation followed by release of the larvae or juveniles.

Overview: Potential Community-Level Impacts of the Red Sea Urchin Fishery. Aside from their economic value, sea urchins are key determinants of community structure. Many authors have documented the importance of urchins as herbivores and the rapid shift in community composition that occurs when urchin density changes (Chapman 1981; Estes, Duggins, and Rathbun 1989; Rowley 1989). Red sea urchins feed preferentially on giant kelp (*Macrocystis pyrifera*) and drift algae (Leighton 1966). "Urchin barrens," where all kelps and macroalgae within a given area are completely grazed, can result from high urchin densities (Tegner and Dayton 1991). Thus harvest pressure has the potential to alter community composition significantly, beyond the direct effect on urchins themselves. For example, the California red urchin fishery may have contributed to kelp recovery from the El Niño of 1957–1959 (Tegner and Dayton 1991; see also Pearse and Hines 1979).

From what we know about the potentially destructive effects of sea-urchin grazing, one might expect that the increase in fishing pressure would result in a lower incidence of massive grazing events, but no examples of decreased incidence of barrens have been identified. The relationship between urchin density and grazing pressure on kelps is complex. Other factors, such as physical disturbance (Ebeling, Laur, and Rowley 1985), the presence of the competing but unfished congener *S. purpuratus* (Leighton 1966; Tegner 1989), and availability of drift algae as an alternative food source (Harrold and Reed 1985) influence the interaction. For example, the kelp forest at Point Loma in southern California experienced an intense grazing event in 1988 in spite of the urchin fishery there; the event probably occurred as a result of storm removal of drift algae (Dayton *et al.* 1992).

The inability to detect a clear biological effect of harvesting urchins results from the confounding effects of direct human harvest on other species in the urchin community. For example, the primary natural predators of urchins are the spiny lobster (*Panulirus interuptus*) and the sheephead fish (*Semicossyphus pulcher*) (Tegner 1980; Tegner and Dayton 1981; Tegner and Levin 1983), both of which are heavily fished. Harvesting multiple species from a food web makes comparisons of urchin density in fished and reserve areas difficult to interpret; to our knowledge, there are no areas where red urchins are the only species harvested in their community. Furthermore, the heavily fished abalone (*Haliotis* spp.) are competitors with urchins for food and space as adults and yet may seek out protection underneath urchin spine canopies as juveniles (Tegner and Dayton 1981; Tegner and Levin 1982). Intense harvest of abalone was implicated in the increase in urchin densities in the 1960s (North and Pearse 1970), and there is anecdotal evidence that current levels of urchin harvest in northern California are benefitting local abalone populations (M. Tegner personal communication). Perhaps because of the omnipresent harvest by humans within the complex community in which urchins occur, we are unaware of any documented changes in the abundance and distribution of kelps or other algal species as a result of the red urchin fishery.

Interestingly, the sea-urchin fishery in the North Pacific may never have developed without overharvest of one of their primary predators, the sea otter (J. Estes personal communication). Current protection of sea otters under the Endangered Species Act has resulted in an impressive increase in their numbers since the early part of this century (Duggins and Estes 1995), which is not good news for urchin fishers. Because of the complex structure of the kelp-urchin-otter community, recovery efforts aimed at individual members can counteract one another. A similar, unexpected result was noted following recovery of predaceous peregrine falcons and subsequent local declines of their seabird prey species (Paine, Wooton, and Boersma 1990). It is clear that, before the impacts of human harvest on marine species embedded in complex communities can be interpreted or predicted, the web of direct and indirect effects must be explored. A community-focused management approach will facilitate such endeavors (see *A Scientific Framework for Management*).

Case Study 3: Sea Otters and the Exxon Valdez *Oil Spill*

The single-hulled tanker *Exxon Valdez* went aground in Prince William Sound, Alaska, on March 24, 1989, resulting in the largest tanker oil spill in U.S. history (EVOS Trustee

Council 1994; Paine *et al.* 1996). We focus on the effects of the oil spill on sea otters, *Enhydra lutris*, because they are important predators in eastern Pacific nearshore habitats, their ecological effects on intertidal and subtidal communities are well documented, and they have been a federally protected species under the Marine Mammal Protection Act since 1972. In part because of their protected status, sea otter data were better than those for any other vertebrate affected by the spill, but as will become clear in this case study, reliable baseline data on the Prince William Sound otter populations are almost nonexistent. We use this case study to illustrate the challenges inherent in assessing damage and recovery in a long-lived, marine species following a major environmental perturbation such as an oil spill.

Threats to Sea Otters. Baseline data from Alaska otter population surveys indicate that there were approximately 10,000 otters in Prince William Sound (PWS) (or 5,800 otters in the western part of the sound where the spill occurred; Garrott, Eberhardt, and Burn 1993) prior to the *Exxon Valdez* oil spill (EVOS). A number of techniques, including surveys from helicopters and boats, estimates of the probability of recovery of sea otter carcasses, direct counts of otters received at rehabilitation centers and analytical models resulted in estimates ranging from 500 to 5,000 individual otters directly killed by the spill (Garrott, Eberhardt, and Burn 1993; Ballachey *et al.* 1994). The large range of estimated mortality is a result of sampling variation among censuses and the numerous assumptions required to estimate a population-level mortality rate from carcass recovery (Ballachey *et al.* 1994; DeGange *et al.* 1994; Garshelis and Estes in press). In 1989, 2,000 to 5,000 otters were estimated to have been killed in the first few months following the spill (Ballachey, Bodkin, and DeGange 1994), and in the period from 1990 to 1991, a large proportion of the carcasses found were of reproductively mature otters (DeGange and Lensink 1990; Monson and Ballachey 1995). The mortality rate of recently weaned juvenile otters was 22% higher in the areas affected by the spill in 1990–91 (Rotterman and Monnett 1991), and although juvenile mortality rates have since declined somewhat, they were still elevated in oiled areas relative to non-oiled areas in 1992–93 (EVOS Trustee Council 1994). Sources of acute mortality to otters may include hypothermia as floating fractions of the oil coat the insulating pelage of the animals, hypoglycemia, and oil-related pathologies (Geraci and Williams 1990; USFWS 1990; Williams and Davis 1990; Lipscomb *et al.* 1993; Lipscomb *et al.* 1994).

Estimates of chronic effects of the oil spill on otter populations are more difficult. Long-term hazards, such as contaminated food sources, and sublethal concentrations of petroleum byproducts may be manifest in high juvenile mortality, reduced reproductive output, reduced growth rates, and increased risk to lactating females (Ralls and Siniff 1990; Monnett and Rotterman 1992a, 1992b; Ballachey 1995). For example, 8–16% of the 10.8 million gallons of spilled oil remains in marine sediments in PWS and is potentially toxic to the marine organisms that are the prey of sea otters (EVOS Trustee Council 1996). Exploration of the effects of the spill on reproductive output of females suggests that pupping rates in oiled areas after the spill were not detectably different from those in the same sites before the spill (Monnett and Rotterman 1992a; Johnson and Garshelis 1995). There is considerable room for error in such estimates, in part because sea otters give birth to young throughout the year, and there has been clear documentation of seasonal variation in pupping rates within populations (Estes 1990; Paine *et al.* 1996). Evaluating the longer-term consequences of the spill is difficult unless a clear demographic framework is used to explore trends in the data (see below). For example, without projected rates of population growth following the spill, it will continue to be debated whether fluctuations in population size are within the range of normal variation observed in the otter populations.

The ecological impact of the *Exxon Valdez* oil spill on otters potentially includes indirect effects of changes in species affected by otter predation. Otters can significantly affect kelp community composition through their direct consumption of prey (such as urchins, clams, mussels, and crabs) and their indirect effects on species in turn eaten by otter prey (Estes, Smith, and Palmisano 1978; Simenstad, Estes, and Kenyon 1978; Bodkin 1988; Duggins 1988; Ebeling and Laur 1988; Laur, Ebeling, and Coon 1988; Duggins, Simenstad, and Estes 1989; Estes, Duggins, and Rathbun 1989; Estes and Duggins 1995). However, the magnitude of indirect effects is likely to be low relative to direct effects for several reasons. First, the relationships between otter population sizes and prey abundances are inconsistent from place to place (Estes, Duggins, and Rathbun 1989). Second, bottom substrata in PWS are primarily soft sands and glacial silts (J. Estes personal communication), so the predicted relationships among kelps, urchins, and otters do not apply to many areas receiving spilled oil. Finally, even in areas with kelps, quantifying indirect effects is extremely difficult because of the confounding direct effects of the oil spill on urchins and kelps (EVOS Trustee Council 1994; Hoffman and Hansen 1994; Jewett *et al.* 1995). Disentangling the direct effects of the spill on constituent species and the indirect effects resulting from changes in the strength of interactions among community members is probably nearly impossible. Nevertheless, the possibility of indirect effects resulting from a disaster such as EVOS illustrates important complexities involved both in assessing damage and in estimating rates of recovery, as we discussed in the red urchin case study.

Attempted Solutions: Steps Towards Recovery of Sea Otter Populations. Most of the sea otters brought into rehabilitation centers alive were carefully cleaned of oil, fed, and kept under observation until they were deemed strong enough to be re-released into the wild (Zimmerman, Gorbics, and Lowry 1994). Approximately 360 live sea otters were taken into rehabilitation centers, 18 pups were born in the centers, and 123 sea otters died in rehabilitation (Estes 1991; Zimmerman, Gorbics, and Lowry 1994). After rehabilitation, 197 sea otters were released in clean-water habitats near where they had been collected; post-release recapture based on a subset of radio-tagged individuals was relatively low (Estes 1991; Monnett and Rotterman 1992b).

In situ response activities immediately following the spill involved otter population surveys to identify areas of high density that should be protected (Zimmerman, Gorbics, and Lowry 1994). Most efforts to contain spilled oil were futile. Protective measures included using booms to exclude oil spreading from adjacent waters and limiting human impacts related to clean-up activities. Aside from general clean up of intertidal sites using the controversial detergent Corexit and the microbe-enhancing fertilizer, Inipol, (Stone 1992); the only potential otter food sources cleaned were selected intertidal mussel beds (Babcock *et al.* 1995). Subtidal clam populations in oiled sites did not differ significantly in hydrocarbon concentrations than clam from non-oiled sites (Doroff and Bodkin 1994), and so were not targeted by recovery efforts. The only recovery actions aimed directly at sea otters were captures of otters for temporary placement in rehabilitation centers (EVOS Trustee Council 1994; B. Ballachey personal communication).

Overview: Evaluation of Sea Otter Population Recovery. Because of high initial mortality of transplanted otters in previous projects (USFWS 1990), managers are relying on natural processes to restore sea otter populations in PWS to their previous levels. In other words, the "restoration" phase of the sea otter recovery plan involves monitoring natural recovery of sea otter populations in oiled and non-oiled areas. To track the progress of the otters, a four-year study is currently being conducted at a subset of the originally oiled sites (B. Ballachey personal communication). The study targets three potential mechanisms affecting recovery of otters: (1) direct recruitment limitation due to injury to oiled otters, (2) continued hydrocarbon exposure from oil-contaminated prey, and (3) limitation of the distribution or abundance of prey species due to the oil spill. Demographic characteristics of oiled and non-oiled otter populations will be monitored. Ultimate recovery of sea otters will be determined by comparison of the community composition of prey species in nonoiled and oiled sites (EVOS Trustee Council 1996; B. Ballachey personal communication). Managers assume that, when there is no longer a significant difference in the abundance and size distributions of prey in oiled and nonoiled sites, the otters will have regained their equilibrium population size. Application of this approach has a number of possible difficulties, including the challenge of accounting for spatially variable invertebrate communities (USFWS 1990), lack of baseline biological similarity between control and oiled sites (Paine *et al.* 1996), and identifying the equilibrium status of control and oiled sea otter populations from prey-community composition.

The final Environmental Impact Statement from the EVOS states that the projected recovery for the sea otters is 7–35 years after the population begins to increase (EVOS Trustee Council 1994). "Recovery" is defined as the population size that would have existed in the absence of a spill, according to regulations under the Comprehensive Environmental Response, Compensation and Liability Act, or CERCLA—the Superfund Act (USFWS 1990). In a separate study, from information from 508 collected sea otter carcasses, a demographic model was developed to reconstruct the population dynamics of the PWS sea otter population before the spill and to estimate population-recovery trajectories following the spill (Udevitz, Ballachey, and Bruden 1996). Age and sex were determined for as many of the retrieved carcasses as possible, and age-specific reproductive rates were calculated. Because of the paucity of data on the PWS population, a number of assumptions were made: (1) survival rate of otters was density-dependent, (2) there was no age- or sex-related bias in mortality due to the oil spill, (3) the intrinsic rate of population increase (lambda) was equal to 1 in the year preceding the EVOS, (4) the population age structure was stable at the time of the spill, and (5) adult survival and reproductive rates were not affected by the oil spill or by the subsequent status of the population. Using what the authors state are "optimistic" assumptions about mortality rates for young age classes, the model projects that recovery of the PWS sea otter population will occur in 10 to 23 years (Udevitz, Ballachey, and Bruden 1996). Whether the assumptions required for making these recovery projections are biologically reasonable can only be determined from the actual rates of sea otter population growth.

Because the estimated time of recovery for the PWS sea otter population is one to three decades (EVOS Trustee Council 1996; Udevitz, Ballachey, and Bruden 1996), a four-year study of otter recovery seems short-sighted. Ideally, the architects of the recovery study will use the data collected to do a demographic analysis of oiled and nonoiled populations, from which the projected rate of intrinsic population increase could be estimated (see Estes 1990; Udevitz, Ballachey, and Bruden 1996) and the life-history stages most critical to population growth identified.

Unfortunately, the data necessary for a full demographic analysis can be prohibitively time-consuming and costly to obtain for species of conservation concern, and the statistical power to detect change in wildlife populations is often weak (Wiens and Parker 1995). Indeed, Ralls and her colleagues (1996) report that the scientific advisory group charged

Table 6.2
Methods used to assess population status of marine mammals. Population models are listed in increasing order of data required for their implementation. References cited are examples only, and do not represent an exhaustive list of applications of each method.

Data available	Population model	Assumptions	References
multiyear estimate of N	density-independent e.g., exponential options: environmental, demographic stochasticity continuous or discrete growth	no migration constant birth, death no age, size structure	1
multiyear estimate of N and information about density-dependent functions	density-dependent e.g., logistic: Ricker, Beverton-Holt options: time lags, discrete population growth, variance in carrying capacity	constant carrying capacity linear density dependence no age, size structure no migration	2
multiyear estimate of N of offspring and adults	stock-recruitment options: density-independent or density-dependent	no size structure equilibrium stock size no migration	3
age-specific reproduction and survival; information about density-dependent functions and spatial variation in vital rates	age-structured population e.g., Leslie matrix options: density-independence or density-dependence, environmental or demographic stochasticity, spatially explicit	no migration no time lags stable age distribution	4

[1]Allen 1976 (baleen whales), Estes 1990, Estes, Jameson, and Doak 1996 (sea otters)
[2]Lankester and Beddington 1986 (gray whales), Cooke 1986 (gray whales), Gerrodette and DeMaster 1990 (gray whales, northern fur seals)
[3]Eberhardt 1981 (northern fur seals), Ragen 1995 (northern fur seals)
[4]Beddington and Cooke 1984 (general fisheries), Brault and Caswell 1993 (killer whales), York 1994 (northern sea lions), Ragen 1995 (northern fur seals), Udevitz, Ballachey, and Bruden 1996 (sea otters), Kirkwood, Beddington, and Rossouw 1994 (general fisheries)

with revising the federal recovery plan for the threatened southern sea otter chose not to use population viability analysis to help establish new delisting criteria because of the restrictive assumptions required by such an approach. In Table 6.2, we provide a list of possible approaches to assessing the population status of a species, ranging from methods that require few data to those that require many. All share an underlying demographic framework, and a number have been applied to sea otter or other marine mammal populations. Demographic approaches such as these can be used to explore the population consequences of alternate management applications, illustrated in a recent analysis of the effects of controlled harvest of sea otters in Washington State (Estes, Jameson, and Doal 1996). Continued vigilance in monitoring post-spill otter population sizes is the only way to evaluate the ultimate effects of the oil spill and the course of recovery.

Case Study 4: Mitigation of Seagrass Losses in the Florida Keys

Seagrasses are marine angiosperms inhabiting soft-bottom nearshore environments throughout the world. Seagrass meadows provide important ecological functions, such as stabilizing sediments, offering refuge and nursery habitat for invertebrates and fish, providing settlement substrata for epiphytic algae and invertebrates, enhancing nutrient cycling into the water column and sediments, and facilitating filtration of the water column by increasing sedimentation rates (den Hartog 1970; Phillips 1984; Thayer, Fonseca, and Kenworthy 1984; Durako, Phillips, and Lewis 1987). Dramatic losses of seagrass habitats have occurred worldwide, because of a number of direct impacts (e.g., boats running aground, anchors and fishing gear uprooting plants, and dredging activities) and indirect effects (e.g., reduction in light and nutrient quality due to eutrophication and high concentrations of suspended solids and competition from invasive species) (Taylor and Saloman 1968; Posey 1988; Fonseca, Kenworthy, and Thayer 1996). In this case study, we illustrate problems associated with reliance on mitigation projects to compensate for destruction of marine habitats.

Threats to Seagrasses: The Florida Keys Bridge-Replacement Project. From 1912 to 1935, the Florida Keys were connected to the mainland by an overseas railroad. A massive hurricane in 1935 destroyed a large section of the railroad, prompting the state of Florida to purchase the track and construct U.S. Highway 1 over the railroad right-of-way. Because of the structural deterioration that occurred on the highway over the next 40 years, the state of Florida authorized the Keys Bridge Replacement Program in 1976, at an estimated cost of $175 million (Lewis *et al.* 1994). A total of 93.3 acres of seagrass were damaged or lost as a result of the replacement of 37 bridges in the Florida Keys from 1977 through 1983. Of those, 47.6 acres (primarily of three species, *Thalassia testudinum*, turtle grass; *Syringodium filiforme*, manatee grass; and *Halodule wrightii*, shoal grass) were directly destroyed by dredging, filling with excess sediment, and shading by construction equipment and the new bridges. An additional 45.7 acres were killed by boat propeller cuts and propeller wash during the construction period (Lewis 1987; Lewis *et al.* 1994). Loss of seagrass habitat leads to the obvious direct effect of changing a vegetated substratum to a sandy bottom, but the resulting cascade of indirect effects can be even more dramatic because of the myriad fish, invertebrate, and turtle species dependent upon seagrasses for food, refuge, and physical support (Phillips 1984; Thayer, Fonseca, and Kenworthy 1984).

Attempted Solutions: Restoration and Mitigation of Seagrass Habitat. The Florida Department of Environmental Regulation (now the Department of Environmental Protection) required that the Florida Department of Transportation mitigate the seagrass losses, through a combination of direct transplantation in numerous areas and re-opening/enhancement of coastal lagoons at three sites in the Keys. Of the 93.3 acres damaged or lost, 30.5 acres were deemed not restorable because of permanent loss of suitable substrata (i.e., rock bottom was all that remained). The remaining 62.8 acres of seagrasses amenable to restoration attempts were distributed over 22 different sites, making coordination of recovery efforts a challenge.

One of the most critical aspects of seagrass-bed restoration success is site selection, and no standardized local, regional, or national criteria presently exist for choosing a site for transplanting seagrasses (reviewed by Fonseca, Kenworthy, and Thayer 1996). One simple and very useful criterion is to determine whether seagrasses existed at the site in the past, indicating that at least the minimum habitat requirements were provided at one time. The partial success of seagrass mitigation after the Keys bridge replacement project may be due in part to the wise limitation of transplants to areas that were known to have supported healthy seagrass beds before the construction occurred.

A second determinant of seagrass mitigation success is the method used to transplant individuals into the site (Fonseca, Kenworthy, and Thayer 1996). Seagrass transplanting techniques have a long and checkered past and a mostly abysmal success rate (Thom 1990; Fonseca, Kenworthy, and Thayer 1996). Restoration methods were chosen on the basis of two previous studies designed to distinguish levels of success among alternate seagrass transplanting techniques (Derrenbacker and Lewis 1983). Diligent work by those involved have yielded over 12 years of follow-up data on the success and rates of spread of seagrasses in the two original transplant sites (Lewis *et al.* 1994). These data represent some of the best anywhere for evaluating the fate of a seagrass restoration project.

In spite of the enviable and unusual advantage of having pilot data available, the subsequent success of seagrass transplant projects as part of the Keys bridge replacement project was very low. Slightly more than 47.5 acres of seagrasses were planted from 1983–1984 at a cost of $150,000. Most of the transplants were to sites where the seagrass damage had originally occurred, and two additional compensatory sites were planted to make up for seagrass acreage lost from sites that had been irreparably damaged (Lewis *et al.* 1994). One year after transplant, only 34.6 acres of seagrasses had survived; nine of 22 sites had failed completely. The primary sources of transplant failure were (1) inadequate sediment depth to support root growth, (2) bioturbation by burrowing invertebrates, (3) poor water quality (i.e., depth and suspended sediment load), and (4) human disturbance (i.e., foot traffic and boat scars) (Lewis 1987; Lewis *et al.* 1994). These sources of failure are common to many other seagrass transplant studies (Thom 1990; Fonseca, Kenworthy, and Thayer 1996), but they were thought to be minimal risks because seagrasses had previously existed at most of the sites.

Ironically, in spite of transplant failures, more acreage of seagrasses now exists almost ten years later in the originally damaged sites, resulting in a greater than acre-for-acre mitigation success (Lewis *et al.* 1994). In spite of excellent pre-planning, the transplant projects provided little, if any, effective mitigation; the recolonization of seagrasses occurred as a result of natural revegetation and as a byproduct of simultaneously conducted mangrove restoration projects. Mitigation for mangrove acreage destroyed by the Keys bridge replacement project involved increasing tidal circulation in mangrove lagoons by means of aluminum culverts or larger circulation cuts placed through sand bars forming the outer edges of lagoons (Lewis *et al.* 1994). Restoring tidal connections to a number of mangrove lagoons resulted in natural revegetation by both mangroves and, unexpectedly, seagrasses (Figure 6.2).

Overview: Remaining Issues in Evaluating "Success" of Seagrass Mitigation. The Florida Keys bridge replacement project illustrates the importance of post-transplantation monitoring, an all-too-often neglected task in mitigation projects (reviewed by Fonseca, Kenworthy, and Thayer 1996; see also Williams and Davis 1996). If monitoring had not been thorough and long-term, the initial transplant failures would have been missed, as would the final lesson that natural re-vegetation was the ultimate source of success. These important data resulting from the careful execution of the Keys seagrass mitigation project should be a model to others planning further seagrass habitat restoration.

The primary regulatory forces underlying mandatory restoration projects in U.S. marine habitats are the Army Corps of Engineers (COE), the NMFS, the U.S. Fish and Wildlife Service (administering the Endangered Species Act), and state-level natural-resource agencies. The COE can require restoration as mitigation for damage to marine habitats resulting from an activity it has permitted, such as dredging a channel for navigation. The ESA has been used in a handful of cases to require restoration of marine

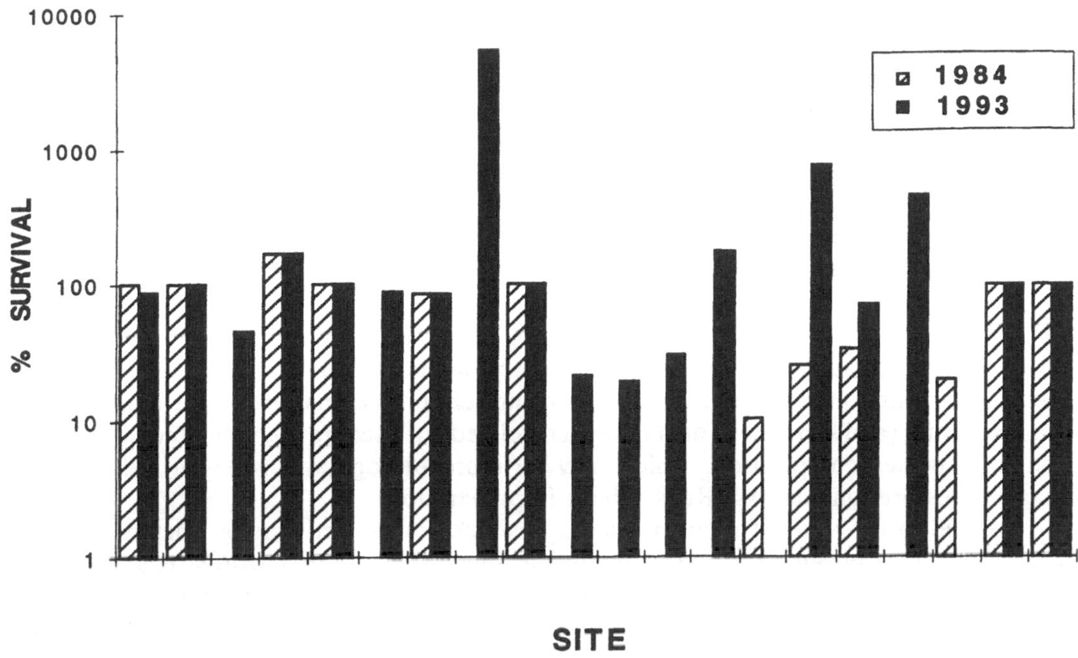

Figure 6.2
Percent survival of seagrasses at 20 sites in the Florida Keys bridge replacement mitigation project from 1983–1993. Seagrasses were transplanted in 1983. Percent acreage surviving was monitored the following year (hatched bars) and then again in 1993 (dark bars). At sites lacking bars in 1984, no transplanted seagrasses survived; all increases in acreage in 1993 at those sites are from natural revegetation, most of which resulted from opening of lagoons to mitigate mangrove losses (see text; data are from Lewis et al. *1994.)*

habitats, such as a salt-marsh restoration project for mitigation of habitat loss for the endangered light-footed clapper rail in San Diego (Roberts 1993; Holloway 1994). However, the extensive clapper rail restoration project, begun over ten years ago, has not succeeded in providing effective habitat for the bird. Perhaps the most important lesson we have learned thus far in the field of wetland restoration is that it often doesn't work.

The future successes of marine restoration projects are dependent on good data with which managers can assess the success of individual projects and the relative effectiveness of alternate restoration techniques. To a large extent, these critical data are not available. For example, in the early 1980s, only 17% of seagrass mitigation projects permitted by the COE included post-mitigation monitoring (Fonseca, Kenworthy, and Thayer 1996); resources since then have not allowed the COE to continue keeping such records. Furthermore, the NMFS Habitat Conservation District offices historically tracked the success of mitigation projects required by the COE, but since the late 1980s, money for monitoring has virtually disappeared (A. Mager personal communication). Even relatively simple data such as the number of permits issued, acreage affected, and success of projects for different habitat types are no longer tallied for any habitats except wetlands (V. Coles personal communication). In other words, data available are inadequate to quantify the extent to which mitigation compensates for loss of habitat, and if trends from wetland mitigation are any indication (Roberts 1993), mitigation generally doesn't work. Indeed, for the majority of cases, NMFS has stopped accepting seagrass mitigation projects in exchange for habitat destruction because of their low perceived success rate (A. Mager personal communication).

Criteria for determining when "recovery" of a restored or created seagrass bed has occurred are completely unstandardized. Two commonly used indicators of successful transplant recovery are the survival of planting units (i.e., adult shoots occurring singly or in clusters, seedlings, etc.) and the percent cover of seagrass within the restoration target area. For example, the NMFS inserts guidelines in its permits that a project is to be deemed successful when approximately 80% vegetative cover has been achieved after what is determined to be a reasonable amount of time (A. Mager personal communication). Upon reviewing 138 seagrass transplant projects across the U.S. and using the survival of planting units as a criterion of success, Fonseca and his colleagues (1996) found that 58 projects were successful; only seven projects achieved 100% cover, and the overall average percent cover attained was 42%. To further confuse matters, these indicators are inconsistently (and sometimes negatively) correlated with seagrass-associated invertebrate and fish species composition, which may be more biologically relevant indicators of seagrass restoration success (Ruckelshaus in preparation).

Related to the issue of recovery in seagrass restoration projects is whether different seagrass species provide equivalent biological functions; in other words, is a seagrass a seagrass? The Florida Keys mitigation project transplanted seagrass species different from those that were destroyed (i.e., *Thalassia testudinum* was destroyed and *Halodule wrightii* and *Syringodium filiforme* were transplanted), a practice not uncommon in such projects (Fonseca, Kenworthy, and Thayer 1996). Some argue that, because many seagrass communities undergo a relatively long-term successional progression (Williams 1990; Lewis *et al.* 1994; Fonseca, Kenworthy, and Thayer 1996), success is more likely if early successional, "weedy" species are used in transplant projects. The practical result is then that a number of seagrass species are used interchangeably in assessing damage, transplant success, and ultimate revegetation patterns. However, seagrass species do not always provide functional equivalency (Ruckelshaus in preparation). Until such equivalency is demonstrated, the time lag in achieving restoration "success" should be factored in to account for the natural rate of successional processes (Fonseca, Kenworthy, and Thayer 1996).

Case Study 5: The introduced Asian clam Potamocorbula amurensis

More than 150 different non-indigenous species have been identified in San Francisco Bay; indeed, almost all of the dominant invertebrate species in the Bay are nonindigenous taxa (Nichols *et al.* 1986). San Francisco Bay is characterized by a great diversity of habitats, spanning highly estuarine mudflats, rocky intertidal areas, and hypersaline ponds. The high success rates of invading taxa may be due in part to this diversity of habitats and their human-mediated disturbance (Nichols and Thompson 1985). Heavy shipping activity and aquaculture practices in the bay surely contribute to the impressive numbers of exotic species. We have selected a recent invader, the Asian clam *Potamocorbula amurensis*, to illustrate the potential ecological effects of nonindigenous species on native communities.

Threats Resulting From the Asian Clam. The Asian clam *Potamocorbula amurensis* is native to eastern Siberia, but the extent of its geographic distribution is not well known (Kimmerer, Gartside, and Orsi 1994). It first appeared in two northern sections of

San Francisco Bay in 1986 (Carlton *et al.* 1990), initially discovered through a monitoring program run by the California Department of Water Resources. In approximately two years, the Asian clam had spread throughout the estuary and is now the numerically dominant benthic organism in the northern portion of the bay, reaching densities of almost 10,000 per m^2 (Carlton *et al.* 1990). The size distribution of early population samples suggests that the introduction event probably occurred in the mid 1980s, via larval dispersal in ballast water (Carlton *et al.* 1990). The Asian clam apparently invaded immediately after a short period of high freshwater input from 1986 through 1988, when the benthic community was relatively depauperate. This period was followed by an unusually dry one that may have allowed the clam to become established (Nichols, Thompson, and Schemel 1990).

After the Asian clam was introduced, the estuarine species found in northern San Francisco Bay were unable to maintain high abundances. The mechanism by which they were excluded is still unclear, but the Asian clam has been shown to be a voracious consumer of bacteria (Alpine and Cloern 1992, Werner and Hollibaugh 1993), zooplankton (Kimmerer, Gartside, and Orsi 1994), and phytoplankton (Cole, Thompson, and Cloern 1992). Its influence as a consumer in the estuary system has been seen in a persistent and widespread decrease in phytoplankton biomass (Cole, Thompson, and Cloern 1992), which dropped from an average production of 106 g C m^{-2} to a mere 20 g C m^{-2} beginning in 1988 (Alpine and Cloern 1992). Clearance rates determined by laboratory feeding measurements suggest that, at population densities present in the early 1990s, the Asian clam can filter the entire water column from one to almost 12 times per day (Werner and Hollibaugh 1993). This rate exceeds that of the phytoplankton population in northern San Francisco Bay (Alpine and Cloern 1992). The Asian clam's persistence in the Bay in spite of the reduced phytoplankton levels suggests that the species receives a substantial proportion of its nutrition from zooplankton and bacterioplankton, and the current high densities of *P. amurensis* may be affecting recruitment of benthic invertebrates via predation on the larvae.

Other potential community-level impacts of the Asian clam include its role as a prey item (*Potamocorbula amurensis* is readily consumed by the European green crab, *Carcinus maenas*, which was introduced into San Francisco Bay several years later; Cohen, Carlton, and Fountain 1995; Grosholz and Ruiz 1995) and potential changes in habitat caused by bioturbation due to the extensive feeding activity of the clam. As far as we are aware, no one has documented population responses of potential predators of *P. amurensis* or quantified changes in habitat characteristics due to the clam.

Attempted Solutions: Steps Towards Reducing the Number of Invading Species. The many shortcomings of current national and international policies addressing introduced species, both terrestrial and aquatic, are illustrated in a report from the U.S. Congressional Office of Technology Assessment (U.S. Congress 1993). The study points out that the U.S. lacks a comprehensive approach toward invasive species. The current system is a patchwork of narrow regulations addressing only the most specific cases rather than the broader issues (U.S. Congress 1993). The regulation of aquaculture practices and the transport of marine organisms is largely done at the state level rather than the federal one, which can lead to conflict when neighboring states differ in policy regarding potential invaders. For example, Virginia and Maryland have opposing views on the proposed introduction of the Pacific oyster, *Crassostrea gigas*, into the Chesapeake Bay (Krantz 1991; U.S. Congress 1993). The difference in opinion is largely a result of economics:

Virginia has potentially more to gain from the introduction than does Maryland, where there is still a viable oyster fishery based on indigenous species.

No introduced marine organism, once established, has ever been eradicated (Carlton 1993a). Thus the problem of nonindigenous species is better addressed by focusing attention and effort on slowing the influx of potential invaders than on mitigating the effects of current exotic species. One such practice that has been highly endorsed is the exchange of ballast in open water. Many planktonic organisms would still be transported, but because of the different habitat types (i.e., pelagic deep water dumped into shallow coastal ports and vice versa) the likelihood of survival and establishment would be reduced. U.S. law instituted in 1993 requires ballast water exchange before entry into the Great Lakes; this is the first such ballast law anywhere (Carlton 1993a). At present, there has not yet been a marine introduction in the United States with an economic impact similar in magnitude to that of the freshwater zebra mussel in the Great Lakes, but unless specific measures are taken to stem the influx of nonindigenous marine organisms, it may be just a matter of time and opportunity before massive ecological impacts occur (Race 1982; Harbison and Volovik 1993; Grosholz and Ruiz 1995).

Overview: What We Can Learn From Invasions. Clearly there is a dire need for national and international measures aimed at stemming continued influx of nonindigenous species into marine habitats (see, for example, Futch and Willis 1992). In addition to encouraging steps towards reducing introductions, the scientific community can take better advantage of ongoing invasions (Lodge 1993). The fates of nonindigenous species can be used to explore basic ecological and evolutionary issues, such as models of community assembly and stability, maintenance of biodiversity, and the genetic consequences of founder events.

Genetic markers provide one way to address the timing and mechanism of recent invasions and to identify cryptically invasive species (Goff *et al.* 1992; Carlton and Geller 1993; Carlton 1996). By characterizing the genetic structure of the introduced populations, one can make inferences about potential sources of colonists, the number of individuals included in the initial population, and rates of population growth in the early stages of invasion. Duda (1994) used starch gel electrophoresis to describe the population-genetic structure of the Asian clam in San Francisco Bay. Individuals were collected from five different sites within the Bay; within-site genetic diversity was high and among-site diversity was relatively low. The high within-subpopulation genetic diversity of the Asian clam could indicate multiple introduction events or a single introduction from one or more highly diverse source population(s). Maintenance of relatively high genetic diversity after a founding event also indicates rapid population growth rates in the early years of colonization (Nei, Maruyama, and Chakraborty 1975; Lande 1980; Barton and Charlesworth 1984). Early population-growth trajectories of the clam (Carlton *et al.* 1990) support the possibility that high growth rates early in the invasion contributed to the maintenance of genetic diversity. The low genetic diversity between the subpopulations of the Asian clam could be the result of either high dispersal within the Bay or the recent timing of the introduction. We would be wise to continue to take advantage of ongoing invasions by studying their biological causes and consequences, as long as we focus with equal intensity on adoption of measures aimed at staving off further introductions (e.g., International Council for the Exploration of the Sea's Code of Practice for introductions and transfers of marine taxa, Carlton 1991).

THE FUTURE OF MARINE SPECIES CONSERVATION

In order to learn from our past successes and mistakes in managing populations of marine species, we must begin to treat management options as hypotheses (Walters and Holling 1990; Hilborn 1992a; Underwood 1995), modifying management strategies according to experiments and observations that evaluate the adequacy of alternate actions. Such management approaches are being used in marine species conservation—for example, the original recovery plan developed for the southern sea otter in 1982 was revised in 1995 so that the status of the population and successes of alternate management tools could be incorporated into an improved strategy for delisting the otter (Ralls, DeMaster, and Estes 1996). In addition, as we discussed in the first case study, the Snake River salmon recovery plan has explicitly incorporated "active adaptive management" (Walters and Holling 1990) into the delisting procedures (NMFS 1995). A recent discussion of the benefits of rethinking our marine fisheries policy concludes that these issues are not simple academic pursuits or "jobs vs. environment" tradeoffs; Safina (1995) estimated that the economic benefits of wise fishing practices could bring in $8 billion dollars annually in the U.S. and restore 300,000 jobs. In this final section, we outline ways to approach the management of marine taxa and discuss general questions emerging from attempts to minimize population declines suffered by marine species. We end by providing a list of more specific research needs.

A Scientific Framework for Management

In spite of the diverse set of circumstances surrounding the case studies we chose, a strikingly similar set of questions remains: (1) How do we best choose marine species for conservation and evaluate their status? (2) How do we most effectively choose areas for management? (3) How can we best define marine conservation goals? Below, we illustrate in turn some broader implications of these issues.

How Do We Best Choose Marine Species for Conservation and Evaluate Their Status? Perhaps even more so than in terrestrial habitats, a broad-scale approach to managing marine environments makes more sense than does one focused on single species. There appears to be a growing consensus among marine scientists and managers that an ecosystem approach to management will most effectively encompass the dispersal biology of marine species (NRC 1994; 1995; NMFS 1995). Because the ecological processes that occur during the recruitment phase of a marine organism's life history are notoriously difficult to quantify (Connell 1985; Gaines and Roughgarden 1985), managers of marine species should be all the more convinced that focusing efforts on habitats or ecosystems rather than on a number of single species is a more viable strategy. Indeed, in the first national research agenda on marine biodiversity (NRC 1995), a panel of marine scientists recommended that biodiversity be measured and studied according to an "integrated regional-scale research strategy" (p. 3) in order to account for the openness of marine systems. We agree, *provided that the "ecosystem" approach is grounded in monitoring the population status of one or more indicator species within each region.*

Developing criteria that can be used in identifying the appropriate species, population, stock, or variety on which to focus management efforts is one of the biggest challenges facing marine conservation managers. Indicator species should represent a critical position in the biological community, so that changes in the status of their populations reflect broader community-wide effects. For this reason, there has been a recent resurgent interest in using "keystone" species as indicators of ecosystem health (see Soulé and Simberloff 1986; Mills, Soulé, and Doak 1993; Paine 1995; Power *et al.* 1996; Simberloff in preparation). This approach will provide one or a few species on which to focus regular population assessments (rather than having to census many species) and increase the probability that a number of other species will benefit from a protection/restoration plan designed for the keystone species. Even though the term "keystone species" was coined in a study of a marine intertidal community (Paine 1966, 1969), they are not widely used as management foci in marine habitats. Keystone species must be empirically identified on a case-by-case basis; a predator critical in one habitat may not play such a role generally (e.g., Underwood and Denley 1984; Mills, Soulé, and Doak 1993; Paine 1995). Identifying trophic links among species takes a lot of work: field experiments, natural history observations, and carefully constructed hypothesis testing are necessary in all cases so that sloppy reasoning does not influence our views of marine community structure (as was observed in the lobster-urchin fishery interaction in the NW Atlantic; see Elner and Vadas 1990).

Even limiting population assessments to a few indicator species will not be simple. Ironically, perhaps the most striking data emerging from the NMFS annual assessment of the status of U.S. marine resources are the number of stocks or species for which the status (increasing, decreasing, or stable in number) is unknown. For example, even for high-profile species such as marine mammals and sea turtles, the status of over 80% of the stocks under the management of NMFS was not known in the latest evaluation, and almost 30% of other fish and shellfish populations were categorized as having unknown status (NMFS 1994). Many overview reports by marine experts conclude that a major problem blocking effective management of marine resources is the low quality and quantity of data (Noss 1990; Ludwig, Hilborn, and Walters 1993; NRC 1994; 1995; Rosenberg and Restrepo 1994; Olver, Shuter, and Minns 1995), and all of the case studies we review further illustrate this problem. We feel hopeful about alternative, less data-intensive methods for estimating population growth rates, such as using non-stage-structured population projections (e.g., Estes 1990, Gerrodette and DeMaster 1990; Estes, Jameson, and Doak 1996; J. Estes and D. Doak personal communication), but much more work is needed in exploring the practical usefulness of such approaches.

How Do We Best Choose Areas for Management? Managing large pieces of marine habitat is advisable for both biological and political reasons. Protecting large areas is perhaps more easily justified in the ocean than in terrestrial environments. As we have discussed, marine taxa differ from terrestrial taxa in at least two fundamental ways: (1) many marine species have long-distance dispersal capabilities, and (2) a number of marine organisms undergo dramatic morphological and habitat shifts during their life histories. The dispersal biology of marine taxa dictates that the distributions of many species will fall across a number of national political boundaries, and that organisms in the sea will experience numerous, widely separated habitats. Because of these biological characteristics of marine species, it is unlikely that isolated management of small areas will work. A promising means of focusing management on large spatial scales is through a link between marine ecological and oceanographic sciences in studies of "large marine ecosystems" (Sherman, Alexander, and Gold 1993). These studies provide important information about the connectivity between local, small-scale patterns and processes and those at larger, oceanographic scales (e.g., Olu *et al.* 1996). For example, Ray (1991) used maps of species distributions to define boundaries of communities using correlations among ranges of 86 species of invertebrates, fishes, birds, and mammals of the Bering, Chukchi, and Beaufort Seas. Once the distribution of habitat types and species in the ocean is known, the controversial issue of which areas to focus on remains. For example, it has been argued that because many marine species have large ranges but small breeding areas, if we give areas with high endemicity the highest priority for conservation, we may unintentionally neglect potentially vulnerable species with apparently widespread distributions when key habitats are narrow and limited (Vermeij 1993).

Coordinated management of ocean basins includes a focus on the interconnection among populations of species. A number of marine species are patchy in their distributions (Palumbi 1994; Knowlton 1993), and in order to estimate adequately their status or likelihood of recovery, we must consider the occupancy, turnover and interconnectedness of a number of populations. Attempts to estimate the actual size of interbreeding populations for managment (e.g., Avise 1989; Waples 1991a, 1994; Shepard and Brown 1993; Hedrick and Hedgecock 1994) and experimental approaches designed to explore the ecological and genetic consequences of patchiness for marine species (e.g., Paine 1988; Quinn, Wolin, and Judge 1989; Dybdahl 1994; Hellberg 1994; Planes, Galzin, and Bonhomme 1996; Ruckelshaus 1996) can be instrumental in assessing the recovery potential of species in addition to helping in reserve design.

A full discussion of the benefits of coordinated national and international efforts at conserving marine species is beyond the scope of this chapter, but managers and scientists alike agree that improving the institutional structure that underlies management decisions concerning marine species is a necessary step in improving conservation efforts (NRC 1994, 1995). An example of a fruitful application of a large-scale approach to managing marine habitats is occurring in the Florida Keys National Marine Sanctuary (FKNMS). Through a program named SEAKEYS, a number of academic, state, and federal biologists are coordinating research efforts over a large area, and are beginning interdisciplinary collaborations that involve automated environmental monitoring stations, detailed mapping of the distribution of habitat types in the region, and oceanographic studies of water circulation patterns throughout the area (Ogden *et al.* 1994). Similarly, the multination Great Barrier Reef Marine Park Authority was established in 1975 to manage the entire 345,000 km^2 park. Up to 25% of the park is closed to some type of fishing, and the authority has been very successful in balancing biological and human needs in the area (Craik 1996; W. Craik personal communication).

Marine protected areas allow large-scale management in marine conservation, although their effectiveness and logistical feasibility is currently a hotly contested topic (Bohnsack 1996; Florida Sportsman 1995; J. Bohnsack and J. Ogden personal communication). Marine protected areas in the U.S. include private and public reserves, national parks, national estuarine research reserve (NERR) sites, and national marine sanctuaries. The degree of protection offered species within marine protected areas is highly variable, ranging from restricted fishing on one species for a few days annually to complete prohibition of harvesting any species (Ballantine 1991; Roberts and Polunin 1991; Bohnsack 1996). Potential direct benefits to target species include (1) increases in abundance, mean size and age, and reproductive output; (2) enhanced recruitment and fishery yields both within and outside protected areas ("spillover" and "replenishment"); and (3) maintenance of the genetic diversity of stocks. Possible benefits of reserves to communities include increased species diversity, enhanced habitat quality and complexity, and increased community stability (Carr and Reed 1993; Roberts 1995; Roberts *et al.* 1995; Bohnsack 1996). With the exception of ample data documenting the benefits described under (1) above, very few good data address the extent to which reserves provide these services.

Many of the data pertaining to reserve effectiveness are inadequate to address the question of whether reserves *cause* changes in species density, abundances, or diversity. The majority of reserve studies involve comparison of species characteristics within and adjacent to the reserve at a given point in time; no replicate "protected" treatments address the generality of the patterns, and few studies provide good "before and after" data within protected and nonprotected sites. Because of poor experimental design, it is not possible to rule out the possibility that the best sites were chosen for reserve status—e.g., those with more diverse microhabitat types and greater abundance or diversity of organisms—or that fish or invertebrates immigrate from fished areas into protected ones. The evidence in support of claimed benefits is thoroughly reviewed by Roberts and Polunin (1991), Rowley (1994), and Bohnsack (1996).

Marine protected areas are invaluable as relatively unaffected sites for gathering baseline data on the biology of marine communities, providing a necessary assessment and inventory of marine biological diversity in a number of locations (Agardy 1994). Reserves can provide sites for biological research addressing such basic issues as understanding functional linkages between target and external communities. Furthermore, reserves can be used to evaluate the relative successes of artificial enhancement tools, such as artificial reefs, mooring buoys, and hatchery or aquaculture techniques in enhancing fish populations (Tegner 1993; Bohnsack 1996). Also, research in marine protected areas provides us with the only means of separating the direct effects of human-induced stresses and other causes of population fluctuations (as we discussed in the red urchin case study).

There are no standardized criteria by which marine protected areas should be chosen or designed *de novo*. As we have noted, the scale of dispersal and larval or gamete connectedness among habitats should be used to help define the units to be managed (Fairweather 1991; Tegner 1993; Glynn *et al.* 1994). For example, New Zealand reserves that successfully protect populations of the spiny lobster *Jasus edwardsii* do not have a detectable impact on its more mobile congener *J. verreauxi* (MacDiarmid and Breen 1992). The community context of reserve siting is also important; biological criteria should include endemicity, diversity, and the specific ecological needs of focal species (Soulé and Simberloff 1986).

The National Biological Service (NBS), coordinates with a number of other private and public groups (e.g., The Nature Conservancy, National Heritage Programs) to inventory diversity patterns in terrestrial habitats in the U.S., with the goal of identifying "hotspots" of diversity for protection. These databases are used to develop standardized criteria for protected-area site selection and design (e.g., areas of high local endemicity, diversity, focal species). The NBS is not involved in cataloging diversity of marine life (R. Hall personal communication), but a number of states have programs that are beginning to catalog and map the spatial distributions of marine resources (such as Florida's Marine Resource Inventory); we encourage such efforts.

The now-famous SLOSS debate (single large or several small reserves), although considered by some to be largely irrelevant when sites are chosen for protection (Soulé and Simberloff 1986), may still be an important consideration when a system of reserves is designed where new habitat is to be created (i.e., when size and shape are controllable variables, such as in seagrass and coral reef mitigation projects) (e.g., McNeill and Fairweather 1993).

How Can We Best Define Marine Conservation Goals? As we saw in each of the case studies we discussed, evaluating the success of management alternatives is very difficult without clearly defined goals. Without good biologically based models of expected changes in community composition or species abundances following implementation of regulations limiting perceived threats to marine species, the relative value of attempted solutions cannot be assessed. For example, the effectiveness of marine reserves can only be evaluated in relation to their objectives. Is the goal of a reserve to increase species diversity in general, or to increase the density or abundance of a few target species? Depending on the aims of protection, monitoring and evaluation should take into account the fact that the ecological and demographic characteristics of nontarget species are expected to change in response to protection of a few species from harvest. For example, harvested fish are often large, predatory species whose increase in number may have dramatic effects on the community residing in the reserve. The variable success of reserves in enhancing density or abundances of fish and invertebrates (Cole, Ayling, and Creese 1990; Armstrong *et al.* 1993) is probably due to a number of factors, including the biology of target species (e.g., dispersal mode, habitat preference, diet breadth, age- and size-specific reproduction) and the ecological interactions among species within a protected area (e.g., response to changes in prey, predator populations) (Duran and Castilla 1989, Cole, Ayling, and Creese 1990; Francour 1993; Russ and Alcala 1996a).

In Figure 6.3, we illustrate the ambiguities in evaluating the success of reserves by plotting the frequency of occurrence and density of fish and invertebrate species in reserves after protected status is imposed as a function of their density or frequency of occurrence before protected status. This figure is based on the few studies we could find with good data before and after reserve establishment. The data on frequencies of occurrence are promising: most species increase in percent occurrence after protective status is implemented. Nevertheless, one could interpret the density data to indicate that reserves are ineffective, because so many key species actually exhibit declines in density following reserve establishment. Alternatively, the data could reflect other potential consequences of protected status, such as changes in species diversity or a shift to larger size distributions in target species (Cole, Ayling, and Creese 1990). Clearly, there are few compelling data indicating *general* reserve success regarding the many proposed benefits. Until the goals of a reserve are made explicit and the data collected are adequate to judge the attainment of those goals, the effectiveness of reserves cannot easily be evaluated. Finally, managers of marine species and habitats should be aware that any one reserve cannot be expected to provide more than a few benefits. With clearly articulated goals, the design and evaluation of marine protected areas will be much more straightforward.

Research Needs. In order to address the general threats facing marine taxa, a number of outstanding research questions must be explored. Next we outline a few specific questions that have emerged from our initial foray into the complicated topic of marine species conservation. For all of these threats, we as biologists need to become more involved in the process of reducing the frequency with which they occur and the magnitudes of their effects. We acknowledge that such involvement often entails complicated political and social questions, and we limit our discussion here to biological issues.

Overfishing. If marine reserves are to be defended as a management option to counter the effects of overfishing, answers to important issues must be provided. For example, questions of whether "replenishment" of stocks occurs outside of reserve areas (e.g., via increase in the number of larvae exported) or whether spillover occurs, require carefully designed experiments, with replicated sites within protected areas, multiple control sites outside of reserves, and good before-and-after data collection. Only with these data can it be determined whether larvae produced within reserves successfully

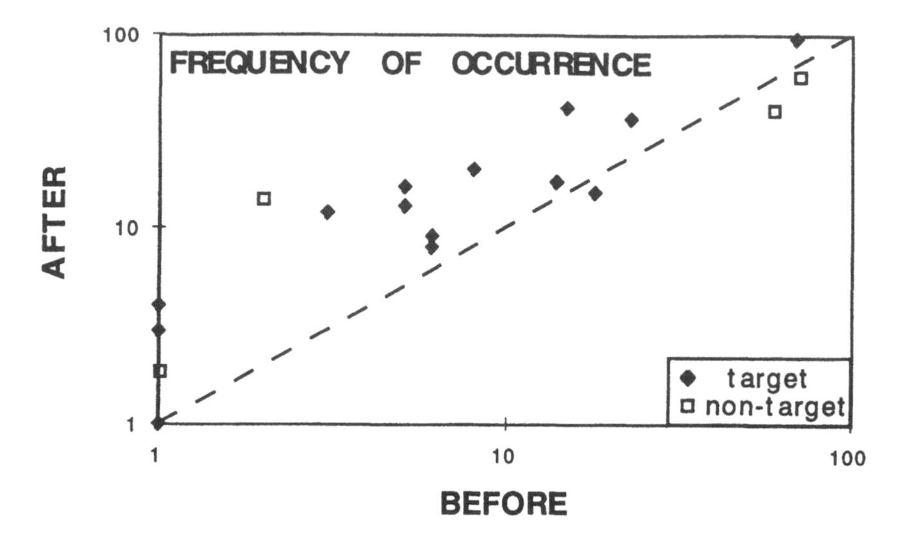

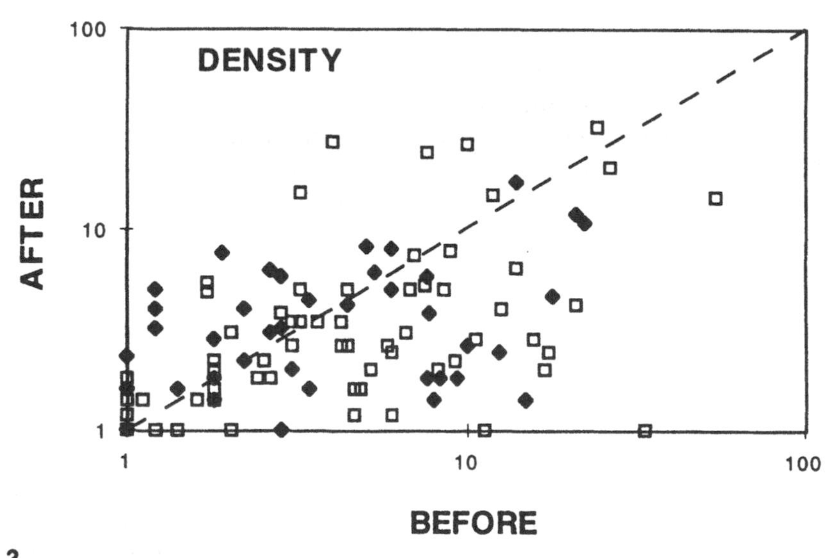

Figure 6.3
Changes in fish and invertebrate species in marine protected areas. The top graph shows the percent frequency of occurrence of fish and invertebrate species in reserves after protection from fishing as a function of their initial abundances. The bottom graph depicts the density of fish and invertebrates after protection from harvest as a function of their initial densities. Filled diamonds indicate species that are targeted by fishers and thus are expected to benefit most from reserves; open squares indicate nontarget species. The dashed line in each graph indicates the expected relationship if there is no effect of protective status offered by reserves. Note log scales. (Data for frequency of occurrence are from three reserves: Davis 1977, Dry Tortugas, Florida; Clark, Causey, and Bohnsack 1989, Florida Keys; and Bennett and Atwood 1991, South Africa. Data for density are from five reserves: Davis 1977, Dry Tortugas, Florida; Duran and Castilla 1989, Chile; Cole, Ayling, and Creese 1990, New Zealand; and Russ and Alcala 1996a, two sites in the Philippines.)

recruit into populations outside them (Rowley 1994) or whether there is a net increase in the number of adults in the total population outside of reserves, as suggested by some models (Man, Law, and Polunin 1995) and experimental studies (Russ and Alcala 1996b).

Are hatcheries a viable management tool? Hatchery supplementation programs generally have not increased the abundances of reef fishes, turtles, or invertebrates (Bohnsack 1996) or of pelagic fishes (Meffe 1992). Research is needed to address questions such as: Under what circumstances (if any) are hatchery supplementation programs biologically sensible? If hatcheries are used, how can deleterious effects on wild populations be minimized?

Finally, we need better ecological and genetic data, especially regarding recruitment patterns and dispersal capabilities of species, so that interbreeding, independently evolving populations can be defined and used as focal points for management. For example, Gall *et al.* (1992) report no detectable allozyme differences between Sacramento River winter-run chinook and other chinook stocks, in spite of their very different life histories. Is each separate chinook stock worth conserving? If so, management strategies will necessarily be different because of differences among stocks in spawning time and the seasonality of river and marine residence.

Pollution. The difficulty in documenting clear biological effects of a discrete and dramatic pollutant introduction such as oil in the *Exxon Valdez* spill illustrates the importance of good baseline data in marine habitats identified as sensitive or at risk from pollutant inputs. The value of understanding trophic relationships among species in the Prince William Sound intertidal and subtidal communities prior to the spill cannot be overstated (Paine *et al.* 1996). Furthermore, greater understanding of the environment-specific effects of pollutants (e.g., synergistic effects in the presence of other contaminants and their biological impacts under different environmental conditions) may help direct attention to specific habitat types or species particularly in need of protection (see, e.g., Bennett, Ostrach, and Hinton 1995; Hoffman 1995). In addition, the cumulative effects of long-term exposure of marine species to pollutants below levels that are deemed acute in their effects must be given greater attention. Finally, ocean currents and tides provide long-distance transport of pollutants in addition to organisms. The nature and extent of threats to marine species stemming from remote sources of toxins need to be assessed (e.g., Estes *et al.*, in press).

Habitat loss. Restoration represents in some cases our only hope of reclaiming some of the marine habitats that have been so dramatically altered that natural processes cannot be relied upon to repair them. The biological equivalence of restored and natural habitats must be experimentally established before the success of restoration projects can be evaluated (reviewed by Fonseca, Kenworthy, and Thayer 1996). Biological criteria for determining "success" must be developed, and the spatial and temporal scales over which recovery occurs need to be defined. Another important but unanswered question is how best to incorporate successional processes into restoration.

Transplantation of corals as a way to restore damaged reef habitats is a relatively new approach that is gaining increased attention. Although it can be successful in replenishing coral acreage on a small scale (Harriott and Fisk 1988; Yap, Licuanan, and Gomez 1990; Yap, Alino, and Gomez 1992; Bohnsack 1996), coral-reef restoration has not yet been implemented on as broad a scale as seagrass transplantation, and the same questions apply.

Introduced species. Even more than in terrestrial habitats, there is a dire need for basic natural history and taxonomic information so that the diversity of life in the sea can be described, including the origins of so-called cryptogenic species (NRC 1995; Carlton 1996). Ecological and genetic effects of introduced marine taxa are just beginning to be addressed (NRC 1995), as we discussed in the Asian clam case study. Possibilities for control of the rate of spread of marine nonindigenous species are virtually unexplored. Are there characteristics of introduced species or habitats that make them particularly likely to become invasive or be invaded? How can molecular genetic techniques best be used to pinpoint sources and times of invasion? Answers to questions such as these will go a long way towards anticipating potential invasions before they occur by slowing the rates of unintentional introductions into marine waters.

Global climate change. The numerous physical changes that are hypothesized to be associated with an increase in the global temperature point to the need for studies that incorporate the cumulative effects of changes in the physical environment on marine biota. Limits to the ocean's ability to buffer changes in atmospheric gases and process nutrients through biological and chemical cycles are not well understood. In addition, the capabilities of marine species to respond evolutionarily to changes in ocean conditions have not been addressed. Will species in the sea adapt, or shift distributions in response to physical changes in their environment? Or will they suffer extinction as the rate of environmental change outstrips their rates of adaptive evolution (Lynch and Lande 1993)?

Finally, the relevance of models predicting terrestrial species responses to global change should be explored carefully for marine taxa. For instance, if there are more long-lived organisms at the top of marine than terrestrial food webs, how does their susceptibility to changes in ocean conditions affect our ability to forecast changes in species abundance and diversity following predicted changes in the marine environment?

CONCLUSIONS

The historical notion that the sea represents a resilient, boundless source of harvestable goods is no longer defensible. Marine organisms are vulnerable to the direct and indirect consequences of human activities, as the case studies in this chapter illustrate for a diverse set of taxa. The challenge facing marine scientists is to provide decision-makers with sound ecological and genetic data with which to evaluate the severity of threats to species and to recognize that lessons from terrestrial conservation may not apply to marine systems. Marine taxa differ fundamentally from terrestrial taxa: they tend to be more widely dispersed. The difference in dispersal biology between marine and terrestrial species results in life histories occurring in multiple, sometimes distant habitats and many species being more broadly distributed than protected areas can encompass. Those involved in implementing marine conservation strategies must treat application of alternate policies as experiments whose results can be incorporated into an improved conservation plan. At present, we are severely limited by our ignorance regarding such fundamental issues as the status of many marine species and the value of marine reserves in combating species declines.

ACKNOWLEDGMENTS

Many people have helped us in pursuit of the details we present in this chapter. B. Ballachey, J. Bohnsack, J. Carlton, P. Dayton, M. Dethier, D. Doak, J. Estes, M. Fonseca,

J. Grubich, P. Haaker, P. Kalvass, D. Levitan, J. Ogden, R.T. Paine, K. Ralls, D. Simberloff, and M. Tegner provided important guidance and data. We thank Peter Kareiva for the opportunity to write this chapter and for his skilled editorial hand. Two anonymous reviewers helped with clarifying our presentation of material, and A.B. Thistle smoothed over rough edges of the draft. Both authors were supported by funds provided by the Department of Biological Science at Florida State University during preparation of this chapter.

LITERATURE CITED

Agardy, M.T. 1994. Advances in marine conservation: the role of marine protected areas. *Trends in Ecology and Evolution* 9:267–270.

Allee, W.C. 1931. *Animal aggregations. A study in general sociology.* Chicago: University of Chicago Press.

Allen, K.R. 1976. A more flexible model for baleen whale populations. *Reports of the International Whaling Commission* 26:247–263.

Alpine, A.E. and J.E. Cloern. 1992. Trophic interactions and direct physical effects control phytoplankton biomass and production in an estuary. *Limnology and Oceanography* 37:946–955.

Armstrong, D.A., T.C. Wainwright, G.C. Jensen, P.A. Dinnel, and H.B. Anderson. 1993. Taking refuge from bycatch issues: Red king crab (*Paralithodes camtschaticus*) and trawl fisheries in the eastern Bering Sea. *Canadian Journal of Fisheries and Aquatic Science* 50:1993–2000.

Avise, J.C. 1989. A role for molecular genetics in the recognition and conservation of endangered species. *Trends in Ecology and Evolution* 4:279–281.

Babcock, M.M., P.M. Harris, S.D. Rice, R.J. Bruyere, and D.R. Munson. 1995. *Recovery monitoring and restoration of oiled mussel beds in Prince William Sound, Alaska.* Annual Report to the *Exxon Valdez* Trustee Council. National Oceanic and Atmospheric Administration, Auke Bay Laboratory, Juneau, Alaska. 21 pp.

Ballachey, B.E. 1995. *Biomarkers of damage to sea otters in Prince William Sound, Alaska following potential exposure to oil spilled from the* Exxon Valdez *oil spill. Exxon Valdez* oil spill state/federal Natural Resource Damage Assessment final report (Marine Mammal Study No. 6–1). Anchorage, Alaska: U.S. Fish and Wildlife Service.

Ballachey, B.E., J.L. Bodkin, and A.R. DeGange. 1994. An overview of sea otter studies. In *Marine mammals and the Exxon Valdez*, ed. T.R. Loughlin, 47–60. New York: Academic Press.

Ballantine, W.J. 1991. *Marine reserves for New Zealand.* Leigh Laboratory Bulletin No. 25, University of Auckland, New Zealand.

Baltz, D.M. 1991. Introduced fishes in marine systems and inland seas. *Biological Conservation* 56:151–177.

Barry, J.P., C.H. Baxter, R.D. Sagarin, and S.G. Gilman. 1995. Climate-related, long-term faunal changes in a California rocky intertidal community. *Science* 267:672–675.

Barton, N.H. and B. Charlesworth. 1984. Genetic revolutions, founder effects and speciation. *Annual Review of Ecology and Systematics* 15:133–164.

Beddington, J.R. and J.G. Cooke. 1982. Harvesting from a predator-prey complex. *Ecological Modeling* 14:155–177.

Beddington, J.R. and J.G. Cooke. 1984. Estimating the response of a population to exploitation from catch and effort data. In: *Mathematical Ecology Proceedings Vol. 54*, eds. S.A. Levin and T.G. Hallam, 247–261. Trieste, Italy.

Bell, P.R.F. 1992. Eutrophication and coral reefs: some examples in the great barrier reef lagoon. *Water Research* 26:553–568.

Bennett, B.A. and C.G. Attwood. 1991. Evidence for recovery of a surf-zone fish assemblage following the establishment of a marine reserve on the southern coast of South Africa. *Marine Ecology Progress Series* 75:173–181.

Bennett, W.A., D.J. Ostrach, and D.E. Hinton. 1995. Larval striped bass condition in a drought stricken estuary: Evaluating pelagic food-web limitation. *Ecological Applications* 5:680–692.

Bodkin, J.L. 1988. Effects of kelp forest removal on associated fish assemblages in central California. *Journal of Experimental Marine Biology and Ecology* 117:227–238.

Bohnsack, J.A. 1996. Maintenance and recovery of reef fishery productivity. In *Management of reef fisheries*, eds. N.V.C. Polunin and C.M. Roberts, 283–238. London: Chapman & Hall.

Brault, S. and H. Caswell. 1993. Pod-specific demography of killer whales (*Orcinus orca*). *Ecology* 74:1444–1454.

Breen, P.A., W. Carolsfeld, and K.L. Yamanaka. 1985. Social behavior of juvenile red sea urchins, *Strongylocentrotus franciscanus* (Agassiz). *Journal of Experimental Marine Biology and Ecology* 92:45–61.

Breitburg, D.L. 1992. Settlement patterns and presettlement behavior of the naked goby, *Gobiosoma bosci*, a temperate oyster reef fish. *Marine Biology* 109:213–221.

Brown, B.E., and J.C. Ogden. 1993. Coral bleaching. *Scientific American* 268:64–70.

Brown, L. R., D. Denniston, C. Flavin, H. French, H. Kane, N. Lenssen, M. Renner, D. Roodman, M. Ryan, A. Sachs, L. Starke, P. Weber, and J. Young. 1995. *State of the world*. New York: Norton.

Butman, C.A., J.T. Carlton, and S.R. Palumbi. 1995. Whaling effects on deep-sea biodiversity. *Conservation Biology* 9:462–464.

Camhi, M. 1995. Industrial fisheries threaten ecological integrity of the Galapagos Islands. *Conservation Biology* 9: 715–724.

Carlton, J.T. 1985. Transoceanic and interoceanic dispersal of coastal marine organisms: The biology of ballast water. *Oceanography and Marine Biology Annual Review* 23:313–371.

Carlton, J.T. 1987. Patterns of transoceanic marine biological invasion in the Pacific Ocean. *Bulletin of Marine Science* 41:452–465.

Carlton, J.T. 1991. An international perspective on species introductions: the ICES Protocol. In *Introductions and transfers of marine species: achieving a balance between economic development and resource protection*, 31–34. South Carolina: Sea Grant Consortium.

Carlton, J.T. 1992a. Introduced marine and estuarine mollusks of North America: An end-of-the-20th-century perspective. *Journal of Shellfish Research* 11:489–505.

Carlton, J.T. 1992b. Dispersal of living organisms into aquatic ecosystems as mediated by aquaculture and fisheries activities. In *Dispersal of living organisms into aquatic ecosystems*, eds. A. Rosenfield and R. Mann, 13–45. College Park: Maryland Sea Grant Publications.

Carlton, J.T. 1993a. Biological invasions and biodiversity in the sea: the ecological and human impacts of nonindigenous marine and estuarine organisms. Keynote address in Nonindigenous Estuarine and Marine Organisms (NEMO), Proceedings of the Conference and Workshop. Seattle, Washington.

Carlton, J.T. 1993b. Neoextinctions of marine invertebrates. *American Zoologist* 33:499–509.

Carlton, J.T. 1996. Biological invasions and cryptogenic species. *Ecology* 77:1653–1655.

Carlton, J.T. and J.B. Geller. 1993. Ecological roulette: The global transport of nonindigenous marine organisms. *Science* 261:78–82.

Carlton, J.T., J.K. Thompson, L.E. Schemel, and F.H. Nichols. 1990. Remarkable invasion of San Francisco Bay (California, USA) by the Asian clam *Potamocorbula amurensis*. I. Introduction and dispersal. *Marine Ecology Progress Series* 66: 81–94.

Carlton, J.T., D.M. Reid, and H. van Leeuwen. 1994. *The role of shipping in the introduction of non-indigenous aquatic organisms to the coastal waters of the United States (other than the Great Lakes) and an analysis of control options*. The National Biological Invasions Shipping Study (NABISS). National Sea Grant Program. 390 pp.

Carr, M.H. 1989. Effects of macroalgal assemblages on the recruitment of temperate zone reef fishes. *Journal of Marine Biology and Ecology* 126: 59–76.

Carr, M.H. and D.C. Reed. 1993. Conceptual issues relevant to marine harvest refuges: Examples from temperate reef fishes. *Canadian Journal of Fisheries and Aquatic Science* 50:2019–2028.

Castilla, J.C. and R. Duran. 1985. Human exploitation from the intertidal zone of central Chile: The effects on *Concholepas concholepas* (Gastropoda). *Oikos* 45:391–399.

Chapman, A.R.O. 1981. Stability of sea urchin dominated barren grounds following destructive grazing of kelp in St. Margaret's Bay, Eastern Canada. *Marine Biology* 62:307–311.

Chapman, J.W. 1988. Invasions of the northeast Pacific by Asian and Atlantic gammaridean amphipod crustaceans, including a new species of *Corophium*. *Journal of Crustacean Biology* 8:364–382.

Chapman, D.W. and K.L. Witty. 1993. *Habitats of weak salmon stocks of the Snake River basin and feasible recovery measures*. Draft appendix to U.S. Department of Energy Bonneville Power Administration Critical Habitat comments, from Don Chapman Consultants, Inc., Boise, Idaho.

Chiappone, M. and K.M. Sullivan. 1994. Patterns of coral abundance defining nearshore hard bottom communities of the Florida Keys. *Florida Scientist* 57:108–125.

Clark, K.B. 1994. *Ascoglossan* (= *Sacoglossa*) molluscs in the Florida Keys: Rare marine invertebrates at special risk. *Bulletin of Marine Science* 54:900–916.

Clark, J.R., B. Causey, and J.A. Bohnsack. 1989. Benefits from coral reef protection: Looe Key Reef, Florida. In *Coastal Zone '89. Vol. 4. Proceedings of the sixth symposium on coastal and ocean management*, eds. O.T. Magoon, H. Converse, D. Miner, L.T. Tobin, and D. Clark, 3076–3086. New York: American Society of Civil Engineers.

Cohen, A.N., J.T. Carlton, and M.C. Fountain. 1995. Introduction, dispersal and potential impacts of the green crab *Carcinus maenas* in San Francisco Bay, California. *Marine Biology* 122:225–237.

Cole, R.G., T.M. Ayling, and R.G. Creese. 1990. Effects of marine reserve protection at Goat Island, northern New Zealand. *New Zealand Journal of Marine and Freshwater Research* 24:197–210.

Cole, B.E., J.K. Thompson, and J.E. Cloern. 1992. Measurement of filtration rates by infaunal bivalves in a recirculating flume. *Marine Biology* 113:219–225.

Connell, J.H. 1985. Variation and persistence of rocky shore populations. In *The ecology of rocky coasts*, eds. P.G. Moore and R. Seed, 467–492. New York: Hudder and Stoughton.

Cooke, J.G. 1986. On the net recruitment rate of gray whales with reference to interspecific comparisons. *Reports of the International Whaling Commission* 36:363–366.

Cooper, S.R. 1995. Chesapeake Bay watershed historical land use: Impact on water quality and diatom communities. *Ecological Applications* 5:703–723.

Craik, W. 1996. *Harvest refugia in the Great Barrier Reef Marine Park*. Townsville, Queensland: Great Barrier Reef Marine Park Authority.

Davis, W.J. 1993. Contamination of coastal versus open ocean surface waters: A brief meta-analysis. *Marine Pollution Bulletin* 26:128–134.

Dayton, P.K., M.J. Tegner, P.E. Parnell, and P.B. Edwards. 1992. Temporal and spatial patterns of disturbance and recovery in a kelp forest community. *Ecological Monographs* 62:421–445.

Dayton, P.K., S.F. Thrush, M.T. Agardy, and R.J. Hofman. 1995. Environmental effects of marine fishing. *Aquatic Conservation: Marine and Freshwater Ecosystems* 5:1–28.

DeGange, A.R., A.M. Doroff, and D.H. Monson. 1994. Experimental recovery of sea otter carcasses at Kodiak Island, Alaska, following the *Exxon Valdez* oil spill. *Marine Mammal Science* 10:492–495.

DeGange, A.R. and C.J. Lensink. 1990. *Distribution, age and sex composition of sea otter carcasses recovered during the response to the T/V* Exxon Valdez *oil spill*. Biological Report 90(12): 124–129. Anchorage, Alaska: U.S. Fish and Wildlife Service.

den Hartog, C. 1970. *The sea-grasses of the world*. Amsterdam: North-Holland.

Denny, M.W. and M.F. Shibata. 1989. Consequences of surf-zone turbulence for settlement and external fertilization. *American Naturalist* 134:859–889.

Derrenbacker, J.A., and R.R. Lewis. 1983. Seagrass habitat restoration, Lake Surprise, Florida Keys. In *Proceedings of the 9th annual conference of wetlands restoration and creation*, ed. R.H. Stovall, 132–154. Hillsborough Community College, Tampa, Florida.

Doroff, A.M. and J.L. Bodkin. 1994. Sea otter foraging behavior and hydrocarbon levels in prey. In *Marine mammals and the* Exxon Valdez, T.R. Loughlin, 193–208. New York: Academic Press.

Downing, N. and C. Roberts. 1993. Has the gulf war affected coral reefs of the northwestern gulf? *Marine Pollution Bulletin* 27:149–156.

Druehl, L.D. 1972. Marine transplantations. *Science* 179:12.

Duda, T.F., Jr. 1994. Genetic population structure of the recently introduced Asian clam, *Potamocorbula amurensis*, in San Francisco Bay. *Marine Biology* 119:235–241.

Duggins, D.O. 1988. The effects of kelp forests on nearshore environments: Biomass, detritus and altered flow. In *The community ecology of sea otters*, eds. G.R. Van Blaricom and J.A. Estes, 192–201. Berlin: Springer-Verlag.

Duggins, D.O., S.A. Simenstad, and J.A. Estes. 1989. Magnification of secondary production by kelp detritus in coastal marine ecosystems. *Science* 245:170–173.

Durako, M.J., R.C. Phillips, and R.R. Lewis III. 1987. *Proceedings of the symposium on subtropical-tropical seagrasses of the Southeastern United States. Florida marine research publications No. 42*. St. Petersburg, Florida: Florida Department of Natural Resources.

Duran, L.R. and J.C. Castilla. 1989. Variation and persistence of the middle rocky intertidal community of central Chile, with and without human harvesting. *Marine Biology* 103:555–562.

Dybdahl, M.F. 1994. Extinction, recolonization and the genetic structure of tidepool copepod populations. *Evolutionary Ecology* 8:113–124.

Ebeling, A.W. and D.R. Laur. 1988. Fish populations in kelp forests without sea otters: Effects of severe storm damage and destructive sea urchin grazing. In *The community ecology of sea otters*, eds. G.R. Van Blaricom and J.A. Estes, 169–191. Berlin: Springer-Verlag.

Ebeling, A.W., D.R. Laur, and R.J. Rowley. 1985. Severe storm disturbances and reversal of community structure in a southern California kelp forest. *Marine Biology* 84:287–294.

Ebert, T.A. 1993. Growth and mortality of subtidal red sea urchins (*Strongylocentrotus franciscanus*) at Sam Nichols Island, California, U.S.A.: problems with models. *Marine Biology* 117:79–89.

Eckman, J.E., D.O. Duggins, and A.T. Sewell. 1989. Ecology of understory kelp environments. I. Effects of kelps on flow and particle transport near bottom. *Journal of Experimental Marine Biology and Ecology* 129:173–187.

Elner, R.W. and R.L. Vadas, Sr. 1990. Interference in ecology: The sea urchin phenomenon in the northwestern Atlantic. *American Naturalist* 136:108–125.

Emlen, J.M. 1991. Heterosis and outbreeding depression: A multi-locus model and an application to salmon production. *Fisheries Research* 12:187–212.

EPA (Environmental Protection Agency). 1975. *DDT: a review of scientific and economic aspects of the decision to ban its use as a pesticide*. Springfield, Virginia: U.S. Environmental Protection Agency.

Estes, J.A. 1990. Growth and equilibrium in sea otter populations. *Journal of Animal Ecology* 59:385–401.

Estes, J.A. 1991. Catastrophes and conservation: Lessons from sea otters and the *Exxon Valdez*. *Science* 254:1596.

Estes, J.A. and D.O. Duggins. 1995. Sea otters and kelp forests in Alaska: Generality and variation in a community ecological paradigm. *Ecological Monographs* 65:75–100.

Estes, J.A., N.S. Smith, and J.F. Palmisano. 1978. Sea otter predation and community organization in the western Aleutian Islands, Alaska. *Ecology* 59:822–833.

Estes, J.A., D.O. Duggins, and G.B. Rathbun. 1989. The ecology of extinctions in kelp forest communities. *Conservation Biology* 3:252–264.

Estes, J.A., R.J. Jameson, and D. Doak. 1996. A preliminary analysis of the Washington sea otter population and the impacts of exploitation. Draft Memo to the U.S. National Biological Service.

Estes *et al.* in press. Marine Pollution Bulletin.

EVOS (*Exxon Valdez* Oil Spill) Trustee Council. 1994. Final Environmental Impact Statement for the *Exxon Valdez* oil Spill Restoration Plan. Anchorage, Alaska.

EVOS (*Exxon Valdez* Oil Spill) Trustee Council. 1996. Mechanisms of impact and potential recovery of nearshore vertebrate predators. Project No. 95025. Briefing.

Fairweather, P.G. 1991. Implications of "supply-side" ecology for environmental assessment and management. *Trends in Ecology and Evolution* 6:60–63.

Fleming, I.A. 1994. Captive breeding and the conservation of wild salmon populations. *Conservation Biology* 8:886–888.

Florida Sportsman Magazine. 1995. *Florida's marine waters: "One big replenishment zone."* A report providing the case against proposed replenishment zones in the Florida Keys National Marine Sanctuary. December, 1995.

Fonseca, M.S., W.J. Kenworthy, and G.W. Thayer. 1996. *Draft guidelines for mitigation and restora-*

tion of seagrass in the United States and adjacent waters. National Marine Fisheries Service, NOAA Coastal Ocean Program Decision Analysis Series.

Francis, R.C. 1990. Climate change and marine fisheries. *Fisheries* 15:7–9.

Francour, P. 1993. Ichthyofauna of the natural reserve of Scandola (Corsica, northwestern Mediterranean): Analysis of the pluriannual reserve effect. *Marine Life* 3:83–93.

Frank, K.T. and W.C. Leggett. 1994. Fisheries ecology in the context of ecological and evolutionary theory. *Annual Review of Ecology and Systematics* 25:423–441.

Futch, C. R. and S.A. Willis. 1992. Characteristics of the procedures for marine species introductions in Florida. In *Introductions and transfers of marine species. Achieving a balance between economic development and resource protection*, ed. R. DeVoe, 59–60. South Carolina Sea Grant Consortium.

Gaines, S. and J. Roughgarden. 1985. Larval settlement rate: A leading determinant of structure in an ecological community of the marine intertidal zone. *Proceedings of the National Academy of Science* 82:3707–3711.

Gall, G.A.E., D. Bartley, and B. Bentley. 1992. Geographic variation in population genetic structure of chinook salmon from California and Oregon. *Fishery Bulletin* 90:77–100.

Garrott, R. A., L. L. Eberhardt, and D. M. Burn. 1993. Mortality of sea otters in Prince William Sound following the *Exxon Valdez* oil spill. *Marine Mammal Science* 9:343–359.

Garshelis, D. and J. Estes. In press. Sea otter mortality from the *Exxon Valdez* oil spill: Evaluation of an estimate from boat-based surveys. *Marine Mammal Science*.

Geraci, J.R. and T.D. Williams. 1990. Physiologic and toxic effects on sea otters. In *Sea mammals and oil: Confronting the risks*, eds. J.R. Geraci and D.J. St. Aubin, 211–222. New York: Academic Press.

Gerrodette, T. and D.P. DeMaster. 1990. Quantitative determination of optimum sustainable population level. *Marine Mammal Science* 6:1–16.

GESAMP (Group of Experts on the Scientific Aspects of Marine Pollution). 1993. *IMO FAO UNESCO and IAEA UN UNEP Reports of the twenty-third session.* London.

Gharrett, A.J. and W.W. Smoker. 1993. A perspective on the adaptive importance of genetic infrastructure in salmon populations to ocean ranching in Alaska. *Fisheries Research* 18:45–58.

Gleason, D.F. and G.M. Wellington. 1995. Variation in UVB sensitivity of planula larvae of the coral *Agaricia agaricites* along a depth gradient. *Marine Biology* 123: 693–703.

Glynn, P.W., S.B. Colley, C.M. Eakin, D.B. Smith, J. Corex, N.J. Gassman, H.M. Guzman, J.B. DelRosario, and J.S. Feingold. 1994. Reef coral reproduction in the eastern Pacific: Costa Rica, Panama and Galapagos Islands (Ecuador): II. *Poritidae. Marine Biology* 118:191–208.

Goff, L.J., L. Liddle, P.C. Silva, M. Voytek, and A.W. Coleman. 1992. Tracing species invasion in *Codium*, a siphonous green alga, using molecular tools. *American Journal of Botany* 79:1279–1285.

Grosholz, E.D. 1996. Contrasting rates of spread for introduced species in terrestrial and marinesystems. *Ecology* 77:1680–1685.

Grosholz, E.D. and G.M. Ruiz. 1995. Spread and potential impact of the recently introduced European green crab, *Carcinus maenas*, in central California. *Marine Biology* 122:239–247.

Gulland, J.A. and S. Garcia. 1984. Observed patterns in multispecies fisheries. In *Exploitation of marine communities*, ed. R. M. May, 155–190. Berlin: Springer-Verlag.

Guzman, H.M. and I. Holst. 1993. Effects of chronic oil-sediment pollution on the reproduction of the Caribbean reef coral *Siderastrea siderea. Marine Pollution Bulletin* 26:276–282.

Guzman, H.M., J.B.C. Jackson, and E. Weil. 1991. Short-term ecological consequences of a major oil spill on Panamanian subtidal reef corals. *Coral Reefs* 10:1–12.

Guzman, H.M., K.A. Burns, and J.B.C. Jackson. 1994. Injury, regeneration and growth of Caribbean reef corals after a major oil spill in Panama. *Marine Ecology Progress Series* 105:231–241.

Hallegraeff, G.M. 1993. A review of harmful algal blooms and their apparent global increase. *Phycologia* 32:79–99.

Hallegraeff, G.M. and C.J. Bolch. 1992. Transport of diatom and dinoflagellate resting spores in ships ballast water: implications for plankton biogeography and aquaculture. *Journal of Plankton Research* 14: 1067–1084.

Hallock, P., F.E. Muller-Karger, and J.C. Halas. 1993. Coral reef decline. *Research and Exploration* 9: 358–378.

Harbison, R. and S.P. Volovik. 1993. The ctenophore, *Mnemiopsis leidyi*, in the Black Sea: a holoplanktonic organism transported in the ballast water of ships. In *Nonindigenous estuarine and marine organisms (NEMO). Proceedings of the conference and workshop*, 25–35. Seattle, Washington.

Hard, J.J., R.P. Jones, Jr., M.R. Delarm, and R.S. Waples. 1992. *Pacific salmon and artificial propagation under the Endangered Species Act*. U.S. Department of Commerce, NOAA Technical Memo NMFS-NWFSC-2.

Harriott, V.J. and D.A. Fisk. 1988. *Coral transplantation as a reef management option. Proceedings of the 6th International Coral Reef Symposium*. Vol. II. Townsville, Queensland.

Harrold, C. and D.C. Reed. 1985. Food availability, sea urchin grazing, and kelp forest community structure. *Ecology* 66: 1160–1169.

Hawkins, J.P. and C.M. Roberts. 1992. Effects of recreational SCUBA diving on fore-reef slope communities of coral reefs. *Biological Conservation* 62:171–178.

Hedgecock, D. 1986. Is gene flow from pelagic larval dispersal important in the adaptation and evolution of marine invertebrates? *Bulletin of Marine Science* 39:550–564.

Hedgecock, D. and F.L. Sly. 1990. Genetic drift and effective population sizes of hatchery-propagated stocks of the Pacific oyster *Crassostrea gigas*. *Aquaculture* 88:21–38.

Hedrick, P.W. and D. Hedgecock. 1994. Effective population size in winter-run chinook salmon. *Conservation Biology* 8:890–892.

Hedrick, P. W., D. Hedgecock, and S. Hamelberg. 1995. Effective population size in winter-run chinook salmon. *Conservation Biology* 9:615–624.

Hellberg, M.E. 1994. Relationships between inferred levels of gene flow and geographic distance in a philopatric coral, *Balanophylia elegans*. *Evolution* 48:1829–1854.

Hilborn, R. 1992a. Institutional learning and spawning channels for sockeye salmon (*Oncorhynchus nerka*). *Canadian Journal of Fisheries and Aquatic Science* 49:1126–1136.

Hilborn, R. 1992b. Hatcheries and the future of salmon in the northwest. *Fisheries* 17:5–8.

Hilborn, R. and J. Winton. 1993. Learning to enhance salmon production: Lessons from the Salmonid Enhancement Program. *Canadian Journal of Fisheries and Aquatic Science* 50:2043–2056.

Hilborn, R., C.J. Walters, and D. Ludwig. 1995. Sustainable exploitation of renewable resources. *Annual Review of Ecology and Systematics* 26:45–67.

Hindar, K., N. Ryman, and F. Utter. 1991. Genetic effects of cultured fish on natural fish populations. *Canadian Journal of Fisheries and Aquatic Science* 48:945–956.

Hinrichsen, D. 1990. *Our common seas: coasts in crisis*. London: Earthscan Publications.

Hoffman, A. and P. Hansen. 1994. *Injury to demersal rockfish and shallow reef habitats in Prince William Sound, 1989–1991. Exxon Valdez* oil spill state/federal Natural Resource Damage Assessment final report (Subtidal Study No. 6, Fish/Shellfish 17). Anchorage, Alaska: Alaska Department of Fish and Game, Division of Sport Fish.

Hoffman, R.J. 1995. The changing focus of marine mammal conservation. *Trends in Ecology and Evolution* 10: 462–465.

Holloway, M. 1994. Nurturing nature. *Scientific American* 270:98–108.

Holmes, B. 1994. Biologists sort the lessons of fisheries collapse. *Science* 264:1252–1253.

Hutchings, P. 1992. Ballast water introductions of exotic marine organisms into Australia. Current status and management options. *Marine Pollution Bulletin* 25:5–8.

Ilan, M. 1990. Sexual reproduction and settlement of the coral reef sponge *Chalinula* sp. from the Red Sea. *Marine Biology* 105:25–31.

Jackson, J.B.C., J.D. Cubit, B.D. Keller, V. Batista, K. Burns, H.M. Caffey, R.L. Caldwell, S.D. Garrity, C.D. Getter, C. Gonzalez, H.M. Guzman, K.W. Kaufmann, A.H. Knap, S.C. Levings, M.J. Marshall, R. Steger, R.C. Thompson, and E. Weil. 1989. Ecological effects of a major oil spill on Panamanian coastal marine communities. *Science* 243:37–44.

Jefferson, T. A., and B. E. Curry. 1994. A global review of porpoise (Cetacea: Phocoenidae) mortality in gillnets. *Biological Conservation* 67:167–183.

Jewett, S.C., T.A. Dean, R.O. Smith, M. Stekoll, L.J. Haldorson, D.R. Laur, and L. McDonald. 1995.

The effects of the Exxon Valdez oil spill on shallow subtidal communities in Prince William Sound, Alaska 1989–93. Exxon Valdez Oil Spill State/Federal Natural Resource Damage Assessment Final Report (Restoration Project 93047, Subtidal Study No. 2A). Anchorage, Alaska: Alaska Department of Fish and Game, Habitat and Restoration Division.

Johnson, C.B. and D.L. Garshelis. 1995. Sea otter abundance, distribution and pup production in Prince William Sound following the *Exxon Valdez* oil spill. In *Exxon Valdez oil spill: fate and effects in Alaskan waters*, eds. P. G. Wells, J. N. Butler, and J. S. Hughes, 894–932. Philadelphia, Pennsylvania: American Society for Testing and Materials.

Johnson, M.W. 1930. Notes on the larval development of *Strongylocentrotus franciscanus. Publications of the Puget Sound Biological Station (Seattle)* 7:401–411.

Jones, J.B. 1992. Environmental impact of trawling on the seabed: A review. *New Zealand Journal of Marine and Freshwater Research* 26:59–67.

Kareiva, P.M., J.G. Kingsolver, and R.B. Huey, eds. 1993. *Biotic interactions and global change.* Sunderland, MA: Sinauer Associates.

Kato, S. and S.C. Schroeter. 1985. Biology of the red sea urchin, *Strongylocentrotus franciscanus*, and its fishery in California. *Marine Fisheries Review* 47:1–20.

Kimmerer, W.J., E. Gartside, and J.J. Orsi. 1994. Predation by an introduced clam as the likely cause of substantial declines in zooplankton of San Francisco Bay. *Marine Ecology Progress Series* 113:81–93.

Kirkwood, G.P., J.R. Beddington, and J.A. Rossouw. 1994. Harvesting species of different lifespans. In *Large-scale ecology and conservation*, eds. P.J. Edwards, N.R. Webb, and R.M. May, 199–227. Oxford: Blackwell Scientific.

Knowlton, N. 1993. Sibling species in the sea. *Annual Review of Ecology and Systematics* 24:189–216.

Krantz, G.E. 1991. Present management position on *Crassostrea virginica* in Maryland with comments on the possible introduction of an exotic oyster, *Crassostrea gigas*. In *Introductions and transfers of marine species: achieving a balance between economic development and resource protection*, 121–126. South Carolina: Sea Grant Consortium.

Lande, R. 1980. Genetic variation and phenotypic evolution during allopatric speciation. *American Naturalist* 116:463–479.

Lankester, K. and J.R. Beddington. 1986. An age structured population model applied to the gray whale (*Eschrichtius robustus*). *Reports of the International Whaling Commission* 36:353–358.

Laur, D.R., A.W. Ebeling, and D.A. Coon. 1988. Effects of sea otter foraging on subtidal reef communities off central California. In *The community ecology of sea otters*, eds. G.R. Van Blaricom and J.A. Estes, 151–167. Berlin: Springer-Verlag.

Laws, R.M. 1977. Seals and whales of the southern ocean. *Philosophical Transactions of the Royal Society of London B* 279:81–96.

Leighton, D.L. 1966. Studies of food preferences in algivorous invertebrates of southern California kelp beds. *Pacific Science* 20:104–113.

Lenihan, H.S. and J.S. Oliver. 1995. Anthropogenic and natural disturbances to marine benthic communities in Antarctica. *Ecological Applications* 5:311–326.

Lesser, M.P. 1996. Acclimation of phytoplankton to UV-B radiation: Oxidative stress and photoinhibition of photosynthesis are not prevented by UV-absorbing compounds in the dinoflagellate *Prorocentrum micans. Marine Ecology Progress Series* 132:287–297.

Levitan, D.R. 1991. Influence of body size and population density on fertilization success and reproductive output in a free-spawning invertebrate. *Biological Bulletin* 181:261–268.

Levitan, D.R. 1992. Community structure in times past: influence of human fishing pressure on algal-urchin interactions. *Ecology* 73:1597–1605.

Levitan, D.R., M.A. Sewell, and F.-S. Chia. 1992. How distribution and abundance influence fertilization success in the sea urchin *Strongylocentrotus franciscanus. Ecology* 73:248–254.

Lewis, R.R., III. 1987. The restoration and creation of seagrass meadows in the southeastern United States. In *Proceedings of the symposium on subtropical-tropical seagrasses of the southeastern United States*. Florida Marine Research Publications No. 42, eds. M.J. Durako, R.C. Phillips, and R.R Lewis III, 139–152. St. Petersburg, Florida: Florida Department of Natural Resources.

Lewis, R.R., III, C.R. Kruer, S.F. Treat, and S.M. Morris. 1994. *Wetland mitigation evaluation report. Florida Keys bridge replacement.* Tallahassee, Florida: State of Florida Department of Transportation.

Lipscomb, T.P., R.K. Harris, R.B. Moeller, J.M. Pletcher, R.J. Haebler, and B.E. Ballachey. 1993. Histopathologic lesions in sea otters exposed to crude oil. *Veterinary Pathology* 30:1–11.

Lipscomb, T.P., R.K. Harris, A.H. Rebar, B.E. Ballachey and R.J. Haebler. 1994. Pathology of sea otters. In *Marine Mammals and the* Exxon Valdez, ed. T.R. Loughlin, 265–280. New York: Academic Press.

Lodge, D.M. 1993. Biological invasions: Lessons for ecology. *Trends in Ecology and Evolution* 8:133–137.

Ludwig, D., R. Hilborn, and C. Walters. 1993. Uncertainty, resource exploitation, and conservation: Lessons from history. *Science* 260:17–36.

Lynch, M. and R. Lande. 1993. Evolution and extinction in response to environmental change. In *Biotic interactions and global change*, eds. P.M. Kareiva, J.G. Kingsolver and R.B. Huey, 234–250. Sunderland, MA: Sinauer Associates.

MacDiarmid, A.B. and P.A. Breen. 1992. Spiny lobster population change in a marine reserve. In *Proceedings of the second international temperate reef symposium*, ed. C.H. Battershill, 633–636. Auckland, New Zealand.

McCauley, J.E. and A.G. Carey, Jr. 1967. Echinoidea of Oregon. *Journal of the Fisheries Research Board of Canada* 24:1365–1401.

McNeill, S.E. and P.G. Fairweather. 1993. Single large or several small marine reserves? An experimental approach with seagrass fauna. *Journal of Biogeography* 20:429–440.

Man, A., R. Law, and N.V.C. Polunin. 1995. Role of marine reserves in recruitment to reef fisheries: a metapopulation model. *Biological Conservation* 71:197–204.

Martin, C. 1995. The collapse of the northern cod stocks: Whatever happened to 86/25? *Fisheries* 20:6–8.

May, R.M. 1984. *Exploitation of marine communities.* Berlin: Springer-Verlag.

May, R.M., J.R. Beddington, C.W. Clark, S.J. Holt, and R.M. Laws. 1979. Management of multispecies fisheries. *Science* 205:267–277.

Meffe, G.K. 1992. Techno-arrogance and halfway technologies: Salmon hatcheries on the Pacific coast of North America. *Conservation Biology* 6:350–354.

Meinesz, A. and B. Hesse. 1991. Introduction of the tropical green alga *Caulerpa taxifolia* and its invasion of the northwestern Mediterranean. *Oceanologica Acta* 14:415–426.

Meinesz, A., J. de Vaugelas, B. Hesse, and X. Mari. 1993. Spread of the introduced tropical green alga *Caulerpa taxifolia* in Mediterranean waters. *Journal of Applied Phycology* 5:141–147.

Messieh, S.N., T.W. Rowell, D.L. Peer, and P.J. Cranford. 1991. The effects of trawling, dredging and ocean dumping on the eastern Canadian continental shelf seabed. *Continental Shelf Research* 11:1237–1263.

Meyer, T.L., R.A. Cooper, and K.J. Pecci. 1981. The performance and environmental effects of a hydraulic clam dredge. *Marine Fisheries Review* 43:14–22.

Mills, L.S., M.E. Soulé, and D.F. Doak. 1993. The keystone species concept in ecology and conservation. *Bioscience* 43:219–23.

Monnett, C. and L.M. Rotterman. 1992a. *Mortality and reproduction of female sea otters in Prince William Sound, Alaska. Exxon Valdez* Oil Spill State/Federal Natural Resource Damage Assessment Final Report (Marine Mammal Study No. 6–13). Anchorage, Alaska: U.S. Fish and Wildlife Service.

Monnett, C. and L.M. Rotterman. 1992b. *Mortality and reproduction of female sea otters oiled and treated as a result of the* Exxon Valdez *oil spill. Exxon Valdez* Oil Spill State/FederalNatural Resource Damage Assessment Final Report (Marine Mammal Study No. 6–14). Anchorage, Alaska: U.S. Fish and Wildlife Service.

Monnett, C., L.M. Rotterman, C. Stack, and D. Monson. 1990. Postrelease monitoring of radio-instrumented sea otters in Prince William Sound. In *Sea otter symposium*, eds. K. Bayha and J. Komendy. Proceedings of a symposium to evaluate the response effort on behalf of sea otters after the T/V *Exxon* Valdez oil spill into Prince William Sound, Anchorage, AK, 17–19 April 1990. U.S. Fish and Wildlife Service Biological Report 90(12).

Monson, D.H. and B. Ballachey. 1995. *Age distributions of sea otters found dead in Prince William Sound, Alaska following the* Exxon Valdez *oil spill. Exxon Valdez* Oil Spill State/Federal Natural Resource Damage Assessment Final Report (Marine Mammal Study No. 6–15). Anchorage, Alaska: U.S. Fish and Wildlife Service.

Moreno, C.A., J.P. Sutherland, and H.F. Jara. 1984. Man as a predator in the intertidal zone of southern Chile. *Oikos* 42:155–160.

Moreno, C.A., K.M. Lunecke, and M.I. Lopez. 1986. The response of an intertidal *Concholepas concholepas* (Gastropoda) population to protection from man in southern Chile and the effects on benthic sessile assemblages. *Oikos* 43:359–364.

Munro, J.L., J.D. Parrish, and F.H. Talbot. 1987. The biological effects of intensive fishing on coral reef communities. In *Human impacts on coral reefs: facts and recommendations*, ed. B. Salvat, 41–49. *Musée National d'Histoire Naturelle et Ecole Pratique des Hautes Etudes, Athenée de Tahiti, Centre de l'Environment*, Moorea (French Polynesia).

Nehlsen, W. 1994. Salmon stocks at risk: Beyond 214. *Conservation Biology* 8:867–869.

Nehlsen, W., J.E. Williams, and J.A. Lichatowich. 1991. Pacific salmon at the crossroads: Stocks at risk from California, Oregon, Idaho, and Washington. *Fisheries* 15:4–21.

Nei, M., T. Maruyama, and R. Chakraborty. 1975. The bottleneck effect and genetic variability in populations. *Evolution* 29:1–10.

Nichols, F.H. and J.K. Thompson. 1985. Persistence of an introduced mudflat community in south San Francisco Bay, California. *Marine Ecology Progress Series* 24:83–97.

Nichols, F.H., J.C. Cloern, S.N. Louoma, and D.H. Peterson. 1986. The modification of an estuary. *Science* 231:567–573.

Nichols, F.H., J.K. Thompson, and L.E. Schemel. 1990. Remarkable invasion of San Francisco Bay (California, USA) by the Asian clam *Potamocorbula amurensis*. II. Displacement of a former community. *Marine Ecology Progress Series* 66:95–101.

Nickelson, T.E., M.F. Solazzi, and S.L. Johnson. 1986. Use of hatchery coho salmon (*Oncorhynchus kisutch*) presmolts to rebuild wild populations in Oregon coastal streams. *Canadian Journal of Fisheries and Aquatic Science* 32:2443–2449.

NMFS (National Marine Fisheries Service). 1993. *Our living oceans*. Report on the Status of U.S. Living Marine Resources. NOAA Technical Memorandum NMFS-F/SPO–15. Washington, D.C.: U.S. Department of Commerce, National Oceanic and Atmospheric Administration.

NMFS (National Marine Fisheries Service). 1994. *Fisheries of the United States*. Washington, D.C.: U.S. Department of Commerce, National Oceanic and Atmospheric Administration.

NMFS (National Marine Fisheries Service). 1995. *Proposed recovery plan for Snake River salmon*. Washington, D.C.: U.S. Department of Commerce, National Oceanic and Atmospheric Administration.

Norse, E. 1993. *Global marine biological diversity*. Washington, D.C.: Island Press.

North, W.J. and J.S. Pearse. 1970. Sea urchin population explosion in Southern California waters. *Science* 167:209.

Noss, R.F. 1990. Indicators for monitoring biodiversity: A hierarchical approach. *Conservation Biology* 4:355–364.

NRC (National Research Council). 1994. *Improving the management of U.S. marine fisheries*. Washington, D.C.: National Academy Press.

NRC (National Research Council). 1995. *Understanding marine biodiversity. A research agenda for the nation*. Washington, D.C.: National Academy Press.

Ogden, J.C., J.W. Porter, N.P. Smith, A.M. Szmant, W.C. Jaap, and D. Forcucci. 1994. A long-term interdisciplinary study of the Florida Keys seascape. *Bulletin of Marine Science* 54:1059–1071.

Olu, K., A. Duperret, M. Sibuet, J.P. Foucher, and A. Fiala-Medioni. 1996. Structure and distribution of cold seep communities along the Peruvian active margin: Relationship to geological and fluid patterns. *Marine Ecology Progress Series* 132:109–125.

Olver, C.H., B.J. Shuter, and C.K. Minns. 1995. Toward a definition of conservation principles for fisheries management. *Canadian Journal of Fisheries and Aquatic Science* 52:1584–1594.

Ortega, S. 1987. The effect of human predation on the size distribution of *Siphonaria gigas* (Mollusca: Pulmonata) on the Pacific Coast of Costa Rica. *The Veliger* 29:251–255.

Paine, R.T. 1966. Food web complexity and species diversity. *American Naturalist* 100:65–75.

Paine, R.T. 1969. A note on trophic complexity and community stability. *American Naturalist* 103:91–93.

Paine, R.T. 1988. Habitat suitability and local population persistence of the sea palm, *Postelsia palmaeformis*. *Ecology* 69:1787–1794.

Paine, R.T. 1993. A salty and salutary perspective on global change. In *Biotic interactions and global change*, eds. P.M. Kareiva, J. Kingsolver, and R. Huey, 347–355. Sunderland, MA: Sinauer Associates.

Paine, R.T. 1995. A conversation on refining the concept of keystone species. *Conservation Biology* 9: 962–964.

Paine, R.T., J.L. Ruesink, A. Sun, E.L. Soulanille, M.J. Wonham, C.D.G. Harley, D.R. Brumbaugh, and D.L. Secord. 1996. Trouble on oiled waters: lessons from the *Exxon Valdez* oil spill. *Annual Review of Ecology and Systematics*, in press.

Paine, R.T., J.T. Wootton, and P.D. Boersma. 1990. Direct and indirect effects of peregrine falcon predation on seabird abundance. *Auk* 107:1–9.

Palumbi, S.R. 1992. Marine speciation on a small planet. *Trends in Ecology and Evolution* 7:114–118.

Palumbi, S.R. 1994. Genetic divergence, reproductive isolation, and marine speciation. *Annual Review of Ecology and Systematics* 25:547–572.

Parfit, M. 1995. Diminishing returns: exploiting the ocean's bounty. *National Geographic* 188:2–37.

Pearse, J.S. and A.H. Hines. 1979. Expansion of a central California kelp forest following the mass mortality of sea urchins. *Marine Biology* 51:83–91.

Pennington, J.T. 1985. The ecology of fertilization of echinoid eggs: the consequences of sperm dilution, adult aggregation, and synchronous spawning. *Biological Bulletin* 51:417–430.

Peterson, C.H., R.T. Barber, and G.A. Skilleter. 1993. Global warming and coastal ecosystem response: how Northern and Southern hemispheres may differ in the Eastern Pacific ocean. In *Earth system responses to global change. Contrasts between North and South America*, eds. H.A. Mooney, E.R. Fuentes, and B.I. Kronberg, 17–34. New York: Academic Press.

Phillips, R.C. 1984. *The ecology of eelgrass meadows in the Pacific Northwest: A community profile.* U.S. Fish and Wildlife Service, FWS/OBS–84/24.

Planes, S., R. Galzin, and F. Bonhomme. 1996. A genetic metapopulation model for reef fishes in oceanic islands: the case of the surgeonfish, *Acanthurus triostegus*. *Journal of Evolutionary Biology* 9:103–117.

Posey, M.H. 1988. Community changes associated with the spread of an introduced seagrass, *Zostera japonica*. *Ecology* 69:974–983.

Power, M.E., D. Tilman, J. Estes, B.A. Menge, W.J. Bond, L.S. Mills, G. Daily, J.C. Castilla, J. Lubchenco and R.T. Paine. 1996. Challenges in the quest for keystone. *Bioscience* 46:609–620.

Quinn, J.F., C.L. Wolin, and M.L. Judge. 1989. An experimental analysis of patch size, habitat subdivision, and extinction in a marine intertidal snail. *Conservation Biology* 3:242–251.

Quinn, J.F., S.R. Wing, and L.W. Botsford. 1993. Harvest refugia in marine invertebrate fisheries: models and applications to the red sea urchin, *Strongylocentrotus franciscanus*. *American Zoologist* 33:537–550.

Race, M.S. 1982. Competitive displacement and predation between introduced and native mud snails. *Oecologia* 54:337–347.

Ragen, T.J. 1995. Maximum net productivity level estimation for the northern fur seal (*Callorhinus ursinus*) population of St. Paul Island, Alaska. *Marine Mammal Science* 11:275–300.

Ralls, K. and D.B. Siniff. 1990. Sea otters and oil: ecologic perspective. In *Sea mammals and oil: confronting the risks*, eds. J.R. Geraci and D.J. St. Aubin, 199–210. New York: Academic Press.

Ralls, K. and B. Taylor. 1996. How viable is population viability analysis? In *Enhancing the ecological basis of conservation: heterogeneity, ecosystem function and biodiversity*, eds. S.T.A. Pickett, R.S. Ostfeld, M. Shachak, and G.E. Likens, 228–235. New York: Chapman and Hall.

Ralls, K., D. DeMaster, and J.A. Estes. 1996. Developing a delisting criterion for the southern sea otter under the U.S. Endangered Species Act. *Conservation Biology* 10:1528–1537.

Randall, J.E. 1987. Introduction of marine fishes to the Hawaiian Islands. *Bulletin of Marine Science* 41:490–502.

Raven, J.A. 1994. Carbon fixation and carbon availability in marine phytoplankton. *Photosynthesis Research* 39:259–273.

Ray, G.C. 1991. Coastal-zone biodiversity patterns. *Bioscience* 41:490–498.

Ray, G.C., B.P. Hayden, A.J. Bulger, and M.G. McCormick-Ray. 1992. Effects of global warming on the biodiversity of coastal-marine zones. In *Global warming and biological diversity*, eds. R.L. Peters and T.E. Lovejoy, 91–104. New Haven: Yale University Press.

Ricker, W.E. 1972. Hereditary and environmental factors effecting certain salmonid populations. In *The stock concept in Pacific salmonids*, eds. R.C. Simon and P.A. Larkin, 27–160. University of British Columbia, Vancouver: Institute of Fisheries.

Roberts, C.M. 1995. Rapid build-up of fish biomass in a Caribbean marine reserve. *Conservation Biology* 9:815–826.

Roberts, C.M. and N.V.C. Polunin. 1991. Are marine reserves effective in management of reef fisheries? *Reviews in Fish Biology and Fisheries* 1:65–91.

Roberts, C.M., W.J. Ballantine, C.D. Buxton, P. Dayton, L.B. Crowder, W. Milon, M.K. Orbach, D. Pauly, and J. Trexler. 1995. *Review of the use of marine fishery reserves in the U.S. southeastern Atlantic*. NOAA Technical Memorandum NMFS-SEFSC–376. Miami, Florida: U.S. Department of Commerce, National Oceanic and Atmospheric Administration.

Roberts, L. 1993. Wetlands trading is a loser's game, say ecologists. *Science* 260:1890–1892.

Rosenberg, A.A. and V. R. Restrepo. 1994. Uncertainty and risk evaluation in stock assessment advice for U.S. marine fisheries. *Canadian Journal of Fisheries and Aquatic Science* 51:2715–2720.

Rotterman, L.M. and C. Monnett. 1991. *Mortality of sea otter weanlings in eastern and western Prince William Sound, Alaska, during the winter of 1990–91. Exxon Valdez* Oil Spill State/Federal Natural Resource Damage Assessment Final Report (Marine Mammal Study No. 6–18). Anchorage, Alaska: U.S. Fish and Wildlife Service.

Rowley, R.J. 1989. Settlement and recruitment of sea urchins (*Strongylocentrotus* spp.) in a sea urchin barren ground and a kelp bed: are processes regulated by settlement or post-settlement processes? *Marine Biology* 100:485–494.

Rowley, R.J. 1994. Marine reserves in fisheries management. *Aquatic Conservation: Marine and Freshwater Ecosystems* 4:233–254.

Ruckelshaus, M.H. 1996. Estimates of genetic neighborhood parameters from pollen and seed dispersal distributions in a marine angiosperm (*Zostera marina* L.). *Evolution* 50:856–864.

Ruckelshaus, M.H. in preparation. Seagrass habitat attributes and habitat quality for resident invertebrate and fish species.

Russ, G.R. and A.C. Alcala. 1989. Effects of intense fishing pressure on an assemblage of coral reef fishes. *Marine Ecology Progress Series* 56:13–27.

Russ, G.R. and A.C. Alcala. 1996a. Marine reserves: Rates and patterns of recovery and decline in abundance of large predatory fish. *Ecological Applications*, in press.

Russ, G.R. and A.C. Alcala. 1996b. Do marine reserves export adult fish biomass: evidence from Apo Island, central Philippines. *Marine Ecology Progress Series* 132:1–9.

Safina, C. 1995. The world's imperiled fish. *Scientific American* 273:46–53.

Sale, P.F., G.E. Forrester, and P.S. Levin. 1994. The fishes of coral reefs: Ecology and management. *Research and Exploration* 10:224–235.

Schroeder, R.E. 1987. Effects of patch reef size and isolation on coral reef fish recruitment. *Second international symposium on Indo-Pacific marine biology*, 441–451. Guam: University of Guam.

Schroeter, S.C., J.D. Dixon, J. Kastendiek, R.O. Smith, and J.R. Bence. 1993. Detecting the ecological effects of environmental impacts: A case study of kelp forest invertebrates. *Ecological Applications* 3:331–350.

Shepard, S.A., and L.D. Brown. 1993. What is an abalone stock: Implications for the role of refugia in conservation. *Canadian Journal of Fisheries and Aquatic Science* 50:2001–2009.

Sherman, K., L.M. Alexander, and B.D. Gold. 1993. *Large marine ecosystems*. Washington D.C.: AAAS Press.

Simberloff, D. In preparation. Flagships, umbrellas and keystones: Is single-species management passe in the landscape era?

Simenstad, C.A., J.A. Estes, and K.W. Kenyon. 1978. Aleuts, sea otters and alternate stable state communities. *Science* 200:403–411.

Sloan, N.A. 1986. World jellyfish and tunicate fisheries, and the northeast Pacific echinoderm fishery. *Canadian Special Publications in Fisheries and Aquatic Science* 92:23–33.

Soulé, M. and D. Simberloff. 1986. What do genetics and ecology tell us about the design of nature reserves? *Biological Conservation* 35:19–40.

SRSRT (Snake River Salmon Recovery Team). 1994. Final recommendations to National Marine Fisheries Service, May, 1994.

Stanley, D.J. and A.G. Warne. 1994. Worldwide initiation of Holocene marine deltas by deceleration of sea-level rise. *Science* 265:228–231.

Steele, J.H. 1985. Comparison of marine and terrestrial ecological systems. *Nature* 313:355–358.

Steele, J.H. 1991. Marine functional diversity. *BioScience* 41:470–474.

Stone, R. 1992. Oil-cleanup method questioned. *Science* 257:320–321.

Suchanek, T.H. 1993. Oil impacts on marine invertebrate populations and communities. *American Zoologist* 33:510–523.

Suchanek, T.H. 1994. Temperate coastal marine communities: Biodiversity and threats. *American Zoologist* 34:100–114.

Taylor, E.G. 1991. A review of local adaptation in Salmonidae, with particular reference to Pacific and Atlantic salmon. *Aquaculture* 98:185–207.

Tegner, M.J. 1980. Multispecies considerations of resource management in Southern California kelp beds. In *Proceedings of the workshop on the relationship between sea urchin grazing and commercial plant/animal harvesting*, eds. J.D. Pringle, G.J. Sharp, and J.F. Caddy, 125–143. Canadian Technical Report of Fisheries and Aquatic Sciences 954.

Tegner, M.J. 1989. The feasibility of enhancing red sea urchin, *Strongylocentrotus franciscanus*, stocks in California: An analysis of the options. *Marine Fisheries Review* 51:1–22.

Tegner, M.J. 1993. Southern California abalones: Can stocks be rebuilt using marine harvest refugia? *Canadian Journal of Fisheries and Aquatic Science* 50:2010–2018.

Tegner, M.J. and P.K. Dayton. 1977. Sea urchin recruitment patterns and implications of commercial fishing. *Science* 196:324–326.

Tegner, M.J. and P.K. Dayton. 1981. Population structure, recruitment, and mortality of two sea urchins (*Strongylocentrotus franciscanus* and *S. purpuratus*) in a kelp forest. *Marine Ecology Progress Series* 5:255–268.

Tegner, M.J. and P.K. Dayton. 1991. Sea urchins, El Niños, and the long-term stability of southern California kelp forest communities. *Marine Ecology Progress Series* 77:49–63.

Tegner, M.J. and L.A. Levin. 1982. Do sea urchins and abalone compete in California kelp forest communities? In *International echinoderms conference, Tampa Bay, Florida*, ed. J. M. Lawrence, 265–271. Rotterdam: Balkema.

Tegner, M.J., and L.A. Levin. 1983. Spiny lobsters and sea urchins: analysis of a predator-prey interaction. *Journal of Experimental Marine Biology and Ecology* 73:125–150.

Thayer, G.W., M.S. Fonseca, and W.J. Kenworthy. 1984. *The ecology of eelgrass meadows of the Atlantic Coast: A community profile*. Washington, D.C.: U.S. Fish and Wildlife Service FWS/OBS-84/02.

Thom, R.M. 1990. A review of eelgrass (*Zostera marina* L.) transplanting projects in the Pacific Northwest. *Northwest Environmental Journal* 6:121–137.

Thomas, C.D. 1990. What do real populations tell us about minimum viable population sizes? *Conservation Biology* 4:324–327.

Thompson, G.G. 1991. *Determining minimum viable populations under the Endangered Species Act*. NOAA Technical Memorandum NMFS F/NWC-198.

Tupper, M. 1995. Effects of habitat on settlement, growth and postsettlement survival of Atlantic cod (*Gadus morhua*). *Canadian Journal of Fisheries and Aquatic Science* 52:1834–1841.

U.S. Congress. 1993. *Harmful non-indigenous species in the United States*. OTA-F-565. Congressional Office of Technology Assessment. Washington, D.C.: U.S. Government Printing Office.

USFWS (United States Fish and Wildlife Service). 1990. *The sea otter (*Enhydra lutris*): behavior, ecology and natural history*. Biological Report 90(14). Washington, D.C.: U.S. Fish and Wildlife Service.

Udevitz, M.S., B.E. Ballachey, and D.L. Bruden. 1996. *A population model for sea otters in western Prince William Sound*. Restoration Project 93043-3 Sea Otter Demographics. Anchorage, Alaska: U.S. National Biological Service.

Underwood, A.J. 1995. Ecological research (and research into) environmental management. *Ecological Applications* 5:232–247.

Underwood, A.J. and E.J. Denley. 1984. Paradigms, explanations and generalizations in models for the structure of intertidal communities on rocky shores. In *Ecological communities: conceptual issues and the evidence*, eds. D.R. Strong, D. Simberloff, L.G. Abele, and A. Thistle, 151–180. Princeton: Princeton University Press.

Upton, H.F. 1992. Biodiversity and conservation of the marine environment. *Fisheries* 17:20–25.

Vermeij, G.J. 1993. Biogeography of recently extinct marine species: Implications for conservation. *Conservation Biology* 7:391–397.

Walters, C.J. and C.S. Holling. 1990. Large-scale management experiments and learning by doing. *Ecology* 71:2060–2068.

Waples, R.S. 1991a. Pacific salmon, *Oncorhynchus* spp., and the definition of "species" under the Endangered Species Act. *Marine Fisheries Review* 53:11–22.

Waples, R.S. 1991b. Genetic interactions between hatchery and wild salmonids: Lessons from the Pacific Northwest. *Canadian Journal of Fisheries and Aquatic Science* 48 (suppl. 1):124–133.

Waples, R. 1994. Genetic considerations in recovery efforts for Pacific salmon. *Conservation Biology* 8:884–886.

Watson, M.F. and R.F.G. Ormond. 1994. Effect of an artisanal fishery on the fish and urchin populations of a Kenyan coral reef. *Marine Ecology Progress Series* 109:115–129.

Werner, I. and J.T. Hollibaugh. 1993. *Potamocorbula amurensis*: comparison of clearance rates and assimilation efficiencies for phytoplankton and bacterioplankton. *Limnology and Oceanography* 38:949–964.

Westbroek, P., B. Buddemeier, M. Coleman, D.J. Kek, D. Fautin, and L. Stal. 1994. Strategies for the study of climate forcing by calcification. *Bulletin de l'Institut Océanographique* 13:37–60.

WHO (World Health Organization). 1989. *DDT and its derivatives—environmental aspects*. Environmental Health Criteria 83. Geneva: World Health Organization.

Wiens, J.A. and K.R. Parker. 1995. Analyzing the effects of accidental environmental impacts: approaches and assumptions. *Ecological Applications* 5:1069–1083.

Williams, S.L. 1990. Experimental studies of Caribbean seagrass bed development. *Ecological Monographs* 60:449–469.

Williams, S.L. and C.A. Davis. 1996. Population genetic analyses of transplanted eelgrass (*Zostera marina*) beds reveal reduced genetic diversity in southern California. *Restoration Ecology* 4:163–180.

Williams, T.M. and R.W. Davis. 1990. *Sea otter rehabilitation program: 1989* Exxon Valdez *oil spill*. London: International Wildlife Research.

Yap, H.T., W.Y. Licuanan, and E.D. Gomez. 1990. Studies on coral reef recovery and coral transplantation in the northern Phillippines: Aspects relevant to management and conservation. In *Proceedings of the first ASEAMS symposium on southeast Asian marine science and environmental protection*, ed. H.T. Yap, 117–127. UNEP Regional Seas Reports and Studies 116. Nairobi: United Nations Environmental Programme (UNEP).

Yap, H.T., P.M. Alino, and E.D. Gomez. 1992. Trends in growth and mortality of three coral species (Anthozoa: *Scleractinia*), including effects of transplantation. *Marine Ecology Progress Series* 83:91–101.

Yodzis, P. 1994. Predator-prey theory and management of multispecies fisheries. *Ecological Applications* 4:51–58.

York, A.E. 1994. The population dynamics of northern sea lions, 1975–1985. Marine Mammal *Science* 10:38–51.

Zann, L.P. 1982. Changing technology in subsistence fisheries. In *Proceedings of the Seminar/ Workshop on Utilization and Management of Inshore Marine Ecosystems of the Tropical Islands, November 24–30, 1979*, ed. P. Helfrich, 69–71. Sea Grant Cooperative Report, University of Hawaii, Hawaii.

Zimmerman, S.T., C.S. Gorbics, and L.F. Lowry. 1994. Response activities. In *Marine mammals and the* Exxon Valdez, ed. T.R. Loughlin, 23–46. New York: Academic Press.

CHAPTER

7

Reptilian Extinctions Over the Last Ten Thousand Years

TED J. CASE,
DOUGLAS T. BOLGER,
and ADAM D. RICHMAN

The fossil record of the earth shows that faunal and floral extinctions increased dramatically during certain periods. These "paleo" upheavals like those at the end of Permian and Cretaceous have long provided the punctuations that geologists and paleontologists use to divide the geological periods. A challenging question in conservation science is whether the processes affecting extinction rates today are helpful in interpreting extinction in the past and, conversely, whether prehistoric extinctions are useful for understanding recent extinctions.

One overriding pattern true of extinctions in historical time that may not be true for prehistoric extinctions is that they are concentrated on islands. Diamond (1984) summarized the modern extinctions of birds and mammals from compilations in International Union for the Conservation of Nature (IUCN) Redbooks. For birds, 171 species and subspecies have gone extinct since about 1600, and over 90% of these have occurred on islands. For mammals, out of 115 documented historical extinctions, 36% of these have occurred on islands. The smaller proportion of island extinctions for mammals is in part simply a reflection of their poor representation on islands relative to birds. Many islands (e.g., New Zealand, Hawaii, Fiji, the Mascarenes, and the Seychelles) with large numbers of bird species and many avian extinctions simply have no native mammals except for bats.

Unfortunately, the IUCN Redbook is not yet complete for reptiles (Honegger 1975, 1981). Here, we attempt to tally the historical and Holocene (Recent) prehistorical extinctions and compare the emerging pattern with that for birds and mammals. We find that, as with birds, the proportion of island compared to continental extinctions is very high. This pattern is in part an unsurprising consequence of island populations. They are small and isolated; thus, they cannot recover from local extirpation following environmental perturbations or long-term climatic changes by immigration from other areas (MacArthur and Wilson 1967; Leigh 1981; Gilpin and Soulé 1986).

A growing body of evidence for birds and mammals suggests that over the last few thousand years, the most important agent of directed change in the environment is not climatic change but human disturbance and alteration of habitats (see Diamond 1984 for review). Most extinctions of entire species in recorded history are attributable to some aspect of human intervention. For example, paleontological investigations in the West Indies and Pacific link the extinction of numerous species of vertebrates with human colonization of these islands in recent prehistory (Steadman, Pregill, and Olson 1984; Olson and James 1982; Steadman and Olson 1985). For birds and mammals, the major mechanisms are habitat destruction, human hunting, effects of introduced taxa, particularly predators, and trophic cascades (i.e., secondary extinctions caused by previous extinctions; Diamond and Case 1986). Here we look for the generality of these findings by evaluating the evidence for the human impact on Holocene reptilian extinctions.

ISLAND REPTILES AND THE PREHISTORICAL LEGACY

Evidence for extinctions of reptiles in historical time is more fragmentary than for birds or mammals. For example, while we have a skeletal specimen of the dodo from Mauritius residing in a museum, the contemporaneous giant skinks also from Mauritius are known only from subfossils. Careful taxonomy and biogeographic documentation of reptiles lagged somewhat behind that for birds and mammals; consequently, early extinctions of reptiles may have gone without detection. Because reptiles are not as generally conspicuous or noisy as birds, they often pass unnoticed even when they are relatively plentiful. Thus, we must rely more on subfossil evidence for inferred extinctions rather than accurate taxonomic descriptions of extant species. It is not often easy to pin an exact date on a species' extinction, and therefore we are forced to rely on an accumulation of evidence rather than a single survey. For these reasons, choosing the year 1600 as a starting point for historical extinctions, as was done with birds and mammals, is rather arbitrary, and we will review all extinctions dating over the Holocene (or Recent), about the last 10,000 years. In what follows we use the term prehistoric to refer to extinctions that occurred prior to the arrival of Europeans to the locality being discussed, and thus the exact dates delimiting this period vary from place to place.

In the rest of this chapter we will focus more closely on the big questions raised in this introduction. What are the geographic patterns in reptilian extinctions? Are extinctions less common on continents and on large islands than on small islands as predicted by theory and demonstrated for bird extinctions? We will also explicitly examine the effect of the presence of man on island extinction rates and shed some light on the mechanisms by which humans impact reptile populations. Specifically, we will consider the effect of human-introduced predators and competitors. How much of a role do they play relative to habitat destruction and is the evolutionary "predator naïveté" of island species important?

A WORLD TOUR OF HOLOCENE REPTILIAN EXTINCTIONS

Continents

Unfortunately, it is nearly impossible to even begin to make Holocene tallies for South America, Africa, and Eurasia. The extant fauna is not completely known, let alone those species that have failed to survive. One lizard species, *Tetradacylus eastwoodae*, which was a very narrow endemic in the northern Transvaal of South Africa, has not been seen since its description in 1913. Its native forest habitat was converted to a pine plantation.

This represents the only historical extinction for continental Africa (Siegfried and Brooke 1995; Branch 1988).

As climates changed at the end of the Pleistocene and plant communities shifted, North American reptiles underwent local extirpations, range contractions or range expansions. These sometimes led to drastic changes in the species associations of reptiles (Van Devender 1977, 1987; Van Devender and Mead 1978), but surprisingly few extinctions. The mammalian megafauna was severely depleted in North America; in the Pleistocene eight families, 46 genera and 191 species or mammals became extinct, yet only reptiles became extinct out of 229 identified Pleistocene taxa (Holman 1995). While it is sometimes difficult to distinguish late Pleistocene from Holocene fossils, it appears that only two Holocene reptilian extinctions have occurred in the continential United States out of perhaps 130 fossil species: two large tortoises, *Geochelone wilsoni* and *G. crassicutata* possibly survived into the Holocene (Moodie and Van Devender 1979; Gehlbach 1965; Estes 1983; Holman 1995). A horned lizard (*Phrynosoma josecitensis*) in Nuevo Leon Mexico (in the northeastern part of the county) may also have become extinct at this time (Brattstrom 1955; Estes 1983). A largish rattlesnake (*Crotalus potterensis*) has now been synonomized with the extant *Crotalus virdis* (Holman 1995).

The situation is similar for mainland Australia. The largest varanid lizard in the world, *Megalania*, which dwarfed the extant Komodo dragons, went extinct in the Late Pleistocene, probably sometime after the entry of the Aborigines in Australia 30,000 to 50,000 years ago; exactly how recently is uncertain, but a date of 20 thousand B.P. would not be unreasonable (Hecht 1975). The largest Australian boid known, the Australian *Wonambi naracoortensis*, became extinct sometime during the same period, as did two species of Australian meiolanid horned tortoises (Molnar 1984a, 1984b; Gaffney 1992; Gaffney and MacNamara 1990) and two crocodiles (*Pallimnarchus pollens*, *Quinkana fortirostrum*). In Australia it seems that the reptile extinctions occurred prior to the Holocene but again in association with the pre-Holocene entry of aborigines into that continent.

Islands

Islands can be grouped into three categories with respect to human settlement histories and thus to their possible influence on reptile extinction. The three histories are discussed in separate subsections.

Islands first colonized in prehistory by aboriginal people and then later colonized by Europeans. Many birds and mammals became extinct on these islands during the aboriginal period and are known only as subfossils (Martin and Klein 1984). This pattern also holds for reptiles. In New Zealand (including Chatham and Stewart island), 43 species of native reptiles are now known from the Holocene period. Using the taxonomy of Doughtery *et al.* (1990), one species of tuatara is completely extinct (*Sphenodon diversum*). Three species of lizards are probably extinct (*Cyclodina northlandi*, larger than any extant form and known only from subfossil deposits in Northland, North Island; the skink *Leiolopisma gracilocorpus* and the gecko *Hoplodactylus delcourti* known only from a solitary museum specimen; Bauer and Russell 1986; Worthy 1987a; Hardy 1977). *Hoplodactylus delcourti* is the largest known gecko in the world, with a snout-vent length of 370 mm. Eleven reptiles today are found only on the off-lying landbridge islands (five skinks, four geckos, and two species of tuatara, *Sphenodon punctatus* and *S. guntheri*). These 11 species include all the relatively large extant species (Hardy and Whitaker

1979). Evidence for a mainland distribution for tuatara as recently as 1,000 years ago, and for some of the other species as well (Cassels 1984; Crook 1973), suggests that many of these present landbridge island distributions are relictual indicating mainland extinctions. Only one skink species, *Leiolopisma fallai* of the Three Kings Islands (which are not landbridge islands but have been isolated for much longer, is regarded as a non-relictual island endemic (Towns 1974; Robb 1986; Patterson and Daugherty 1990). In the case of two large skinks, *Cyclodina macgregori, C. alani*, as well as the large gecko, *Hoplodactylus duvaucelii*, main island subfossils indeed establish these species as formerly occurring on the North Island as recently as 1000 A.D. (Worthy 1987a).

The time of disappearance of tuataras and most of the lizards from the main islands coincides with the date for human occupation of New Zealand and the subsequent introduction of the Polynesian rat (*Rattus exulans*). On islands where the rat is present, the tuatara is either absent or not successfully recruiting young (Crook 1973; Cree, Daugherty, and Hay 1995). Tuatara populations have become extinct on at least four islands inhabited by Polynesian rats in the last 150 years (Cree, Daugherty, and Hay 1995). Three largest of six species of frog (*Leiopelma*) have gone extinct in New Zealand in the Holocene (Worthy 1987b), and the largest surviving frog (*L. hamiltoni*) occurs only on two rat-free islands.

Shifting to the Carribean region, the pattern is not dissimilar. The large herbivorous iguanine *Cyclura* has gone extinct on a number of Caribbean islands in the recent past (Pregill 1981, 1986). The giant *Cyclura pinguis*, which probably became extinct on Puerto Rico in Holocene times, survives on the off-lying small island of Anegada. Two species of iguanids in the curly-tailed lizard genus *Leiocephalus, L. eremitus* and *L. herminieri*, occupied small islands in the Caribbean and became extinct in the last 100 years (Pregill 1992). *Leiocephalus eremitus* is known only from the type specimen, a female 63 mm snout-vent (sv), which is moderate to large for the genus. *Leiocephalus herminieri* was very large (up to 140 mm sv) and is thought to have occupied Martinique although some confusion exists regarding the collecting localities of the few known museum specimens (Pregill 1992). Six other relatively large species for the genus (reaching 200 mm sv) are known only from fossil material and probably became extinct during aboriginal occupation on Hispaniola, Jamaica, Puerto Rico, and the Barbuda bank, but other smaller species survive in the Bahamas, on Hispaniola, and Cuba (Pregill 1992).

The extinction of large endemic forms occurs in other reptile groups in the Caribbean. For example, the giant gecko (*Aristelliger titan*) disappeared from Jamaica sometime before European settlement (Hecht 1951). More recently, the very large legless lizard Celestus (*Diploglossus occiduus*) vanished from Jamaica. The last specimens were collected around 50 years ago. Pregill *et al.* (1991) found ample fossil material dating no more than 800 years B.P., when presumably it was much more common. The giant anole, (*Anolis roosevelti*) is known only from a few specimens from tiny Culebra Island off the east coast of Puerto Rico and has not been seen since 1932 (Pregill 1981). Finally in the Caribbean, we have Holocene fossils of giant tortoises (*Geochelone*) from the Bahamas (e.g., San Salvador), Mono Island, and Curaçao. While these tortoise extinctions can probably be attributed to human hunting, early Holocene losses of native populations of pond turtles (Emydidae) and crocodiles in the Bahamas may be attributed to the interior lakes of these low islands becoming more saline as sea levels rose (Olson, Pregill, and Hilgartner 1990).

The Canary Islands in the East Atlantic are home to the largest lacertid lizards in the world and in the recent past were occupied by even larger species. In 1974, the large *Gallotia* (*Lacerta*) *simonyi*, long thought to be extinct, was rediscovered on Hierro (Böhme and Bings 1977). Hierro is the smallest major island of the Canaries and the most distant from the African mainland. Before the Spanish arrived in the fourteenth century, all the islands were occupied by an aboriginal people, the Guanches, whose ancestors probably arrived around 2,000 to 4,000 years ago (Schwidetzky 1976; Mercer 1980).

In Figure 7.1, we compare the extant lacertids to the fauna that presumably existed before the arrival of humans. We restrict our attention to the five relatively mesic western islands where fossil forms have been collected. Four of these islands originally were inhabited by two or three *Gallotia* species in the late Pleistocene and Holocene. There was a small species (*G. galloti*) sympatric with a larger species, (*G. simonyi* or *G. stehlini*) and/or a still larger *G. goliath* (Mertens 1942; Bravo 1953; Arnold 1973; Marrero Rodriguez and Garcia Cruz 1978; Hutterer 1985; López-Jurado 1985). The exception is Gran Canaria where no *G. galloti* exists today and none is evident in existing fossil deposits. At least on Tenerife, an even larger species, *G. maxima*, existed from probably the Pliocene to the early Pleistocene. It is still unclear whether the largest "species" *G. maxima* evolved into *G. goliath* or are in fact the same species (Izquierdo, Medina, and Hernandez 1989). In any event, this very large form seems to have disappeared before the islands were colonized by Guanches (Bravo 1953).

The other large *Gallotia* species in the Canary Islands became extinct or were extirpated more recently and these extinctions were contemporaneous with human colonization. Fossils of the now extirpated *G. simonyi* on Gomera have been found at one 500-year-old, pre-Hispanic site (Hutterer 1985). Elsewhere fossil lizards are found in association with abundant human artifacts (Böhme *et al.* 1981; Bings 1985). There are also a few historical references to the presence of gigantic lizards on Hierro, Gomera, and Gran Canaria (see review in Machado 1985b). We know that the early Canary Island aborigines hunted and ate lizards (Hooton 1925; Schwidetsky 1976; Bings 1985), but lizards were not a major portion of their diet. Indirect anthropogenic influences such as the introduction of rats, goats, pigs, and especially dogs by the aborigines may have been more important in the large lizards' eventual demise. Machado (1985b) attributes the exceptional survival of the large *G. stehlini* on Gran Canaria to the absence of any smaller-sized lacertid competitors in the face of introduced predators. He speculates that the tenuous survival of *G. simonyi* on Hierro might be due to the apparent absence of Guanche dogs.

On Madagascar, as with continents, known reptilian extinctions are rare. The extant herpetofauna is still incompletely known and new reptile species are still being discovered at a prodigious rate (Glaw and Vences 1992; Raxworthy and Nussbaum 1993a, 1993b). The sole documented extinctions out of 274 described indigenous species are two species of giant tortoise that are known only from subfossils and probably became extinct around the same time as the giant elephant birds, after the arrival of humans in about 500 A.D. (Dewar 1984). Some authorities, however, believe the giant tortoises actually went extinct before human contact (Paulian 1984). Three smaller species still survive on Madagascar, and other giant tortoises survived until European settlement in the Seychelles and still survive today on Aldabra (Arnold 1976). The absence of fresh water on Aldabra restricted permanent human settlement, a gratuitous benefit to the tortoises. In the nearby Mascarenes (Mauritius, Reunion, and Rodrigues), at least six species of large tortoises also became extinct shortly after human contact (Cheke 1987; Arnold 1980).

Giant tortoises also occurred on Sicily, Malta, and the Balearic Islands in the Mediterranean, but their time of extinction is unclear. It may have occurred earlier than the arrival of humans (Reese 1989). The giant lizard *Lacerta siculimelitensis* (220 mm sv) inhabited Malta and Sicily but became extinct sometime toward the end of the Pleistocene (Böhme and Zammit-Maempel 1982). It is not known if this lizard survived until the Holocene

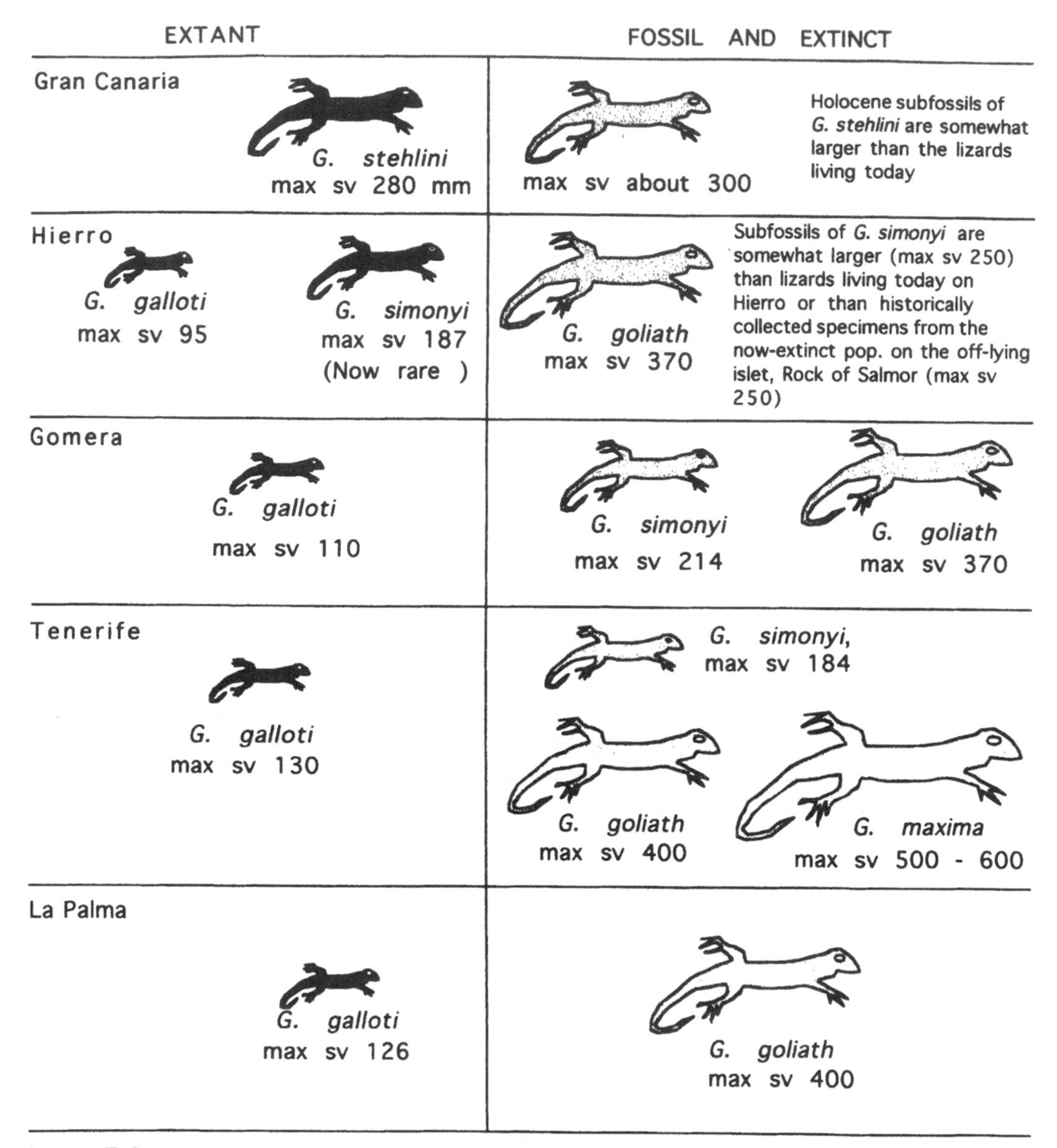

Figure 7.1
Body sizes (maximum snout-vent length, sv) of extant and extinct lizards of the genus Gallotia (Lacerta) *on the Canary Islands. Size data were gathered for extant species during field work by TJC in the Canaries during 1980 and were supplemented by measurements of preserved specimens at the British Museum and records in Machado (1985a) and López-Jurado (1989) for Hierro, and Thorpe (1985) for Lanzarote and Fuerteventura. Extinct species' maximum sizes are based on estimates in Mertens (1942), Bravo (1953), Marrero Rodriguez and Garcia Cruz (1978), Böhme* et al. *(1981), Hutterer (1985), Izquierdo, Medina, and Hernandez 1989, and López-Jurado (1985). The length of the lizards is drawn roughly to scale.*

This figure illustrates the preferential extinction of larger forms, probably due to human influences, and the downward size shifts of surviving large species on Hierro and Gran Canaria. The extinct and extant species were all contemporaneous in the late Pleistocene and Holocene with the possible of exception of G. maxima.

or if it was contemporaneous with humans. Recent excavations on nearby Cyprus, however, suggest an earlier date for human influence and a later date for the extinction of that island's megafauna than was previously thought (Reese 1989). Here the megafauna and humans are known to have been contemporaneous about 10,000 years ago.

On Tonga in the South Pacific, Pregill and Dye (1989) found subfossils about 2,000 years old of an extinct large iguanine in the genus *Brachylophus* on the island of Lifuka. The length of these lizards is estimated to be about twice that of the extant *Brachylophus* on nearby Fiji. The fossils are directly associated with human artifacts and bear distinctive marks that testify to their use as human food. Also extinct on Tonga is a very large skink (175 mm sv), *Eugongylus microlepus*, known only from two specimens collected in 1838 (Rinke 1986; Pregill 1993). In New Caledonia, fossils of now extinct large varanid lizards, giant meiolaniid turtles, and large terrestrial crocodiles (*Mekosuchus inexpectatus*) occur near or with sediments containing human artifacts (Gaffney, Balouet, and DeBroin 1984; Rich 1982; Gifford and Shutler 1956).

Islands with a colonial period but no aboriginal history. Here, unique island reptile species survived to be described as living species only too often to meet their demise shortly thereafter. For example, Rodrigues Island in the Indian Ocean experienced a period of intensive European settlement in the late seventeenth century. At that time large numbers of some spectacular endemic geckos were found. *Phelsuma edwardnewtonii* was a large diurnal species, bright green with blue spots. It was described as being so tame that it inhabited houses and would eat fruits from the owners' hands (Leguat 1708). However, the species was devastated, apparently by rats and cats, on the main island around the mid-nineteenth century. It survived for a short time on small outlying islets but finally disappeared from these too, as they became infested with rats. An even larger species, *P. gigas*, reaching nearly one half meter in total length vanished from the main island prior to the disappearance of *P. edwardnewtonii*. It, too, survived on uninhabited off-lying cays only to disappear later when rats were introduced (Vinson and Vinson 1969).

Many of the other endemic reptile species on Mauritius became extinct after conversion of habitat to agriculture, overgrazing, and the introduction of rats, cats, and other predators in the seventeenth century. The presence of these "missing species" was only confirmed from fossil deposits of quite recent age (Arnold 1980). A huge skink, *Leiolopisma mauritiana* (about 300 mm sv), inhabited Mauritius. This species is known only from subfossils, and the cause and chronology of its extinction is not understood. Some species survived on satellite islands that lie on the same island bank and were connected in times of lowered sea level. Most important in this respect is rat-free and uninhabited Round Island, where four species survive that have gone extinct on Mauritius. This includes the three largest lizards known from Mauritius (*Phelsuma guentheri*, *Leiolopisma telfairii*, and *Nactus sepensinsula*) and one snake (*Casarea dussumieri*), a member of a distinct group of primitive boas, the Bolyerinae. The only other species in the Bolyerinae, *Bolyeria multocarinata*, also may occur today on Round Island, although it has not been seen since 1975 (North, Bullock, and Dulloo 1994). Because Round was connected to Mauritius less than 12,000 years ago, *Bolyeria* probably became extinct on Mauritius as well, although confirming fossil evidence is so far lacking. The small skink, *Scelotes bojerii*, occurs on Round Island and surrounding islets but was known from Mauritius and Reunion in the last century and was once thought to be extinct on both islands. It was rediscovered, however, in the Macabé forest of Mauritius where it appears to be very rare (Vinson 1973; Arnold 1980).

Several of these reptiles were imperiled on Round Island because of recent habitat loss due to severe overgrazing by introduced rabbits and goats (North, Bullock, and Dulloo 1994). An eradication program was inititated and the last goat was shot in 1978 and rabbits were eradicated in 1986. Comparisons of lizard numbers between 1982 and 1989 showed many increases attributable to increases in habitat and a presumed increase in food availability. The indigenous snake *Caseria*, which feeds strictly on lizards, seems to have increased its abundance as well (North, Bullock, and Dulloo 1994).

The Cape Verde Islands off the west coast of Africa are the home of the second largest living skink, *Macroscinus coctei* (320 mm sv), the "endproduct" of a small adaptive radiation of *Mabuya* skinks dating back to the Cretaceous, when these islands were probably formed. The *Macroscinus* skinks have been described by residents as tame and easy to catch (Greer 1976). Perhaps this is why this giant species, prior to becoming completely extinct, was in this century restricted to two tiny (total area 10 km^2) uninhabited islands in the archipelago, Branco and Razo (Mertens 1956; Greer 1976). These islands were severed from the other larger islands on the bank about 10,000 years ago with rising sea levels. Thus, the skinks' absence from the larger adjacent islands suggests a recent extirpation. The islands were first colonized in the late fifteenth century by the Portuguese, who left no records of a broader range for this species. Interestingly, the geckos in the genus *Tarentola* are divided into three endemic species on the Cape Verdes (Joger 1984). The largest species, *T. borneensis* (max. sv 127 mm), inhabits the same two islands as *Macroscinus*, whereas the substantially smaller species (*T. capoverdianus* and *T. darwini*, max. sv 70–80 mm) occur on most of the remaining islands in the Cape Verdes (Mertens 1956; Greer 1976).

None of the endemic reptile species of the Galapagos have become extinct, but population densities have declined and island extirpations (and subspecies extinctions) have occurred in association with introduced cats, rats, dogs, and pigs (Honegger 1975, 1981; Steadman *et al.* 1991). The land iguana (*Conolophus subcristatus*) is extirpated on Baltra and James islands although the cause of extinction on James Island is not known; the species was abundant in Darwin's time but existed only as subfossils during the California Academy of Science expedition of 1905–1906. Although feral dogs are not now present, they were in the nineteenth century and perhaps drove the local extinction. Land iguanas on Santa Cruz Island were thought to have been exterminated by feral dogs before 1906, but small populations remained at Conway Bay, Cerro Colorado, and East Tortuga Bay. These populations persisted until they were heavily attacked by feral dogs in the 1970s. Today, only captive individuals remain. Giant tortoises (*Geochelone elephantopus*) have become extinct on Barrington and Floreana and are rare on all the other major islands except Isabela, Duncan, and Santa Cruz (Steadman 1986; Kramer 1984; Thorton 1971).

Islands with no permanent human settlement to date. Usually these islands are too small or too bleak and isolated to support human settlement (e.g., tiny islets in the Caribbean and Pacific, Malpelo Island off South America, most of the desert islands in the Sea of Cortez or off arid Australia, and polar islands). The high-latitude islands are too cold to support any reptiles. The desert islands are near shore and usually support few endemics, so local extirpations do not result in the extinction of a species. Yet these islands are extremely important for calibrating the magnitude of natural extinctions apart from the effects of human disturbance.

Richman, Case, and Schwaner (1988) examined data from some relatively undisturbed arid landbridge islands to estimate extinction rates of reptiles in the absence of human disturbance. Landbridge islands were formed as a consequence of rising sea levels at the end of the Pleistocene. They are convenient in this regard because one may estimate the rate of extinction for a particular taxon using: (a) information on the number of species in the taxon of interest on the island today; (b) an estimate of the number of species on the island at its time of isolation, as determined by counting the average number of species on the adjacent mainland today in a similar-sized area; and, (c) an estimate of the time elapsed since island isolation. These data are then fit to an *a priori* model that describes the dynamics or "relaxation" of species loss over time (Diamond 1972).

Richman, Case, and Schwaner (1988) estimated the relaxation rate for reptilian faunas of two landbridge island groups, one off Baja California and the other near South Australia. If relaxation rates on landbridge islands are to provide a valid estimate of a natural background rate of extinction, it is essential to evaluate the importance of human-related effects on these islands. The islands of Baja California are arid, extreme environments, and the establishment of human settlements or introduced animals have been severely limited as a result (Bahre 1983). Most of the islands of South Australia are similarly uninhabited, though a few of the largest islands have been settled for some time. However, in these instances initial surveys of the resident faunas began quite early. Thus, species lists for the islands used in this analysis have not been impoverished by anthropogenic extinctions.

Compared to other vertebrate taxa, reptiles present particular advantages for partialling out the contribution of extinction to observed relaxation rates. Terrestrial reptiles are generally poor overwater dispersers and thus rarely recolonize these islands subsequent to their isolation from the mainland. In addition, they are relatively resistant to extinction compared to warm-blooded vertebrates (Wilcox 1980; Case and Cody 1987), presumably because of their lower metabolic requirements and often higher population densities. Thus, the observed disparity between current island censuses and estimates of species number at the time of island isolation may be attributed largely to extinctions occurring in the absence of confounding immigration events.

Conclusions from these relaxation studies are as follows:

1. Even in the absence of much habitat disturbance or regional climatic change, a substantial number of extinctions may occur, and the rate of extinction declines with increasing area. Figure 7.2 shows a significant negative correlation between extinction risk as measured by the extinction rate parameter k_2 and increasing island area. The rate of species loss is empirically unlike radioactive decay, where there is a constant half-life independent of initial abundance. Instead, species loss in better described by nonlinear models of faunal relaxation. The constant k_2 is the rate parameter from an equation describing faunal relaxation, $dS/dt = -k_2S^2$. Greater extinction rates correspond to higher levels of k_2. This pattern is a general one. Case and Cody (1987) calculated extinction rates based on Baja island mammals and found the regression of k_2 with island area to have about the same slope (but higher overall magnitude) as that for reptiles.

In addition, experimental support for declining extinction rates with increasing island area comes from the study of Schoener and Schoener (1983) who introduced *Anolis sagrei* or *Leiocephalus* spp. onto 30 very small islands in the Bahamas having no lizards naturally. Small island populations quickly became extinct while larger island populations survived until the end of the study (five years) and even beyond (T. Schoener personal communication, 1990).

2. In spite of widely different faunas with little taxonomic overlap even at the family level, the two island groups (Australia and Baja California) display the same pattern and magnitude of extinction rate as a function of island area, i.e., the two regressions are not

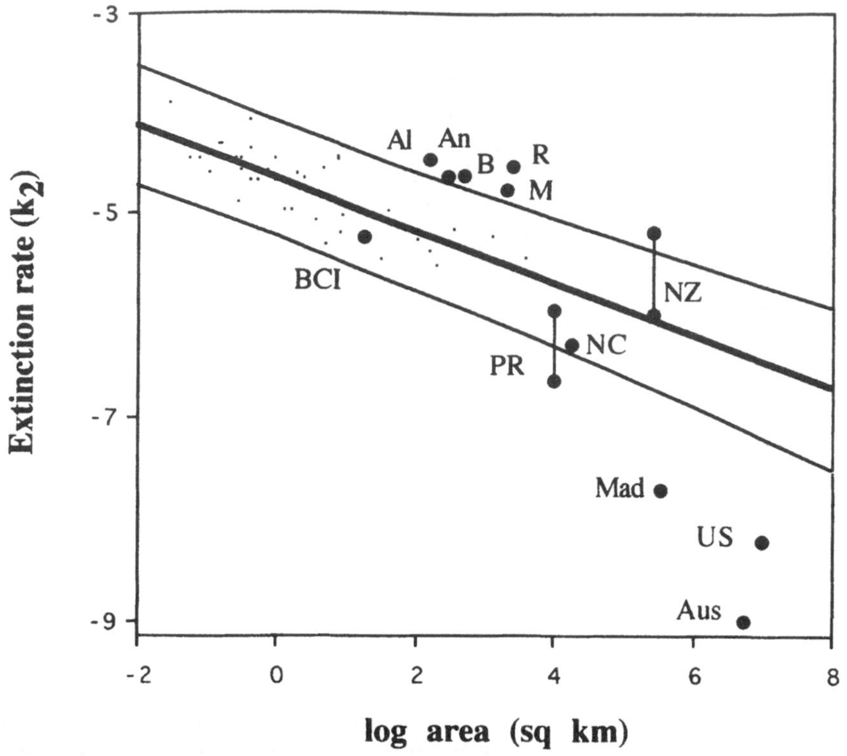

Figure 7.2

The extinction rate for recent reptile faunas (log of the relaxation parameter k^2) is plotted against the log of area for selected areas discussed in the text: An (Antigua), B (Barbuda), PR (Puerto Rico), US (continental United States). The small unlabeled points represent landbridge islands off Baja California and Australia. The regression line shown and its 95% prediction intervals are calculated for only these landbridge islands. Major reference sources are in Richman et al. *(1988) with additions and changes as follows:*

Place	Reference	Spp. now	Spp. extinct	Time (yr)
New Zealand (NZ) (two major islands)	Robb 1986; Bauer and Russel 1986; Worthy 1991; Patterson and Daugherty 1990	30	13	10,000
			8	1,000
Madagascar (Mad)	Blanc 1972, 1984; Glaw and Vences 1992	274	2	2,000
New Caledonia (NC)	Gaffney, Balouet, and DeBroin 1984; Sadlier 1986, 1988	42	3	3,000
Mauritius (M)	Cheke 1987; Tirvengadum and Bour 1985	4	9	10,000
Reunion (R)	Cheke 1987; Tirvengadum and Bour 1985	2	3	10,000
Aldabra (Al)	Gardner 1986; Arnold 1976	2	6	10,000
Australia (Aus)	Molnar 1984b; Gaffney 1992	28	6	20,000
Barro Colorado (BCI)	Myers and Rand 1969	66	2	77
Antigua	Pregill *et al.* 1994	6	5	3,500
Barbuda	Pregill *et al.* 1994	5	4	3,500
Puerto Rico	Pregill 1981	30	4	3,500
		30	4	17,000

(continued)

Figure 7.2. ***(continued)***

*The two extinction rates for New Zealand are based solely on the main two islands and do not include their satellites. The second calculation is based on the roughly 1,000-year period since the islands were first occupied by Maoris and includes complete extinctions (*Leiolopisma gracilocorpus, Cyclodina northlandi, *and* Hoplodactylus delcourti*) plus five extirpated species that can be identified and dated from major island subfossils yet still survive today on satellite islands. We assume here that only one species of tuatara occupied the two major islands. The first calculation includes the assumption that all the endemic island species on satellite landbridge islands also were present on the two main islands 10,000 years ago when these islands would have been connected to the main islands, and have since gone extinct on the major islands. In all cases, species whose incidence on an island was probably human-aided are excluded from the species counts. This excludes the following species for all these sites:* Hemidactylus frenatus, Hemidactylus mercatorius, Lepidodactylus lugubris, Gehyra mutilata, Cryptoblepharus boutonii, *and* Typhlops braminus *(see Case and Bolger 1991 and Darlington 1957).*

significantly different (test for coincidence of regression lines, $p < 0.5$). Similarly, Figure 7.2 plots the known reptilian extinctions for Barro Colorado Island (BCI). Since its inception about 80 years ago as a reserve in the Panama Canal, BCI has lost two reptile species (Myers and Rand 1969). This translates to about 3% of the initial fauna compared to 23% for bird species over the same time (Willis 1974). Where plotted in Figure 7.2 the point for BCI falls roughly on the regression line for the undisturbed arid islands. Since these islands all have relatively minimal human impact and little climatic change over the interval measured, we believe that these extinctions primarily represent "background" extinction rates in the absence of significant human intervention.

Worldwide, a few islands have both a reasonable historical record documenting environmental changes and abundant fossil records documenting past extinctions. With these data we are not inferring extinctions but actually have the "smoking gun" in subfossil form. Richman, Case, and Schwaner (1988) calculated the extinction rates for these islands assuming that the original species number is the present species number (minus all species introduced by humans) plus the number of extinct species (or forms) as determined by subfossil evidence. If anything, this gives a conservative estimate of extinction since undiscovered fossils may include new extinct forms. Of course, there are many islands with no known extinctions simply because geological conditions are not favorable for their deposition or discovery, or researchers have not yet looked.

Results are superimposed as points on Figure 7.2. It is apparent that the effects of disturbance are most telling on the smallest islands; the per species extinction rate for Antigua, Barbuda, Mauritius, and Reunion is approximately ten times that on islands of similar size in Australia or Baja California. It is impossible from these data alone, however, to disentangle the causative role of introduced predators, competitors, and the like from simple habitat destruction.

Significantly, the now-familiar trend of decreasing extinction risk with increasing area is preserved even for these disturbed areas, where extinction rates are calculated on the basis of known fossils. The observed elevation of the extinction parameter decreases with area, with no elevation in risk for the very large "islands" of Australia, the continental United States, and Madagascar. Indeed, these points lie far below the predicted extinction rate based on relatively undisturbed landbridge islands. The low number of extinct reptiles recorded in the continental U.S. contrasts with the large numbers of Holocene extinctions of reptiles in the nearby West Indies (Etheridge 1964; Pregill 1981, 1992; Pregill, Steadman, and Watters 1994). Although data for Madagascar are probably much less complete because of its larger area, fewer fossil digs, and a reptilian fauna that is still incompletely documented, the calculated extinction rate, 0.4% (or two species in 2,000 years) yields an extinction rate roughly similar to that of the continental U.S. and substantially lower than that for the next smaller islands of New Zealand and New Caledonia. This low extinction rate for reptiles is all the more surprising given the tremendous amount of habitat destruction on Madagascar; approximately 80–90% of the original vegetation has been cleared (Jolly, Oberlé, and Albignac 1984) and along with other human impacts has resulted in the extinction of at least 13 of the 75 native mammal species (approximately 17%; Jolly, Oberlé, and Albignac 1984).

For completeness, it is desirable to compare extinctions on the continents of Africa, Eurasia, and South America with those from North America and Australia, but unfortunately the fossil record for these areas is not well known for the Holocene and late Pleistocene. Based on present knowledge, the record for South America is not unlike that of North America in that practically all fossils known for the past 10,000 years are referable to extant taxa (Baez and Gasparini 1979).

It probably is the case that many undocumented large-island and continental extinctions simply await discovery, but we see no reason that there should be any particular bias against fossil discovery on mainlands compared to islands. Moreover, abundant evidence for mammalian extinctions on continents occurs worldwide over this same time period and from the same deposits. The greater number of mammalian extinctions is probably at least partly due to overhunting by Pleistocene/Holocene humans (Martin and Klein 1984). Certainly, it is often difficult to distinguish taxa at the level of species from fossil material alone, but this problem befalls islands as well as mainlands.

Conventional explanations for island extinctions emphasize the extreme vulnerability of native island species to introduced predators, and this effect cannot be denied (see Predation section). Mainlands and large islands have endemic predators with which the fauna has presumably coevolved. The prey have probably evolved better defenses and the predators' populations are in turn kept in check by higher-order predators and parasites. Perhaps equally important, low extinction rates are expected given large area and thus increased opportunity for immigration after local extirpation.

3. Reptiles have lower rates of extinction than birds or mammals. Case and Cody (1987) demonstrated that for the same islands in the Sea of Cortez, mammals have extinction rates about an order of magnitude higher than those for reptiles. Additionally, Schoener (1983), based on a wide review of the literature found that species turnover rates for reptiles generally fall below those of birds and mammals and most arthropod systems. Lizard populations should be expected to be more resistant to extinction because their lower metabolic rate should allow higher densities than either birds or mammals and thus larger population sizes.

We have made a tally of reptile extinctions over the last 10,000 years, both historic and prehistoric. This count is only an approximation because of the previously mentioned paucity of Pleistocene fossils from Asia, South America, and mainland Africa and also because of the difficulty in deciding whether fossils finds in certain taxa such as turtles and tortoises represent one or several extinct species. This minimum estimate is likely to be revised upwards in the future but presently there is evidence for 63 Holocene (late Pleistocene/Holocene for Australia) species extinctions, eight (or 13%) of which are continental species. There are 23 reptile species that have become extinct since 1600 (not counting subspecies and recently rediscovered species like *Tiliqua adelaidensis* and *Dromicus ornatus*; Honegger 1981; Henderson 1992; Groombridge 1992); seven of these species are giant tortoises from islands in the Indian Ocean and for which the taxonomy may be confused. None of these historical extinctions are continental species. Adding about eight subspecies extinctions (Honnegger 1981) produces 31 reptilian extinctions compared to 115 mammalian (64% continental) and 171 avian extinctions (10% continental) since 1600 (Diamond 1984). We conclude cautiously, given the caveats above that Recent reptile extinctions are primarily on islands and that reptiles have been less extinction prone during the Holocene than birds and mammals. [Note that the lower number of extinct reptiles relative to birds and mammals is not simply due to a lower number of species available for extinction. There are approximately 6,550 reptile, 9,672 bird and 4,327 described mammal species (Groombridge 1992). We do not have accurate tallies for each of these groups as to the number of species exclusively on islands compared to mainlands, but there are certainly more island reptile species than island mammals and we suspect that the number of bird species on islands is roughly comparable or greater than that of reptiles.]

Terrestrial mollusks are one of the few invertebrate taxa for which some compilation of extinctions has been attempted, although not comprehensively. In Hawaii alone, over 20 terrestrial snail species have gone extinct (Hadfield, Miller, and Carwile 1989). This vulnerability is probably caused by the limited geographical range of many of the endemic species and again the introduction of exotic species, chiefly rats, carnivorous snails, and ants (Cowie 1992).

PREDATION

One of the most important factors influencing lizard abundance on islands is the number and variety of predators. On predator-free islands, lizards can achieve extremely high densities. For example, on small rat-free islands off New Zealand densities reach 1,390 lizards per acre or nearly one lizard every 3 m^2 (Crook 1973; Whitaker 1968, 1973). In other parts of the world, one finds this pattern repeated. Up to 2,074 diurnal lizards per acre have been reported for rat-free Cousin Island in the Seychelles (Brooke and Houston 1983), and 1,214 per acre for San Pedro Martir in the Sea of Cortez, Mexico (Wilcox 1981; Case unpublished data).

That predation can have a large impact on lizard densities is tested by the introduction of lizard predators to some islands and not to others. Although strictly illegal in most places today, this "experiment" was conducted historically many times with rats, cats, dogs, and mongooses. The mongoose is one of the most potent predators on diurnal ground-foraging lizards. Mongooses have been introduced to various islands around the world with the hope of controlling rats and other vertebrate pests. Although their success in this regard has been mixed, their impact on native reptiles (as well as bird populations), particularly ground-foraging forms, like skinks, teiids, lacertids and snakes, has been devastating. In Puerto Rico, reptiles and insects, not rats, form the bulk of the mongoose diet (Pimentel 1955).

One of us (TJC) attempted to quantify the impact of the mongoose on diurnal lizard abundance on islands in the South Pacific by censusing lizards on islands with and without the mongoose (Case and Bolger 1991). Lizards were counted along two to three transects per island of about 1 km each. There is nearly a 100-fold increase in diurnal lizard abundance on islands without mongooses compared to islands with mongooses (Figure 7.3).

The same qualitative pattern is evident in the West Indies. Nearly 50 years after the

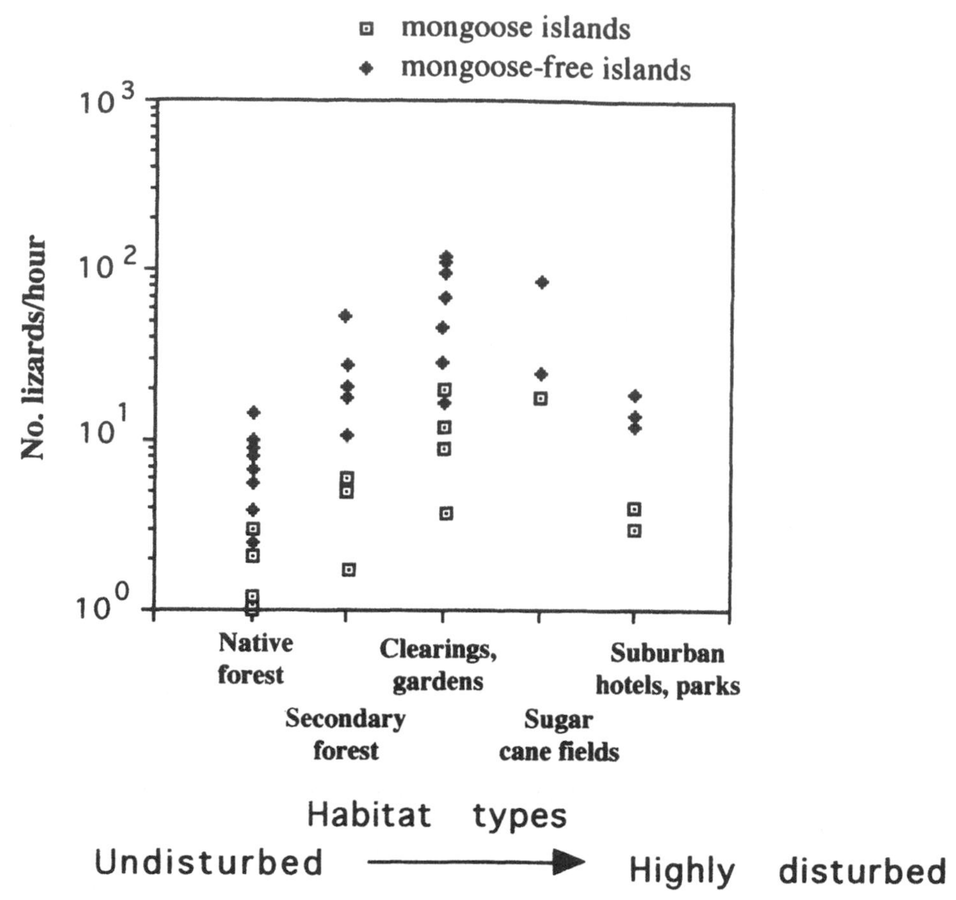

Figure 7.3
Crude lizard censuses (expressed as the average number of diurnal lizards seen per hour; also see Case 1975) in natural and human-modified habitats on mongoose-inhabited and mongoose-free islands in the tropical Pacific. All censuses were conducted during sunny days from 1984–1988 by TJC. No attempt was made to capture any of the lizards so that a constant search speed could be maintained. Nearly all lizards seen were skinks and include both native and introduced species. Each point represents the average of two to four censuses. The islands and the number of habitats censused are as follows. Mongoose-inhabited islands in Hawaii: Hawaii (2), Oahu (2), Molokai (2), and Maui (1). Fiji: Vite Levu (5), Rabi (2), and Vanua Levu (1). Mongoose-free islands are New Caledonia (3), Kauai (3), Efaté (2), Espiritu Santo (2), Tahiti (3), Moorea (1), Roratonga (1), and Atiu (2); in Fiji; Kadavu (4), Taveuni (3), and Ovalau (1).

introduction of the mongoose to Jamaica, Barbour (1910; p. 273) noticed the "almost complete extinction of many species which were once abundant . . . true ground inhabiting forms have, of course, suffered most . . . snakes have perhaps suffered more than lizards." This effect on lizard abundance is also seen today on small cays off Jamaica and elsewhere in the West Indies. Where the mongoose is absent, terrestrial lizards and snakes are much more common (Barbour 1930; Schmidt 1928; Pregill 1986; Mittermaier 1972; Henderson 1992).

The actual role of mongoose in causing extinctions and extirpations relative to other exotic predators and habitat destruction in the West Indies is difficult to sort out because of the vagueness of historical records, particularly regarding largely cryptic species. Henderson (1992) documents at least 26 instances of extirpation in the West Indies since about 1830. At least 12 of these occurred on islands before the introduction of mongoose or on islands still free of mongoose (although in both cases other exotic predators are or were present). That leaves about 14 reptiles, all ground-dwelling forms, that became locally extinct shortly after the introduction of mongoose.

Interestingly none of these cases occur on the Greater Antilles, where mongoose was introduced at a relatively early date. This pattern suggests that large island area along with the fact that mongoose densities appears to be higher on smaller islands buffers the Greater Antilles reptiles from extinction relative to smaller islands (also see Conclusions and Summary).

An example of the pattern on small islands is illustrated by St. Lucia in the Lesser Antilles. Three reptile species have been extirpated in historical time coincident with the introduction of mongoose: one skink (*Mabuya mabuya*) and two colubrid snakes (*Clelia clelia* and *Liophis ornatus*; Corke 1987). *Liophis ornatus* once thought to be extinct survives only on the tiny offshore island of Maria Major along with the ground-foraging lizard, *Cnemidophorus vanzoi*, which is curiously absent from St. Lucia. It seems unlikely that this lizard was not once present on St. Lucia at an earlier time in that the islands are so close, yet no specimens were ever deposited in museum collections. Similarly, the colubrid snake, *Alsophis antillensis*, once occupied Barbuda and Antigua but today can be found only on mongoose-free offshore cays (Pregill *et al.* 1988).

The last major experiment with rat control by mongoose introduction took place in Mauritius around 1900, after most of the large endemic species had already become extinct. Today, the only surviving ground-dwelling species on Mauritius, the skink *Scelotes bojerii*, is extremely rare and until recently, was thought to be extinct. The other surviving species are relatively common but are arboreal (three endemic *Phelsuma* day geckos) or are widely distributed non-endemic species (the skink *Cryptoblepharus boutonii* and the house gecko *Hemidactylus frenatus*) of continental origin whose introductions here and elsewhere have been man-aided (Cheke 1984, 1987).

Domestic cats and dogs also have had devastating effects on island species. We have already mentioned the role of dogs in the local extinction of land iguanas in the Galapagos. Dogs have also reduced populations of marine iguanas as well, but to date no extirpations are known. Because they are more arboreal than dogs or mongoose, feral cats and tree rats (*Rattus rattus*) affect prey species that the mongooses and dogs are less likely to capture. Gibbons and Watkins (1982) suggest that cats may have been even more damaging than mongooses to highly arboreal Fijian lizards and in particular to the now rare endemic Fijian iguanas. Today substantial populations can be found only on small islands lacking both mongooses and cats. The combination of cats and mongooses on the two largest islands of Fiji, Viti Levu and Vanua Levu, has apparently resulted in the local extinction of the ground foraging skinks *Emoia nigra* and *E. trossula*. These are the two largest skinks in Fiji and they have not been seen on these islands in over 100 years (Zug 1992), although they survive quite well on mongoose-free islands in the archipelago (e.g., Ovalau, Rotuma, and Taveuni). Interestingly, nearly all these islands have rats (*Rattus rattus* and *R. exulans*). The ground-nesting banded rails (*Rallus phillippensis*) and ground doves also are absent from the mongoose-inhabited islands and are presumed to be extinct (Gibbons 1984).

The large herbivorous iguanine *Cyclura carinata* was nearly extirpated on Pine Cay in the Caicos islands (West Indies) during the three years following construction of a hotel and tourist facility (Iverson 1978). Predation by cats and dogs introduced during the hotel construction resulted in the decline from about 5,500 adults to approximately five. Iverson also presents evidence suggesting that population declines of *Cyclura* elsewhere in the Turks and Caicos Banks stems directly from cat and dog predation.

Thomson (1922) noted that New Zealand lizards became much less common after the mid-nineteenth century and he attributed this decline to loss of cover and predation by cats. Today reptile density and species numbers are almost invariably higher on predator-free islands than on mainland New Zealand or islands with exotic mammalian predators (Whitaker 1982; Carmichael, Gillingham, Keall 1989). Off-lying islands with *Rattus exulans*, the Polynesian rat, support smaller populations of lizards and tuataras than do islands without rats (Crook 1973, Whitaker 1973). The only exceptions occur in relatively predator-proof habitat at some mainland sites, such as deep boulder banks where local lizard densities may exceed one per m^2 (Whitaker 1982; Towns 1972).

If introduced predators reduce reptile densities low enough, extinction follows, particularly on smaller islands. The role of predators in causing many of the extinctions documented here is circumstantial but voluminous. Tuataras are the last remaining representative of a widespread Mesozoic order of reptiles known as the Ryncocephalia. Today it is found on uninhabited landbridge islands off New Zealand, but subfossils, less than a thousand years old, are found on both of the main islands (Cassels 1984). In all, ten species of lizards (about one third of the New Zealand lizard fauna), in addition to the tuatara, are restricted to small off-lying islands formerly connected to the main islands (Robb 1986; Newman 1982, pp. 303–7; Cassels 1984). Predation by exotic animals, predominantly rats, is thought to be responsible for this pattern of extinctions. Whitaker (1973) found that small islands off New Zealand with Polynesian rat have fewer lizard species (all natives) for their size than islands without rats. McCallum (1986) documents the changes to the herpetofauna following the colonization of Lizard Island by the Polynesian rat in 1977. Two lizard species appeared to go locally extinct, and overall lizard densities dropped by at least one order of magnitude. Norway rats colonized Whenuakura Island in 1983–84, and by 1985 the previously thriving tuatara population had disappeared, as had nearly all the lizards.

The pattern of endemic lizards being restricted or at least much more common on smaller rat-free islands off-lying larger rat-infested islands is repeated in the Mascarenes (Vinson and Vinson 1969), Seychelles (Gardner 1986), Canary Islands (Klemmer 1976), Cape Verdes (Greer 1976), Norfolk Islands (Cogger, Sadlier, and Cameron 1983) and Lord Howe Islands (Cogger 1971). On most of these islands, rats arrived so early historically that we do not have adequate pre-rat reptile records or census data. In the case of Lord Howe Island, however, the numerical decline of the only two native lizards a gecko, *Phyllodactylus guentheri*, and a skink, *Leiolopisma lichenigerum*, on the main island seems to have occurred after the arrival of rats in 1918 (Cogger 1971).

In the Seychelles, all populations of the largest extant skink, *Mabuya wrightii*, are on rat-free islands that usually also have nesting seabirds (Cheke 1984; Gardner 1986). When Lantz visited Marianne in 1877, both seabird colonies and *M. wrightii* were present while rats were not found (Cheke 1984). Subsequently rats were introduced and today neither *M. wrightii* or breeding seabirds are present.

Mammals are not the only taxa implicated in causing reptile extinctions or extirpations. The introduced brown tree snake, *Boiga irregularis*, which has become infamous for decimating populations of endemic birds on Guam (Savidge 1988; Engbring and Fritts 1988), has also severely impacted the lizard fauna there. Juvenile snakes prey predominantly on lizards and are suspected of being a major factor in the possible extirpation of three species of skinks and two geckos (Savidge 1988; Rodda and Fritts 1992). They have also apparently reduced the numbers of forest populations of some other geckos (*Gehyra oceanica*, *G. mutilata*, and *Lepidodactylus lugubris*), species that are usually abundant in these habitats in the absence of the tree snake (Case, Bolger, and Petren 1994). The brown tree snake originates from New Guinea, the Solomons, northern Australia, and Indonesia where it is not particularly common and where it coexists with a rich land bird and reptilian fauna. For example, McCoy (1980) describes it as uncommon in the Solomons. The havoc that it is causing on Guam could stem from its high densities due to release from its own predators, prey naïveté or some degree of both.

Most human-introduced predators were brought to islands shortly after they were colonized, either by aborigines or Europeans. Pregill (1986) correlated the settlement time of islands with the extinction times of a number of insular reptiles. The overall picture is quite convincing. The arrival of humans on an island is closely associated with increased reptile extinction, especially of large endemic species. The inference is that habitat destruction and predation by humans and/or their entourage of introduced animals is responsible for these extinctions.

Another feature of these extinctions and extirpations is that they seem to occur most often within the endemic component of the fauna and on islands with high levels of endemism. The fossil species going extinct on the islands in Figure 7.2 are typically forms that are endemic at the species or genus level on the island or archipelago. This trend is explored further in a set of seven islands in Table 7.1 that have both endemic and nonendemic species as well as good fossil evidence for extinction. For each island the proportion of species becoming locally extinct on the island is broken down into three categories: (1) species endemic to the island (or its immediate satellites); (2) species that are not endemic; and, (3) species that are endemic to the island group (e.g., a species endemic to Aldabra and the Comoros would be counted in category 3 but not in category 1 or 2). Endemic species have significantly higher extinction rates than nonendemics ($p = 0.011$ for comparison of group (1) to (2) and $p = 0.012$ for group (1) to (3); Mann-Whitney U tests).

Endemic species have been isolated on islands lacking mammalian predators for long periods and have presumably become relatively defenseless to introduced predators. Few attempts have been made, however, to quantify this supposition, although Shallenberger (1970) measured the flushing distance of insular and mainland iguanid lizards and found that a human can get up to ten times closer to several insular varieties.

A similar behavioral naïveté, this time in birds, is apparent when one compares the effects of introduced predators on islands that previously had no similar predators to those that did. On Hawaii, Midway, Lord Howe, New Zealand, and other islands, introduced rats have led to the extinction of many native bird species. Yet on others, like Fiji, Tonga, Samoa, Marquesas, Rennell, the Solomons, Aldabra, Christmas (Indian Ocean), and the Galapagos, the introduction of rats was not accompanied by a wave of avian extinctions. Atkinson (1985) points out that the extinction-resistant islands all have native rats or land crabs. Atkinson argues that land crabs fill an ecological niche very similar to that of the rat. Birds on predator-free islands are easy prey for the rats, whereas on other islands birds have presumably acquired a more effective predator-avoidance behavior.

Rats are similarly implicated in reptile extinctions or extirpations on many of the same islands as for birds (except, of course, those where no native reptiles occur). Rats have had the greatest effect in terms of lizard densities and numbers of extinctions in

Table 7.1

Terrestrial reptile extinctions on selected islands for endemic and non-endemic species over roughly the last 10,000 years.

The table's entries give the number of species in one of three categories becoming locally extinct on the island during the Holocene divided by the number of species in that category that were initially present on the island. Human introduced species are excluded. Species are grouped into three categories: endemic to the island or its satellites, endemic to the region, or non-endemic to the island or region. Species that are "endemic to the island" occur only on the island in question (and possibly its satellites. Species that are "endemic to the island or region" include species confined to the island plus species endemic to the island and nearby islands or archipelagoes. The total number of species in each category on the island appears in parentheses. For example, in Recent times, Antigua had a total of seven species that were not endemic to the island or nearby islands. One of those species became extinct. Aldabra is excluded because of confusion regarding the species status (and thus endemism) of its extinct forms (Arnold 1976). Overall, extinction rates are significantly higher in the endemic species component of the fauna compared to the non-endemic.

	Extinction Probabilities		
Island	**Non Endemic**	**Endemic**	**Endemic to Island or Region**
Tenerife	— (0)	1.00 (1)[a]	0.33 (6)[a]
Mauritius	— (0)	0.88 (8)	0.69 (13)
Reunion	0.00 (2)	1.00 (1)	1.00 (3)
Rodrigues	— (0)	1.00 (4)	1.00 (4)
Puerto Rico	0.00 (4)	0.13 (30)	0.13 (30)
Antigua	0.14 (7)[b]	1.00 (1)[b]	1.00 (4)
Jamaica	0.00 (6)	0.18 (28)	0.18 (28)
Means	0.035 (4)	0.74 (7)	0.62 (7)

[a]This includes *Gallotia maxima,* so far only known from Tenerife, although it is not clear if it is distinct from *Gallotia goliath.* We follow the taxonomy in Machado *et al.* (1985) for *Gallotia* and Joger (1984) for *Tarentola Hemidactylus turcicus* is considered as introduced.

[b]The extinct non-endemic on Antigua is *Boa constrictor* and the extinct endemic is the Boidae sp. (Pregill *et al.* 1994). *Hemidactylus mabouia* is considered as introduced.

the Mascarenes, Seychelles, Lord Howe, Norfolk, and New Zealand but interestingly seem to have had little effect in most of the Central Pacific, for example, Fiji, New Caledonia, Tonga, Samoa, the Solomons, the Galapagos, and Australia.

Why should the reptile fauna in these different places exhibit such varying susceptibilities to rat introductions? This question needs further study. Invading rats freed from their continental predators and parasites can reach high densities on islands and often invade forest habitats; in more continental faunas they are nearly restricted to man-modified habitats. This factor however does not readily explain the apparent differences between islands lacking rat predators. One likely factor is the high frequency of introduced reptile species on islands where rats have not had a big impact. Introduced reptiles generally come from mainland areas where they have had a long coevolutionary history with predators. In Fiji, for example, 44% (11/25) of its present-day reptile fauna is probably introduced (Case and Bolger 1991; Zug 1992; Zug and Ineich 1995) and rats have had little apparent effect (Case and Bolger 1991). Contrast this with New Zealand, with only one introduced reptile (*Lampropholis delicata* from Australia), and a reptile fauna severely impacted by rat introduction. Introduced mongoose has played a large role in the Central Pacific in affecting overall lizard densities but it has not greatly affected the number of introduced reptile species on islands. After partialling out differences between islands in their area and maximum elevation, islands in the mid-Pacific with mongooses do not have significantly fewer species of introduced reptiles than mongoose-free islands (Case and Bolger 1991). (A similar analysis cannot be done for the native species because about half the islands have none at all.)

In the Seychelles and the Comoros, rat-free islands have approximately equal numbers of introduced lizard species as do rat-infested islands, although densities may be very different (Cheke 1984; Evans and Evans 1980; Brooke and Houston 1983). Because nearly all introduced species come originally from predator-rich continental areas, they may be less susceptible to introduced predators than the sympatric endemic predator-naive species.

Another contributing factor to the apparent vulnerability of island endemics could be the lack of recolonization sources. When a population of nonendemics or regional endemics becomes locally extinct on an island, the island can potentially be recolonized from individuals still surviving on other nearby islands. For a single-island endemic, however, extirpation and extinction are synonymous.

COMPETITION FROM AND AMONG INTRODUCED SPECIES

Unlike birds, lizards have not been able to colonize on their own the remotest islands of the world, such as those in the mid-Pacific (e.g., Hawaii and the Marquesas). For the most part, reptiles reached these islands when the Polynesians and Melanesians inadvertently began spreading a set of geckos and skinks throughout much of the Pacific about 4,000 years ago (Dye and Steadman 1990; Case and Bolger 1991; Pregill 1993). Additions to this set of aboriginal introductions have occurred more recently during European settlement. The reptilian faunas of somewhat less-isolated islands (e.g., Guam, Fiji, Vanuatu) today are a mixture of native and introduced species. These introductions, although unconscious, poorly documented, and not as well controlled as a manipulative experiment, can be used to partially sort out competitive relationships among species because of the huge sample sizes involved (i.e., literally hundreds of island and mainland locations).

Case and Bolger (1991) reviewed this literature for reptiles and found no documented case in which a native reptile species is reduced to extinction by the introduction of a reptilian competitor. We are aware of only one example where an introduced species seemed to numerically supplant a native species. South Florida has only two native anoles, *Anolis carolinensis* and *A. distichus*. In recent years it has been a beachhead for at least six introduced anoles from the more anole-rich Greater Antilles. Most of these introductions are still highly localized in urban areas, but *Anolis sagrei* is successfully displacing the native *A. carolinensis* as the most common anole in urban areas, penetrating agricultural and even native habitats (Wilson and Porras 1983; Salzburg 1984).

Case and Bolger (1991) also found evidence that native species-rich faunas seem to resist invasion by exotics, and that the densities of resident introduced species may decline after the introduction of new competing species. The mechanisms behind both these effects, however, are not well understood.

A striking example of historical competitive displacement seems likely among introduced lizards in Hawaii. Until aproximately 1940 one of the most common skinks in Hawaii was *Emoia impar*. This species was recently taxonomically separated from the very similar *Emoia cyanura* (Ineich and Zug 1991; Bruna, Fisher, and Case 1995). Museum collections reveal that *E. impar* was far more common in Hawaii than *E. cyanura*. *Emoia impar* was probably a Polynesian introduction as evidenced by an analysis of its mt-DNA that reveals it to be identical in cytochrome b sequence to *E. impar* elsewhere in Polynesia (Bruna, Fisher, and Case 1996). *Emoia impar* is still one the most common skinks in Fiji, Samoa, the Marquesas, and nearly everywhere else in the eastern Pacific where it occurs (Oliver and Shaw 1953; McKeown 1978; Jones 1979). It is also common in subfossil deposits on Hawaii from the Polynesian period (G.K. Pregill personal communiation). This pattern changed after *Lampropholis delicata* was accidentally introduced to Hawaii from southeast Australia (McKeown 1978; Case and Bolger 1991). Today *E. impar* is rare, whereas *Lampropholis* is the most frequently seen ground-dwelling skink on the islands. The only other place where *Lampropholis delicata* was introduced in the Pacific is New Zealand where it seems to thrive in suburban settings on the North Island (Robb 1986). On the many "control" islands in the tropical Pacific lacking *Lampropholis delicata*, *E. cyanura* and/or *E. impar* are still very common.

Another example of competitive affects has also occurred in Hawaii. After World War II, a new gecko appeared in Hawaii, the common house gecko *Hemidactylus frenatus*, native to Asia and the Indo-Pacific. It subsequently increased in numbers in urban/suburban habitats, while three other Polynesian-introduced geckos, the fox or Polynesian gecko (*Hemidactylus garnotti*) the mourning gecko, (*Lepidodactylus lugubris*), and the stump-toed gecko (*Gehyra mutilata*), formerly occupying this niche, became scarce in these habitats (Oliver and Shaw 1953). Today, the most common association on lighted building walls is the house gecko, alone or sometimes in association with the smaller and typically less abundant mourning gecko (Table 7.2). The pattern is complicated by two additional factors, however. The house gecko has spread beyond Oahu to other Hawaiian islands but this spread has been recent and the situation is still dynamic. Secondly, based on our studies in progress in Hawaii and Fiji, it is apparent that climatic factors also impinge on the competitive interaction between house geckos and mourning geckos (Case, Bolger, and Petren 1994). The competitive displacement goes more slowly in the more mesic habitats on the windward sides of islands. Today one can still find substantial numbers of mourning geckos on building walls in Hilo (on Hawaii) for example, although they are typically far outnumbered by house geckos. This same leeward/windward difference is also evident on islands in Fiji.

Frogner (1967) found that the house gecko could displace the mourning gecko from favored shelter sites in laboratory experiments and that they would eat juvenile *Lepidodactylus*. The reverse is not true, however, in that hatchling house geckos are larger than the largest prey taken by *Lepidodactylus* in the field. Laboratory experiments have shown that *H. frenatus* is behaviorally dominant to both the smaller *L. lugubris* and the equivalently sized *H. garnotii* (Bolger and Case 1992). However replicated experiments under more natural conditions reveal that the competitive superiority of *H. frenatus* is attributable to its greater foraging success, particularly in urbanized conditions (Petren, Bolger, and Case 1993; Petren and Case 1996).

Elsewhere in the Pacific where the house gecko has yet to invade, for example, most of the Societies, Tuamotus, and Marquesas, most of the Cooks, and many islands of Fiji, *G. mutilata* or *G. oceanica* with *Lepidodactylus lugubris*, and/or *Hemidactylus garnotii* have remained dominant in the "human building" niche (Table 7.2). This appears to be rapidly changing, however, on the main Fijian island, Viti Levu. Although unrecorded until recently, the house gecko has been in the Nadi area on the west for at least 20 years (Pernetta and Watling 1979; D. Watling personal communication) and now is the only gecko common in towns along the west. It appeared in the major port city of Suva on

Table 7.2
Gecko introductions and faunal affects in the guild of gecko species occupying human structures (a,b).

Archipelago	Island	Historical house geckos	Recent invaders (a)	House geckos today (b)
Fiji	w. Viti Levu	GO LL (c)	HF in 1960 (d)	HF:LL 30:1 (n=90)
	e. Viti Levu	GO, LL (c)	HF in 1982 (e)	HF:LL:GO 30:20:1 (n=several hundred)
	Other islands	GO, LL (c)	none	LL:GO 7:3 (n=several hundred)
Western Samoa	Upolu	GO, LL, GM (f)	HF to Upolu in 1960's (g)	Upolu—HF dominant. Other islands—LL and GO (g)
	Savai'i	GO, GM, LL (f)	None (except local invasion of HF at Saleloga Wharf)	LL:GO:GM 10:10:1 (n=41)
American Samoa		GO, LL (f)	HF to Tuitilla mid-1960's (h)	Tuitilla—HF dominant Other islands—LL, GO, GM (h)
Vanuatu	Espiritu Santo	GO, LL (i)	HF post 1971	HF:LL 17:3 (n=99)
	Efate (Port Vila)	GO, LL (i)	HF post 1971	HF:LL:GO 85:15:1 (n=99)
	north Efate	GO, LL	none	LL:GO 2:1 (n=32)
	Emao	GO, LL	none	LL:GO 1:1 (n=33)
Hawaii	Oahu	GM, LL, HG (j)	HF 1951 (k)	HF:LL:GM:HG 30:1:0:0 (n=31)
	e. Hawaii	GM, LL, HG (j)	HF post 1965 (l)	HF:LL:GM:HG 10:1:0:0 (n=37)
	w. Hawaii	GM, LL, HG (j)	HF post 1965 (l)	HF:LL:GM:HG 2:1:0:0 (n=72)
Societies	Tahiti, Papeete port area	GO, GM, HG, LL (m)	mid 1980's	HF:LL:GO 30:1:2 (n=36)
	Tahiti, elsewhere	GO, GM, HG, LL (m)		LL:GO:GM 27:12:1 (n=40)
	Moorea	GO, GM, HG, LL (m)		LL:GO:GM 12:9:1 (n=22)
Mainland Mexico (San Blas and Mazatlan)		GM (n)	HF post 1963 (n)	HF dominant (n)

(a) In most cases the exact date of the invasion is unknown: date given is the date of the last survey that did not find the invader.
(b) Except where noted these are personal observations by the authors. GM = *Gehyra mutilata*; GO = *Gehyra oceanica*; LL = *Lepidodactylus lugubris*; HF = *Hemidactylus frenatus*; HG = *Hemidactylus garnotii*
(c) Pernetta and Watling, 1979
(d) Watling personal communication, 1986
(e) Gibbons personal communication, 1984
(f) Burt and Burt 1932
(g) Zug 1994
(h) Amerson *et al.* 1982
(i) Medway and Marshall 1975
(j) Stejneger 1899
(k) Hunsaker and Breese 1967
(l) Jones 1979
(m) I. Ineich 1987
(n) N. Scott 1987

the southeast windward side in about 1983 and already has become the most frequent gecko on walls at the University of the Pacific in Suva with the concomitant decline of the previous resident geckos on the same walls (Bolger 1991, Case, Bolger, and Petren 1994). Today the area around Suva is a mosaic with H. frenatus already dominant in some areas but absent in others, and instead the other geckos are found in high numbers. In areas where *H. frenatus* is present but not common, its numbers have been increasing (Bolger 1991).

The house gecko fauna also changed rapidly in Vanuatu (New Hebrides). In 1971, the Royal Society did not find a single *H. frenatus* in Vanuatu (Medway and Marshall 1975). Today it is virtually the only urban gecko seen in the major city of Port Vila on Efate (although it is still restricted to the Port Vila area on Efate) and is by far the most common gecko in the town of Santo on Espiritu Santo (Table 7.2). Despite much recent work on the geckos of the Society Islands, Ineich (1987) and Ineich and Blanc (1988) did not find any *Hemidactylus frenatus*. In 1989 we recorded the presence of this species on Tahiti for the first time. It was restricted to the wharf area of Papeete where it had become the most common gecko on buildings. A second census of Papeete was performed in 1991; *H. frenatus* had now spread about ten km beyond where we found it localized in 1989 (Bolger and Case personal observation). Even more recently, our co-workers found *H. frenatus* in downtown Avarua (Rarotonga, Cook Islands) in 1994 (E. Bruna and R. Fisher personal observation; specimens at the California Academy of Science).

Bermuda has no native lizards other than a single endemic skink. Wingate (1965) documents the introduction of *Anolis grahami* from Jamaica in 1905. After about 35 years the lizard had spread throughout the main island. Sometime around the early 1940s a second larger and ecologically dissimilar anole (*Anolis leachi*) was introduced from the Barbuda Bank in the Lesser Antilles; the exact circumstances are unknown. The rate of spread of this second species was considerably slower than that of *A. grahami*, but today the range of *A. leachi* is encompassed within that of *A. grahami*. *Anolis leachi* is much larger and appears to be behaviorally and numerically dominant. The presence of *A. leachi* apparently causes *A. grahami* to perch substantially lower in vegetation (Wingate 1965; Schoener 1975). The two species are allotopic on a fine scale but broadly sympatric at a larger geographic scale. Finally, a third anole, *A. extremus*, from Barbados, was introduced sometime prior to 1945. *Anolis extremus* is similar in size and habitat to *A. grahami* has not spread yet into the range of *A. leachi* but is sympatric with *A. grahami* which it resembles in body size and habits. In spite of this ecological similarity, Wingate (1965) found no obvious numerical displacement as for the previous size-dissimilar species pair.

Losos, Marks, and Schoener (1993) provide a review of this and other *Anolis* introductions in the Carribean. They find no documented extinctions of natives by invaders nor invaders by ecologically similar natives. However, they do find that in those cases where an introduced species manages to become widespread (seven cases in total), no resident species was ecological similar to the invader. Moreover, the twelve instances of failed or only marginally successful anole invasions occurred in situations where there was at least one ecological similar resident anole species.

CONCLUSIONS AND SUMMARY

A worldwide survey of Holocene (Recent) reptile extinctions yielded several conclusions:

1. Humans are implicated either directly or indirectly in many extinction, extirpations, and population declines.
2. In the absence of much environmental change, either climatically or due to humans, a background extinction rate still exists, and the magnitude of local extinction decreases with increasing island area.
3. In the presence of humans, this background rate is exaggerated, sometimes by an order of magnitude for small islands, but the effect of island area is, if anything, accentuated because very large islands and mainlands have lower extinction rates than predicted from landbridge island extrapolations.
4. Island extinctions are more common than mainland extinctions. This apparent resilience of continental faunas may in part be an artifact of the difficulties in finding fossils spread throughout larger areas. Yet, native island species seem more vulnerable to introduced predators. Perhaps equally important, low extinction rates are expected given large area and thus increased opportunity for immigration after local extirpation.
5. Species becoming extinct are usually those with relatively large body size and a long history of island isolation, resulting in endemic status.
6. Predators, chiefly mongooses, rats, cats and dogs are often implicated in the extinction of reptiles.
7. Introduced reptiles do not seem to seriously affect native reptiles, at least in terrestrial habitats, although they sometimes have dramatic impacts on the densities of other introduced species.
8. Reptile extinction rates are often lower than those calculated for mammals and birds.
9. Overall loss of habitat due to human induced changes in landscapes, probably interacts with the introduction of exotic predators and competitors, to elevate reptile extinction rates (Atkinson 1989). Assuming that a minimum absolute population size is necessary for a population's long-term persistence, the decline of total area due to habitat destruction combined with decreases in population density within remaining core habitat areas caused by the depredations of exotics, will likely bring population sizes below necessary thresholds for persistence.

ACKNOWLEDGMENTS

At various stages during the development of this paper, many people contributed ideas and suggestions and useful unpublished information: Nick Arnold, Aaron Bauer, Emilio Bruna, Harold Cogger, Ronald Crombie, Robert Fisher, John Gibbons, Mike Gilpin, Jon Losos, Gerald McCormack, Ken Petren, Greg Pregill, Tom Schoener, Dick Watling, Ernest Williams, Rhys Walkley, Tony Whitaker, Richard Zweifel, and George Zug. We thank them and we thank Allison Alberts, Jared Diamond, Peggy Fiedler, Mike Gilpin, Greg Pregill, Tom Schoener, and Tony Whitaker for improving the presentation.

LITERATURE CITED

Amerson, A.B., W.A. Whistler, and T.D. Schwaner. 1982. *Wildlife and wildlife habitat of American Samoa. II. Accounts of flora and fauna.* Washington D.C.: U.S. Department of Interior, Fish and Wildlife Service.

Arnold, E.N. 1973. Relationships of the palaearctic lizards assigned to the genera *Lacerta*, *Algyroides*, and *Psammodromus* (Reptilia: *Lacertidae*). *Bulletin of the British Museum of Natural History* 25:291–366.

Arnold, E.N. 1976. Fossil reptiles from Aldabra Atoll, Indian Ocean. *Bulletin of the British Museum of Natural History* 29:83–116.

Arnold, E.N. 1980. Recently extinct populations from Mauritius and Reunion, Indian Ocean. *Journal of the Zoological Society of London* 191:33–47.

Atkinson, I. 1985. Effects of rodents on islands. In *Conservation of island birds*, ed. P.J. Moors, 35–81. Cambridge: International Council for Bird Preservation.

Atkinson, I. 1989. Introduced animals and extinctions. In *Conservation for the twenty-first century*, ed. D. Western and M. Pearl, 54–75. Oxford: Oxford University Press.

Baez, A.M. and Z.B. Gasparini. 1979. An evaluation of the fossil record. In *The South American herpetofauna: its origin, evolution, and dispersal*, ed. W.E. Duellman, 29–54. University of Kansas Museum of Natural History Monograph.

Bahre, C.J. 1983. Human impact: The Midriff Islands. In *Island biogeography in the Sea of Cortez*, eds. T.J. Case and M.L. Cody, 290–306. Berkeley: University of California Press.

Barbour, T. 1910. Notes on the herpetology of Jamaica. *Bulletin of the Museum of Comparative Zoology 52 (no. 15)*.

Barbour, T. 1930. Some faunistic changes in the Lesser Antilles. *Proceedings of the New England Zoological Club* 11:73–85.

Bauer, A.M. and A.P. Russell. 1986. *Hoplodactylus delcourti* n. sp. (Reptilia: Gekkonidae), the largest known gecko. *New Zealand Journal of Zoology* 13:141–148.

Bings, W. 1985. *Zur früheren Verbreitung von* Gallotia simonyi *auf Hierro, mit Vorschlägen zur Wiederansiedlung*. *Bonn. Zoolgische Beitr.* 36:417–427.

Blanc, C.P. 1972. Les reptiles de Madagascar et des iles voisines. In *Monographiae biological*, ed. J. I. Schlitz, 501–614. vol 21. The Hague: Junk.

Blanc, C.P.1984. The reptiles. In *Madagascar (key environments)*, eds. A. Jolly, P. Oberlé, and R. Albignac, 105–114. Oxford: Pergamon Press.

Böhme, W. and B. Bings. 1977. *Nachtrage zur Kenntnis der kanarischen Rieseneidechsen* (Lacerta simonyi-*Gruppe*). *Salamandra* 13:105–111.

Böhme, W., W. Bischoff, H. Nettmann, S. Rykena, and J. Freundlich. 1981. *Nachweis von* Gallotia simonyi *(Steindachner, 1889) (Reptilia: Lacertidae) aus einer fruhmittelalterlichen Fundschicht auf Hierro, Kanarische Inseln*. *Bonn. Zoolgische Beitr.* 32:157–166.

Böhme, W. and G. Zammit-Maempel. 1982. *Lacerta siculimelitensis* sp. n. (Sauria:Lacertidae), a giant lizard from the late Pleistocene of Malta. *Amphibia-Reptilia* 3:257–268.

Bolger, D.T. 1991. Community perturbation: introduced species and habitat fragmentation. Ph.D. dissertation, University of California, San Diego.

Bolger, D.T. and T.J. Case. 1992. Intra-specific and inter-specfic interference behavior among sexual and asexual geckos. *Animal Behavior* 44:21–30.

Branch, W.R., ed. 1988. *South African Red Data Book—reptiles and amphibians*. South African Scientific Programs Report. Pretoria: CSIR.

Brattstrom, B.H. 1955. Pleistocene lizards from the San Josito Cavern, Mexico, with description of a new species. *Copeia* 1955:133–134.

Bravo, T. 1953. Lacerta maxima *n. sp. de la fauna continental extinguida en el Pleistoceno de las Islas Canarias*. *Inst. Invest. Geol. Lucas Mallada* 9:7–34.

Brooke, M.L. and D.C. Houston. 1983. The biology and biomass of the skinks *Mabuya sechellensis* and *Mabuya wrightii* on Cousin Island, Seychelles (Reptilia: Scincidae). *J. Zool. London* 200:179–195.

Bruna, E.M., R.M. Fisher, and T.J. Case. 1995. Cryptic species of Pacific skinks (*Emoia*): support form mitochondrial DNA sequences. *Copeia* 1995:981–983.

Bruna, E.M., R.M. Fisher, and T.J. Case. 1996. Morphological and genetic evolution appear decoupled in Pacific skinks (Squamata: Scincidae: *Emoia*). *Proceedings of the Royal Society of London, B*, in press.

Burt, C.E. and M.D. Burt. 1932. Herpetological results of the Whitney south sea expedition. VI. Pacific island amphibians and reptiles in the collection of the American Museum of Natural History. *Bulletin of the American Museum Natural History* 63:461–597.

Carmichael, C.K., J.C. Gillingham, and S.N. Keall. 1989. Feeding ecology of the tuatara (*Sphenodon punctatus*) on Stephens Island based on niche diversification. *New Zealand Journal of Zoology* 16:269.

Case, T.J. 1975. Species numbers, density compensation, and colonizing ability of lizards on islands in the Gulf of California. *Ecology* 56:3–18.

Case, T.J. and D.T. Bolger. 1991. The role of introduced species in shaping the abundance and distribution of island reptiles. *Evolutionary Ecology* 5:272–290.

Case, T.J., D.T. Bolger, and K. Petren. 1994. Invasions and competitive displacement among house geckos in the tropical Pacific. *Ecology* 75:464–477.

Case, T.J. and M.L. Cody. 1987. Island Biogeographic theories: Tests on islands in the Sea of Cortez. *Ameican Scientist* 75:402–411.

Cassels, R. 1984. The role of prehistoric man in the faunal extinctions of New Zealand and other Pacific islands. In *Quaternary extinctions*, eds. P.S. Martin and R.G. Klein, Chapter 34. Tucson: University of Arizona Press.

Cheke, A.S. 1984. Lizards of the Seychelles. In *Biogeography and ecology of the Seychelles Islands*, ed. D.R. Stoddart, Chapter 19. The Hague: Dr W. Junk.

Cheke, A.S. 1987. An ecological history of the Mascarene Islands with particular reference to extinctions and introductions of land vertebrates. In *Studies of Mascarene Island Birds*, eds. A.W. Diamond and A.S. Cheke, 5–89. Cambridge: Cambridge University Press.

Cogger, H.G. 1971. The reptiles of Lord Howe Island. *Proceedings of the Linnean Society of New South Wales* 96:23–38.

Cogger, H.G., R. Sadlier, and E. Cameron. 1983. *The terrestrial reptiles of Australia's Island Territories*. Australian National Parks and Wildlife Service, Special Publication No. 11.

Corke, D. 1987. Reptile conservation on the Maria Islands (St. Lucia, West Indies). *Biological Conservation* 40:263–279.

Cowie, R.H. 1992. Evolution and extinction of the Partulidae, endemic Pacific island land snails. *Philosophical Transactions of the Royal Society of London B* 335:167–191.

Cree, A., C.H. Daugherty, and J.M. Hay. 1995. Reproduction of a rare New Zealand reptile, the tuatara, *Shenodon punctatus*, on rat-free and rat-inhabited islands. *Conservation Biology* 9:373–383.

Crook, I.G. 1973. The tuatara, *Sphenodon punctatus*, on islands with and without populations of the Polynesian rat, *Rattus exulans*. *Proceedings of the New Zealand Ecological Society* 20:115–120.

Darlington, P.J. 1957. *Zoogeography, the geographical distribution of animals*. New York: John Wiley.

Daugherty, C.H., A. Cree, J.M. Hay, and M.B. Thompson. 1990. Neglected taxonomy and continuing extinctions of tuatara (*Sphenodon*). *Nature* 347:177–179.

Dewar, R.E. 1984. Extinction in Madagascar: The loss of the subfossil fauna. In *Quaternary extinctions*, eds. P.S. Martin and R.G. Klein, Chapter 26. Tucson: University of Arizona Press.

Diamond, J.M. 1972. Biogeographic kinetics: Estimation of relaxation times for avifaunas of Southwest Pacific Islands. *Proceedings of the National Academy of Sciences* 69:3199–3203.

Diamond, J.M. 1984. Historic extinctions: A rosetta stone for understanding prehistoric extinctions. In *Quaternary extinctions*, eds. P.S. Martin and R.G. Klein, Chapter 38. Tuscon: University of Arizona Press.

Diamond, J.M. and T.J. Case. 1986. Overview: introduction, extinctions, exterminations, and invasions. In *Community ecology*, eds. J.M. Diamond and T.J. Case, 65–79. New York: Harper and Row.

Dye, T. and D.W. Steadman. 1990. Polynesian ancestors and their animal world. *American Scientist* 78: 207–215.

Engbring, J. and T.H. Fritts. 1988. Demise of an insular avifauna: the brown tree snake on Guam. *Transactions of the W. Section Wildlife Society* 24:31–37.

Estes, R. 1983. *Handbuch der Paläoherpetologie, Sauria terrestria Amphisbaenia*. Teil 10A. Stuttgart, Germany: Gustav Fischer Verlag.

Etheridge, R. 1964. Late Pleistocene lizards from Barbuda British West Indies. *Bulletin of the Florida State Museum* 9:43–75.

Evans, P.G.H. and J.B. Evans. 1980. The ecology of lizards on Praslin Island, Seychelles. *Journal of Zool. London* 191:171–192.

Frogner, K.J. 1967. Some aspects of the interaction between the gecko species *Hemidactylus frenatus* and *Lepidodactylus lugubris* in Hawaii. M.S. thesis, University of Hawaii, Honolulu.

Gaffney, E.S. 1992. *Ninjemys*, a new name for "*Meiolania*" *oweni* (Woodward), a horned turtle from the Pleistocene of Queensland. *American Museum Novitates* 3049:1–10.

Gaffney, E.S. and G. McNamara. 1990. A meiolaniid turtle from the Pleistocene of Northern Queensland. *Mem. Queensland Museum* 28:107–113.

Gaffney, E.S., J.C. Balouet, and F. DeBroin. 1984. New occurences of extinct meiolaniid turtles in New Caledonia. *American Museum Novitates* 2800:1–6.

Gardner, A.S. 1986. The biogeography of the lizards of the Seychelles Islands. *Journal of Biogeography* 13:237–253.

Gehlbach, F.R. 1965. Amphibians and reptiles from the Pliocene and Pleistocene of North America: A chronological summary and selected bibliography. *Texas Journal of Science* 17:56–70.

Gibbons, J.R.H. 1984. Iguanas of the south Pacific. *Oryx* 18:82–92.

Gibbons, J.R.H. and I.F. Watkins. 1982. Behavior, ecology, and conservation of South Pacific banded iguanas, *Brachylophus*, including a newly discovered species. In *Iguanas of the world, their behavior, ecology, and conservation*, eds. G.M. Burghardt and A.S. Rand, Chapter 23. Park Ridge, NJ: Noyes Publishers.

Gifford, E.W. and D. Shuter, Jr. 1956. Archaeological excavations in New Caledonia. *Anthropological Records* 18:1–148.

Gilpin, M.E. and M.E. Soulé. 1986. Minimum viable populations: Processes of species extinction. In *Conservation biology*, ed. M.E. Soulé, 19–34. Sunderland, MA: Sinauer Associates.

Glaw, F. and M. Vences. 1992. *A field guide to the amphibians and reptiles of Madagascar.* Leverkusen, Germany: Moos-Druck.

Greer, A.E. 1976. On the evolution of the giant Cape Verde scincid lizard *Macroscincus coctei. Journal of Natural History* 10:691–712.

Groombridge, B. 1992. *Global biodiversity: Status of the Earth's living resources.* World Conservation Monitoring Centre. New York: Chapman and Hall.

Hadfield, M.G., S.E. Miller, and A.H. Carwile. 1989. Recovery plan for the Oahu tree snails of the genus *Achatinella*. Washington D.C.: U.S. Department of Interior, Fish and Wildlife Service.

Hardy, G.S. 1977. The New Zealand Scincidae (Reptilia: *Lacertilia*) a taxonomic and zoogeographic study. *New Zealand Journal of Zoology* 4:221–325.

Hardy, G.S. and A.H. Whitaker. 1979. The status of New Zealand's endemic reptiles, and their conservation. *Forest and Bird* 13:34–39.

Hecht, M.K. 1951. Fossil lizards of the West Indian genus *Aristelliger* (Gekkonidae). *American Museum Novitates* 1538:1–33.

Hecht, M.K. 1975. The morphology and relationships of the largest known terrestrial lizard, *Megalania prisca* Owen, from the Pleistocene of Australia. *Proceedings of the Royal Society Victoria* 87:239–249.

Henderson, R.W. 1992. Consequences of predator introductions and habitat destruction on amphibians and reptiles in the Post-Columbus West Indies. *Carribean Journal of Science* 28:1–10.

Holman, J.A. 1995. *Pleistocene amphibians and reptiles in North America.* Oxford: Oxford University Press.

Honegger, R.E. 1975. *Red data book. Vol 3. Amphibia and Reptilia.* Morges, Switzerland: International Union for the Conservation of Nature and Natural Resources (IUCN).

Honegger, R.E. 1981. List of amphibians and reptiles either known or thought to have become extinct since 1600. *Biological Conservation* 19:141–158.

Hooton, E.A. 1925. Ancient inhabitants of the Canaries. *Harvard African Studies* 7:1–401.

Hunsaker, D. and P. Breese. 1967. Herpetofauna of the Hawaiian Islands. *Pacific Science* 21:168–172.

Hutterer, R. 1985. Neue Funde von Rieseneidechsen (Lacertidae) auf der Insel Gomera. *Bonn. Zoolgische Beitr.* 36:365–394.

Ineich, I. 1987. Recherches sur le peuplement et l'evolution des retiles terrestres de polynesie française. Ph.D., Université des Sciences et techniqes du Languedoc, Academie de Montpellier, France.

Ineich, I. and C.P. Blanc. 1988. Distribution des reptiles terrestres en Polynesie Orientale. *Atoll Research Bulletin* 318:1–75.

Ineich, I. and G. Zug. 1991. Nomenclatural status of *Emoia cyanura* (Lacertilia, Scincidae). *Copeia* 1991:1132–1136.

Iverson, J.B. 1978. The impact of feral cats and dogs on populations of the West Indian rock iguana, *Cyclura carinata. Biological Conservation* 14:63–74.

Izquierdo, I., A.L. Medina, and J.J. Hernandez. 1989. Bones of giant lacertids from a new site on El Hierro (Canary Islands). *Amphibia-Reptilia* 10:63–69.

Joger, U. 1984. Die Radiation der Gattung Tarentola in Macaronesien (Reptilia: Sauria: Gekkonidae). *Cour. Forsch. Inst. Senckenberg* 71:91–111.

Jolly, A., P. Oberlé, and R. Albignac, eds. 1984. *Madagascar (key environments)*. Oxford: Pergamon.

Jones, R.E. 1979. Hawaiian lizards—their past, present and future. *Bulletin of the Maryland Herpetological Society* 15:37–45.

Klemmer, K. 1976. The amphibia and reptilia of the Canary Islands. In *Biogeography and ecology in the Canary Islands*, ed. G. Kunkel, Chapter 15. The Hague: Junk.

Kramer, P. 1984. Man and other introduced animals. In *Evolution of the Galapagos*, ed. R.J. Berry, 253–258. London: Academic Press.

Leguat, F. 1708. Voyages et aventures de Francois Leguat & de ses compagnons en deux iles désertes des Indes Orientales. Volumes 1 and 2. London: David Mortier.

Leigh, E.G. 1981. Average lifetime of a population in a varying environment. *Journal of Theoretical Biology* 90:213–239.

López-Jurado, L.F. 1985. Los reptiles fósiles de la Isla de Gran Canaria (Islas Canarias). *Bonn. Zoolgische Beitr.* 36:355–364.

López-Jurado, L.F. 1985. A new canarian lizard subspecies from Hierro (Canarian Archipelago). *Bonn. Zoolgische Beit.* 40:265–272.

Losos, J.B., J.C. Marks, and T.W. Schoener. 1993. Habitat use and ecological interactions of an introduced and a native species of *Anolis lizard* on Grand Cayman, with a review of the outcomes of anole introductions. *Oecologia* 95:525–532.

MacArthur, R.H. and E.O. Wilson. 1967. *The theory of island biogeography*. Princeton: Princeton University Press.

Machado, A. 1985a. New data concerning the Hierro giant lizard and the lizard of Salmor (Canary Islands). *Bonn. Zoolgische Beitr.* 36:429–470.

Machado, A. 1985b. Hypothesis on the reasons for the decline of the large lizards in the Canary Islands. *Bonn. Zoolgische Beitr.* 36:563–575.

Machado, A., L.F. López-Jurado, and A. Martin. 1985. Conservation status of reptiles in the Canary Islands. *Bonn. Zoolgische Beitr.* 36:585–606.

Marrero Rodriguez, A. and C.M. Garcia Cruz. 1978. Nuevo yacimiento de restos subfósiles de dos vertebrados extintos de la Isla de Tenerife (Canarias), Lacerta maxima *Bravo, 1953 y Canariomys bravoi Crus. et Pet., 1964. Vieraea* 7:165–174.

Martin, P.S. and R.G. Klein, eds. 1984. *Quaternary extinctions*. Tucson: University of Arizona Press.

McCallum, J. 1986. Evidence of predation by kiore upon lizards from the Mokohinau Islands. *New Zealand Journal of Ecology* 9:83–87.

McCoy, M. 1980. *Reptiles of the Solomon Islands*. Wau, Papua New Guinea: Wau Ecology Institute Handbook No. 7.

McKeown, S. 1978. *Hawaiian reptiles and amphibians*. Honolulu: Oriental.

Medway, L. and A.G. Marshall. 1975. Terrestrial vertebrates of the New Hebrides: origin and distribution. *Philosophical Transactions of the Royal Society of London B* 272:423–465.

Mercer, J. 1980. *The Canary Islands, their prehistory, conquest, and survival*. London: Rex Collins.

Mertens, R. 1942. Lacerta goliath *n. sp., cine ausgestorbene Rieseneidechse von den Kanaren. Senckenbergiana* 25:330–339.

Mertens, R. 1956. *Dei Eidechsen der Kapverden. Commentat. Biol.* 15:1–16.

Mittermeier, R.A. 1972. Jamaica's endangered species. *Oryx* 11:258–262.

Molnar, R. 1984a. Cainozoic reptiles from Australia (and some amphibians). In *Vertebrate zoology and evolution in Australasia*, eds. M. Archer and G. Clayton, 337–341. Perth: Hesperian.

Molnar, R. 1984b. A checklist of Australian fossil reptiles. In *Vertebrate zoology and evolution in Australasia*, ed. M. Archer and G. Clayton, 405–406. Perth: Hesperian.

Moodie, K.B. and T.R.V. Devender. 1979. Extinction and extirpation in the herpetofauna of the Southern High Plains with emphasis on *Geochelone wilsoni* (Testitudinae). *Herpetologica* 35:198–206.

Myers, C.W. and A.S. Rand. 1969. Checklist of amphibians and reptiles of Barro Colorado Island, Panama, with comments on faunal change and sampling. *Smithsonian Contributions to Zoology No. 10.*

Newman, D.G., ed. 1982. New Zealand herpetology: Proceedings of a symposium held at Victoria University of Wellington, January 1980. *New Zealand Wildlife Service Occasional Publications No. 2.*

North, S.G., D.J. Bullock, and M.E. Dulloo. 1994. Changes in the vegetation and reptile populations on Round Island, Mauritius, following eradication of rabbits. *Biological Conservation* 67:21–28.

Oliver, J.A. and C.E. Shaw. 1953. The amphibians and reptiles of the Hawaiian Islands. *Zoologica* 38:65–95

Olson, S.L. and H.G. James. 1982. Prodromus of the fossil avifauna of the Hawaiian Islands. *Smithsonian Contributions to Zoology No. 365.*

Olson, S.L., G.K. Pregill, and W.B. Hilgartner. 1990. Studies on fossil and extant vertebrates from San Salvador (Watling's) Island, Bahamas. *Smithsonian Contributions to Zoology* 508: 1–15.

Patterson, G.B. and C.H. Daugherty. 1990. Four new species and one new subspecies of skinks, genus *Leiolopisma* (Reptilia: Lacertilia: Scincidae) from New Zealand. *Journal of the Royal Society of New Zealand* 20:65–84.

Paulian, R. 1984. Madagascar: a micro-continent between Africa and Asia. In, *Madagascar (key environments)*, eds. A. Jolly, P. Oberlé, and R. Albignac, 1–26. Oxford: Pergamon Press.

Pernetta, J.C. and D. Watling. 1979. The introduced and native terrestrial vertebrates of Fiji. *Pacific Science* 32:223–244.

Petren, K., D.T. Bolger, and T.J. Case. 1993. Mechanisms in the competitive success of an invading sexual gecko over an asexual native. *Science* 259:354–358.

Petren, K. and T.J. Case. 1996. An experimental demonstration of exploitation competition in an ongoing invasion. *Ecology* 77:118–132.

Pimentel, D. 1955. Biology of the Indian mongoose in Puerto Rico. *Journal of Mammology* 36:62–68.

Pregill, G.K. 1981. Late Pleistocene herpetofaunas from Puerto Rico. *University of Kansas Museum of Natural History Miscellaneous Publication No.71.*

Pregill, G.K. 1986. Body size of insular lizards: A pattern of Holocene dwarfism. *Evolution* 40:997–1008.

Pregill, G.K.1992. Systematics of the West Indian lizard genus *Leiocephalus* (Squamata: Iguania: Tropiduridae). *University of Kansas Museum of Natural History Miscellaneous Publication No.* 84:1–69.

Pregill, G.K. 1993. Fossil lizards from the late Quaternary of 'Eua, Tonga. *Pacific Science* 47:101–114.

Pregill, G.K., R.I. Crombie, D.W. Steadman, L.K. Gordon, F.W. Davis, and W.B. Hilgartner. 1991. Living and late Holocene fossil vertebrates, and the vegetation of the Cockpit Country, Jamaica. *Atoll Research Bulletin* 353:1–19.

Pregill, G.K. and T. Dye. 1989. Prehistoric extinction of giant iguanas in Tonga. *Copeia* 1989:505–508.

Pregill, G.K., D.W. Steadman, S.L. Olson, and F.V. Grady. 1988. Late Holocene fossil vertebrates from Burman Quarry, Antiqua, Lesser Antilles. *Smithsonian Contributions to Zoology No. 463.*

Pregill, G.K., D.W. Steadman, and D.R. Watters. 1994. Late Quaternary vertebrate faunas of the Lesser Antilles: Historical components of Caribbean biogeography. *Bulletin of the Carnegie Museum of Natural History* 30:1–51.

Raxworthy, C.J. and R.A. Nussbaum. 1993a. Four new species of *Amphiglossus* from Madagascar (Squamata: Scincidae). *Herpetologica* 49:326–341.

Raxworthy, C.J. and R.A. Nussbaum. 1993b. A new Madagascan *Phelsuma*, with a review of *Phelsuma trilineata* and comments on *Phelsuma cepidiana* in Madagascar (Squamata: Gekkonidae). *Herpetologica* 49:342–349.

Reese, D.S. 1989. Tracking the extinct pygmy hippopotamus of Cyprus. *Field Museum of Natural History Bulletin* 60:22–29.

Rich, P.V. 1982. Kukwiede's revenge: a look into New Caledonia's distant past. *Hemisphere* 26:166–171.

Richman, A., T.J. Case, and T.D. Schwaner. 1988. Natural and unnatural extinction rates of reptiles on islands. *American Naturalist* 131:611–630.

Rinke, D. 1986. The status of wildlife in Tonga. *Oryx* 20:146–151.

Robb, J. 1986. *New Zealand amphibians and reptiles in colour*. 2nd ed. Auckland: William Collins.

Rodda, G.H. and T.H. Fritts. 1992. The impact of the introduction of the colubrid snake *Boiga irregularis* on Guam's lizards. *Journal of Herpetology* 26:166–174.

Sadlier, R.A. 1986. A review of the Scincid lizards of New Caledonia. *Records of the Australian Museum* 39:1–66.

Sadlier, R.A. 1988. *Bavayia validiclavis* and *Bavayia septuiclavis*, two new species of gekkonid lizard from New Caledonia. *Records of the Australian Museum* 40:365–370.

Salzburg, M.A. 1984. *Anolis sagrei* and *Anolis cristatellus* in Southern Florida: A case study in interspecific competition. *Ecology* 65:14–19.

Savidge, J.A. 1988. Food habits of *Boiga irregularis*, an introduced predator on Guam. *Journal of Herpetology* 22:275–282.

Schmidt, K.P. 1928. Amphibians and land reptiles of Porto Rico, with a list of those reportedfrom the Virgin Islands. *New York Academy of Science* 10:1–160.

Schoener, T.W. 1975. Presence and absence of habitat shift in some widespread lizard species. *Ecological Monographs* 45:233–258.

Schoener, T.W. 1983. Rates of species turnover decreases from lower to higher organisms: A review of the data. *Oikos* 41:372–377.

Schoener, T.W. and A. Schoener. 1983. The time to extinction of a colonizing propaqule of lizards increases with island area. *Nature* 302:332–334.

Schwidetzky, I. 1976. The prehispanic population of the Canary Islands. In *Biogeography and ecology in the Canary Islands*, ed. G. Kunkel, Chapter 2. The Hague: Junk.

Shallenberger, E.W. 1970. Tameness in insular animals: A comparison of approach distances of insular and mainland iguanid lizards. Ph.D. dissertation, University of California, Los Angeles.

Siegfried, W.R. and R.R. Brooke. 1995. Anthropogenic extinctions in the terrestrial biota of the Afrotropical region in the last 500,000 years. *Journal of African Zoology* 109:5–14.

Steadman, D.W. 1986. Holocene vertebrate fossils from Isla Floreana, Galapagos. *Smithsonian Contributions to Zoology No. 413.*

Steadman, D.W. and S.L. Olson. 1985. Bird remains from an archaeological site on Henderson Island, South Pacific: man-caused extinctions on an "uninhabited" island. *Proceedings of the National Academy of Sciences* 82:6191–6195.

Steadman, D.W., G.K. Pregill, and S.L. Olson. 1984. Fossil vertebrates from Antigua Lesser Antilles: evidence for late Holocene human-caused extinctions in the West Indies. *Proceedings of the U.S. National Museum* 81:4448–4451.

Steadman, D.W., T.W. Stafford Jr., D.J. Donahue, and A.J.T. Jull. 1991. Chronology of Holocene vertebrate extinction in the Galapagos Islands. *Quaternary Research* 36:126–133.

Stejneger, L. 1899. The land reptiles of the Hawaiian Islands. *Proceedings of the U.S. National Museum No. 1174.*

Tirvengadum, D.D. and R. Bour. 1985. Checklist of the herpetofauna of the Mascarene Islands. *Atoll Research Bulletin* 292:49–60.

Thomson, G.M. 1922. *The naturalisation of animals and plants in New Zealand.* Cambridge: Cambridge University Press.

Thorpe, R.S. 1985. Body size, island size and variability in the Canary Island lizards of the genus *Gallotia*. *Bonn. Zoologische Beitr.* 36:481–487.

Thorton, I. 1971. *Darwin's islands: a natural history of the Galapagos*. New York: American Museum of Natural History Press.

Towns, D.R. 1972. Ecology of the black shore skink, *Leiolopisma suteri* (Lacertilia: Scincidae), in boulder beach habitats. *New Zealand Journal of Zoology* 2:389–407.

Towns, D.R. 1974. Zoogeography and the New Zealand scincidae. *Journal of the Royal Society of New Zealand* 4:217–226.

Van Devender, T.R. 1977. Holocene woodlands in the Southwestern Deserts. *Science* 198:189–192.

Van Devender, T.R. 1987. Holocene vegetation and climate in the Puerto Blanco Mountains, southwestern Arizona. *Quaternary Research* 27:51–72.

Van Devender, T.R. and J.M. Mead. 1978. Early and Late Pleistocene amphibians and reptiles in Sonoran Desert packrat middens. *Copeia* 1978:464–475.

Vinson, J. 1973. A new skink of the genus *Gongylomorphus* from Macabé forest (Mauritius). *Revue Agric. Sucr. Ile Maurice* 52:39–40.

Vinson, J. and J. Vinson. 1969. The Saurian fauna of the Mascarene Islands. *Mauritius Inst. Bull.* 6:203–320.

Whitaker, A.H. 1968. The lizards of the Poor Knights Islands, New Zealand. *New Zealand Journal of Zoology* 11:623–651.

Whitaker, A.H. 1973. Lizard populations on islands with and without Polynesian rats, *Rattus exulans*. *Proceedings of the New Zealand Ecological Society* 20:121–130.

Whitaker, A.H. 1982. Interim results from a study of *Hoplodactylus maculatus* (Boulenger) at Turakirae Head, Wellington. In *New Zealand herpetology*, ed. D.G. Newman, 363–374. New Zealand Wildlife Service Occasional Publication No. 2. Auckland: New Zealand Wildlife Service.

Wilcox, B.A. 1980. Insular ecology and conservation. In *Conservation biology: an evolutionary-ecological perspective*, eds. M.E. Soulé and B.A. Wilcox, Chapter 6. Sunderland, MA: Sinauer Associates.

Wilcox, B.A. 1981. Aspects of the biogeography and evolutionary ecology of some island vertebrates. Ph.D. dissertation, University of California, San Diego.

Willis, E.O. 1974. Populations and local extinction of birds in Barro Colorado Island, Panama. *Ecological Monographs* 44:153–169.

Wilson, L.D. and L. Porras. 1983. The ecological impact of man on the South Florida herpetofauna. *University of Kansas Museum of Natural History Special Publication No. 9.*

Wingate, D.W. 1965. Terrestrial herpetofauna of Bermuda. *Herpetologica* 21:202–218.

Worthy, T.H. 1987a. Osteological observations on the larger species of the skink *Cyclodina* and the subfossil occurrence of these and the gecko *Hoplodactylus duvaucelii* in the North Island, New Zealand. *New Zealand Journal of Zoology* 14:219–229.

Worthy, T.H. 1987b. Paleoecological information concerning members of the frog genus *Leiopelma*: Leiopelmatidae in New Zealand. *Journal of the Royal Society of New Zealand* 17:409–420.

Worthy, T.H. 1991. Fossil skink bones from Northland, New Zealand, and description of a new species of *Cyclodina*, Scincidae. *Journal of the Royal Society of New Zealand* 21: 329–348.

Zug, G.R. 1992. *The lizards of Fiji: Natural history and systematics*. Honolulu: Bernice P. Bishop Museum Press.

Zug, G.R. and I. Ineich. 1995. A new skink (*Emoia*: Lacertilia: Reptilia) from the forest of Fiji. *Proceedings of the Biological Society of Washington* 108:395–400.

CHAPTER 8

Conservation Status of the World's Fish Fauna: An Overview

ROBERT A. LEIDY
and PETER B. MOYLE

In the five years since we first reviewed the status of the world's fishes (Moyle and Leidy 1992), there has been an explosion of new information on the conservation of aquatic organisms and their ecosystems. Notwithstanding this surge of interest, many aquatic ecosystems remain poorly understood because conservation biology remains primarily focused on the loss of biotic diversity in terrestrial environments. Loss of diversity in aquatic environments has received comparatively little attention, even though the physical, chemical, and biological degradation of aquatic environments is widely recognized as a major problem, usually in the context of the spread of human disease, loss of fisheries, or degraded water quality for drinking, irrigation, or recreation. Yet aquatic habitats support an extraordinary array of species, many of which are being lost as their habitats deteriorate.

Ironically, while interest in the conservation of aquatic organisms has increased over the last five years, so too has the rate and extent of degradation of aquatic environments upon which these organisms depend. The root cause of the greatest loss of aquatic biodiversity remains the rapid expansion of human populations. The earth's human population has increased by approximately 500 million to about 5.7 billion people between 1990–1995, and it continues to increase at a rate of more than 86 million people annually (United Nations Population Fund 1995). The implications of increasing human populations for the conservation of aquatic biodiversity are not trivial, as about 60% of the world's population lives within 100 km of the ocean and most inland cities center on lakes and rivers that eventually drain into the ocean (The World Resources Institute *et al.* 1996).

General forms of aquatic degradation due to increasing human populations include physical conversion and degradation of natural habitats, various forms of water pollution, development of freshwater resources, introduction of exotic organisms, and overfishing. For example, it is estimated that 950 million people rely on fish as their primary source of protein; however, more than two-thirds of the earth's marine fish stocks are either overfished or reaching their limit to sustain a fisheries (World Resources Institute *et al.* 1996; Food and Agriculture Organization 1993, 1995; see chapter 6). Human demands on the world's freshwater resources are also increasing. While fresh water covers about 1% of the continents, increased human demands for freshwater are rapidly outstripping supply in many regions of the world (Postel 1996). Engelman and LeRoy (1993) estimate that the scarcity of fresh water is already a problem in 20 countries, and that by the year 2025 the number of people in water-scarce countries could be as high as 2.42 billion (see World Resources Institute *et al.* 1996). Clearly, the threat from increasing water demands on the world's freshwater biotas, particularly in regions of projected chronic scarcity (e.g., Africa, western Asia), will be severe.

Our currently inadequate appreciation of the conservation status and ecological functioning of aquatic ecosystems and their constituent fishes is the result of a number of factors. From a logistical and technological perspective, most marine habitats are too remote or too deep for easy study; most freshwater habitats are too turbid, deep, or physically dynamic to monitor easily. In addition to these methodological difficulties, there are experimental design constraints, specifically, the inherent but confounding variability in spatial and temporal scale, especially in aquatic environments such as rivers and nearshore marine environments where disturbance is a major force structuring fish communities. Also, many aquatic organisms are microscopic (e.g., the planktonic component) or extremely small (e.g., many aquatic invertebrates), therefore being difficult to identify. Thus, the organisms we know best in aquatic habitats are fish because of their relatively larger size, abundance, economic importance, and comparative ease of capture and identification (Karr 1981). Fortunately, fishes are appropriate indicators of trends in aquatic biodiversity because their enormous variety reflects a wide range of environmental conditions. Trends observed in fish also have been noted similarly for other less documented groups, such as freshwater molluscs and crustaceans (Williams *et al.* 1993).

With the exception of only a very few regions (e.g., North America and Europe), the distribution, ecology, and status of most fish species remain poorly understood. Even in the better known regions, new fish taxa continue to be described, although individual species abundances may not be well known. Examination of trends in freshwater fish faunas from different parts of the world indicate that most fauna are in serious decline and in need of immediate protection. Species most likely to be threatened with immediate extinction are either: (1) specialized for life in large rivers, although the number of threatened species in small rivers is increasing; (2) of lacustrine or inland sea habitats; (3) found in arid, freshwater environments; (4) dependent on coastal estuaries and reefs and other shallow-water marine habitats; or (5) endemics restricted to very small areas, such as springs or caves.

The purposes of this chapter are to review the status of aquatic environments as reflected by fishes and to make recommendations for worldwide fish conservation. To do this, we examine the following questions:

(1) Why are aquatic environments so vulnerable to degradation?
(2) How much diversity as measured by fishes exists in aquatic environments?
(3) How appropriate are fish as indicators of aquatic biodiversity?
(4) What are the taxonomic and ecological characteristics of threatened fishes?
(5) What is the status of fish faunas from different regions of the world and what are some of the reasons for their decline?
(6) What actions can be taken to protect aquatic biodiversity?

VULNERABILITY OF AQUATIC ENVIRONMENTS TO DEGRADATION

Several reasons exist why aquatic environments are so vulnerable to degradation. These include the complex properties of water itself, the interactions between aquatic and terrestrial environments, and the proximity of human populations to aquatic systems. Productivity of various aquatic environments is driven largely by the capacity of water as a solvent and its tendency to ionize dissolved substances. As a result, inland and nearshore marine environments are not only affected by internal biogeochemical processes, but they are strongly affected by processes in adjacent terrestrial environments. Such aquatic habitats consequently are extremely productive, and humans have exploited this productivity for food, especially fish. However, the great assimilative properties of water also have resulted in the use of inland and marine aquatic environments as seemingly endless "sinks" for cultural wastes produced in terrestrial biomes. In many areas, the capacity of aquatic systems to absorb these wastes has been exceeded and aquatic wastelands have developed, increasingly unfit for most forms of life.

Some energy and material flow from aquatic to terrestrial environments, but it is much less than what flows in the opposite direction. Thus, cutting rainforests adjacent to streams in the Pacific Northwest of North America, for example, causes major shifts in biotic communities of the streams through such processes as siltation, increased growth of algae in exposed reaches, and reduced amounts of large woody debris in the channel. In many instances, the effects of logging are felt far downstream as coastal lagoons and estuaries fill with sediment derived from the upper watershed. The effects of terrestrial changes on aquatic environments are particularly severe in areas where human populations are dense, because urbanization and agricultural development cause major alterations of aquatic communities. The fact that aquatic environments are the recipients of nearly every form of human waste has resulted in their rapid and continuous degradation, often with long-term consequences to the health of fish populations (Kime 1995). It is an indication of their resilience that so many aquatic habitats still retain much of their native biota, although the most altered environments are often dominated by introduced species.

The concentration of people around freshwater systems has resulted in a much greater degree of degradation to these systems than to most open marine systems (World Resource Institute 1994; Abramovitz 1996). Increasing demand for freshwater resources generated by continued population growth, urbanization, industrialization, and irrigation will likely result in further declines of freshwater biotas. The loss of diversity we have seen in freshwater systems is starting to spread seaward, especially into shallow coastal areas and coral reefs that harbor a majority of marine species. The ability of freshwater environments to absorb so much abuse has buffered coastal marine environments from secondary effects of human civilization, but this buffering capacity is rapidly being lost. For example, the World Resources Institute *et al.* (1996), determined that about one-half of the earth's coastal ecosystems were threatened by human-related activities. In Europe and Asia, development-related activities threatened 86% and 69% of coastlines, respectively (World Resources Institute *et al.* 1996).

DIVERSITY IN AQUATIC ENVIRONMENTS

More than one-half of all vertebrate species are fish. Approximately 24,600 fish species have been described, but the total is likely to reach at least 28,500 (Nelson 1994) when the world's ichthyofauna is better known. This taxonomic diversity is spread over 482 described families. Eighty percent (384) of the fish families contain less than 50 species, 67 (14%) of which contain just one species (Table 8.1). Such statistics reflect the many adaptations of fish to specialized conditions and ways of life. Fish families with only a few extant species include ancient lineages such as the coelacanth (Coelacanthidae, 1 species), lungfishes (three families, 9 species), and sturgeons (Acipenseridae, 25 species). In contrast, the few highly speciose families, such as the minnow and carp family (Cyprinidae), with over 2000 species, are widespread, often exhibiting rapid evolutionary responses to local conditions. One of the most extraordinary speciation events on record is the evolution in less than 1,000 years of over 900 highly specialized species of cichlids in the Great Lakes of east Africa (Owen *et al.* 1990).

Diversity of fishes is high in part because fish represent an ancient lineage of vertebrates and in part because fish occupy an extraordinary array of aquatic environments. Because over two-thirds of the earth's surface is covered with water and because the average ocean depth is approximately 4,000 m, it might be expected that a large majority of aquatic species would be found in the open ocean. This is not the case, however, because the surface and deep waters of the ocean are relatively uniform habitats of low productivity and contain relatively few fish species (Angel 1993). The greatest diversity of aquatic life is distributed overwhelmingly along continental shelves, in reefs associated with islands, or in freshwater. Thus, 41% (ca. 11,500 species) of all fish species are exclusively freshwater; 1% (ca. 250 species) are diadromous, moving on a regular basis between the ocean and freshwater; 46% (ca. 13,000 species) are shallow water marine species; 12% (ca. 3,400) are deepsea fishes; and 1% (250) are epipelagic, open ocean fishes (Berra 1981; Nelson 1994).

Diversity of fishes in shallow water marine environments and freshwater is caused by the same evolutionary processes that have created high diversity in terrestrial habitats—i.e., opportunities for speciation following events such as the rise of mountain ranges that isolate regions, creation of isolated island systems, or sea level fluctuations that isolate bays. Considering that freshwater covers approximately 1% of the continents and that the continental shelves (to 200 m depth) have an area less than 10% of the continents, the diversity of fish species is surprisingly high.

Perhaps because of this great diversity of fishes, ecological knowledge of most species is limited. Indeed, many species remain to be discovered or described, especially in tropical freshwaters, tropical reefs, and deepsea environments (Greenwood 1992). For most described species, little is known about their life histories, environmental requirements, or present status. For example, 25% of the 188 species described from the African freshwater family Mormyridae are known only from the locality from which they were originally collected (Stiassny 1996). Thus any estimate of conservation needs of fishes based on species lists is likely to be very conservative because most poorly known species are not included on lists of potentially endangered species.

FISH AS INDICATORS OF AQUATIC BIODIVERSITY

The main reason for using fish to monitor aquatic biodiversity is that we know more about them than other aquatic organisms. As discussed previously, they are also relatively easy to collect and identify. However, there are other compelling arguments for using fish.

First, they are enormously diverse, with different species reflecting different environ-

Table 8.1

Summary of the numbers of recognized extant families, genera, and species in the 57 orders of fishes that contain living representatives (adapted from Nelson 1994).

	Number of Families								
Total Number of Families	1 Species	2–50 Species	51–100 Species	101–500 Species	501–2,000 Species	> 2,000 Species	Freshwater Species	Species Using Freshwater	Marine Species
482	67	317	44	48	5	1	9,966	10,465	14,460

mental conditions (Moyle and Cech 1996). Second, fish often have major effects on the distribution and abundance of other organisms in the waters they inhabit. In lakes, for example, plankton-feeding fish can causes changes to the the plankton communities, which in turn causes cascading changes in the abundance of other organisms in the food web (Carpenter and Kitchell 1988). In streams, algae-feeding fish alter algal communities, except where their populations are kept low by predators (Power 1990, 1992). Goulding (1980) presents evidence that certain species of characins in the Amazon Basin may be important dispersers of tree seeds in seasonally-flooded lowland forests of the Amazon. And the presence of kelp forests off the Southern California coast is determined in part by the ability of fish to control invertebrates that graze on the kelp (Foster and Schiel 1985). When kelp forests are large and widespread, many species of fish are more abundant than they would be without the forests.

Third, fishes often form an important ecological link between aquatic and terrestrial environments. For example, anadromous fishes often provide resources important to the survival and reproduction of terretrial wildlife, especially as a source of food (Willson and Halupka 1995).

WORLDWIDE STATUS OF FISHES

The International Union for the Conservation of Nature (IUCN) has developed a list of fish species known or suspected to be globally threatened with extinction (1996), which is the best overview of fish conservation problems. The IUCN list puts threatened fishes into the categories extinct, extinct in the wild, critically endangered, endangered, vulnerable, data deficient, and lower risk, this latter category supporting three subcategories (Table 8.2). For the critically endangered, endangered, and vulnerable categories, the IUCN list includes semi-quantitative evaluations to assess extinction risks based upon such information as rates of decline, known distribution, or estimates of population viability (Hudson and Mace 1996). However, while this list represents a major step toward improving the basis for determining the conservation status of fishes, it is essentially a snapshot in time, a point on a graph with a downward trend.

The IUCN list is very conservative as tool for indicating trends in biodiversity loss in aquatic systems for the following reasons:

(1) Some widely distributed "species" may actually represent groups of species. For example, numerous species of freshwater fish have been described in recent years from the Mediterranean countries of Europe. Most taxa had previously been lumped with common species from northern Europe (Crivelli 1996).

(2) By focussing on the global status of species, the list ignores local and regional trends in species loss. For example, Holcik (1996) considers 21 species of fish to be threatened in Slovakia, while the IUCN list includes only seven of them.

(3) The IUCN lists primarily freshwater species because the majority of information on fishes exists for freshwater taxa. Marine species are not only more poorly known, but tools normally employed for assessing extinction risk may not be appropriate for some marine species that have high reproductive potential, fast maturation, and wide distribution. Such fish are nevertheless vulnerable to extinction through a combination of human influences (e.g., overexploitation) and sudden changes in oceanographic conditions (including those caused by human-induced climate change). Some species of fish may be "passenger pigeon species," where the theoretical minimum viable population size may be a very large number.

Table 8.2
IUCN categories for species at risk (Modified from IUCN 1996).

Category	Definition
Extinct (E)	A taxon is **Extinct** when there is not reasonable doubt that the last individual has died.
Extinct in the Wild (EW)	A taxon is *Extinct in the Wild* when it is known only to survive in cultivation, in captivity or as a naturalized population (or populations) well outside the past range. A taxon is presumed extinct in the wild when exhaustive surveys in known and/or expected habitat, at appropriate times (diurnal, seasonal, annual), throughout its historic range have failed to record an individual. Surveys should be over a time frame appropriate to the taxon's life cycle and life form.
Critically Endangered (CR)	A taxon is *Critically Endangered* when it is facing an extremely high risk of extinction in the wild in the immediate future, as defined by specified criteria.
Endangered (EN)	A taxon is *Endangered* when it is not Critically Endangered but is facing a very high risk of extinction in the wild in the near future, as defined by specified criteria.
Vulnerable (VU)	A taxon is *Vulnerable* when it is not Critically Endangered or Endangered but is facing a high risk of extinction in the wild in the medium-term future, as defined by specified criteria.
Lower Risk (LR)	A taxon is *Lower Risk* when it has been evaluated, does not satisfy the criteria for any of the categories Critically Endangered, Endangered or Vulnerable. Taxa included in the Lower Risk category can be separated into three subcategories: 1. *Conservation-Dependent (cd).* Taxa which are the focus of a continuing taxon-specific or habitat-specific conservation program targeted toward the taxon in question, the cessation of which would result in the taxon qualifying for one of the threatened categories above within a period of five years. 2. *Near-Threatened (nt).* Taxa which do not qualify for Conservation Dependent, but which are close to qualifying for Vulnerable. 3. *Least Concern (lc).* Taxa which do not qualify for Conservation Dependent or Near Threatened.
Data-Deficient (DD)	A taxon is *Data-Deficient* when there is inadequate to make a direct, or indirect, assessment of its risk of extinction based on its distribution and/or population status. A taxon in this category may be well studied, and its biology well known, but appropriate data on abundance and/or distribution are lacking. Data Deficient is therefore not a category of threat or Lower Risk. Listing of taxa in this category indicates that more information is required and acknowledges the possibility that future research will show that threatened classification is appropriate. It is important to make positive use of whatever data are available. In many cases great care should be exercised in choosing between DD and threatened status. If the range of a taxon is suspected to be relatively circumscribed, if a considerable period of time has elapsed since the last record of the taxon, threatened status may well be justified.
Not Evaluated (NE)	A taxon is *Not Evaluated* when it has not yet been assessed against the criteria.

(4) The IUCN list considers only species and therefore ignores the high diversity of evolutionary lines below the species level. The U.S. Endangered Species Act of 1973 attempted to circumvent this problem by extending protection to distinct populations of vertebrates. More recently, attempts have been made to apply this level of protection to runs of Pacific salmon (*Oncorhynchus* spp.), many which are in danger of extinction in California, Oregon, and Washington, through the development of the concept of Evolutionary Significant Unit ("ESU"; Waples 1995). An ESU is a reproductively isolated population that contains enough genetic distinctiveness from past evolutionary events to have potential for its own evolutionary trajectory, independent of other similar populations. Unfortunately, our knowledge of ESUs for the vast majority of fishes (as well as for most organisms) is totally lacking.

(5) The status of fish species in many environments (e.g., deepsea, tropical rivers) is basically unknown.

Despite these shortcomings, we have used the IUCN list as the main source of information for this chapter, supplementing it with more detailed local information where appropriate. To demonstrate the breadth of taxa that are threatened and the diversity of causes of their declines, we sequentially consider conservation problems by reviewing selected taxa, then by habitat, and finally, by region.

THE CONSERVATION STATUS OF FISHES BY TAXA

Inspection of the IUCN list reveals that virtually any fish taxon can be threatened with extinction as the direct or indirect result of human activity. Not surprisingly, most at-risk species are members of speciose families such as the Cyprinidae, Cichlidae, and Gobiidae; however, in some cases entire taxonomic groups of fishes are at risk (Table 8.3). For example, 25 of the 27 sturgeons and paddlefishes (order Acipenseriformes) are threatened, as is the one species of coelacanth (order Coelacanthiformes). In the following sections we present an assessment of the conservation status of selected groups of fishes to demonstrate the breadth of taxa that are threatened and the diversity of causes of their declines.

Lampreys

The lampreys (Petromyzontiformes) are ancient, jawless vertebrates with complex life cycles, found in temperate regions of the northern and southern hemispheres (Moyle and Cech 1996). Of the 41 species currently recognized, IUCN (1996) considers nine (23%) as extinct, threatened, or near threatened, and another 7% as having insufficient data about which to make a determination of their status. Threatened lamprey species are mostly small, cryptic, non-predatory brook lampreys that require clean, flowing streams for their survival. Thus developed rivers within the industrialized regions of the northern hemisphere, such as the United States, Europe, and portions of Asia, contain the greatest number of threatened lamprey species. The single species thought to be extinct (*Lampetra minima*) is a small, predatory lamprey that was eradicated when the lake system it inhabited was treated with piscicides to kill off unwanted species (Miller, Williams, and Williams 1989). Two non-parasitic lamprey species within the genus *Mordacia*, one from Chile

Table 8.3
Species numbers for orders of fishes by IUCN status categories.[1]

Class/Order	IUCN Status Categories[2] EX (a)	CR (b)	EN (c)	VU (d)	LR (e)	DD (f)	Total Listed Species[3]	Total Described Species (%)[4]	Listed Families (%)[5]	Listed Genera (%)[6]
Class: Cephalaspidomorphi										
Petromyzontiformes	1	0	1	2	6	3	13	41 (32)	1 (100)	5 (83)
Class: Chondrichthyes										
Hexanchiformes	0	0	1	0	0	0	1	5 (20)	1 (50)	1 (25)
Orectolobiformes	0	0	0	0	0	1	1	31 (3)	1 (14)	1 (7)
Carchariniformes	0	1	1	2	0	0	4	208 (2)	1 (14)	1 (4)
Lamniformes	0	0	1	3	0	0	4	16 (25)	3 (43)	4 (40)
Squaliformes	0	0	0	1	0	0	1	74 (1)	1 (25)	1 (4)
Pristiophoriformes	0	0	4	0	0	0	4	5 (80)	1 (100)	1 (50)
Rajiformes	0	0	1	0	0	1	2	456 (<1)	2 (17)	2 (3)
Class: Sarcopterygii										
Coelacanthiformes	0	0	1	0	0	0	0	1 (100)	1 (100)	1 (100)
Class: Actinopterygii										
Acipenseriformes	0	6	11	8	2	0	27	27 (100)	2 (100)	6 (100)
Osteoglossiformes	0	0	1	0	2	1	4	217 (2)	2 (33)	3 (10)
Clupeiformes	0	0	2	2	1	8	13	357 (14)	2 (40)	7 (8)
Cypriniformes	18	41	36	114	75	69	353	2,662 (13)	4 (80)	126 (45)
Characiformes	0	0	1	0	0	6	7	1,343 (<1)	1 (10)	4 (1)
Siluriformes	1	8	7	22	7	24	69	2,405 (3)	14 (41)	35 (8)
Esociformes	0	0	0	1	1	0	2	10 (20)	1 (50)	2 (50)
Osmeriformes	1	4	1	11	5	8	30	236 (13)	5 (38)	11 (15)
Salmoniformes	4	4	5	10	6	14	43	66 (65)	1 (100)	10 (91)
Percopsiformes	0	1	0	3	0	0	4	9 (44)	1 (33)	3 (50)
Gadiformes	0	1	0	2	0	0	3	482 (<1)	2 (17)	3 (4)
Ophidiiformes	0	0	0	7	0	1	8	355 (2)	1 (20)	5 (5)
Batrachoidiformes	0	0	0	5	0	0	5	69 (7)	1 (100)	2 (11)
Lophiiformes	0	1	0	0	0	0	1	297 (<1)	1 (6)	1 (2)
Atheriniformes	1	6	5	31	23	28	94	285 (33)	2 (25)	22 (47)
Beloniformes	0	2	3	8	3	2	18	191 (9)	2 (40)	7 (18)

(continued)

Table 8.3 *(continued)*

Class/Order	IUCN Status Categories[2]						Total Listed Species[3]	Total Described Species (%)[4]	Listed Families (%)[5]	Listed Genera (%)[6]
	EX (a)	CR (b)	EN (c)	VU (d)	LR (e)	DD (f)				
Cyprinodontiformes	15	18	20	25	10	11	99	807 (12)	4 (50)	32 (36)
Gasterosteiformes	0	2	0	37	3	15	57	257 (22)	3 (27)	12 (17)
Synbranchiformes	0	0	1	0	0	4	5	87 (6)	2 (67)	4 (33)
Scorpaeniformes	1	3	1	5	3	2	15	1,271 (<1)	2 (8)	4 (1)
Perciformes	50	55	29	136	70	53	393	9,293 (4)	31 (21)	81 (5)
Pleuronectiformes	0	0	1	1	0	0	2	570 (<1)	1 (9)	2 (2)
Tetraodontiformes	0	0	0	3	0	5	8	339 (2)	2 (22)	6 (6)
Totals	91	153	134	439	217	256	1,290 (5)	22,464[7]	99 (21)	405 (10)

[1]Species data compiled from IUCN (1996). Classification of fishes follows Nelson (1994).
[2]EX = extinct, CR = critically endangered, EN = endangered, VU = vulnerable, LR = lower risk, DD = data deficient.
[3]Species totals for IUCN status categories from columns a through f.
[4]Total number of described species followed by the percentage of listed speices out of total described species.
[5]Total number of families within an IUCN status category followed by the percentage of listed families out of all described families within an order.
[6]Total number of genera within an IUCN status category followed by the percentage of listed general out of all described genera within an order.
[7]Total number of species for listed orders only. The total number of all described species is 24,600 (Nelson 1994).

and the other from Australia, are also of conservation concern (IUCN 1996). As often is the case with other anadromous fishes (e.g., salmon, sturgeon), threatened lamprey species have suffered population declines due to the blockage of migration routes from the construction of dams and water diversions on large rivers, as well as various forms of pollution and stream alteration.

Sharks and Skates

The sharks, skates, and their kin (*Elasmobranchi*) are a highly successful group (800+ species) of largely marine predators (Nelson 1994). There has been little concern for their status until recently, when alarming declines in the numbers of many species have been noted, largely due to overfishing for other taxa (Manire and Gruber 1990). Elasmobranchs are particularly vulnerable to the effects of overfishing because most species grow and mature slowly, and exhibit delayed reproduction and low fecundity (Hoenig and Gruber 1990). Adult survivorship is critical to the stability of populations and, therefore, increased mortality of adults by overfishing can have drastic effects on population structure and persistence (Dayton *et al.* 1995). However, often there are too little data on the magnitude of the fisheries or the status of individual stocks (Vas 1995).

One of the more serious threats to marine communities in general, and elasmobranchs in particular, is "bycatch" mortality from modern fishing methods (Dayton *et al.* 1995). Bycatch refers to organisms that are caught in addition to the target species. Bycatch not only directly affects populations of the non-target species, but may have indirect affects on the abundance and distribution of other organisms in the food web. International concern over the conservation status of sharks led to the establishment of the Shark Specialist Group (SSG) under the auspices of the IUCN in 1991 (Anon 1991). The SSG recently recommended the addition of 15 cartilaginous fishes to the 1996 IUCN *Red List of Threatened Animals*, and the specialist group is drafting an international *Action Plan for Chondrichthyan Fishes* (Fowler 1996).

Of the six species of sawfish (Pristidae), four are considered endangered by the IUCN (1996) because of population declines of at least 50% over the last ten years. One endangered sawfish of particular interest is *Pristis microdon*, a species known to ascend rivers in Australia and New Guinea.

Other elasmobranchs listed as endangered by the IUCN (1996) include the dusky shark (*Carcharhinus obscurus*) and the giant freshwater stingray (*Himantura chaophraya*), the latter being restricted to several rivers in Australia, Indonesia, Papua New Guinea, and Thailand. Finally, the critically endangered Ganges shark (*Glyphis gangeticus*) currently consists of a single subpopulation of less than 250 individuals, and is one of only seven shark species that ascend into freshwater (IUCN 1996).

Coelacanth

The coelacanth (*Latimeria chalumnae*) is the last remaining representative of the line of lobefinned fishes (Sarcopterygii) that gave rise to terrestrial vertebrates. It is a large (to 95 kg), slow-growing, live-bearing fish, which is largely confined to the deep waters off the Comoro Islands. A recent population estimate for coelacanths around Grand Comoro Island, the center of its range, is only 210 fish (Bruton 1995). The total number of adult individuals is likely to be less than 1,000. The main threat to the continued existence of the coelocanth is hook-and-line fisheries for other deep-dwelling fish as well as some demand for this lobefinned fish as a scientific curiosity.

Sturgeons and Paddlefishes

Perhaps more than any other group of fishes, the sturgeons and paddlefishes are in urgent need of protection and recovery. Of the 27 species described worldwide, almost all species (25/27) may be considered threatened (IUCN 1996). This is alarming if for nothing more than their place in the history of vertebrate evolution—sturgeons are the last remnants of the Chondrostei, the ancestral group of all modern ray-finned fishes with a fossil history dating back to the Jurassic (Grande and Bemis 1991).

Five species of sturgeon are critically endangered and near extinction by virtue of their very small, fragmented populations (IUCN 1996). These taxa include the Yangtze sturgeon (*Acipenser dabryanus*) from China; the Atlantic/Baltic sturgeon (*A. sturio*) from Europe and the Mediterranean region; and the central Asian Syr-Dar shovelnose (*Pseudoscaphirhynchus fedtschenkoi*) from Kazakhstan; the small Amu-Dar shovelnose (*P. hermanni*) from Turkmenistan and Uzbekistan; and the large Amu-Dar shovelnose (*P. kaufmanni*) from Tajikistan, Turkmenistan, and Uzbekistan.

The European sturgeon (*A. sturio*) is thought to persist only as two reproductively viable populations, one from the Garonne-Dordogne-Gironde River system in France and the other in River Rioni, a tributary to the Black Sea. These two populations number less than 600 and 1,000 individuals, respectively (Debus 1996). A host of interrelated factors are responsible for the decline of the European sturgeon, including the disruption of spawning migrations from dam construction, deterioration of spawning grounds from pollution, and the overfishing of juveniles and adults in coastal estuaries (Debus 1996).

Birstein (1993) recently provided a thorough review of the status of sturgeons and paddlefish in the former Soviet Union and China, two countries that account for 16 threatened species of sturgeon (IUCN 1996). The drying of the Aral Sea through agricultural water diversions, construction of canals and barriers to migrating fish, and pesticide pollution on tributary streams, is largely responsible for the decline of three species from the genus *Pseudocaphirhynchus*, as well as a fourth species, the ship sturgeon, *A. nudiventris*. Another species of sturgeon, the beluga (*Huso huso*), is one of the largest freshwater fish in the world, reaching lengths of 6 m and weights of up to 3,200 kg. The beluga has experienced drastic population declines during the last 50 years due to overfishing, pollution, and the construction of large dams that block migrations of the Volga, Don, and Danube rivers and their tributaries (Birstein 1993).

The two species of threatened paddlefish are *Polyodon spatula*, from the Mississippi River system of North America, and *Psephurus gladius*, from the Yangtze River system of China. Mississippi paddlefish are in trouble because pollution, dams, and harvest have reduced populations and blocked spawning migrations. Chinese paddlefish have declined for similar reasons and are facing an additional population decline of at least 80% in the next 10 years (or three generations), as the result of habitat changes caused by immense Three Gorges Dam.

Order Cypriniformes: Family Cyprinidae (minnows)

The Cypriniformes consist of about 2,700 species or 11% of the world's total fish species (Nelson 1994). The IUCN (1996) lists 190 and 18 species as threatened with extinction, or extinct, respectively. Another 77 species are considered "lower risk", while 69 species are "data deficient." It is likely that with additional studies, many of the species listed in these latter two categories would also qualify as threatened, as will many others not now listed.

Within the Cypriniformes, the Cyprinidae contain the largest number of listed species (n = 157, or 8% of the total species in the family), followed by the Balitoridae (n = 17, or 14%), Catostomidae (n = 13, or 19%) , and Cobitidae (n = 4, or 4%) (IUCN 1996). The Cyprinidae in North America contain about 50 genera and 270 species. The IUCN (1996) lists 45 species, or 24% of the total North American minnow species. The majority of threatened North American cyprinid genera are fishes confined to desert or Mediterranean climates within Mexico and the United States.

Threatened European and South African cyprinids inhabitat primarily riverine habitats that are impacted by dams, water diversions, the introduction of exotic fishes, and various forms of water pollution (Kirchhofer and Hefti 1996; Crivelli and Maitland 1995; Skelton 1990). Europe contains about 36 threatened cyprinids. Selected genera with the most listed threatened taxa include *Barbus* (14 species), *Leuciscus* (8), and *Paraphoxinus* (4) (IUCN 1996). Approximately 26% of all fishes in southern Africa are cyprinids, so it is not surprising that many threatened fishes are also from this family (Skelton 1990). For example barbs (*Barbus*) and redfins (*Pseudobarbus*) are two groups of cyprinids from South Africa, with ten and seven threatened species, respectively (IUCN 1996).

Trout and Salmon

Members of the Salmonidae (cisco, whitefish, trout, salmon, char, grayling), a group with a circumpolar distribution in the temperate regions of the world, are becoming increasingly threatened. Of the approximately 66 described species worldwide, 19 and 14 taxa, or 29% and 21% are considered threatened and data-deficient, respectively (IUCN 1996). An additional four species are listed as extinct. The Salmonidae is probably the best studied family of fishes in the world because of their economic and cultural importance in North America, Europe, and Japan. Salmonid studies have given us a good appreciation for the importance of the conservation of genetic and life history variation within species. However, because of lingering taxonomic confusion within several subfamilies of the Salmonidae (e.g., Coregoninae, Salmoninae), the conservation status of some taxa remains unclear.

Concern over the status of "stocks" (a concept roughly equivalent to populations) of the seven species of Pacific salmon (*Oncorhynchus*) has led to the publication of numerous reports and books on their conservation (e.g., National Research Council 1996; Hedgecock, Siri, and Strong 1994). Nehlsen, Williams, and Lichatowich (1991) identified 214 native Pacific salmon stocks whose persistence is threatened: 101 are considered at a high risk of extinction; 58 at a moderate risk of extinction; and 54 of special conservation concern. An additional 106 populations were identified as extinct, 67 of which were from the Columbia River Basin of Washington and Idaho alone (Nehlsen, Williams, and Lichatowich 1991). The main reasons for the decline of anadromous fishes in general, including Pacific salmonids, are habitat loss and degradation from water quality degradation, dam construction, diversions, overfishing of native fish in mixed stock fisheries, and interbreeding between native and hatchery stocks (Moyle 1994; Nehlsen 1994).

Similarly, populations of several European salmonids are of conservation concern. Most notably, some local stocks of brown trout (*Salmo trutta*), particularly sea-migrating as well as lake populations, have experienced ongoing population declines (Ruhle 1996). Other threatened salmonids in Europe include the world's largest salmonid, the Danube huchen (*Hucho hucho*), which occupies the Danube River of Central Europe, and several salmonid species in the southern Mediterranean region. Unfortunately, local populations of a number of endemic trout species are disappearing before their taxonomy and genetic relationships have been adequately described. For example, local endemics such as the marbled trout (*Salmo marmoratus*), that occur in the rivers of Italy, Slovenia, Croatia, Bosnia-Herzegovina, Montenegro, and Albania are being displaced by, or hybridizing with, introduced brown trout (Crivelli 1995).

Other threatened salmonids include members of the genus *Coregonus* (e.g., cisco, whitefish, vendace), primarily from the Great Lakes of the United States and Canada, but also from Russia and Northern Europe. Three species of cisco from the North American Great Lakes, the longjaw cisco (*C. alpenae*), deepwater cisco (*C. johannae*), and blackfin cisco (*C. nigripinnis*), are now extinct, largely as the result of commercial overfishing during the late 19th century. Another unusual species that is threatened with extinction is the long-finned charr (*Salvethymus svetovidovi*). This plantivorous charr is endemic to Russia's Lake El'gygtgyn, an ultraoligotrophic lake formed approximately 3.5 million years ago from a meteorite crater, which served as a refugium during the Pleistocene glaciation (Chereshnev and Skopets 1990; Nelson 1994). Similarly, the Lake Ohrid trout, *Acantholingua ohridana*, is a monotypic genus of trout endemic to Lake Ohrid in Albania and Macedonia that is listed as vulnerable.

Freshwater Smelts

The order Osmeriformes contains two groups (making up the suborder Osmeroidei) that are largely found in freshwater—i.e., the southern hemisphere "smelts" (Galaxidae, Retropinnidae, and Lepidogalaxidae) and the northern hemisphere smelts (Osmeridae, Salangidae, Sundasalangidae). The 72 species in these families are all small in size but most are prized by humans as food. Perhaps because of this favor they find with humans, 30 (42%) have made it to the 1996 IUCN list.

Galaxids are small fishes characteristic of streams in Australia, New Zealand, southern South America, and South Africa, a distribution reflecting their ancient ties to the original southern continent, Gondwana. Most are at risk because of alterations of their stream habitats, combined with the introduction of predatory salmonids, especially brown trout (*Salmo trutta*), into their stream habitats. Galaxids currently persist in stream refugia from which brown trout are absent. In Australia, four species are critically endangered because of populations declines of at least 80% in the last ten years (IUCN 1996).

Of the three species of osmerids on the IUCN list, the delta smelt (*Hypomesus transpacificus*) is the most endangered. This species is confined to the Sacramento-San Joaquin Estuary of California. It has shown a major decline in its population due to massive changes in the Estuary, including increased water diversions (Moyle *et al.* 1992) and invasion by the inland silverside (*Menidia beryllina*), an egg and larval predator. An additional threat to its existence has been the recent invasion of the wakasagi (*H. nipponensis*), introduced from Japan, with which it hybridizes (Moyle unpublished data). The delta smelt therefore is a classic example of a species whose exact cause of decline is hard to delimit because so many different negative factors are acting on it at once (Bennett and Moyle 1996).

Pupfishes and Splitfins

The pupfishes (Cyprinodontidae) and splitfins (Goodeidae), members of the Cyprinodontiformes, contain about 100 and 40 species, respectively (Nelson 1994). Splitfins primarily inhabit small isolated springs, pools, and streams in the desert regions of Nevada and west central Mexico (Mesa Central and Rio Lerma basin)—aquatic habitats that are particularly vulnerable to human activities such as groundwater pumping for agricultural uses and other water development schemes, habitat conversion, and the introduction of exotic fishes (Minckley and Deacon 1991). Human-caused extinctions have occurred in several genera within the Goodeidae including *Ameca* (1 extinct species), *Characodon* (1), *Empetrichthys* (1), and *Skiffia* (1). Several additional splitfin species representing seven genera are threatened with extinction in Mexico (IUCN 1996).

The cyprinodonts are found in fresh, brackish, and coastal marine habitats in North and Central America and parts of South America, as well as northern Africa and the Mediterranean Anatolian region (Nelson 1994). Alarmingly, approximately 40% of the species in this family are either extinct, extinct in the wild, or threatened with extinction (IUCN 1996). As in the splitfins, pupfishes are often restricted to isolated habitat in arid environments such as springs and pools, intermittent streams, and small lakes that are particulary susceptable to human perturbations. For example, most species within the genus *Cyprinodon* of the deserts of the United States and Mexico are threatened with extinction (Minckley and Deacon 1991). One species, the Devil's Hole pupfish (*Cyprinodon diabolis*), has the smallest geographic distribution of any vertebrate, its entire population restricted to a tiny thermal spring in the Nevada desert (Minckley, Hendrickson, and Bond 1986).

Recent extinctions of Mexican fishes include cyprinids from the genera *Evarra* (3 extinct species), *Notropis* (3), and *Stypodon* (1); cyprinodont species from the genera *Cyprinodon* (6), and *Megupsilon* (1); and the genus *Priapella* (1) within the family Poeciliidae (IUCN 1996; Miller, Williams, and Williams 1989). Similarly, the most critically endangered fishes of Mexico are within the families Cyprinidae, Cyprinodontidae, Goodeidae, and Poeciliidae.

Other cyprinodonts threatened with extinction include several species within the genera *Aphanius* and *Valencia* (toothcarps) from the Mediterranean regions of Spain, Turkey, and northern Africa; members of the genus *Orestrias* from lakes on the continental divide of Chile, Bolivia and Peru (e.g., Lake Titicaca); and *Pantanodon madagascariensis* from the island of Madagascar (IUCN 1996).

Pipefishes and Seahorses

The IUCN (1996) lists 37 species of pipefish and seahorse as threatened, or approximately 43% of the 87 described taxa. An additional 11 species (13%) have insufficient data to determine their conservation status. The vast majority of the listed taxa are from the Indian Ocean and northwest and western central Pacific Ocean, perhaps reflecting the family's importance in the commercial medicinal trade of the Old World. Though largely a marine family, three freshwater forms are also threatened, most notably the critically endangered river pipefish, *Syngnathus watermayeri*, from South Africa (IUCN 1996).

Cichlids

The Perciformes are the largest order of fishes, containing approximately 9,300 described species. Within the Perciformes are the Cichlidae, which conform to at least 1,500 species, with well over half of these occurring in Africa (Nelson 1994). Of enormous scientific interest are the diverse species flocks of cichlids that evolved in the Great Rift Lakes of East Africa—of interest because of their extraordinary rates of speciation and as an example of the how an entire native freshwater fish community that exhibited extreme trophic radiation may virtually disappear in a few decades because of human activities

(Kaufman 1992). Of the more than 300 haplochromine cichlids endemic to Lake Victoria, about two-thirds, or 200+ species, have gone extinct over the last 30 years (Lowe-McConnell 1993). The dramatic and tragic loss of cichlids in Lake Victoria is largely attributable to predation pressures from the introduction of the Nile perch (*Lates niloticus*) and exotic tilapias interacting with other factors such as increased fishing pressures and habitat alteration (e.g., pollution) (Ogutu-Ohwayo 1990; Witte *et al.* 1992a, 1992b; Lowe-McConnell 1993).

The recent species changes have had other dramatic effects on the ecology of Lake Victoria. For example, severe deoxygenation now occurs in 25–50 m depth range, with local upwellings of hypoxic water resulting in fish kills. Eutrophication and associated algal blooms are also common (Kaufman 1992; Bootsma and Hecky 1993; Goldschmidt, Witte, and Wanink 1993). In addition, the introduced Nile perch and Nile tilapia (*Oreochromis niloticus*) now account for over 80% of the lake's biomass; native cichlids comprise less than 1% of the total catch (Stiassny 1996). As in the case of Lake Victoria, introduction of the Nile perch into two nearby lakes, Lake Kyoga and Lake Nabugabo, has resulted in the depletion and extinction of many native cichlid species (Ogutu-Ohwayo 1993). In addition, Lake Tanganyika, which has more fish families (n = 24) than any of the Great Rift Lakes and contains over 287 fish species of which 220+ are endemic (76%), including over 172 endemic cichlid species, is threatened threatened with sediment pollution from deforestation within its watershed (Lowe-McConnell 1993; Hori *et al.* 1993; Cohen *et al.* 1993).

This pattern of high diversity and endemism of fishes is similar for Lake Malawi, the third largest lake in Africa. Lake Malawi supports up to 500 cichlid species, almost all of which are endemic, as well as a species flock of clariid catfish, and endemic cyprinids (Lowe-McConnell 1993; Reinthal 1993). However, the Lake Malawi fish fauna is relatively intact compared to Lake Victoria, and as such, there is still time to implement conservation measures to protect the entire lake ecosystem. Reinthal (1993), in recognizing the importance of managing the entire ecosystem rather than these particular species or groups of species, has recommended an ecosystem approach to the conservation of Lake Malawi fishes. Such recommendations include the establishment of an international park between the countries of Malawi, Tanzania, and Mozambique.

A situation similar to that of Lake Victoria could develop for the 16+ species of native cichlids in Lake Nicaragua, Central America, the largest tropical lake outside Africa. The introduction into Lake Nicaragua of several species of African tilapia (genus *Oreochromis*) in the 1980s to increase the fishery, corresponded with declines in native cichlid yields (McKaye *et al.* 1995). There is concern that in addition to the loss of the native cichlid fauna, introduced tilapia will also affect the planktonic community and primary productivity of Lake Nicaragua (McKaye *et al.* 1995).

HABITAT UTILIZATION BY THREATENED FISHES

To discern general patterns of habitat utilization by fishes that are threatened, we categorized the percentage of fish species according to habitat types for the various IUCN status categories (Table 8.4). Species most likely to be threatened with immediate extinction occur primarily in lacustrine or inland sea habitats, as well as in rivers. Interestingly, most threatened species utilize multiple habitat types, likely reflecting their complex life histories. Anadromous fishes such as salmon and sturgeon, for example, exhibit life

Table 8.4

Percentage of fish species occurring in generalized habitat types for various IUCN status categories.

Status Category[2]	Marine	Estuarine	Large Riverine	Small Riverine	Lacustrine	Spring/Cave
Extinct	1	1	13	12	62	12
Extinct in the Wild	0	0	9	27	36	27
Critically Endangered	8	6	20	22	36	8
Endangered	16	12	23	22	19	8
Vulnerable	19	4	21	29	20	5
Data-Deficient	15	11	28	34	11	1
Lower Risk (near threatened)	11	14	27	40	6	1
Lower Risk (conservation dependent)	11	5	28	56	0	0
Lower Risk (least concern)	7	8	40	28	14	3

[1]Data compiled from IUCN (1996)
[2]Status categories defined in Table 2.

histories that require time spent in marine, estuarine, riverine, and in some cases, lacustrine habitats.

Human activities have disrupted natural flow patterns and ecological processes on large rivers throughout the world, with adverse affects on their native fish faunas (Moyle and Leidy 1992; Petts 1984). For example, 77% of the 139 largest river systems (discharges of 350 cubic meters per second or greater) in the northern third of the world have been strongly or moderately affected by fragmentation through the construction of dams, reservoir operation, interbasin water transfers, and irrigation practices (Dynesius and Nilsson 1994; see chapter 4). However, the percentage of threatened and extinct species occurring in small rivers is also increasing, an indication of the degradation of smaller tributaries as the direct and indirect effects of human impacts in larger rivers spread upstream. Further support that small river fishes are increasingly threatened also is evidenced by the large percentage of species listed within the "data-deficient" and "near-threatened" categories. The largest percentages of "extinct" and "extinct in the wild" fishes are for lacustrine or spring/cave habitats, which is not surprising given the insular nature of these environments. Other threatened groups of fishes include freshwater species in arid environments, species dependent upon coastal estuaries and reefs or other shallow-water marine habitats, and endemic species restricted to very small areas, such as springs or caves.

REGIONAL STATUS OF FISH FAUNA

Clearly, freshwater fishes and their habitats continue being degraded and lost more rapidly than marine fishes and habitats, although there is a growing list of exceptions (e.g., enclosed marine environments such as the Aral Sea, near coastal coral reef environments). This pattern may be changing, however, as exploitation of oceanic resources, marine pollution, and coastal habitat alteration increase (Hughs 1994; see chapter 6). Marine communities, with the exception of certain near coastal environments and commercially important fish species, are monitored much less closely than freshwater communities.

In this section we summarize briefly what is known about the status of freshwater fishes in various regions of the world. Specifically, we discuss fish faunas of selected regions that have adequate information on the status of their fish faunas: (1) Europe; (2) Africa; (3) Asia; (4) Australia; (5) Central and South America; (6) North America; and (7) oceanic islands and other insular environments. We conclude with a brief overview of the status of fishes in marine environments.

Western Europe

The status of freshwater fishes in several European countries' fresh water has been recently reviewed by Crivelli and Maitland (1995) and Kirchhofer and Hefti (1996). Below we present an overview of the conservation status of fishes in several European countries to illustrate the problems faced by fishes.

British Isles and Ireland. The native fish faunas of the British Isles are relatively depauparate compared to continental Europe (56 vs. 221 species, respectively). This is largely because the islands of Great Britain and Ireland are relatively small, the geography relatively homogeneous, and the opportunities for speciation much reduced compared to the remainder of Europe. Maitland and Lyle (1991, 1996) and Sweetman, Maitland, and Lyle (1996) identify nine species that are either extinct or threatened in the United

Kingdom. Five of these species are anadromous: the European sturgeon, Allis shad (*Alosa alosa*), Twaite shad (*A. fallax*), Arctic charr (*Salvelinu salpinus*), houting (*Coregonus oxyrinchus*), and powan (*C. lavaretus*), while the remaining four species are entirely freshwater (Winfield, Fletcher, and Cragg-Hine 1994). In addition to these anadromous species, an additional anadromous smelt, *Osmerus eperlanus*, may be considered "vulnerable" in Ireland (Quigley and Flannery 1996). The IUCN (1996) lists only two freshwater fishes as threatened within the United Kingdom and Ireland. The reason for the discrepancy between the IUCN (1996) listing and others may be due to the fact that the IUCN lists most of the above anadromous forms as lacking sufficient data to make an assessment of risk possible.

France. Of the 49 native freshwater fish species known to occur in France, 25 (51%) are considered threatened by the IUCN (IUCN 1996). The construction of large dams and diversions that disrupt spawning migrations has been the primary cause for declines in anadromous species. For example, the European sturgeon frequented the lower reaches of many of the larger rivers in France (e.g., Seine, Loire, Rhone, Saone, and Doubs rivers) until the latter half of the 19th century at which time most runs experienced precipitous declines and ultimately extinction (Keith and Allardi 1996). Today sturgeon only occur in the Garonne and Dorgogne Rivers where they are extremely rare (Keith and Allardi 1996). Other diadromous species that are threatened include the sea lamprey (*Petromyzon marinus*), river lamprey, (*Lampetra fluviatilis*), Allis shad (*Alosa alosa*), Atlantic salmon (*S. salar*), sea trout (*S.* trutta *trutta*), grayling (*Thymallus thymallus*), and eel (*Anguilla anguilla*).

Operation of hydroelectric dams on several large rivers in France has been identified as a contributing factor in the decline of France's fish fauna, including the zingel (*Zingel asper*), pike (*Esox lucius*), and grayling (Keith and Allardi 1996). Organic and chemical pollution also have been identified as a primary factor in the decline of native fishes, as 37% of the stream sections in the French Mediterranean basin (134,360 km^2 or 24% of the area of France) studied by Changuex and Pont (1995) were either detrimental to fish reproduction or survival or totally unsuitable for most native fishes. The introduction of exotic fishes has also led to the decline of two endemic freshwater fishes. It is thought that population declines in the endemic cyprinid *Chondrostoma toxostoma* may be the result of competition with *C. nasus*, a species that is spreading through artificial connections between adjacent drainages. The introduced *Gambusia affinis* has been implicated in the decline of the endemic cyprinodont *Aphanius fasciatus* in lagoon environments (Changeux and Pont 1995).

Spain and Portugal. The Iberian peninsula exhibits a high degree (62%) of endemism in its freshwater fishes, particularly in the Cyprinidae (90% endemic), Cobitidae (> 90% endemic) and Cyprinodontidae (100% endemic) (Almaca 1995; Elvira 1995a). These three families account for approximately 90% of the endemic freshwater fishes occurring within Spain, Portugal, and adjacent areas (Blanco and Gonzalez 1992). Elvira (1996) lists 28 threatened species and subspecies of freshwater fish in Spain, 17 (61%) of which are endemic taxa. Notably, of approximately twenty species and subspecies of endemic cyprinids, 12 taxa (60%) may be considered threatened, or are insufficiently known to determine their conservation status (Elvira 1995a, 1996). Given the great habitat degradation within the Iberian penninsula, a reasonable estimate of the percentage of threatened freshwater species out of the total freshwater fish fauna in Portugal and Spain is forty percent.

Overall threats to Spanish and Portuguese freshwater fishes include habitat destruction, especially the filling of coastal wetlands and sand extraction, dam construction, pollution, overfishing (*i.e.*, *Alosa fallax rhodanensis*), the introduction of exotic fish predators (most notably *Micropterus salmoides*, *Lepomis cyanellus*, *Oncorhynchus mykiss*, *Gambusia holbrooki*, and *Fundulus heteroclitus*), stream desiccation through small water diversions, use of native fishes as bait, and hybridization between endemic and exotic forms (Elvira 1990; Almaca 1995; Elvira 1995b, 1996). Loss of coastal wetlands and the introduction of exotic *Gambusia holbrooki* and *Fundulus heteroclitus* are thought to be responsible for the localized extinction and overall population declines of the endemic toothcarps *Valencia hispanica* and *Aphanius iberus* (Elvira 1990).

Italy. Bianco (1995) lists 71 species as occurring in Italian freshwaters, of which 44 species (61%) are native. As in several other Mediterranean countries, Italy has a highly endemic freshwater fish fauna (61%), especially within the Cyprinidae (14 species). Of the 44 natives, 25 are endemic to Italy and the western Balkans, while three are Mediterranean endemics and the remainder are found in other non-Mediterranean countries (Bianco 1995). The IUCN (1996) lists nine freshwater fishes as either endangered or vulnerable, and another nine species as "near-threatened." These three categories represent 40% of the total native fish fauna. An additional 13 species lack the data to make an assessment of possible risk, although it is likely that with additional data some of these fishes would qualify as threatened. Two anadromous species, the beluga and possibly the European sturgeon, may be considered extinct in Italy (IUCN 1996). Thus 84% of the native fish fauna is either extinct or at risk of extinction.

Italy has been subjected to a large number of introductions of exotic fishes as well as the deliberate human transfer of native endemic forms between drainages. This transfer of native fishes during the past 20 years between geographically isolated drainage basins has had negative effects on the receiving fish faunas through interspecific hybridization and the resultant loss of genetic identity of formerly isolated forms, as well as through population declines and local extinctions as a result of predation and competition (Bianco 1990). Bianco (1995) found that only 36% of the original native fishes have not been contaminated by introductions of conspecific populations or introduced to one or more localities outside their original range.

An illustration of the range and complexity of threats facing Italian freshwater fishes through faunal transfers may be found in the Cyprinidae. While *Rutilus rubilio* has been eliminated from its native habitats in central Italy by transfers of *R. aula* and other lacustrine cyprinids, *R. rubilio* itself has been introduced into rivers in southern Italy where it has eliminated the endemic cyprind *Alburnus albidus* (Bianco 1990). In addition, *A. albidus* has hybridized with the introduced *Leucisans cephalus* (Bianco 1982). Similarly, the transfer of the endemic *Barbus plebejus plebejus* into the rivers of central Italy has resulted in massive hybridization with *B. fucini* (Bianco 1995). Bianco (1995) has predicted that continued interbasin transfers will eventually result in the zoogeographic and genetic homogenization of the entire Italian fish fauna.

Central Europe

The IUCN (1996) lists 11 species as threatened in Hungarian and Romanian waters, or about 17% of the total native faunas. Six of the most endangered species in Hungary and Romania are the Russian sturgeon (*Acipenser gueldenstaedti*), ship sturgeon, beluga, sterlet (*A. ruthenus*), stellate sturgeon (*A. stellatus*), and European sturgeon. The European mudminnow, *Umbra krameri*, which occurs in Hungary, is considered threatened throughout Europe (IUCN 1996).

Slovakia contains a relatively rich native fish fauna (approximately 67 taxa) for its small size (49,000 km^2), in part because of its shared drainages with the Danube and Tisza Rivers (Ponto-Caspian-Aralian Province) and the Vistula River (Baltic Province) (Holcik 1996). The IUCN (1996) lists seven species as threatened. With the exception of population declines and the extinction of several sturgeon species (four forms are now considered extinct and another two critically endangered), the native fish fauna of Slovakia remained relatively intact prior to the 1960's, at which time increased pollution, overfishing, and the construction of dams and channelization of rivers resulted in declines of other species (Holcik 1996). One species considered critically endangered by Holcik (1996) is the wild form of the common carp (*Cyprinus carpio*). Interestingly, the common carp has been widely introduced throughout the world, especially in North America, where it is considered a pest and has been implicated in the decline (and in some cases) the extinction of native fishes. Other critically endangered species in Slovakia include two species of cyprinid (*Gobio kessleri* and *G. uranoscopus*), the weatherfish (*Misgurnus fossilis*), the European mudminnow, and the Danubian huchen, the largest salmonid in the world (Holcik 1996).

The Czech Republic contains three hydrographic provinces that drain into the North, Baltic, and Black Seas. Its fish fauna is similar to, but less diverse than, Slovakia. The IUCN (1996) lists six out of approximately 53 species (11%) as threatened, with another 12 and four species listed as "data-deficient" and "near-threatened" respectively. Combining the IUCN assessment with data from a recent review by Lusk (1996), realistically places the number of threatened species in the Czech Republic at 14, while another 11 taxa known to historically occur in waters of the Czech Republic are now considered extinct within its borders. Thus, 47% of the Czech fish fauna is either threatened or has been extirpated. Three of the most threatened fishes in the Czech Republic are members of the Percidae: the striped ruffe or yellow pope (*Gymnocephalus schraetser*), that occurs primarily in the Danube River basin and has experienced drastic population declines in several countries within its range; and the streber (*Zingel streber*) and zingel (IUCN 1996). These percids along with several members of the Cyprinidae have experienced population declines due in large part to channelization, dam construction, and flow regulation schemes (Jurajda and Penaz 1996).

The Mediterranean Region of Slovenia, Croatia, Montenegro, and Albania. The countries of Slovenia, Croatia, Montenegro, and Albania cover a relatively small geographic area. However, the complex hydrogeography of the region has resulted in a fish fauna that exhibits the highest degree of endemism in Europe. Notwithstanding this diversity, the taxonomy and ecology of many local endemics remains poorly understood, and under increasing development pressures, many forms may disappear before being adequately described.

Slovenia is a small country (20,000 km^2), however it contains 25 and 61 native fish taxa in its Adriatic and Danube River drainages, respectively (Povz 1995). The IUCN (1996) lists five species as threatened and another 13 as "data-deficient." In marked contrast to the IUCN estimate of threatened status is that of (Povz 1996) who, using the pre-1996 IUCN Red Data Book status categories, lists 23 taxa as endangered, eight as vulnerable, and four as rare, for a total of 35 taxa. This would compare to approximately 23 taxa listed for all IUCN (1996) status categories. This discrepancy between the two estimates probably reflects in part, the lack of thorough information on status for certain species, as well as differences in the interpretation of status categories.

Povz (1996) identified pollution, primarily in the form nutrient and pesticide inputs from agriculture, channelization, regulation of flows, and the introduction of exotic fishes as the primary causes of species declines in Slovenia. For example, it is estimated that only 5% of the total length of rivers in Slovenia remain unaffected by pollution and channelization, and other forms of hydrologic or hydraulic regulation (Povz 1996).

Croatia contains approximately 63 and 50 native species and subspecies within the Danube River and Adriatic basins, respectively (Mrakovcic, Misetic, and Povz 1995). The IUCN lists 20 species, or approximately 17% of the total native fish fauna as threatened. However, an additional 18% (n = 21 species) of the fish fauna does not have sufficient data to determine the degree of threat.

The fish fauna of Montenegro is split between catchments that flow into either the Black or Adriatic seas. Many of the endemic fishes are known from a series of shallow, lowland lakes (e.g., Lake Skadar, Lake Sasko) and their tributaries (Maric 1995). Commercial and sport overfishing, fishing with non-selective gear, and municipal and industrial wastes have contributed to declines in the catches of once common commercial species as well as non-commercial taxa. For example, the sturgeons *Acipenser naccarii*, and *A. sturio* have not been recorded in Lake Skadar for 15 years, and catches and size distribution of two Salmonidae (*Salmo marmoratus* and *Salmothymus obtusirostris zetensis*) have also declined (Maric 1995). *Salmo obtusirostris zetensis*, along with the endemic cyprinid *Phoxinellus stimphalicus montenegrinus*, are now the two most endangered fish in Montenegro.

Very little information on the taxonomy, ecology, and status of Albanian fishes is published. Albania supports 54 species of native freshwater fish, 43 of which are endemic to the northern Mediterranean region (Rakaj and Flloko 1995). The IUCN (1996) lists seven species as threatened and an additional 16 and six species as "data-deficient" and "near-threatened" status, respectively. Many of the species with insufficient data on their taxonomy and status are members of the Cyprinidae (12 of 16 total listed species), especially members of the genus *Barbus* (IUCN 1996). Another Mediterranean endemic cyprinid, *Chondrostoma scodrensis*, occurs in Lake Skadar and is considered critically endangered, with estimates of at least an 80% decline in its population over the last ten years or three generations (IUCN 1996). Conservatively assuming that one-third of the "data-deficient" and "near-threatened" species warrant listing as threatened means that approximately 25% of the fish fauna is threatened.

Greece. The complex geological history of Greece, combined with its arid climate, has effectively restricted the movement of fish between drainages, leading to high species diversity (n = 100) and endemism (48%) (Economidis 1995). Native fishes of Greece have persisted through thousands of years of human occupancy of the region. However, over the last thirty years the fish fauna has been more seriously impacted by human activities than it has in the previous 3,000 years. Economidis (1995) identifies water extraction, overfishing, tourism, agricultural activities, industrial development, and the introduction of exotic species as having negative effects on native fish communities. In two instances, the development of large karstic limestone springs (one within Sperchios River drainage in central Greece, the other on the island of Lefkas in the Ionian Sea) caused the springs to go completely dry, with the concomitant loss of two endemic fishes originally described from these springs, *Pungitius hellenicus* and *Economidichthys pygmaeus* (Economidis and Miller 1990; Economidis 1995). Of the 22 endangered, threatened, and rare Greek fishes reviewed by Economidis (1995), water extraction for human uses was listed as a contributing factor for population declines in 15 of the species. The IUCN (1996) lists 16 fishes, as threatened, with an additional 26 species for which there is insufficient data to determine extinction risk. Approximately one-quarter to one-third of the fish fauna of Greece can be regarded as at risk of extinction in the near future.

Turkey. Turkey shares a fish fauna that is characteristic of Europe and Asia, and is an example of where information on the conservation status of the fish fauna is incomplete. Balik (1995) lists 109 taxa (88 species and 21 subspecies) as occurring in the freshwaters of Turkey. While the IUCN (1996) categorizes 18 species as threatened, another 37 taxa lack the data to make an assessment of risk. Again, a conservative assumption of one-third of the "data-deficient" taxa as warranting listing as threatened combined with those already listed, represents 28 percent of the total fish fauna.

One area of Turkey where there is some information of the status of its freshwater fishes is the fast developing southern Mediterranean region of Anatolia (Balik 1995). The Anatolia region contains 20 and ten endemic species and subspecies, respectively, with representatives primarily from the Cyprinidae (n= 20) (Balik 1995). Several of these endemic cyprinids inhabit a series of large inland freshwater lakes that are being degraded through increased inputs of industrial and domestic sewage, overfishing, the introduction of exotic fishes, and reclamation. For example, the draining of Lake Amik has resulted in the extinction of the endemic cyprinid *Tor canis* (Balik 1995). Two other critically endangered cyprinids, *Phoxinellus egridiri* and *P*. handlirschi, occur only in Lake Egirdir where they have not been found recently (Balik 1995). The introduced zander, *Stizostedion lucioperca*, a large predatory percid that feeds primarily on small fish, is thought to be responsible for the possible extirpation of both cyprinids.

Africa

Africa contains approximately 2800 + species of freshwater fish, or 10% to the world's total (Daget *et al.* 1991). The IUCN (1996) lists 154 African species as threatened, or approximately 6% of the African fish fauna. This certainly greatly underestimates the number of threatened species, as much of Africa's freshwater fish fauna is poorly known (Stiassny 1996). In addition, only six of 57 countries (Cameroon, Kenya, Madagascar, South Africa, Tanzania, and Uganda) account for over 86% of the fishes considered threatened, reflecting in part a better understanding of the conservation status of their fish faunas compared to the remainder of Africa. However, these are also areas dominated by arid or Mediterranean climates. Countries in Africa that contain a large number of threatened species include South Africa, Cameroon, and the island of Madagascar. South Africa has 27 threatened species, and an additional 14 species that are considered "near-threatened" (IUCN 1996). Cyprinids within the genus *Barbus* account for about 30% of the threatened species. Two species, the Twee River redfin (*B. erubescens*) and the Treue River barb (*B. trevelyani*) are "critically endangered" having experienced drastic population declines over the last ten years (IUCN 1996).

East Africa. Eighty-one threatened species (53%) are from a single family, the Cichlidae, in Lake Victoria (Kenya, Tanzania, and Uganda). Approximately 200+ cichlid species from Lake Victoria are listed as extinct or extinct in the wild, while many of those remaining out of the approximately 300 original species are considered critically endangered, and many more will likely go extinct. Arguably, the catastrophic declines and extinctions of cichlids in Lake Victoria represnt the most dramatic example of human-caused extinctions within an ecosystem. Causes of the decline are attributable to a host of interacting physical and biological factors, but introduction of the predatory Nile perch, combined with intensive fishing pressure and habitat degradation are the primary factors (Hughes 1986, Kaufman 1992).

Selected threatened South African fishes include two riverine species of catfish within the family Bagridae, Knysna seahorse (*Hippocampus capensis*), freshwater river pipefish (*Syngnathus watermayeri*), Saint Lucia mullet (*Liza luciae*), and several species of goby (IUCN 1996). Skelton (1990) identified threats to southern African fishes, which includes South Africa, including stream channelization, water diversions and extraction, agricultural development, dams and weirs, water pollution, and the introduction of exotic fishes.

Cameroon has 26 threatened fish species, 25 of which are cichlids. Eleven of the most critically endangered cichlid species belong to the genera *Konia*, *Myaka Sarotherodon*, and *Stomatepia*, and inhabit Lake Barombi, while the remaining 14 species are from the genus *Tilapia* in lakes Ejagham and Bermin (IUCN 1996).

Madagascar. The island of Madagascar off the east coast of Africa is noteworthy for its long isolation and resultant highly unusual and endemic flora and fauna; about 84% of the island's freshwater fishes are endemic (Stiassny 1996). The IUCN (1996) lists 13 threatened species in Madagascar, another nine as "data deficient", and 14 of "least conservation concern" in terms of population risk. However, given continuing trends of habitat degradation many species in the latter category of fishes may soon also be threatened. Deforestation and the introduction of exotic fishes have been identified as the greatest threats to the fishes of Madagascar (Reinthal and Stiassny 1991, Stiassny and Raminosoa 1994, Stiassny 1996). Threatened freshwater fishes in Madagascar include the zona (*Rheocles wrightae*), an atherinid inhabiting the River Manambola of eastern Madagascar; the cyprinodont, *Pantanondon madagascariensis*; several riverine species of cichlid from the genera *Paretroplus*, *Ptychochromis*, *Ptychochromoides*, and *Oxylapia*; two gudgeons in the genus *Typhleotris*; and a goby, *Glossogobius ankarensis* (IUCN 1996).

Asia

Of the 51 Asian countries listed by the IUCN (1996), nine countries (Indonesia, China, Malaysia, Philippines, Russia, Sri Lanka, Thailand, Turkey, and Ukraine) account for 71% of the fish species listed as threatened. Indonesia is the world's fourth most populous nation (estimated 190 million), and its burgeoning population along with deforestation and rice cultivation has caused population declines in many of its native freshwater fishes. Indonesia, therefore, has the greatest number of listed threatened fish species in Asia (n = 60), with an additional 21 species having insufficient data to determine their conservation status (IUCN 1996). Many of Indonesia's threatened fishes are members of the Atheriniformes (rainbowfishes), and include the endemic Celebes rainbowfish of Lakes Malili, Towuti, Matano, and Poso on the island of Sulawesi and several cyprinids from Lakes Toba, Tawar, and Bratran on the island of Sumatra (Nelson 1994, IUCN 1996). Other Indonesian freshwater fishes that are endangered include the freshwater sawfish (*Pristis microdon*), giant freshwater stingray (*Himantura chaophraya*), and Asian bonytongue (*Scleropages formosus*) (IUCN 1996).

China. China is the most populous country in the world with well over one billion people, yet it has only 28 threatened fish species. However, if one considers China's size and population relative to other countries such as the United States which is only a little larger in terms of area, but with one-fifth the population, it has surprisingly few listed threatened fishes—i.e., 26 for China compared to 123 in the United States. This clearly is the result of much better information and concern for the status of fishes in the United States, combined the high degree of development-related impacts to aquatic habitats in the United States. However, the number of threatened fishes and aquatic habitats in China is likely to increase dramatically in the near future with increasing economic growth and wide-scale alteration of river systems.

Eleven of China's threatened fishes are cyprinids, and nine of these inhabit lakes Dianchi and Fuxian (IUCN 1996). Four other endangered large river fishes that have experienced significant population declines over the last ten years include the Yangtze sturgeon, Amur sturgeon (*A. schrencki*), Chinese sturgeon (*A. sirensis*), the kaluga (*Huso dauricus*), and the Chinese paddlefish (IUCN 1996). Water diversions and recently proposed dam building schemes on the Yangtze, Pearl, and Amur Rivers may result in the eventual extinction of remaining populations of these fishes in China and neighboring Russia (Birstein 1993).

Phillipines. The IUCN (1996) lists 26 threatened fishes in the Phillipines, almost all of which are lacustrine endemics threatened by introduced species. For example, Lake Lanao on the island of Mindanao alone contains 19 of 26 listed threatened fish species, 14 of which are considered "critically endangered" (IUCN 1996). Threatened fishes of Lake Lanao belong to the family Cyprinidae whose members are largely restricted in their distribution in the Phillipines to the islands of Mindanao, Palawan, and Mindoro (Berra 1981). Many of Lake Lanao endemic fishes became extinct due to pollution and the introduction of exotic fishes before they were adequately described (Kornfield and Echelle 1984). Critically endangered species in Lake Lanao include *Cephalakompsus pachycheilus*, the bagangan (*Mandibularca resinus*), the bitungu (*Ospatulus truncatus*), and eleven members of the genus *Puntius*, which has experienced extensive speciation (IUCN 1996). In addition, the dwarf pygmy goby (*Pandaka pygmaea*) from the River Malabon is critically endangered and the goby, *Sicyopus axillimentus*, is considered "vulnerable" (IUCN 1996).

Other Asian countries with a relatively large number of threatened fish species include Malaysia (14 species), Thailand (14), Russia (13), Ukraine (12), and Sri Lanka (8) (IUCN 1996). The largest group of threatened fishes in Malaysia belong to the freshwater family Belontidae (gouramies), including six species in the genus *Betta* (six threatened species) and the ornate paradisefish (*Parosphromenus harveyi*) (IUCN 1996). Three other threatened fish are riverine catfish from the family Clariidae (= airbreathing catfishes), including three members in the genus *Encheloclarias*, two of which are critically endangered. Another threatened species that also is found in several other Southeast Asian countries is the Asian bonytongue (*Scleropages formosus*) an osteoglossid mouthbrooder that is a locally important food fish (IUCN 1996, Berra 1981).

Sri Lanka. Sri Lanka is a densely populated tropical country that also has a rich endemic fish fauna in its fresh waters (Pethiyagoda 1991). Most of these fishes are small but have value in the aquarium trade, as food, and as control for mosquitoes. Despite the dense populations and the civil wars that have periodically ravaged the country, the endemic biota seems largely intact. Of the 88 species, only 18% are in need of special protection. The reasons for this are complex, but are related to strong religous traditions that protect life, a thousand year history of protected areas, a British colonial rule that also favored limited conservation, a history of academic study of the fishes, and until recently, a non-industrial economy.

Despite this history of moral and political protection, many of the endemic fishes of Sri Lanka are now in danger of extinction as the result of large water projects, deforestation, and other symptoms of a rapidly expanding population and westernization of the economy (Senanayake and Moyle 1982; Pethiyagoda 1991). Some of the species may have been saved through translocation (Wikramanayake 1990) but long-term survival depends on protection of watersheds, especially in rain forest areas. The persistence of its endemic fish fauna will be a major test of whether or not Sri Lanka's ancient conservation ethic will survive its clash with western economic development.

Australia

While Australia covers an area of approximately 7,682,000 km^2, the majority of the continent is either desert or scrubland that receives less than 20 cm of annual rainfall. Aridity and geographic isolation in part explain why the vast majority of Australian freshwater fishes have evolved from marine ancestors. There are approximatly 195 native species and subspecies of freshwater fish, representing 39 families described from Australian waters, and an additional 22 undescribed taxa (Wager and Jackson 1993). Of the described taxa, only four species may be considered exclusively freshwater fishes: the Australian lungfish *(Neoceratodus foresteri*), northern saratoga (*Scleropages leichhardti*), southern saratoga (*S. jardinii*), and salamanderfish (*Lepidogalaxias salamandroides*); two of these may be considered as threatened or vulnerable taxa.

The IUCN (1996) lists 31 freshwater (n = 27) and marine (n = 8) taxa as either critically endangered, endangered, or vulnerable, with an additional 21 taxa described as near threatened. These likely underestimated the total number of threatened fishes because the categories do not include wide ranging pelagic marine species (i.e., sharks) that frequent Australian waters, nor some of the 19 other taxa for which there was insufficient information to determine their risk status. A conservative estimate, therefore, is that approximately 52 species, or 27 percent of the fish fauna is threatened, depending upon whether strictly marine taxa are also included, although estimates of threatened fishes range as high as 34% of the total fish fauna (Pollard *et al.* 1990). The IUCN estimate is somewhat lower than estimates provided by Wager and Jackson (1993), who estimated 73 taxa, or 34% of the total freshwater fish fauna as either extinct in the wild, endangered, vulnerable, poorly known, or rare. However, it is reasonable to assume that some of the taxa listed by Wager and Jackson (1993) as poorly known or rare would upon further study not warrant threatened status, as likely would also be the case with the data deffícient species currently on the IUCN list.

Many factors have been identified in the decline of Australian fishes, most related to the development of surface and groundwater water resources. Thus, groundwater mining for agricultural use and livestock grazing reduces flows in springs inhabited by native Australian fishes (Rinne *et al.* 1995). Likewise, dams and diversions modify of the natural hydrologic and hydraulic patterns in riverine environments (Weger and Jackson 1993). For example, alteration of the timing and amount of flows has reduced the extent and duration of floodplain inundation. This has deterimental effects on species that require the floodplain for reproduction and rearing of their young, such as the murray cod (*Maccullochella peelii*), golden perch (*Macquaria ambigua*), silver perch (*Bidyanus bidyanus*), common galaxias (*Galaxias maculatus*), and Hyrtl's tandan *(Neosilurus hyrtli*) (Weger and Jackson 1993).

Another threat facing Australian freshwater fishes is the introduction of exotic fishes and the transfer of native fishes into basins from which they were historically absent. Approximately 20 species of exotic fish have been successfully introduced into Australian waters (Wager and Jackson 1993). Brown trout and mosquitofish (*Gambusia holbrooki*) have been implicated in the decline of several natives species (Arthington and Lloyd 1989; Arthington 1991). In the case of *G. holbrooki*, the inferred mechanisms for the decline of native species are competition for food and space, agressive behavior, and egg predation. Similarly, the translocation of the native climbing galaxias (*Galaxias brevipinnis*) into Lake Pedder following its flooding for a hydroelectric project, combined with predation from exotic brown trout, has been implicated in the near extinction of the pedder galaxias (*G. pedderensis*) (Weger and Jackson 1993). According to the IUCN (1996), *G. pedderensis* is critically endangered, has experienced a population reduction of at least 80% over the last ten years (three generations). With the implementation of drastic conservation measures, this galaxias will likely go extinct within the next ten years.

The center of diversity for galaxids is Australia with 20 species, followed by New Zealand with 13 species, South America with three species, with one species from South Africa (Cohen 1994). Of the 28 species from Australia and New Zealand, 16 species or 57% are considered threatened with extinction (IUCN 1996). Several Australian species such as the swan galaxias (*Galaxias fontanus*), barred galaxias (*G. fuscus*), clarence galaxias (*G. johnstoni*), and pedder galaxias are considered critically endangered because of population declines of 80% or greater over the last ten years (IUCN 1996). Other groups of Australian fishes that are threatened include several endemic desert fishes in the Lake Eyre, Western Plateau, and Bulloo-Bacannia regions. Up to 40% of the fish fauna in these regions are considered to be endangered, vulnerable, or restricted (Jackson 1992; Rinne *et al.* 1995).

Central and South America

Central and South America contain somewhere between 2,700 to 4,250 species of freshwater fish (Moyle and Cech 1996). This represents between 11–17% of total fish species, approximately 40% of all freshwater fish species. The IUCN (1996) lists 30 species of threatened fish for all of Central and South America, with an additional 33 species having insufficient data to determine their conservation status. This is a surprisingly small number given the large geographic area and diversity of fishes. The low number of threatened species likely reflects a lack of information on the distibution, ecology, and conservation status of most fishes, as well as the fact that many aquatic habitats and their fish communities are relatively intact compared to other more industrialized regions of the world. Although aquatic habitats within the Amazon River basin are experiencing degradation through activities such as deforestation, construction of dams, and arsenic pollution from gold mining, much of the basin contains fully functioning riverine systems.

Brazil, which includes the largest portion of the Amazon River basin, holds 12 threatened fish species, six of which are riverine cyprinodonts within the genus *Cynolebias*, (pearlfish, family Aplocheilidae). Interestingly, although Brazil's freshwater environments support one of the greatest diversities of fishes on earth, half the species listed as threatened are marine forms. Threatened marine fishes include two serranid species, *Anthias salmopunctatus* and *Mycteroperca microlepis*, that, like other grouper species, have experienced population declines due to overfishing. Other threatened marine taxa found in Brazilian coastal waters include the brownstripe grunt (*Anisotremus moricandi*), which also occurs in Columbia, Panama, and Venezuela, a butterflyfish (*Chaetodon obliquus*), and a pomacentrid (*Stegastes santipaulae*) (IUCN 1996). Another fish of conservation interest is the Pirarucu (*Arapaima gigas*) one of world's largest scaled freshwater fishes. This species reaches lengths of 2.5 m, inhabits large rivers in Brazil, Guyana and Peru (Nelson 1994).

Chile contains only three threatened fish species although another 21 and 5 species are listed as "data-deficient" and "near-threatened", respectively (IUCN 1996). There are several species of catfish (Siluriformes) of conservation concern, including five species of *Trichomycterus* some of which inhabit torrential, high-gradient streams (Nelson 1994). Other threatened freshwater fishes include two cyprinodonts from the genus *Orestias* that inhabit high altitude lakes and rivers along the Andean divide (Parenti 1984). Other interesting taxa of conservation concern are the nonparasitic anadromous lamprey *Mordacia lapicida*, (family Petromyzontidae) and several genera within the family Percichthydae (temperate perches) (IUCN 1996; Nelson 1994). The only listed threatened marine fish is the Easter Island butterflyfish (*Chaetodon litus*) in the southwest Pacific Ocean.

Other groups of Central and South American marine fishes that are threatened include five species of toadfish (family Batrachoididae) from Columbia, Venezuela, and Belize and two species of blennies from Columbia and Venezuela (IUCN 1996). Freshwater threatened fishes of conservation concern include two poecilid species, one (*Gambusia aestiputeus*) from San Andres Island and the other, *Poecilia vetiprovidentiae*, from Isla de Provivencia, both off the coast of Columbia (IUCN 1996). Another freshwater fish, the catfish *Rhizosomichthys totae* endemic to Lake Tota in Columbia, is now extinct (IUCN 1996).

North America

North America contains approximately 1,174 species of freshwater fish (Williams and Miller 1990). The IUCN (1996) lists 222 threatened species in Canada (n = 13), the United States (n = 123), and Mexico (n = 86), or about 21% of the total fish fauna. However, because many additional species are listed as "data-defficient" or "near-threatened", a more reliable percentage estimate of threatened fishes, conservatively assuming a third of the taxa in these categories are actually threatened, is probably 31%. For the United States alone, the IUCN (1996) assessment indicates that about 15% of the fauna is threatened. However, the IUCN estimate is significantly lower than that compiled by the America Fisheries Society (AFS) in 1989, which listed 254 taxa, or about 32%, as imperiled (Williams *et al.* 1989). The differences may be largely attributable to various definitions of the "threatened" status categories between the IUCN and AFS listings, as well the fact that the IUCN lists taxa only to species. For example, the AFS list used the same "endangered" and "threatened" categories contained in the Endangered Species Act of 1973, as well as an additional "special concern" category, while the IUCN threatened list includes critically endangered, endangered, or vulnerable fishes. The AFS inclusion of a "special concern" category may result in listing of more species than the IUCN.

Johnson (1995) compared changes in the status categories on the AFS lists for 1979 (Deacon *et al.* 1979) and 1989 (Williams *et al.* 1989) and found that the number of imperiled fishes in the United States increased 38% in ten years! States with the most threatened fishes include California (42), Tennessee (40), and Nevada (39), while regions with the highest percentage of threatened fishes include the Southwest (48%), Northwest (19%) and Southeast (10%) (Warren and Burr 1994; Johnson 1995). To illustrate the status of North American fish communities, we present three regional examples: California; the southeastern United States, and Mexico.

California. California, because of its size, its aridity, and high degree of endemism (60%) in its fishes, represents the problems facing freshwater fishes throughout the western United States. According to Warren and Burr (1994), California contains more endangered fishes that any other state. Of the 116 taxa native to the state, eight (7%) are extinct, 15 (13%) are formally recognized as threatened or endangered by state or federal governments, 27 (23%) qualify for such formal listing, and 22 (19%) may qualify in the near future if present population trends continue (Moyle *et al.* 1995). These numbers represent an increase in the percentage of fishes listed in the first three categories of 19% between 1989 and 1992. Today only 28% of the native fish fauna of California can be regarded as "secure."

Threatened or endangered fish taxa in California are most likely to be endemic forms with limited ranges, especially those confined to desert springs (e.g., members of the Cyprinodontidae = pupfish), small streams (e.g., various members of the family Cyprinidae), or fishes that inhabit large rivers and their estuaries (e.g., chinook salmon, *O. tshawytscha*) (Moyle 1995). Not suprisingly, these fishes occur most frequently in arid basins, where competition for water from humans is most severe. In the lower Colorado River, for example, all native fishes are either extinct or endangered. In the arid Los Angeles region, all native fishes need formal protection (Moyle and Williams 1990). However, California's endangered fishes are not limited only to the most arid regions of the state. For example, in extreme northwestern coastal California where rainfall may exceed 200 cm per year, most anadromous fish, including coho salmon (*Oncorhynchus kisutch*) and steelhead trout (*O. mykiss mykiss*), are in serious decline.

Southeastern United States. The center for freshwater species diversity in temperate North America is the southeastern United States, where over 60% of North America's freshwater fish fauna north of Mexico resides (Walsh, Burkhead and Williams 1995). Of the approximately 485 native species of freshwater fish in the southeast, 93 taxa or 19% may be considered threatened (Walsh, Burkhead and Williams 1995). The IUCN (1996) lists about 35 species of darters (family Percidae), or about 25% of the 150 species of darters from the region as threatened or near threatened. A similar trend of endangerment is found for the Cyprinidae and madtom catfishes within the family Ictaluridae (IUCN 1996).

In the southeastern state of Alabama, 31 out of a total 303 freshwater fish species are of conservation concern (Lydeard and Mayden 1995). Within the region many species are becoming increasingly restricted in their distributions, as peripheral populations disappear. The primary causes of decline of native populations of southeastern fishes are siltation and sedimentation that fill in pool habitats of bottom-dwelling species, and habitat fragmentation and alteration of natural hydrologic and hydraulic regimes (Lydeard and Mayden 1995; Walsh, Burkhead, and Williams 1995).

Mexico. Mexico contains about 375 species of purely freshwater fishes of which 60% are endemic, and another 125 species inhabiting estuarine and coastal lagoon habitats (Miller 1986; Espinosa-Perez *et al.* 1993; Lyons *et al.* 1995). The IUCN (1996) lists 86 threatened fish species in Mexico, with an additional five and eight species listed as "near-threatened" and "data-deficient", respectively. Alarmingly, 19 species are listed as extinct or extinct in the wild, far more than any other country (IUCN 1996). Six families account for 76% of the listed taxa: the Cyprinidae (19 threatened species), Cyprinodontidae (13), Poeciliidae (11), Goodeidae (10) Catostomidae (7), and Cichlidae (5). Recent extinctions of Mexican fishes include cyprinids from the genera *Evarra* (3 extinct species), *Notropis* (3), and *Stypodon* (1); cyprinodont species from the genera *Cyprinodon* (6) and *Megupsilon* (1); several genera within the Goodeidae including *Ameca* (1), *Characodon* (1), *Empetrichthys* (1), and *Skiffia* (1); and the genus *Priapella* (1) within the family Poeciliidae (IUCN 1996, Miller, Williams, and Williams 1989).

Many of Mexico's extinct and threatened fishes are desert endemics once confined to small springs, lakes, pools, and streams (Williams *et al.* 1985). One of the most remarkable ecosystems for several threatened fishes lies within the small valley of Cuatro Cienegas in the Chihuahuan Desert of northen Mexico. The 500 square-mile valley is characterized by a labyrith of springs, pools, and small lakes, and their associated wetlands, fed by water from subterranean channels. These aquatic habitats support over 60 endemic aquatic species, including 16 fishes, eight of which are endemic (Grall 1995). The cyprinodont Sardinilla Cuatro Cienegas (*Lucania interioris*) is considered critically endangered because of its extremly limited distribution and because its population is estimated to number less than 250 individuals. No subpopulation is estimated to contain more than 50 mature individuals (IUCN 1996). Other threatened fishes of Cuatro Cienegas include the poecilids *Gambusia longispinis*, and *Xiphophorus gordoni*, and the cichlid *Cichlasoma minckleyi*.

Two marine species, the giant sea bass (*Stereolepis gigas*) and totoaba (*Totoaba macdonaldi*), are also critically endangered (IUCN 1996). The giant sea bass, which matures slowly and has low fecundity, has declined primarily from overfishing. The totoaba, the largest member of the Sciaenidae (with recorded weights of over 100 kg and lengths of over 2 meters), is endemic to the Gulf of Mexico (Flanagan and Hendrikson 1976). Decline of the totoaba is attributed to several interating factors including alteration of spawning and nursery grounds from decreased spring water inputs from the Colorado River in the upper Gulf of Mexico, overfishing of reproducing adults, and the incidental take of juveniles in the shrimp fishery (Cisneros-Mata, Montemayor-Lopez, and Roman-Rodriguez 1995).

Oceania and Other Insular Environments

Oceanic islands vary greatly in the diversity of their fishes in terms of number of species and degree of endemism depending largely on their age, size, location, and genesis. While marine fish communities surrounding islands, especially in the subtropical and tropical seas may exhibit a high fish species diversity, with a few exceptions (e.g., the larger islands of New Zealand, Papua New Guinea, Phillipines, Sri Lanka, Madagascar), the freshwater fish fauna is not diverse in terms of the number of species, although there may be a large porportion of endemic forms.

Oceania is defined as the islands of the central and south Pacific, including Micronesia, Melanesia, Polynesia, and Australasia. Exclusive of Australia which is discussed above, Oceania contains about 40 threatened and another 15 listed as data deficient marine fish species, many of which are wide ranging in in the Pacific Ocean and other oceans of the world. For example there are as many as a dozen species of threatened elasmobranchs that range throughout various portions of Oceania. There are also about 25 threatened freshwater fish species mostly represented by saltwater dispersant species (IUCN 1996).

Papua New Guinea and New Zealand contain the majority of the threatened freshwater forms, with 13 and eight species respectively (IUCN 1996). Families of threatened freshwater fishes found in New Guinea include the Pristidae (sawfishes), Dasyatidae (*i.e.*, giant freshwater stingray), Plotosidae (eeltail catfishes), Atherinidae (especially the genera

Craterocephalus [= hardyheads]), Melanotaeniidae (rainbowfish), and Pseudomugilidae (blue-eyes), Syngnathidae (pipfishes), Terapontidae (grunters), and Eleotridae (gudgeons or mogurnda) (IUCN 1996). An assemblage of six threatened fishes occur in Lake Kutubu, including the Kutubu tandan (*Oloplotosus torobo*), a hardyhead (*Craterocephalus lacustris*), Adamson's grunter (*Hephaestus adamsoni*), and three species of gudgeon (*Mogurnda* spp.) (IUCN 1996). Another 15 species have insufficient data to determine their conservation status.

Threatened fishes of New Zealand include five freshwater galaxid species (genera *Galaxias* and *Neochanna*) and the Tarndale bully (*Gobiomorphus alpinus*), and two marine members of the Pegasidae, the big-bellied seahorse (*Hippocampus abdominalis*) and spiny pipehorse (*Solegnathus spinosissimus*) (IUCN 1996). Other islands with threatened fishes include French Polynesia with three marine species—a grouper (*Pseudanthias regalis*), a labrid (*Xyrichtys virens*), and a puffer (*Canthigaster rapaensis*). Guam contains the other threatened species from the region—i.e., the yellow-crowned butterflyfish (*Chaetodon flavocoronatus*).

Other oceanic islands with several listed threatened fishes include St. Helena (seven threatened species), Japan (7), Taiwan (6), and Cuba (4) (IUCN 1996). The volcanic island of St. Helena in the southeast Atlantic Ocean is 308 km^2 in size. Threatened fishes of St. Helena are all marine species and include representatives of the families Modidae (morid cods), Scorpaenidae (rockfishes), Pomacentridae (damselfishes), Labridae (wrasses), and Callionymidae (dragonets). There are two critically endandered species, the deepwater jack (*Pontinus nigropunctatus*) and the sculpin (*Physiculus helenaensis*), populations of each species are thought to total less than 50 mature individuals (IUCN 1996).

Similar to other continental regions within the north temperate zone (e.g., United States and Europe), Japan and Taiwan contain several species of threatened salmonids. Japan has two threatened salmonids, one the Satsukimasu salmon (*Oncorhynchus ishikawai*) inhabits the River Nagara of central Japan and the Kirikuchi char (*Salvelinus japonicus*) of the River Totsukawa, both of which are estimated to have experienced a population reductions of at least 50% over the last ten years or three generations, and continue to decline (IUCN 1996). Similarly, the habitat of critically endangered *Oncorhynchus formosanus* of Taiwan is estimated to be severely fragmented and may only exist at a single location (IUCN 1996). Interestingly, Taiwan lying on the Tropic of Cancer represents the southernmost distribution of salmonids (Berra 1981).

MARINE FISHES

About 60% of all fishes are marine, but a majority (78%) of marine fishes inhabit the narrow band of shallow water habitat on the continental shelves or associated with reefs around islands. In shallow water habitats, all fish are subject to such pervasive human influences as pollution and fishing. Nevertheless, the IUCN (1996) lists only 156 exclusively marine species (most of these from the family Pegasidae [seahorses]) as being threatened (up from just 20 in 1994), which presumably reflects mainly a lack of information rather than the reality of the situation (Vincent and Hall 1996). The majority of the marine fishes recognized as at risk are found in enclosed basins, such as the Black and Aral seas, or in nearshore habitats that are exceptionally vulnerable to degradation (Micklin 1988). The number of marine taxa at risk of extinction is increasing rapidly as coastal development fragments marine and salt marsh habitats, as pollution becomes more pervasive, as introduced species become more pervasive, and as modern technology increases our ability to exploit the ocean's fish, invertebrates, and mineral resources. A 1996 conference of experts on marine conservation recommended that 120 additional marine species be added to the 1996 IUCN list (Vincent and Hall 1996).

Over-exploitation of fishes is increasingly a major threat to ocean ecosystems and the species they contain. The 1992 collapse of the Atlantic cod (*Gadus morhua*) fishery off Newfoundland and Labrador is a spectacular example of the decline of a major fishery due to overexploitation (Hutchings and Meyers 1994). Fishing for cod continued even after predictions of its collapse. Attempts were made to blame poor ocean conditions for growth and survival of young fish as the ultimate cause of the collapse, but the evidence pointed overwhelmingly to overfishing (Hutchings and Meyers 1994). However, while the Atlantic cod populations off Canada collapsed because of the direct effects of exploitation, those off Norway collapsed because of the overexploitation of a principal prey species, the Atlantic herring (*Clupea harengeus*) (Hamre 1994). When stocks of an alternate prey species, the capelin (*Malotus villosus*) were reduced due to natural factors, hungry cod became cannibalistic on their own young. This was followed by starvation of large cod, as well as of seabirds and seals. Thus the fisheries for cod, herring, and capelin all collapsed.

Similarly, shark fisheries off the Galapagos Islands are not only threatening 27 species of sharks, but significantly altering the abundances of other fishes upon which the sharks prey (Camhi 1995). Tropical reef fisheries are driving large species, especially predatory groupers (Serranidae), to local extinction (Roberts 1995a, 1995b). The removal of fish of all kinds from coral reefs in Jamaica had led to the collapse of the reef ecosystem, which itself depends upon the intense activites of fish (e.g., grazing on algae) to maintain ecosystem functions (Hughes 1994). Thus exploitation can have many unexpected indirect effects on marine ecosystems and on non-target species and its effects may not always be reversible.

In many instances, fishing literally alters the physical environment in which fish live. Repeated bottom trawling plows the bottom as thought it were an agricultural field, drastically altering the structure of the benthic community. Fishing on coral reefs with sodium cyanide for the international aquarium business and Asian restuarants is increasingly common in Indo-Pacific reef habitats, especially in the Phillipines and Indonesia (Ariyoshi 1997). Cyanide fishing kills coral, invertebrates, and fish, subsequently destroying much of the structure that enables the reef to support such a diversity of life. Cyanide fishing for the humphead wrasse (*Cheilinus undulatus*) has contributed to its listing as a threatened species by the IUCN (1996). The wrasse sells for as much as $225 per plate in Asian restuarants (Ariyoshi 1997).

A relatively new but growing threat to marine fishes is the invasion of exotic species of fish and invertebrates, especially in bays and estuaries, which can completely change inshore ecosystems, presumably to the detriment of native species. The principal invaders are the hundreds of species and millions of individuals transported and released from ballast water of ships (Carlton 1985, 1987, 1993). In some estuaries, such as San Francisco Bay, most of the invertebrates and some of the fish are non-native species, with the number of exotic species growing on a near daily basis (Cohen and Carleton 1995).

CONCLUSIONS

It is clear that aquatic biodiversity, as exemplified by fish, is rapidly being lost. In freshwater systems the losses are probably much more severe than losses in surrounding terrestrial environments, and losses are beginning to be sustained in marine environments as well, despite the enormous buffereing capacity of the oceans for human insults. In freshwater systems, fish losses appear to be highest in (1) industrialized countries, (2) in regions with arid or Mediterranean climates, (3) in tropical regions with large human populations, (4) in regions with high endemism, (5) in small, isolated bodies of water such as desert spring, and (6) in big rivers. In regions such as California or South Africa, where several of these conditions apply simultaneously, two-thirds or more of the native fish fauna may be in danger of extinction within the next 50 years. It is also evident that regions with intensively studied fish faunas have more threatened species than regions with poorly known fish faunas, suggesting that much of the loss of aquatic biodiversity is going unrecorded. Indeed, the estimate of Moyle and Leidy (1992) that 20% of the freshwater fishes of the world is at risk of extinction is probably very conservative.

In marine environments, there are no documented extinctions of a single fish species yet, but losses of fish species are most likely to occur in (1) enclosed areas such as inland seas, bays, and estuaries, (2) shallow water areas close to large urban areas, (3) areas subjected to intense, poorly regulated, multi-species fisheries, and (4) tropical reefs. Even more so than for freshwater fishes, our ignorance of the status of most marine fishes, even exploited species, is profound. Today it is likely that at least 5% of marine fishes are at some risk of extinction, not including stocks of widely distributed fish driven to extinction by fishing and other factors.

Protecting Aquatic Biodiversity

Reversing the decline aquatic biodiversity will not be easy and is only likely to happen if humanity recognizes it is its best interests to have healthy aquatic environments. The following are some suggestions, in no particular order, for foundations for global policies leading towards protection of aquatic biodiversity.

(1) Religious and moral leaders and institutions should take major responsibility for convincing the people of the world of their responsibilities towards conservation. Conservation of aquatic systems is likely to be sustained in the long run only if such conservation is a major part of the ethical systems under which people live. The major religions of the world arguably all have such ethics built into them, but they are typically not emphasized (Callicott 1994).

(2) Recognize watersheds as the fundamental unit of the landscape for all conservation efforts. Practically speaking, an aquatic biodiversity protection strategy will necessarily involve work at many levels (species, groups of species, watersheds of different sizes, landscapes, bioregions) (e.g., Moyle and Yoshiyama 1994). However, watersheds have the advantage of being natural (physically defined) units that are small enough to be easily recognizable by humans yet large enough to protect ecosystem processes (Naiman 1992). In the short term some kind of triage system may have to implemented in some regions of the world, so limited resources available for conservation can be focussed on watersheds where long-term protection of the biota is most likely to occur (e.g., Moyle and Randall 1996).

(3) Recognize that our present economic system greatly undervalues water in the natural environment by externalizing the cost of replacing ecological services and other values (Moyle and Moyle 1995; see chapter 16). There is a need to end or greatly reduce subsidies for water development and use (especially for

irrigated agriculture) in order to encourage more efficient use of developed water in order to provide more water to meet environmental goals.

(4) Make ecosystem management, as defined by the Ecological Society of America (1996), the basis for management of watersheds and landscapes. Fundamental to ecosystem management is the idea of sustainability, which boils down to making sure that future generations are no worse off in terms of natural resources and environments than the present generation.

(5) Increase research, inventory, and monitoring activities in aquatic environments, in order to determine where to place limited funds for conservation and rehabilitation of aquatic systems. Naiman *et al.* (1995a,b) present an agenda for the United States that would be a good model for all nations to follow.

(6) Develop international laws and regulations to greatly reduce the introduction non-native species through international trade and travel. The most severe problem is the worldwide spread of marine organisms through ballast water transport, although there are growing problems with the introduction of aquarium and aquaculture fishes.

(7) Develop new management strategies for the conservation of marine fishes and habitats (Suchanek 1994). In particular, there is growing evidence that large-scale marine reserves can be effective for conservation of marine fishes.

(8) Adopt as the fundamental principle of fisheries management that aquatic ecosystems must be managed to sustain all native fish stocks for the indefinite future. Oliver, Shuter, and Mirns (1995) suggests a series of supporting principles to guide management actions under this general principle:
- "The sustainability of a fish stock requires protection for the specific physical and chemical habitats utilized by the individual members of that stock."
- "The sustainability of a fish stock requires maintenance of its supporting native community."
- "Vulnerable, threatened, and endangered species must be rigidly protected from all anthropogenic stresses."
- "Exploitation of populations or stocks undergoing rehabilitation will delay, and may preclude, full rehabilitation."
- "Harvest must not exceed the regeneration rate of a population or its individual stocks."
- "Direct exploitation of spawning aggregations increases the risk to sustainability of fish stocks."

(9) Develop aquatic conservation strategies that recognize that there is a direct relationship between aquatic biodiversity, habitat complexity (including natural flow regimes of rivers) and water quality (e.g., Kirchhofer 1995). Such strategies should include serious consideration of reversing previous environmental damage, such as removing dams from rivers (Shuman 1995) restoring river floodplains (Gore and Shields 1995) or cleaning polluted ocean waters to allow kelp forests to grow again.

(10) Manage aquaculture operations to reduce their negative influences on natural systems, by restricting the release of cultured fish and other organisms, as well as their diseases, providing stricter controls over the movement of cultured fish among regions, and reducing the their effects on natural habitats through pollution and alteration of natural environments for aquaculture facilities.

ACKNOWLEDGMENTS

The authors would like to especially thank B. Groombridge at the World Conservation Monitoring Centre, Cambridge, U.K. for providing preliminary copies of the 1996 IUCN Red List of Threatened Animals.

LITERATURE CITED

Abramovitz, J.N. 1996. Imperiled waters, impoverished future: The decline of freshwater ecosystems. *Worldwatch paper 128.*

Almaca, C. 1995. Freshwater fish and their conservation in Portugal. In *Conservation of endangered freshwater fish in Europe*, eds. A. Kirchofer and D. Helti, 125–127. Basel, Switzerland: Birkhauser Verlag.

Angel, M.V. 1993. Biodiversity of the pelagic ocean. *Conservation Biology* 7:760–772.

Anon. 1991. Elasmobranch conservation: the establishment of the IUCN shark specialist group. *Aquatic Conservation* 1:193–194.

Ariyoshi, R. 1997. Halting a coral catastrophe. *Nature Conservancy*. January/February: 20–25.

Arthington, A.H. 1991. Ecological and genetic impacts of introduced and translocated freshwater fishes in Australia. *Canadian Journal of Fisheries and Aquatic Sciences* 48:33–43.

Arthington, A.H. and L.N. Lloyd. 1989. Introduced poeciliids in Australia and New Zealand. In *Evolution of livebearing fishes (Poeciliidae)*, ed. G.K. Meffee and F.F. Snelson. New Jersey: Prentice Hall Publishers.

Balik, S. 1995. Freshwater fish in Anatolia, Turkey. *Biological Conservation* 72:213–223.

Bennett, W.A. and P.B. Moyle. 1996. Where have all the fishes gone? Interactive factors producing fish declines in the Sacramento-San Jooaquin estuary. In *San Francisco Bay: the ecosystem*, ed. J.T. Hollibaugh, 519–541. San Francisco: American Association for the Advancement of Science.

Berra, T.M. 1981. *An atlas of distribution of the freshwater fish families of the world.* Lincoln, Nebraska: University of Nebraska Press.

Bianco, P.G. 1982. Hybridization between *Alburnus albidus* (C.) and *Leuciscus cephalus cabeda* R. in Italy. *Journal of Fish Biology* 21:593–603.

Bianco, P.G. 1990. Vanishing freshwater fishes in Italy. *Journal of Fish Biology* 37 (Supplement A):235–237.

Bianco, P.G. 1995. Mediterranean endemic freshwater fishes of Italy. *Biological Conservation* 72:159–170.

Birstein, V.J. 1993. Sturgeons and paddlefishes: threatened fishes in need of conservation. *Conservation Biology* 7:773–787.

Blanco, J.C. and J.L. Gonzalez. 1992. *Libro rojo de los vertebrados de Espana.* Madrid: Coleccion Tecnica, ICONA.

Bootsma, H.A. and R.E. Hecky. 1993. Conservation of the African Great Lakes: A limnological perspective. *Conservation Biology* 7:644–656.

Brown, L.R., P.B. Moyle and R.M. Yoshiyama. 1994. Historical decline and current status of coho salmon in California. *North American Journal of Fisheries Management* 14:237–261.

Bruton, M.N. 1995. Threatened fishes of the world: *Latimeria chalumanae* Smith 1939 (Latimeriidae). *Environmental Biology of Fishes* 43:104.

Callicott, J.B. 1994. Conservation values and ethics. In *Principles of Conservation Biology*, eds. G.K. Meffe and C.R. Carroll, 24–49. Sunderland, MA: Sinauer Associates.

Camhi, M. 1995. Industrial fisheries threaten ecological integrity of the Galapagos Islands. *Conservation Biology* 9:715–724.

Carlton, J.T. 1985. Transoceanic and interoceanic dispersal of coastal marine organisms: The biology of ballast water. *Oceanography and Marine Biology Annual Review* 23:313–371.

Carlton, J.T. 1987. Patterns of transoceanic marine biological invasions in the Pacific Ocean. *Bulletin of Marine Science* 41:452–465.

Carlton, J.T. and J. Geller. 1993. Ecological roulette: The global transport and invasion of nonindigenous marine organisms. *Science* 261:239–241.

Carpenter, S.R., and J.F. Kitchell. 1988. Consumer control of lake productivity. *BioScience* 38:764–769.

Changeux, T. and D. Pont. 1995. Current status of the riverine fishes of the French Mediterranean basin. *Biological Conservation* 72:137–158.

Chereshnev, I.A. and M.B. Skopets. 1990. *Salvethymus svetovidovi* gen. et *sp. nova*—a new endemic fish of the subfamily Salmoninae from Lake El'gygytgyn (Central Chukotka). *Journal of Ichthyology* 30:87–103.

Cisneros-Mata, M.A., G. Montemayor-Lopez, and M.J. Roman-Rodriguez. 1995. Life history of *Totoaba macdonaldi*. *Conservation Biology* 9:806–814.

Cohen, A.N. and J.T. Carlton. 1995. *Nonindigenous aquatic species in a United States estuary: a case study of the biological invasions of the San Francisco Bay and Delta*. A report for the United States Fish and Wildlife Service, Washington, D.C. and the National Sea Grant College Program, Connecticut Sea Grant.

Cohen, A.S., R. Bills, C.Z. Cocquyt, and A.G. Caljon. 1993. The impact of sediment pollution on biodiversity in Lake Tanganyika. *Conservation Biology* 7:667–677.

Crivelli, A.J. 1995. Are fish introductions a threat to endemic freshwater fishes in the northern Mediterranean region? *Biological Conservation* 72:311–319.

Crivelli, A.J. and P. S. Maitland. 1995. Special issue: Endemic freshwater fishes of the Northern Mediterranean region. *Biological Conservation* 72.

Crivelli, A.J. 1996. *The freshwater fish endemic to the northern Mediterranean region: an action plan for their conservation*. Arles: Tour du Valet Publications.

Daget, J., J.P. Gosse, G.G. Teugels, and D.F.E. Thys van den Audenaerde. 1991. *Check-list of the freshwater fishes of Africa. Volumne IV*. ISBN-MRAC-ORSTOM.

Dayton, P.K., S.F. Thrush, M. Tundi Agardy, and R.J. Hofman. 1995. Environmental effects of marine fishing. *Aquatic Conservation: Marine and Freshwater Ecosystems* 5:205–232.

Deacon, J.E., G. Kobetich, J.D. Williams, S. Contreras, *et al.* 1979. Fishes of North America endangered, threatened, or of special concern: 1979. *Fisheries* 4:30–44.

Debus, L. 1996. The decline of the European sturgeon *Acipenser sturio* in the Baltic and NorthSea. In *Conservation of endangered freshwater fish in Europe*, eds. A. Kirchofer and D. Helti, 147–156. Basel, Switzerland: Birkhauser Verlag.

Dynesius, M. and C. Nilsson. 1994. Fragmentation and flow regulation of river systems in the northern third of the world. *Science* 266:753–762.

Ecological Society of America. 1996. The report of the Ecological Society of America Committee on the Scientific Basis for Ecosystem Management. *Ecological Applications* 6:665–691.

Economidis, P.S. 1995. Endangered freshwater fishes of Greece. *Biological Conservation* 72:201–211.

Economidis, P.S. and P.J. Miller. 1990. Freshwater gobies from Greece (Teleostei: Gobiidae). *Journal Zoology of London* 221:125–170.

Elvira, B. 1990. Iberian endemic freshwater fishes and their conservation status. *Journal of Fish Biology* 37 (Supplement A):231–232.

Elvira, B. 1995a. Conservation status of endemic freshwater fish in Spain. *Biological Conservation* 72:129–136.

Elvira, B. 1995b. Native and exotic freshwater fishes in Spanish river basins. *Freshwater Biology* 33:103–108.

Elvira, B. 1996. Endangered freshwater fish of Spain. In *Conservation of endangered freshwater fish in Europe*, eds. A. Kirchofer and D. Helti, 55–61. Basel, Switzerland: Birkhauser Verlag.

Engelman, R. and P. LeRoy. 1993. *Sustaining water: population and the future of renewable water supplies*. Population Action International, Washington, D.C.

Espinosa-Perez, H., P. Fuentes-Mata, M.T. Gaspar-Dillanes, and V. Arenas. 1993. Notes on Mexican ichthyofauna. In *Biological diversity of Mexico: origins and distribution*, eds. R. Bye, A. Lot, and J. Fa, 229–251. New York: Oxford University Press.

Flanagan, C.A. and J.R. Hendrikson. 1976. Observations on the commercial fishery and reproductive biology of the totoaba, *Cynoscion macdonaldi*, in the northern Gulf of California. *Fishery Bulletin* 74:531–544.

Food and Agricultural Organization of the United Nations (FAO). 1995. *The state of the world fisheries and aquaculture*. FAO, Rome.

Food and Agricultural Organization of the United Nations (FAO). 1993. *Global fish and shellfish production in 1993*. FAO Fisheries Department, Rome.

Foster, M.S. and D.R. Schiel. 1985. *The ecology of giant kelp forests in California: a community profile*. U.S. Fish and Wildlife Service Biological Report 85(7.2).

Fowler, S. 1996. Shark specialist group. *Species* 26–27:88–89.

Goldschmidt, T., F. Witte, and J. Wanink. 1993. Cascading effects of the introduced Nile perch on the detritivorous/phytoplanktivorous species in the sublittoral areas of Lake Victoria. *Conservation Biology* 7:686–700.

Gore, J.A. and F.D. Shields. 1995. Can large rivers be restored? *BioScience* 45:142–152.

Goulding, M. 1980. *The fishes and the forest. Explorations in Amazonian natural history*. Berkeley: University of California Press.

Grall, G. 1995. *Cuatro Cienegas,* Mexico's desert aquarium. *National Geographic* 188:84–97.

Grande, L. and W.E. Bemis. 1991. Osteology and phylogenetic relationships of fossil and recent paddlefishes (Polyodontidae) with comments on the interrelationships of Acipenseriformes. *Journal of Vertebrate Paleontology* 11(suppl. 1):1–121.

Greenwood, P.H. 1992. Are the major fish faunas well-known? *Netherlands Journal of Zoology* 42:131–138.

Hamre, J. 1994. Biodiversity and exploitation of the main fish stocks in the Norwegian-Barents ecosystem. *Biodiversity and Conservation* 3:473–492.

Hedgecock, D., P. Siri, and D.R. Strong. 1994. Conservation biology of endangered Pacific salmonids: introductory remarks. *Conservation Biology* 8:863–864.

Holcik, J. 1996. Vanishing freshwater fish species of Slovakia. In *Conservation of endangered freshwater fish in Europe*, eds. A. Kirchofer and D. Helti, 79–88. Basel, Switzerland:Birkhauser Verlag.

Hoenig, J. and S.H. Gruber. 1990. Life history patterns in the elasmobranchs: implications for fisheries management. *National Oceanic and Atmospheric Administration Technical Report, National Marine Fisheries Service* 90:1–28.

Hori, M., M. Mukwaya Gashagaza, M. Nshombo, and H. Kawanabe. 1993. Littoral fish communities in Lake Tanganyika: Irreplaceable diversity supported by intricate interactions among species. *Conservation Biology* 79:657–666.

Hudson, E. and G. Mace. 1996. Marine fish and the IUCN Red List of Threatened Animals. Report of a workshop held in collaboration with WWF and IUCN at the Zoological Society of London, April 29th–May 1st, 1996.

Hughes, N.F. 1986. Changes in the feeding biology of the Nile perch, *Lates niloticus* (L.) (Pisces: Centropomidae), in Lake Victoria, East Victoria, East Africa since its introduction in 1960, and its impact on the native fish community of the Nyanza Gulf. *Journal of Fish Biology* 29:541–548.

Hughes, T.P. 1994. Catastrophes, phase shifts, and large-scale degradation of a Caribbean coral reef. *Science* 265:1547–1551.

Hutchings, J.A. and R.A. Myers. 1994. What can be learned from the collapse of a renewable resource: Atlantic cod, *Gadus morhua*, of Newfoundland and Labrador. *Canadian Journal of Fisheries and Aquatic Sciences* 51:2126–2146.

International Union for the Conservation of Nature. 1996. *Red list of threatened animals*. Cambridge: World Conservation Monitoring Centre.

Jackson, P. 1992. Australian threatened fishes—1991 supplement. *Australian Society of Fisheries Biology* 21:20–23.

Johnson, J.E. 1995. Imperiled freshwater fishes. In *Our living resources, a report to the nation on the distribution, abundance, and health of U.S. plants, animals, and ecosystems*, eds. E.T. LaRoe, G.S. Farris, C.E. Puckett, P.D. Doran, and M.J. Mac, 142–144. Washington, D.C.: U.S. Government Printing Office.

Jurajda, P. and M. Penaz. 1996. Endangered fishes of the River Morava (Czech Republic). In *Conservation of endangered freshwater fish in Europe*, eds. A. Kirchofer and D. Helti, 35–54. Basel, Switzerland: Birkhauser Verlag.

Kaufman, L. 1992. Catestrophic change in species-rich freshwater ecosystems: The lessons of Lake Victoria. *BioScience* 42:846–858.

Karr, J.R. 1981. Assessment of biotic integrity using fish communities. *Fisheries* 7:2–8.

Keith, P. and J. Allardi. 1996. Endangered freshwater fish: the situation in France. In *Conservation of endangered freshwater fish in Europe*, eds. A. Kirchofer and D. Helti, 35–54. Basel, Switzerland: Birkhauser Verlag.

Keresztessy, K. 1996. Threatened freshwater fish in Hungary. In *Conservation of endangered freshwater fish in Europe*, eds. A. Kirchofer and D. Helti, 73–77. Basel, Switzerland: Birkhauser Verlag.

Kime, D.E. 1995. The effects of pollution on reproduction in fish. *Reviews in Fish Biology and Fisheries* 5:52–96.

Kirchhofer, A. 1995. Morphological variability in the ecotone—an important factor for the conservation of fish species in Swiss rivers. *Hydrobiologia* 303:103–110.

Kirchhofer, A. and D. Hefti, eds. 1996. *Conservation of endangered freshwater fish in Europe*. Basel: Birkhauser.

Kornfield, I. and A.A. Echelle. 1984. Who's tending the flock? In *Evolution of fish species flocks*, eds. A.A. Echelle and I. Kornfield, 251–254. Maine: Orono Press.

Lowe-McConnell, R.H. 1993. Fish faunas of the African Great Lakes: origins, diversity, and vulnerability. *Conservation Biology* 7:634–643.

Lusk, S. 1996. The status of the fish fauna in the Czech Republic. In *Conservation of endangered freshwater fish in Europe*, eds. A. Kirchofer and D. Helti, 89–98. Basel, Switzerland: Birkhauser Verlag.

Lydeard, C. and R.L. Mayden. 1995. A diverse and endangered aquatic ecosystem of the southeast United States. *Conservation Biology* 9:800–805.

Lyons, J., S. Navarro-Perez, P.A. Cochran, E. Santana C., and M. Guzman-Arroyo. 1995. Index of biotic integrity based on fish assemblages for the conservation of streams and rivers in west-central Mexico. *Conservation Biology* 9:569–584.

Maitland, P.S. and A.A. Lyle. 1991. Conservation of freshwater fish in the British Isles: the status and biology of threatened species. *Aquatic conservation, marine and freshwater ecosystems* 1:25–54.

Maitland, P.S. and A.A. Lyle. 1996. Threatened freshwater fishes of Great Britain. In *Conservation of endangered freshwater fish in Europe*, eds. A. Kirchofer and D. Helti, 9–21. Basel, Switzerland: Birkhauser Verlag.

Manire, C.A. and S.H. Gruber. 1990. Many sharks may be headed toward extinction. *Conservation Biology* 4:10–11.

Maric, D. 1995. Endemic fish species of Montenegro. *Biological Conservation* 72:187–194.

Mayden, R.L., B.M. Burr, L.M. Page, and R.R. Miller. 1992. The native freshwater fishes of North America. In *Systematics, historical ecology, and North American freshwater fishes*, ed. R.L. Mayden, 827–863. Stanford: Stanford University Press.

McKaye, K.R., J.D. Ryan, J.R. Stauffer, Jr., L.J. Lopez Perez, G.I. Vega, and E.P. Van den Berghe. 1995. African tilapia in Lake Nicaragua. *BioScience* 45:406–411.

Micklin, P.P. 1988. Dessication of the Aral Sea: A water management diaster in the Soviet Union. *Science* 241:1170–1176.

Miller, R.R. 1986. Composition and derivation of the freshwater fish fauna of Mexico. *Anales de la Escuela nacional de Ciencias Biologicas, Mexico, Distrito Federal* 30:121–153.

Miller, R.R., J.D. Williams, and J.E. Williams. 1989. Extinctions of North American fishes during the past century. *Fisheries* 14:22–38.

Minckley, W.L. and J.E. Deacon. 1991. *Battle against extinction: Native fish management in the American west.* Tucson: University of Arizona Press.

Minckley, W.L., D.A. Hendrickson, and C.E. Bond. 1986. Geography of western North American freshwater fishes: Description and relationships to intercontinental tectonism. In *The zoogeography of North American freshwater fishes*, eds. C.H. Hocutt, and E.O. Wiley, 519–614. New York: John Wiley and Sons.

Moyle, P.B. 1995. Conservation of native freshwater fishes in the Mediterranean-type climate of California, USA: A review. *Biological Conservation* 72:271–279.

Moyle, P.B. 1994. The decline of anadromous fishes in California. *Conservation Biology* 8:869–870.

Moyle, P.B. and J.J. Cech. 1996. *Fishes: An introduction to ichthyology*. 3rd ed. New Jersey: Prentice Hall.

Moyle, P.B., B. Herbold, D.E. Stevens, and L.W. Miller. 1992. Life history and status of delta smelt in the Sacramento-San Joaquin Estuary, California. *Transactions of the American Fisheries Society* 121:67–77.

Moyle, P.B. and R.A. Leidy. 1992. Loss of biodiversity in aquatic ecosystems: Evidence from fish faunas. In *Conservation biology: the theory and practice of nature conservation, preservation, and management*, eds. P.L. Fiedler and S.K. Jain, 127–170. New York: Chapman & Hall.

Moyle, P.B. and P.R. Moyle. 1995. Endangered fishes and economics: Intergenerational obligations. *Environmental Biology of Fishes* 43:29–37.

Moyle, P.B. and P.J. Randall. 1996. Biotic integrity of watersheds. In *Sierra Nevada Ecosystem Project: final report to Congress. Vol. 2. Assessments and scientific basis for management options*, 975–985. University of California, Davis: Centers for Water and Wildland Resources.

Moyle, P.B. and R.M. Yoshiyama. 1994. Protection of aquatic biodiversity in California: A five-tiered approached. *Fisheries* 19:6–18.

Moyle, P.B., R.M. Yoshiyama, E. Wikramanayake, and J.E. Williams. 1995. *Fish species of special concern in California*. Sacramento: California Department of Fish and Game.

Moyle, P.B. and J.E. Williams. 1990. Biodiversity loss in the temperate zone: Decline of the native fish fauna of California. *Conservation Biology* 4: 275–284.

Mrakovcic, M., S. Misetic, and M. Povz. 1995. Status of freshwater fish in Croation Adriatic river systems. *Biological Conservation* 72:179–185.

Naiman, R.J. 1992. *Watershed management*. New York; Springer-Verlag.

Naiman, R.J., J.J. Magnuson, D.M. Knight, and J.A. Stanford. 1995a. *The freshwater imperative: A research agenda*. Washington, D.C.: Island Press.

Naiman, R.J., J.J. Magnuson, D.M. McKnight, J.A. Stanford, and J.R. Karr. 1995b. Freshwater ecosystems and their management: A national initiative. *Science* 270:584–585.

National Research Council. 1996. *Upstream: Salmon and society in the Pacific Northwest*. Washington, D.C.: NRC Press.

Nehlsen, W. 1994. Salmon stocks at risk: Beyond 214. *Conservation Biology* :867–869.

Nehlsen, W., J.E. Williams, and J.A. Lichatowich. 1991. Pacific salmon at the crossroads: Stocks at risk from California, Oregon, Idaho, and Washington. *Fisheries* :4–21.

Nelson, J.S. 1994. *Fishes of the World*. New York: John Wiley and Sons, Inc.

Ogutu-Ohwayo, R. 1993. The effects of predation by Nile perch, *Lates niloticus* L., on the fish of Lake Nabugabo, with suggestions for conservation of endangered endemic cichlids. *Conservation Biology* 7:701–711.

Ogutu-Ohwayo, R. 1990. The decline of the native species of Lakes Victoria and Kyoga (E. Africa) and the impact of introduced species, especially the Nile perch, *Lates niloticus*, and the Nile tilapia, *Oreachromis niloticus*. *Environmental Biology of Fishes* 27:81–96.

Oliver, C.H., B.J. Shuter, and C.K. Minns. 1995. Toward a definition of conservation principles for fisheries management. *Canadian Journal of Fisheries and Aquatic Sciences* 52:1584–1594.

Owen, R.B., R. Crossley, T.C. Johnson, D. Tweddle, I. Kornfield, S. Davidson, D.H. Eccles, and D.E. Engstrom. 1990. Major low levels of Lake Malawi and their implications for speciation rates in cichlid fishes. *Proceedings of the Royal Society of London Biologists* 240:519–553.

Parenti, L.R. 1984. A taxonomic revision of the Andean killifish genus *Orestias* (Cyprinodontiformes, Cyprinodontidae). *Bulletin of the American Museum of Natural History* 178:107–214.

Pethiyagoda, R. 1991. *Freshwater fishes of Sri Lanka*. Colombo: Wildlife Heritage Trust.

Petts, G.E. 1984. *Impounded rivers: Perspectives for ecological management*. United Kingdom: John Wiley and Sons.

Pollard, D.A., B.A. Ingram, J.H. Harris, and L.F. Reynolds. 1990. Threatened fishes in Australia—an overview. *Journal of Fish Biology* 37 (Supplement):67–78.

Postel, S. 1996. Dividing the waters: Food security, ecosystem health and the new politics of scarcity. *Worldwatch paper 132*. Washington, D.C.: Worldwatch Institute.

Postel, S.L., G.C. Dailey, and P.R. Ehrlich. 1996. Human appropriation of renewable fresh water. *Science* 271:785–788.

Power, M.E. 1990. Effects of fish in river food webs. *Science* 250:811–814.

Power, M.E. 1992. Habitat heterogeneity and the functional significance of fish in river food webs. *Ecology* 73:1675–1688.

Povz, M. 1995. Status of freshwater fishes in the Adriatic catchment of Slovenia. *Biological Conservation* 72:171–177.

Povz, M. 1996. The red data list of freshwater lampreys (*Cyclostomata*) and fishes (Pisces) of Slovenia. In *Conservation of endangered freshwater fish in Europe*, eds. A. Kirchhofer and D. Hefti, 63–72. Basel, Switzerland.

Quigley, D.T.G. and K. Flannery. 1996. Endangered freshwater fish in Ireland. In *Conservation of endangered freshwater fish in Europe*, eds. A. Kirchofer and D. Helti, 27–34. Basel, Switzerland: Birkhauser Verlag.

Rakaj, N. and A. Flloko. 1995. Conservation status of freshwater fish of Albania. *Biological Conservation* 72:195–199.

Reinthal, P. 1993. Evaluating biodiversity and conserving Lake Malawi's cichlid fish fauna. *Conservation Biology* 7:712–718.

Reinthal, P.N. and M.L.J. Stiassny. 1991. The freshwater fishes of Madagascar: A study of an endangered fauna with recommendations for a conservation strategy. *Conservation Biology* 5:231–243.

Rinne, J.N., W. Ivantsoff, L.E.M. Crowley, and J. Lobon-Cervia. 1995. Conservation of desert fishes: Spain, Australia, and the United States. In *Biodiversity in managed landscapes*, 377–400. Oxford: Oxford University Press.

Roberts, C.M. 1995a. Rapid build-up of fish biomass in a Caribbean marine reserve. *Conservation Biology* 9:815–826.

Roberts, C.M. 1995b. Effects of fishing on the ecosystem structure of coral reefs. *Conservation Biology* 9:988–995.

Ruhle, C. 1996. Decline and conservation of migrating brown trout (*Salmo trutta* f. *lacustris* L.) of Lake Constance. In *Conservation of endangered freshwater fish in Europe*, eds. A. Kirchofer and D. Helti, 203–211. Basel, Switzerland: Birkhauser Verlag.

Senanayake, F.R. and P.B. Moyle. 1982. Conservation of the freshwater fishes of Sri Lanka. *Biological Conservation* 22:181–195.

Shuman, J.R. 1995. Environmental considerations for assessing dam removal alternatives for river restoration. In *Regulated rivers: Research and management, Vol.2*, 249–261. New York: John Wiley and Sons.

Skelton, P.H. 1990. The conservation and status of threatened fishes in southern Africa. *Journal of Fish Biology* 37 (Supplement): 87–95.

Stiassny, M.L.J. 1996. An overview of freshwater biodiversity: With some lessons from African fishes. *Fisheries* 21:7–13.

Stiassny, M.L.J. and N. Raminosoa. 1994. The fishes of the inland waters of Madagascar, biological diversity in African fresh- and brackish water fishes. Geographical overviews. *Annual Mus. r. Afr. Cent. Zool.* 275:133–149.

Suchanek, T.H. 1994. Temperate coastal marine communities: Biodiversity and threats. *American Zoologist* 34:100–114.

Sweetman, K.E., P.S. Maitland, and A.A. Lyle. 1996. Scottish natural heritage and fish conservation in Scotland. In *Conservation of endangered freshwater fish in Europe*, eds. A. Kirchofer and D. Helti, 23–26. Basel, Switzerland: Birkhauser Verlag.

United Nations Population Fund. 1995. *The state of world population 1995*. New York: United Nations Population Fund.

Vas, P. 1995. The status and conservation of sharks in Britain. *Aquatic Conservation: Marine and Freshwater Ecosystems* 5:67–79.

Vincent, A.C.J. and H.J. Hall. 1996. The threatened status of marine fishes. *Trends in Ecology and Evolution* 11:360–361.

Wager, R. and P. Jackson. 1993. *The action plan for freshwater fishes*. Brisbane: Australian Nature Conservation Agency, Endangered Species Program.

Walsh, S.J., N.M. Burkhead, and J.D. Williams. 1995. Southeastern freshwater fishes. In *Our living resources, a report to the nation on the distribution, abundance, and health of U. S. plants, animals, and ecosystems*, eds. E.T. LaRoe, G.S. Farris, C.E. Puckett, P.D. Doran, and M.J. Mac, 144–147. Washington, D.C.: U.S. Government Printing Office.

Waples, R.S. 1995. Evolutionary significant units and the conservation of biological diversitry under the endangered species act. In *Evolution and the aquatic ecosystem: defining units in population conservation*, ed. J.F. Nielsen, 8–27. Bethesda, MD: American Fisheries Society.

Warren, M.L., Jr. and B.M. Burr. 1994. Status of freshwater fishes of the United States: an overview of an imperiled fauna. *Fisheries* 19:6–18.

Wikramanayake, E.D. 1990. Conservation of endemic rain forest fishes of Sri Lanka: Results of a translocation experiment. *Conservation Biology* 4:32–37.

Williams, J.E. and R.R. Miller. 1990. Conservation status of the North American fish fauna in freshwater. *Journal of Fish Biology* 37:79–85.

Williams, J.E., M.L. Warren, Jr., K.S. Cummings, J.L. Harris, and R.J. Meves. 1993. Conservation status of freshwater mussels of the United States and Canada. *Fisheries* 18:6–22.

Williams, J.E., J.E. Johnson. D.A. Hendrikson, S. Contreras-Balderas, J.D. Williams, M. Navarro-Mendoza, D.E. McAllister, and J.E. Deacon. 1989. Fishes of North America endangered, threatened, or of special concern: 1989. *Fisheries* 14:2–20.

Williams, J.E., D.B. Bowman, J.E. Brooks, A.A. Echelle, R.J. Edwards, D.A. Hendrickson, and J.J. Landye. 1985. Endangered aquatic ecosystems in North American deserts with a list of vanishing fishes of the region. *Journal of the Arizona-Nevada Academy of Science* 20:1–62.

Willson, M.F. and K.C. Halupka. 1995. Anadromous fish as keystone species in vertebrate communities. *Conservation Biology* 9:489–497.

Winfield, I.J., J.M. Fletcher, and D. Cragg-Hine. 1994. *Status of rare fish: A literature review of freshwater fish in the UK*, R&D Report 18. Bristol, UK: Institute of Freshwater Ecology, National Rivers Authority.

Witte, F., T. Goldschmidt, P.C. Goudswaard, W. LigtVoet, M.J.P. Van Oijen, and J.H. Wanink. 1992a. Species extinction and concomitant ecological changes in Lake Victoria. *Netherlands Journal of Zoology* 42:214–232.

Witte, F., T. Goldschmidt, J. Wanink, M. van Oijen, K. Goudswaard, E. Witte-Maas, and N. Bouton. 1992b. The destruction of an endemic species flock: Quantitative data on the decline of the haplochromine cichlids of Lake Victoria. *Environmental Biology of Fishes* 34:1–28.

World Resources Institute. 1994. *World resouces*. Oxford, UK: Oxford University Press.

World Resources Institute, United Nations Environment Programme, United Nations Development Programme, and World Bank. 1996. *World resources*. Oxford, UK: Oxford University Press.

CHAPTER

9

Challenges in Insect Conservation: Managing Fluctuating Populations In Disturbed Habitats

CHERYL B. SCHULTZ
and GARY C. CHANG

The vast majority of the world's species are insects (at least 80%; Mawdsley and Stork 1995). Their importance is overwhelming by almost any measure. For example, insects and other arthropods contribute substantially to standing biomass; 1,000 kg/ha is an estimate for the United States (Pimentel *et al.* 1980). In most terrestrial and freshwater ecosystems they play critical roles as prey, predators, herbivores and pollinators (Free 1970; Debach and Rosen 1991; Kellert 1993; Lloyd and Barrett 1996). Indeed, in one of the first issues of the Society for Conservation Biology's journal, E.O. Wilson (1987) called insects "the little things that run the world." Because they comprise the majority of the earth's biodiversity, insects should be considered pivotal in conservation efforts (Kim 1993). Unfortunately, an alarmingly small percent of our conservation literature focusses on insect issues. For example, in 1993, 1994 and 1995, the journals *Ecological Applications*, *Conservation Biology*, and *Biological Conservation* published 1,070 articles with only 62 related to insect issues and still fewer related to conservation of declining insect populations. Thus, only 6% of our conservation literature is aimed at 80% of our planet's biodiversity. This neglect of insect conservation cannot be justified on the basis of insects not being endangered. In Britain where the biodiversity is relatively well-documented, approximately 22,500 insect species occur; 43 insects are believed to have gone extinct between 1900 and 1987 (Hambler and Speight 1996). The number of insect species believed extinct in Britain is over eight times that of number of extinct vertebrates, and over three times that of flowering plants (Hambler and Speight 1996). Thus our goal is this chapter is to highlight some of the problems associated with insect conservation. While the issues we address are not unique to insects, they are particularly problematic in insect conservation.

ECOSYSTEM FUNCTIONS PERFORMED BY INSECTS

As diverse and abundant as insects are, it is not surprising that they perform many ecosystem functions. Although we focus mainly on insects' trophic roles in pollination and biological control, it is worth noting that they are also important as food for other animals, including humans (Bodenheimer 1951). In fact, insects are still a part of human diets in several non-Western societies, in some cases serving as a major source of protein (Berenbaum 1995). One hundred grams of *Usta terpsichore* caterpillars contains 28.2 g of protein, comparable to chicken and lean ground beef, 31.6 g and 27.4 g, respectively (Berenbaum 1995). In addition, insects are "ecosystem engineers" (*sensu* Lawton and Jones 1995). Ants and termites, for example, turn soil while constructing their nests. A colony of termites, *Macrotermes bellicosus*, in Uganda annually excavates up to 230 g of soil per square meter (Kalule-Sabiti 1980), while the ant community (primarily *Chelaner* spp., *Pheidole* sp., and *Iridomyrmex* sp.) in Australian shrub steppe turns 350–420 kg of soil/ha/yr (Briese 1982). Such nest building activity subsequently alters soil properties, usually increasing amounts of available nutrients and rates of water infiltration (Holldobler and Wilson 1990; Eldridge 1994).

Insects As Pollinators

As can be found in all introductory biology texts, insects play a major role in most terrestrial ecosystems as pollinators (Free 1970). Pollination services of insects translate into conventional human benefits; indeed, the value of insect-pollinated crops in the United States was $24 billion in 1988, exceeding the estimated 14 billion dollars in products lost that year to insect damage (Metcalf and Metcalf 1993). Bees, wasps, flies, beetles, moths, butterflies, and even thrips are major pollinators of flowering plants (Endress 1994). Yet the significance of insect pollinators to global biodiversity is underrated.

Increasing plant diversity correlates with reliance on animal pollination (Endress 1994). It may be that animal pollination is crucial to maintaining high plant diversity. For example, Janzen (1971) observed that euglossine bees and other long-distance foraging pollinators allow low populations of plants to persist. Distributions of insect pollinators may limit the distributions of the plants they pollinate. For example, the rare Polish orchid *Dactylorhiza fuchsii* is pollinated by a longhorn beetle, *Alosterna tabacicolor* (Gutowski 1990). While losing the orchid would not significantly affect populations of the beetle, loss of the beetle would jeopardize the orchid. To emphasize the importance of pollinators vividly, consider the following analogy. Some plants depend entirely on a single insect species to distribute their male gametes (Thompson 1994); should the pollinators of such plants be eliminated, the entire plant species has been effectively castrated.

Although the importance of pollination for crops has long been recognized, only recently have ecologists realized the extent to which pollinators may be a limiting resource for plants in general (see Bierzychudek 1981). From a conservation perspective, pollination limitation can compound other threats to endangered plant species. In Sweden, habitat fragmentation has reduced flowering plant and pollinator diversity and abundance, visitation rates, and, for at least the maiden pink (*Dianthus deltoides*), lowered seed set (Jennersten 1988). Similarly, a network of insects and flowers maintains the rare orchid *Spiranthes diluvialis* in Colorado and Utah (Sipes and Tepedino 1995). When specialized pollinators are replaced by generalists in a system, hybridization can jeopardize plant species (Levin, Francisco-Ortega, and Jansen 1996).

One insect that has captured the attention of environmentalists and conservation biologists is the honeybee. The European honey bee, *Apis mellifera*, was brought into the United States by early European settlers and first reported to have established in 1622 (Free 1982). *Apis mellifera* is the only species of honey bee occuring in North America (Borror, Triplehorn, and Johnson 1989). A thorough economic analysis considering crop output, lower production costs, and other societal values placed the annual social gains in the United States due to the pollinating activities of *A. mellifera* between $1.6 and $5.7 billion (Southwick and Southwick 1992). Thus any threat to honey bee is considered a major problem. The unintentional death of bees due to nontarget effects of pesticides initially became an issue in the United States in the 1870s (Johansen 1977) with the earliest arsenic-based insecticides (e.g., Paris green), other inorganic insecticides, such as those based on copper and sulfur, and botanically-derived nicotine and pyrethrum insecticial compounds (Shaw 1941).

Honey bee poisoning was exacerbated during the chemical intensification of agricultural practices after World War II (Johansen 1977). The development of DDT and other synthetic organic pesticides increased the amount of chemicals used in pest control (Perkins 1982) as well as the area treated with pesticides (Johansen 1977). An increase in aerial spraying of crops also exposed honey bees to pesticides via drift (Martin and McGregor 1973). Clearly, bee poisoning is a problem for bee keepers, but conservation concerns extend beyond *A. mellifera*. For example, although *A. mellifera* was moved to safety during spraying for pest grasshoppers in Wyoming, over 660 native bee species were potentially subjected to the pesticide (Tepedino 1979).

One way of conserving pollinators is by using only highly selective pesticides that do not kill bees. An example of such an approach can be found in New Brunswick, where over 20 species of native bees are key pollinators of commecially harvested blueberries, *Vaccinium angustifolium* and *V. myrtilloides* (Kevan 1975). In 1952, spraying of DDT was begun in nearby forests to control spruce budworm, *Choristoneura fumiferana*, a voracious forest pest. Due to the toxicity of DDT, in 1969 it was replaced by fenitrothion, a less persistent insecticide but one that is more toxic to bees (Kevan 1975). With the introduction of fenitrothion, native bees and the blueberry crops declined. When fenitrothion spraying ceased, bee populations recovered, but so did spruce budworm populations. An alternative pesticide, trichlorfon, was then employed against spruce budworm. Trichlorfon proved to be less toxic to the bees, and the bees and the blueberry crop have recovered (Kevan and Baker 1984).

Insects as Population Regulators

Insects utilize a diverse range of foods, and regulate the populations of the resources upon which they feed in their role as consumers. While it is difficult to demonstrate rigorously the importance of insect natural enemies (Luck, Shepard, and Kenmore 1988), they have long been successfully employed by practitioners of biological control (Hagen and Franz 1973; Huffaker and Messenger 1976; Debach and Rosen 1991; Van Driesche and Bellows 1996). One line of evidence regarding the importance of insects as stabilizing factors comes from the literature on exotic weeds and exotic pest insects. Weed and insect pest species are often not problematic in their regions of origin (but see chapter 11). The explanation is usually thought to be because they are controlled by their natural enemies. For example, the Yellow starthistle, *Centaurea solstitialis*, is a weed from Eurasia that has invaded the western United States. In southern Europe where at least 42 insect species have been found to feed on it, the yellow starthistle does not reach high population densities (Clement 1990). In contrast to the situation in Eurasia, however, only eight endemic insect species were found to feed on the weed in northern Idaho, and none prevent it from achieving high densities (Johnson, McCaffrey, and Merickel 1992).

Deliberate introductions of natural enemies that successfully control pest species are high-profile examples of classical biological control (Caltagirone 1981; Murdoch, Chesson, and Chesson 1985). McEvoy, Cox, and Coombs (1991) conducted a large-scale study (i.e., 42 sites throughout western Oregon over 12 years) of a classical biological control of an alien weed, the Tansy ragwort, *Senecio jacobaea*. The ragwort is native to Europe but not to the United States, where it was first recorded in 1913. It is a biennial or short-lived perennial that displaces native plants and poisons livestock. Three insect natural enemies of the ragwort, the cinnabar moth (*Tyria jacobaeae*), the ragwort flea beetle (*Longitarsus jacobaeae*), and the ragwort seedfly (*Hylemya seneciella*), were located in France and Italy and imported into the United States during the 1960s. Together, these three herbivorous insects are capable of reducing ragwort populations to less than one percent of their previous size (McEvoy, Cox, and Coombs 1991). The three insects exploit the ragwort in different fashions; *T. jacobaeae* is a defoliator, *L. jacobaeae* consumes roots and shoots, and *H. seneciella* feeds on immature seeds (McEvoy, Cox, and Coombs 1991). The insects' compatibility as control agents is explained by the differences in where they feed on the plant, as well as seasonal differences in activity. *Tyria jacobaeae* larvae feed in the summer, while *L. jacobaeae* feeds during the rest of the year (James, McEvoy, and Cox 1992). Thus, together they can reduce ragwort populations further than any one species could alone (James, McEvoy, and Cox 1992).

Finally, although the "prey" of decomposers are already dead, these insects are strikingly important to ecosystems. Waterhouse (1974) recounts Australia's battle against ever accumulating cow dung. Cows are not native to Australia and their dung qualitatively (and quantitatively) differs from marsupial dung. Appoximately 250 species of Australian scarab beetles utilize and thus dispose of, marsupial dung, but not one of these beetles substantially uses the larger, moister cow dung. Since a single cow can potentially cover 5–10% of an acre with dung in one year, several aesthetic, agricultural, environmental, and public health concerns mounted until scarab beetles from Africa and Europe were imported into Australia, beginning in 1967. Four species were initially chosen to import, the most important of these being *Onthophagus gazella* (Waterhouse 1974). These and other subsequently imported beetles successfully thwarted the dung menace in Australia (Hughes, Jones, and Gutierrez 1984).

CHALLENGES FACING INSECT CONSERVATION

Challenge 1: Unraveling Basic Biology

Over the last decade, conservation biologists have developed many new theories and strategies for managing declining species and ecosystems (e.g., population viability theory, metapopulation theory). However, while general theories may be useful in assessing conservation priorities, they do not replace knowledge regarding the biology and natural history of declining species. Natural history data are the building blocks for sound conservation for any taxa.

Many insects, along with many plants and other animals, have complex life cycles and require a wide range of habitats to complete their life cycle. Experiences with several rare insects highlight the importance of understanding the idiosyncracies of the target species before implementing conservation management plans. In Britain, for example, despite the establishment of several reserves, populations of *Maculinea arion*, the large blue butterfly, declined for many years (Thomas 1995). The butterfly's disappearance in many seemingly appropriate habitats was a mystery until biologists began scrutinizing several aspects of the larva's microhabitat and its habitat needs. It was discovered that *Maculinea arion* larvae rely on the red ant, *Myrmica sabuleti*, to avoid predation. *Maculinea arion* larvae spend up to 11 months of the year in *M. sabuleti* nests where the butterfly larvae supply ants with sugar while ants provide food and protection for the butterfly larvae. However, in the 1950s, rabbits that grazed *M. arion* habitats declined due to an outbreak of myxomytosis. Subsequently, *M. sabuleti* populations declined, which resulted in declines in *M. arion* populations. By the time biologists figured out why *M. arion* populations were declining, there was only one butterfly colony left in England. By manipulating former *M. arion* sites to mimic historic grazing intensities, biologists have begun to restore *M. sabuleti* populations and *M. arion* is now being reintroduced into historic sites.

Challenge 2: Monitoring Population Trends

Unlike most vertebrates, insect populations fluctuate by orders of magnitude over very short time scales. Ito (1980) assembled population time series for 67 animal species. Insects fluctuate by many orders of magnitude more than birds, mammals, or fish, with the maximum population size as much as 35,000 times the minimum population for one moth, *Zeiraphera diniara* (Figure 9.1; Ito 1980). The large swings in population size

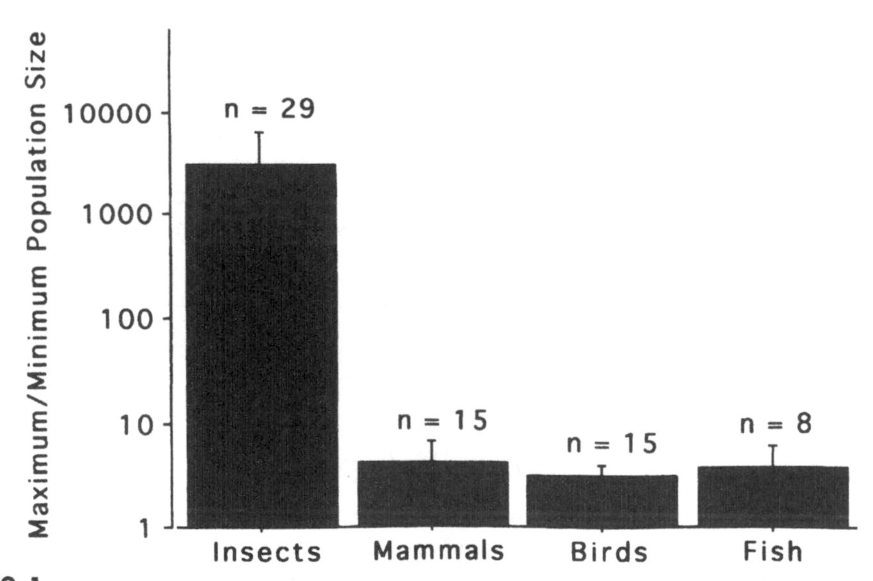

Figure 9.1
Ratio of maximum to minimum population size in 67 animal species with at least ten years of census data. Sample sizes above the bars indicate the number of species in each group. Data are from Ito (1980).

characteristic of insects pose a significant challenge to predicting population changes in rare insect populations.

Over the last decade, conservation biologists have achieved significant advances in theories predicting population change, namely population viability theory (see chapter 1). However, most viability theory has been developed to address changes in populations of long-lived vertebrates such as the Northern spotted owl, *Strix occidentalis* (McKelvey, Noon, and Lamberson 1993) or the African elephant, *Loxodonta africana* (Armbruster and Lande 1993), while few advances have been made in applying viability theory to rare insects and other organisms with one or more generation(s) per year. For example, one use of viability theory involves the analysis of linear population projection matrices as a means of learning whether current demographic trends imply long-run population decline (Caswell 1989). Projection matrices are needed in this effort because with long-lived animals there can be so much inertia that a population could be increasing because of past events yet possess a birth and death schedule that inevitably implies certain extinction. However, this matrix approach relies on the assumption that a few years of data aptly capture the demographic rates for the study organisms (see chapter 2). In insects, year-to-year fluctuations are typically so severe that the approach is unlikely to work. To determine whether insects are increasing or decreasing, alternative methods are needed.

Ludwig (1996) developed one promising approach for evaluating survival probabilities for insect populations. Ludwig's model addresses populations with two characteristics that are common and problematic in rare insect populations—i.e., small population size and the tendency to have large changes in population size between successive generations. He constructed a discrete model with non-overlapping generations that investigated the probability of extinction as a function of initial population size and carrying capacity. Ludwig noted that populations with large swings from year to year are likely to reach carrying capacity from time to time. Carrying capacity and a populations' proximity to carrying capacity also influence survival probability. Thus, density-independent models, such as most matrix models, may be inappropriate for evaluating the risk of extinction in highly varying populations.

Extrapolating from a diffusion approximation, Ludwig estimated survival probabilities and found that a large percentage of small populations are likely to go extinct rather quickly but that remaining populations tend to survive a long time. Such an approach may prove useful in generating insight into how to manage rare insect species, particularly those that are now comprised of many small populations scattered across the landscape.

Challenge 3: Reducing Fragmentation-induced Isolation

Land use over the last 200 years has resulted in the fragmentation of nearly all habitats around the world (Groom and Schumaker 1992). Fragmentation threatens numerous rare species through effects such as exposed edges, reduced habitat area, and increased distance between patches. Indeed, for many insect species, isolation of habitat fragments is the key problem facing rare species (Thomas 1995). In Britain, for example, despite the establishment of dozens of butterfly reserves in the 1960s and 1970s, the long distances between suitable habitats have resulted in butterfly populations that are still declining (Prendergast and Eversham 1995). Rare butterfly species tend to be relatively sedentary and individuals are reluctant to move beyond the edge of their primary habitat (Tables 9.1 and 9.2). Isolation results in population decline because butterfly species with low dispersal abilities rarely recolonize patches that have gone extinct. Half of all British

Table 9.1.
Dispersal behavior of British butterflies. Data are from Tables 1.1 and A1 in Pollard and Yates (1993).

	Mobility		
Abundance	**Sedentary**	**Intermediate**	**Wide-Ranging**
Common	8	4	7
Rare	12	8	0

Table 9.2.
Dispersal behavior in some European and North American butterflies. 'Low' habitat sensitivity refers to the willingness of butterflies to travel beyond hostplant patches into inhospitable habitat. 'High' habitat sensitivity refers to the reluctance of butterflies to travel into inhospitable habitat. Data are from Brooke, Lees, and Lawman 1985; Cappucino and Kareiva 1985; Fahrig and Paloheimo 1987; Mattoni 1990; New 1992; Pollard and Yates 1993; Shreeve 1992; Thomas 1983; Thomas 1984a and 1984b; Thomas 1985; Thomas et al. *1986; Thomas 1991; Warren, Pollard, and Bibby 1986; Watt* et al. *1977; and Watt, Han, and Tabashnik 1979.*

	Habitat Sensitivity	
Abundance	**High**	**Low**
Common	0	9
Rare	16	0

butterflies are so sedentary that once a population becomes isolated, rates of patch extinction exceed rates of colonization, even when habitat loss in one place is matched by habitat regeneration elsewhere (Thomas 1991).

Once isolated, butterfly populations may also experience severe inbreeding. In a study of the endangered French butterfly *Parnassius mnemosyne*, Descimon and Napolitano (1993) detected reduced genetic diversity in small patches and patches far from the center of the species' distribution. They observed a decline in gene flow in peripheral colonies of this montane butterfly using electrophoretic data. Similarly, Dempster, King, and Lakhani (1976) identified distinct dispersal and non-dispersal morphologies in the rare British swallowtail, *Papilio machaon*. Dempster and colleagues observed that swallowtails with wing and thorax characteristics of the isolated population in Wicken Fen were weaker fliers than those with characteristics of the less isolated populations in Norfolk Broads. These researchers hypothesized that in isolated populations the dispersal morph dwindled because dispersers emigrated out of the population while no dispersers arrived to maintain the trait.

Challenge 4: Managing at the "Right" Scale

How large is large enough? This question has plagued conservation biology since its inception and the question has long been debated acrimoniously in the context of island biogeography and the SLOSS controversy (Simberloff 1974, 1988). In contrast to vertebrates, which are often reported to require several thousand of hectares of habitat (e.g., grizzly bears, panthers, and spotted owls), conservation biologists have suggested that plants and invertebrates can survive on "postage-stamp" parcels in urban and agricultural areas (Simberloff 1982; Hafernik 1992). Indeed, in the SLOSS debate, plant and invertebrate biologists frequently prefer the option of "several small" preserves (Figure 9.2). In contrast, bird and mammal biologists typically suggest that because very large reserves are required to maintain key species and only a limited amount of land is available, the "single large" option is preferred.

Of course, small areas are only adequate to an insect species if its potentially diverse and often unknown habitat requirements are met in a habitat fragment. In addition, ecological processes needed to maintain critical habitats may require much larger expanses of area. In fire-maintained ecosystems, for example, fire may eliminate the insect fauna while encouraging the regeneration of the flora. To maintain both plants and insects in such a community, conservation plans must consider a large enough area to introduce fire on the appropriate fire cycle and patch size so as not eliminate insects of concern. Thus the challenge to insect conservation is formulating broad management strategies at the right scale for rare insects.

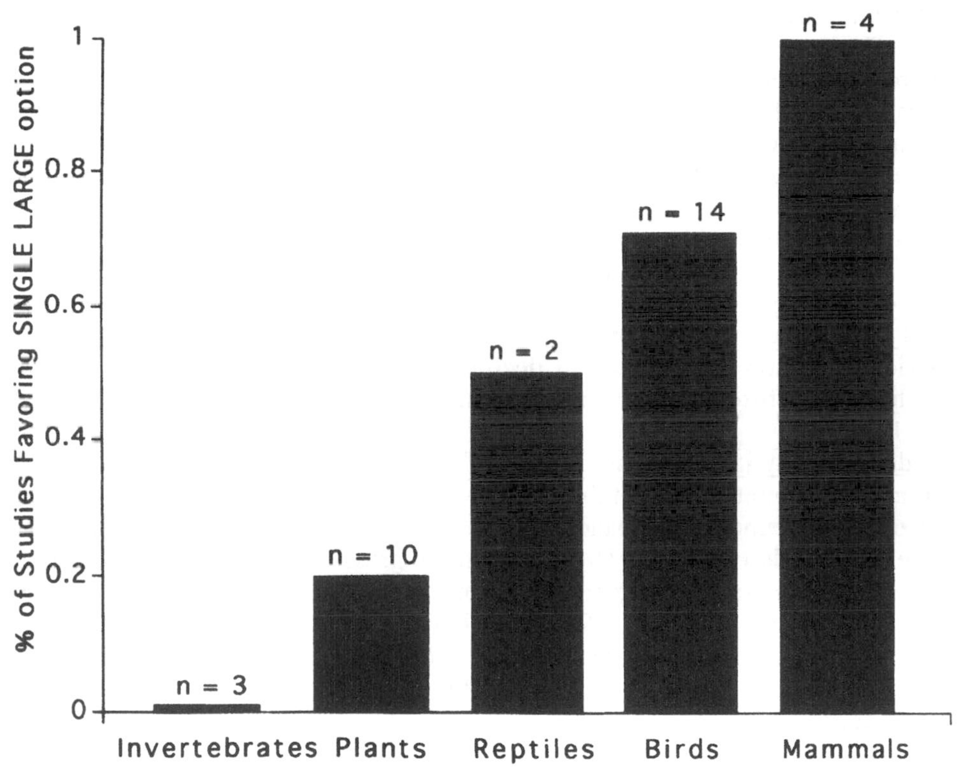

Figure 9.2
Influence of taxonomic bias on argument favored in the SLOSS (Single Large or Several Small) debate. Sample sizes above the bars indicate the number of studies in each group. Data are from Abele and Patton 1976; Bassett and Miers 1984; Brown 1978; Dawson 1984; Deshaye and Morrisett 1989; Diamond 1975; Forman et al. *1976; Gilpin and Diamond 1980; Higgs and Usher 1984; Janzen 1983; Jarvinen 1982; King 1984; Lahti and Ranta 1986; Loman and von Schantz 1991; Miller, Bratton, and White 1987; Nilsson 1978; Picton 1979; Quinn and Robinson 1987; Quinn, Wolin and Judge 1989; Simberloff 1982; Simberloff and Abele 1976; Simberloff and Gotelli 1984; Soulé and Simberloff 1986; Temple and Cary 1988; Terbourgh 1976; Williams 1984; Woolhouse 1987 and Wright and Hubbell 1983.*

Challenge 5: Managing Semi-Natural Habitats

Frequently conservation biologists assume that rare species require natural environments, implying that natural habitats are not influenced by humans. However, humans have influenced their surroundings for millennia (e.g., Bergland 1991). If maintaining species is our conservation objective, the key is to maintain these habitats as they were managed in pre-modern times. The role of pre-modern human influences on their surrounding ecosystems is particularly poignant in insect conservation in the temperate zone. Many habitats that depend on traditional management methods are declining. In Britain, 44% of threatened beetles and 25% of threatened butterflies are declining due to changes traditional management practices, and an additional 13% of threatened beetles and 35% of threatened butterflies are declining due to lack of habitat management (Mawdsley and Stork 1995).

In northwest Europe, the demise of managed heathland has begun to attract a lot of attention (Webb 1990; Blackstock *et al.* 1995). Heathland butterfly communities have thrived for millenia in habitats with low intensity farming (Erdhardt and Thomas 1991; Webb and Thomas 1994). Heathland habitats were managed by regular burning, cutting and grazing (Webb and Thomas 1994). In lowland Wales, Blackstock *et al.* (1995) noted that overall 44% of the managed heathland has disappeared since the 1920s due to agricultural intensification while some specific heath types have declined by more than 95% since the 1920s. In Scotland, 30% of semi-natural habitats have been modified (Usher 1995).

Similarly, semi-natural habitats are disappearing in other temperate regions. In South Africa, Samways (1995) noted that 64% of historical fires in grasslands were set by humans. Today, fire prevention policies restrict fires. The changes in the plant community that result from the elimination of fires adversely effects numerous insect species in the South Africa (Samways 1995). In Japan, decline in traditional agricultural and silvicultural practices has led to forest succession, resulting in the decline in many butterfly species (Sibatani 1990).

In addition, many insect species that spend most of their lives in unmanaged habitats require managed habitat for part of their life cycle. More than half of the British butterflies that are frequently found in woodlands or forests require semi-natural grasslands for some part of their life cycle (Oates 1995). Because these grasslands revert to woodlands if left unmanaged, active habitat management is clearly critical to the survival of many British butterfly species (Oates 1995). As Webb and Thomas (1994, p. 143) describe the situation British insects are facing, conservationists are now dealing with species whose "... dynamics were moulded over perhaps 5,000 generations by a recently obsolete form of land management."

Challenge 6: Making Insect Conservation and Pest Management Compatible

The phrase "insect management" must come to embrace both pest control and the preservation of insect diversity. Approximately 70% of the Earth's terrestrial surface is utilized in agriculture and forestry, compared to 2.8% that has been set aside as nature reserves (Western 1989). While agricultural land may not be as diverse uncultivated areas, it is not merely crops and pests. For example, the Philippines' extensively studied rice ecosystem contains 687 taxa from different kingdoms (Schoenly *et al.* 1996); of these, just 25 insects are considered pests (Cohen *et al.* 1994). The remainder probably constitute desirable biodiversity. The potential for incorporation of agricultural land in conservation strategies is starting to be recognized, but remains largely untapped (Gall and Orians 1992). Agriculturalists and conservationists are beginning to realize that both insect control and preservation will be important in global sustainability. The challenge is integrating the two seemingly disparate activities (Gall and Orians 1992; Paoletti *et al.* 1992; Van Hook 1994).

Individual farmers must frequently choose among insect pest control options. Decisions at this level can have potent conservation consequences. The adverse environmental impacts of pesticides have received much attention since the pivotal publication of Rachel Carson's *Silent Spring* (1962). Pesticides have often been found to impact non-target organisms (Edwards 1993), although it is not clear how responsible they might be in insect species extirpations. Resurgence of primary pests and secondary pest outbreaks (Edwards 1993), and the development of pesticide resistant strains (McKenzie 1996) provide compelling ecological and economic reasons for curbing reliance on chemical control tactics. These issues, along with misuse, result in an estimated annual $8.123 billion environmental and social cost to pesticides in the United States (Pimentel *et al.* 1992).

Yet chemicals remain a primary option in pest control. Over 200 chemicals are regisitered for insecticidal use in the United States (Metcalf 1994), and some unregistered compounds (DDT among them) are used in other countries (Metcalf and Metcalf 1993). The estimated amount of all pesticides annually used in the United States is 434 million kg spread over 148 million treated ha (Pimentel *et al.* 1991). Ninety-nine million kg of these pesticides are directed at insects (Pimentel *et al.* 1991). It is also worth noting that pesticide use is more concentrated in households than on crops. On agricultural land the ratio of kg of pesticide to ha treated is 2.8:1 versus 13.8:1 in households; 358 million ha of agricultural land stay untreated, while essentially all four million ha of household land is treated (Pimentel *et al.* 1991). In the tropics, insecticides are additionally relied upon to control populations of insect vectors of diseases, such as mosquitoes (*Culicidae*) that transmit malaria (Metcalf and Metcalf 1993).

Ways to Make Pest Control More Conservation Sensitive. While pesticides serve important roles in society, they tend to be misused. Specific actions on particular crops that could be taken to reduce pesticide use are presented by Pimentel *et al.* (1993); in several crops, effectively monitoring pest populations to optimize the timing of pesticide application would reduce usage. Sweden, Denmark, and the Netherlands have legislated 50% reductions in total pesticide use (Matteson 1995) as has Ontario (Surgeoner and Roberts 1993). Sweden has attained its goal, although through some controversial tactics such as substituting more toxic pesticides (Matteson 1995). Some pesticides are effective against certain pests, yet are less damaging to humans and other components of the environment (Coats 1994; Metcalf 1994).

An environmental pest-management rating system that incorporates toxicity to mammals, fish, birds, honeybees and environmental persistence data has been developed for commonly used insecticides (Metcalf 1994; Table 9.3). Scores range from a low of three to a possible high of 15, with lower scores reflecting lower toxicity and persistence (Metcalf 1994).

Different pesticides cause different effects in the field. For example, in pears, conventional spraying with azinphosmethyl has been compared to a diflubenzuron-based management program designed to kill pestiferous pear psylla, *Cacopsylla pyricola*, and codling moth, *Cydia pomonella*, while minimizing harm to predacious arthropods (Booth and Riedl 1996). Pear psylla natural enemies, including various Dermaptera, Hemiptera,

Table 9.3
Pest-Management Rating of Widely Used Insecticides[a]

Insecticide	Mammalian Toxicity	Nontarget Toxicity				Environmental Persistence	Overall Rating
		Fish	Bird	Bee	Average		
Aldicarb	5	3	5	5	4.3	3	12.3
Aldrin	4	4	4	4	4.0	5	13.0
Azinphos-methyl	4	3	2	4	3.0	3	10.0
B.t.	1	1	1	1	1.0	1	3.0
Carbaryl	2	1	1	4	2.0	2	6.0
Carbofuran	5	2	5	5	4.0	3	12.0
Carbophenothion	4	2	4	4	3.3	2	9.3
Chlordane	2	3	2	2	2.3	3	7.3
Chlorpyrifos	3	3	3	5	3.7	3	9.7
Cryolite	1	1	1	2	1.3	5	7.3
Cyfluthrin	2	5	1	5	3.7	2	7.7
Cypermethrin	2	5	2	5	4.0	2	8.0
DDT	3	4	2	2	2.7	5	10.7
Demeton	5	2	5	2	3.0	2	10.0
Diazinon	3	2	5	4	3.7	3	9.7
Dicofol	2	1	2	1	1.3	4	7.3
Dieldrin	4	4	3	4	3.7	5	12.7
Diflubenzuron	1	1	1	4	2.0	4	7.0
Dimethoate	3	1	4	5	3.3	2	8.3
Disulfoton	5	3	5	2	3.3	3	11.3
Endosulfan	4	4	2	2	2.7	3	9.7
Endrin	5	5	5	2	4.0	5	14.0
EPN	4	2	3	4	3.0	4	11.0
Fenvalerate	2	4	2	5	3.7	2	7.7
Flucythrinate	3	5	1	5	3.7	3	9.7
Heptachlor	4	3	4	4	3.7	5	12.7
Imidacloprid	2	1	3	4	2.7	4	8.7
Lindane	3	3	2	4	3.0	4	10.0
Malathion	2	2	1	4	2.3	1	5.3
Methomyl	4	4	3	4	3.7	2	9.7
Methoprene	1	1	1	2	1.3	2	4.3
Methoxychlor	1	3	1	2	2.0	2	5.0
Methyl parathion	4	1	5	5	3.7	2	9.7
Mevinphos	5	3	5	4	4.0	1	10.0
Mexacarbate	4	1	5	5	3.7	2	9.7
Naled	2	2	3	4	3.0	1	6.0
Ovex	1	2	1	1	1.3	4	6.3
Oxydemeton-methyl	3	2	4	2	2.7	2	7.7
Parathion	5	2	5	5	4.0	2	11.0
Permethrin	2	4	2	5	3.7	2	7.7
Phorate, terbufos	5	4	5	2	3.7	3	11.7
Phosphamidon	4	1	5	3	3.0	2	9.0
Resmethrin	1	5	1	4	3.3	1	5.3
Stirofos	1	4	1	4	3.0	1	5.0
TEPP	5	2	5	5	4.0	1	10.0
Toxaphene	3	4	4	1	3.0	4	10.0
Trichlorfon	2	1	2	1	1.3	1	4.3

[a]For an explanation of the ratings, see Section IV.B.

Neuroptera, and Coleoptera, were generally more abundant and perhaps more diverse in the diflubenzuron-based orchards (Booth and Riedl 1996). Similarly, in a study of six crops (wheat, peas, beets, potatoes, onions, and carrots), the general pattern was that reduced pesticide farming practices resulted in increases in the abundance of carabids, staphylinids, and spiders (Booij and Noorlander 1992).

Enlightened pesticide use can also positively influence land beyond farms. Dover (1991) presents an account of the history and efficacy of "conservation headlands" in Britain. Conservation headlands are created by managing the outer six meters of fields with reduced amounts of selective pesticides. Originally devised to enhance game bird populations, conservation headlands also bolster populations of butterflies and probably other insects as well. Hedgerows, which are habitat for insects of conservation interest themselves, may benefit from conservation headlands in two ways: pesticide drift from the main portion of the cultivated field is reduced and the hedgerows may be effectively widened by the headlands.

Chemicals are not our only option in pest control. Among the alternative or complementary tactics to pesticides are biological control, intercropping, pest resistant crops, and hampering pest reproduction by massive releases of sterile males of the pest species or exploiting insect pheromones (Metcalf and Metcalf 1993; Metcalf and Luckmann 1994; Vandermeer 1989, 1995). Thousands of biocontrol efforts have been launched against insects, weeds, and other pests (Van Driesche and Bellows 1996). While the degree of success of each program varies, biocontrol has certainly provided economic returns. Benefit-cost analyses determined a composite 31.6:1 benefit/cost ratio for four biocontrol research projects conducted in Australia by the CSIRO Division of Entomology, versus a composite 2.5:1 ratio for nine non-biocontrol projects conducted by the same organization during the same years of 1960–1975 (Tisdell 1990). Biocontrol is often an environmental improvement over conventional pesticides (Howarth 1991; Carruthers and Onsager 1993; U.S. Congress 1995).

Conservation need not be viewed as in conflict with food production—in fact, conservation may assist food production. For example, biodiversity provides a pool from which effective biocontrol agents may be drawn (Waage 1991), and biodiversity may itself improve pest control in situations where the perpetual presence of natural enemies prevents potential pest populations from outbreaking (Altieri 1991). The brown planthopper, *Nilaparvata lugens*, has emerged as a major rice pest due to pesticide misuse (Cohen *et al.* 1994; Way and Heong 1994; Settle *et al.* 1996). At least 188 natural enemies prey upon *N. lugens*, and the continual presence of at least some of them controls planthopper populations (Way and Heong 1994). When pest species are not abundant, generalist predators feed on detritivores and plankton feeders (Settle *et al.* 1996). Insecticides kill nontarget species in this system, disrupting the dynamics of the rice community (Cohen *et al.* 1994). Via indirect effects, pesticides can completely undo pest suppression provided by natural enemies (Fagan *et al.* unpublished manuscript). Indeed, the greater the pesticide input, the more damage is inflicted by *N. lugens* (Settle *et al.* 1996).

Successfully reducing pesticide use must not lull us into complacency. Alternatives to pesticides can affect the environment in complex and not necessarily positive ways. *Bacillus thuringiensis* (Bt) and its derivatives bridge conventional pesticides and biological control techniques. Bt is a bacterium that produces insect-specific endotoxins that have been used for over 30 years without adverse human health impacts (Krimsky and Wrubel 1996). Bt scores the lowest possible on the environmental pest-management rating system for insecticides (Metcalf and Metcalf 1993; Metcalf 1994; Table 3). However, Bt-based products account for merely 1% of global insecticide sales, due in part to excessive specificity in multiple-pest situations and low residual activity (Krimsky and Wrubel 1996). Such qualities are desirable from a conservation standpoint, but even in this context Bt is imperfect.

The difficulty is that "specific" is a relative term. For example, some Bt endotoxins target only members of the order Lepidoptera (butterflies and moths)—a group with over 11,000 species in the United States and Canada alone (Borror, Triplehorn, and Johnson 1989). Thus Bt would be insufficient for protecting rice being devoured by planthoppers (homopterans) as well as caterpillars, yet potentially excessive if any of the 17 lepidopterans listed for federal protection under the Endangered Species Act live near a spray area. In fact, Bt has been considered detrimental to biodiversity. The monarch butterfly, *Danaus plexippus*, known for its bright coloration, long migrations, and overwintering in large colonies, overwinters in Mexican sites where a defoliating moth pest, *Evita hyalinaria*, can reach high population densities (Leong, Yoshimura, and Kaya 1992). A laboratory study found low susceptibility of adult monarchs to Bt; this combined with untested susceptibility of monarch caterpillars led Leong, Yoshimura, and Kaya (1992) to recommend using caution when spraying. Yet they also point out that without measures to control *E. hyalinaria*, damage to the plants in the vicinity could harm the monarch colony. A similar example involves the cinnabar moth, one of the herbivores that successfully control the tansy ragwort (McEvoy, Cox, and Coombs 1991). Serious lepidopteran forest pests such as gypsy moth, *Lymantria dispar*, and western spruce budworm, *Choristoneura occidentalis*, occur where the ragwort is found. The cinnabar moth is susceptible to Bt in its last stages as a caterpillar, but largely unaffected in its early stages. Thus James, Millerm and Lighthart (1993) suggest timing Bt applications to coincide with the moth's early larval stage.

Classical biological control also has risks (Howarth 1991; Lockwood 1993a, 1993b; U. S. Congress 1995; Secord and Kareiva 1996). Introduced species are sometimes cause for concern among conservationists, but introduced species are, by definition, employed in classical biocontrol. Ironic cases of intentionally introduced biocontrol agents becoming undesirable themselves most frequently involve, but are not limited to, vertebrate agents such as the mongoose, *Herpestes auropunctatus*, and the so-called cane toad, *Bufo marinus* (Howarth 1991). In most of these recognized failures of biocontrol, the agent is a natural enemy of a desirable species in addition to the target pest. The mongoose, for example, has been implicated in the extinctions of seven reptiles: *Jamaica iguana*, *Cyclura collei*; Martinique giant ameiva, *Ameiva major*; Jamaican gaint galliwasp, *Diploglossus occiduus*; Jamaica tree snake, *Alsophis ater*; St. Croix racer, *A. sancticrucis*; Martinique racer, *Dromicus cursor*; and St. Lucia racer, *D. ornatus* (Honegger 1981, and see chapter 7).

Insect examples of excessive consumption of nontarget species may be largely unrecognized (Howarth 1991). One example comes from Guam, where 27 biocontrol agents have been introduced against seven pest Lepidoptera (Nafus 1993). Two of those agents, *Brachymera lasus* and *Trichogramma chilonis*, are both polyphagous parasitoids that attack nontarget endemic butterflies such as *Hypolimnas anomala* and *H. bolina* (Nafus 1993). This is potentially problematic, since Guam is (or was) home to *Euploea eleutho*, *Vagrans egistina*, and *Hypolimnas octoculata*, among other unique and extremely rare butterflies (Nafus 1993).

Additional research on insects will be particularly valuable at the interface of agriculture and conservation. Core assumptions of each side are starting to align, but gaps remain. For example, does conserving insect predators for natural biological control necessarily conserve native species? Research and implementation of alternative pest control techniques in specific crops and growing conditions must be refined. A strategy of critical pluralism may work best. Food producers should be willing to learn a variety of pest control tactics rather than rely on a non-existant single best technology. Scientists cannot be satisfied with implementing a pest control strategy, but must examine why it is working or failing to attain environmentally and economically optimized management.

CASE STUDY OF THE FENDER'S BLUE BUTTERFLY: A RARE BUTTERFLY IN A DECLINING ECOSYSTEM

Background

The Fender's blue butterfly is a rare butterfly that resides in the dwindling native prairies of western Oregon. Native prairies may be the most endangered habitat in western Oregon and Washington. Currently occupying less that one percent of its presettlement range of 300,000 acres, urbanization and invasion by nonnative species further threaten the prairies (Alverson 1993). Many native species endemic to prairies are thought to be relicts of ancestral types that survived in unglaciated refuges while the surrounding areas were covered in ice (Barnosky, Anderson, and Bartlein 1987).

Censuses of Fender's blue butterfly indicate that less than four thousand butterflies survive in a dozen isolated patches in the Willamette Valley (Figure 9.3; Hammond and Wilson 1993). The butterfly is found only in habitats that contain its larval foodplants, the Kincaid's lupine (*Lupinus sulphureus* spp. *kincaidii*) and the spurred lupine (*Lupinus laxiflorus*). The Kincaid's lupine is itself rare and survives only in about three dozen prairie patches in the Willamette Valley (Kuykentall and Kaye 1993). Both the Kincaid's lupine and the Fender's blue butterfly are candidates for listing under the federal Endangered Species Act of 1973 (Anonymous 1995). The spurred lupine is rare in the Willamette Valley, but common in the nearby Coast and Cascade Ranges.

Puzzle 1: Where are the Butterflies in the Winter? While the adult butterfly stage is most easy to observe, the Fender's blue spends the majority of its life as a larva. Larvae are active for a few weeks in June and a month or so in March and April. During the bulk of their life (from July through March), larvae are in winter diapause. Therefore identifying where larvae diapause is critical to formulating management strategies.

On protected lands where the Fender's blue butterfly resides, burning or mowing for the management of weeds is implemented in the late summer and fall when the lupine food plants are senescent. If the larvae diapause in a place exposed to these management activities, such as on the soil surface, management efforts such as burning would potentially risk destroying the larvae. However, if the larvae diapause below the surface, most weed management would have little effect on larvae. Thus the primary threat facing populations on protected land is weed management.

To ascertain specifically where larvae diapause, 47 newly-hatched larvae were transplanted to potted lupine in June 1995 (Schultz 1996). In mid-July when larvae were no longer visible on the plants, the pots were carefully excavated to find the larvae. Thirty-five of the transplanted larvae were located. Of these, all but one were less than 1 cm from the surface and less than 5 cm from the lupine. It appeared that the larvae crawled down the lupine and found a convenient clod of dirt to crawl under. They attached themself to the dirt with a bit of silk, but otherwise had no apparent protection. When larvae were found and placed on a sheet of paper, they immediate 'woke up' and started crawling.

These data are consistent with the other studies indicating that significant soil impacts

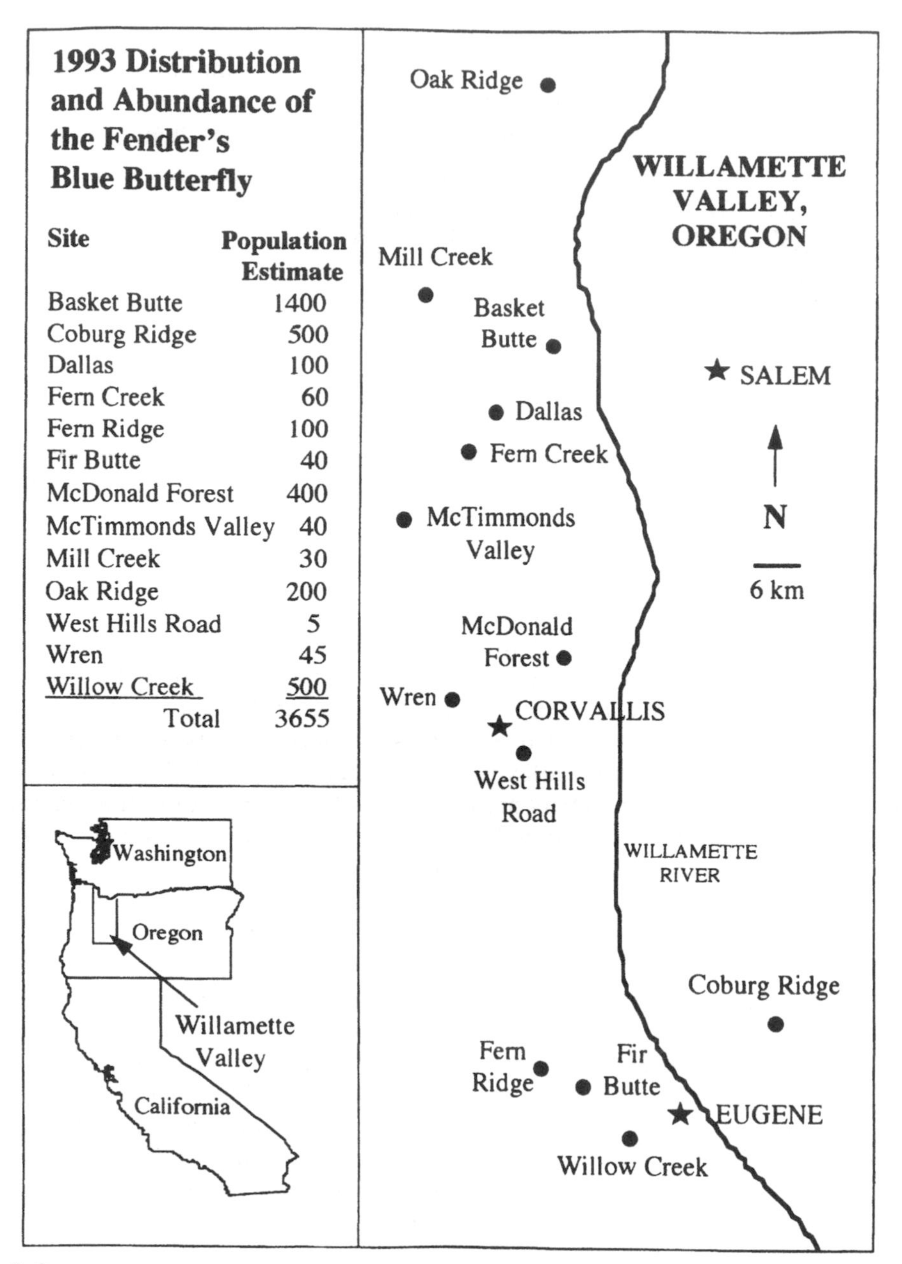

Figure 9.3
Distribution and abundance of the Fender's blue butterfly in 1993, the most recent year in which all sites were surveyed. Data are from Hammond and Wilson 1993.

such as burning negatively impact the Fender's blue butterfly larvae. Wilson and Clark (1995), for example, investigated the effect of burning, mowing and herbicides on native plants and the Fender's blue butterfly and found that all treatments significantly reduced the number of Fender's blue larvae. In mowed areas the larval densities were reduced while in burned areas, the larvae were completely eliminated. Clearly, studies of larval biology can help guide habitat management plans for the Fender's blue butterfly. Specifically, what we have learned strongly suggests that management strategies should focus on reducing weeds in a small percent of the habitat each year so that only a small portion of the butterfly population is effected by management in each weeding cycle.

Puzzle 2: Are Fender's Blue Butterfly Populations Declining? Assessing population size over several years is critical to determining the health of the remaining Fender's blue butterfly population. Fender's blue butterfly populations are small and fluctuate widely from year to year (Figures 9.3 and 9.4). Only four populations support more than a few hundred butterflies and only one population occasionally numbers more than a thousand.

A few of the butterfly populations in the southern portion of the range were monitored closely from 1993–1995 (Schultz 1996). By studying any one population over a number of years, it is difficult to detect the underlying cause of population fluctuations. However, by studying several populations over the same time period, some conclusions about the causes of population change might be inferred. For example, 1993 was one of the wettest years on record in the Willamette Valley (45 cm, April–June weather data; NOAA 1993). Average rainfall during the spring is usually around 15 cm (April–June weather data; NOAA 1993). Because the butterflies do not oviposit during inclement weather, oviposition time was limited and population numbers fell from 1993–1994. In 1994, however,

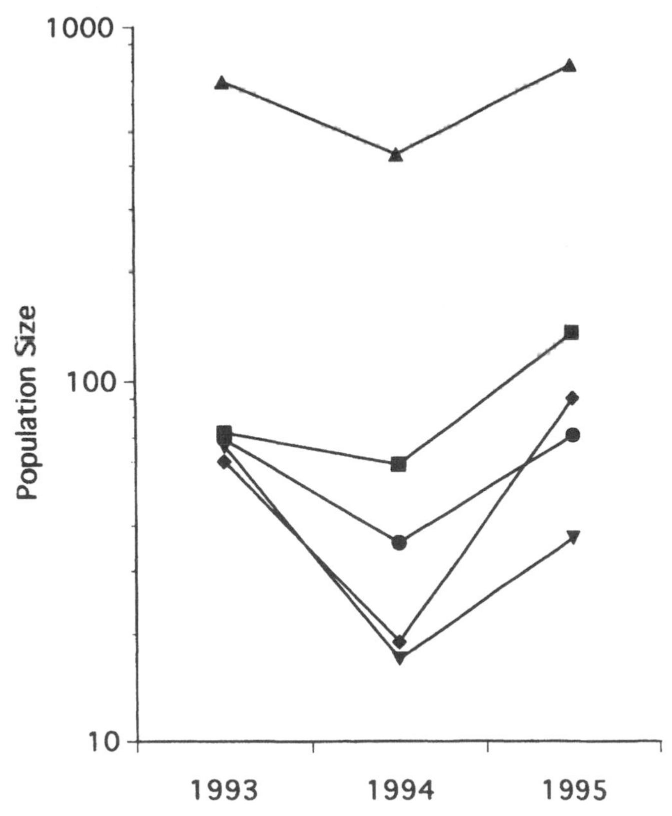

Figure 9.4
Population sizes at five Fender's blue butterfly sites west of Eugene, Oregon from 1993–1995. The central butterfly area at Willow Creek is the largest population. The other areas are Fern Ridge, Fir Butte, and the Bailey Hill and West 18th areas of Willow Creek.

rainfall was near average (12 cm in April–June, NOAA 1994), and the butterfly populations in 1995 rebounded to their previous levels.

One insightful approach into population fluctuations induced by weather and fluctuations induced by other possible events is to simultaneously monitor several populations that experience similar weather patterns. If a local population falls out of synch with its neighbors, other forces may be driving population change. For example, at one site, Coburg, the butterfly population in 1995 did not rebound. On their own, these data are inconsistent with the rainfall trend (Figure 9.3). However, environmental changes occurred at Coburg in 1995, i.e., approximately one quarter of the butterfly's habitat was converted into a tree farm. Other parts of Coburg that were not disturbed exhibited trends similar to those in Figure 9.3, i.e., a drop from 1993–1994 and an increase from 1994–1995.

Population monitoring at a fine scale within a site and at several sites throughout the range also can be invaluable. For example, in 1995 a large corporation began construction of a multi-million dollar semi-conductor chip plant immediately adjacent to Willow Creek, the site supporting the second largest remaining population of the Fender's blue. This industrial plant will significantly increase the air pollution in the vicinity. Given natural fluctuations in the Fender's blue butterfly population, it will be difficult to detect changes that occur due to the effect of the factory if only the Willow Creek population is monitored. However, if several populations are monitored and only Willow Creek declines after the factory begins emitting pollutants, declines might be linked to the increase in air pollutants from the factory. This monitoring scheme is known as BACI, Before-After-Control-Impact (Osenberg and Schmitt 1994). Underwood (1994) suggests that an asymmetric BACI design with several control sites, in this case non-factory sites, and a single impact site, Willow Creek, allows strong inference in detecting temporal variation.

Puzzle 3: Does Fragmentation Hinder Successful Dispersal? Maps of Fender's blue butterfly habitat suggest that its prairie habitat was patchy historically. Before European settlement in the 1850s, the Willamette Valley was a mosaic of upland and wetland prairies. The lupines required by the Fender's blue grow only in the drier soils of the upland prairies (Kaye and Kuykentall 1993). Settlement records indicate that upland prairie patches were common, rarely with more than ½ km between patches (Alverson 1993). Today, however, lupine patches in upland prairie are uncommon and the average distance between patches is much greater, frequently more than 3–30 km (Hammond and Wilson 1993).

In 1994 and 1995, Fender's blue butterfly dispersal was quantified in lupine patches, outside lupine patches in an abandoned, weedy field, and at the boundary between habitats (Schultz, in review). Observations at the edge of lupine patches indicate that the butterfly's behavior is biased toward the lupine, but they nonetheless leave lupine at an observable rate. By mapping butterfly flight paths and recording the time in flight, movement rates for the Fender's blue were calculated in lupine habitat and outside of lupine habitat. The butterflies spread 3 m^2/s in lupine and 15 m^2/s out of lupine, on average.

In addition, the amount of time a butterfly flies was assessed by quantifying time spent in different activities throughout the day and by estimating longevity with a limited mark-recapture-release study in 1994 (Schultz 1995). Adult butterflies spend about 2 hours/day flying on sunny days and live about 10 days. With known diffusion rates and time spent flying, it was estimated that a butterfly flying in lupine spreads up to ¾ km in its lifetime and one flying outside of lupine in disturbed habitats spreads about 2 km in its lifetime. Thus the movement rates calculated from dispersal observations suggest that given the historical habitat distribution, dispersing Fender's blue butterflies had a high probability of arriving at a new patch of lupine. Today, however, it is unlikely that a dispersing butterfly has time to find a new patch of habitat. Therefore, the challenge to Fender's blue butterfly conservation is to restore habitat at distances that increases the probability dispersing butterflies will successfully find new habitat.

Puzzle 4: At What Spatial and Temporal Scales Should Fire be Set in the Prairies? The minimum area required for insect populations should be based on the area needed to preserve important ecological processes. In the case of the upland prairie that is home to the Fender's blue butterfly, Native Americans burned the Willamette Valley prairies on an annual basis to maintain the prairie so as to hunt deer and to prepare the land for a fall growing season (Boyd 1986). Fires, however, eliminate larvae (Wilson and Clark 1995). In August and September, when most of the fires were set historically, Fender's blue larvae were in diapause. Given this historical disturbance regime, the Fender's blue must spread their eggs among several habitat patches to insure that at least some of their offspring survive. Because the Fender's blue has a tendency to disperse out from habitat patches to look for new areas, management plans should focus on strategies that include historical disturbances at the spatial scale of the butterfly's dispersal. Thus while many small populations of Fender's blue butterflies hang on in "postage stamp" areas, a much larger area is required to maintain the species and its habitat if fire is to be reintroduced as a structuring ecological process.

Puzzle 5: How Do We Manage Semi-Natural Habitats? Anthropological evidence indicates that Native Americans began using fire to manage the habitat in the Willamette Valley as early as 4,000–6,000 years ago (Boyd 1986) At that time, the valley began to cool down and trees began to invade the prairies. The Native Americans depended on the valley's game for meat and upon the grasslands for edible bulbs and seeds, and grasses for basketmaking. They began burning the prairies to maintain the plants and animals they required. In the process, they maintained a flora and fauna that included the Kincaid's lupine and the Fender's blue butterfly. As conservation biologists, we must recognize that many species and ecosystems we wish to preserve today exist because native peoples directly manipulated their homelands for thousands of years. If these ecosystems are to survive into our future, we need to continue these structuring ecological processes.

In the case of the Willamette Valley, fire is a key process. Not only does fire prevent tree invasion into the prairies, but after fire native perennial plants such as the Kincaid's lupine grow vigorously (Wilson and Clark 1995). In addition, the Fender's blue butterfly is attracted to these newly burned areas. Wilson and Clark (1995) observed that while larvae are eliminated by burning, the density of Fender's blue eggs increases in the season after burning. The study was conducted at a very small scale, with treatment areas 15 m × 4 m and spaced 2 m apart. Butterflies from unburned areas were close enough to easily disperse into newly burned areas—results that reinforce the conclusion that viewing the prairies of the Willamette Valley as semi-natural habitats requiring human management is essential to maintaining the flora and fauna of this rare prairie ecosystem.

CONCLUSION

As central but declining players in virtually all ecosystems we are trying preserve, insects warrant additional attention from conservation biologists. Insects provide critical services both in natural systems and in highly-managed agricultural systems. Without key insect pollinators, many of our crops would fail, countless wildflowers would disappear, and large segments of many ecosystems would crumble. Without key predators, natural habitats and managed farmlands would be decimated. The scant attention paid to insects in our recent literature indicates the low priority insects are given in traditional conservation planning. However, with several decades of "vertebrate-biased" conservation, insect conservation has a strong foundation on which to build. New approaches in insect conservation must branch out and address special features of insect biology that both drive them to rarity and propel them to problematic overabundance.

Several special challenges cause particularly perplexing problems in designing conservation strategies for insects. Progressing toward solutions to these problems will help address not just declining insect species, but also species across diverse taxa that are caught in similar situations.

1. Unraveling basic biology
Because insect species often have complex life histories and habitat needs, conservation biologists need to pay careful attention to the details of species' natural histories before they draw on general developments in conservation biology. For example, details on "invisible" stages, such as where insects go when they are young and sedentary, may be vital to persistence of many insect species.

2. Monitoring population change
Because insect populations may experience large fluctuations over short time intervals, biologists need to develop new methods to detect population decline. One approach involves integrating traditional population biology with methods from other branches of biology. For example, combining viability theory and key factor analysis with BACI schemes originally designed to detect pollution effects may yield predictions about decline in focal populations or about where to invest restoration efforts (Stewart-Oaten, Bence, and Osenberg 1992).

3. Reducing fragmentation-induced isolation
Insects are particularly vulnerable to rampant fragmentation. Conservation plans must be sensitive to the behavior of insects both in their preferred habitat and in disturbed intervening areas. However, as we move to integrate conservation planning on nature reserves with agricultural concerns about maintaining key insect pollinators and predators, we must recognize both the limitations and the advantages of today's patchwork of habitats. While fragmentation may isolate rare insects, it may also provide nearby habitat for pollinators and generalist predators. The challenge will be to balance these needs.

4. Managing at the "right" scale
While insects may currently linger in "postage stamp" preserves, these preserves are frequently too small to maintain the ecological processes on which the insects depend. In terms of adequate size and scale of reserve, insect conservation biologists can learn much from vertebrate biologists who have long argued that small reserves are often too small to maintain ecological processes. Unfortunately, scaling reserves to ecological processes may not be possible today given the scattering of remaining habitat remnants. The trick, then, may be to learn how to scale down the processes in a manner meaningful to key insect species.

5. Managing semi-natural habitats
As conservation biologists we must acknowledge that many species survive today because of, not in spite of, human intervention over the last several millennia. For species to

persist, we need to rediscover the land management methods of pre-modern times. Unlike "natural" processes, human-maintained processes may be amenable to relatively small scales. Grazing and coppicing, for example, can be accomplished at scales much smaller than the scales over which fire raced across the tallgrass prairies. However, regardless of the true "appropriate" size, scale will be a critical component in plans to recreate pre-modern land management methods.

6. Making insect conservation and pest management compatible
To our food supply, insects are both our friends and our enemies. In agricultural landscapes, conservation biologists will need to learn how to maintain insect species that pollinate our crops and that prey on crop herbivores. This means not just managing insects in crop lands, but also those in the shelterbelts, the hedgerows and the undisturbed natural areas that are intermingled with the farms. Neither farm edges nor preserve boundaries form impenetrable barriers. We need to manage areas to encourage dispersal of rare species between undisturbed fragments while promoting movement of beneficial species between farms and preserves. Insects are critical to making agriculture succeed; in their management lies the future of our food supply.

Answering vexing questions in insect conservation will contribute to conservation of numerous taxa that are caught in similar predicaments. Integrating these challenges across the diversity of disturbed landscapes must become a high priority for conservation biologists—both for those who study insects and for those who study the systems that depend on the services insects provide.

ACKNOWLEDGMENTS

We thank Peter Kareiva and Peggy Fiedler for improving this manuscript. William Fagan and William Settle kindly provided unpublished work. The case study of the Fender's blue butterfly would not have been possible without support from many people and agencies. Paul Hammond, Mark V. Wilson, Peter Kareiva, Nathan Rudd, Cindy Hartway, Andrea Gambaro, Stacy Philpott, Samantha Cross, Katrina Dlugosch, Jon Jay, Rich van Buskirk, and volunteers from the Nature Conservancy and Peter Kareiva's lab all contributed to the study of the Fender's blue. The Nature Conservancy, the city of Eugene, and the U.S. Army Corps of Engineers gave us permission to work on their land. Financial support was provided by the National Science Foundation, U.S. Fish and Wildlife Service, the U.S. Bureau of Land Management, The Nature Conservancy, Sigma Xi, and the Northwest Scientific Association.

LITERATURE CITED

Abele, L.G. and W.K. Patton. 1976. The size of coral heads and the community biology of associated decapod crustaceans. *Journal of Biogeography* 3:5–47.

Altieri, M.A. 1991. Increasing biodiversity to improve insect pest management in agroecosystems. In *The biodiversity of microorganisms and invertebrates: its role in sustainable agriculture*. ed. D.L. Hawksworth, 165–182. Wallingford, UK: CAB International.

Alverson, E.R. 1993. *Assessment of proposed wetland mitigation areas in west Eugene*. Report to the Environmental Protection Agency.

Anonymous. 1995. Endangered and threatened wildlife and plants: Review of plant and animal taxa that are candidates for listing as endangered or threatened species. *Federal Register* 61(40):7596–7613.

Armbruster, P. and R. Lande. 1993. A population viability analysis for African Elephant: How big should reserves be? *Conservation Biology* 7:602–610.

Barnosky, C.W., P.M. Anderson, and P.J. Bartlein. 1987. The northwestern U.S. during deglaciation: Vegetational history and paleoclimatic implications. In *North America and adjacent oceans during the last deglaciation: The Geology of North America*, eds. W.F. Ruddiman and H.E. Wright, Jr., 289–321. Boulder, CO: Geological Society of America.

Bassett, C. and K.H. Miers. 1984. Scientific reserves in state forests. *Journal of the Royal Society of New Zealand* 14:29–35.

Berenbaum, M.R. 1995. *Bugs in the system.* Reading, MA: Addison-Wesley Publishing.

Berg, C.C. and J.T. Wiebes. 1992. *African fig trees and fig wasps.* New York: Koninklijke Nederlandse Akademie van Wetenschappen.

Berglund, B.E. 1991. The cultural landscape during 6000 years in southern Sweden—the Ystad Project: The project—background, aims and organization. *Ecological Bulletins* 41: 13–17.

Bierzychudek, P. 1981. Pollinator limitation of plant reproductive effort. *American Naturalist* 117:838–840.

Blackstock, T.H., J.P. Stevens, E.A. Howe and D.P. Stevens. 1995. Changes in the extent and fragmentation of heathland and other semi-natural habitats between 1920–22 and 1987–88 in the Llyn Peninsula, Wales, UK. *Biological Conservation* 72: 33–44.

Bodenheimer, F.S. 1951. *Insects as human food.* The Hague: Dr. W. Junk.

Booij, C.J.H. and J. Noorlander. 1992. Farming systems and insect predators. *Agriculture Ecosystem and Environment* 40:125–135.

Booth, S.R. and H. Riedl. 1996. Diflubenzuron-based management of the pear pest complex in commercial orchards of the Hood River Valley in Oregon. *Journal of Economic Entomology* 89:621–630.

Borror, D.J., C.A. Triplehorn, and N.F. Johnson. 1989. *An introduction to the study of insects.* 6th ed. San Diego, CA: Harcourt, Brace, and Jovanovich.

Boyd, R. 1986. Strategies of Indian burning in the Willamette valley. *Canadian Journal of Anthropology* 5:65–86.

Briese, D.T. 1982. The effect of ants on the soil of a semi-arid saltbrush habitat. *Insectes Sociaux* 29:375–386.

Brooke, M.D.L., D.R. Lees, and J.M. Lawman. 1985. Spot distribution in the meadowbrown butterfly, *Maniola jurtina* L. (Lepidoptera: Satyridae): South Welsh populations. *Biological Journal of the Linnaen Society* 24:337–348.

Brown, J.H. 1978. Mammals on mountaintops: Nonequilibrium insular biogeography. *American Naturalist* 105:467–478.

Caltagirone, L.E. 1981. Landmark examples in classical biological control. *Annual Review of Entomology* 26:213–232.

Cappuccino, N. and P. Kareiva. 1985. Coping with a capricious environment: A population study of a rare peirid butterfly. *Ecology* 66:152–161.

Carruthers, R.I. and J.A. Onsager. 1993. Perspective on the use of exotic natural enemies for biological control of pest grasshoppers (Orthoptera: Acrididae). *Environmental Entomology* 22:885–903.

Carson, R. 1962. *Silent spring.* Boston: Houghton Mifflin Co.

Caswell, H. 1989. *Matrix population models.* Sunderland, MA: Sinauer Associates.

Clement, S.L. 1990. Insect natural enemies of yellow starthistle in southern Europe and the selection of candidate biological control agents. *Environmental Entomology* 19:1882–1888.

Coats, J.R. 1994. Risks from natural versus synthetic insecticides. *Annual Review of Entomology* 39:489–515.

Cohen, J.E., K. Schoenly, K.L. Heong, H. Justo, G. Arida, A.T. Barrion, and J.A. Litsinger. 1994. A food web approach to evaluating the effect of insecticide spraying on insect pest population dynamics in a Philippine irrigated rice ecosystem. *Journal of Applied Ecology* 31:747–763.

Collins, N.M. and J.A. Thomas, eds. 1991. *The conservation of insects and their habitats.* London: Academic Press.

Compton, S.G., S.J. Ross, and I.W.B. Thornton. 1994. Pollinator limitation of fig tree reproduction on the island of Anak Krakatau (Indonesia). *Biotropica* 26:180–186.

Dawson, D.G. 1984. Principles of ecological biogeography and criteria for reserve design. *Journal of the Royal Society of New Zealand* 117:629–638.

Debach, P. and B. Bartlett. 1951. Effects of insecticides on biological control of insect pests of citrus. *Journal of Economic Entomology* 44:372–383.

Debach, P. and D. Rosen. 1991. *Biological control by natural enemies*. New York: Cambridge University Press.

Decsimon, H. and M. Napolitano. 1993. Enzyme polymorphim, wing pattern variability and geographical isolation in and endangered butterfly species. *Biological Conservation* 66:117–123.

Dempster, J.P. 1991. Fragmentation, isolation and mobility of insect populations. In *The conservation of insects and their habitat*, eds. M. Collins and J.A. Thomas, 71–108. London: Academic Press.

Dempster, J.P., M.L. King, and K.H. Lakhani. 1976. The status of the swallowtail butterfly in Britain. *Ecological Entomology* 1:71–84.

Deshaye, J. and P. Morisset. 1989. Species-area relationships and the SLOSS effect in asubarctic archipelago. *Biological Conservation* 48:265–276.

Diamond, J. 1975. The island dilemma: Lessons of modern biogeographic studies for the design of nature reserves. *Biological Conservation* 7:129–146.

Dover, J.W. 1991. The conservation of insects on arable farmland. In *The conservation of insects and their habitats*, eds. N.M. Collins and J.A. Thomas, 293–318. London: Academic Press.

Edwards, C.A. 1993. The impact of pesticides on the environment. In *The pesticide question: Environment, economics, and ethics*, eds. D. Pimentel and H. Lehman, 13–46. New York: Chapman & Hall.

Eldridge, D.J. 1994. Nests of ants and termites influence infiltration in a semi-arid woodland. *Pedobiologia* 38:481–492.

Endress, P.K. 1994. *Diversity and evolutionary biology of Tropical flowers*. Cambridge: Cambridge University Press.

Erdhardt, A. and J.A. Thomas. 1991. Lepidodoptera as indicators of change in semi-natural grasslands of lowland and upland Europe. In *The conservation of insects and their habitats*, eds. N.M. Collins and J.A. Thomas, 213–236. London: Academic Press.

Fagan, W.F., A.L. Hakim, H. Ariawan, Yuli, and W. Settle. Unpublished manuscript. Intraguild predation in tropical rice agroecosystems.

Fahrig, L. and J.E. Paloheimo. 1987. Interpatch dispersal of the cabbage butterfly. *Canadian Journal of Zoology* 65:616–622.

Forman, R.T.T., A.E. Galli, and C.F. Leck. 1976. Forest size and avian diversity in New Jersey woodlots with some land use implications. *Oecologia* 26:1–8.

Free, J.B. 1970. *Insect pollination of crops*. New York: Academic Press.

Free, J.B. 1982. *Bees and mankind*. Boston: George Allen and Unwin.

Gaston, K.J. and B.H. McArdle. 1993. All else is not equal: Temporal population variability and insect conservation. In *Perspectives on insect conservation*, eds. K.J. Gaston, T.R. New, and M.J. Samways, 171–184. Andover: Intercept, Ltd.

Gall, G.A.E. and G.H. Orians. 1992. Agriculture and biological conservation. *Agriculture, Ecosystem and Environment* 42:1–8.

Gilpin, M.E. and J.M. Diamond. 1980. Subdivision of nature reserves and the maintenance species diversity. *Nature* 285:567–568.

Groom, M.J. 1994. Quantifying the loss of species due to tropical deforestation. In *Principles of Conservation Biology*, eds. G.K. Meffe and C.R. Carroll, 121–122. Sunderland, MA: Sinauer Associates.

Groom, M.J. and N. Schumaker. 1992. Evaluating landscape change: Patterns of worldwide deforestation and local fragmentation. In *Biotic interactions and global change*, eds. P.M. Kareiva, J.G. Kingsolver and R.B. Huey, 24–44. Sunderland, MA: Sinauer Associates.

Gutowski, J.M. 1990. Pollination of the orchid *Dactylorhiza fuchsii* by longhorn beetles in primeval forests of northeastern Poland. *Biological Conservation* 51:287–297.

Hafernick, J.E., Jr. 1992. Threats to invertebrate biodiversity: Implications for conservation strategies. In *Conservation biology: the theory and practice of nature conservation, preservation and management*, eds. P.L. Fiedler and S.K. Jain, 171–194. New York: Chapman & Hall.

Hagen, K.S. and J.M. Franz. 1973. A history of biological control. In *History of entomology*, eds. R.F. Smith, T.E. Mittler, and C.N. Smith, 433–476. Palo Alto: Annual Reviews, Inc.

Hambler, C. and M.R. Speight. 1996. Extinction rates in British nonmarine invertebrates since 1900. *Conservation Biology* 10:892–896.

Hammond, P.C. and M.V. Wilson. 1993. *Status of the Fender's blue butterfly*. Report to the U.S. Fish and Wildlife Service.

Higgs, A.J. and M.B. Usher. 1984. Should nature reserves be large or small? *Nature* 285:568–569.

Holldobler, B. and E.O. Wilson. 1990. *The ants*. Cambridge, MA: Harvard University Press.

Honegger, R.E. 1981. List of amphibians and reptiles either known or thought to have become extinct since 1600. *Biological Conservation* 19:141–158.

Howarth, F.G. 1991. Environmental impacts of classical biological control. *Annual Review of Entomology* 36:485–509.

Huffaker, C.B. and P.S. Messenger, eds. 1976. *Theory and practice of biological control*. San Francisco: Academic Press.

Hughes, R.D., R.E. Jones, and A.P. Gutierrez. 1984. Short-term patterns of population change: the life system approach to their study. In *Ecological entomology*, eds. C.B. Huffaker and R.L. Rabb, 309–357. New York: John Wiley and Sons.

Hunter, A. 1995. Ecology, life history, and phylogeny of outbreak and nonoutbreak species. In *Population dynamics: New approaches and synthesis*, eds. N. Cappuccino and P.W. Price, 41–63. San Diego: Academic Press.

Ito, Y. 1980. *Comparative ecology*. Cambridge: Cambridge University Press.

James, R.R., J.C. Miller, and B. Lighthart. 1993. *Bacillus thuringiensis* var. *kurstaki* affects a beneficial insect, the cinnabar moth (Lepidoptera: Arctiidae). *Journal of Economic Entomology* 86: 334–339.

James, R.R., P.B. McEvoy, and C.S. Cox. 1992. Combining the cinnabar moth (*Tyria jacobaeae*) and the ragwort flea beetle (*Longitarsus jacobaeae*) for control of ragwort (*Senecio jacobaea*): an experimental analysis. *Journal of Applied Ecology* 29: 589–596.

Janzen, D.H. 1971. Euglossine bees as long-distance pollinators of tropical plants. *Science* 171:203–205.

Janzen, D.H. 1983. No park is an island: Increase in interference from the outside as park size decreases. *Oikos* 41:402–410.

Jarvinen, O. 1982. Conservation of endangered plant populations: Single large or several small reserves. *Oikos* 38:301–307.

Jennersten, O. 1988. Pollination in *Dianthus deltoides* (Caryophyllaceae): Effects of habitat fragmentation on visitation and seed set. *Conservation Biology* 2:359–366.

Johansen, C.A. 1977. Pesticides and pollinators. *Annual Review of Entomology* 22:177–192.

Johnson, J.B., J.P. McCaffrey, and F. Merickel. 1992. Endemic phytophagous insects associated with yellow starthistle in northern Idaho. *Pan-Pacific Entomologist* 68: 169–173.

Kalule-Sabiti, J.M.N. 1980. The ecological role of ants (Formicidae) in the ecosystem. *African Journal of Ecology* 18:113–121.

Kellert, S.R. 1993. Values and perceptions of invertebrates. *Conservation Biology* 7:845–855.

Kevan, P.G. 1975. Forest application of the insecticide Fenitrothion and its effect on wild bee pollinators (Hymenoptera: Apoidea) of lowbush blueberries (*Vaccinium* spp.) in southern New Brunswick, Canada. *Biological Conservation* 7:301–309.

Kevan, P.G. and H.G. Baker. 1984. Insects on flowers. In *Ecological entomology*, eds. C.B. Huffaker and R.L. Rabb, 607–631. New York: Wiley-Interscience.

Kim, K.C. 1993. Biodiversity, conservation and inventory: Why insects matter. *Biodiversity and Conservation* 2:191–214.

King, C.M. 1984. Open discussion on biological reserve design. *Journal of the Royal Society of New Zealand* 14:39–44.

Krimsky, S. and R.P. Wrubel. 1996. *Agricultural biotechnology and the environment: Science, policy, and social issues*. Urbana: University of Illinois Press.

Kuykentall, K. and T. Kaye 1993. *Lupinus sulphureus* spp. *kincaidii* survey and reproductive studies. Report to the Bureau of Land Management and the Oregon Department of Agriculture.

Lahti, R. and E. Ranta. 1986. Island biogeography and conservation: A reply to Murphy and Wilcox. *Oikos* 47:388–389.

Lawton, J.H. and C.G. Jones. 1995. Linking species and ecosystems: Organisms as ecosystem engineers. In *Linking species and ecosystems*, eds. C.G. Jones and J.H. Lawton, 141–150. New York: Chapman & Hall.

Leong, K.L.H., M.A. Yoshimura, and H.K. Kaya. 1992. Low susceptibility of overwintering monarch butterflies to *Bacillus thurigiensis* Berliner. *Pan-Pacific Entomology* 68:66–68.

Levin, D.A., J. Francisco-Ortega, and R.K. Jansen. 1996. Hybridization and the extinction of rare plant species. *Conservation Biology* 10:10–16.

Lloyd, D.G. and S.C.H. Barrett, eds. 1996. *Floral biology: studies on floral evolution in animal pollinated plants*. New York: Chapman & Hall.

Lockwood, J.A. 1993a. Environmental issues involved in biological control of rangeland grassshoppers (Orthoptera: Acrididae) with exotic agents. *Environmental Entomology* 22:503–518.

Lockwood, J.A. 1993b. Benefits and costs of controlling rangeland grasshoppers (Orthoptera: Acrididae) with exotic organisms: Search for a null hypothesis and regulatory compromise. *Environmental Entomology* 22:904–914.

Loman, J. and T. von Schantz. 1991. Birds in a farmland—more species in small than large habitat islands. *Conservation Biology* 5: 176–188.

Luck, R.F., B.M. Shepard, and P.E. Kenmore. 1988. Experimental methods for evaluating arthropod natural enemies. *Annual Review of Entomology* 33:367–391.

Ludwig, D. 1996. The distribution of population survival times. *American Naturalist* 147:506–526.

Martin, E.C. and S.E. McGregor. 1973. Changing trends in insect pollination of commercial crops. *Annual Review of Entomology* 18:207–226.

Matteson, P.C. 1995. The "50% pesticides cuts" in Europe: A glimpse of our future? *American Entomologist* 41:210–220.

Mattoni, R.H.T. 1990. The endangered El Segundo blue butterfly. *Journal of Research on the Lepidoptera* 29: 277–304.

Mawdsley, N.A. and N.E. Stork. 1995. Species extinctions in insects: Ecological and biogeographical considerations. In *Insects in a changing environment*, eds. R. Harrington and N.E. Stork, 322—371. London: Academic Press.

McKenzie, J.A. 1996. *Ecological and evolutionary aspects of insecticide resistance*. Austin, TX: R.G. Landes Co.

McEvoy, P., C. Cox, and E. Coombs. 1991. Successful biological control of ragwort, *Senecio jacobaea*, by introduced insects in Oregon. *Ecological Applications* 1:430–442.

Metcalf, R.L. 1980. Changing role of insecticides in crop protection. *Annual Review of Entomology* 25: 219–256.

Metcalf, R.L. 1994. Insecticides in pest management. In *Introduction to insect pest management*, ed. R.L. Metcalf and W.H. Luckmann, 245–314. New York: John Wiley and Sons, Inc.

Metcalf, R.L. and W.H. Luckmann, eds. 1994. *Introduction to Insect Pest Management*. New York: John Wiley and Sons, Inc.

Metcalf, R.L. and R.A. Metcalf. 1993. *Destructive and useful insects: Their habits and control*. New York: McGraw-Hill.

Miller, R.I., S.P. Bratton, and P.S. White. 1987. A regional strategy for reserve design and placement based on an analysis of rare and endangered species' distribution patterns. *Biological Conservation* 39:255–268.

Murdoch, W.W. 1994. Population regulation in theory and practice. *Ecology* 75:271–287.

Murdoch, W.W., J. Chesson, and P.L. Chesson. 1985. Biological control in theory and practice. *American Naturalist* 125:344–366.

Myers, N. 1994. Global biodiversity II: Losses. In *Principles of conservation biology*, eds. G.K. Meffe and C.R. Carroll, 110–140. Sunderland, MA: Sinauer Associates.

Nafus, D.M. 1993. Movement of introduced biological control agents onto nontarget butterflies, *Hypolimnas* spp. (Lepidoptera: Nymphalidae). *Environmental Entomology* 22:265–272.

National Oceanic and Atmospheric Administration (NOAA). 1993. *Climatological data annual summary: Oregon*. Vol. 99.

National Oceanic and Atmospheric Administration (NOAA). 1994. *Climatological data annual summary: Oregon*. Vol. 100.

New, T.R. 1991. *Butterfly conservation*. Oxford: Oxford University Press.

Nilsson, S.G. 1978. Fragmented habitat, species richness and conservation practice. *Ambio* 7:26–27.

Oates, M.R. 1995. Butterfly conservation within the management of grassland habitats. In *The ecology and conservation of butterflies*, ed. A.S. Pullin, 98–112. London: Chapman & Hall.

O'Neill, R.V. 1976. Ecosystem persistence and heterotrophic regulation. *Ecology* 57:1244–1253.

Orians, G.H. and P. Lack. 1992. Arable lands. *Agriculture, Ecosystems and Environment* 42:101–124.

Osenberg, C.W. and R.J. Schmitt. 1994. Detecting environmental impacts in marine environments. *Ecology* 4:1–2.

Paoletti, M.G., D. Pimentel, B.R. Stinner, and D. Stinner. 1992. Agroecosystem biodiversity: Matching production and conservation biology. *Agriculture, Ecosystems and Environment* 40:3–23.

Perkins, J.H. 1982. *Insects, experts, and the insecticide crisis*. New York: Plenum Press.

Peterken, G.F. 1991. Ecological issues in the management of woodland nature reserves. In *The scientific management of temperate communities for conservation*, eds. I.F. Spellerberg, F.B. Goldsmith, and M.G. Morris, 245–272. London: Blackwell Scientific Publications.

Picton, H. 1979. The application of insular biogeography theory to the conservation of large mammals in the northern Rocky Mountains. *Biological Conservation* 15:73–79.

Pimentel, D., H. Acquay, M. Biltonen, P. Rice, M. Silva, J. Nelson, V. Lipner, S. Giordano, A. Horowitz, and M. D'Amore. 1992. Environmental and economic costs of pesticide use. *BioScience* 42:750–760.

Pimentel, D., E. Garnick, A. Berkowitz, S. Jacobson, S. Napolitano, P. Black, S. Valdes-Cogliano, B. Vinzant, E. Hudes, and S. Littman. 1980. Environmental quality and natural biota. *BioScience* 30:750–755.

Pimentel, D., L. McLaughlin, A. Zepp, B. Lakitan, T. Kraus, P. Kleinman, F. Vancini, W.J. Roach, E. Graap, W.S. Keeton, and G. Selig. 1991. Environmental and economic effects of reducing pesticide use. *Bioscience* 41:402–409.

Pimentel, D., L. McLaughlin, A. Zepp, B. Lakitan, T. Kraus, P. Kleinman, F. Vancini, W.J. Roach, E. Graap, W.S. Keeton, and G. Selig. 1993. Environmental and economic impacts of reducing U.S. agricultural pesticide use. In *The pesticide question: Environment, economics, and ethics*, eds. D. Pimentel and H. Lehman, 223–278. New York: Chapman & Hall.

Pollard, E. and T.J. Yates. 1993. *Monitoring butterflies for ecology and conservation: the British monitoring scheme*. London: Chapman & Hall.

Prendergast, J.R. and B.C. Eversham. 1995. Butterfly diversity in southern Britain: Hotspot Losses since 1930. *Biological Conservation* 72:109–114.

Quinn, J.F. and G. Robinson. 1987. The effects of experimental subdivision on flowering plant diversity in California annual grassland. *Journal of Ecology* 75:837–856.

Quinn, J.F., C.L. Wolin, and M.L. Judge. 1989. An experimental analysis of patch size, habitat subdivision and extinction in a marine intertidal snail. *Conservation Biology* 3:242–251.

Robertson, P.A., S.A. Clarke, and M.S. Warren. 1995. Woodland management and butterfly diversity. In *Ecology and Conservation of Butterflies*, ed. A.S. Pullin, 113–134. London: Chapman & Hall.

Samways, M.J. 1994. *Insect conservation biology*. London: Chapman & Hall.

Samways, M.J. 1995. Southern hemisphere insects: Their variety and the environmental pressures upon them. In *Insects in a changing environment*, eds. R. Harrington and N.E. Stork, 298–321. London: Academic Press.

Schoenly, K., J.E. Cohen, K.L. Heong, J.A. Litsinger, G.B. Aquino, A.T. Barrion, and G. Arida. 1996. Food web dynamics of irrigated rice fields at five elevations in Luzon, Philippines. *Bulletin of Entomological Research* 86: 451–466.

Schultz, C.B. 1995. Status of the Fender's blue butterfly (*Icaricia icarioides fenderi*) in Lane County, Oregon: A Year of Declines. Report to the U.S. Fish and Wildlife Service.

Schultz, C.B. 1996. Status of the Fender's blue butterfly (*Icaricia icarioides fenderi*) in Lane County, Oregon: Population Ups and Downs. Report to the U.S. Fish and Wildlife Service.

Schultz, C.B. In review. The role of dispersal behavior in designing reserves for a rare Oregon butterfly. *Conservation Biology*.

Secord, D. and P. Kareiva. 1996. Perils and pitfalls in the host specificity paradigm. *BioScience* 46:448–453

Settle, W.H., H. Ariawan, E.T. Astuti, W. Cahyana, A.L. Hakim, D. Hindayana, A.S. Lestari, *et al.* In press. Managing rice pests through complexes of generalist natural enemies: an alternative to the single species approach. *Ecology.*

Shaffer, M. 1987. Minimum viable populations: coping with uncertainty. In *Viable populations for conservation*, ed. M.E. Soulé, 69–86. Cambridge: Cambridge University Press.

Shaw, F.R. 1941. Bee poisoning: A review of the more important literature. *Journal of Economic Entomology* 34:16–21.

Shreeve, T.G. 1992. Monitoring butterfly numbers. In *The ecology of butterflies in Britain*, ed. R.L.H. Dennis, 120–138. Oxford: Oxford University Press.

Sibatani, A. 1990. Decline and conservation of butterflies in Japan. *Journal of Research on the Lepidoptera* 29:305–315.

Simberloff, D. 1974. Equilibrium theory of island biogeography and ecology. *Annual Review of Ecology and Systematics* 5:165–182.

Simberloff, D. 1982. Big advantages of small refuges. *Natural History* 91:6–14.

Simberloff, D. 1988. The contribution of population and community biology to conservation science. *Annual Review of Ecology and Systematics* 19:473–511.

Simberloff, D. and L.G. Abele. 1976. Island biogeography theory and conservation practice. *Science* 191:285–6.

Simberloff, D. and N. Gotelli. 1984. Effects of insularization on plant species richness in the prairie-forest ecotone. *Biological Conservation* 29:27–40.

Sipes, S.D. and V.J. Tepedino. 1995. Reproductive biology of the rare orchid, *Spiranthes diluvialis*: Breeding system, pollination, and implications for conservation. *Conservation Biology* 9:929–938.

Soulé, M.E. and D. Simberloff. 1986. What do genetics and ecology tell us about the design of nature reserves? *Biological Conservation* 35:19–40.

Southwick, E.E. and L. Southwick, Jr. 1992. Estimating the economic value of honey bees (Hymenoptera: Apidae) as agricultural pollinators in the United States. *Journal of Economic Entomology* 85:621–633.

Stewart-Oaten, A., J. Bence, and C. Osenberg. 1992. Assessing effects of unreplicated perturbations: no simple solutions. *Ecology* 73:1396–1404.

Surgeoner, G.A. and W. Roberts. 1993. Reducing pesticide use by 50% in the province of Ontario: challenges and progress. In *The pesticide question: environment, economics, and ethics*, eds. D. Pimentel and H. Lehman, 206–222. New York: Chapman & Hall.

Temple, S.A. and J.R. Cary. 1988. Modeling dynamics and habitat-interior bird populations in fragmented landscapes. *Conservation Biology* 2:340–346.

Tepedino, V.J. 1979. The importance of bees and other insect pollinators in maintaining floral species composition. *Great Basin Naturalist Memoirs* 3:139–150.

Terborgh, J. 1976. Island biogeography and conservation: Strategy and limitations. *Science* 193:1029–1030.

Thomas, C.D. 1985. The status and conservation of the butterfly *Plebejus argus* L. (Lepidoptera:Lycaenidae) in north west Britain. *Biological Conservation* 33:29–51.

Thomas, J.A. 1983. The ecology and conservation of *Lysandra bellargus* (Lepidoptera: Lycaenidae) in Britain. *Journal of Applied Ecology* 20:59–83.

Thomas, J.A. 1984a. The behavior and habitat requirements of *Maculinea nausithous* (the dusky large blue butterfly) and *M. teleius* (the scarce large blue) in France. *Biological Conservation* 28:325- 347.

Thomas, J.A. 1984b. The conservation of butterflies in temperate countries: Past efforts and lessons for the future. In *The biology of butterflies*, eds. R.I. Vane-Wright and P.R. Ackery, 333–354. London: Academic Press.

Thomas, J.A. 1991. Rare species conservation: Case studies of European butterflies. In *The scientific management of temperate communities for conservation*, eds. I.F. Spellerberg, F.B. Goldsmith, and M.G. Morris, 149–157. London: Blackwell Scientific Publications.

Thomas, J.A. 1995. The colony and conservation of Maculinea arion and other European species of large blue butterfly. In *Ecology and conservation of butterflies*, ed. A.S. Pullin, 180–197. London: Chapman & Hall.

Thomas, J.A., C.D. Thomas, D.J. Simcox, and R.T. Clarke. 1986. The ecology and declining status of the silver-spotted skipper butterfly (*Hesperia comma*) in Britain. *Journal of Animal Ecology* 23:365–380.

Thompson, J.N. 1994. *The coevolutionary process*. Chicago: The University of Chicago Press.

Tisdell, C.A. 1990. Economic impact of biological control of weeds and insects. In *Critical issues in biological control*, eds. M. Mackauer, L.E. Ehler, and J. Roland, 301–316. Andover, UK: Intercept.

Underwood, A.J. 1994. On beyond BACI: Sampling designs that might reliably detect environmental disturbances. *Ecology* 4:3–15.

U.S. Congress, Office of Technology Assessment. 1995. *Biologically based technologies for pest control, OTA-ENV-636*. Washington, D.C.: U.S. Government Printing Office.

Usher, M.B. 1995. A world of change: Land-use patterns and arthropod communities. In *Insects in a changing environment*, eds. R. Harrington and N.E. Stork, 372–399. London: Academic Press.

Vandermeer, J. 1989. *The ecology of intercropping*. New York: Cambridge University Press.

Vandermeer, J. 1994. The ecological basis of alternative agriculture. *Annual Review of Ecology and Systematics* 26:201–224.

Van Driesche, R.G. and T.S. Bellows, Jr. 1996. *Biological control*. New York: Chapman and Hall.

Van Hook, T. 1994. The conservation challenge in agriculture and the role of entomologists. *Florida Entomologist* 77: 42–73.

Waage, J.K. 1991. Biodiversity as a resource for biological control. In *The biodiversity of microorganisms and invertebrates: its role in sustainable agriculture*, ed. D.L. Hawksworth, 149–163. Wallingford, UK: CAB International.

Warren, M.S. 1993. A review of butterfly conservation in central southern Britain: I. Protection evaluation and extinction on prime sites. *Biological Conservation* 64: 25–35.

Warren, M.S., E. Pollard, and T.J. Bibby. 1986. Annual and long-term changes in a population of the wood white butterfly *Leptidea sinapsis*. *Journal of Animal Ecology* 55:707–719.

Waterhouse, D.F. 1974. The biological control of dung. *Scientific American* 230:100–109.

Watt, W.B., F.S. Chew, L.R.G. Snyder, A.G. Watt, and D.E. Rothschild. 1977. Population structure in pierid butterflies: I. Numbers and movements of some montane species. *Oecologia* 27:1–22.

Watt, W.B., D. Han, and B.E. Tabashnik. 1979. Population structure of pierid butterflies II:A "native" population of *Colias philodice eriphyle* in Colorado. *Oecologia* 44:44–52.

Way, M.J. and K.L. Heong. 1994. The role of biodiversity in the dynamics and management of insect pests of tropical irrigated rice—a review. *Bulletin of Entomological Research* 84:567–587.

Webb, N.R. 1990. Changes on the heathlands of Dorset England between 1978 and 1987. *Biological Conservation* 51:273–286.

Webb, N.R. and J.A. Thomas. 1994. Conserving insect habitats in heathland biotopes: a question of scale. In *Large-scale ecology and conservation biology*, eds. P.J. Edwards, R.M. May and N.R. Webb, 129–152. London: Blackwell Scientific Publications.

Western, D. 1989. Conservation without parks: Wildlife in the rural landscape. In *Conservation for the twenty-first century*, eds. D. Western and M. Pearl, 158–165. New York: Oxford University Press.

Williams, G.R. 1984. Has island biogeography theory any relevance to the design of biological reserves in New Zealand? *Journal of the Royal Society of New Zealand* 14:7–10.

Wilson, E.O. 1987. The little things that run the world. *Conservation Biology* 1:344–346.

Wilson, E.O. 1985. The biological diversity crisis. *BioScience* 35:700–706.

Wilson, M.V. and D.L. Clark. 1995. *Effects of fire and fire fighting techniques on Fender's blue butterfly and Kincaid's lupine*. Report to Western Oregon Refuges, U.S. Fish and Wildlife Service.

Wood, G.W. 1979. Recuperation of native bee populations in blueberry fields exposed to drift of fenitrothion from forest spray operations in New Brunswick. *Journal of Economic Entomology* 72:36–39.

Woolhouse, M.E.J. 1987. On species richness and nature reserve design: An empirical study of U.K. woodland avifauna. *Biological Conservation* 40:167–178.

Wright, S.J. and S.P. Hubbell. 1983. Stochastic extinction and reserve design: A focal species approach. *Oikos* 41:466–476.

CHAPTER

10

An Australian Perspective on Plant Conservation Biology in Practice

Stephen D. Hopper

Conservation biology has emerged as an important scientific discipline in Australia, reflected by the appearance of the new journal *Pacific Conservation Biology*, and the publication of a number of recent symposium volumes (e.g., Saunders *et al.* 1987, 1990, 1993, 1995; Saunders and Hobbs 1992; Moritz and Kikkawa 1994; Bradstock *et al.* 1995; Hopper *et al.* 1996). My purpose here is to provide an overview of current Australian plant conservation practice and the scientific knowledge that has delivered tangible outcomes. The emphasis on plants is deliberate because some exciting insights have recently unfolded.

Although national in scope, I will focus on Western Australia because the southwest contains one of the world's richest floras, different in many respects even from those in eastern Australia (Hopper 1992a, 1995; Hopper *et al.* 1996). Plant conservation biologists have been particularly active in their recent attempts to understand and help manage this remarkably diverse and highly endemic flora in the face of dramatic threatening processes. I outline their approaches and suggest worthy areas for future work. A brief overview of the reasons why an Australian perspective, of the nation's flora and vegetation, and of post-colonial land use precedes the discussion of conservation strategies.

WHY AN AUSTRALIAN PERSPECTIVE?

An Australian perspective on plant conservation biology might be illuminating from two viewpoints. Firstly, as the world's driest, flattest, oldest, and most insular continent, it contains a flora and fauna like no other, with a unique evolutionary history. Australia is recognised as one of the world's megadiverse countries (Common and Norton 1992). This raises the possibility that geohistorical, ecosystem, and evolutionary processes might be different from those commonly seen elsewhere (Dodson and Westoby 1985; Westoby 1988; Clark 1990; Hopkins, How and Saunders 1990; Hobbs 1992; Hopper 1992a; Pate and Hopper 1993; Cowling and Witkowski 1994). If so, management for conservation may need to be applied in novel ways, and ample material exists for the elucidation of new principles for conservation theory.

Second, the science of conservation biology as practised in Australia has arisen in a cultural and environmental context different from those of Europe and North America. For example, Australia has endured rapid onslaught of western technological land use, with European colonisation dating back a mere two centuries (Hobbs and Hopkins 1990; Saunders, Hopkins, and How 1990). Although approximately the same size as the US or the European continent, Australia has a much smaller and highly urbanized population, mainly confined to a narrow, temperate coastal fringe adjacent to the vast arid inland and tropical north. This has allowed the persistence of significant components of the biota. Moreover, it is not uncommon for Australian governmental agencies as well as universities to have active conservation research programs, providing direct links between academic research and land managers. For all these reasons, an account of how plant conservation biology has been practised in Australia may provide a case study of interest in helping refine improvements to the interchange between theory and practice so urgently needed across the globe.

AUSTRALIA'S FLORA AND VEGETATION

It is not surprising that the earliest landfall by European scientists on Australia's southeastern seaboard was named Botany Bay. Joseph Banks and Daniel Solander, botanists on James Cook's voyage in 1770, were so enthralled by their first encounter with Australian plants, many new to science, that Cook named the bay on which the Sydney airport now sits to record the enthusiasm of his colleagues.

The Australian flora differs dramatically at the species and generic level from that elsewhere, due to the Gondwanan origins of significant components and a long period of insular evolution in the Tertiary (Barlow 1981). An estimated 25,000 species of vascular plants (10% of the global total) occur in Australia, with at least 85% endemism. At the level of plant family, Australia shows very little endemism, primarily because the evolution of most families predated the fragmentation of Gondwana.

Cosmopolitan weeds introduced from other continents number in excess of 3,000 species in Australia (an estimated 15% of the flora; Humphries *et al.* 1991). Native plant communities are especially vulnerable to weed invasion wherever disturbance occurs, particularly bulldozing, grading or deep ploughing that strips the topsoil of its vital native seed store and associated microbial symbionts. This is particularly so in the southwest, where high vagility and colonizing ability has been of much less selective value to native plants than limited dispersal and sophisticated adaptations to living in species-rich communities under challenging environmental conditions. Disturbance associated with European agriculture has not been an evolutionary force on the flora, resulting in poor competitive ability of native plants against introduced species.

Within Australia, as elsewhere, plant species are distributed in distinctive communities, some vast in distribution, others highly localized (see Groves 1994 for a review). From the earliest analysis of the composition of the native Australian vascular flora by Hooker (1860), three key floristic elements have been identified:

(1) the bulk of species (15,000+) are endemic, evergreen, sclerophyllous plants of forests, woodlands, mallee (e.g., lignotuberous multi-stemmed eucalypts, Figure 10.1), shrublands, sedgelands, and grasslands. The continent is especially noteworthy in the dominance of two woody but evergreen genera—*Eucalyptus* (800+ species) and *Acacia* (1,000+ species). Spinifex hummock grasses of the genera *Triodia* and *Plectrachne* are prominent over vast desert areas. Species-rich Gondwanan families in Australia include the Proteaceae (banksias, grevilleas, etc.), Myrtaceae (eucalypts, melaleucas etc.), Epacridaceae (southern heaths), and Resti-

Figure 10.1
Eucalyptus caesia, *a rare mallee with multiple stems to 5 m tall arising from an underground lignotuber, on a granite outcrop in southwestern Australia. Associated shrubs included typically Australian genera such as* Acacia *and* Allocasuarina. *Note the flat horizon, also characteristic of much of the island continent west of the Great Dividing Range.*

onaceae (southern rushes). Occupying about 5% of the continent, the southwest of Western Australia is especially rich in this major Gondwanan element of the Australian flora, with an estimated 8,000 species, 75% endemic to the region itself (Hopper 1979, 1992). In contrast, the extensive central Australian desert region occupying one third of the continent has only 2,000 species (Jessop 1981).

(2) An ancient rainforest element covered only 1% of the Australian landmass at European colonization. Rainforests occur in scattered sites from Tasmania to north Queensland and westwards across the tropical Northern Territory to the Kimberley region of Western Australia, and contain upwards of 2,000 species, many endemic to Australia. The rainforests are living Gondwanan museums—i.e., fragmented and depleted relicts of vegetation that covered much of the continent prior to the onset of late-Tertiary aridity when Australia drifted north away from Antarctica. These rainforest patches differ significantly in composition, with three major floristic groups recognised (Tracey 1981; Webb *et al.* 1984). There are cool-wet temperate rainforests of Tasmania, Victoria, and New South Wales; hot-wet subtropical-tropical rainforests from near Sydney north to Queensland and the wettest parts of the Northern Territory; and hot-dry semi-deciduous or deciduous rainforests and vine thickets extending from the Kimberley across to north Queensland and south into semi-arid New South Wales.

(3) A cosmopolitan element consists of up to 3,000 species occupying coastal habitats, saltlands, wetlands and alpine or mountainous areas. Typically, endemism is lower in this component of the flora.

Two-thirds of Australia is arid (Barker and Greenslade 1982), with less than 300 mm annual rainfall, including the Pilbara region of the northwest where the desert meets the coast. Arid Australia is unusually well vegetated, with the deserts dominated by spinifex hummock grasslands and scattered low shrubs, mallees or small trees. Extensive red sand dune systems are interspersed among low rocky ranges and hills throughout much of the arid zone. Large salt lakes occupy desert basins and support fringing halophytic vegetation of samphires (chenopods) and other salt-tolerant plants. The arid zone flora is relatively recent in origin, with few endemics (Barlow 1981).

Surrounding the arid zone on three sides are broad semi-arid belts in which woodlands, mallee, and sclerophyll shrublands are prevalent (Groves 1994). Again, salt lakes with uncoordinated drainage are common in broad valley floors, particularly in the west and south of the semi-arid zone, but the east is occupied by the bulk of Australia's largest river system, the Murray-Darling River System. The temperate semi-arid zone is rich in Australian endemics, particularly the southwest with its kwongan shrublands (Pate and Beard 1984), and the mallee regions of Western Australia, South Australia, Victoria and New South Wales (Noble and Bradstock 1989; Noble, Joss, and Jones 1991).

The wetter parts of Australia occupy less than a fifth of the continent, south and eastwards of the eastern highlands, across the tropical north, and in a small isolated region of the south-west. Vegetation is complex, matching the topography and geology of these areas. The wettest places with deep fertile soils in eastern and northern Australia are occupied by rainforests, while shallow soils in the highest rainfall areas have stunted woodlands and heaths. Wet sclerophyll forests occupy fire-prone sites, and contain immense hardwoods such as mountain ash (*Eucalyptus regnans*) in the southeast, and karri (*E. diversicolor*) in the southwest. Mountains support alpine vegetation and stunted sclerophyll communities. Less fertile soils on rock outcrops throughout the wetter parts of Australia also have sclerophyll shrublands and stunted woodlands. Freshwater lakes and streams are strongly seasonal, as are coastal estuaries and embayments into which the latter discharge (McComb and Lake 1990).

Coastal floras include mangroves and the usual cosmopolitan plants of dunes and strand (e.g., Cribb and Cribb 1985), while adjacent Australian marine environments are noteworthy in the north for their coral communities of the Great Barrier Reef on the east coast and such places as Ningaloo Reef on the west coast. Temperate marine algal floras are among the richest in the world (Womersley 1981, 1990).

POST-COLONIAL LAND USE

Initial wonderment and scientific curiosity among explorers and colonists did little to save the Australian flora from massive disruption and fragmentation, as a European agricultural economy was imposed from 1788 as fast as natural hindrances (including aboriginal resistance) could be overcome (Saunders, Hopkins, and How 1990; Flannery 1994; Kirkpatrick 1994). In just two centuries, 48% of Australia's landcover has witnessed substantial disturbance, including 20% on which all native vegetation has been cleared, with human impacts evident to varying degrees in the remainder (Graetz, Wilson, and Campbell 1995). Today, 39% of Australia south and east of the arid zone supports intensive land use (cropping, pasture, plantation, urban and industrial infrastructure), while the other 61% has extensive land use (mainly pastoral activities).

The most productive and best watered ecosystems have sustained the greatest impact. For example, only 0.5% of the 2 million ha of lowland native grasslands on rich soils in southeastern Australia persist (McDougall and Kirkpatrick 1994). Of the once extensive Australian temperate woodlands and mallee, a highly fragmented 10% remain; rainforests, originally occupying a mere 1% of the continent, have 25% persisting, as do 40% of the southern and eastern coastal wetlands, and 59% of the forests. Seagrass beds have declined in shallow nearshore areas adjacent to urban and industrial centres.

Of considerable concern is that recent destruction of native Australian vegetation continued through the 1980s and early 1990s at an annual rate approaching that seen in Sudan, Thailand, Venezuela, Bolivia, Mexico and Zaire. The clearing rate was eclipsed only by rates in Indonesia and Brazilian Amazonia (Glanznig 1995). The Australian Inventory for the Estimation of Greenhouse Gas Emissions (1994) indicated that 5,177,000 ha of native vegetation was destroyed between 1983 and 1993, 664,000 ha alone in 1990, with semi-arid eucalypt woodlands, brigalow (*Acacia harpophylla*) scrub, and open forest in Queensland and New South Wales sustaining most of this recent loss. A smaller but significant area was cleared in southwestern Australia, where the richest, highly endemic kwongan and mallee-woodland floras occupy land used for cereal agriculture.

Clearing of land and other threatening processes such as invasion of exotic weeds, diseases, and feral animals have resulted in an estimated 76 extinctions of plant species, with 301 listed nationally as endangered, and a further 708 as vulnerable (Briggs and Leigh 1996). In southern Australia, especially the southwest kwongan and woodlands, a pandemic of plant disease is in progress, with cinnamon fungus (*Phytophthora cinnamomi*) and other pathogens threatening populations of hundreds of susceptible species in the Proteaceae, Epacridaceae, Fabaceae, and other families (Withers *et al.* 1994). Resilience of native vegetation to invasion by exotic pathogens, weeds, and feral animals is unusually low in many communities where any major disturbance other than fire has been a rare event. Collectively, threatened species are concentrated in the southwest and southeast wheatbelts and along the eastern seaboard and hinterland supporting the greatest concentrations of people and their associated ecosystem disturbance.

A recent assessment of the status of plant communities in Australia, although somewhat limited in sampling certain parts of the nation, draws similar conclusions (Specht *et al.* 1995). Poorly conserved plant formations included grasslands, grassy woodland, *Acacia*-dominated formations, dry scrubs, monsoon rainforests, and humid and arid wetlands, whereas moist rainforests, alpine vegetation, and eucalypt forests were assessed to be the best conserved formations.

Australia, therefore, is unlike most other continents, in that it is mainly in temperate lowland woodlands, grasslands, mallee, and heath, not tropical rainforests or temperate forests, where the greatest threats to loss of plant biodiversity lie. Yet most public debate and media comment on conservation in Australia is about forests, perhaps signalling strong empathy for large living organisms, as well as the effective international communication of the very real dire plight of forests elsewhere (e.g., Myers 1992). A clear challenge for Australian conservation biologists is to turn community concern and action towards neglected plant communities where the major biodiversity conservation issues lie.

CONSERVATION STRATEGIES

In 1996, The National Strategy for the Conservation of Australia's Biological Diversity was signed by the Prime Minister, all State Premiers, and Chief Ministers of Territories. The goal of the strategy is "to protect biological diversity and maintain ecological processes and systems." The package of objectives includes:

(1) identification of biodiversity components and threatening processes,
(2) establishment of protected areas,
(3) strengthening of off-reserve conservation,
(4) integrated management in a bioregional planning context,
(5) ensuring survival of threatened species and ecological communities,
(6) working with aboriginal people and all interested sectors of the community,
(7) using *ex situ* approaches to complement conservation of biodiversity in the wild,
(8) improvement of our knowledge and understanding of biodiversity, and
(9) recognition of Australia's international responsibilities.

In many ways, these objectives are similar to those articulated by other nations, and are embraced in the concept of integrated conservation of rare plants articulated by Falk (1990, 1992). This conservation strategy emphasizes "interactions among land conservation, biological management, offsite research and propagation, and (re)introduction and habitat restoration." (Falk 1992, p. 397). However, achieving these objectives requires a combination of sound biological knowledge and local know-how. Emphases and strategies will change from place to place if success is to be assured.

In Australia, Main (1996) succinctly stated the conservation management challenge as the need to maintain all the components of biodiversity through a natural system of patch dynamics while at the same time ensuring that rare and endangered species, and others requiring special treatment, are not disadvantaged. He articulated five key scientific areas pertinent to successful conservation:

(1) *Taxonomy and genetics*: i.e., to enable recognition of elements and establish what is important.
(2) *Biogeography and evolution*: i.e, to understand scales of change and interpret the biota.
(3) *Regeneration and replacement ecology*: i.e., to understand patch dynamics and chance events underpinning population and ecosystem processes.
(4) *Resource recycling*: i.e., to understand how space and nutrients are made available for regeneration and growth.
(5) *Risk assessment for land and water management*: i.e., to determine probabilities that the biota will regenerate, and decide upon management priorities in the face of changing environments.

The following synopsis provides key findings and references to each of these areas of conservation biology relevant to Australian plants.

Taxonomy and Genetics

Accurate definition of the units of conservation is fundamental to achieve objectives of the national biodiversity strategy. Poor resolution of taxa has very real implications when applied to activities such as recovery programs or restoration of communities—the scarce funds allocated for such works potentially can be wasted, or at least diverted from taxa that are in critical need of help (Hopper 1994).

Australia is not as advanced as the United States or European countries in describing its flora. Even for eucalypts, views expressed just two decades ago that most species had been found and described (e.g., Pryor and Johnson 1971) have been demonstrably in error. The often quoted total of "over 500 species" (Chippendale 1988) is now estimated to be 800, with descriptions of substantial numbers of new species appearing across the nation (e.g., Brooker and Kleinig 1983, 1990, 1994). Similarly, the number of recognised Australian species of *Acacia* has risen from 660 to 950 in just two decades (Maslin 1995).

Among less well-known groups, even more impressive rises have been documented. George (1991), for example, described 41 new species of featherflowers (*Verticordia*: Myrtaceae) among the 97 he recognized. In the endemic southwest Australian genus *Synaphea* (Proteaceae), which had not been examined in detail since 1810, George (1995) described as new 41 of the 50 species recognised, and signalled that further work on some species complexes may reveal additional undescribed taxa.

This story is repeated in less dramatic form across diverse groups of the Australian flora, as a perusal of recent issues of the nation's taxonomic journals will reveal. While some of the rise in new taxa is due to better understanding and dissection of broadly based "polymorphic" species, many new taxa are distinct with no living close relatives. Among the most remarkable recent finds include the rare araucarian Wollemi Pine (*Wollemia nobilis*) found in the Blue Mountains just 200 km from the centre of Sydney (Jones, Hill, and Allen 1995). The rainforest relict tree *Eidothea zoexylocarya* (possibly representing a new subfamily of Proteaceae), from Mt. Bartle Frere in north Queensland, was an equally startling discovery (Douglas and Hyland 1995).

In Western Australia, eucalypts new to science such as Rock Mallee (*E. petrensis*) have turned up recently within the confines of outer metropolitan Perth (Brooker and Hopper 1993). Very distinct locally endemic relict species such as the forest tree Rate's Tingle (*E. brevistylis*) and Mount Lesueur Mallee (*E. suberea*) have been discovered in easily accessible sites in the southwest, and new local endemics continue to be discovered and described each year (Hopper *et al.* 1990). Among southwestern orchids, two editions of the recent field guide published eight years apart signalled a doubling in recognised taxa (Hoffman and Brown 1984, 1992).

This surprising rate of recent discovery and description of new species signals that Australia has a taxonomic impediment to conservation of a magnitude comparable to that for some tropical countries (cf. Gentry 1996). Australia is nevertheless fortunate in having herbaria staffed by government botanists in all capital cities, and taxonomy is still taught in some universities. In many cases, however, funding for taxonomic research has declined significantly, as has university teaching in traditional methods for taxonomic description. In line with trends in North America and Europe, Australian students increasingly are directed towards molecular and computer-based studies, which require long periods of research in a laboratory environment and reduce time available for exploration in the field and herbarium.

Australia runs the risk of losing significant undiscovered and undescribed parts of its floristic heritage unless field taxonomy and formal description of new taxa are given the priority they deserve by the relatively few botanists available to do this important work. For example, of the 238 taxa declared in 1989 as Rare Flora in Western Australia under the Wildlife Conservation Act, 60 (25%) were yet to be formally named, but judged to be well-defined entities by taxonomists (Hopper *et al.* 1990). Five years later, in 1994, 53 (19%) of 274 declared taxa were similarly undescribed. Therefore, it is for very sound reasons that the national biodiversity strategy includes among its objectives accelerated research and training in taxonomy.

The description of patterns of genetic variation in the Australian flora is advanced beyond simple chromosome counts in relatively few groups. From an early focus on species of economic importance or of special academic interest, Australian geneticists have only recently directed their attention to conservation concerns (Hopper and Coates 1990; Moritz *et al.* 1994; Hopper 1994, 1996; Frankham 1995). An important contribution has been clearer definition of the units of evolution and conservation in groups where patterns of morphological variation are complex or cryptic. For example, using allozyme markers, Sampson (1988; Sampson, Hopper and James 1989; Sampson, Hopper, and Coates 1990) unequivocally dispelled the hypothesis that the colourful Rose Mallee (*Eucalyptus rhodantha*) was a hybrid of *E. macrocarpa* x *pyriformis*, and provided an understanding of population genetic architecture that has been useful in framing a recovery plan for this endangered species (Sampson, Hopper, and Coates 1990; Kelly and Coates 1995).

Allozymes enabled Coates (1988) to establish that the narrowly endemic subshrub *Acacia anomala* (Chittering Grass Wattle) had both sexually reproducing and strongly clonal populations in two disjunct areas across its 30 km range. He advocated contrasting conservation strategies, in which all four of the known clonal populations needed protection because of their genetic differences, whereas reserving only the two largest of the six sexual populations would adequately conserve their detectable genetic variation.

Other examples where conservation genetic studies have been useful include the identification of a pair of cryptic species within the highly endangered herbaceous rosetted Wongan triggerplant, *Stylidium coroniforme* (Stylidiaceae) (Coates 1992), as well as the elucidation of genetically distinct disjunct populations in the herbaceous Rock Isotome (*Isotoma petraea*, Lobeliaceae; James 1965, 1970, 1982, 1984; James *et al.* 1983; James, Playford and Sampson 1990), in two rare mallee eucalypts of isolated granite outcrops (*Eucalyptus caesia*; Moran and Hopper 1983, Figure 10.1; and *Eucalyptus crucis*, Sampson, Hopper and James 1988), and in the endangered narrowly endemic tree *Banksia cuneata* (Proteaceae; Coates and Sokolowski 1992). The latter examples have enabled the identification of priorities for conservation action where resources are limited and not all populations of an endangered species can be protected.

Genetic studies also have enabled the identification of transilient and hybrid populations where the generation of evolutionary novelty has occurred (Hopper 1995, 1996). Well documented cases of transilience include the Pigeon Rock population of *Isotoma petraea*, from which complex translocation heterozygosity evolved in the southwest of the species' range (James 1965, 1970, 1982, 1984; James *et al.* 1983; James, Playford, and Sampson 1990), and the Yanneymooning Hill population of *Eucalyptus caesia*, from which *E. c.* ssp. *magna* may have originated (Hopper, Campbell, and Moran 1982; Moran and Hopper 1983; Hopper and Burgman 1983).

Natural hybrids may be beneficial or harmful to conserving biodiversity, but they require careful genetic study before managers decide what to do with them (Hopper 1995). Hybrid speciation through allopolyploidy is well documented, and convincing data on diploid hybrid speciation, introgression, and improved gene dispersal are accumulating (Grant 1981; Potts and Reid 1988; Rieseberg and Wendel 1993). From this perspective, conservation of hybrid populations has merit. Conversely, eradication is favored where natural hybridization leads to outbreeding depression and genetic or ecological swamping, threatening small parental populations (e.g., Rieseberg 1991).

These few case studies highlight that a genetic approach has potential, and indeed, is vital, to the clear identification of the units of conservation in components of the Australian flora. We are many decades away from an adequate exploration of this field. An ongoing challenge, therefore, given the rapidity with which populations of some species are disappearing, is the astute selection of case studies that elucidate key principles and have wide applicability (Moran and Hopper 1987; Moritz *et al.* 1994).

A significant observation is that natural rarity and population fragmentation are not necessarily equivalent to reduced genetic variation in Western Australian plants (Moran and Hopper 1987). Geographic range was found by Hamrick and Godt (1989) to be correlated with allozyme variability in plants, but exceptions to the rule were noted (e.g., Conkle 1987). Western Australia has an extraordinary number of rare plants, and conditions have existed continuously for terrestrial life unencumbered by glaciation since the Permian (Hopper *et al.* 1996). Many species (e.g., *Stylidium coroniforme*; Coates 1992) have evolved over long periods the means of retaining heterozygosity in the face of small population size. A compelling hypothesis explaining this in terms of balanced lethals interacting under enforced inbreeding has been developed by James (1992).

Biogeography and Evolution

Having established the base units of conservation, a knowledge of where genes, species, and communities are distributed geographically and how they came to be there over time is essential to ensure conservation reserves adequately represent biota, to identify rare endemics, and to establish a framework for management.

Studies of plant communities in Australia have been conducted for more than a century, but an agreed national descriptive system remains elusive, despite ongoing attempts (e.g., Specht *et al.* 1995). Quantitative floristic community approaches show the greatest promise, having been used effectively for the selection of conservation reserves at a regional scale across the nation. Representative examples are, for Tasmania, Kirkpatrick 1983; New South Wales, Pressey *et al.* 1993; southeast Australian lowland grasslands, McDougall and Kirkpatrick 1994; and Western Australia's kwongan, Griffin, Hopkins and Hnatiuk 1983; mallee, Burgman 1988; Nullarbor Plain, McKenzie *et al.* 1989; Kimberley vine thickets, McKenzie, Johnson, and Kendrick 1991; and Swan Coastal Plain, Gibson et al 1994.

Arguably the most effective biogeographic studies in terms of conservation outcomes have been those where floristic community data have been analysed together with the distribution of threatened species. This approach seeks to locate large conservation reserves with representative communities on areas richest in threatened, endemic species (Kirkpatrick 1983; Kirkpatrick and Brown 1991; Burbidge, Hopper, and van Leeuwen 1990). A sound biogeographic understanding of threatened species is an essential prerequisite. Knowledge of Australian threatened plants has been compiled nationally for more than two decades (Briggs and Leigh 1996), building initially on assessments of distribution and abundance from taxonomists, and becoming more accurate and refined as the biogeography of threatened flora has been studied directly.

In Western Australia, remarkable revisions of conservation status have occurred as plant geographers have intensified exploration recently. For example, 72 species listed as presumed extinct (not recorded in the previous 50 years) in published compilations had been rediscovered in the two decades leading up to 1989 (Hopper *et al.* 1990), and more have turned up subsequently (e.g., *Eucalyptus rameliana*, Hopper 1992b, 1994; *Stylidium merrallii*, Kenneally and Lowrie 1994). This trend signals the need for caution in ascribing conservation status where field survey data are limited.

A systematic process to record biogeographical data on threatened and poorly known plants of Western Australia was established by the author in the 1980s to ensure that plants legally protected as Rare Flora under the Wildlife Conservation Act had been adequately searched for and found to be of conservation concern (Hopper *et al.* 1990). This system continues in use, and has been adopted nationally. At present in Western Australia, some 274 taxa are declared as Rare Flora, another 200 have been adequately surveyed to establish they are not under immediate threat but need to be monitored, and another 1,500 are poorly known and possibly under threat (Coates and Atkins 1996). Such numbers are among the highest globally.

Careful documentation in the 1980s of the distribution of 1,386 populations of declared Rare Flora in Western Australia (Hopper *et al.* 1990) established that only 28% were on conservation reserves, thus the majority were on land not primarily set aside for conservation. Such lands included road reserves (22%), private property (16%), and vacant Crown land not set aside for any designated use (13%). These statistics remain much the same today (Coates and Atkins 1996). It is clear that effective conservation of threatened species requires broadly-based community support and action because conservation land managers cannot achieve biodiversity goals on their own.

To facilitate a community approach, apart from a color book on endangered plants (Hopper *et al.* 1990), published compilations of district level biogeographic and taxonomic information on threatened and poorly known flora were prepared in Western Australia, commencing with forest and wheatbelt areas nearest Perth (Kelly *et al.* 1990). Since then, three other area-based treatments have been published, and seven are in preparation, covering the entire South West Botanical Province where threatened species are concentrated (Coates and Atkins 1996).

Direct community involvement in acquiring biogeographical data also has occurred in projects such as the *Banksia Atlas* (Taylor and Hopper 1988, 1991), the survey of rare and poorly known eucalypts of Western Australia (Kelly, Napier and Hopper 1995), and the floristic survey of the southern Swan Coastal Plain (Gibson *et al.* 1994). Such projects increased survey efforts manyfold, leading to the discovery of new rare taxa in Western Australia (e.g., *Banksia seminuda* ssp. *remanens*), and had the added benefit of encouraging stewardship in local communities through personal involvement in worthwhile conservation research.

A combination of the above biogeographic approaches enables the following synoptic overview (from Hopper *at al.* 1996). The flora of Western Australia is distributed across three biogeographic regions:—the south-west with a Mediterranean climate (Hobbs 1992), the tropical north or Kimberley (Burbidge, McKenzie, and Kenneally 1991), and the intervening arid zone of desert, Pilbara, pastoral and goldfields country (Beard 1980). Highest endemism in many plant groups is seen in the southwest; some large genera have more than 90% of their southern Western Australian species endemic (e.g., *Banksia*; spider orchids, *Caladenia*; beard heaths, *Leucopogon*), and even whole genera are endemic to this region (e.g., *Dryandra* and *Synaphea* of the Proteaceae; George 1995, 1996).

Endemic plant families, in contrast, are few indeed, with the monotypic Albany pitcher plant (*Cephalotus follicularis*, Cephalotaceae) belonging to one. The southwest is also where the State's few cryptogamic endemics occur (George 1992; Orchard 1994; Wyatt and Stoneburner 1989). Examples include the lichens *Xanthoparmelia gerhardii* from the Stirling Range, *X. nana* from the Darling Scarp near Perth, *Degelia flabellata* from the Porongurup Range, and the moss *Pleurophascum occidentale* from south coast granite peaks near Albany and Esperance.

The flora of Western Australia's deserts and tropical north is similar to that of adjacent States, with fewer endemics than the southwest (e.g., grasses with 20% endemic W.A. species), and arid plant genera such as *Atriplex* (Chenopodiaceae), with 31%, and *Solanum*

(Solanaceae), with 37%). Extremes in the Kimberley are seen in predominantly tropical genera such as *Cyperus* (Cyperaceae, only 6% WA endemics) and *Triumfetta* (Tiliaceae, 61%).

Eucalypts illustrate these statewide trends well, with endemism as occurring in the arid and tropical northern half of Western Australia (N of 26°) at only 32% (29 species) among the 92 species treated by Brooker and Kleinig (1994). In comparison, some 88% (252) of the 285 listed by Brooker and Kleinig (1990) as occurring in the southern half are endemic in W.A.

The flora of the southwest is dominated by speciose genera of woody and herbaceous perennials in families such as the Myrtaceae, Proteaceae, Fabaceae, Mimosaceae, Epacridaceae, Orchidaceae, and Restionaceae, which are all Gondwanan or significantly so in relevant groups dominant in Western Australia. There are few relicts of the Tertiary rainforest flora extant, and only a modest diversity of families and genera remains (Lamont, Hopkins and Hnatiuk 1984). But the flora has run riot at the species level. In the southwest, *Acacia* has at least 400+ species, *Eucalyptus* 285 or more, *Grevillea* 150, *Stylidium* 130, *Leucopogon* 115, *Dryandra* 92, and *Caladenia* 91. This pattern is in stark contrast to the rainforest flora of Queensland, where there is great diversity in families and genera but relatively few species within genera (Webb and Kikkawa 1990; Webb, Tracey and Williams 1984). It is clear that explosive speciation has occurred in the southwest in response to the environmental perturbations of the late Tertiary and Quarternary. Moreover, the Transitional Rainfall Zone of the wheatbelt is where such speciation has been concentrated for the majority of woody perennial taxa (Hopper 1979; Lamont, Hopkins, and Hnatiuk 1984).

For almost a century, it has been recognized that there are exceptionally rich floras in the Stirling Range—Fitzgerald River National Park region near the south coast, and in the Lesueur National Park area to the north of Perth. For example Fitzgerald River National Park has 1750+ named species (Chapman and Newbey 1987), and Lesueur National Park 820+ (Griffin, Hopper, and Hopkins 1990). More than 110 species have been found in quadrats 10m x 10m near Mt. Lesueur, a level of species richness at this scale with few parallels globally. Indeed, Lamont, Downes and Fox (1977, p. 440) recorded species diversity statistics in kwongan near Eneabba that were "as high as, if not higher than, those for the most diverse communities previously recorded, including the richest of tropical rainforests."

Another notable feature is the high turnover of species across the landscape. Adjacent quadrats on laterite in Lesueur National Park may have as few as 60% of their species in common. Moving as little as 0.5 km away within the same vegetation type can reduce the similarity to less than 40% (Griffin, Hopper, and Hopkins 1990). This striking replacement of species across small distances in seemingly uniform habitat is seen throughout the wheatbelt or transitional rainfall zone of Western Australia (Hopper 1979; Burgman 1988). For conservation managers, the pattern signals an unparalleled need to develop fine-scale local knowledge of plants, particularly where reconstruction of native vegetation in degraded sites is contemplated.

At the regional level in the southwest, there are major differences in the concentration of genera, and recurrent examples of allopatric replacement series of closely related narrowly endemic species within genera. For example, genera of Myrtaceae (*Calothamnus*, *Melaleuca*, *Darwinia*) and Proteaceae (*Dryandra*, *Hakea*) show bimodal concentrations of species in the two major kwongan nodes near the west and south coasts. Others are concentrated in the southern kwongan and mallee (e.g., *Adenanthos* [Proteaceae], *Eucalyptus*, *Banksia*, *Leucopogon*), or the Mt. Lesueur-Kalbarri northern kwongan *Grevil-*

lea, *Conostylis* [Haemodoraceae], *Lechenaultia* [Goodeniaceae], or are concentrated inland through the wheatbelt (e.g., *Acacia*, *Verticordia* [Myrtaceae]).

It is unusual to find species that occur throughout much of the Transitional Rainfall Zone. The majority of known examples have good means of seed dispersal (e.g., the annual composites *Quinettia urvillei*, *Millotia tenuifolia*; the orchids *Spiculaea ciliata*, *Caladenia flava*, *Leporella fimbriata*; and the mistletoes *Amyema miquelli*, and *Nuytsia floribunda*). It is much more common, however, for species to be confined to smaller ranges within the Transitional Rainfall Zone, often in allopatric replacement series with congeners (e.g., *Eucalyptus wandoo*, *E. redunca* and related species; Brooker and Hopper 1991). Indeed, the Transitional Rainfall Zone flora has Australia's highest concentration of rare and endangered local endemics, and is exceptional in this respect from a global perspective as well (Hopper *et al.* 1990). It seems likely that the Transitional Rainfall Zone flora has had to cope with small population sizes induced by recurrent late Tertiary-Quaternary climatic and sea-level oscillations in a heterogeneous soil mosaic. Consequently, the southwest has much to teach about rare species and their roles in ecosystem function (Pate and Hopper 1993), as well as diverse genetic system responses to inbreeding imposed by small population size (James 1992).

Although most of the southwestern flora is concentrated in the Transitional Rainfall Zone, a few genera such as *Agonis* (Myrtaceae) favor the wetter conditions of the High Rainfall Zone. A recent account of the flora of the karri (*Eucalyptus diversicolor*) forest and adjacent communities (Hopper, Keighery, and Wardell-Johnson 1992) listed 1947 taxa, the most important families being Fabaceae (165 taxa), Orchidaceae (152), Myrtaceae (125) and Proteaceae (115), with the largest genera *Acacia* (64), *Stylidium* (55), *Caladenia* (46), *Leucopogon* (41), *Eucalyptus* (26) and *Drosera* (24). Most groups of wetland monocotyledons, including genera of Cyperaceae, Xyridaceae, Juncaginaceae, Restionaceae and Orchidaceae, are species-rich in the karri southern forest region.

It is in this highest rainfall extreme southcoastal area that the few relicts of rainforest affinities are concentrated (e.g., *Cephalotus follicularis*, *Podocarpus drouynianus* (Podocarpaceae), *Eucalyptus brevistylis* (Ladiges, Humphries, and Brooker 1987; Wardell-Johnson and Coates 1996). Moreover, there are several rare taxa persisting in mesic sites with rainfall more or less year-round whose closest affinities may be with eastern Australian summer rainfall taxa (e.g., the banksias described by Hopper 1989).

Evolutionary studies provide an essential understanding of how communities and taxa change through time. They highlight the difference between conservation, where the aim is maintenance of ecological and evolutionary processes, and preservation, where the goal is to retain for as long as possible what is present at a selected point in time. With dramatically different evolutionary components in its flora, Australia highlights the value of an evolutionary approach. This is nowhere more evident than the interface between rainforest and sclerophyll eucalypt communities (Webb and Kikkawa 1990; Werren and Kershaw 1991a, 1991b), the former being relictual remnants of the great Tertiary Gondwanan rainforests, and the latter being derivative recently-evolved lineages arising from the late Tertiary onset of aridity and attendant fire. This single insight enables managers to understand that not all components of the flora can tolerate fire and seasonalaridity, with considerable caution called for whenever relictual rainforest taxa occur where fire might be frequent.

In the Kimberley, conservation of vine thickets depends upon a clear recognition of the potent impact of fire (McKenzie, Johnson, and Kendrick 1991). In the southwest, a more subtle situation occurs with very few rainforest relicts clearly identifiable in the flora. However, forest eucalypts restricted to narrow ranges in the highest rainfall southcoastal uplands have been shown to be relictual in phylogenetic and ecological studies (Ladiges, Humphries, and Brooker 1987; Wardell-Johnson and Coates 1996). They and their dependent invertebrate fauna appear less tolerant of frequent fire than their more recently-evolved congeners in adjacent habitats (Hopper *et al.* 1996). This observation leads directly into the next arena of conservation biology pertinent to an Australian perspective—i.e., understanding regeneration and recruitment.

Regeneration and Replacement Ecology

Fire orchestrates the tempo and mode of much in Australian vegetation. The emergence of concepts such as resprouters and obligate seeders in response to fire has been a noteworthy Australian contribution with clear implications for management (Gill 1981; Bell, Hopkins, and Pate 1984; Pate and Hopper 1993). Obligate seeders are killed by canopy fire and recruit only from germination, so they are vulnerable to local extirpation should fire intervals exceed the time required for them to flower and set seed. Resprouters can tolerate more frequent fires because regrowth of vegetative buds protected beneath bark or underground enables survival. Many threatened plants in Western Australia are perennial shrubs and obligate seeders (Hopper *et al.* 1990).

Dixon, Roch, and Pate (1995) demonstrated that smoke plays a key role in securing germination of 45 of 94 Western Australian species normally difficult to germinate. This exciting find, building on the work of Brown (1993) in South African fynbos, has led to a better understanding of the life history of many native plants, as well as made them available for use inbushland restoration and horticulture. Another significant discovery is that seed banks of kwongan species of Restionaceae and Epacridaceae are short-lived after wildfire, irrespective of parental fire response as a resprouter or an obligate seeder (Meney, Nielssen, and Dixon 1994). This counter-intuitive finding needs further elaboration, but suggests that post-fire recruitment in the first year is critical for a diversity of species.

Westoby (1988) noted that factors other than fire in the abiotic environment, notably shade, moisture, acidity, salinity, and soil nutrients, also play similarly striking roles in controlling recruitment. There is a growing literature in these fields (e.g., Pate and McComb 1981; Dodson and Westoby 1985; Saunders, Hopkins and How 1990). Thus, elucidation of critical recruitment factors for threatened species in Australia is underway (Hopper *et al.* 1990; Cropper 1993; Coates and Atkins 1996).

Grazing by feral animals or introduced stock has caused the decline of critically endangered species such as *Darwinia carnea* (Myrtaceae) and *Stylidium coroniforme*. Simple fencing has yielded dramatic recovery for populations of these species (Coates and Atkins 1996). In contrast, light grazing on native grassland pasture in Tasmania favoured the unpalatable rare daisy *Leucochrysum albicans* (Gilfedder and Kirkpatrick 1994a, 1994b).

Competition with invasive weeds (Humphries *et al.* 1991) and dieback disease (Withers, Cowling, and Wills 1994) also account for significant declines in threatened Australian flora. To a lesser extent, so does rising saline groundwater in overcleared agricultural landscapes (MacFarlane, George, and Farrington 1993; George, MacFarlane, and Speed 1995).

The life histories of Western Australia's declared Rare Flora are illuminating in terms of assessing recruitment prospects. Most (176) of the 238 declared taxa in 1989 are woody long-lived perennials such as mallee eucalypts. However, at least 60 of these woody perennials are short-lived disturbance opportunists and obligate seeders after fire (e.g., species of *Acacia* and *Grevillea*). Of the 59 terrestrial perennial herbs declared as Rare Flora, 30 are orchids, being tuberous and flowering best in open recently-burnt communities. Three species are aquatics of ephemeral (vernal) ponds.

Spring flowering occurs in 160 of the 238 taxa, and more than a third (97, 41%) have flowers likely to be pollinated by birds and/or mammals (Figure 10.2). This is almost three times the proportion of the southwestern flora at large that is vertebrate-pollinated (15%, Keighery 1982). Bird pollination provides one mechanism for increasing gene flow and hybridity in the face of small population size and inbreeding (James and Hopper 1981; Hopper and Burbidge 1986; Sampson, Hopper, and James 1989). Reproductive failure in small remnant populations on road verges and farmland has been documented in *Eucalyptus rhodantha*, *B. cuneata*, *B. tricuspis* and *Banksia goodii* (Sampson 1988; Sampson *et al.* 1989, 1996; Coates and Sokolowski 1992; Lamont *et al.* 1993). Reduced abundance of vertebrate pollinators due to loss of habitat is implicated in this process because either inbreeding has increased and/or seed set is negligible.

Among the insect-pollinated taxa, orchids in genera such as *Caladenia* and *Drakaea* have an almost species-specific relationship with male thynnid flower wasps who are sexually deceived by the floral odors and anatomy (Stoutamire 1974, 1983). Such relationships are vulnerable to disruption through habitat fragmentation and competition from introduced honey bees.

Figure 10.2
*A western pigmy possum (*Cercatetus concinnus*) feeding on nectar and pollinating* Eucalyptus caesia. *Pollination by birds and mammals is common in Australia, particularly the southwest. Ensuring such ecological processes persist is an important challenge for conservation biologists.*

Perhaps the most bizarre species of all in the endangered Australian flora is the underground orchid (*Rhizanthella gardneri*), an epiparasite that grows, flowers, and fruits beneath the soil surface, drawing nutrient through mycorrhizal connections to broombush *Melaleuca uncinata* (Dixon, Pate, and Kuo 1990). With a related species (*R. slateri*) growing adjacent or in rainforests on the east coast, the Western Australian underground orchid may be a Tertiary rainforest relict that has evaded aridity through its epiparasitic subterranean lifestyle. It has the largest seeds among orchids, and fruits may be dispersed by marsupials in search of hypogean fungi. With such unusual life histories awaiting investigation, and community interest rapidly developing in revegetation and restoration, the regeneration and replacement ecology of Australian plants is a major field in need of elucidation.

Resource Recycling and Risk Assessment

Although recognised as major areas requiring study, both resource recycling and risk assessment have received little attention from Australian conservation biologists. Yet they are critical to successful management for conservation. Persistence of ecosystems is intimately linked to cycling of water and essential nutirents. Risk assessment provides useful decision-making tools to help conservation managers set achievable targets and work towards attaining them.

For example, Western Australian soils in the main differ from the younger fertile soils on the east coast and elsewhere, in being old, highly weathered, and nutrient deficient. Phosphorous and potassium are limiting, not nitrogen, except for the first year or so after fire when most nitrogen literally goes up in smoke (Pate and Hopper 1993). A diversity of microbial symbioses exists among the vascular flora and soil microbes, partnerships that increase nutrient extraction and water uptake (Lamont 1982).

Risk assessment procedures illuminate a key issue for managers in a time of changing environments, including potential global climate change; should present communities be encouraged where they are growing now, even if reproductive failure is in evidence, or should new communities be synthesized based on best predictions of changed circumstances? To a large extent, the latter is already happening through global agriculture and forestry operations, but it is not a mainstream operation among conservation managers. Without active intervention, many fragmented relict communities face a dismal future. This has certainly happened in Kings Park, Perth, where more than a century of urban activity has led to one third of the flora comprising introduced exotics, and the transition to communities dominated by particularly invasive trees and herbs is well underway. Active management to alter this track commenced only recently.

Formal risk assessment in conservation biology has been explored by Burgman, Ferson and Akcakaya (1993), and applied to the endangered tree *Banksia cuneata* in Western Australia (Burgman and Lamont 1992). This approach is useful in highlighting the kinds of data required to predict outcomes of differing management regimes. Interestingly, Burgman and Lamont (1992) concluded that the most effective plan for minimizing the risk of extinction of *B. cuneata* over 50 years was to provide water to seedlings in the summer drought following fire. Alternatively, the next best thing was to burn populations at infrequent intervals (ca. 25 years) when the canopy seed store is high, but before plants senesce. Burning at more frequent intervals raised the probability of extinction because this obligate seeder needed time to build up a seed store.

Apart from managing populations in the wild, another form of risk management is to enlist modern horticultural techniques to ensure germplasm remains alive and is propagated for restoration and recovery programs. Dixon (1994) outlined pertinent activities in the botanic garden at Kings Park, including conventional cutting and seed propagation, grafting, whole seed germination (symbiotic and asymbiotic), excised seed embryo culture, and tissue culture of shoot apices and inflorescence sections. Experiments on slow growth and on cryostorage have been increasingly successful with a diversity of genera (e.g., Touchall, Dixon, and Tan 1992; Touchall and Dixon 1993).

Trial recovery operations from cultivated stock are underway with several endangered Western Australian species, most notably Corrigin Grevillea (*G. scapigera*; Bunn and Dixon 1992; Rossetto 1995; Rossetto *et al.* 1996) and *Eucalyptus rhodantha* (Sampson, Hopper, and Coates 1990; Kelly and Coates 1995). Involvement of local communities is a key part of the strategy. Astute sampling of germplasm is needed to ensure adequate genetic variation is included in stock for return to the wild (Brown and Briggs 1991). However, the success of these recovery activities may well require some decades to elapse before a reasonable judgement could be made, given the importance of rare threshold events such as prolonged drought in controlling regeneration and survivorship. The educational value of such high profile activities should not be underestimated, however.

CONCLUSION

Being the world's oldest, flattest, driest, and most isolated continent with the most recent and rapid transformation of ecosystems by European humans, Australia provides a unique insight into the management and mismanagement of a remarkable and highly vulnerable flora. This "island continent" contains 25,000 species, about 10% of the world's flora, with endemism at 85%. Today's eucalypt and acacia communities are arid-adapted and fire-prone derivatives from Gondwanan rainforests, the fragmented and depleted relicts of which occupied just 1% of the continent. Covering only 5% of Australia, the southwest is notably rich in species (8000) and endemism (75%), while the vast arid zone of central Australia has about 2,000 species.

Changes wrought on the land following European colonization have profoundly altered most vegetation, with Western Australia in particular having more species of threatened plants than other States or most countries of the world. More than 3,000 environmental weed species are established, and plant diseases are a major threat in species-rich heath (kwongan). Areas of conservation biology where exciting insights have recently unfolded include taxonomic and genetic studies, revealing a hitherto unsuspected floristic diversity in need of description; biogeographic and evolutionary investigations, providing first insights into the conservation status of the majority of species and communities; and a clear identification of recently evolved and relictual taxa. The southwest of Australia has a remarkably high level of turnover of plant species across the landscape, requiring caution in the application of conservation management approaches across broad regions. Regeneration and recruitment work has highlighted the importance of fire and smoke-stimulated germination, the ephemeral nature of some seed banks, and the impact of fragmentation on mating systems, pollination and seed set. Risk assessment has barely been touched, save for some interesting quantitative modelling and the use of modern propagation and germplasm storage technology to help in recovery programs of critically endangered taxa.

Protection and restoration of remnant vegetation, especially involving local communities as active stewards, remains the most important and effective priority for conservation biologists in Australia. The flora is poorly documented and much remains to be done to elucidate its biology. Consequently, reconstruction of native plant communities is a field in its infancy, with far greater risks of ecosystem failure than there is in caring for intact remnant vegetation. Conservation biologists must cherish what remains, and work with local people best placed to look after remnants to maximise opportunities for wild plant life.

ACKNOWLEDGMENTS

My understanding of plant conservation has benefited considerably from collaboration and discussion with numerous colleagues, most of whom are cited in the reference list. For enhancing my appreciation of the special aspects of the Western Australian flora, I am particularly grateful to S.H. James, D.J. Coates, K.W. Dixon, M.I.H. Brooker, A.P. Brown, G.J. Keighery, A.R. Main, and others. Needless to say, the responsibility for the above interpretation of Australian plant conservation biology is mine alone.

LITERATURE CITED

Australian Inventory for the Estimation of Greenhouse Gas Emissions: Carbon Dioxide From the Biosphere. 1994. Australian Government Publishing Service, Canberra.

Barker, W.R. and P.J.M. Greenslade, eds. 1982. *Evolution of the flora and fauna of arid Australia.* Adelaide: Peacock Publications.

Barlow, B.A. 1981. The Australian flora: Its origin and evolution. *Flora of Australia* 1:25–75.

Beard, J.S. 1980. A new phytogeographic map of Western Australia. *Western Australian Herbarium Research Notes* 3:37–58.

Bell, D.T., A.J.M. Hopkins, and J.S. Pate. 1984. Fire in the kwongan. In *Kwongan: Plant life of the sandplains*, ed. J.S. Pate and J.S. Beard, 178–204. Perth: The University of Western Australia Press.

Bradstock, R.A., T.D. Auld, D.A. Keith, R.T. Kingsford, D. Lunney, and D.P. Sivertsen, eds. 1995. *Conserving biodiversity: Threats and solutions.* Sydney: Surrey Beatty and Sons.

Briggs, J.D. and J.H. Leigh. 1996. *Rare or threatened Australian plants—1995 revised edition.* Canberra: CSIRO.

Brooker, M.I.H. and S.D. Hopper. 1991. A taxonomic revision of *Eucalyptus wandoo*, *E. redunca*, and allied species (*E.* series *Levispermae* Maiden—Myrtaceae) in Western Australia. *Nuytsia* 8:1–189.

Brooker, M.I.H. and S.D. Hopper. 1993. New series, subseries, species and subspecies of *Eucalyptus* (Myrtaceae) from Western Australia and from South Australia. *Nuytsia* 9:1–68.

Brooker, M.I.H. and D.A. Kleinig. 1983. *Field guide to Eucalypts*, Vol. 1, South-eastern Australia. Melbourne: Inkata Press.

Brooker, M.I.H. and D.A. Kleinig. 1990. *Field guide to Eucalypts*, Vol. 2, Western Australia (southern part, below 26 latitude), northern South Australia, Eyre Peninsula, and Kangaroo Island, New South Wales (west and north of the Darling River). Melbourne: Inkata Press.

Brooker, M.I.H. and D.A. Kleinig. 1994. *Field guide to eucalypts*, Vol. 3, Northern Australia. Inkata Press: Melbourne.

Brown, A.H.D. and J.D. Briggs. 1991. Sampling strategies for genetic variation in ex situ collections of endangered plant species. In *Genetics and conservation of rare plants*, eds. D.A. Falk and K.E. Holsinger, 99–119. New York: Oxford University Press.

Brown, N.H.C. 1993. Promotion of germination of fynbos seeds by plant-derived smoke. *New Phytologist* 122:1–9.

Bunn, E. and K.W. Dixon. 1992. *In vitro* propagation of the rare and endangered *Grevillea scapigera* (Proteaceae). *HortScience* 27:261–262.

Burbidge, A.A., N.L. McKenzie, and K.F. Kenneally. 1991. *Conservation reserves in the Kimberley Western Australia.* Como, W.A.: Department of Conservation and Land Management: Como.

Burbidge, A.A., S.D. Hopper, and S. van Leeuwen, eds. 1990. *Nature conservation, landscape and recreation values of the Lesueur area.* EPA Bulletin 424. Perth: Environmental Protection Authority.

Burgman, M.A. 1988. Spatial analysis of vegetation patterns in southern Western Australia: Implications for reserve design. *Australian Journal of Ecology* 13:415–429.

Burgman, M.A. and B.B. Lamont. 1992. A stochastic model for the viability of *Banksia cuneata* populations: environmental, demographic and genetic effects. *Journal of Applied Ecology* 29:719–727.

Burgman, M.A., S. Ferson, and H.R. Akcakaya. 1993. *Risk assessment in conservation biology.* London: Chapman & Hall.

Chapman, A. and K.R. Newbey. 1987. *A biological survey of the Fitzgerald area, Western Australia.* Report to the Australian Heritage Commission.

Chippendale, G.M. 1988. *Eucalyptus, Angophora* (Myrtaceae). *Flora of Australia* 19. Canberra: Australian Government Publishing Service.

Clark, R.L. 1990. Ecological history for environmental management. In *Australian ecosystems 200 years of utilization, degradation and reconstruction*, eds. D.A. Saunders, A.J.M. Hopkins, and R.A. How. *Proceedings of the Ecological Society of Australia* 16:1–21.

Coates, D.J. 1988. Genetic diversity and population genetic structure in the rare Chittering grass wattle *Acacia anomala. Australian Journal of Botany* 36:273–86.

Coates, D.J. 1992. Genetic consequences of a bottleneck and spatial genetic structure in the triggerplant *Stylidium coroniforme* (Stylidiaceae). *Heredity* 69:512–520.

Coates, D.J. and K. Atkins. 1996. Threatened flora of Western Australia: A focus for conservation outside reserves. In *Conservation outside nature reserves*, eds. P. Hale and D. Lamb, in press.

Coates, D.J. and R.E.S. Sokolowski. 1992. The mating system and patterns of genetic variation in *Banksia cuneata* A.S. George (Proteaceae). *Heredity* 69:11–20.

Common, M.S. and T.W. Norton. 1992. Biodiversity: Its conservation in Australia. *Ambio* 21:258–265.

Conkle, M.T. 1987. Electrophoretic analysis of variation in native Monterey cypress (*Cupressus macrocarpa* Hartw.). In *Conservation and management of rare and endangered plants*, ed. T.S. Elias, 249–256. Sacramento: California Native Plant Society.

Cowling, R.M. and E.T.F. Witkowski. 1994. Convergence and non-convergence of plant traits in climatically and edaphically matched sites in Mediterranean Australia and South Africa. *Australian Journal of Ecology* 19:220–232.

Cribb, A.B. and J.W. Cribb. 1985. *Plant life of the Great Barrier Reef and adjacent shores.* St. Lucia: University of Queensland Press.

Cropper, S.C. 1993. *Management of endangered plants.* East Melbourne: CSIRO.

Dixon, K.W. 1994. Towards integrated conservation of Australian endangered plants—the Western Australian model. *Biodiversity and Conservation* 3:148–159.

Dixon, K.W., J.S. Pate, and J. Kuo. 1990. The Western Australian fully subterranean orchid *Rhizanthella gardneri.* In *Orchid biology. Reviews and perspectives*, ed. J. Arditti, 37–62. Portland: Timber Press.

Dixon, K.W., S. Roche, and J.S. Pate. 1995. The promotive effect of smoke derived from burnt native vegetation on seed germination of Western Australian plants. *Oecologia* 101:185–192.

Dodson, J.R. and M. Westoby, eds. 1985. Are Australian Ecosystems Different? *Proceedings of the Ecological Society of Australia* 14.

Douglas, A.W. and B.P.M. Hyland. 1995. *Eidothea.* Flora of Australia 16:127–129.

Ehrlich, P.R. and E.O. Wilson. 1991. Biodiversity studies: Science and policy. *Science* 253:758–62.

Falk, D.A. 1990. Integrated strategies for conserving plant genetic diversity. *Annals of the Missouri Botanical Garden* 77:38–47.

Falk, D.A. 1992. From conservation biology to conservation practice; strategies for protecting plant diversity. In *Conservation biology: the theory and practice of nature conservation, preservation and management*, eds. P.L. Fiedler and S.K. Jain, 397–431. New York: Chapman & Hall.

Flannery, T.F. 1994. *The future eaters. An ecological history of the Australasian lands and people.* Chatswood, New South Wales: Reed Books.

Frankham, R. 1995. Conservation genetics. *Annual Review of Genetics* 29:305–327.

Gentry, A.H. 1996. Species expirations and current extinction rates: A review of the evidence. In *Biodiversity in managed landscapes: theory and practice*, eds. R.C. Szaro and D.W. Johnston, 17–26. Oxford University Press: New York.

George, A.S. 1991. New taxa, combinations and typifications in *Verticordia* (Myrtaceae: Chamelaucieae). *Nuytsia* 7:231–394.

George, A.S., ed. 1992. *Flora of Australia* Vol. 55, *Lichens—Introduction, Lecanorales 1*. Canberra: Australian Biological Resources Study.

George, A.S. 1995. *Synaphea. Flora of Australia* 16:271–315.

George, A.S. 1996. New taxa and a new infrageneric classification in *Dryandra* R. Br. (Proteaceae: Grevilleoideae). *Nuytsia* 10:313–408.

George, R.J., D.J. McFarlane, and R.J. Speed. 1995. The consequences of a changing hydrologic environment for native vegetation in southwestern Australia. In *Nature conservation 4. The role of networks*, eds. D.A. Saunders, J.L. Craig, and E.M. Mattiske, 9–22. Chipping Norton, NSW: Surrey Beatty and Sons.

Gibson, N., B. Keighery, G. Keighery, A. Burbidge, and M. Lyons. 1994. *A floristic survey of the southern Swan Coastal Plain*. Unpublished report for the Australian Heritage Commission prepared by the Department of Conservation and Land Management and the Conservation Council of Western Australia (Inc.).

Gilfedder, L. and J.B. Kirkpatrick. 1994a. Culturally induced rarity? The past and present distributions of *Leucochrysum albicans* in Tasmania. *Australian Journal of Botany* 42:405–416.

Gilfedder, L. and J.B. Kirkpatrick. 1994b. Climate, grazing and disturbance, and the population dynamics of *Leucochrysum albicans* at Ross, Tasmania. *Australian Journal of Botany* 42:417–30.

Gill, A.M. 1981. Coping with fire. In *The biology of Australian plants*, eds. J.S. Pate and A.J. McComb, 65–87. Perth: The University of Western Australia Press.

Glanznig, A. 1995. Native vegetation clearance, habitat loss and biodiversity decline: an overview of recent native vegetation clearance in Australia and its implications for biodiversity. *Biodiversity Series, Paper No. 6*. Canberra: Biodiversity Unit, Department of the Environment, Sport, and Territories.

Graetz, R.D., M.A. Wilson, and S.K. Campbell. 1995. Landcover disturbance over the Australian continent: a contemporary assessment. *Biodiversity Series, Paper No. 7*. Canberra: Biodiversity Unit, Department of the Environment, Sport and Territories.

Grant, V. 1981. *Plant speciation*. 2nd ed. New York: Columbia University Press.

Griffin, E.A., A.J.M. Hopkins, and R.J. Hnatiuk. 1983. Regional variation in mediterranean-type shrublands near Eneabba, south-western Australia. *Vegetatio* 52:103–127.

Griffin, E.A., S.D. Hopper, and A.J.M. Hopkins. 1990. Flora. In *Nature conservation, landscape and recreation values of the Lesueur area*, eds. A.A. Burbidge, S.D. Hopper, and S. van Leeuwen, 39–69. EPA Bulletin 424. Perth: Environmental Protection Authority.

Groves, R.H., ed. 1994. *Australian vegetation*. 2nd edition. Cambridge: Cambridge University Press.

Hamrick, J.L. and M.J.W. Godt. 1989. Allozyme diversity in plant species. In *Plant population genetics, breeding and genetic resources*, eds. A.H.D. Brown, M.T. Clegg, A.L. Kahler, and B.S. Weir, 43–63. Sunderland, MA: Sinauer Associates.

Hobbs, R.J., ed. 1992. *Biodiversity of mediterranean ecosystems in Australia*. Sydney: Surrey Beatty and Sons.

Hobbs, R.J. and A.J.M. Hopkins. 1990. From frontier to fragments: European impact on Australia's vegetation. In *Australian ecosystems 200 years of utilization, degradation and reconstruction*, eds. D.A. Saunders, A.J.M. Hopkins, and R.A. How. *Proceedings of the Ecological Society of Australia* 16:93–114.

Hoffman, N. and A. Brown. 1984. *Orchids of south-west Australia*. 1st Edition. Nedlands: University of Western Australia Press.

Hoffman, N. and A. Brown. 1992. *Orchids of south-west Australia*. 2nd Edition. Nedlands: University of Western Australia Press.

Hooker, J.D. 1860. Introductory essay, *Botany of the Antarctic Voyage of H.M. Discovery ships 'Erebus' and 'Terror' in the years 1839–1843, III. Flora Tasmaniae*. London: Reeve.

Hopkins, A.J.M., R.A. How, and D.A. Saunders. 1990. Managing Australia's environment: directions for the future.In *Australian ecosystems 200 years of utilization, degradation and reconstruction*, eds. D.A. Saunders, A.J.M. Hopkins, and R.A. How. *Proceedings of the Ecological Society of Australia* 16:579–593.

Hopper, S.D. 1979. Biogeographical aspects of speciation in the south west Australian flora. *Annual Review of Ecology and Systematics* 10:399–422.

Hopper, S.D. 1989. New subspecies of *Banksia seminuda* and *B. occidentalis* (Proteaceae) from the south coast of Western Australia. *Nuytsia* 7:15–24.

Hopper, S.D. 1992a. Patterns of diversity at the population and species levels in south-west Australian mediterranean ecosystems. In *Biodiversity of mediterranean ecosystems in Australia*, ed. R.J. Hobbs, 27–46. Chipping Norton, Sydney: Surrey Beatty and Sons.

Hopper, S.D. 1992b. In the footsteps of Giles. *Landscope* 7:28–34.

Hopper, S.D. 1994. Plant taxonomy and genetic resources: foundations for conservation. In *Conservation biology in Australia and Oceania*, eds. C. Moritz and J. Kikkawa, 269–285. Chipping Norton, Sydney: Surrey Beatty and Sons.

Hopper, S.D. 1995. Evolutionary networks: Natural hybridization and its conservation significance. In *Nature conservation 4: the role of networks*, eds. D.A. Saunders, J.L. Craig, and E.M. Mattiske, 51–66. Chipping Norton, NSW: Surrey Beatty and Sons.

Hopper, S.D. 1996. The use of genetic information in establishing reserves for nature conservation. In *Biodiversity in managed landscapes*, eds. R.C. Szaro and D.W. Johnston 253–260. New York: Oxford University Press.

Hopper, S.D. and A.H. Burbidge. 1986. Speciation of bird-pollinated plants in south-western Australia. In *The dynamic partnership: birds and plants in southern Australia*, eds. H.A. Ford and D.C. Paton, 20–23. Adelaide: Govt. Printer

Hopper, S.D. and M.A. Burgman. 1983. Cladistic and phenetic analyses of phylogenetic relationships among populations of *Eucalyptus caesia*. *Australian Journal of Botany* 1:35–49.

Hopper, S.D. and D.J. Coates. 1990. Conservation of genetic resources in Australia's flora and fauna. In *Australian Ecosystems: 200 years of utilization, degradation and restoration*, eds. D.A. Saunders, A.J.M. Hopkins, and R.A. How. *Proceedings of the Ecological Society of Australia* 16: 567–577.

Hopper, S.D., N.A. Campbell, and G.F. Moran. 1982. *Eucalyptus caesia*, a rare mallee of granite rocks from south-western Australia. In *Species at risk. Research in Australia*, ed. R.H.Groves and W.D.L. Ride, 46–61. Canberra: Australian Academy of Science.

Hopper, S.D., S.D. van Leeuwen, S. Brown, and S.J. Patrick. 1990. *Western Australia's endangered flora*. Perth: Department Conservation and Land Management.

Hopper, S.D., G.J. Keighery, and G. Wardell-Johnson. 1992. Flora of the karri forest and other communities in the Warren Botanical Subdistrict of Western Australia. In *Research on the impact of forest management in south-west Western Australia*, ed. Anon, 1–32. Western Australian Department of Conservation and Land Management Occasional Paper 2/92.

Hopper, S.D., J.A. Chappill, M.S. Harvey, and A.S. George, eds. 1996. *Gondwanan heritage: Past, present and future of the Western Australian biota*. Chipping Norton, NSW: Surrey Beatty and Sons. In press.

Hopper, S.D., M.S. Harvey, J.A. Chappill, A.R. Main, and B.Y. Main. 1996. The Western Australian biota as Gondwanan Heritage. In *Gondwanan heritage: past, present and future of the Western Australian biota*, eds. S.D. Hopper, J.A. Chappill, M.S. Harvey, and A.S. George, 1–46. Chipping Norton, NSW: Surrey Beatty and Sons. In press.

Humphries, S.E., R.H. Groves, and D.S. Mitchell. 1991. *Plant invasions. The incidence of environmental weeds in Australia. Kowari 2*. Canberra: Australian National Parks and Wildlfie Service.

James, S.H. 1965. Complex hybridity in *Isotoma petraea* I. The occurrence of interchange heterozygosity, autogamy and a balanced lethal system. *Heredity* 20:341–53.

James, S.H. 1970. Complex hybridity in *Isotoma petraea* II. Components and operation of a possible evolutionary mechanism. *Heredity* 25:53–77.

James, S.H. 1982. The relevance of genetic systems in *Isotoma petraea* to conservation practice. In *Species at risk. Research in Australia*, eds. R.H. Groves and W.D.L. Ride, 63–71. Canberra: Australian Academy of Science.

James, S.H. 1984. The pursuit of hybridity and population divergence in *Isotoma petraea*. In *Plant biosystematics*, ed. W.F. Grant, 169–177. Toronto: Academic Press.

James, S.H. 1992. Inbreeding, self-fertilization, lethal genes and genomic coalescence. *Heredity* 68:449–456.

James, S.H. and S.D. Hopper. 1981. Speciation in the Australian flora. In *The biology of Australian plants*, eds. J.S. Pate and A.J. McComb, 361–381. Perth: University of Western Australia Press.

James, S.H., J.F. Sampson, and J. Playford. 1989. Complex hybridity in *Isotoma petraea* VII. Assembly of the genetic system in the O6 Pigeon Rock population. *Heredity* 64:289–295.

James, S.H., J. Playford, and J.F. Sampson. 1990. Complex hybridity in *Isotoma petraea* VIII. Variation for seed aborting lethal genes in the O6 Pigeon Rock population. *Heredity* 66:173–180.

James, S.H., A.P. Wylie, M.S. Johnson, S.A. Carstairs, and G.A. Simpson. 1983. Complex hybridity in *Isotoma petraea* V. Allozyme variation and the pursuit of hybridity. *Heredity* 51:653–63.

Jessop, J., ed. 1981. *Flora of central Australia*. Syndey: Reed Books.

Jones, W.G., K.D. Hill, and J.M. Allen. 1995. *Wollemia nobilis*, a new living Australian genus and species in the Araucariaceae. *Telopea* 6:173–176.

Keighery, G.J. 1982. Bird-pollinated plants in Western Australia. In *Pollination and evolution*, eds. J.A. Armstrong, J.M. Powell, and A.J. Richards, 77–90. Sydney: Royal Botanic Gardens.

Kelly, A.E. and D.J. Coates. 1995. *Rose mallee recovery plan*. Wildlife Management Program No. 14. Como, W.A.: Department of Conservation and Land Management.

Kelly, A.E., A.C. Napier, and S.D. Hopper. 1995. Survey of rare and poorly known eucalypts of Western Australia. *CALM Science Supplement* 2:1–206.

Kelly, A.E., D.J. Coates, I. Herford, S.D. Hopper, M. O'Donogue, and L. Robson. 1990. *Declared rare flora and other plants in need of special protection in the Northern Forest Region*. Wildlife Management Program No. 5. Perth: Department of Conservation and Land Management.

Kenneally, K.F. and A. Lowrie. 1994. Rediscovery of the presumed extinct triggerplant *Stylidium merrallii* (Stylidiaceae) with an amended description of the species and its conservation status. *The Western Australian Naturalist* 19:269–277.

Kirkpatrick, J.B. 1983. An iterative method for establishing priorities for the selection of Nature Reserves: an example from Tasmania. *Biological Conservation* 25:127–134.

Kirkpatrick, J.B. 1994. *A Continent transformed. Human impact on the natural vegetation of Australia*. Melbourne: Oxford University Press.

Kirkpatrick, J.B. and M.J. Brown. 1991. Planning for species conservation. In *Nature conservation: cost effective biological surveys and data analysis*, eds. C.R. Margules and M.P. Austin, 83–89. Melbourne: CSIRO.

Ladiges, P.Y., C.J. Humphries, and M.I.H. Brooker. 1987. Cladistic and biogeographic analysis of Western Australian species of *Eucalyptus* L'Herit. Informal Subgenus *Monocalyptus* Pryor & Johnson. *Australian Journal of Botany* 35:251–81.

Lamont, B.B. 1982. Mechanisms for enhancing nutrient uptake in plants, with particular reference to mediterranean South Africa and Western Australia. *The Botanical Review* 48:597–689.

Lamont, B.B., S. Downes, and J.E.D. Fox. 1977. Importance-value curves and diversity indices applied to species-rich heathland in Western Australia. *Nature* 265:438–441.

Lamont, B.B., A.J.M. Hopkins, and R.J. Hnatiuk. 1984. The flora—composition, diversity and origins. In *Kwongan plant life of the sandplain*, eds. J.S. Pate and J.S. Beard, 27–50. Nedlands: University of Western Australia Press.

Lamont, B.B., P.G.L. Klinkhamer, and E.T.F. Witkowski. 1984. Population fragmentation may reduce fertility to zero in *Banksia goodii*—a demonstration of the Allee effect. *Oecologia* 94:446–450.

Lindsay, A.M. 1985. Are Australian soils different? *Proceedings of the Ecological Society of Australia* 14:83–97.

Main, A.R. 1996. Conservation. In *Gondwanan heritage: Past, present and future of the Western Australian biota*, eds. S.D. Hopper, J.A. Chapill, M.S. Harvey, and A.S. George. Chipping Norton, NSW: Surrey Beatty and Sons. In press.

Maslin, B.R. 1995. Systematics and phytogeography of *Acacia*: An overview. *Institute of Foresters Australia Newsletter* 36:2–5.

McComb, A.J. and P.S. Lake. 1990. *Australian wetlands*. North Ryde, NSW: Angus and Robertson.

McDougall, K. and J.B. Kirkpatrick. 1994. *Conservation of lowland native grasslands in south-eastern Australia.* Sydney: World Wide Fund for Nature Australia.

McFarlane, D.J., R.J. George, and P. Farrington. 1993. Changes in hydrologic balance. In *Reintegrating fragmented landscapes towards sustainable production and nature conservation*, eds. R.J. Hobbs and D.A. Saunders, 146–186. New York: Springer-Verlag.

McKenzie, N.L., L. Belbin, C.R. Margules, and G.J. Keighery. 1989. Selecting representative reserve systems in remote areas: A case study from the Nullarbor region, Australia. *Biological Conservation* 50:239–261.

McKenzie, N.L., R.B. Johnston, and P.G. Kendrick, eds. 1991 *Kimberley rainforests of Australia.* Chipping Norton, NSW: Surrey Beatty and Sons.

Meney, K.A., G.M. Nielssen, and K.W. Dixon. 1994. Seed bank patterns in Restionaceae and Epacridaceae after wildfire in kwongan in southwestern Australia. *Journal of Vegetation Science* 5:5–12.

Moran, G.F. and S.D. Hopper. 1983. Genetic diversity and insular population structure of the rare granite rock species *Eucalyptus caesia.* Benth. *Australian Journal of Botany* 31:161–72.

Moran, G.F. 1987. Conservation of the genetic resources of rare and widespread eucalypts in remnant vegetation. In *Nature conservation: The role of remnants of native vegetation*, eds. D.A. Saunders, G.W. Arnold, A.A. Burbidge, and A.J.M. Hopkins, 151–162. Sydney: Surrey Beatty and Sons.

Moritz, C. and J. Kikkawa, eds. 1994. *Conservation biology in Australia and Oceania.* Chipping Norton, Sydney: Surrey Beatty and Sons.

Moritz, C., D.J. Coates, W. Sherwin, T. Clancy, and C.J. Limpus. 1994. Population ecology and genetics. In *Conservation biology in Australia and Oceania*, eds. C. Moritz and J. Kikkawa, 359–362. Chipping Norton, Sydney: Surrey Beatty and Sons.

Myers, N. 1992. *The primary source: Tropical forests and our future.* 2nd edition. New York: Norton and Company.

Noble, J.C. and R.A. Bradstock, eds. 1989. *Mediterranean landscapes in Australia: Mallee ecosystems and their management.* CSIRO: East Melbourne.

Noble, J.C., P.J. Joss, and G.K. Jones, eds. 1991. *The mallee lands: A conservation perspective.* East Melbourne: CSIRO.

Orchard, A.E. 1994. *Flora of Australia* Volume 54, *Lichens—Lecanorales 2, Parmeliaceae.* Canberra: Australian Biological Resources Study.

Pate, J.S. and J.S. Beard. 1984. *Kwongan plant life of the sandplain.* Nedlands: University of Western Australia Press.

Pate, J.S. and S.D. Hopper. 1993. Rare and common plants in ecosystems, with special reference to the south-west Australian flora. In *Biodiversity and ecosystem function*, eds. E.D. Schultze and H.A. Mooney, 293–325. New York: Springer Verlag.

Pate, J.S. and A.J. McComb. 1981. *The biology of Australian plants.* Perth: University of Western Australia Press.

Potts, B.M. and J.B. Reid. 1988. Hybridization as a dispersal mechanism. *Evolution* 42:1245–55.

Pressey, R.L., C.J. Humphries, C.R. Margules, R.I. Vane-Wright, and P.H. Williams. 1993. Beyond opportunism: Key principles for systematic reserve selection. *Trends in Ecology and Evolution* 8:124–128.

Pryor, L.D. and L.A.S. Johnson. 1971. *A classification of the eucalypts.* Canberra: Australian National University Press.

Rieseberg, L.H. 1991. Hybridization in rare plants: Insights from case studies in *Cercocarpus* and *Helianthus.* In *Genetics and conservation of rare plants*, eds. D.A. Falk and K.E. Holsinger, 171–181. New York: Oxford University Press.

Rieseberg, L.H. and J.F. Wendel. 1993. Introgression and its consequences in plants. In *Hybrid zones and the evolutionary process*, ed. R.G. Harrison, 70–109. New York: Oxford University Press.

Rossetto, M. 1995. Integrated conservation of the rare and endangered *Grevillea scapigera* A.S. George. Ph.D. dissertation, University of Western Australia, Nedlands.

Rossetto, M., K.W. Dixon, K. Atkins, and D.J. Coates. 1996. *Corrigin grevillea recovery plan.* Wildlife Management Program No. 24. Como, W.A.: Department of Conservation and Land Management. In press.

Sampson, J.F. 1988. The population genetic structure of *Eucalyptus rhodantha* Blakely & Steedman and its allies *Eucalyptus crucis* Maiden and *Eucalyptus lane-poolei* Maiden. Ph.D. dissertation, University of Western Australia, Nedlands.

Sampson, J.F., S.D. Hopper, and S.H. James. 1988. Genetic diversity and the conservation of *Eucalyptus crucis*. Maiden. *Australian Journal of Botany* 36:447–60.

Sampson, J.F. 1989. The mating system and population genetic structure in a bird-pollinated mallee, *Eucalyptus rhodantha*. *Heredity* 63:383–93.

Sampson, J.F., S.D. Hopper, and D.J. Coates. 1990. *Eucalyptus rhodantha*. Wildlife Management Program No. 4. Perth: Department of Conservation and Land Management.

Sampson, J.F., S.D. Hopper, and S.H. James. 1990. Temporal variation in *Eucalyptus rhodantha* pollen pool allele frequencies. *Heredity* 65:189–99.

Sampson, J., D.J. Coates, and S.J. van Leeuwen. 1996. Mating system variation in animal-pollinated rare and endangered plant populations in Western Australia. In *Gondwanan heritage: past, present and future of the Western Australian biota*, eds. S.D. Hopper, J.A. Chapill, M.S. Harvey, and A.S. George. Chipping Norton, NSW: Surrey Beatty and Sons. In press.

Saunders, D.A. and R.J. Hobbs, eds. 1992. *Reintegrating fragmented landscapes*. New York: Springer-Verlag.

Saunders, D.A., J.L. Craig, and E.M. Mattiske, eds. 1995. *Nature Conservation 4. The role of networks*. Chipping Norton, NSW: Surrey Beatty and Sons.

Saunders, D.A., R.J. Hobbs, and P.R. Ehrlich, eds. 1993. *Nature Conservation 3. The reconstruction of fragmented ecosystems*. Chipping Norton, NSW: Surrey Beatty and Sons.

Saunders, D.A., A.J.M. Hopkins, and R.A. How, eds. 1990. *Australian ecosystems. 200 years of utilization, degradation and reconstruction*. Proceedings of the Ecological Society of Australia 16. Chipping Norton, NSW: Surrey Beatty and Sons.

Saunders, D.A., G.W. Arnold, A.A. Burbidge, and A.J.M. Hopkins, eds. 1987. *Nature Conservation: the role of remnants of native vegetation*. Chipping Norton, NSW: Surrey Beatty and Sons.

Specht, R.L., A. Specht, M.B. Whelan, and E.E. Hegarty. 1995. *Conservation atlas of plant communities in Australia*. Lismore: Centre for Coastal Management, Southern Cross University.

Stoutamire, W.P. 1974. Australian terrestrial orchids, thynnid wasps, and pseudocopulation. *American Orchid Society Bulletin* 43:13–18.

Stoutamire, W.P. 1983. Wasp-pollinated species of *Caladenia* (Orchidaceae) in south-western Australia. *Australian Journal of Botany* 31:383–394.

Taylor, A. and S.D. Hopper. 1988 (1991, reprinted with amendments). *The* Banksia *atlas*. Australian Flora and Fauna Series Number 8. Canberra: Australian Government Publishing Service.

Touchall, D.H. and K.W. Dixon. 1993. Cryopreservation of seed of Western Australian native species. *Biodiversity and Conservation* 2:594–602.

Touchell, D.H., K.W. Dixon, and B. Tan. 1992. Cryopreservation of shoot-tips of *Grevillea scapigera* (Proteaceae): A rare and endangered plant from Western Australia. *Australian Journal of Botany* 40:305–310.

Tracey, J.G. 1981. Australia's rain forests: Where are the rare plants and how do we keep them? In *The biological aspects of rare plant conservation*, ed. H. Synge, 165–178. London: John Wiley and Sons.

Wardell-Johnson, G. and D.J. Coates. 1996. Links to the past: Local endemism in four species of forest eucalypts in south-western Australia. In *Gondwanan heritage: Past, present and future of the Western Australian Biota*, eds. S.D. Hopper, J.A. Chapill, M.S. Harvey, and A.S. George. Chipping Norton, NSW: Surrey Beatty and Sons. In press.

Wasson, R.J. 1982. Landform development in Australia. In *Evolution of the flora and fauna of arid Australia*, eds. W.R. Barker and P.J.M. Greenslade, 23–33. Adelaide: Peacock Publications.

Webb, L.J. and J. Kikkawa, eds. 1990. *Australian tropical rainforests. Science—values—meaning*. Melbourne: CSIRO.

Webb, L.J., J.G. Tracey, and W.T. Williams. 1984. A floristic framework of Australian rainforests. *Australian Journal of Ecology* 9:169–198.

Werren, G. and P. Kershaw, eds. 1991a. *The rainforest legacy Australian Rainforests. Study Vol. 2—Flora and fauna of the Rainforests*. Canberra: Australian Government Publishing Service.

Werren, G. and P. Kershaw. 1991b. *The rainforest legacy Australian Rainforests. Study Vol. 3—Rainforest history, dynamics, and management*. Canberra: Australian Government Publishing Service.

Westoby, M. 1988. Comparing Australian ecosystems to those elsewhere. *BioScience* 38:549–556.

Wills, R.T. 1993. The ecological impact of *Phytophthora cinnamomi* in the Stirling Range National Park, Western Australia. *Australian Journal of Ecology* 18:145–159.

Withers, P.C., W.A. Cowling, and R.T. Wills. 1994. Plant diseases in ecosystems. Threats and impacts in south-western Australia. *Journal of the Royal Society of Western Australia* 77:97–186.

Womersley, H.B.S. 1981. Aspects of the distribution and biology of Australian marine macro-algae. In *The biology of Australian plants*, eds. J.S. Pate and A.J. McComb, 294–306. Perth: University of Western Australia Press.

Womersley, H.B.S. 1990. Biogeography of Australasian marine macroalgae. In Biology of *Marine plants*, eds. M.N. Clayton and R.J. King, 367–381. Cheshire: Longman.

Wyatt, R. and A. Stoneburner. 1989. *Pleurophascum occidentale*: A new moss from Western Australia. *The Bryologist* 92:299–301.

HABITAT DEGRADATION AND ECOLOGICAL RESTORATION: HUBRIS, HEGEMONY, AND HEALING

Perhaps it is a given that no habitat on Earth remains uninfluenced by humans. Such impacts run the gamet from the obvious decimation by a forest clearcut to the subtle but insidous effects of long-distance atmospheric inputs from acid rain. The notion of a broad, pristine wilderness is chimaeric. Habitat degradation by humans has existed for millenia, but we are still at an embryonic stage in the understanding of its affects and of their remedy. Jane Hamilton, contemporary American novelist, has perhaps best captured our sense of loss and plaguing bewilderment:

> "I can't . . . visualize the endless prairie, the vast tracks of woodland. I can't hold it in my mind long enough to know absolutely what we've lost. And so the loss is magnified, knowing, as I do, that my powers are poor, and that our world has become diminished beyond all measure" (J. Hamilton. 1994. *A Map of the World.* Anchor Books, Doubleday, New York, p. 154).

Our response to the losses caused by habitat degradation has not always been constructive, in part, because too many conservation biologists and environmental activitists engage in the maudlin (and unproductive) pasttime of romanticizing about times past when smaller human population sizes and simpler technologies wreaked relatively less havoc on their supporting ecosystems. Such emotions are predicated on the assumption that the causes and consequences of habitat degradation were simpler and fundamently easier to remedy than during our more recent prehistory. Hamilton (*loc. cit.*) insightfully suggests that "[t]here has never been a time of simple light."

When we look at degraded habitats there is a tendency to pick out some singular cause of doom and ruin—e.g., prairies disappearing because of the absence of fire, marshes lost because of excessive drainage, and so forth. In reality, habitat degradation always involves a tesselation of causes; consequently successful restoration of habitat will inevitably require a synthesis that addresses the many different threats to ecosystem integrity. The contributors to this section try to escape the narrow thinking of "one problem/one cause" and look at the challenge of restoration from many perspectives.

In the leadoff chapter of section three, Parker and Reichard point out that "biotic resistence" to exotic invasion is composed of at least two parts. The first is the determination of what makes a species a good invader. This task goes beyond the long lists of auteocological attributes compiled by earlier researchers that, 30 years ago, represented tremendous insight. The second part to understanding biotic resistence is the determination of what characteristics make which communities invasible—a task infinitely more intractable, especially because baseline data are generally not available. Parker and Reichard make the point that the intellectual tools we have for understanding invasions, that is to say, the analytical models currently in vogue, have soft spots. However, empirical work is accumulating that suggests it is species-rich, not species-poor communities that are susceptible to invasion—a conclusion curiously at odds with the earlier studies and current management convention. Smith and Fancy's contribution on Hawaiian forest birds lends creedance to this emerging hypothesis.

Invasion of oceanic islands by exotic species and the havoc wreaked by such organisms is indeed old news, as discussed by Maunder and his colleagues, as well as by Smith and Fancy. The latter conservation biologists champion the cause of Hawaii's avifauna, while Maunder and his colleagues report that the reknown island floras of such notable sites as Rapa Nui, Ascension, and Islas Canarias have been irreparably devastated. Interestingly, Maunder *et al.* point out that populations (or more accurately, scattered individuals) exist in degraded habitats as the "living dead," reminding us that because a higher proportion of dioecious species exists on oceanic islands than on mainland floras, a relatively greater proportion of island plant species are susceptible to a process of "living extinction." The notorious demise of the Dusky seaside sparrow (*Ammodramus maritimus nigrescens*) illustrates the vertebrates' nearly parallel "living dead" dilemma, as does the unhappy fate of many of the Hawaiian honeycreepers.

Ironically, just because a species is officially labelled "extinct" we should not consider it a permanent appellation. Maunder and colleagues illustrate in their contribution that extinct species have a disconcerting (and sometimes embarrassing) habit of turning up. Also, many plant species that are extinct in the wild can be found in living collections. These *ex situ* collections may be the last gasp of a species, or at least the "genetic memory" of a formerly more widespread taxon. Conservationists need to be ever so creative with depleted gene pools, particularly ones that have endured either intra- and interspecific (or both) hybridization that often occurs *ex situ*, particularly if the ultimate (long-term) conservation goal is to restore or thwart the loss or degradation of indigenous habitats. The point here is that *ex situ* collections are critically important for the restoration of degraded ecosystems, but the process of collection and management imposes new evolutionary pressures that demand consideration in the restoration process.

Restoration, or in many instances habitat rehabilitation, is by no means a new activity. Maunder, Culham, and Hankamer provide an example from the eighteenth century of habitat restoration with the explicit purpose of modifying, i.e., "ameliorating," the landscape. Eighteenth century motives differed profoundly from those of contemporary conservation biology, but the notion of "improvement," in whatever guise, is central to all efforts, past and present. What is not central is the restoration of ecological *processes*, which is only now being appreciated fully in conservation today. Restoration is a complex endeavor that should be integrated not only with the establishment of the ecological processes that sustain the targeted ecosystem (e.g., sediment deposition, flooding, etc.), but also with the processes of invasion by exotic species that degrade existing ecosytems. This is absolutely necessary because successful restoration of damaged ecosystems cannot be accomplished without consideration of controlling invasive species. Thus the tasks of exotic control and restoration are integrally linked, and no restoration attempt can succeed without a thorough understanding of the mechanisms of invasive species.

Given that restoration efforts will play an increasingly important role in conservation biology, Knapp and Dyer provide a persuasive rationale for worrying about genetics in restoration ecology. These authors argue quite cogently why we have to worry at all about genetics in conservation, specifically ecological restoration. In actuality, it's pretty basic: success and failure of a restoration project will depend in great part on proper matching of adapted germplasm to the restoration site. Potential complications of using a poorly adapted population include a familiar litany, e.g., increased mortality, genetic contamination of extant, local populations, and the loss of interspecific interactions that affect ecosystem processes such as pollination or seed dispersal. In short, the "proper" genetic complement is part of the solution. Given how expensive ecological restoration is, and how profoundly important it is to ensure that restoration truly restores (and not simply substitutes), we need to make every effort at the onset to insure that our efforts will succeed.

Again, why? Because conservation biology cannot be solely concerned with saving what remains. There isn't enough political will, scientific knowledge, money, or viable habitat left to make it so. We need to start investing in ecological restoration, in a very big and creative way. Zedler's final chapter in this section points in such a direction. Combining conservation science with legislative mandates (and adequate funding), Zedler has shown that degraded wetlands, easily the most visible ecosystems in the restoration arena, present some of the greatest challenges. Restoration and recovery of the Salt marsh bird's beak (*Cordylanthus maritimus* ssp. *maritimus*), a rare, annual facultative hemiparasite, in San Diego Bay, illustrates clearly the integration of the scientific contributions in this section. Habitat for this rare species is high marsh and upland ecotone, a seasonally inundated/saturated landscape that has seen degradation and conversion from the earliest settlements along the California coast. Zedler and her students have documented significant negative effects on host taxa through competition from exotic species. But the mechanisms of competition between the exotic and the host species remain unknown in this ecosystem, as would be predicted by Parker and Reichard. Perhaps even more interesting, exotic vertebrates have reduced the population sizes of the small native mammals that create habitat patches for colonization by the Salt marsh bird's beak, illustrating that exotic competitors can have negative effects on a single targeted species at several ecosystem levels.

In summary, the contributions provided herein illustrate individually and as a suite of arguments that conservation biology remains a highly synthetic science of an unprecedented magnitude. This should come as no surprise, however, as the challenges to maintain and restore biotic integrity at any temporal or spatial scale are more profound than ever before.

CHAPTER 11

Critical Issues in Invasion Biology for Conservation Science

INGRID M. PARKER
and SARAH H. REICHARD

Modern conservation biology had its origins in the management of game species and later of rare species, focusing attention on understanding the biological origins and causes of population decline, rarity, and endangerment. But while anthropogenic global change has resulted in the decline of some species, others have thrived and proliferated, accompanied by sometimes dramatic impacts on both single populations and whole ecosystems (Office of Technology Assessment 1993). Although some have long recognized invasive, non-native species as a force capable of irreversibly transforming the natural world (Elton 1958; Baker 1965), these scientists were in large part acting in isolation from those doing traditional conservation biology. In fact, even recent books on conservation biology often include only a cursory treatment, if any, of the problem of nonnative species (e.g., Soulé 1986; Fiedler and Jain 1992; Given 1994; Jordan 1995).

One reason for this exclusion is perhaps that invasion issues have been seen as the province of agriculture, i.e., pest management and weed science. The most high profile invasions are, of course, those species that cause serious economic damage, and 50% to 75% of major crop weeds are imports (Office of Technology Assessment 1993). Until recently species that invaded outside of the agricultural setting were often neglected in discussions on the topic (Parker and Kareiva 1996). As a result, there has been confusion over how much of our theoretical understanding of invasions pertains only to "weeds," that is, crop weeds or ruderal species, and how much is generalizable to all invading species. This is a particularly important point, because from a conservation point of view we are often most concerned about the species that behave least like the traditional weed species. Unlike agricultural pests, the pernicious invaders of natural areas are not necessarily dependent on frequent disturbance, nutrient inputs, and open space. Rather they are characterized by an aggressive ability to expand and displace.

With the development of ideas such as ecosystem management that advocate conserving whole ecosystems instead of single populations (LaRoe 1993), nonnative species are being recognized as a problem inherent, rather than peripheral, to conservation biology. This shift requires us to place next to the science of rarity a new focus on the science of aggression. In this chapter we review a few of the questions commonly addressed in invasion biology; although our selection is not by any means exhaustive, it is meant to provide the reader an introduction to the types of issues important in understanding why invasions occur and what determines their impacts. Our second objective is to provide a sense of the wide variety of approaches that can be used to address these questions. Our final objective is to present the limitations of the state of the field at this point in time, and suggest avenues for important future research.

DOES "BIOTIC RESISTANCE" EXIST?

The first and perhaps most important question in invasion biology is: What allows new species to invade in the first place? The question actually has two parts: (1) What makes a species a good invader?, and (2) What makes a community invasible? Early work on the characteristics promoting invasion led to many different and often divergent lists of traits, with little agreement among them (Baker 1965; Forcella and Wood 1984; Dritschilo *et al.* 1985; Lawton and Brown 1986; Moulton and Pimm 1986; Pimentel 1986; Bergelson and Crawley 1989; Perrins, Williamson, and Fitter 1992). This led to the discouraging view held by many ecologists that invasions were highly idiosyncratic, and that statistical summaries would never provide us with general rules that could be used for making decisions about future introductions (Crawley 1989; Kareiva *et al.* 1991). However, sophisticated recent applications of statistical models to improved databases of both failed and successful introductions have generated practical risk assessment protocols that show surprisingly high rates of success in prediction (Reichard 1994; Rejmánek and Richardson 1996; Reichard and Hamilton in press).

The second part of the question, "What makes communities invasible?," has been a focus for both theory and experimentation. The two properties most commonly thought to influence invasibility are species richness and disturbance, which are sometimes lumped together in the term "biotic resistance" (Simberloff 1986). Disturbed, species-poor communities are thought to be more invasible because they offer less biotic resistance. Island or species-poor communities may be less likely to have enemies capable of controlling a potential invader, or may have fewer competitors capable of excluding invaders through the preemption of limiting resources (Elton 1958). Mathematical models based on the structure of Lotka-Volterra interactions have been used to introduce a new species to a stable model assemblage (Drake 1983, 1988; Post and Pimm 1983; Drake 1988; Case 1990, 1991). The most general result is that the probability of successful integration into the community is negatively related to the number of resident species (Case 1990). In contrast, the level of "impact" of the invader is positively related to the number of residents, meaning that when a larger community is invaded, the probability of extinction is higher for resident species. These analytical models provide support for the biotic resistance hypothesis. However, the weakness of the modeling approach is that these "communities" seldom exceed 20 species, while natural communities are very rarely that small. If the relationship between invasibility and species number is an asymptotic one, then species number over the range seen in nature could be irrelevant to invasibility.

Statistical models also have been used to test the theory of biotic resistance. Moulton and Pimm (1983, 1986) found a declining rate of invasion success for birds in Hawaii with increasing species richness, where the increase was provided primarily by previous invaders. However, Simberloff and Boecklen (1991) criticized this result by pointing to two alternative explanations: habitat destruction over time may have resulted in fewer possibilities for invaders, or fewer invasions may have succeeded as one by one the best invaders were already incorporated into the resident community. Simberloff (1989) tested the hypothesis that insect biological control agents should have higher establishment rates on species-poor islands than mainland sites, but the evidence was weak. However, studies by Reichard (1996) indicate that woody plant species of continental origin are significantly more likely to invade Hawaii than are those originating from other islands. In contrast, invasion success in North America is independent of origin.

The biotic resistance hypothesis is very amenable to empirical study, and recent work has begun to shed new light on the issue. A study of riparian zones in France, Oregon, and Washington demonstrated a positive, not negative, relationship between total species richness and the percentage of nonnative plant species at several different scales (Planty-Tabacchi *et al.* 1996). Similarly, forest patches in New Zealand were more likely to contain the non-native herb *Hieracium lepidulum* (hawkweed) when they had higher species richness and greater numbers of species within the same guild (Wiser *et al.* 1996). In an experimental manipulation of host community diversity of intertidal macroalgae, reintroduction was slowest in low diversity treatments (Allison 1996). Using the native herb *Eschscholzia californica* (California poppy) as an "invader," Robinson, Quinn, and Stanton (1995) introduced individuals into plots of California grassland and obtained higher establishment rates in the plots with greatest species richness. Although these two manipulative experiments did not use nonnative invaders, they represent the type of study that we need to address the biotic resistance issue. Interestingly, these studies all suggest that for plants and sessile organisms, species diversity may in fact be associated with higher invasibility.

In conclusion, empirical tests of the biotic resistence hypothesis have so far provided mixed results, with species-rich communities only sometimes proving to be less invasible. Many more studies are needed from a wide variety of systems before we will be able to formulate a synthetic view of the relationship between species number and invasibility. Why might greater invasibility accompany high species richness in some cases? One possibility is that some species-poor communities may be made up of a few, very competitive species, which are often highly successful invaders themselves. Another possible explanation is suggested by the intermediate disturbance hypothesis (Connell 1978), i.e., a high level of local species turnover could be responsible for high resident species diversity and could also provide opportunities for the establishment (and sometimes coexistence) of an introduced invader.

The connection between ecological disturbance and invasion is both important and complex. Many have noted a relationship between nonnative species and disturbed habitats (Fox and Fox 1986; Orians 1986; Hobbs 1989; Hobbs and Huenneke 1992), and several studies have demonstrated an empirical correlation between the abundance of particular invaders and levels of disturbance (Cowrie and Werner 1993; Parker, Mertens, and Schemske 1993; Knops, Griffin, and Royalty 1995). In some cases this relationship may be explained by a reduction in percent cover or levels of competition (Crawley 1986), or may reflect an increase in the availability of specific resources. Although a sudden increase in nutrient input can produce a similar result without any physical disruption of the soil (Huenneke *et al.* 1990), several studies have documented the rapid spread of nonnatives over open ground (Hobbs and Atkins 1991; De Pietri 1992; D'Antonio 1993). Once the space is occupied there is potential for an invader to exclude native species, a topic to which we now turn our attention.

IMPACT OF NONNATIVE SPECIES ON NATIVE SPECIES AND ECOSYSTEMS

It is likely that most nonnative species have little to no effect on species in the invaded community or on the structure and function of that community (Simberloff 1981). Williamson and Brown (1986) estimate that only 10% of invaders become established and of those only 10% become pests affecting native species and ecosystems. While their estimates seem a bit low in comparison to other studies (e.g., Reichard 1994), they demonstrate that relatively few species actually appear to be of concern to conservation efforts.

Some nonnative species can, however, interact with native species in a number of ways detrimental to natives. Nonnative species may compete with native species for limited resources, alter ecosystem processes, and increase disturbances such as fire frequency and intensity. They may also hybridize with closely related native species (Thompson 1991; Abbott 1992). Non-native plant species may also affect the use of native communities by wildlife (Neill 1983; Olson and Knopf 1986).

Competition

The type of interaction between nonnative and native plants that is usually cited is competition. Competition is the direct interference of one organism with another and is most commonly exploitative, with the effect of the interaction occurring through the reduction of the available pool of resources. These resources can include light, nutrients, water, and biotic agents such as pollinators and seed dispersers. The simplest result of such competition is the replacement of one species by another. Non-native species are generally assumed to be more competitive than the native species with which they interact.

Papers on the competition of plants with one another are common (e.g., Fowler 1986; Gaudet and Keddy 1995; Wilson and Tilman 1995) but despite the potential importance of competition as a conservation concern, few studies actually document competition between native and non-native invasive plants, especially in an analytical or quantitative sense. Furthermore, few studies have explicitly documented which resources are limited and are thus the object of competition. Such information is critical if we are to understand the impact of non-natives, manage the resources that may be limited, and convince those that develop policy that there is a legitimate concern.

To review the current level of research on the competition of non-native invasive species on native species and ecosystems, we examined the last five to six volumes (representing the years 1990–1995) of 17 international, national, and regional journals (Table 11.1). While this sample is not exhaustive in depth or breadth, it represents a substantial portion of the ecological literature over the last six years, including applied journals as well as basic research journals, and including publications from Europe, Great Britain, Australia, and New Zealand, in addition to North America. Every publication within the volumes that focused on an aspect of native/non-native plant competition was examined to determine if the limiting resource was identified (e.g., light, water, nutrients, pollinators, or seed dispersers), the type of analyses used (e.g., spatial or temporal observance of changes in species density or abundance; transects or plots to document relative density or abundance; addition, removal, or replacement experiments; greenhouse studies such as relative nutrient use or resource manipulations; and site comparisons for resource attributes), and outcome of competition (Table 11.2). Studies could include more than one resource or method.

Many of the papers discussed factors relating to establishment of the non-native species. The factor most frequently cited as facilitating invasion into an area was disturbance (e.g., Beatty and Licari 1992; McClaran and Anable 1992; Parker, Mertens, and Schemske 1993; Knops, Griffin, and Royalty 1995). As previously discussed, disturbance decreases vegetation cover and increases resource availability. Other factors shown to provide a competitive advantage over native species in the establishment phase include

Table 11.1
Journals used for competition study review.

Journal	Years
American Midland Naturalist	1990–1994
Australian Journal of Ecology	1990–1994
Biological Conservation	1990–1995
Bulletin of the Torrey Botanical Club	1990–1995
Conservation Biology	1990–1995
Ecological Applications	1991–1995
Ecological Monographs	1990–1995
Ecology	1990–1995
Journal of Applied Ecology	1990–1995
Journal of Ecology	1990–1995
Journal of Vegetation Science	1990–1995
Madroño	1990–1995
New Zealand Journal of Botany	1990–1995
Northwest Science	1990–1994
Oecologia	1990–1995
Rhodora	1990–1994
Weed Technology	1990–1995

decreased salinity levels in salt marshes, also interpretable as a form of disturbance (e.g., Zedler, Gautier, and McMaster 1983; Mesléard *et al.* 1993), and seed biology and other reproductive factors (Shafroth, Auble, and Scott 1995). Thus, competition between species was not necessarily the objective of the study nor the cause of the non-native species success in an area. Instead, a more rapid or efficient establishment of the seedling stage resulted in later dominance of an area.

While most of the articles in Table 11.2 documented the results of competitive interaction between native and non-native species, only four specified the resource that was limited to the native. Light was intercepted by nonnatives in two of the studies (Woods 1993; Shafroth, Auble, and Scott 1995) and ground water was also found to be restricted by the presence of non-natives (Melgoza, Nowak, and Tausch 1990; Shafroth, Auble, and Scott 1995). Sallabanks (1993) discovered that competition for seed dispersers occured in two closely related and now sympatric species of *Crataegus* (hawthorn). American robins, the primary seed dispersers of hawthorn species in the Pacific Northwest, were preferentially attracted to the larger, pulpier, and more numerous fruits of a non-native species than the native. In addition, two studies revealed that a competitive interaction between native and non-native species existed on the root level but did not specify the exact nature of the interaction (Dillenburg *et al.* 1993; Jones 1993). More investigations into limiting resources could increase our understanding of the apparent greater competitive ability of nonnative species, and might also provide clues for managing resources to favor native species.

Competitive interaction can be determined in a number of ways and the papers addressing competition of native and nonnative species varied widely in their approaches. The simplest method is to infer superior competitive ability of nonnatives by finding a concentration of non-native species in areas lacking natives (e.g., Allen 1991; Parker, Mertens, and Schemske 1993; Knops, Griffin, and Royalty 1995), often with changes in relative species abundance over time (e.g., McClaran and Anable 1992; Webb and Kaunzinger 1993; Burdon and Chilvers 1994). The inference is that the nonnatives must be more competitive in the area than natives because they have increased in number over

Table 11.2
Summary of articles documenting competition between native and nonnative plant species.

Citation	Methods Used	Summary of Results
Melgoza, Nowak, and Tausch 1990	Removal experiments Site resource comparisons	Nonnative grass competes with native species for soil water and negatively affects their productivity.
Zedler, Paling, and McComb 1990	Greenhouse salinity and addition experiments	*Typha orientalis* outcompetes a native marsh species when salinity levels were lower, as near street drains. It is less competitive in more undisturbed higher salinity areas.
Allen 1991	Transects or plots	In New Zealand native species diversity declined slightly but not significantly as *Berberis darwinii* increased basal area. No resources were specified.
Burgess, Bowers, and Turner 1991	Aerial surveys and transects	The Desert Laboratory grounds in Arizona were surveyed and 52 nonnative species found. Apparent competitive interactions were discussed but no actual studies done.
Beatty and Licari 1992	Transects or plots Spatial variation	Fennel *(Foeniculum vulgare)* was less competitive in chaparral vegetation than in grassland. Disturbance appeared to assist fennel invasion into grassland but not in chaparral. No resources specified.
McClaran and Anable 1992	Transects or plots Spatial/temporal variation	Native grasses do not tolerate grazing; nonnative grass increases in abundance as grazing intensity increases; no resources specified.
De Pietri 1992	Transects or plots Site resource comparisons	Nonnative rose established more quickly on bare areas than native but provided native seedlings with protection from grazing and more friable soil beneath; no resources specified.
Hester and Hobbs 1992	Transects or plots Site resource comparisons Removal experiments	Native species increase when nonnatives are removed from plots; no resources specified.
Dillenburg, Whigham, Teramura, and Forseth 1993	Addition experiments	An nonnative vine *(Lonicera japonica)* had a more negative effect on growth of a native tree than did a native vine; the competitive interaction was on the root level but resources were not specified.
Jones 1993	Addition experiments	*Sapium sebiferum* did not show competitive advantage over two native species at room temperature but when pots were heated it increased in interspecific competitive ability perhaps due to root activities.
Mesleard, Ham, Boy, von Wijck, and Grillas 1993	Replacement experiments Greenhouse salinity experiments	Nonnative grass increased in abundance at low salinity but was less competitive than the native grass when salinity levels were higher; it is expected that the two species will thus not compete in salt marshes which are spatially varied in salinity.

(continued)

Table 11.2 *(continued)*

Citation	Methods Used	Summary of Results
Parker, Mertens and Schemske 1993	Transects or plots Spatial variation	Densities of two nonnatives species were positively correlated with each other and negatively correlated with native species. The two nonnatives were more common in disturbed areas. No resources were specified.
Sallabanks 1993	Comparative attractions of frugivores	*Crataegus monogyna* had more and larger fruit than a native *Crataegus,* traits apparently attractive to robins. Fruit of the native species appears to be selected less frequently.
Webb and Kaunzinger 1993	Temporal variation	*Acer platanoides* has increased in number of individuals in an eastern hardwood forest. No resources were specified.
Woods 1993	Transects or plots Spatial changes	At 30% cover *Lonicera tartarica* depressed the diversity and abundance of native species in mesic but not dry sites. Light is the limiting resource.
Burdon and Chilvers 1994	Transects or plots Temporal variation	As individuals of *Pinus radiata* increase in size and abundance in a native eucalypt forest, eucalypt juveniles decrease in number; no resources specified.
Huenneke and Thomson 1994	Transects or plots Temporal variation Site resource comparisons Replacement experiments	Sites supporting nonnative teasel were similar to those of an endangered thistle in measure characteristics. Greenhouse density experiments showed a negative effect of teasel on the thistle but field density studies were inconclusive; resources not specified.
Robinson, Yurlina, and Handel 1994	Temporal variation	Over a 112-year period nonnative species have increased from 19% of the flora to 33% while native species declined in abundance. Native species rare in early surveys have a greater likelihood of being absent now. No resources were specified.
Bruce, Cameron, and Harcombe 1995	Temporal variation	In comparisons of 19 stands of known age *Sapium sebiferum* was most common in younger (<15 years) stands. Native species appear to become more established as the stands aged; no resources were specified.
Knops, Griffen, and Royalty 1995	Transects or plots Spatial variation	Nonnative species were positively associated with disturbed areas in nature reserve and natives were positively associated with forests and woodlands; no resources specified.
Shafroth, Auld, and Scott 1995	Experimental resource manipulations	Nonnative Russian-olive succeeds under similar light and water conditions optimal for native cottonwood and reproductive strategies favor early site establishment of Russian-olive.

time while the natives decreased. These studies may or may not include some additional analyses of the physical characteristics of the sites and usually find patterns in the distributions of the species that suggest competitive interactions. Such studies of temporal or spatial variation of relative abundance suggest competition, but they should include a discussion of other factors that may also result in the same distribution patterns. For instance, diseases affecting native species or the loss of a seed dispersal agent may cause changes in abundance of the natives over time, and site factors such as disturbance or changes in soil chemistry within an area can affect species distribution over space. Some of the studies (e.g., De Pietri 1992; Hester and Hobbs 1992; Huenneke and Thomson 1994) did do further site examinations to determine what additional factors could be contributing to the patterns discovered.

A number of the papers used other, more experimental, methods to determine competitive interactions, including addition experiments in which one species is added to another to detect interactions (e.g., Zedler, Gautier, and McMaster 1983; Dillenburg, *et al.* 1993; Jones 1993). Others used a similar type of experiment that involves removal of a species from a plot or pot to determine changes in the remaining species (Melgoza, Nowak, and Tausch 1990; Hester and Hobbs 1992). Finally, the most complicated studies used replacement series experiments, which vary the ratio of two different species while maintaining a constant total density of plants, and can be used to assess the abundance at which a negative interaction occurs (Mesléard *et al.* 1993; Huenneke and Thomson 1994).

Studies that revealed the strongest evidence of competition between native and nonnative species were generally those that combined more than one method of analysis. For instance, Huenneke and Thomson (1994) used spatial and temporal patterns of variation in distribution to suggest that the nonnative teasel (*Dipsacus sylvestris*) was affecting an endangered thistle (*Cirsium vinaceum*). They then examined the sites in which the species occurred for physical differences that might contribute to the patterns, as well as performed both field and greenhouse density experiments. From these tests they determined that teasel had the same habitat requirements as the thistle but germinated under lower light, indicating that teasel would be able to germinate under thistle but not *vice versa*. The greenhouse experiment suggested that the thistle was reduced by teasel but that the invader was unaffected, while the field experiments were equivocal but suggested interference. Such studies provide a clear indication of the importance and nature of the competitive interaction.

While the studies varied in methods, all but two (Zedler, Paling, and McComb 1990 and Mesléard *et al.* 1993) found a negative response of native species to the presence of nonnatives, with no effect of natives on nonnatives. This underscores the need to better understand and document the actual types and magnitudes of competitive interactions. The accumulation of solid evidence of competitive abilities in certain types of species will allow inferences to be made about the impacts of unstudied nonnative species with similar traits, helping to set management priorities for control. There should also be greater effort to clarify the resources that are limited by the presence of the non-native species; too often assumptions are made. Documentation of competition for biotic resources would also be helpful to conservation efforts. Sallabanks (1993) provided support for the idea that frugivores may be dispersing the seeds of nonnative species more than those of natives. It has also been suggested that non-native species may compete with natives for pollinators (Breytenbach 1986), but our literature study revealed no evidence for such an interaction, nor are we aware of any other articles which do so.

Our literature survey revealed that currently untapped but valuable information on the nature of competition between native and nonnative species could be gained even without performing new studies. We uncovered a number of state-of-the-art competition studies that involved both native and nonnative species but were not designed for the purpose of evaluating the impact of invasions specifically (DiTommaso and Aarssen 1991; Gaudet and Keddy 1995; Gerry and Wilson 1995; Wilson and Tilman 1995). Because results generally were not separated on the basis of native vs. nonnative in these studies, we could not extract generalizations from them. It would be simple for those already doing plant competition experiments to examine the comparative competitive ability of natives and nonnatives, strengthening our knowledge about such competitive interactions. In addition, these studies would be more likely to provide information on negative results, or situations in which native species outcompete nonnatives.

Hybridization

When previously isolated related species become sympatric hybridization can occur; plants often have few barriers to interspecific reproduction. Hybridization between a native and an introduced species may have evolutionary consequences and may lead to weakening of genetic stability within populations of the native species (Abbott 1992). In some cases initial sterility of hybrids between native and nonnative species may be overcome by later chromosome doubling, as was the case for *Spartina anglica* (Thompson 1991). We do not know how prevalent this negative impact is; however, while most interspecific hybrids found in the British Isles are highly or completely sterile, some degree of fertility was found for 49% of the hybrids between a native and nonnative species and 62% of hybrids between two non-native species (Stace 1991).

Ecosystem Processes

The normal functioning of ecosystems and the interactions of organisms within can be altered by the inclusion of some nonnative species. Changes in processes such as nutrient cycling, soil formation, hydrologic cycles, and geomorphic processes can irrevocably change the character of the ecosystem, which in turn may facilitate the entry of additional non-native species. Below is a brief summary of some of the ecosystem processes that may be affected.

Nitrogen Fixation. A nonnative species with the capacity to "fix" atmospheric nitrogen is clearly at a competitive advantage if nitrogen is limited within the community and can further alter the community by releasing nitrogen into the immediately surrounding soil. The association between many species of the legume family (Fabaceae) and species of the bacterium *Rhizobium*, for example, can have a significant effect on soil fertility. In addition, a number of other vascular plant families with invasive members (e.g., Casurinaceae, Betulaceae, Myricaceae, Myrsinaceae, Elaeagnaceae) have associations with species of the bacteria genus *Frankia*, also with the same effect. A classic study of *Myrica faya* (Myricaceae), an nonnative invasive plant in Hawaii, found that it altered succession on lava flows by increasing nitrogen levels in this nitrogen-limited ecosystem (Vitousek *et al.* 1987). Nonnative *Acacia* (Fabaceae) species, especially *A. saligna* and *A. cyclops*, also have impacted native species where they invade, especially in the unique South African fynbos ecosystem. These *Acacia* species increase the size of the nutrient pools as well as alter the spatial distribution of nutrients within the invaded ecosystem (Stock, Wienand, and Baker 1995). Not only do they fix nitrogen through microbial associations, but they also produce more leaf litter and have foliar nitrogen levels three times that of native species, further enhancing the nitrogen status of the ecosystem (Witkowski 1991).

Hydrologic Cycles. A few nonnative species can alter the hydrologic cycle of ecosystems. In the American southwest *Tamarix ramossissima* (saltcedar) has demonstrated the ability to cause dramatic changes in invaded ecosystems. The species requires moisture for seed germination (Horton 1977) so the seedlings generally establish in wetlands and riparian areas within the desert matrix where water-loving native species occur. Once established the primary root grows downward 30 feet or more until it reaches the water table and begins to form secondary roots. Saltcedar has the highest evapotranspiration rate of any species studied in the region, with annual water losses that may total as much as $2.1 m^3/m^2$ (Carman and Brotherson 1982). These high rates of evapotranspiration can result in a lowering of the water table, making water unavailable to native species and completely altering the structure of wetland communities. In one dramatic case, the removal of a substantial saltcedar invasion at Eagle Borax Spring in Death Valley National Monument that had resulted in the disappearance of the marsh led to its return and renewed use by migratory birds (Neill 1983). In addition, along the rivers it invades, saltcedar can also increase sedimentation rates by creating "ponds" and thus reducing current velocity (Blackburn, Knight, and Schuster 1982).

Mycorrhizal fungi. Many plants are significantly better competitors for soil resources when mycorrhizal then when not mycorrhizal; the presence of vesicular-arbuscular mycorrhizal (VAM) fungi can alter the outcome of competitive interactions among plants. In the Intermountain West native bunchgrasses are mycorrhizal but introduced non-native plants are both mycorrhizal and non-mycorrhizal (Goodwin 1992). When a site is invaded by non-mycorrhizal species, VAM fungi populations decline, leaving native bunchgrasses at a competitive disadvantage (Goodwin 1992).

Disturbance Regimes

One impact of nonnative plants with important conservation implications is a change in disturbance frequency or intensity in native ecosystems. Perhaps the most serious disturbance of this kind is fire. Some nonnative species have evolved in fire-affected areas and have morphological characteristics that exacerbate fires as well as adaptations for recovering quickly, while the native species lack these adaptations (Bock and Bock 1992). In Hawaii, for example, nonnative species (especially grasses) that are adapted to fire have resulted in wide-scale burning in ecosystems that have never known fire as a natural disturbance (D'Antonio and Vitousek 1992). The effect has been devastating to the already threatened Hawaiian flora. Native species are intolerant of fire, leaving diverse woodlands to be replaced by nonnative grasslands. Some native species, including the candidate endangered species, *Pittosporum terminalioides*, may be driven to extinction by the introduction of fire in Hawaiian ecosystems (Hughes, Vitousek, and Tunison 1991). In California chaparral ecosystems where fire is already part of the natural disturbance regime, nonnative grasses have increased the frequency of fires (Zedler, Gautier, and McMaster 1983). Native species are eventually unable to recover because seed stocks are depleted and the non-native species increase in dominance. The result is a shift in community structure with a greater abundance of annual rather than perennial grasses (Jackson 1985).

Use by Wildlife

Nonnative species may alter community structure in ways incompatible with use by wildlife for shelter or food. Along many western river systems tall native cottonwood trees (*Populus sargentii*) have been replaced with Russian-olive (*Elaeganus angustifolia*). The replacement impacts native cavity-nesting birds that now have fewer potential nest sites (Olson and Knopf 1986). And, on the southern coasts of Florida, between one-third and one-half of all undeveloped beaches are lined with ironwood (*Casuarina equisetifolia*), changing the structure from beach and mangrove thickets to forests unsuitable for nests of threatened sea turtles and American crocodiles (Johnson 1994).

FACTORS AFFECTING RANGE EXPANSION

Theory

Once a species has been introduced to a new area it must expand from its initial focus. This physical spread is determined by two processes: (1) an increase in population numbers, described by demographic rates, and (2) a simultaneous increase in the area occupied by the population, driven by the dispersal of propagules. Traditionally invasions have been modeled with "reaction-diffusion" equations, which couple a growth term with dispersal represented by simple diffusion (Fisher 1937; Skellam 1951):

$$\delta N/\delta t = rN\ (1 - N/K) + D\ \delta 2N/\delta x2, \quad (1)$$

where N is the population size, K is the carrying capacity of the environment, D is a constant dispersal rate, and r is the intrinsic rate of increase. One basic result from Fisher-type models is that a constant speed of invasion (a "travelling wave") is attained, with that speed (c) given by the expression

$$c = 2\ \sqrt{(rD)} \quad (2)$$

The form of these equations has increased in sophistication and in recent years, range-expansion models have incorporated more realistic conditions and biological details, such as directed dispersal (Holmes 1993), density-dependent dispersal (Pablo and Vasquez 1991), Allee effects (Lewis and Kareiva 1993), and discrete dispersal phases (Kot and Schaffer 1986; Hardin, Takac, and Webb 1990; Kot 1992).

While theory has advanced rapidly, so have studies analyzing empirical records of spread. Most simply, these studies use observed range expansion rates to fit models, for example, the European starling (*Sturnus vulgaris*), English sparrow (*Passer domesticus*), House finch (*Carpodacus mexicanus*; Okubo 1988), the collared turtle dove (*Streptopelia decaocto*; Hengeveld 1989), the ungulate Himalayan thar (*Hemitragus jemlahicus*; Caughley 1970), California sea otter (*Enhydra lutris*; Lubina and Levin 1988), cholla (*Opuntia imbricata*; Allen *et al.* 1991), and *Mimosa pigra* (Lonsdale 1993). The calculated rate of spread can be used in some cases to gain insights into processes that may not be amenable to direct empirical study. For example, Caughley (1970) evaluated whether the spread of Himalayan thar in New Zealand was better explained by random dispersal or population pressure, concluding that thar disperse at an innately determined rather than disperse away from centers of high density. Lubina and Levin (1988) dismissed water currents as a factor controlling variation in the spread of the sea otter, suggesting instead the importance of either movement behavior or mortality rates. Lonsdale (1993) quantified the role of dispersal by water in the invasion of a floodplain weed.

Others have used direct data on demography and dispersal to make independent estimates of predicted rates of spread, and then compared these estimates to observed rates. Such a comparison has been made for organisms as diverse as gray squirrels (*Sciurus carolinensis*; Okubo *et al.* 1989), cereal leaf beetles (*Oulema melanopus*; Andow *et al.* 1990, 1993), and Neolithic farmers (Ammerman and Cavalli-Sforza 1984). Results in the majority of cases correspond surprisingly well, when one considers the level of variation that might be expected in demographic and dispersal parameters and the simplicity of the models. In one case, Andow *et al.* (1990, 1993) found that the predicted rate of spread for the cereal leaf beetle was much slower than the observed rate of spread. From this incongruency they concluded that long-distance dispersal events, not included in the model, must be important in explaining the dynamics of this system.

One common focus of recent theory using models of range expansion is understanding how the details of dispersal contribute to rates of spread. Because the tails of the distributions determine what happens at the very edge of an invading front, the shape of the dispersal curve, not simply the mean dispersal distance, can be very important to the asymptotic wave speed (Kot, Lewis, and van den Driessche in press). Another way to formulate the problem is to evaluate the contribution of occasional long-distance propagules, which have also been found to be important in analytical models (Goldwasser, Cook, and Silverman 1994; Lewis 1997). Earlier simulation models addressed an analogous issue, i.e., the effect on spread of additional "foci" of introduction (Auld and Coote 1980; Moody and Mack 1988). Simulating the increase of diameter in simple circles of infestation, Moody and Mack (1988) concluded that control efforts would be most effective when directed at isolated peripheral patches under most situations. Auld and Coote (1980) used cellular automata models to investigate the spread of a weed from farm to farm, finding faster spread when multiple farms were infected simultaneously.

A SIMULATION OF SPREAD IN *CYTISUS SCOPARIUS*

Although the results of analytical and simulation models point to factors that might be important in the spread of invaders, we have a very limited idea of the relative importance of different factors in determining the magnitude and variation in rates of spread of actual invading populations. Here we present a model designed to evaluate the importance of details of dispersal within the context of an actual invading population. The model is an extension of that presented in Parker (1996), and is based on demographic rates collected from over 3,000 individuals of Scotch broom (*Cytisus scoparius*), an invasive shrub on the west coast of North America. Demographic data were taken from three populations in each of two distinct habitat types: urban fields and glacial outwash prairies, a rare ecosystem in Washington state. *Cytisus scoparius* has very different stage-specific growth, mortality, and fecundity rates in these two habitats (Parker 1996). Stage structure was included by dividing the life cycle into four stages: seeds, seedlings, juveniles, and adult plants.

The model has the structure of a cellular automata, with invasions occurring into a square field or grid of 10,000 $1 m^2$ "bins." At each time step, the production of new adults, juveniles, seedlings and seeds is calculated from field-collected demographic rates for each bin. Seeds are then redistributed among neighboring bins with a dispersal function (see below). The lattice has absorbing boundaries—that is, it is modeled as a field or prairie surrounded by uninhabitable land, such as forest, agricultural field, or asphalt. Density-dependence is included simply as follows: bins are "saturated" once they reached a set number of adult plants, at which point densities are held constant. To summarize spread, the model tabulates the increase in number of bins occupied per year (i.e., area covered by plants). The square root of that increase in bin occupancy is then used to measure "invasion rate." The square root transformation allows the simulation model to be compared directly with analytical theory, which emphasizes the linear, or radial, spread of a range. Invasion rate is calculated only over the time interval that produces a linear rate of expansion. Early in the invasion there is a lag time before bins fill, and late in the invasion spread slows down as the edges of the habitable area are reached.

This model was used to investigate the relative importance of dispersal and demography under three scenarios: (1) spread from a single founder, contrasting three empirically-derived dispersal distributions, (2) spread from a single founder, systematically varying the proportion of long-distance propagules, and (3) spread from varying spatial arrangements of founders. In each case the results of the model were analyzed with ANOVA to reveal significant effects.

1. The consequences of dispersal mechanism for rate of spread.
Cytisus scoparius has two dispersal mechanisms, ballistic dispersal and secondary ant dispersal (Bossard 1990). Experiments were used to tease apart the contribution of each of these mechanisms to the overall dispersal distribution (Parker 1996). The simulations were then run using one of three distributions: ballistic only (humped distribution with short tail), ants only (highly leptokurtic distribution with longer tail), and ballistic + ants (humped distribution with long tail).

Before discussing the results, we mention that the conclusions drawn from the model depend on the threshold used for identifying an occupied bin (Figure 11.1). Densities were fractions rather than integers, and if one accepts as a threshold anything greater than zero seedlings, then infinitesimally small densities produce an "occupied bin" and spread is entirely driven by the dispersal distribution. When this artificial measure is used, ballistic dispersal results in the slowest spread, ant dispersal is intermediate, and ballistic + ant dispersal results in the most rapid spread (Figure 11.1). What is interesting about this result is that it does not follow from the relative means of the three distributions, which increase from ant, then to ballistic, finally to ballistic + ant dispersal (Parker 1996). Rather, the maximum distance of the three distributions is the important factor, illustrating the critical nature of the tails of dispersal distributions in this case as in other invasion studies.

However, when a more relevant definition of occupancy was used, i.e., full saturation of bins, the importance of variation in demography became apparent. Rate of spread was much higher for prairie populations than for urban populations (Figure 11.2; ANOVA, $F = 96.0$, $P < 0.0001$), and the effect of dispersal distribution was small by comparison ($F = 0.11$, $P = 0.90$). Interestingly, the ranking of the three dispersal types changed between prairie and urban habitats (Figure 11.2). Ballistic dispersal consistently resulted in the fastest spread in urban populations, while ballistic + ant dispersal resulted in the fastest spread in prairie populations. This suggests a potentially complex interaction between dispersal distribution and demography. However, the main result from this use of the model is that dispersal by ants, which extends the tail of the distribution, has little effect on the rate of spread relative to the large differences in demographic parameters among habitats. More generally, three contrasting dispersal distributions, quite different in both mean and shape, result in similar predictions for spread at the scale of a local field or prairie.

2. The effect of varying the proportion of long-distance propagules.
To move away from the specifics of a single experiment, we asked more generally how varying the proportion of seeds traveling relatively long distances might affect rates of

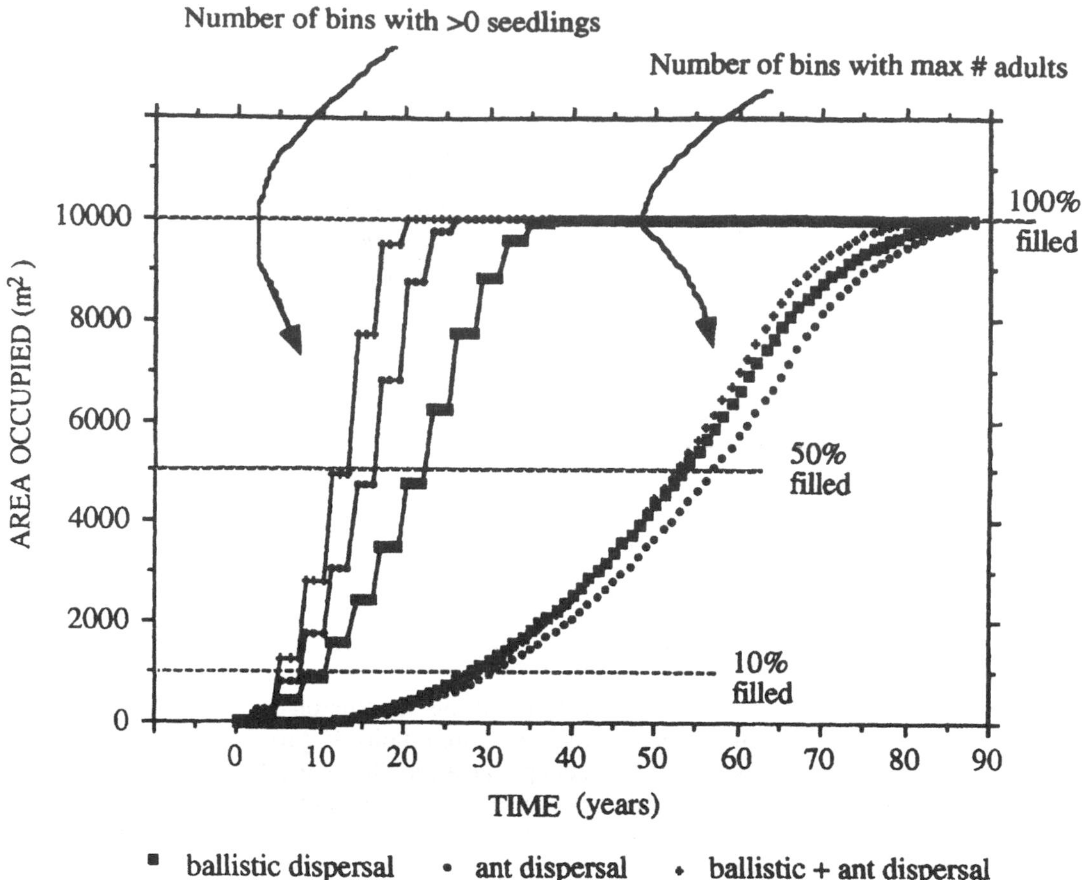

Figure 11.1
An example of the output of one run of a simulation model of the invasion of Cytisus scoparius *for each of three dispersal types estimated from the field (Parker 1996). Ballistic dispersal is represented by filled squares, ant dispersal by open squares, and ballistic + ant dispersal by crosses. The results shown are for one set of demographic parameters taken from a prairie population. The number of bins occupied is presented for two criteria: bins with greater than zero seedlings in them (including fractions of individuals), and bins with the maximum allowable density in them (five adults).*

spread in the *C. scoparius* system. Instead of the three empirical dispersal distributions, dispersal was modelled in two parts: first, seeds were redistributed according to the ballistic distribution; then a proportion of them (0, 0.00001, 0.0001, 0.001, 0.01, 0.02, or 0.05) were "dispersed" to a randomly chosen bin between 5 and 10 m away from the source. This random long-distance dispersal was similar to the action of ants moving groups of seeds to a nest site or seed cache, but it allowed a systematic evaluation of the importance of this component of dispersal.

Again the difference in rate of spread between prairie and urban populations was much greater than differences among runs, due to the variance in proportion of seeds going out to the tail of the distribution (Figure 11.3). Even with 5% of seeds (equivalent to as many as 200 seeds from every filled bin in each time step) going relatively long distances, urban populations did not approach the spread rates achieved by prairie populations with no long-distance dispersal.

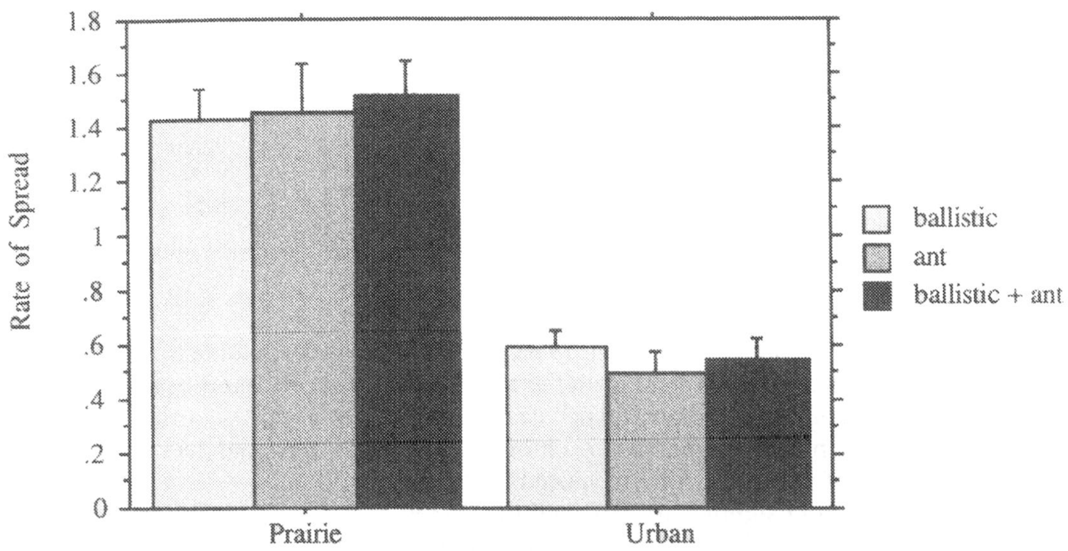

Figure 11.2
Rate of spread for Cytisus scoparius *(increase in the square root of bins occupied by the maximum density of adults, over time) from simulations incorporating demographic rates for prairie and urban populations of the invasive shrub. Simulations utilized one of three empirically derived dispersal distributions: ballistic dispersal, ant dispersal, or ballistic + ant dispersal.*

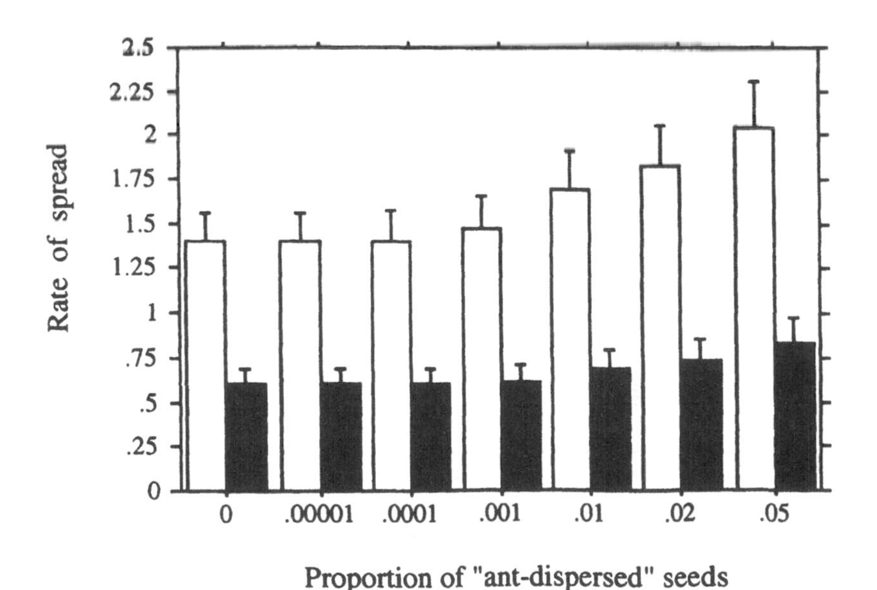

Figure 11.3
The effect on rate of spread (the increase in the square root of bins occupied over time) of changing the proportion of long-distance dispersers in prairie populations (open bars) and urban populations (filled bars) of Cytisus scoparius. *The proportion of seeds dispersed between five and ten meters, extending the maximum dispersal distance beyond the 6m achieved by ballistic dispersal, was increased systematically from 0.00–0.05.*

Table 11.3
Analysis of variance of the effect of populations type (prairie/urban) and the number of initial foci (1, 2, 4) on rate of spread (the increase in square-root of bins occupied over time).

Source	DF	Sum Sq.	F	P-Value
Population Type	1	9.9	52.2	0.0001
# Foci	2	3.2	8.5	0.0044
Pop Type * Foci	2	0.8	2.2	0.15
Residual	13	2.5		

3. Spread of populations from varying spatial arrangements of founders.
Populations were allowed to expand from the same number, but different spatial arrangement, of initial founders. The "one focus" case placed all four plants in the center bin, the "two foci" case place two plants in each of two bins, and the "four foci" case placed one plant in each of four bins equally spaced on the lattice.

The number of foci significantly affected the spread rate (Table 11.3), and even with the same initial number of founding adults, the spread rate was almost twice as fast for four foci as it was for a single focus (Figure 11.4). However, again the variance due to demographic rates was much greater than that due to number of foci (Table 11.3). Although the effect of additional foci appeared to be more important in the fast-growing prairie populations, the proportional increase was about the same, and an analysis of variance revealed no significant interaction between population type and the number of foci (Table 11.3).

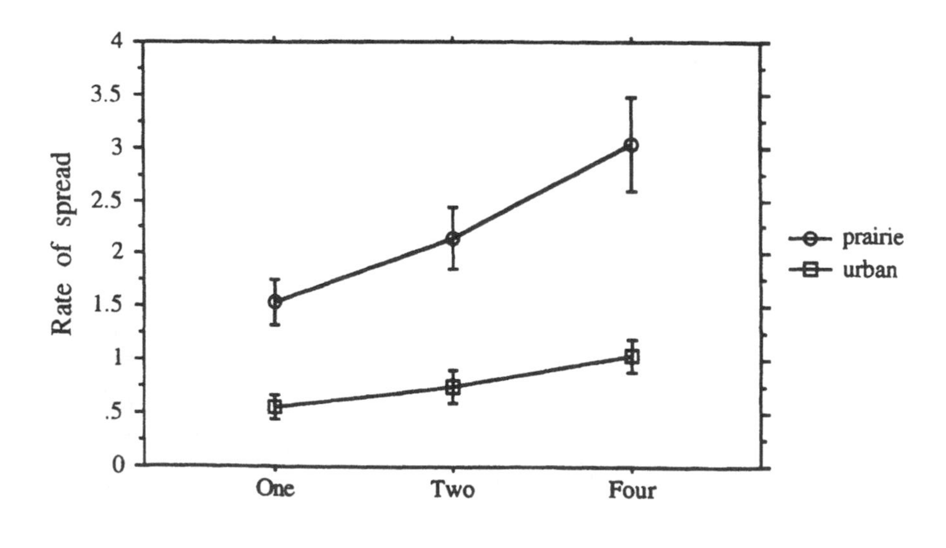

Figure 11.4
The effect of the number of initial foci, or founder patches, on rate of spread of the invasive shrub Cytisus scoparius *for prairie populations (circles) and urban populations (squares). Rate of spread, measured as the increase in the square root of bins occupied over time, was generated from simulation models incorporating field-estimated demographic rates and dispersal (ballistic + ants).*

Summary of Model Results and Implications

The results presented here both provide insight into the specifics of *C. scoparius* invasion and shed some light on the factors important to range expansion in general. Two generalizations were upheld. First, as expected, both population growth rate and dispersal can contribute to variation in rates of spread. Second, increasing the number of independent sources for population growth (foci) can greatly increase the rate of spread, as has been suggested previously by both models (Moody and Mack 1988) and empirical case histories (Selleck, Coupland, and Frankton 1962).

Although these generalizations did not furnish any big surprises, applying the models to the particular case of *C. scoparius* did result in some unexpected patterns. Specifically, although recent theory has shown that the details of dispersal can be very important in determining rates of spread, neither field-estimated variation in dispersal nor artificially generated variation (e.g., ant-only dispersal, which does not occur in the field) had an overriding effect on how fast space was filled. The proportion of seeds going beyond the primary dispersal distribution also had a surprisingly small effect. Possibly dispersal would have increased in importance if the tail of the distribution had been much longer, similar to the effect of having multiple foci. However, even in the case of multiple foci, variation in demographic parameters such as that observed between prairie and urban populations completely dominated variation in rates of spread.

The importance of demographic variation underscores environmental heterogeneity as a key factor influencing the invasion process (Kareiva, Parker, and Pascual 1996), a factor that is rarely incorporated into either analytical or simulation models of range expansion (but see Shigesada, Kawasaki, and Teramoto 1986). In addition, the observation that urban or old-field populations of *C. scoparius* have lower rates of spread due to lowered demographic rates in a less favorable environment also has consequences for nonnative species management. The economic and logistic feasibility of different approaches to control is very much influenced by how fast a species will come back, and how vigilant one must be in attacking occasional new infestations. Hand-pulling and other time-intensive control strategies, though ineffective in prairie ecosystems, might work quite well in urban parks. An awareness of the potential importance of spatial variation in invasion rates cautions us against blanket generalizations and can help inform our decisions regarding the management of nonnative species.

CONCLUSIONS

The study of invasions has only very recently been recognized as a significant subfield of conservation biology. Nonnative species are now widely cited both in academic and popular sources as serious threats to native populations, the "integrity" of communities, and ecosystem functions (Temple 1990; Hedgpeth 1993; Reichard and Campbell 1996). However, invasion biology is still in its early stages as a scientific field of inquiry, and like many young fields it appears to be in a transition of moving from more descriptive and anecdotal evidence to more rigorous experimental and empirical work. The debate over biotic resistance in communities as a determinant of invasibility has, until recently, focused almost exclusively on analytical theory and statistical tests. New empirical work suggests that the widely accepted idea that invasion success declines with species richness may not be the rule in natural systems. There is room for much more productive work in this area.

In a similar way, models of range expansion have run ahead of our empirical understanding of what controls rates of spread at different scales for invading species. Models have provided us with a theoretical context for range expansion research and point explicitly to what types of factors should be studied. But as shown by work on *Cytisus scoparius*, patterns in natural variation may produce a different view of what factors are most important in field populations. In particular, demographic rates vary so much between invasible habitats in this species that demography dominates invasion rates even under very different dispersal scenarios.

Given current common acceptance of the negative impacts of invasive species, we remained surprised by the paucity of studies rigorously demonstrating these impacts. At a recent conference convened specifically to address ecological and management issues for invasive broom species (*Cytisus* spp., *Genista* spp.), the speaker responsible for reviewing their ecological impacts was forced to admit that no data were available on the subject. Clearly if we are to demonstrate to policy-makers the importance of controlling old and prohibiting new nonnative species, we need better documentation of their effects on native species and ecosystems. In addition, we argue that accumulating the results of experiments addressing the mechanisms underlying competition between nonnative and native species will provide us with improved understanding of what controls the impact of successful invaders.

ACKNOWLEDGMENTS

The authors would like to thank Peter Kareiva, Martha Groom, Jennifer Ruesink, and Douglas Schemske for discussions that contributed to parts of this paper, and P. Fiedler and P. Kareiva for helpful comments on the manuscript. The work presented here was supported by NSF dissertation improvement grant DEB–9411702 and NSF training grant BIR–9256532 in Mathematical Biology to IMP, and grant HO-STEW–010195 from The Nature Conservancy to SHR.

LITERATURE CITED

Abbott, R.J. 1992. Plant invasions, interspecific hybridization and the evolution of new plant taxa. *Trends in Ecology and Evolution* 7:401–405.

Allen, L.J.S., E.J. Allen, C.R.G. Kunst, and R.E. Sosebee. 1991. A diffusion model for dispersal of *Opuntia imbricata* (cholla) on rangeland. *Journal of Ecology* 79:1123–1135.

Allen, R.B. 1991. A preliminary assessment of the establishment and persistence of *Berberis darwinii* Hook., a naturalised shrub in secondary vegetation near Dunedin, New Zealand. *New Zealand Journal of Botany* 29:353–360.

Allison, G.W. 1996. Does high diversity facilitate or suppress species invasions? Some lessons from a diversity manipulation and reintroductions. *Bulletin of the Ecological Society of America* 77:9.

Ammerman, A.J. and L.L. Cavalli-Sforza. 1984. *The neolithic transition and the genetics of populations in Europe*. Princeton: Princeton University Press.

Andow, D., P. Kareiva, S.A. Levin, and A. Okubo. 1990. Spread of invading organisms. *Landscape Ecology* 4:177–188.

Andow, D.A., P.M. Kareiva, S.A. Levin, and A. Okubo. 1993. Spread of invading organisms: patterns of spread. In *Evolution of insect pests: patterns of variation*, eds. K. Kim and B.A. McPheron, 219–242. New York: John Wiley and Sons.

Auld, B.A. and B.G. Coote. 1980. A model of a spreading plant population. *Oikos* 34:287–292.

Baker, H.G. 1965. Characteristics and modes of origin of weeds. In *The genetics of colonizing species*, eds. H.G. Baker and G.L. Stebbins, 147–168. New York: Academic Press.

Beatty, S.W. and D.L. Licari. 1992. Invasion of fennel (*Foeniculum vulgare*) into shrub communities on Santa Cruz Island, California. *Madroño* 39:54–66.

Bergelson, J. and M. Crawley. 1989. Can we expect mathematical models to guide biological control programs?: A comment based on case studies of weed control. *Comments on Theoretical Biology* 1:197–216.

Blackburn, W., R.W. Knight, and J.L. Schuster. 1982. Saltcedar influence of sedimentation in the Brazos River. *Journal of Soil and Water Conservation* 37:298–230.

Bock, J.H. and C.E. Bock. 1992. Vegetation responses to wildfire in native versus exotic Arizona grassland. *Journal of Vegetation Science* 3:439–446.

Bossard, C.C. 1990. Secrets of an ecological interloper: Ecological studies on *Cytisus scoparius* in California. Ph.D. dissertation, University of California, Davis.

Breytenbach, G.J. 1986. Impacts of alien organisms on terrestrial communities with emphasis on communities of the south-western Cape. In *The ecology and management of biological invasions in South Africa*, eds. I.A.W. MacDonald, F.J. Kruger, and A.A. Ferrar, 229–238. Cape Town: Oxford University Press.

Bruce, K.A., G.N. Cameron, and P.A. Harcombe. 1995. Initiation of a new woodland type on the Texas coastal prairie by the Chinese tallow tree (*Sapium sebiferum* (L.) Roxb.). *Bulletin of the Torrey Botanical Club* 122:215–225.

Burdon, J.J. and G.A. Chilvers. 1994. Demographic changes and the development of competition in a native Australian eucalypt forest invaded by exotic pines. *Oecologia* 97:419–423.

Burgess, T.L., J.E. Bowers, and R.M. Turner. 1991. Exotic plants at the desert laboratory, Tuscon, Arizona. *Madroño* 38:96–114.

Carman, J.G. and J.D. Brotherson. 1982. Comparison of sites infested and not infested with saltcedar (*Tamarix pentandra*) and Russian olive (*Elaeagnus angustifolia*). *Weed Science* 30:360–364.

Case, T.J. 1990. Invasion resistance arises in strongly interacting species-rich model competition communities. *Proceedings of the National Academy of Sciences* 97:9610–9614.

Case, T.J. 1991. Invasion resistance, species build-up and community collapse in metapopulation models with interspecies competition. *Biological Journal of the Linnean Society* 42:239–266.

Caughley, G. 1970. Liberation, dispersal and distribution of Himalayan thar (*Hemitragus jemlahicus*) in New Zealand. *New Zealand Journal of Science* 13:220–239.

Connell, J.H. 1978. Diversity in tropical rainforests and coral reefs. *Science* 199:1302–1310.

Cowrie, I.D. and P.A. Werner. 1993. Alien plant species invasive in Kakadu National Park, tropical northern Australia. *Biological Conservation* 63:127–135.

Crawley, M.J. 1986. The population biology of invaders. *Philosophical Transactions of the Royal Society of London B* 314:711–731.

Crawley, M.J. 1989. Chance and timing in biological invasions. In *Biological invasions: A global perspective*, eds. J.A. Drake, H.A. Mooney, F. d. Castri, R.H. Groves, F.J. Kruger, M. Rejmanek and M. Williamson, 407–423. New York: John Wiley and Sons.

D'Antonio, C.M. 1993. Mechanisms controlling invasion of coastal plant communities by the alien succulent *Carpobrotus edulis*. *Ecology* 74:83–95.

D'Antonio, C.M. and P.M. Vitousek. 1992. Biological invasions by exotic grasses, the grass/fire cycle and global change. *Annual Review of Ecology and Systematics* 23:63–87.

De Pietri, D.E. 1992. Alien shrubs in a national park: can they help in the recovery of natural degraded forest? *Biological Conservation* 62:127–130.

Dillenburg, L.R., D.F. Whigham, A.H. Teramura, and I.N. Forseth. 1993. Effects of below- and aboveground competition from the vine *Lonicera japonica* and *Parthenocissus quinquefolia* on the growth of the tree host *Liquidambar styraciflua*. *Oecologica* 93:48–54.

Di Tommaso, A. and L.W. Aarssen. 1991. Effect of nutrient level on competition intensity in the field for three coexisting grass species. *Journal of Vegetation Science* 2:513–522.

Drake, J.A. 1983. Invasibility in Lotka-Volterra interaction webs. In *Current trends in food web theory*, eds. D. DeAngelis, W.M. Post, and G. Sugihara, 83–90. Oak Ridge, Tennessee: Oak Ridge National Laboratories.

Drake, J.A. 1988. Models of community assembly and the structure of ecological landscapes. In *Proceedings of the international conference on mathematical ecology*, eds. L. Gross, T. Hallam, and S. Levin, 585–604. Singapore: World Press.

Dritschilo, W., D.E. Carpenter, J.L. Hastings, O. Meyn, D. Moss, and M.N. Weinstein. 1985. *Implications of data on introduced species in California for field releases of recombinant DNA organisms*. Los Angeles: School of Public Health, University of California, Los Angeles.

Elton, C.S. 1958. *The ecology of invasions by animals and plants*. London: Methuen.

Fiedler, P.L. and S.K. Jain, eds. 1992. *Conservation biology: The theory and practice of nature conservation, preservation, and management*. New York: Chapman & Hall.

Fisher, R.A. 1937. The wave of advance of advantageous genes. *Annals of the Eugenics Society (London)* 7:355–369.

Forcella, F. and J. Wood. 1984. Colonization potentials of alien weeds are related to their native distributions: implications for plant quarantine. *Journal of the Australian Institute of Agricultural Science* XX:35–41.

Fowler, N. 1986. The role of competition in plant communities in arid and semi-arid regions. *Annual Review of Ecology and Systematics* 17:89–110.

Fox, M.D. and B.J. Fox. 1986. The susceptibility of natural communities to invasion. In *Ecology of biological invasions: an Australian perspective*, eds. R.H. Groves and J.J. Burdon, 57–66. Canberra: Australian Academy of Sciences.

Gaudet, C.L. and P.A. Keddy. 1995. Competition performance and species distribution in shoreline plant communities, a comparative approach. *Ecology* 76:280–291.

Gerry, A.K. and S.D. Wilson. 1995. The influence of initial size on the competitive responses of six plant species. *Ecology* 76:272–279.

Given, D.R. 1994. *Principles and practice of plant conservation*. Portland: Timber Press.

Goldwasser, L., J. Cook, and E.D. Silverman. 1994. The effects of variability on metapopulation dynamics and rates of invasion. *Ecology* 75:40–47.

Goodwin, J. 1992. The role of mycorrhizal fungi in competitive interactions among native bunchgrasses and alien weeds: A review and synthesis. *Northwest Science* 66:251–260.

Hardin, D.P., P. Takac, and G.F. Webb. 1990. Dispersion population models discrete in time and continuous in space. *Journal of Mathematical Biology* 28:1–20.

Hedgpeth, J. 1993. Foreign invaders. *Science* 261:34–35.

Hengeveld, R. 1989. *Dynamics of ecological invasions*. Cambridge: Cambridge University Press.

Hester, A.J. and R.J. Hobbs. 1992. Influence of fire and soil nutrients on native and non-native annuals at remnant vegetation edges in the Western Australian wheatbelt. *Journal of Vegetation Science* 3:101–108.

Hobbs, R.J. 1989. The nature and effects of disturbance relative to invasions. In *Biological invasions: a global perspective*, eds. J.A. Drake, H.A. Mooney, F. d. Castri, R.H. Groves, F.J. Kruger, M. Rejmanek, and M. Williamson, 389–405. New York: Wiley and Sons.

Hobbs, R.J. and L. Atkins. 1991. Interactions between annuals and woody perennials in western Australian nature reserves. *Journal of Vegetation Science* 2:643–654.

Hobbs, R.J. and L.F. Huenneke. 1992. Disturbance, diversity, and invasion: Implications for conservation. *Conservation Biology* 6:324–337.

Holmes, E.E. 1993. Are diffusion models too simple? A comparison with telegraph models of invasion. *The American Naturalist* 142:779–795.

Horton, J.S. 1977. *The development and perpetuation of the permanent tamarisk type in the phreatophyte zone of the Southwest*. General Technical Report, U.S.D.A.

Huenneke, L.F., S.P. Hamburg, R. Korste, H.A. Mooney, and P.M. Vitousek. 1990. Effects of soil resources on plant invasion and community structure in Californian serpentine grassland. *Ecology* 71:478–491.

Huenneke, L.F. and J.K. Thomson. 1994. Potential interference between a threatened endemic thistle and an invasive nonnative plant. *Conservation Biology* 9:416–425.

Hughes, F., P.M. Vitousek and T. Tunison. 1991. Alien grass invasion and fire in the seasonal submontane zone of Hawaii. *Ecology* 72:743–746.

Jackson, L.E. 1985. Ecological origins of California's Mediterranean grasses. *Journal of Biogeography* 12:349–361.

Johnson, A.P. 1994. Coastal impacts of non-indigenous species. In *An assessment of invasive, non-*

indigenous species in Florida's public lands, eds. D.C. Schmitz and T.C. Brown, 119–126. Tallahassee, Florida: Florida Department of Environmental Protection.

Jones, R.H. 1993. Influence of soil temperature on root competition in seedlings of *Acer rubrum*, *Liquidambar styraciflua* and *Sapium sebiferum*. *American Midland Naturalist* 130:116–126.

Jordan, C.F. 1995. *Conservation*. New York: John Wiley and Sons.

Kareiva, P., M.J. Groom, I.M. Parker, and J. Ruesink. 1991. *Risk analysis as a tool for making decisions about the introduction of non-indigenous species into the United States*. Washington D.C.: United States Office of Technology Assessment.

Kareiva, P., I.M. Parker, and M. Pascual. 1996. Can we use experiments and models in predicting the invasiveness of genetically engineered organisms? *Ecology* 77:1670–1675.

Knops, J.M.H., J.R. Griffin, and A.C. Royalty. 1995. Introduced and native plants of the Hastings Reservation, central coastal California: A comparison. *Biological Conservation* 71:115–123.

Kot, M. 1992. Discrete-time travelling waves: Ecological examples. *Journal of Mathematical Biology* 30:413–436.

Kot, M., M.A. Lewis, and P. van den Driessche. In press. *Dispersal data and the spread of invading organisms*.

Kot, M. and W.M. Schaffer. 1986. Discrete-time growth-dispersal models. *Mathematical Biosciences* 80:109–136.

LaRoe, E.T. 1993. Implementation of an ecosystem approach to endangered species conservation. *Endangered Species Update* 10:3–12.

Lawton, J. and K. Brown. 1986. The population and community ecology of invading insects. *Philosophical Transactions of the Royal Society of London B* 314:607–617.

Lewis, M. 1997. Invasion and the importance of rare long-distance dispersal events. In *Spatial processes in ecology*, eds. D. Tilman and P. Kareiva. Princeton: Princeton University Press.

Lewis, M.A. and P. Kareiva. 1993. Allee dynamics and the spread of invading organisms. *Theoretical Population Biology* 43:141–158.

Lonsdale, W.M. 1993. Rates of spread of an invading species—*Mimosa pigra* in northern Australia. *Journal of Ecology* 81:513–521.

Lubina, J.A. and S.A. Levin. 1988. The spread of a reinvading species: Range expansion of the California sea otter. *American Naturalist* 133:526–543.

McClaran, M.P. and M.E. Anable. 1992. Spread of introduced Lehmann lovegrass along a grazing intensity gradient. *Journal of Applied Ecology* 29:92–98.

Melgoza, G., R.S. Nowak, and R.J. Tausch. 1990. Soil water exploitation after fire: Competition between *Bromus tectorum* (cheatgrass) and two native species. *Oecologia* 83:7–13.

Mesléard, F., L.T. Ham, V. Boy, C. van Wijck, and P. Grillas. 1993. Competition between an introduced and an indigenous species: The case of *Paspalum paspalodes* (Michx) Schribner and *Aeluropus littoralis* (Gouan) in the Camargue (southern France). *Oecologia* 94:204–209.

Moody, M.E. and R.N. Mack. 1988. Controlling the spread of plant invasions: The importance of nacent foci. *Journal of Applied Ecology* 25:1009–1021.

Moulton, M. and S. Pimm. 1986. Species introduction to Hawaii. In *Ecology of biological invasions of North America and* Hawaii, eds. H.A. Mooney and J.A. Drake, 231–249. New York: Springer-Verlag.

Moulton, M.P. and S.L. Pimm. 1983. The introduced Hawaiian avifauna: Biogeographic evidence for competition. *American Naturalist* 121:669–690.

Neill, W. 1983. The tamarisk invasion of desert riparian areas. Educational Bulletin of the Desert Protective Council, vol. 83–84. Spring Valley, CA: Educational Foundation of the Desert Protective Council.

Office of Technology Assessment. 1993. *Harmful non-indigenous species in the United States*. Washington, D.C.: U.S. Congress.

Okubo, A. 1988. Diffusion-type models for avian range expansion. In *Acta XIX congressis internationalis ornithologici*. 1, ed. H. Ouellet, 1038–1049. Ottowa, Canada: University of Ottowa Press.

Okubo, A., P.K. Maini, M.H. Williamson, and J.D. Murray. 1989. On the spatial spread of the grey squirrel in Britain. *Proceedings of the Royal Society of London B* 23:113–125.

Olson, T.E. and F.L. Knopf. 1986. Agency subsidization of a rapidly spreading exotic. *Wildlife Society Bulletin* 14:492–493.

Orians, G.H. 1986. Site characteristics favoring invasions. In *Ecology of biological invasions of North America and Hawaii*, eds. H.A. Mooney and J.A. Drake, 133–148. New York: Springer-Verlag.

Pablo, A.P. and J.L. Vasquez. 1991. Travelling waves and finite propagation in a reaction-diffusion equation. *Journal of Differential Equations* 93:19–61.

Parker, I.M. 1996. Ecological factors affecting rates of population growth and spread in *Cytisus scoparius*, an invasive exotic shrub. Ph.D. dissertation, University of Washington.

Parker, I.M. and P. Kareiva. 1996. Assessing the risks of genetically engineered organisms: Acceptable evidence and reasonable doubt. *Biological Conservation* 78:193–203.

Parker, I.M., S.K. Mertens, and D.W. Schemske. 1993. Distribution of seven native and two exotic plants in a tallgrass prairie in southeastern Wisconsin: The importance of human disturbance. *The American Midland Naturalist* 130:43–55.

Perrins, J., M. Williamson, and A. Fitter. 1992. A survey of differing views of weed classification: Implications for regulation of introductions. *Biological Conservation* 60:47–56.

Pimentel, D. 1986. Biological invasions of plants and animals in agriculture and forestry. In *Ecology of biological invasions of North America and Hawaii*, eds. H.A. Mooney and J.A. Drake, 149–162. New York: Springer-Verlag.

Planty-Tabacchi, A., E. Tabacchi, R.J. Naiman, C. Deferrari, and H. Décamps. 1996. Invasibility of species-rich communities in riparian zones. *Conservation Biology* 10:598–607.

Post, W.M. and S.L. Pimm. 1983. Community assembly and food web stability. *Mathematical Biosciences* 64:169–192.

Reichard, S. 1996. Biogeographic and taxonomic patterns in the invasive and non-invasive flora of Hawaii. *Bulletin of the Ecological Society of America* 77:371.

Reichard, S. and F. Campbell. 1996. Invited but unwanted. *American Nurseryman* 184:39–45.

Reichard, S.E. 1994. Assessing the potential of invasiveness in woody plants introduced to North America. Ph.D. dissertation, University of Washington.

Reichard, S.H. and K. Hamilton. In press. Predicting invasions of woody plants introduced into North America. *Conservation Biology*.

Rejmánek, M. and D.M. Richardson. 1996. What attributes make some plant species more invasive? *Ecology* 77:1655–1661.

Robinson, G.R., J.F. Quinn, and M.L. Stanton. 1995. Invasibility of experimental habitat islands in a California winter annual grassland. *Ecology* 76:786–794.

Robinson, G.R., M.E. Yurlina, and S.N. Handel. 1994. A century of change in the Staten Island flora: ecological correlates of species losses and invasions. *Bulletin of the Torrey Botanical Club* 121:119–129.

Sallabanks, R. 1993. Fruiting plant attractiveness to avian seed dispersers: Native vs. invasive *Crataegus* in western Oregon. *Madroño* 40:108–116.

Selleck, G.W., R.T. Coupland, and C. Frankton. 1962. Leafy spurge in Saskatchewan. *Ecological Monographs* 32:1–29.

Shafroth, P.B., G.T. Auble, and M.L. Scott. 1995. Germination and establishment of the native plants cottonwood (*Populus deltoides* Marshall subsp. *monilifera*) and the exotic Russian-olive (*Elaeagnus angustifolia* L.). *Biological Conservation* 9:1169–1175.

Shigesada, N., K. Kawasaki, and E. Teramoto. 1986. Traveling periodic waves in heterogeneous environments. *Theoretical Population Biology* 30:143–160.

Simberloff, D. 1981. Community effects of introduced species. In *Biotic crises in ecological and evolutionary time*, ed. M.H. Nitecki, 53–81. New York: Academic Press.

Simberloff, D. 1986. Introduced insects: A biogeographic and systematic perspective. In *Ecology of Biological Invasions of North American and Hawaii*, eds. Mooney, H.A. and J.A. Drake, 3–26. New York: Springer-Verlag.

Simberloff, D. 1989. Which insect introductions succeed and which fail? In *Biological invasions: A global perspective*, eds. J.A. Drake, H.A. Mooney, F.D. Castri, R.H. Groves, F.J. Kruger, M. Rejmanek, and M. Williamson, 61–75. New York: Wiley and Sons.

Simberloff, D. and W. Boecklen. 1991. Patterns of extinction in the introduced Hawaiian avifauna: A reexamination of the role of competition. *The American Naturalist* 138:300–327.

Skellam, J. 1951. Random dispersal in theoretical populations. *Biometrika* 38:196–218.

Soulé, M.E., ed. 1986. *Conservation biology: The science of scarcity and diversity*. Sunderland, MA: Sinauer Associates.

Stace, C.A. 1991. New flora of the British Isles. Cambridge: Cambridge University Press.

Stock, W.D., K.T. Wienand, and A.C. Baker. 1995. Impacts of invading N_2-fixing *Acacia* species on patterns of nutrient cycling in two Cape ecosystems: Evidence from soil incubation studies and ^{15}N natural abundance values. *Oecologia* 101:375–382.

Temple, S.A. 1990. The nasty necessity: Eradicating exotics. *Conservation Biology* 4:113–114.

Thompson, J.D. 1991. The biology of an invasive plant: What makes *Spartina anglica* so successful? *Bioscience* 41:393–401.

Vitousek, P.M., L.R. Walker, L.D. Whiteaker, D. Mueller-Dombois, and P. Matson. 1987. Biological invasion by *Myrica faya* alters ecosystem development in Hawaii. *Science* 238:802–804.

Webb, S.L. and C.K. Kaunzinger. 1993. Biological invasion of the Drew University (New Jersey) Forest Preserve by Norway maple (*Acer platanoides* L.). *Bulletin of the Torrey Botanical Club* 120:343–349.

Williamson, M. and K. Brown. 1986. The analysis and modelling of British invasions. *Philosophical Transactions of the Royal Society of London B* 314:505–522.

Wilson, S.D. and D. Tilman. 1995. Competitive responses of eight old-field plant species in four environments. *Ecology* 76:1169–1180.

Wiser, S.K., R.B. Allen, P.W. Clinton, and K.H. Platt. 1996. Invasibility of species-poor forest by a perennial herb over 25 years. *Bulletin of the Ecological Society of America* 77:488.

Witkowski, E.T.F. 1991. Effects of invasive alien acacias on nutrient cycling in the coastal lowlands of the cape fynbos. *Journal of Applied Ecology* 28:1–15.

Woods, K.D. 1993. Effects of invasion by *Lonicera tatarica* L. on herbs and tree seedlings in four New England forests. *American Midland Naturalist* 130:62–74.

Zedler, J.B., E. Paling, and A. McComb. 1990. Differential responses to salinity help explain the replacement of native *Juncus kraussii* by *Typha orientalis* in western Australian salt marshes. *Australian Journal of Ecology* 15:57–72.

Zedler, P.H., G.R. Gautier, and G.S. McMaster. 1983. Vegetation change in response to extreme events: the effect of a short fire interval between fires in a California chaparral and coastal scrub. *Ecology* 64:809–818.

CHAPTER 12

Challenges and Approaches for Conserving Hawaii's Endangered Forest Birds

THOMAS B. SMITH
and STEVEN G. FANCY

Although the Hawaiian Islands exhibit a rich diversity of birds, with especially high rates of endemism, the avian diversity is only a small fraction of what the archipelago once supported. In fact, it is estimated that over two thirds of Hawaii's forest bird species have gone extinct since human contact (Freed, Conant, and Fleisher 1987; Jacobi and Atkinson 1994). Few faunas on Earth and no other island avifauna have experienced as many recent extinctions or include as many endangered species. In total, 21 of the 59 historically known species and subspecies of Hawaiian birds have gone extinct in the last 150 years and most of those remaining are threatened (Pyle 1990). Interestingly, it is not only contact with "modern" western Europeans that caused extinctions, but as fossils suggest, it is also contact with Polynesians that began 15 centuries ago (Olsen and James 1982). Current threats to Hawaiian birds are especially acute (Table 12.1), and we risk losing perhaps the most celebrated example of adaptive radiation in any vertebrate, the Hawaiian honeycreepers, finches in the subfamily Drepanidinae (Freed, Conant, and Fleisher 1987). Honeycreepers are believed to be derived from a single colonization event (Johnson, Martin, and Ralph 1989) and are exceedingly diverse in morphology, coloration, and habits (Figure 12.1). In addition to the honeycreepers, threatened forest species from three other passerine families occur in the archipelago: the Melaphagidae (honeyeaters), Corvidae (crows), and Muscicapidae (old world flycatchers and solitaires).

While many of the threats to Hawaii's endangered birds are shared with birds on other islands (Temple 1985; Johnson and Stattersfield 1990) and birds faunas worldwide (Groombridge 1992), the severity of threats, especially from disease, is greatly magnified in the Hawaiian Islands (Figure 12.2). The introduction of the mosquito (*Culex quinquefasciatus*) provided a vector for avian diseases, such as avian pox and malaria, which have had devastating effects on forest bird populations (Warner 1968; van Riper *et al.* 1986; Atkinson *et al.* 1996). Introductions of feral ungulates (pigs, *Sus scrofa*; goats, *Capra hircus*; sheep, *Ovis* spp.; and cattle, *Bos taurus*) and exotic plant species have severely altered native plant communities (Tomich 1969; Jacobi 1990; Hobdy 1993; Jacobi and

Table 12.1
Population size and trend for critically-endangered species of Hawaiian forest birds (Modified from Ellis et al. *1992). Declining populations result primarily from the combined effects of disease, introduced species and habitat loss (see Figure 12.2).*

Species	Island	Estimated size	Population trend
Hawaiian goose *(Branta sandvicenses)*	Maui	<200	stable?
Hawaiian goose *(Branta sandvicenses)*	Hawaii	<339	declining
Hawaiian goose *(Branta sandvicenses)*	Kauai	100	increasing
Short-eared owl *(Asio flammeus)*	Lanai	±50	declining
Short-eared owl *(Asio flammeus)*	Oahu	±200	declining
'Alala *(Corvus hawaiiensis)*	Hawaii	13	declining
Oloma'o *(Myadestes lanaiensis)*	Molokai	0–10	declining
Kama'o *(Myadestes myadestinus)*	Kauai	0–10	declining
O'ahu 'Elepaio *(Chasiempis sandwichensis)*	Oahu	200–500	declining
Kaua'i Nuku-pu'u *(Hemignathus lucidus)*	Kauai	0–10	declining
Maui Nuku-pu'u *(Hemignathus lucidus)*	Maui	<10	declining?
O'ahu creeper *(Paroreomyza maculata)*	Oahu	0–10	declining
Maui Akepa *(Loxops coccineus)*	Maui	0–10	declining
Molaka'i I'iwi *(Vestiaria coccinea)*	Molokai	<50	declining
O'ahu I'iwi *(Vestiaria coccinea)*	Oahu	<50	declining
Lana'i Apapane *(Himatione sanguinea)*	Lanai	<100	declining
Po'ouli *(Melamprosops phaeosoma)*	Maui	<50	declining

Atkinson 1994). Introduced mammalian predators, including the Polynesian rat (*Rattus exulans*), black rat (*Rattus rattus*), Norway rat (*Rattus norwegicus*), mongoose (*Herpestes auropunctatus*), and domestic cat (*Felis catus*) have contributed to declines and extinctions of many species (Atkinson 1977; Ebenhard 1988; Scott *et al.* 1986; Snetsinger *et al.* 1994). There have been more than 125 introductions of at least 47 species of alien birds to the six main islands in the past 100 years alone (Moulton and Pimm 1983; Pratt 1994). In addition to being reservoirs for pox and malaria, some of these alien species have had deleterious effects on the fitness of native bird populations due to competition (Mountainspring and Scott 1985; see also chapter 11). Threats from still additional introductions loom on the horizon. For example, the introduction of a temperate, cold-hardy mosquito, capable of vectoring diseases at higher elevations or of the brown tree snake (*Boiga irregularis*), a highly efficient nest predator, which has decimated native bird populations on the island of Guam, would likely lead to the extinction of many more species.

Despite the severity of threats, island archipelagos such as Hawaii still offer unique opportunities for conservation biologists to examine and test conservation approaches. In this chapter we review the risks facing Hawaii's forest avifauna and the conservation approaches available to biologists. We then examine two case studies of conservation projects that highlight some of the challenges that conservation biology must face as a science if its "practice" is to be improved.

CURRENT CONSERVATION CHALLENGES AND APPROACHES

The conservation of Hawaii's forest birds depends on the protection and management of native forest above 1,500 m where mosquitoes are uncommon. In addition to land acquisition, other important steps towards protecting native forests should include providing

Figure 12.1
Three species of Hawaiian honeycreepers, (a) adult male Palila and (b) Iiwi in mamane (Sophora chrysophylla) *and (c) a mosquito taking a blood meal from an Apapane* (Himatione sanguinea). *Introduced Japanese White-eye (d), believed to be an important competitor with native birds.*

economic incentives to landowners to promote conservation. Unfortunately, landowners currently can obtain a lower tax rate by introducing cattle and clearing trees than by leaving forests intact. Increased public awareness and education of politicians and landowners about the uniqueness and vulnerability of native species is essential.

Exotic plant species, such as banana poka (*Passiflora mollissima*), firetree (*Myrica faya*), guava (*Psidium* spp), Christmas berry (*Schinus terebinthifolius*), pines (*Pinus* spp.) and eucalyptus (*Eucalyptus* spp.) are choking out the native vegetation on which the native avifauna depends. Few economical solutions exist to halt this onslaught. For example, banana poka, originally introduced as an ornamental, is now estimated to cover more than 40,000 ha across the archipelago (Gillis 1992). Recently, considerable concern has been raised by the spread of *Miconia calvescens*, a fast growing neotropical tree species that threatens a quarter of the Tahitian flora (Gillis 1992). Control of alien plant species is expensive and has been successful in only a few localized areas.

In higher elevational areas, cattle ranching, introduction of exotic grasses and vegetation damage by feral pigs have radically changed the understory of native forests. Feral pigs are now the most significant modifier of native forests (Anderson and Stone 1993); their rooting destroys the shrub layer, spreads alien weeds into new areas, and facilitates the spread of mosquitoes by providing breeding pools (Tomich 1969; Scott *et al.* 1986).

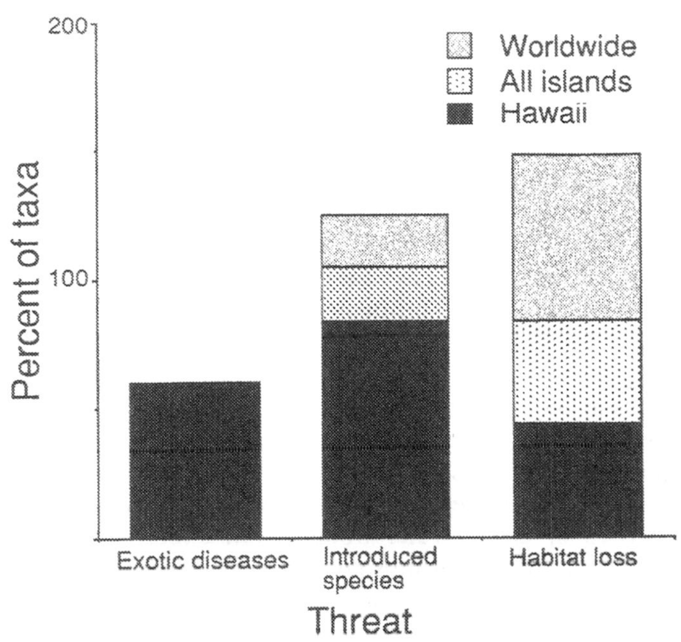

Figure 12.2
A comparison of three of the major threats to endangered and threatened birds among Hawaiian forest birds (modified from Ellis et al. *1992), endemic island birds (modified from Johnson and Stattersfleld 1990) and birds worldwide (modified from Groombridge 1992). The severity of the threat from introduced disease, particularly avian malaria is unique to Hawaii when compared to other island and mainland bird faunas.*

Today, exotic grasses from Africa, Asia, and North America, many of which are favored by cattle ranchers, dominate the understory of various regions and are believed to have also radically changed the native invertebrate populations on which many native forest species depend. Introductions of grasses have meant not only an increase in the incidence of destructive fires in native forests, but also an increase in the frequency of certain exotic grasses that are adapted to frequent burning and are particularly detrimental to native understory plants (Hughes, Vitousek, and Tunison 1991). Although there are few practical solutions to the control of introduced grasses, fencing and eliminating feral ungulate populations have been successful where implemented. For example, Hawaii Volcanoes National Park on Hawaii, and Haleakala National Park and the Hanawi Natural Area Reserve on Maui, have seen rapid recovery of native vegetation following fencing and removal of feral ungulates. But still greater protection and management of high-elevation native forests will be critical to the survival and recovery of most Hawaiian forest birds.

Finding solutions to exotic diseases will be one of the greatest challenges of the next decade. The introduction of *Culex* mosquitoes, followed by the introduction of domestic fowl, game birds, and cage birds, is thought to be responsible for the establishment of avian malaria and pox in Hawaii (Warner 1968; van Riper *et al.* 1986; Atkinson *et al.* 1996). Currently, avian disease is the single most important factor preventing the recovery of several species of endangered Hawaiian birds. Experiments with captive birds have shown that most species of Hawaiian honeycreepers show less resistance to avian malaria than introduced species. In a dramatic experiment, Atkinson *et al.* (1996) found that a single bite from a mosquito infected with *Plasmodium relictum* caused the death of nine of ten Iiwi (*Vestiaria coccinea*). The greatest impact of disease appears to be in mid-elevation forests between 900 m and 1,500 m, where the distributions of mosquitoes and native bird species overlap most (van Riper *et al.* 1986; Atkinson *et al.* 1996). Epizootics of avian malaria have been found at elevations as high as 1,200 m during the periods of seasonal warming that allow mosquitoes to occur at higher elevations (C. Atkinson unpublished data). The introduction of a cold-hardy mosquitoes that could spread avian malaria and pox to high-elevation forests would likely lead to the extinction of many of the remaining species of Hawaiian forest birds. This highlights the need of strict enforcement of importation and quarantine laws to the islands. This is clearly a challenge when one considers that in 1991 alone, 4.3 million airline passengers visited the island bringing with them 10.4 million pieces of baggage (Gillis 1992).

Totally eliminating mosquitoes from Hawaii is not feasible, but it may be possible to control mosquitoes through a combination of pig control and forest restoration. Recent studies (C. Atkinson unpublished data; D. LaPointe unpublished data) indicate that mosquitoes do not typically lay their eggs in the pools of water that collect in intact forest, but rather prefer stagnant water in tree ferns hollowed out by pigs, or artificial structures such as discarded tires or cattle water troughs. An experiment is now underway to determine if it is possible to control mosquitoes by eliminating potential breeding sites and using larvicides in pools of water. The restoration of forests, especially involving fencing and ungulate control also holds promise for controlling avian disease, since mosquitoes occur more often near the edges of forests or in fragmented forests, and are uncommon in intact native forests.

Perhaps the best hope for the control of avian disease is the evolution of resistance to disease by native birds. The occurrence of breeding populations of several species of native birds at lower elevations where mosquitoes are common suggests that some individuals have in fact developed resistance to avian malaria. On the island of Oahu breeding populations of Common Amakihi (*Hemignathus virens*) occur below 1,300 m where mosquitoes are abundant. Similarly, on the island of Hawaii populations of Apapane (*Himatione sanguinea*) and Omao (*Mydestes obscurus*) co-occur with mosquitoes. Of the seven Omao captured below 1,000 m all had antibodies for malaria and five had survived past pox infections (S. Fancy unpublished data). If resistant individuals can be identified, management approaches—such as screening and translocating resistant individuals, and selected breeding for resistant individuals for release—might speed the development of resistance in threatened populations. One innovative approach under development, which may prove more effective than traditional blood smears, is the use of the Polymerase Chain Reaction (PCR) to detect unique DNA sequences of *Plasmodium* from the blood of birds. In this way it may be possible to identify resistant individuals that harbor the *Plasmodium* parasite but are clinically healthy (Feldman, Freed, and Cann 1995; Cann *et al.* 1996). Although additional research is necessary, data suggest that an integrated approach using both diagnostic techniques has the potential of better identifying resistant individuals for selective captive propagation and translocations.

Because the Hawaiian avifauna evolved in the absence of mammalian predators, the introduction of rats, cats, and mongooses has had devastating effects on bird populations (Atkinson 1977; Griffin *et al.* 1989). Although Norway and Polynesian rats also prey on birds, the black rat remains the greatest threat because of its arboreal habits and greater abundance in native forests (National Biological Service unpublished data). Feral cats are found in both dry and wet forests throughout Hawaii, where they probably have a greater effect on native birds than historically realized (Ebenhard 1988). Trapping and poisoning is now used in localized areas around nests of endangered species to control mammalian predators, but no approach has yet been fully evaluated in Hawaii.

The most ominous exotic predator threatening Hawaii's avifauna is the brown tree snake, which is native to New Guinea, the Solomon Islands, and Australia (Savidge 1987; Rodda, Fritts, and Conry 1992; Rodda and Fritts 1992). Although this species is not yet established in Hawaii, this voracious nest predator has already been responsible for the extirpation of most of Guam's native terrestrial vertebrates species. On Guam, brown tree snakes reach densities as high as 100 per hectare, and have recently spread to several other islands in the Pacific (T. Fritts personal communication). The risk of the brown tree snake dispersing to Hawaii from Guam is heightened by several factors. First, Guam is a major transport center for civilian and military traffic for the central pacific region. Second, brown tree snakes easily colonize secondary and disturbed habitats and are capable of maintaining high densities in urban areas. Third, once in a port or cargo facility their nocturnal activity causes them to seek shade during the day, accounting for their propensity to be discovered in vehicles, shipping containers and other cargoes moved by sea and air. Since 1981, seven snakes have been found on Oahu arriving on civilian and military aircraft originating from Guam, one involving the discovery of a dead snake in an airplane wheel well. Currently, there are no effective means of controlling brown tree snake populations. If the brown tree snake were to become established in the Hawaiian Islands there would be little hope of preventing mass extinctions of Hawaii's remaining avifauna.

More species of alien birds have been purposefully introduced to Hawaii than to any other place on Earth (Pratt 1994). Most of the species that have become established have filled unoccupied niches in low-elevation habitats and have had a relatively benign influence on native ecosystems, except as potential disease reservoirs (Moulton and Pimm 1983; Pratt 1994). Impacts of introduced species that have become established in higher elevational native forests have been difficult to evaluate, in part because of the difficulties of measuring competition in natural populations. Available data are scarce, mostly coming from one study that reported negative correlations between population sizes of exotic Japanese White-eye (*Zosterops japonicus*) and native Iiwi, and between the Japanese White-eye and Elepaio (*Chasiempis sandwichensis*) (Mountainspring and Scott 1985). The Japanese White-eye (Figure 12.1(d)) is now the most abundant bird in the Hawaiian Islands and presumably competes with many native species (Scott et al. 1986). Other alien species that are rapidly expanding their ranges, but for which little or no information is available on possible competitive effects, include the Red-billed Leiothrix (*Leiothrix lutea*), Melodius Laughing-Thrush (*Garrulax canorus*), Northern Cardinal (*Cardinalis cardinalis*), and Japanese Bush-Warbler (*Cettia diphone*).

CASE STUDIES

Hakalau Forest National Wildlife Refuge

The establishment and management of the Hakalau Forest National Wildlife Refuge provides an example of measures being undertaken in several parks, preserves, and refuges in Hawaii to conserve native birds. Established as a refuge in 1985, Hakalau Forest National Wildlife Refuge was established as a direct result of the Hawaii Forest Bird Surveys (HFBS; Scott *et al.* 1986), in which the forests of all of the main Hawaiian Islands were systematically surveyed to determine the distribution and status of forest birds. The region that is now the Hakalau Forest National Wildlife Refuge was identified by the HFBS as a prime forest bird habitat lacking legal protection, but crucial to the long-term survival of at least three endangered species. The area was threatened by feral pigs, cattle ranching, and planned harvest of koa (*Acacia koa*) trees. Private ranches were initially purchased by The Nature Conservancy and turned over to the U.S. Fish and Wildlife Service once federal funding was obtained. The Nature Conservancy and other private organizations have played a critical role in several land acquisitions in Hawaii because they are able to act quickly to secure and hold valuable conservation lands while government agencies slowly obtain funding through Congress or the State legislature. The Hakalau Forest National Wildlife Refuge has since been expanded to include 13,252 ha.

The current management strategy of Hakalau Forest National Wildlife Refuge is to protect refuge lands from further degradation and to restore them as much as possible to their natural state. By the end of 1994, more than 39 km of ungulate-proof fencing had been constructed around four management units, and more than 181 feral cattle and 525 pigs had been removed from enclosures by staff and public hunting. Construction and maintenance of fencing in remote forested areas is very expensive. Construction costs for the Hakalau management units alone were $17,126–$21,554 per kilometer of fence. Without fencing, however, there is little chance of controlling feral ungulate populations and preventing ingress of additional animals into managed areas. Fencing and pig control are also controversial. Feral pigs are a source of food and recreation for some residents, and many hunting groups oppose fencing and control of pigs anywhere in Hawaii. In 1994, the State of Hawaii was forced by hunting groups to tear down a recently completed fence in one of the State's Natural Area Reserves. To date, complete eradication of feral ungulates has only been possible in sections of two national parks and in a portion of the Hakalau Forest National Wildlife Refuge.

Control of alien plants and restoration of former ranch lands to native forest are part of habitat restoration efforts at the Hakalau Forest National Wildlife Refuge. Banana poka, gorse (*Ulex europaeus*), and blackberry (*Rubus* spp.) are controlled manually and by hand-application of herbicides. However, alien plant control has been successful in few areas, and in some portions of the refuge is considered futile. Reforestation of previously grazed areas will be a slow process, but considerable work has already been accomplished. To date, more than 130,000 koa seedlings have been planted in formerly grazed areas. The fast-growing koa are planted in belts from which they will spread naturally, resulting in an uneven-age forest. In addition, a native plant nursery is being built on site to raise seedlings of several native tree and shrub species to assist in the reforestation efforts.

The National Biological Service and University of Hawaii both have active research programs on the refuge to provide recommendations to the refuge manager on what actions might best benefit native birds. One experiment involves the use of a Before-After-Control-Impact (BACI) design (Stewart-Oaten, Bence, and Osenberg 1986) to examine the effects of introduced predators. The experiment involves removing rats, cats, and mongooses from a 48 ha randomly selected treatment site and comparing it to an adjacent control area to determine the effects of predator removal on native bird populations. Although the response of bird populations have yet to be analyzed, preliminary results from 1996, the first year of the treatment, were nevertheless dramatic: 3,143 rats were trapped or poisoned on the experimental site, yielding a density of 65 rats per hectare, the highest recorded density of rats recorded anywhere in the world (S. Fancy, unpublished data). Other research is aimed at determining the factors limiting bird populations on the refuge. Considering the high cost of management actions, research to determine which actions result in a response by bird populations is an important and pragmatic component of the management strategy.

Recovering Alala Populations

The Alala or Hawaiian crow (*Corvus hawaiiensis*) is a primarily frugivorous forest corvid found only on the island of Hawaii. This endemic crow has experienced drastic declines in numbers and a corresponding contraction of its geographic range since the 1890s. By 1974 the wild population had declined to an estimated 61 individuals (Banko and Banko 1980). The causes of population declines include habitat loss and modification; predation by introduced rats, cats, and mongooses, and by the native Hawaiian hawk (*Buteo solitarius*); avian disease; and shooting (NRC 1992). Based on very low recruitment and the advanced ages of wild Alala, it was concluded that Alala were unlikely to survive in the wild without intensive human intervention.

In response to precipitous population decline of Alalas, a captive propagation program was initiated in the early 1970s. In early 1993, the captive flock held at the Olinda Endangered Species Propagation Facility (Olinda) on Maui consisted of 11 birds that were moderately to highly inbred (NRC 1992). Between 1979 and 1991, only eight (24%) chicks were hatched from 33 fertile eggs, and only six of the eight chicks raised in captivity survived. Because of concerns over the possibility of inbreeding depression, several groups advocated the taking of one or more Alala from the wild to augment the gene pool of the captive flock. However, the U.S. Fish and Wildlife Service was unable to gain access to the wild population after a private landowner believing that the Alala would do best if left alone refused to allow biologists onto their land.

A Population and Habitat Viability Assessment (PHVA) conducted in December 1992 predicted a probability of extinction of 59% or more during the next 100 years (Eliss *et al.* 1992). By early 1993 a population of only 12 individuals of unknown sex and age structure were known to exist on private lands in South Kona. In 1991, a lawsuit was filed by a group of conservation organizations against the U.S. Fish and Wildlife Service and the private landowner to force the federal government to take action to save the Alala before it went extinct. Remarkably, it was this lawsuit that provided the impetus for an intensive, interagency recovery effort involving nest manipulations, captive propagation, and release of captive-reared birds to the wild. In 1993 and 1994, the entire first clutch of eggs was removed from wild pairs for captive propagation. In each case, the breeding pair laid a second clutch, which they were allowed to incubate naturally. During 1993 and 1994, 14 fertile eggs were collected from wild pairs and 12 Alala juveniles were subsequently released to the wild. To increase genetic variation, four of the juveniles hatched from wild eggs were added to the captive flock, and four juveniles hatched in 1994 were released to the wild. Because none of the subsequent nesting attempts by wild pairs produced fledglings in 1993 or 1994, the release of captive raised birds will be essential to the continued existence of a wild population.

Although a lawsuit was required to bring all of the parties together, the Alala recovery program is an example of a coordinated effort by several agencies and private organizations working together. Private ranchers, The Peregrine Fund, U.S. Fish and Wildlife Service, National Biological Service (NBS), and Hawaii Department of Land and Natural Resources all played important roles in the recovery effort. Research to determine the status of each nesting pair and the optimum day for removing eggs was conducted by the USFWS and NBS. Eggs were incubated and hatched by The Peregrine Fund, with funding from the USFWS. The Hawaii Department of Land and Natural Resources managed the captive flock. At approximately two months of age, the captive-reared Alala chicks were moved to a large release (or hacking) facility constructed near the wild pairs; here the chicks learned foraging and social skills for up to two months before being released to the wild. After release, monitoring using radio telemetry and systematic searches were conducted by USFWS and The Peregrine Fund. Although an intensive level of management must be continued for many years to increase the size of the Alala population to self sustaining levels, the success of this program to date had made it a model for the recovery of several other critically endangered species.

In March 1996, The Peregrine Fund began operating the Keauhou Bird Conservation Center on the island of Hawaii, and will assume operations of the Olinda captive propagation facility from the State of Hawaii. As part of efforts to rescue species from the brink of extinction, eggs or nestlings of the Alala, Puaiohi (*Myadestes palmeri*), Poo-uli (*Melamprosops phaeosoma*), Palila (*Loxioides bailleui*) and other endangered species will be hatched and reared at the two captive propagation facilities for subsequent release to the wild. Field research programs led by NBS are currently underway for each of these species to locate nests and obtain information needed for captive propagation and release efforts.

CONCLUSIONS

The many threats that have caused the extinction of more than half of Hawaii's avifauna continue to cause declines in most of the remaining endemic forest birds. Habitat loss and modification, which has led to the extinction or endangerment of species on islands throughout the world, has had the greatest effect on Hawaii's avifauna. However, threats posed by exotic disease, especially avian malaria have played a greater role in causing extinctions and population declines in Hawaii than elsewhere. These threats are especially acute and will require aggressive and immediate management actions to prevent further population declines or extinctions. A priority for conservation is to acquire native forest lands above the elevation where mosquitoes are common, remove exotic species and restore them. The establishment and management of the Hakalau Forest National Wildlife Refuge serves as a model for the acquisition, recovery and management of other federal, state, and private lands in Hawaii. However, increased public awareness and cooperation among private landowners, and governmental and non-governmental organizations will be essential to prevent further avian extinctions. For critically endangered species, intensive management will be required and will involve nest manipulations, captive propagation and release, translocations and intensive predator control. The next decade will determine just how successful these current contrasting approaches are.

ACKNOWLEDGMENTS

The authors thank Carter Atkinson, Peggy Fiedler, James Jacobi, Peter Kareiva, Brenda Larison and Michael Moore for their comments on the manuscript, and Jack Jeffrey for kindly providing photographs of Hawaiian forest birds.

LITERATURE CITED

Anderson, S.J. and C.P. Stone. 1993. Snaring to control feral pigs *Sus scrofa* in a remote Hawaiian rain forest. *Biological Conservation* 63:195–201.

Atkinson, C.T., K.L. Woods, R.J. Dusek, L. Sileo, and W.M. Iko. 1996. Wildlife disease and conservation in Hawaii: Pathogenicity of avian malaria (*Plasmodium relictum*) in experimentally infected Iiwi (*Vestiaria coccinea*). *Parasitology* 111(n. Suppl.):S59-S69.

Atkinson, I.A.E. 1977. A reassessment of factors, particularly *Rattus rattus* L., that influenced the decline of endemic forest birds in the Hawaiian Islands. *Pacific Science* 31:109–133.

Banko, W.E. and P.C. Banko. 1980. *History of endemic Hawaiian birds. Part 1. Population histories - species accounts: Forest birds: 'Alala of Hawaiian Raven/Crow*. Avian Hist. Rep. 6B Coop. National Park Resoure Studies Unit, University of Hawaii, Manoa, Honolulu, 121 pp.

Cann R.L., R.A. Feldman, L. Agullana, and L.A. Freed. 1996. A PCR approach to detection of malaria in Hawaiian birds. In *Molecular genetic approaches in conservation*, eds. T.B. Smith and R.K. Wayne, 202–213. New York: Oxford University Press.

Ebenhard, T. 1988. Introduced birds and mammals and their ecological effects. *Swedish Wildlife Research Viltrevy* 13:5–107.

Ellis, S., C. Kuehler, R. Lacy, K. Hughes, and U.S. Seal. 1992. *Hawaiian forest birds conservation assessment and management plan*. Captive Breeding Specialist Group, IUCN-The World Conservation Union/Species Survival Commission.

Feldman, R.A., L.A. Freed, and R.L. Cann. 1995. A PCR test for avian malaria in Hawaiian birds. *Molecular Ecology* 4:663–673.

Freed, L.A., S. Conant, and R.C. Fleischer. 1987. Evolutionary ecology and radiation of Hawaiian passerine birds. *Trends in Ecology and Evolution* 2:196–203.

Gillis, A.M. 1992. Keeping aliens out of paradise. *BioScience* 42:482–485.

Griffin, C.R., C.M. King, J.A. Savidge, F. Cruz, and J.B. Cruz. 1989. Effects of introduced predators on island birds: contemporary case histories from the Pacific. In *Proceedings of XIX Ornithological Congress, Vol. 1.*, ed. H. Ouellet, 687–698. Ottawa, Canada: University of Ottawa Press.

Groombridge, B., ed. 1992. *Global biodiversity*. A report by the World Conservation Monitoring Centre. New York: Chapman & Hall.

Hobdy, R. 1993. Lana'i—A case study: The loss of biodiversity on a small Hawaiian island. *Pacific Science* 47:201–210.

Hughes, F., P.M. Vitousek, and T. Tunison. 1991. Alien grass invasion and fire in the seasonal submontane zone of Hawaii. *Ecology* 72:743–746.

Jacobi, J.D. 1990. Distribution maps, ecological relationships, and status of native plant communities on the island of Hawaii. Ph.D. dissertation, University of Hawaii, Honolulu.

Jacobi, J.D. and C.T. Atkinson. 1994. *Status and conservation prospects for Hawaii's endemic birds*. Status and Trends Report. Hawaii: National Biological Service.

Johnson, N.K., J.A. Martin, and C.J. Ralph. 1989. Genetic evidence for the origin and relationships of Hawaiian honeycreepers (Aves: Fringillidae). *Condor* 91:379–396.

Johnson, T.H. and A.J. Stattersfield. 1990. A global review of island endemic birds. *Ibis* 132:167–180.

Moulton, M.P. and S.L. Pimm. 1983. The introduced Hawaiian avifauna: Biogeographic evidence for competition. *American Naturalist* 121:669–690.

Mountainspring, S. and J.M. Scott. 1985. Interspecific competition among Hawaiian forest birds. *Ecological Monographs* 55:219–239.

National Research Council (NRC). 1992. *Scientific bases for the preservation of the Hawaiian crow*. National Academy Press, Washington, D.C.

Olson, S.L. and H.F. James. 1982. Fossil birds from the Hawaiian Islands: Evidence for wholesale extinction by man before western contact. *Science* 217:633–635.

Pratt, H.D. 1994. Avifaunal change in the Hawaiian Islands, 1983–1993. *Studies in Avian Biology* 15:103–118.

Pyle, R.L. 1990. Native breeding birds of Hawaii. *Elepaio* 50:99–100.

Pyle, R.L. 1992. Checklist of the birds of Hawaii–1992. *Elepaio* 52:53–62.

Ralph, C.J. and C. van Riper III. 1985. Historical and current factors affecting Hawaiian native birds. *Bird Conservation* 2:7–42.

Rodda, G.H., T.H. Fritts, and P.J. Conry. 1992. Origin and population growth of the brown tree snake *Boiga irregularis* on Guam. *Pacific Science* 46:46–57.

Rodda, G.H. and T.H. Fritts 1992. The impact of the introduction fo the colubrid snake *Boiga irregularis* on Guam's lizards. *Journal of Herpetology* 26:166–174.

Savidge, J.A. 1987. Extinction of an island forest avifauna by a introduced snake. *Ecology* 68:660–668.

Scott, J., M.S. Montainspring, F.L. Ramsey, and C.B. Kepler. 1986. Forest bird communities of the Hawaiian Islands: Their dynamics, ecology, and conservation. *Studies in Avian Biology* 9:1–431.

Snetsinger, T.J., S.G. Fancy, J.C. Simon, and J.D. Jacobi. 1994. Diets of owls and feral cats in Hawaii. *Elepaio* 54:47–50.

Stewart-Oaten, A.J., R. Bence, and C.W. Osenberg. 1992. Assessing the effects of unreplecated perturbations: No simple solutions. *Ecology* 73:1396–1404.

Temple, S.A. 1985. Why endemic islands are so vulnerable to extinction. *Bird Conservation* 2:3–6.

Tomich, P.Q. 1969. Mammals in Hawaii. B.P. Bishop Museum Special Publication 57. Honolulu: Bishop Museum Press.

van Riper, C. III, S.G. van Riper, M.L. Goff, and M. Laird. 1986. The epizoology and ecological significance of malaria in Hawaiian land birds. *Ecological Monographs* 56:327–344.

Warner, R.E. 1968. The role of introduced diseases in the extinction of the endemic Hawaiian avifauna. *Condor* 70:101–120.

CHAPTER 13

Picking Up The Pieces: Botanical Conservation on Degraded Oceanic Islands

MIKE MAUNDER

ALASTAIR CULHAM

and CLARE HANKAMER

"Tropical islands can provide a preview of the environmental situation that is likely to become more prevalent on the world's continents in the future. These islands typically have high population densities, exhibit highly fragmented landscapes, and have already experienced significant extinction events."

(McNeely *et al.* 1995)

"Most of the research about species extinction has been conducted on islands because islands are controlled environments and scientists can get drinks with little umbrellas in them there."

(O'Rourke 1994)

Historical debates about environmental degradation on oceanic islands acted as crucibles for the evolution of modern conservation thought (Grove 1995). These largely colonial debates recognized the link between forest loss and watershed decline and the possibility that habitat loss can result in species loss. Currently, oceanic islands are manifesting very high levels of extinction that demand urgent and innovative approaches to conservation. The paradigms established for continental areas, based primarily on the establishment of protected areas, are not sufficient to ensure the survival of the highly modified biotas and ecologies of many oceanic islands. On such islands the habitats prior to human colonization are largely destroyed, the original ecological processes lost or diverted, and the populations of endemic taxa severely reduced and fragmented. To salvage endemic species and their ecologies, habitat conservation needs to be matched with intensive species management and habitat restoration.

ISLAND ENVIRONMENTS AND SPECIES LOSS

Oceanic island ecosystems have a unique evolutionary history as a result of their geographic isolation and small size (Mueller-Dombois 1981). As oceanic islands were formed, initial plant colonizers originated from continental floras. This colonization process resulted in patterns of taxonomic diversity quite different to those found on the source continents. Disharmonious island floras, differing in composition and proportion of plant families in comparison with the source area, share a number of characteristics (derived from Eliasson 1995):

(1) Adaptive radiation: An original founder taxon gives rise to a variety of species utilizing the available ecological niches on an island. The Asteraceae family is a classic example, e.g., *Argyranthemum* spp. in Macaronesia (Ortega, Jansen, and Santos-Guerra 1996) and the Hawaiian silversword alliance (Baldwin, Kyhos, and Dvorak 1990).

(2) Island woodiness: A number of well known plant families and genera exhibit a pattern of woodiness on islands not observed in continental areas, for instance, *Chenopodium* spp. (Chenopodiaceae) in Hawaii and Archipelago Juan Fernández.

(3) Loss of dispersability: Fruits of island species may show a reduced dispersal ability when compared with putative continental ancestors. For instance, endemic species of *Bidens* (Asteraceae) in Hawaii have reduced or no barbed awns associated with the epizoic dispersal characteristic of continental species.

(4) Changes in reproductive biology: The proportion of dioecious species seldom exceeds 3% on continental areas, but often exceeds 10% on island floras.

Patterns of colonization and radiation have resulted in rich floras, characteristically high in endemism, often derived from a limited number of founder taxa. These unique assemblages of plants on oceanic islands are threatened by a variety of human-induced factors. None of these are unique to islands but are most dramatically manifested on the vulnerable island floras. The introduction of exotic animals and plants by man is cited as the single most destructive force acting on Hawaiian ecosystems (Mueller-Dombois and Loope 1990; see chapters 11 and 12). Table 13.1 illustrates some of the major threats to the biodiversity of six oceanic islands/island archipelagos.

Historical Patterns of Habitat Modification and Species Loss

Island degradation is arguably as old as the human colonization of islands. Archaeological evidence suggests that European Mediterranean island ecosystems were dramatically altered by Neolithic settlers (Broodbank and Strasser 1991), although Schüle (1993) proposed that earlier devastation of these islands occurred as a result of colonizing ungulates (e.g., *Myotragus* and *Hippopotamus*) during the Pleistocene. Many Pacific islands were subject to profound environmental changes and high levels of avian extinctions following Polynesian colonization (Diamond 1986; Cuddihy and Stone 1990). One example, the Pacific island of Rapa Nui (Easter Island), underwent massive environmental degradation as long ago as 1200 to 800 B.P. (Flenley *et al.* 1991). In the Caribbean (Watts 1987), Atlantic islands (Grove 1995), and Mascarenes (Cheke 1987; Gade 1985), patterns of environmental degradation can be traced to European colonial administration

Table 13.1

Threats to the botanical biodiversity of six oceanic islands/island archipelagos

Major threats to biodiversity: 1. Early devastation for timber extraction; 2. Development pressure from tourism industry; 3. Continued loss due to exotic animals; 4. Continued loss due to exotic plants; 5. Erosion; 6. Wildfire. (Sources: Davis, Heywood, and Hamilton 1994, 1995)

Islands	Area (km^2)	Popn. density (per km^2)	No. of native vascular plant spp.	No. of endemics:		No. of threatened taxa[3]	% of island protected area	No. of extinct taxa	Major threats to biodiversity
				Genera	Species				
Ascension Island (South Atlantic)	94	11 (1994)	25	0	11	12	0	4	3,4
Hawaiian Islands (132 islands) (Pacific Ocean)	16,641	>60 (1992)	1,200	32	1,000	856[4]	12	97	2,3,4,6
Islas Canarias (7 islands) (Atlantic Ocean)	7,542	192 (1994)	1,200	20	500	475	ca. 4.0	1	2,3,4
Mauritius incl. Rodrigues (Indian Ocean)	1,865	598 (1995)	685	8	311	319	ca. 2.4	39[5]	1,2,3,4
Rapa Nui (Pacific Ocean)	166	13 (1992)[1]	30	0	4	5	ca. 20	1	1
St. Helena (South Atlantic)	122	43 (1991)	60	10(8)[2]	50	51	ca. 2.0	7	1,3,4,5

[1](Bahn and Flenley 1992)

[2]Figure in parentheses refers to the number of monotypic genera.

[3]Number of known threatened taxa. IUCN categories of threat are given at the world level and do not include categories: 'nt', '?' and 'c' (World Conservation Monitoring Center, Threatened Plants Database 1996).

[4]Figure for the island of Hawaii only.

[5](Page personal communication)

and the subsequent development of unstable and fragile economies based upon plantation products (Grove 1995; Brookfield 1985).

Colonial settlement often focused on the clearance and agricultural development of the lowlands, with the retention, planned or otherwise, of upland habitats, albeit in a modified and fragmented form. Examples of islands with severely modified lowlands include Mauritius (Strahm 1994), the Gulf of Guinea Islands (Fa 1991; Figueiredo 1994), and New Caledonia (Bouchet, Jaffre, and Veillon 1995). While the lowlands of Gran Canaria, Islas Canarias, were converted to plantations, the uplands were devastated by deforestation for firewood and by removal of topsoil. On Gran Canaria less than 1% of the original laurel forest now survives (Bramwell 1990).

Some islands have suffered a complete loss of original habitats. This can result from periods of over-exploitation during the course of centuries, e.g., the Polynesian settlement of Rapa Nui and the European settlement of St. Helena. On Rapa Nui, no original plant communities have survived human settlement and Chilean sheep and cattle ranching. On St. Helena, only fragments of gumwood forest and montane tree fern thicket have survived. The total transformation of an oceanic island can also result from industrial exploitation, for instance phosphate mining on Nauru in the Pacific (Williams and Macdonald 1985).

Case Study 1: Ascension. The Release of a Botanic Garden Plant Collection on an Oceanic Island. The archetypal 'island degradation' case study describes a once forested island now reduced to desert. The opposite is true for Ascension, where the effects of exotic species have transformed a desert island into a forested one (Figure 13.1). This profound modification of the island environment and subsequent loss of an original, albeit sparse, ecosystem has taken place with great rapidity and under conditions of both low human population density and restricted land use. The net effect has been a dramatic increase in total botanical diversity, primary productivity, and the creation of a new 'synthetic' ecosystem.

Ascension was discovered in 1501 by the Portuguese navigator, Juan da Nova. The island was a bare volcanic landscape supporting only 25 vascular plant species. Eleven of these are endemic (Cronk 1980). Since 1815, the island has been a United Kingdom (UK) garrison settlement. It has a transient population of about 1,000 with a tiny area of land devoted to agriculture. However, a large number of exotic plants have been introduced for both agricultural and horticultural purposes. For example, following instructions from the botanist, Sir Joseph Hooker, the British government installed a farm manager with instructions to obtain exotic plants for vegetating the island. In 1847, for instance, 700 packets of seed were sent from Kew, and in 1858 the botanic gardens at Kirstenbosch, South Africa, sent an additionl 200 species to the island. Until 1850, consignments of plants and seed were arriving on Ascension on a monthly basis (Duffey 1964). These represented massive influxes of both botanical and associated invertebrate and fungal diversity, which were introduced, via gardens and plots, into a species-poor ecosystem. This practice of large-scale plant introductions continued until relatively recently. Ascension has lost a minimum of four vascular plant species through the impacts of invasives.

Consequently, the original upland *Marattia purparescens* (Marattiaceae) fern thicket has been transformed into an exotic woodland dominated by an introduced bamboo. In the middle altitudes, introduced *Casuarina equesitifolia* (Casuarinaceae), *Araucaria excelsa* (Araucariaceae) and *Juniperus bermudiana* (Cupressaceae) have formed open woodlands. Ironically, *J. bermudiana* is a threatened species in its native Bermuda.

The xeric lowland lavafields of Ascension have been transformed into an open woodland of *Prosopis juliiflora* (Fabaceae), coupled with introduced grass species. *Prosopis*

Figure 13.1
View from Green Mountain towards George Town, Ascension Island, South Atlantic. On discovery, Ascension had a sparse vegetation cover, with large areas of bare lava field. Following a colonial program of plant introductions an exotic forest is covering the island.

juliiflora was introduced in the early 1980s and now covers all lowland habitat in varying densities (Figure 13.2). Introduced feral donkeys are dispersing the seed. The thicket is now encroaching upon the breeding areas of migratory animals of international importance such, as the green turtle (*Chelonia mydas*), and unique colonies of sooty tern (*Sterna fusca*).

Persistence and Extinction

Oceanic islands are suffering high levels of species loss. It is estimated that approximately one third of the world's known threatened plants are island endemics (Synge 1992). It has been proposed that oceanic islands have sustained heavy, but under-recorded, periods of species loss upon human colonization (Balmford 1996; see also chapter 7). As many islands were subject to high levels of habitat degradation prior to scientific description, the exact proportion of each island's flora that has already gone extinct may never be fully recorded. *Paschalococos disperta* (Arecaceae), an extinct cocoid palm from Rapa Nui, only recently described from sub-fossil fruits, is an example of such an unrecorded extinction (Zizka 1991). On St. Helena, the main phase of extinctions probably occurred in the seventeenth and eighteenth centuries (Figure 13.3). Hooker (quoted in Melliss 1875) estimated that the island originally had over 100 endemic species.

Rodrigues was colonized from 1691 onwards with the first systematic botanical survey undertaken in 1874 (Strahm 1989). Balmford (1996) suggested that those islands with a history of human settlement will manifest lower recorded extinction rates than newly

Figure 13.2
Exotic colonization of lavafields on Ascension, South Atlantic. A bare volcanic island has been transformed by a colonial program of plant introductions. The main tree in the lowlands is Prosopis juliiflora *with a ground cover dominated by* Catharanthus roseus *and introduced African forage grasses.*

colonized areas where extinctions are more likely to be recorded. This may explain why some islands have been subject to high levels of recent habitat clearance with relatively low levels of recent extinction. The initial phase of human settlement may have acted as the first extinction filter for extinction prone species. For example, Bramwell (1990) records only one possible extinction, *Solanum nava* (Solanaceae), from the Islas Canarias, where over 67% (383) of the endemic plants are threatened.

It can be argued that even the documented extinctions do not fully reflect the ongoing loss of botanical (and associated) diversity. Populations, or more accurately, scattered individuals, have the ability to persist over long periods of time (see Table 13.2). A large proportion of an island's threatened taxa may persist as isolated, reproductively redundant individuals, surviving outside of viable habitat fragments. Approximately 22% of the endemics on St. Helena survive outside of viable habitats. A prime example from Rodrigues is *Ramosmania heterophylla* (Rubiaceae), the last individual bush surviving in roadside scrub among agricultural plots (Strahm 1989). This last plant is a functionally sterile male (Owens *et al.* 1993).The persistence of these taxa depends on the longevity of these individuals, the 'living dead' (*sensu* Janzen 1986) (see Table 13.2). Therefore, it could be argued that some of these taxa, without horticultural intervention, are already on a trajectory to extinction, illustrating merely a time-lag between almost complete habitat destruction during the last 200 years or more and eventual extinction.

There are practical difficulties in assessing the status of island endemics. Apart from an imperfect historical perspective, it is difficult to physically locate plants. Many

Figure 13.3
Eroded landscape on St. Helena, South Atlantic. Until the eighteenth century this area supported an open woodland of endemic tree species.

populations will persist on cliff areas protected from introduced browsing species or in thickets dominated by exotics. Search priorities based on the distribution of surviving indigenous habitat may not necessarily locate the scattered plants surviving in disturbed habitats. Accordingly, the assignment of categories of threat such as 'Extinct' should not be seen as a permanent appellation; island species have a disconcerting habit of turning up (see Table 13.2).

Hybridization as a Threat to Island Endemics

There is accumulating evidence that island taxa that have undergone recent adaptive radiation are vulnerable to hybridization. This is particularly the case when habitat disturbance allows contact with previously unencountered but indigenous congenerics (Ortega, Jansen, and Santos-Guerra 1996). It has been proposed that hybridization has played an important role in the evolution of some island genera, for instance *Argyranthemum* spp. (Asteraceae) (Levin, Ortega, and Jansen 1996). Patterns of recent hybridization for *Argyranthemum* spp. (Brochmann 1984; Humphries 1976; Ortega, Jansen, and Santos-Guerra 1996) and for *Lavandula* spp. (Lamiaceae) (Humphries 1979; Ortega, Jansen, and Santos-Guerra 1996) have been documented for Islas Canarias, where a local endemic has been swamped by a more weedy but indigenous congeneric. Similar examples are recorded for *Cercocarpus* spp. (Rosaceae) (Rieseberg and Gerber 1995) and *Lotus* spp. (Fabiaceae) (Liston, Rieseberg, and Mistretta 1990) from the Santa Catalina and San Clemente islands, respectively, off the southern Californian coast. Hybridization has

Table 13.2
Persistence of Highly Endangered Island Endemics

Species	Family	Island	Current wild population	Historical decline	Breeding biology	Ease of propagation and cultivation
Commidendrum rotundifolium	Asteraceae	St. Helena	Extinct in the Wild	Last wild tree 1982–1986, blown from cliff.	Dioecious	Poor seed fertility and difficult from cuttings.
Hibiscus liliiflorus	Malvaceae	Rodrigues	2 trees	One tree killed in 1981, two new trees found in 1983.	Monoecious	Readily from seed and cuttings.
Hyophorbe amaricaulis	Arecaceae	Mauritius	1	Last specimen, pre 1950s, still extant.	Self-incompatible	*In vitro* propagation attempted with no success.
Lachanodes arborea	Asteraceae	St. Helena	65	Population discovered in 1977, 89 trees in 1993, 62 trees remaining in 1995.	Dioecious (?)	Poor seed fertility and difficult from cuttings.
Nesiota elliptica	Rhamnaceae	St. Helena	Extinct in the Wild	Last wild tree died in 1994, self-incompatible. Died from systemic fungal infection.	Self-incompatible	Poor seed fertility and difficult from cuttings.
Pandanus macrocarpus	Pandanaceae	Mauritius	Extinct	Last specimen, pre 1975–1995, died.	?	Vegetative propagation 'air-layering' attempted with no success.
Ramosmania heterophylla	Rubiaceae	Rodrigues	1	Last specimen located in 1980, still extant.	Last plant functionally male.	No prospect of seed production, difficult vegetative propagation.
Ruizia cordata	Sterculiaceae	Réunion	ca. 3	21 individuals located over last decade, ca. 10 genotypes represented *ex situ*.	Monoecious	Readily from seed and cuttings.
Sophora toromiro	Fabaceae	Rapa Nui	Extinct in the Wild	Last wild tree felled for firewood in 1960.	Dioecious and self-fertile.	Readily from seed and cuttings.
Trochetiopsis melanoxylon	Sterculiaceae	St. Helena	2	2 plants discovered in 1980. Still extant on cliff.	Dioecious and self-fertile.	Readily from seed and cuttings.

been implicated as a threat to the last surviving individuals of *Argyroxiphium virescens* (Asteraceae) in Hawaii (Loope and Medeiros 1994). Misguided introductions and translocations can unwittingly create hybrid swarms, for instance, the translocation of *Hibiscadelphus giffordianus* and *H. hualalaiensis* (Malvaceae) into Hawaii Volcanoes National Park created a hybrid swarm (Baker and Allen 1976). On St. Helena, an artificial hybrid between *Trochetiopsis melanoxylon* and *T. erythroxylon* (Sterculiaceae) threatens the integrity of both parental species through introgression (Upson and Maunder 1993; Maunder 1995b) (Figure 13.4).

Introduced congenerics may pose a threat, for instance with *Senecio* spp. (Asteraceae) (Gilmer and Kadereit 1989) and *Arbutus* spp. (Ericaceae) (Salas-Pascual, Acebes-Ginoves, and Del Arco-Aguilar 1993) in Islas Canarias. The Mascarene *Hibiscus* spp. (Malvaceae) are threatened by the widespread planting of inter-fertile exotic congenerics as garden and roadside ornamental plantings (Maunder and Culham in press). Similarly, sub-specific taxa of the Mascarene endemic palm genus, *Dictyosperma* (Arecaceae) are threatened by introduced stock from other Mascarene islands planted for the palm heart trade (Strahm 1983; Page and Maunder in press). Similar concerns have been expressed regarding the introduction of nonindigenous provenances of *Dracaena draco* (Agavaceae) into Madeira and *Phoenix canariensis* (Arecaceae) into Islas Canarias (Santos-Guerra personal communication).

Case Study 2: Macaronesian *Echium* sect. *Simplicia*. *Ex situ* management of island endemics can pose considerable risks. During the nineteenth century the giant

Figure 13.4
Plantation of Trochetiopsis *spp. on St. Helena, South Atlantic. The hybrid between* T. erythroxylon *and* T. melanoxylon *stands out amongst a slower growing plantation of* T. melanoxylon. *The very vigorous hybrid can back cross to the parental species.*

monocarpic *Echium* (Boraginaceae) species, section *Simplicia,* became standard components of the European botanic garden flora (Hooker 1902; Perez 1912a, 1912b). The three wild species, *E. pininana*, *E. simplex* and *E. wildpretii*, are separated in the wild by both ecology and geography. The spectacular nature of the plants encouraged collection managers to create artificial hybrids and to promote unintended hybrids when they occurred (Perez 1912b). One wild taxon, *E. wildpretii* ssp. *trichosiphon*, is 'Critically Endangered' (*sensu* IUCN 1994) as a result of grazing by introduced Barbary goat (*Ammotragus laevis*). A survey of wild and cultivated populations of these three species has revealed widespread intra and inter-specific hybridization in *ex situ* collections (Maunder and Culham in press). Accordingly, extreme caution should be exercised in utilizing botanic garden stock for reintroduction that has been exposed to inter-fertile congenerics.

APPROACHES TO BOTANICAL CONSERVATION

Conservation and Sustainable Development

Economic exploitation has systematically and successively focused on the highest value natural resources available at the time, dropping to the next valuable until the point of economic exhaustion. The history of island exploitation started with lumber (e.g., Caribbean mahogany) and slaves and continues today with plantation crops, marine resources, and tourism. It is sobering to note that among the first recorded extinctions from both the Caribbean and Islas Canarias were the indigenous peoples, the Caribs and Guanchez, respectively.

Any conservation planning on small islands should recognize not only the extreme environmental variability (Stoddardt and Walsh 1992) but also their economic and demographic characteristics. Such characteristics include (derived from Bass and Dalal-Clayton 1995):

(1) economic dependence on larger countries for markets, investment and transport,
(2) geographic isolation,
(3) a relatively small human population and hence limited pool of professional skills,
(4) typically high population densities and accordingly strong local demand on resources,
(5) highly finite land resources,
(6) intimate linkage of all island ecosystems; impacts on one will affect other parts,
(7) high ratio of coastline to land area; accordingly vulnerable to marine and climate influences (Loope 1995),
(8) vulnerability of island ecosystems to other external ecological influences, notably introduction of invasives and pathogens (D'Antonio and Dudley 1995), and
(9) restricted and vulnerable watersheds.

Conservation on islands cannot proceed effectively without taking into account the needs and perspectives of the island inhabitants. It is worth noting that among the greatest concerns expressed by the St. Helenans during the public participation and feedback of the Sustainable Environment and Development Strategy (SEDS) mission was the constant stream of UK government consultants working on the island (Royal Botanic Gardens, Kew and International Institute for Environment and Development 1993). An understanding of local political and social issues is vital both in understanding local environmental issues and in maintaining working relationships and collaborations between different constituencies (see chapter 18). For instance shared concerns can generate viable collaborations (e.g., St. Helena SEDS case study).

The massive development of islands in proportion to their land mass has resulted in some of the most densely populated areas of the world, e.g., Mauritius. Today, the lure of their isolation and apparent tranquillity makes islands one of the most popular tourist destinations. Influx as a result of tourism may swell the human population dramatically. For example, Mauritius received over 400,000 tourists in 1995 (Mauritius Government Tourist Office 1996), which is equivalent to approximately one third of the population. Further, the Galápagos Islands in the Pacific Ocean received 46,000 tourists in 1993, which was equivalent to approximately four times the population (derived from Brandon 1996). Ironically, it is the natural landscape, particularly beaches, that tourists seek. Therefore, island degradation via development will ultimately affect tourism (Coccossis 1987).

On some oceanic islands, e.g., Cabo Verde and St. Helena, the fragility of the volcanic substrate and the vulnerability of the few, condensed watersheds to erosion has led to desertification. Low levels of groundwater then lead to saltwater penetration, e.g., British Virgin Islands (Davidson, Teleki, and Lamb 1988). This is a grave problem and, in some cases, freshwater has to be brought in by ship. Increased demands for freshwater supplies from tourism, as well as the predicted rise in sea level as a result of global warming, will exacerbate this problem for low-lying islands (Bass and Dalal-Clayton 1995). Tropical montane cloud forest plays a vital role in the hydrological cycle of oceanic islands. Hamilton, Juvik, and Scatena (1995) documented the distribution of tropical montane cloud forest, demonstrating that approximately one-fifth of the known cloud forest sites were located on oceanic islands. Evergreen upland forest has been shown to double the mean annual rainfall by cloud water interception on La Gomera, Islas Canarias (Gioda *et al.* 1995). Soil and water conservation then become critical issues inseparable from botanical conservation.

Consequently, conservation planning on islands must take place within the context of sustainable development planning and the application of *Agenda 21*. *Agenda 21* was drawn up to provide a framework for the issues addressed at the United Nations Conference on Environment and Development (UNCED 1992). This 'action plan for UNCED' (Dalal-Clayton *et al.* 1994) is an example of 'international soft law' (Johnson 1993). Although it is not legally binding, it has moral force to support national sustainable development plans, the need for which are highlighted throughout the *Agenda 21* document. This is particularly applicable to those densely populated islands that are experiencing large-scale environmental degradation (e.g., the Comoros, Indian Ocean/Mozambique Channel), or undergoing economic transition driven by new crops or tourism (e.g., Galápagos Islands).

Case Study 3: St. Helena, South Atlantic Ocean—Sustainable Environment and Development Strategy. The Sustainable Environment and Development Strategy (SEDS) for St. Helena was an attempt to review the environmental and development processes underway in a remote South Atlantic Island, as well as an attempt to promote a cross-sectoral and participatory approach to resource management (Maunder *et al.* 1995). The island has an approximate area of 122 km^2. It is a Dependent Territory of the United Kingdom with a population of ca. 5,000. Islanders are fiercely proud of both their island and of their UK nationality. Any discussion on the application of sustainable development on St. Helena needs to take the following local factors into account:

(1) No unmodified natural communities or ecological processes survive. The ecology of all habitats has been profoundly modified through extinction, changes in ecological processes (for instance loss of island sea-bird colonies) and exotic invasion.

(2) The vast proportion of the island (over 60%) is a manmade landscape resulting from centuries of agriculture and over-grazing (Figure 13.5).

(3) Habitat distribution and agricultural usage are highly correlated to altitude and local climate.

(4) St. Helena is a settled island with a resident population; therefore any protected area development must take into account the views of the populace.

The primary concerns of the islanders were identified through questionnaires and interviews. To a large extent the main points of public concern overlapped with biodiversity issues (refer to Table 13.3). The involvement of the island population in natural resource management is fundamental to the conservation of St. Helena's biota.

Within the island government, a SEDS Response Committee has been established to promote cross-disciplinary planning and decision-making. In 1994, a government-based Advisory Committee on Environmental Affairs was created to move the SEDS process on (Bass and Dalal-Clayton 1995). A series of studies were undertaken by the International Institute for Biological Control to initiate an island-wide integrated pest management

Figure 13.5
Degraded rangeland, island of St. Helena, South Atlantic. Approximately 60% of the island's land area is an eroded wasteland dominated by exotic scrub.

Table 13.3
Major Environmental Concerns on St. Helena, South Atlantic (derived from Royal Botanic Gardens, Kew in association with International Institute for Environment and Development 1993)

Concerns expressed by St. Helenans, ranked in order of importance	Biodiversity Linkages
Food security (ranked 1)	Need to increase island food production, linked to integrated pest management and organic agriculture.
Education (ranked 2)	Requirement to increase environmental understanding and involvement of islanders.
Soil Erosion (ranked 2)	Retention of forest cover and promotion of regeneration on the "Crown Wastes"
International marine pollution and over-fishing (ranked 2)	Control of deep sea fishing and traffic in territorial waters.
Water shortages (ranked 3)	Protection and enhancement of watersheds.
No cash for environmental work (ranked 3)	Need to locate UK and international funds for island-led conservation initiatives.
Increase in pests (ranked 6)	Need for integrated pest management to support local agriculture and protect endemic flora, with quarantine as a major priority.
Regeneration of natural vegetation (ranked 7)	Need for Biodiversity Action Plan.

scheme (Fowler 1993). Priority areas for botanical conservation and soil restoration have been identified (Maunder 1995b) and incorporated within a strategic land use plan. The island's first national park has been established on Diana's Peak (Smith and Williams 1996). The St. Helena Nature Conservation Group has been formed and small scale agricultural initiatives are being developed by land owners and the island government. However, lack of external funding is hampering the implementation of many of the SEDS action plans and there are indications that the government will use it mainly as an internal planning exercise (Bass and Dalal-Clayton 1995).

Introduced Pests and Quarantine

The impacts of introduced mammals and plants are well known, as they are easily documented and can have catastrophic impacts. The effects of rodents can be profound: evidence is accumulating that rats are a major inhibitor of palm and broad-leaf forest species regeneration in the Mascarenes (see also their affects on island reptiles in chapter 7). Restoration projects in the Mascarenes will, accordingly, require extensive rodent control programs to ensure regeneration (Carl Jones personal communication).

Among the greatest threats to the long-term survival of St. Helena's unique biota are the exotic plant and pest species. There is no evidence that the rate of introduction is slowing down (Fowler 1993). All the surviving habitats are threatened by invasive plants, most notably the New Zealand flax, *Phormium tenax* (Agavaceae), expanding from abandoned plantation fields.

Introduced insect pests are recognized increasingly as a major concern impacting both crops and endemics. St. Helenans are unable to grow citrus crops because of Mediterranean fruit fly (*Ceratitis capitata*). Similarly, householders have now stopped growing cucurbits because of the Pumpkin fly (*Bactrocera cucurbitae*) (Fowler 1993). A number of endemic species are being adversely affected by introduced pests. For instance, the Vine weevil (*Otiorhynchus* sp.) has caused heavy foliar damage to *Trochetiopsis* spp.

(Sterculiaceae) reintroductions, while as yet unidentified taeniad moths are contributing to the decline of populations of the Cabbage Tree, *Lachanodes arborea* (Asteraceae). An introduced homopteran (*Orthezia insignis*) nearly destroyed the main population of the endemic Gumwood *Commidendrum robustum* (Asteraceae). The rapid introduction of a biological control agent (*Hyperaspis pantherina*) to curtail *O. insignis* populations prevented the certain extinction of the Gumwood in the wild (Fowler 1993). It has been strongly recommended that strict phytosanitary measures be adopted to minimize the risk of new species establishing on St. Helena (Fowler 1993; Maunder 1995b).

The impacts of exotics are perhaps nowhere as dramatic as in the Hawaiian islands (see chapter 12). Nonindigenous species have been identified as an important, if not the most important, current threat to the islands' indigenous flora (Cuddihy and Stone 1990; Mueller-Dombois and Loope 1990; Vitousek, Loope and Stone 1987). For a 50-year period, between 1937 and 1987, Hawaii received an average of eighteen new insect and other arthropod species annually (Anon. 1987). This threat can best be dealt with through the promotion of interagency cooperation (Anon. 1992; U.S. Congress, Office of Technology Assessment 1993), with a particular focus on pre-entry prevention.

Integrated Species Management for Fragmented Non-regenerating Populations

Case Study 4: Mascarene Palms, *Tectiphiala ferox* (Arecaceae). The endemic palm flora of Mauritius and Rodrigues contains a high proportion of threatened taxa. From a total of eleven taxa, six have wild populations estimated to number less than 40 individuals, and two taxa are represented in the wild by single individuals only. One example is the monotypic genus, *Tectiphiala* (Arecaceae), endemic to Mauritius, which was only described in 1978. The wild population is restricted to the heath-forest communities of upland southwest Mauritius (Page and Maunder in press). It is estimated that less than 40 adult individuals survive in fragmented habitat pockets. No natural regeneration has been observed since the discovery of this species. The precise reasons for the lack of reproductive success are not known but are suspected to include rat predation on fruit and seedlings, monkey predation on fruit, and the swamping of regeneration niches by exotic weeds such as *Ravenala madagascariensis* (Strelitziaceae; Traveller's palm). In addition, many adult individuals appear to produce sterile inflorescences.

The conservation of this genus is dependent upon a number of actions (derived from Page and Maunder in press), including the (1) monitoring of all wild specimens; (2) prospecting for new populations and individuals; (3) habitat restoration for extant populations through weeding and fencing; (4) netting and collection of fruit for *ex situ* seed sowing and subsequent replanting of seedlings; (5) establishment of field gene banks and new wild populations; and (6) research on reproductive biology.

The Mauritian Wildlife Foundation has recently propagated this species and reintroductions are underway. Young plants have been introduced to the Mare Longue Reserve and the recent completion of weeding in the Petrin and Florin reserves will make available further reintroduction sites.

Case Study 5: St. Helenan Endemics. A large proportion of the island's endemic angiosperms survive as scattered individuals, i.e., the remnants of once larger populations (Figure 13.6). It is likely that many of these species (e.g., *Trochetiopis erythroxylon*,

Figure 13.6
The last wild population of the endemic, Lachanodes arborea, *St. Helena, South Atlantic. About 60 trees survive on a fence line between pastures. This population has declined from 89 trees in 1993 and shows little evidence of regeneration.*

T. melanoxylon, (Sterculiaceae) *Commidendrum* spp. (Asteraceae) and *Nesiota elliptica* (Rhamnaceae)) have been surviving in this way for over 200 years.

A total of thirteen species have a high chance of extinction if intensive conservation intervention is not undertaken (Table 13.4). In the short term, conservation efforts must focus on the protection of the available individuals and, in parallel, the establishment of secure cultivated stocks prior to attempting to re-establish a wild reproducing population. Importantly, there are a significant number of taxa that survive only because of timely intervention on the island. For instance, three species now 'Extinct in the Wild' survive because of horticultural intervention, and eight 'Conservation Dependent' species have increased in number because of protection from grazing.

The Agriculture and Forestry Department and the Department of Development and Planning have identified the main areas of surviving habitat, and management guidelines have been developed for a series of 'Endemic Management Zones' (Maunder 1995b). In 1996, the main area of surviving tree fern thicket on Diana's Peak, believed to be important in the maintenance of the watershed, was gazetted as the island's first national park (Smith and Williams 1996).

To ensure the survival of the endemic taxa, the following actions are required:

(1) location and salvage (propagation) of all available individuals,
(2) genetic assessment of surviving individuals,
(3) effective control program of exotic species utilizing mechanical, chemical and biological techniques,

Table 13.4
Status of Priority Species, St. Helena, South Atlantic (derived from Maunder 1995b)
EW = 'Extinct in the Wild', CR = 'Critically Endangered' (IUCN 1994). Notes: 1. Heavily reduced or no natural regeneration; 2. Largest proportion of the wild population (>80%) at one site; 3. Disease or pest problems; 4. Contamination of local genotypes; 5. Relictual population(s) or individuals surviving outside of natural vegetation; 6. Breeding biology problems, manifest through low fertility; 7. Hybridization.

Taxon	IUCN Category	Wild Pop.	Cult. Pop.	No. of Populations (with estimated numbers)	Notes
1. *Commidendrum robustum* (Asteraceae)	CR	ca. 2,500	ca. 500	1(2,500), 2(5), 3(1), 4(20)	2,3,4
2. *C. rotundifolium*	EW	0	23	N/A	6
3. *C. spurium*	CR	9	ca. 20	1(6), 2(2), 3(1)	1,2,5,6
4. *Lachanodes arborea* (Asteraceae)	CR	65	10–20	1(65)	1,2,3,5,6
5. *Nesiota elliptica* (Rhamnaceae)	EW	0	5	N/A	6
6. *Nesohedyotis arborea* (Rubiaceae)	CR	ca. 200–250	10	?	
7. *Petrobium arboreum* (Asteraceae)	CR	ca. 200–250	0	N/A	
8. *Phylica polifolia* (Rhamnaceae)	CR	25–30	0	1 (ca. 10), 2(1), 3(6), 4(ca. 12)	1,5
9. *Pladaroxylon leucadendron* (Asteraceae)	CR	ca. 200	0	5 pop.	1,3,6
10. *Sium burchellii* (Apiaceae)	CR	ca. 200	35	4 pop.	
11. *Trochetiopsis erythroxylon* (Sterculiaceae)	EW	0	'000s	N/A	7
12. *T. melanoxylon*	CR	2	'000s	1(2)	1,2,7
13. *Wahlenbergia linifolia* (Campanulaceae)	CR	30–60	0	2 pop.	7(?)

(4) establishment of permanent field gene banks, at middle and high altitudes,

(5) development of horticultural protocols and a training program for the care of endemic taxa, and

(6) applied research on habitat restoration, linked to watershed enhancement.

Ex situ *Management of Island Taxa*

The very low population sizes and the associated high risks of extinction mean that *ex situ* conservation will play an important role in conserving island species. It is only through the establishment of local facilities that *ex situ* conservation can play a real role. Cultivated stocks should be seen as a support for *in situ* conservation (see chapter 5), in terms of genetic management, publicity, and fundraising.

In a small number of cases, stock from international collections has been utilized for conservation programs (refer to *Sophora toromiro* case study). This process presents a number of risks, with the repatriated plants subject to high levels of mortality during and after transit (Owadally, Dulloo, and Strahm 1991). Another risk is the inadvertent introduction of pathogens from cultivated collections, for instance the occurrence of Cucumber Mosaic virus in cultivated populations of the threatened New Zealand endemic, the Chatham Island forget-me-not, *Myosotidium hortensia* (Boraginaceae) (Thomson 1981).

Stocks of threatened species held within botanic gardens, or similar institutes, are characterized by a number of shared genetic and demographic features (derived from Maunder and Culham in press), namely:

(1) numerically small and/or closely related founder stocks for cultivated populations,
(2) fluctuating cultivated population size,
(3) little or no associated ecological/biological information,
(4) poorly documented cultivation history, often no satisfactory horticultural protocols,
(5) stock scattered through a number of collections with varying horticultural/curatorial skills,
(6) long history in cultivation, often over decades or several generations,
(7) held in mixed collections with high risk of pathogen transfer, and
(8) open to artificial selection (horticultural tolerance and merit), genetic drift, inbreeding and hybridization.

Ex situ facilities should focus on the needs of one island, archipelago, or a particular protected area and avoid the tendency to develop mixed exotic collections. Priority should be given to reducing the risks of hybridization and transfer of pathogens between and to islands. Most colonial botanic gardens were developed in lowland areas, often now in restricted urban sites. In islands such as Mauritius, Réunion and Hawaii there is an urgent need to establish upland *ex situ* facilities (Center for Plant Conservation 1994). This would suggest a dual system where lowland old-colonial gardens, with existing and largely undocumented exotic collections (frequent in the Caribbean and Indian Ocean) act primarily as centers for environmental education supplemented by new dedicated facilities at appropriate altitudes that provide field gene banks for indigenous species. The main concern should be the provision of horticultural facilities to propagate and hold adequate samples of threatened plant species. In addition, tools are needed to identify candidate species for *ex situ* conservation. A Conservation Assessment Management Plan (CAMP) workshop is such a tool. Information is gathered on the species' distribution and range. Categories of threat are assessed based upon available information and recommendations are made for further research and species management including, if appropriate, *ex situ* management. A CAMP workshop was convened with the Conservation Breeding Specialist Group of the IUCN to assess the St. Helenan biota, the first time this tool was used to assess an entire island biota (Maunder *et al.* 1994).

For *ex situ* management, particular emphasis should be given to the development of conventional horticultural skills. The vast majority of candidate species will require propagation by conventional vegetative and seed techniques. More expensive capital investments, such as micropropagation facilities, should be developed on a regional level and should not be regarded as obligatory for local facilities. Micropropagation can play a key role in propagating difficult material (Krogstrup, Norgaard, and Hamman 1990) and storing it by cryopreservation. However, *in vitro* material needs care on weaning to conventional cultivation and monitoring against the risk of somaclonal variation. Seed banks can play an invaluable role in maintaining population samples for species with conventional seeds.

The value of *ex situ* conservation to taxon survival can be demonstrated by the number of taxa surviving only in cultivation (Table 13.5). For instance, between six and eight Hawaiian taxa are now thought to survive only in cultivation (G. Ray personal communication). However, these species are all descended from very few founders and not all are secure in cultivation nor represented by large and genetically diverse *ex situ* populations.

Although salvaged from extirpation in the wild, some species are still not secure. One example is the Bastard Gumwood, *Commidendrum rotundifolium* (Asteraceae), which survives only in cultivation on St. Helena. The last wild specimen died in 1986, but served as the founder for all known extant individuals. Between 1983 and 1992, a total of 52 individuals were planted in a domestic garden acting as a field gene bank. A survey of the plantings in February 1995 revealed a total of only 22 plants (57% mortality). Between 1986 and 1992, 84 individuals have been planted at Horse Point Plain, a local nature reserve; five remained in 1993, and one remained in early 1995 (98% mortality) (Maunder 1995b). This suggests that for highly threatened species priority should be placed on the creation of secure field gene banks, holding duplicate samples of available founders prior to utilizing valuable individuals for reintroduction. *Nesiota elliptica* (Rhamnaceae) on St. Helena survives as only four individuals in cultivation derived from one founder (Maunder 1995b).

Forecasting tools are required to catch such taxa earlier in the process of decline. On Rodrigues, prior to the development of *ex situ* propagation programs, a survey was undertaken to identify all potential individuals for inclusion in field gene banks. Indeed, the poor sampling of remnant individuals can have profound impacts on the outcome of a reintroduction, an example being the use of seed from only one or two maternal parents from the wild population of the Hawaiian endemic, *Argyroxiphium sandwicense* subsp. *sandwicense* (Asteraceae) (Rieseberg and Swensen 1996). This resulted in outplants with

Table 13.5
Some Island Species 'Extinct in the Wild'

Taxon	Origin	Notes
Commidendrum rotundifolium (Asteraceae)	St. Helena, South Atlantic	ditto
Dombeya rodriguesiana (Sterculiaceae)	Rodrigues, Indian Ocean	Survives in Forestry Department nursery on Rodrigues.
Hyophorbe amaricaulis (Arecaceae)	Mauritius, Indian Ocean	Last individual tree, a wild specimen incorporated into botanic garden.
Kokia cookei (Malvaceae)	Hawaii, Pacific	In cultivation in Hawaiian botanic gardens.
Lysimachia minoricensis (Primulaceae)	Minorca, Mediterranean	Established in European commercial nursery trade.
Nesiota elliptica (Rhamnaceae)	St. Helena, South Atlantic	Maintained by St. Helena Agriculture and Forestry Section nurseries.
Sophora toromiro (Leguminosae)	Rapa Nui, Pacific	Stock in Chilean private collections, European, Australian and American botanic gardens.
Trochetiopsis erythroxylon (Sterculiaceae)	St. Helena, South Atlantic	ditto
Wahlenbergia larrainii (Campanulaceae)	Archepelago Juan Fernández, Pacific	Grown in the Viná del Mar Botanic Garden, Chile.

poor genetic representation, manifesting an atypical growth form and high levels of self-incompatibility.

In Situ *Conservation and Habitat Restoration*

Many oceanic islands have suffered massive losses of terrestrial habitat, with lowland habitats being particularly prone to clearance. As a matter of priority, protected areas should be established to secure available fragments of habitat. Surviving habitat fragments are vulnerable to invasive species, pests/pathogens, and chemical disturbance even if protected against incursions by wood cutters and livestock. Thus the establishment of protected areas will not guarantee the survival of a habitat and its component biodiversity; at best, it will retard the processes of loss and degradation. A parallel commitment to habitat management and restoration may be needed to secure the original habitat fragments.

For some islands, the main phase of habitat clearance has passed. Some islands such as Mauritius are moving towards an economy with an increased emphasis on tourism and light industry, and accordingly, pressure from agricultural land may decline. Thus, the decline in monoculture crops for export is making available areas of land for potential restoration and rehabilitation projects. The opportunity exists to expand, through restoration, areas of habitat adjacent to retained areas. This will serve both to increase the area of potential occupancy for threatened species and to compensate, to some extent, for inevitable habitat degradation.

Restoration, or more accurately, rehabilitation, of island environments, is not a new topic. J.R. Forster (quoted in Grove 1995), traveling in the eighteenth century proposed exotic plantings to 'improve' the degraded landscape of Rapa Nui:

> "seeds of various plants as yams and coconuts, with Breadfruit and all kinds of trees that love a hilly country would do exceedingly well and cause more moisture by their shade and procure rivulets, to this now poor, parched-up land."

A similar philosophy has led to the complete transformation of Ascension Island (refer to Ascension Island case study).

Restoration of species, where possible, should seek to integrate with local agricultural needs and landscape services while improving and expanding areas of available habitat. Restoration work for indigenous tree fern thicket on St. Helena is working, within the context of watershed security, while weed control for *Lantana camara* (Verbenaceae) in the Mascarenes benefits both conservationist and local grazier. Ongoing studies (e.g., Brown and Lugo 1994; Lugo, Parrotta, and Brown 1993) suggest that exotic plantations, if carefully managed, can provide a crop and encourage the establishment of indigenous species. This may be necessary on highly degraded sites, e.g., Rapa Nui, where an initial nurse crop is needed.

It is apparent from experience in the Mascarenes that restoration is a complex process that must be supported by a full commitment to the control of exotic invasive species (see chapter 11). Habitat restoration on Mauritius involves a number of parallel activities (Jones and Hartley 1995; Strahm personal communication). These include, amongst others, fencing of plots to exclude pigs, deer, etc.; eradication of rats, feral cats and mongoose which are able to climb protective fences; regular weeding of plots; regular monitoring of regeneration; monitoring of threatened bird populations utilizing restored areas; translocation of understorey species (e.g., terrestrial orchids) threatened by pigs and deer, to fenced reserves; and reintroduction of threatened woody plant species.

The establishment of fenced, highly managed, small plots and reserves cannot be regarded as a long-term conservation policy. It is, however, a necessary policy where fragile populations are surrounded by encroaching threats that are beyond the control of the conservation manager. For instance, on Mauritius, feral pigs are destroying colonies of native terrestrial orchids. Accordingly, a number of species are being translocated to fenced plots (Page and Maunder unpublished data). The introduced Barbary goat (*Ammotragus laevis*) is devastating montane communities on Tenerife and La Palma, Islas Canarias. Fenced plots have been established to preserve populations of threatened endemics (e.g., *Echium gentianoides* (Boraginaceae) and *Genista benehoavensis* (Fabaceae)) in the Tabouriente National Park, La Palma, Islas Canarias. The fencing of reserves is an expensive and labor demanding exercise. Installation of 45 miles of fencing around Haleakala National Park, Hawaii, cost $2.4 million, with an annual maintenance of $130,000 (US Office of Technology Assessment 1993).

Off-shore islands, which are theoretically more secure from invasion than main island sites, can be developed as reserve areas (if suitable), in terms of hydrology and climate, as alternatives to maintaining highly managed enclaves. The restoration of off-shore islands has been utilized for conservation purposes in New Zealand (Towns, Daugherty, and Atkinson 1990), Bermuda (Wingate 1985), Mauritius (Jones and Hartley 1995; North, Bullock, and Dulloo 1994). Atkinson (1990) has reviewed the topic and categorized island reserves according to their conservation goals and level of intervention and management. The increasing need for restoration activities on tropical islands will need a commitment to supportive research on seed and plant ecology and planting techniques, as undertaken by Ray and Brown (1994).

Reintroduction

The reintroduction of taxa has been promoted widely as a technique for reinstating wild populations (Falk and Olwell 1992; Maunder 1992). However, for oceanic islands particular problems are associated with species reintroduction. Habitat destruction has, in some cases been complete, so that no available habitat for reintroduction remains, e.g., Rapa Nui and lowland St. Helena. In addition, the greatest period of habitat destruction often preceded any botanical or ecological documentation. Goals for restoration and reintroduction in these cases therefore must be based on an interpretation of historical records, on surviving habitat fragments, and on herbaria specimens. Major ecological processes may have been disrupted or destroyed over historical time, e.g., the loss of massive sea-bird colonies will have changed the nutrient status of St. Helena's terrestrial ecosystems. Additionally, colonial settlement was accompanied by periods of severe soil erosion. Accordingly, reintroduction/restoration is proceeding on a highly modified edaphic environment.

Case Study 6: The Toromiro tree of Rapa Nui. In 1960, the last specimen of the endemic tree, *Sophora toromiro* (Fabaceae) was felled for firewood on Rapa Nui. The species is 'Extinct in the Wild' and survives only in cultivation (Aldén and Zizka 1989; Lobin and Barthlott 1988). There has been a minimum of thirteen unsuccessful reintroduction attempts for *S. toromiro* between 1965 and 1994. These have involved small numbers of plants; in some cases only two or three. Some of these reintroductions were kept secret from local authorities (Christensen and Schlatzer 1993).

A multi-disciplinary research team, The Toromiro Management Group (Maunder 1995a), has subsequently been established, incorporating Chilean government agencies and private growers, and European, American, and Australian botanic gardens and research groups. The work of this group is to ensure sound genetic management of cultivated stocks and to establish secure populations of the Toromiro in cultivation and, importantly, on Rapa Nui.

Until recently, the only known *S. toromiro* material of definite identity was descended from a Thor Heyerdahl collection made in 1958 and subsequently maintained by the Göteborg Botanic Gardens. In addition to the European stocks, there were putative Toromiro trees in Chile, New Zealand, and Australia. A program of genetic screening has been completed using RAPDs (Welsh and McClelland 1990; Williams *et al.* 1990) and Anchored Microsatellites (Charters *et al.* 1996) to assess surviving diversity within the species. Genetic analysis confirms that a number of true toromiro trees do exist in the Chilean, European, and Australian collections (Maunder, Culham, and Wilkinson unpublished data). However, it is likely that all the known trees are derived from a single founder. Plantations have been established in botanic gardens in Chile (Viña del Mar), France (Menton), and Hawaii (Waimea). An experimental reintroduction using stock from Bonn and Göteborg was started in May 1995. This trial is being undertaken to assess logistical and environmental difficulties before using stock from the breeding program.

Networks and Regional Programs

Islands are by definition isolated. However, the many small island states share similar environmental problems. Networks to support plant conservation are developing at the national, regional, and international levels. For example, the Hawaiian islands contain a large number of threatened plants, with 375 (G. Ray personal communication) of the 1,100 indigenous flowering plant species in danger of extinction. A number of these are at very low population levels, with fifteen taxa persisting as single species left in the wild (Center for Plant Conservation 1994). As a response to both the scale and complexity of the problem, the Center for Plant Conservation (Hawaii) is facilitating and working with two groups, the Hawaii Endangered Plant Task Force and the Hawaii Rare Plant Restoration Group, to promote integrated conservation programs (Center for Plant Conservation 1994). A third networking organization is the U.S. Fish and Wildlife Service-appointed Hawaii Pacific Plant Recovery Coordinating Committee. These activities are linking governmental and private land owners, protected area and legal authorities, botanic gardens, research and community groups. An action plan has been produced outlining an integrated conservation program and research requirements (Center for Plant Conservation 1994).

Regional programs are also being promoted by the Indian Ocean Commission of the European Union. This program is funding the completion of the *Flore des Mascareignes* and supporting other plant conservation efforts in the Mascarenes and Seychelles. The South Pacific Biodiversity Conservation Program (Miller and Manou 1995) is another coordinated regional program, which established a series of "locally owned and locally managed" conservation areas in the South Pacific region. Within the Caribbean region, a number of agencies are coordinating environmental activities; examples include United Nations Environment Program, Regional Coordinating Unit for the Caribbean, Organization of Eastern Caribbean States, Natural Resources Management Unit, and the Caribbean Conservation Association.

The Programme of Action for the Sustainable Development of Small Island Developing States (UNESCO 1994) makes clear recommendations for activities at the national and regional level. These include:

(1) "Formulate and implement integrated strategies for the conservation and sustainable use of terrestrial and marine biodiversity, in particular endemic species, including protection from the introduction of certain non-indigenous species and the identification of sites of high biological significance . . ."

(2) "Ratify and implement the Convention on Biological Diversity . . . and other relevant international and regional conventions."

(3) "Promote community support for the conservation of biological diversity and the designation of protected areas by concentrating on educational strategies that increase awareness of the significance of biodiversity conservation . . ."

(4) "Generate and maintain buffer stocks or gene banks . . . for reintroduction into their natural habitat . . . especially in the case of post-disaster restoration and rehabilitation."

(5) "Develop or continue studies and research on biological resources . . ."

(6) "Conduct detailed inventories of existing flora, fauna, and ecosystems to provide basic data needed for the preservation of biodiversity."

(7) "Ensure that the ownership of intellectual property rights is adequately and effectively protected," and

(8) "Support the involvement of NGOs, women, indigenous people and other major groups, as well as fishing communities and farmers, in the conservation and sustainable use of biodiversity. . . ."

On a political level, the conservation activities on the UK's remaining Dependent Territories are supported by the UK-based NGO, the UK Dependent Territories Conservation Forum, which supports the development of conservation NGOs in the Dependent Territories.

Lastly, the Species Survival Commission of the World Conservation Union (IUCN) has a number of specialist groups supporting botanical conservation needs on oceanic islands. The Specialist Groups are voluntary networks of conservationists collecting information and identifying priorities and mechanisms for action. They include an Island Plants Specialist Group (with regional chapters in the Atlantic and Indian Oceans), Invasive Species Specialist Group, and the Re-introduction Specialist Group.

CONCLUSIONS

Island flora face immediate and varied threats. In contrast with continental areas, a large and increasing proportion of the threatened plant taxa have been reduced to dangerously low population numbers where they are highly vulnerable to stochastic and genetic impacts. Botanical conservation on oceanic islands can no longer be seen in isolation. It is inextricably bound to the conservation of watersheds, soil fertility, and ground water supplies. With relatively small land areas and often dense human populations, the response to environmental degradation and species loss is necessarily part biological and part political and economic. The creation of protected areas is fundamental to the conservation of island floras. However, as land comes under increasing pressure for development, "management strategies prohibiting resource use by local people . . . are difficult to implement" (Ishwaran 1993). Community-backed conservation initiatives are essential.

National activities will need investment to support an integrated island program such as the one developed by the Center of Plant Conservation for Hawaii (CPC 1994). This should incorporate reserve management, monitoring, and the development of *ex situ* facilities. *Ex situ* conservation should always complement (not replace) *in situ* conservation. Plant populations should be intensively managed and coordinated between institutions for genetic and demographic security. At least two island endemics "Extinct in the Wild," *Astiria rosea* (Sterculiaceae, Mauritius) and *Streblorrhiza speciosa* (Fabaceae, Phillip Island), have been lost entirely from cultivation. It could be argued that the greatest contemporary challenge for *ex situ* botanical conservation is the conservation of endangered island species, as island floras continue to manifest high levels of extirpation. Only intensive habitat management and *ex situ* salvage of individual specimens will prevent extirpation becoming extinction.

Lessons need to be learned from zoo captive breeding programs to ensure a more comprehensive approach to botanical conservation on oceanic islands through genetic screening, creation of field gene banks, population and habitat viability analysis, and skilled horticultural intervention. Many of these specialist fields of expertise are being utilized, for example, for the Toromiro species recovery program.

At the regional level, island networks and NGOs should foster coordination with regard to invasive species management and quarantine. These species are probably the greatest threat to island floras. In-country training and capacity building will also be important. Regional, issue-based environment strategy planning is seen as the way forward for Small Island Developing States in terms of sustainable environment and development planning (Bass and Dalal-Clayton 1995).

ACKNOWLEDGMENTS

Our thanks go to Harriett Gillett and Charlotte Jenkins of the World Conservation Monitoring Centre, UK; Björn Aldén, Alberto Bordeu, Sabine Glissman-Gough, Wolfram Lobin, Yves Monnier, Cathérine Orliac, Marcia Ricci, Keith Woolliams and Georg Zizka of the Toromiro Management Group; Gary Ray of CPC, Hawaii; Jean Yves Lesouef of Conservatoire Botanique National de Brest, France; Christian Vericel, Réunion; Wendy Strahm, IUCN-SSC; Ehsan Dulloo, Carl Jones, Wayne Page and Pierre Baissac, Mauritius; Colin Parberry, Rodrigues; George Benjamin, St. Helena; Arnoldo Santos-Guerra, Tenerife; Mike Wilson, ODA; Bryan Spooner, Global Impacts; Barry Dalal-Clayton, IIED; and Mike Sinnott, Peter Atkinson, Sarah Higgens and Graham Lyons of RBG Kew for invaluable support.

LITERATURE CITED

Aldén, B. and G. Zizka. 1989. Der Toromiro (*Sophora toromiro*) eine ausgestorbene Pflanze wird wiederentdeckt. *Natur und Museum* 119:145–152.

Anon. 1987. *Hawaii Department of Agriculture, Annual Report.* Hawaii, USA.

Anon. 1992. *The alien pest species invasion in Hawaii: background study and recommendations for interagency planning.* The Nature Conservancy of Hawaii and Natural Resources Defense Council.

Atkinson, I.A.E. 1990. Ecological restoration on islands: Prerequisites for success. In *Ecological restoration of New Zealand Islands,* Conservation Sciences Publication No. 2, eds. D.R. Towns, C.H. Daugherty, and I.A.E. Atkinson, 73–90. Wellington: Department of Conservation.

Bahn, P. and J. Flenley. 1992. *Easter Island earth island.* London: Thames and Hudson Ltd.

Baker, K. and S. Allen. 1976. Studies on the endemic Hawaiian genus *Hibiscadelphus* (Haukauahiwi). *Proceedings from the First Conference in Natural Sciences,* 19–22. Honolulu: University of Hawaii.

Baldwin, B.G., D.W. Kyhos, and J. Dvorak. 1990. Chloroplast DNA evolution and adaptive radiation in the Hawaiian silver sword alliance (Asteraceae-Madiinae). *Annals of the Missouri Botanic Garden* 77:96–109.

Balmford, A. 1996. Extinction filters and current resilience: The significance of past selection pressures for conservation biology. *Trends in Ecology and Evolution* 11:193–196.

Bass, S. and B. Dalal-Clayton. 1995. *Small Island States and Sustainable Development: Strategic Issues and Experience*, 3, Environmental Planning Issues, No 8, Environmental Planning Group, London: International Institute for Environment and Development.

Bouchet, P., T. Jaffre, and J.M. Veillon. 1995. Plant extinction in New Caledonia: Protection of the sclerophyll forests urgently needed. *Biodiversity and Conservation* 4:415–428.

Bramwell, D. 1990. Conserving Biodiversity in the Canary Islands. *Annals of the Missouri Botanic Garden* 77:28–37.

Brandon, K. 1996. *Ecotourism and conservation: a review of key issues*, Environment Department Papers, Biodiversity Series, Global Environment Division, Discussion Paper, World Bank.

Brochmann, C. 1984. Hybridization and distribution of *Argyranthemum coronopifolium* (Asteraceae-Anthemidae) in the Canary Islands. *Nordic Journal of Botany* 4:729–736.

Broodbank, C. and T.F. Strasser. 1991. Migrant farmers and the Neolithic colonisation of Crete. *Antiquity* 65:233–245.

Brookfield, H.C. 1985. Problems of monoculture and diversification in a sugar island: Mauritius. *Economic Geography* 34:25–40.

Brown, S. and A.E. Lugo. 1994. Rehabilitation of tropical lands: A key to sustaining development. *Restoration Ecology* 2:97–111.

Center for Plant Conservation (CPC). 1994. *An action plan for conserving Hawaiian plant diversity.* Missouri: Center for Plant Conservation.

Charters, Y.M., A. Robertson, M.J. Wilkinson, and G. Ramsay. 1996. PCR analysis of oilseed rape cultivars (*Brassica napus* L. ssp. *oleifera*) using 5' anchored simple sequence repeat (SSR) primers. *Theoretical and Applied Genetics* 92:442–447.

Cheke, A.S. 1987. An ecological history of the Mascarene Islands, with particular reference to extinctions and introductions of land vertebrates. In *Studies of Mascarene Island Birds*, ed. A.W. Diamond, 5–89. Cambridge: Cambridge University Press.

Christensen, S.S. and G. Schlatzer. 1993. Comments on the conservation of *Sophora toromiro* Skottsb., from Rapa Nui. *Botanic Gardens Conservation News* 2:22–25.

Coccossis, H.N. 1987. Planning for islands. *Ekistics*, 323/324, March/April–May/June, 84–87.

Cronk, Q.C.B. 1980. Extinction and survival in the endemic vascular flora of Ascension Island. *Biological Conservation* 17:207–219.

Cuddihy, L.W. and C.P. Stone. 1990. *Alteration of native Hawaiian vegetation: Effects of humans, their activities and introductions.* Honolulu: University of Hawaii Press.

Dalal-Clayton, B., S. Bass, B. Sadler, K. Thomson, R. Sandbrook, N. Robins, and R. Hughes. 1994. *National Sustainable Development Strategies: Experience and Dilemmas*, Environmental Planning Issues. No. 6, Environmental Planning Group. London: International Institute for Environment and Development.

Davidson, S., G.C. Teleki, and B. Lamb. 1988. *Caribbean environmental programming strategy, final report: volume 3, background information.* Quebec: Canadian International Development Agency.

Davis, S.D., V.H. Heywood, and A.C. Hamilton, eds. 1994. *Centres of plant diversity: a guide and strategy for their conservation, volume 1, Europe, Africa, South West Asia and the Middle East.* Oxford: WWF and IUCN, IUCN Publications Unit.

Davis, S.D., V.H. Heywood, and A.C. Hamilton, eds. 1995. *Centres of plant diversity: A guide and strategy for their conservation, volume 2, Asia, Australasia and the Pacific.* Oxford: WWF and IUCN, IUCN Publications Unit.

D'Antonio, C.M. and T.L. Dudley. 1995. Biological invasions as agents of change on islands versus mainlands. In *Islands: Biological diversity and ecosystem function,* Ecological Studies 115, eds. P.M. Vitousek, L.L. Loope, and H. Andersen, 103–121. Berlin: Springer-Verlag.

Diamond, J. 1986. The environmentalist myth. *Nature* (London) 324:19–20.

Duffey, E. 1964. The terrestrial ecology of Ascension Island. *Journal of Applied Ecology* 1:219–251.

Eliasson, U. 1995. Patterns of Diversity in Island Plants. In *Islands: Biological diversity and ecosystem function,* Ecological Studies 115, eds. P.M. Vitousek, L.L. Loope, and H. Andersen, 36–41. Berlin: Springer-Verlag.

Fa, J.E. 1991. *Conservación de los ecosistemas forestales de Guinea Ecuatorial*, Gland and Cambridge: IUCN.

Falk, D.A. and P. Olwell. 1992. Scientific and policy considerations in restoration and reintroduction of endangered species. *Rhodora* 94:287–315.

Figueiredo, E. 1994. Diversity and endemism of angiosperms in the Gulf of Guinea islands. *Biodiversity and Conservation* 3:785–793.

Flenley, J.R., A.S.M. King, J. Jackson, C. Chew, J. Teller, and M.E. Prentice. 1991. The late Quaternary vegetation and climatic history of Easter Island. *Journal of Quaternary Science* 6:85–115.

Fowler, S. 1993. Report on a visit to St. Helena: 6–25th June, 1993. Unpublished report. Silwood Park, UK: International Institute of Biological Control.

Gade, D.W. 1985. Man and nature on Rodrigues, tragedy of an island commons. *Environmental Conservation* 12:207–216.

Gilmer, K. and J.W. Kadereit. 1989. The biology and affinities of *Senecio teneriffae* Schultz Bip., an annual endemic from the Canary Islands. *Botanische Jahrbuch* 11:263–273.

Gioda, A., J. Maley, R.E. Guasp, and A.A. Baladón. 1995. Some low elevation fog forest of dry environments: applications to African paleoenvironments. In *Tropical montane cloud forests*, Ecological Studies 110, eds. L.S. Hamilton, J.O. Juvik, and F.N. Scatena, 156–163. New York: Springer-Verlag.

Grove, R.H. 1995. *Green imperialism: Colonial expansion, tropical island edens, and the origins of environmentalism 1600–1860*, 24. Cambridge: Cambridge University Press.

Hamilton, S.L., J.O. Juvik, and F.N. Scatena. 1995. The Puerto Rico Tropical Cloud Forest Symposium: Introduction and Workshop Synthesis. In *Tropical montane cloud forests,* Ecological Studies 110, eds. L.S. Hamilton, J.O. Juvik, and F.N. Scatena, 1–23. New York: Springer-Verlag.

Hooker, J. D. 1902. *Echium wildpretii. Curtis Botanical Magazine* 128:Tab 7847.

Humphries, C.J. 1976. Evolution and endemism in *Argyranthemum* Webb ex Schultz Bip. (Compositae-Anthemidae). *Botanica Macaronesia* 1:25–50.

Humphries, C.J. 1979. Endemism and Evolution in Macaronesia. In *Plants and islands*, ed. D. Bramwell, 171–199. London: Academic Press.

Ishwaran, N. 1993. Protected Areas and Sustainable Development in Small Islands. *Insula* 2:31–33.

IUCN 1994. *IUCN Red List Categories*. Gland: IUCN Species Survival Commission.

Janzen, D.H. 1986. The future of tropical ecology. *Annual Review of Ecology and Systematics* 17:305–324.

Johnson, S.P. (Series ed.). 1993. *The earth summit: The United Nations conference on environment and development (UNCED)*, International Environmental Law and Policy Series, 125–503. London: Graham & Trotman/Martinus Nijhoff.

Jones, C. and J. Hartley. 1995. A Conservation project on Mauritius and Rodrigues: An overview and bibliography. *Dodo* 31:40–65.

Krogstrup, P., J.V. Norgaard, and O. Hamman. 1990. Micropropagation of threatened endemic and indigenous plant species from the island of Rodrigues. *Botanic Gardens Micropropagation News* 1:8–11.

Levin, D.A., J.F. Ortega, and R.K. Jansen. 1996. Hybridization and the extinction of rare plant species. *Conservation Biology* 10:10–16.

Liston, A., L.H. Rieseberg, and O. Mistretta. 1990. Ribosomal evidence for hybridization between island endemics of *Lotus. Biochemical Systematics and Ecology* 18:239–244.

Lobin, W. and W. Barthlott. 1988. *Sophora toromiro* (Leguminosae); the lost tree of Easter Island. *Botanic Gardens Conservation News* 1:32–34.

Loope, L.L. 1995. Climate change and island biological diversity. In *Islands: Biological diversity and ecosystem function*, Ecological Studies 115, eds. P.M. Vitousek, L.L. Loope, and H. Andersen, 123–132. Berlin: Springer-Verlag.

Loope, L.L. and A.C. Medeiros. 1994. Impacts of biological invasions on the management and recovery of rare plants, in Haleakala National Park, Maui, Hawaii. In *Restoration of endangered species: Conceptual issues, planning and implementation*, eds. M.L. Bowles and C.J. Whelan, 139–142. Cambridge: Cambridge University Press.

Lugo, A.E., J. Parrotta, and S. Brown. 1993. Loss in species caused by tropical deforestation and their recovery through management. *Ambio* 22:106–109.

Maunder, M. 1992. Plant reintroductions: an overview. *Biodiversity and Conservation* 1:51–61.

Maunder, M., ed. 1995a. *Report of the meeting of the Toromiro Management Group*. October 1994, at University of Bonn Botanischer Garten. Royal Botanic Gardens, Kew: Conservation Projects Development Unit.

Maunder, M. 1995b. *Endemic plants: Options for an integrated conservation strategy*. An unpublished report submitted to the Overseas Development Administration and the Government of St. Helena. Royal Botanic Gardens, Kew: Conservation Projects Development Unit.

Maunder, M. and A. Culham. 1997. Practical aspects of threatened species management, Genetic aspects of conserving small populations, York Conference, *British Ecological Society* [in press].

Maunder, M., U.S. Seal, A. Culham, and P. Pearce-Kelly. 1994. Conservation assessment and management plan for St. Helena. *Botanic Gardens Conservation News* 2:44–48.

Maunder, M., T. Upson, B. Spooner, and T. Kendle. 1995. Saint Helena: Sustainable development and conservation of a highly degraded island ecosystem. In *Islands: Biological diversity and ecological function*, Ecological Studies 115, eds. P.M. Vitousek, L.L. Loope, and H. Andersen, 205–217. Berlin: Springer-Verlag.

McNeely, J.A., M. Gadgil, C. Levèque, C. Padoch, and K. Redford. 1995. Human influences on biodiversity. In *Global biodiversity assessment*, UNEP, eds. V.H. Heywood and R.T. Watson, 782. Cambridge: Cambridge University Press.

Melliss, J.C. 1875. *St. Helena: A physical, historical and topographical description of the island, including its geology, fauna, flora and meteorology*. London: L. Reeve and Co.

Miller, K. and S.M. Manou. 1995. *National biodiversity planning*. Washington D.C., Nairobi, Gland: World Resources Institute, United Nations Environment Programme and the World Conservation Union.

Mueller-Dombois, D. 1981. Island Ecosystems: What is unique about their ecology? In *Island ecosystems: biological organization in selected Hawaiian communities*, US/IBP Synthesis Series 15, eds. D. Mueller-Dombois, K.W. Bridges, and H.L. Carson, 485–501. Pennsylvania: Hutchinson Ross Publishing Company.

Mueller-Dombois, D. and L.L. Loope. 1990. Some unique ecological aspects of oceanic island ecosystems. In *Botanical research and management in Galápagos*, Proceedings of the Workshop on Botanical Research and Management in Galápagos, 11–18 April, 1987, eds. J.E. Lawesson, O. Hamann, G. Rogers, G. Reck, and H. Ochoa. *Monographs in Systematic Botany-Missouri Botanical Garden* 32:21–27.

North, S.G., D.J. Bullock, and M.E. Dulloo. 1994. Changes in the vegetation and reptile populations on Round Island, Mauritius, following eradication of rabbits. *Biological Conservation* 67:21–28.

O'Rourke, P.J. 1994. *All the trouble in the world: The lighter side of famine, pestilence, destruction and death*, 162. London: Picador.

Ortega, J.F., R.K. Jansen, and A. Santos-Guerra. 1996. Chloroplast DNA evidence of colonization, adaptive radiation, and hybridization in the evolution of the Macaronesian flora. *Proceedings of the National Academy of Science USA* 93:4085–4090.

Owadally, A.W., M.E. Dulloo, and W. Strahm. 1991. Measures that are required to help conserve the flora of Mauritius and Rodrigues in *ex situ* collections. In *Tropical botanic gardens: their role in conservation and development*, eds. V.H. Heywood and P.W. Wyse-Jackson, 95–117. London: Academic Press.

Owens, S.J., A. Jackson, M. Maunder, P. Rudall, and M. Johnson. 1993. The breeding system of *Ramosmania heterophylla*—dioecy or heterostyly? *Botanical Journal of the Linnean Society* 113:77–86.

Page, W. and M. Maunder. 1997. A conservation review of the Mascarene palms. *Principes*. In press.

Perez, G.V. 1912a. *Echium pininana*. *Revue Horticole*, pp. 350–351.

Perez, G.V. 1912b. Les Echium des Iles Canaries. *Revue Horticole*, pp. 440–442.

Ray, G.J. and B.J. Brown. 1994. Seed ecology of woody species in a Caribbean dry forest. *Restoration Ecology* 2:156–163.

Rieseberg, L.H. and D. Gerber. 1995. Hybridization in the Catalina Island mountain mahogany *(Cercocarpus traskiae)*: RAPD evidence. *Conservation Biology* 9:199–203.

Rieseberg, L.H. and S.M. Swensen. 1996. Conservation genetics of endangered island plants. In *Conservation Genetics: Case Histories from Nature*, eds. J.C. Avise and J.L. Hamrick, 305–334. New York: Chapman & Hall.

Royal Botanic Gardens, Kew, in association with International Institute for Environment and Development. 1993. *Report on the sustainable environment and development strategy and action plan for St. Helena*. Under assignment from the Overseas Development Administration for the Government of St. Helena.

Salas-Pascual, M., J.R. Acebes-Ginoves, and M. Del Arco-Aguilar. 1993. *Arbutus x androsterilis*, a new interspecific hybrid between *A. canariensis* and *A. unedo* from the Canary Islands. *Taxon* 42:789–792.

Schüle, W. 1993. Mammals, vegetation and the initial human settlement of the Mediterranean islands: A palaeoecological approach. *Journal of Biogeography* 20:399–412.

Smith D. and N. Williams. 1996. *Diana's Peak National Park of St. Helena, The Management Plan for 1996–2001*, Agriculture and Forestry Department, unpublished report.

Stoddardt, D.R. and R.P.D. Walsh. 1992. Environmental variability and environmental extremes as factors in the island ecosystem. *Atoll Research Bulletin* 356:1–71.

Strahm, W. 1983. Rodrigues—can its flora be saved? *Oryx* 17:122–125.

Strahm, W. 1989. *Plant Red Data Book for Rodrigues*. Konigstein: IUCN/Koeltz Scientific Books.

Strahm, W. 1994. Regional Overview: Indian Ocean Islands. In *Centres of plant diversity: a guide and strategy for their conservation, Volume 1, Europe, Africa, South West Asia and the Middle East*, eds. S.D. Davis, V.H. Heywood, and A.C. Hamilton, 265–270. Oxford: WWF and IUCN, IUCN Publications Unit.

Synge, H. 1992. Threatened species on islands: Plants. In *Global biodiversity: status of the Earth's living resources*, A report compiled by the World Conservation Monitoring Centre, in collaboration with The Natural History Museum London, IUCN, UNEP, WWF and WRI, ed. B. Groombridge, 244–245. London: Chapman & Hall.

Thomson, A.D. 1981. New plant disease record in New Zealand: Cucumber mosaic virus in *Myosotidium hortensia*. *New Zealand Journal of Agricultural Research* 24:401–2.

Towns, D.R., C.H. Daugherty, and I.A.E. Atkinson, eds. 1990. *Ecological restoration of New Zealand islands*, Conservation Sciences Publication No. 2, Wellington: Department of Conservation.

UNCED. 1992. *Agenda 21*, United Nations Conference on Environment and Development (UNCED), United Nations General Assembly, New York.

UNESCO. 1994. *Report of the global conference on the sustainable development of small island developing states*, Chapter IX, Biodiversity Resources, 28. Bridgetown, Barbados, 25 April–6 May 1994.

Upson, T. and M. Maunder. 1993. Status of the endemic flora and preliminary recovery programmes, Volume 3. In *Report on sustainable environment and development strategy and action plan for St. Helena*, Royal Botanic Gardens, Kew in association with International Institute for Environment and Development under assignment from the Overseas Development Administration for the Government of St. Helena.

United States Office of Technology Assessment. 1993. *Harmful non-indigenous species in the United States*. Washington DC, USA.

Vitousek, P.M., L.L. Loope, and C.P. Stone. 1987. Introduced species in Hawaii: Biological effects and opportunities for ecological research. *Trends in Ecology and Evolution* 2:224–227.

Watts, D. 1987. *The West Indies: Patterns of development, culture and environmental change since 1492*. Cambridge: Cambridge University Press.

Welsh, J. and M. McClelland. 1990. Fingerprinting genomes using PCR with arbitrary primers. *Nucleic Acids Research* 18:7213–7218.

Williams, J.G.K., A.R. Kubelik, K.J. Livak, J.A. Rafalski, and S.V. Tingey. 1990. DNA polymorphisms amplified by arbitrary primers are useful as genetic markers. *Nucleic Acids Research* 18:6531–6535.

Williams, M. and B. Macdonald. 1985. *The Phosphateers*. Victoria: Melbourne University Press.

Wingate, D.B. 1985. The restoration of Nonsuch Island as a living museum of Bermuda's precolonial terrestrial biome. *ICBP Technical Publication No. 3*, Cambridge, UK.

Zizka, G. 1991. *Flowering plants of Easter Island*, Scientific Reports and Research Activities, *Palmarum Hortus Francofurtensis*, 3, Frankfurt: Palmengarten.

CHAPTER

14

When do Genetic Considerations Require Special Approaches to Ecological Restoration?

ERIC E. KNAPP
and ANDREW R. DYER

Conservation biology cannot be concerned solely with preserving what remains. With many habitat types reduced to as little as 1% (e.g., tallgrass prairie east of the Mississippi River, USA) or even 0.1% (e.g., Central Valley riparian forest, California, USA) of their original area (see references in Noss *et al.* 1995), protecting what is left often represents "too little, too late." Even habitat types that remain relatively common often occur in isolated patches that are too small for long-term conservation of viable populations of all organisms, particularly those of large carnivores and ungulates (Schonewald-Cox 1983). For these reasons, effective preservation of biodiversity may require investment in ecological restoration to increase the size as well as the connectivity of available habitat (Jordan *et al.* 1988). Restoration will be especially vital for restoring native diversity to many of the world's most fertile and productive communities, where habitat destruction resulting from human activities has been concentrated (Janzen 1988).

Two of the most widely discussed dimensions of restoration involve the application of ecological understanding to the re-establishment of natural vegetation (see chapter 15), and methods of sampling genetic variation for seed stock used in reintroduction (Guerrant 1992; chapter 5). Another important but often less appreciated aspect of restoration involves the proper matching of genotypes to the environment. In this chapter, we explore the implications of population differentiation and local adaptation to ecological restoration and present some strategies for reintroducing genetically appropriate populations. We then discuss the consequences of non-local introductions on existing native populations, and conclude by describing some community-level genetic processes that may require consideration when attempting restoration. The examples we use are weighted toward plants since the practice of restoration thus far has been concerned mainly with the manipulation of vegetation. However, many of the points should be equally applicable to animals.

GENETIC DIFFERENTIATION AND ADAPTATION: HOW IMPORTANT IS IT?

Extensive genetic differences are common among populations of many species, and can be found across a multitude of spatial scales, ranging from meters to broad geographic regions. Striking differences in metal tolerance and morphological characteristics have been documented across mine tailing soil-type boundaries only 1–2 m wide in the grass species *Agrostis tenuis* (Bradshaw 1959) and *Anthoxanthum odoratum* (Antonovics and Bradshaw 1970). Such fine-scale genetic differentiation can occur in response to the biotic as well as the abiotic environment. For example, Turkington and Harper (1979) found genetically based differences in competitive ability of the clover *Trifolium repens* within a single pasture characterized by a mosaic of different associated grass species. Clover genotypes grew most vigorously when planted back into the sward of origin, or into pure stands of the grass species that dominated the sward from which the clover plants originated. Local biotic differentiation may also arise in response to pathogen virulence, as reported by Parker (1985) for plant-fungal interactions. Plants of the species *Amphicarpaea bracteata* transplanted back into the local environment became infected with the fungus *Synchytrium decipiens*, whereas plants from as little as 30 m away did not.

Genetic differences are also common across more gradual environmental gradients, although these changes may require larger spatial scales to be detected. Environmental factors commonly associated with such regional genetic differentiation include aspect, elevation, and climate. For example, populations of Douglas fir (*Pseudotsuga menziesii*) collected from west-facing and east-facing sides of mountains in coastal Oregon were found to produce seeds that differed in size and timing of germination, and seedlings that differed in the ratio of resource allocation to roots and shoots, when planted in a common garden (Sorensen 1983). Genetic differences in these traits were also recorded with changing elevation on the same slope. In the Chinook salmon (*Oncorhynchus tshawyscha*), a species in which certain populations will likely depend on restoration efforts for survival, genetic differentiation for allozymes has been reported among populations occupying different watersheds (Bartley and Gall 1990).

Degree of differentiation among populations depends on the amount of gene flow. Populations lacking movement of genes between them will tend to differentiate over time as a result of localized selective pressures, genetic drift, or both. The great variability among species used in restoration in the patterns and magnitude of such genetic differentiation can be attributed, in part, to different mating systems, degree of environmental heterogeneity, and differences in how populations are distributed across the landscape. In general, populations of self-pollinating plant species show greater differentiation than populations of outcrossing species (Loveless and Hamrick 1984). High levels of differentiation are also common among populations of widespread species found in multiple environments, and in species whose populations are spatially isolated from each other (Loveless and Hamrick 1984).

Selection is commonly implicated as the evolutionary force responsible for genetic differentiation because patterns of differentiation often follow detectable environmental gradients. From the standpoint of successfully matching germplasm to a restoration planting, differentiation due to selection and the subsequent adaptation of populations to their local environment is likely to be more critical than differentiation resulting from genetic drift. However, it should not be universally assumed that the local population represents the best adaptive "solution" to environmental variability at a particular site. The possibility exists that all populations are equally well adapted, or that a species contains no genetic variation, with phenotypic differences entirely the result of plasticity. In addition, cases where the local population is not the best adapted have also been reported (e.g., Rapson and Wilson 1988; Rice and Mack 1991). Maladaptation can result from limited sampling of a species' genetic variability during colonization events, where a propagule(s) arrives at a site and forms a new population by chance and not due to genetic superiority.

Since adaptation, or lack thereof, cannot be assumed, collecting empirical data on the fitness of different plant stock is a valuable first step towards restoring genetically appropriate populations. Unfortunately, even this direct experimental approach can yield misleading results if the temporal scale is too short for adequately gauging plant responses to all possible environmental variability. For example, Millar and Libby (1989) describe cases in which initial monitoring of common garden plantings of commercial conifer species indicated that certain non-local populations became established and grew more rapidly than local populations. As a result, these apparently superior non-local genotypes were widely planted. While the introduced populations often grew well initially, they eventually succumbed to infrequent environmental extremes that favored the local populations over the long term (Millar and Libby 1989).

Unfortunately, testing plant populations over appropriate spatial and temporal scales is in many cases not feasible. In the absence of data, however, it is prudent to assume that local populations are the best adapted. The body of literature on among-population genetic differentiation does not suggest that maladaptation is a common phenomenon. Furthermore, the likelihood of detecting maladaptive differentiation is greatest at fine spatial scales, across which germplasm transfers are not usually as problematic. The greatest concern about germplasm transfer in restoration is across regional scales where adaptation to different environments within the region is expected.

Adaptation and Restoration: What is Native?

The degree to which restorationists are concerned with the genetic provenance of their source materials is highly variable (Jordan *et al.* 1988). The literature abounds with cases where so-called "native" species were used, but closer examination reveals that the source of the "native" species was often a commercially bred variety or other non-local population. For example, several "native" grass species planted in Saskatchewan, Canada, were reported to exhibit inferior initial establishment compared to commonly grown exotic grasses (Kilcher and Looman 1983), yet the "native" species were actually derived from collections made as far away as Kansas (USA). Padgett and Crow (1994) describe a wetland mitigation project in New Hampshire (USA) where planting was accomplished using commercial stock of various native species obtained from Connecticut and Wisconsin. Another example of an uncritical definition of "native" involves the 'Lassen' strain of antelope bitterbrush (*Purshia tridentata*), which originated from a single native population in Lassen County, California, yet has been promoted for restoring depleted range lands, burned areas, mined lands, and other disturbed sites in the western United States as far east as Colorado and as far north as the Canadian border (Soil Conservation Service 1986). Soil Conservation Service literature claims that the 'Lassen' population, selected for its forage quality and upright growth habit, is "probably adapted wherever antelope bitterbrush occurs naturally" (Soil Conservation Service 1986). It is not surprising then, that while 'Lassen' bitterbrush often does establish well in many locations, its vigor commonly declines before maturity (S. Monsen [U.S.D.A. Forest Service Shrub Sciences Laboratory, Provo, UT], personal communication).

Using non-local sources that provide a poor genetic match to the environment may result in the failure of restoration efforts. This brings up the frequently asked but hard to answer question: "How local is local?" While much of the interest in this question is in reference to plant species, current discussion about restoring populations of the endangered Florida panther (*Felis concolor* ssp. *coryi*) by introducing panthers from Texas (*F. concolor* ssp. *stanleyana*) (Hedrick 1995) provides a good wildlife example of the dilemma. Genetic deterioration in the Florida subspecies, resulting from small population sizes and inbreeding, might be reversed by hybridization with panthers from other sources (Hedrick 1995). However, the introduction of foreign genes is controversial (see Maehr and Caddick 1995), and raises other questions. For example, would successful restoration with hybrid populations of panthers containing genetic combinations that did not evolve in Florida really contribute to restoration of the "native" ecosystem? Perhaps more importantly, assuming that animals containing genes foreign to the "native" population do thrive, would the failure to maintain genetic purity of the Florida subspecies matter if the ecological role of the panther is filled by the restoration? How such questions are resolved will hinge on the motives for restoration, and will certainly vary depending on the situation.

It is safe to say that the development of strategies for collecting and utilizing germplasm is complicated by the need to recognize genetic differences among populations. Whenever possible, emphasis should be placed on restoring not just the native species, but the native genotypes. The genetic differentiation commonly found among populations also suggests that for each species, the unit of restoration should be the population.

REINTRODUCING ADAPTED (AND ADAPTABLE) POPULATIONS

Obtaining appropriate germplasm is one of the greatest challenges facing the restorationist. Many decisions are influenced more by what is available in the quantities needed, and on the price, than on the potential for the propagules to be well adapted to the restoration site. In the restoration of ecosystems, a great deal of environmental variability (and therefore variability in selection pressures) will be encountered both at the regional scale (e.g., regional edaphic differences and climatic factors such as rainfall, temperature, and number of cooling/heating degree days) and the local scale (e.g., local edaphic heterogeneity, aspect, and local moisture gradients). Given the potential complexity of resulting patterns of adaptation, data on which to base decisions about the collection of genetically appropriate germplasm for restoration are rarely available. Research is needed to provide, at the very least, broad recommendations on the scale over which germplasm can be transferred successfully.

Strategies for Matching Genotype and Environment

Recommendations for obtaining germplasm with the best chance of being well-adapted include collecting propagules within a certain distance of the restoration site. For example, Linhart (1995) suggested that, where possible, seed of herbaceous plants should be collected within 100 m and seed of woody plants should be collected within 1 km. Without additional information, such broad recommendations can provide a general guideline for how proximal collections of germplasm should be made. Unfortunately, patterns of environmental and genetic heterogeneity rarely follow such simple contours. In addition, the desired native species might not be found within spatial scales such as those recommended by Linhart (1995). This is especially true for areas where habitat degradation is the worst, and hence the urgency for restoration the greatest.

More sophisticated strategies have been developed for species in which the genetic structure is at least partially understood. For example, the U.S. Forest Service uses "seed zones" to delineate boundaries of similarly adapted tree populations (Millar and Libby 1991; Ledig 1992). Seed zone boundaries are drawn on the basis of natural patterns of genetic diversity and structure, but topographic, climatic, and edaphic divisions are often used when genetic information is incomplete or lacking (Ledig 1992). Replanting following logging or wildfire is done using native tree species grown from seed collected within the same local seed zone. Restorationists may find the concept of seed zones useful for making decisions about germplasm transfer. However, applying seed zones developed for commercially harvested trees to other groups of plants may be problematic due to the variation among species in patterns and magnitude of within-species genetic differentiation (Knapp and Rice 1996). For instance, commercially harvested trees tend to have high rates of outcrossing, fairly continuous distributions, and typically shed pollen over long distances. These characteristics all promote gene flow and relative genetic homogeneity among populations. The potential for gene flow may be much more limited with many other plant species used in restoration.

Even the crudest recommendations for the appropriate scales of germplasm transfer are not possible for many species because the most rudimentary information on mating systems and other parameters linked to gene flow is lacking. Until such data are gathered, careful use of germplasm collections made as locally as possible and/or from natural environments as similar as possible to the one being restored should be a goal.

In the absence of basic genetic or life history data, matching of germplasm and environment can be approximated by using existing regional environmental data, such as horticultural zones and soil maps. Numerous studies have attributed patterns of ecotypic variation to environmental and climatic variables including day length, annual precipitation, elevation, and soil type (e.g., Clausen *et al.* 1940; McMillan 1959; Clary 1975), providing evidence for the potential of this approach. In addition, a strong relationship between climate and patterns of genetic variation has been demonstrated for purple-needlegrass (*Nassella pulchra*) (Knapp and Rice in preparation), a perennial grass species commonly used in restoration projects in California. Measurements of morphological traits (leaf shape, plant height, growth habit) and phenological traits (time to heading) in a common garden planting of populations collected from diverse environments indicated that trait mean values were associated with climatic differences among the collection sites, but were not associated with the geographic distance between populations. Coastal populations, under the moderating influence of the Pacific Ocean, generally were shorter, more prostrate, produced more seed, and had seeds of smaller size than populations from the interior, where winters are colder and summers much hotter and drier. These data suggest that the transfer of *N. pulchra* across coastal to interior gradients might result in the greatest risk of planting a poorly adapted source. Germplasm obtained from populations growing in climates more similar to the restoration site would likely be superior, even if it originated a greater distance away.

The Role of Environment on the Fitness Differential between Local and Non-Local Populations

Data obtained in the previously mentioned study of *N. pulchra* populations (Knapp and Rice, in preparation) suggest that the fitness differential between local and non-local germplasm depends on the environment at the planting site. Very little mortality was observed in any of the local and non-local populations after several growing seasons at Davis, California (Figure 14.1a), despite the diversity of environments from which the populations were originally collected. In contrast, a separate planting on a more xeric interior site with poor soil and competition with exotic weeds (at Winters, California [20 km west of Davis]) revealed dramatic differences in mortality between local and non-local populations. After three years, nearly 80% of the plants from the non-local sources had died, whereas mortality of plants from the local sources was less than 20% (B. Young and J. Anderson, unpublished data) (Figure 14.2). Although the same populations were not planted at each site, populations in the Winters planting were actually collected in closer proximity to each other than the populations grown at the Davis site. This suggests that environmental differences between the two sites likely played a dominant role in the mortality differences between local and non-local sources. In contrast to the conditions at the Winters site, conditions in the Davis common garden were highly favorable for plant growth, with deep, fertile soils, and no interspecific competition from weeds. It is

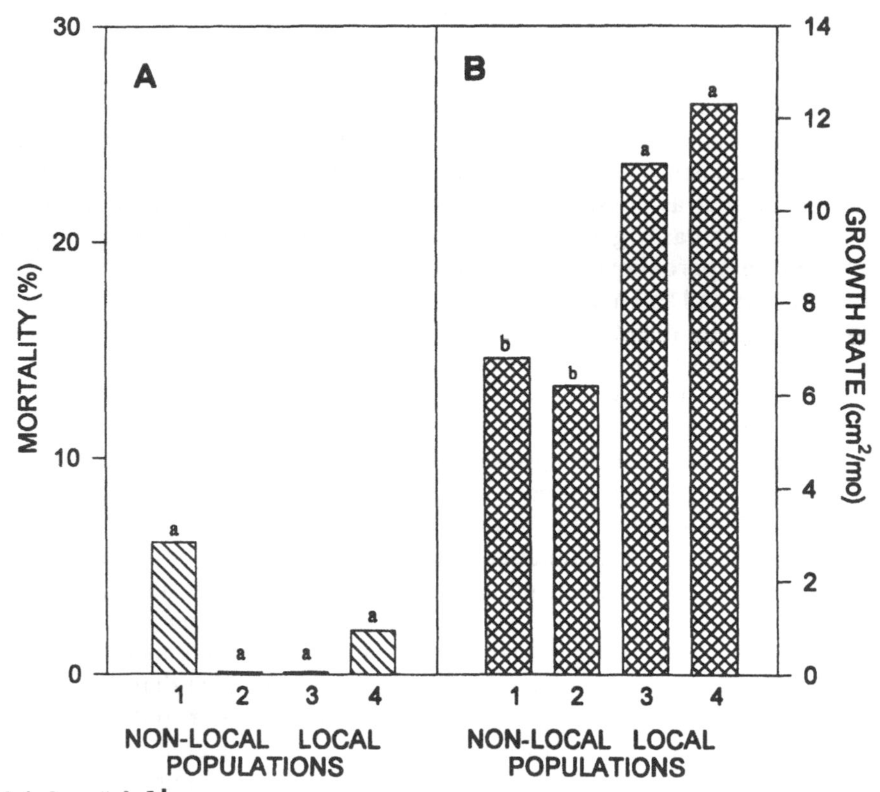

Figure 14.1a, 14.1b

Survival (A) and growth rates (B) of four Nassella pulchra *populations in a common garden at Davis, California. Seed of population 1 was collected in coastal San Luis Obispo County (Central California), and seed of population 2 was collected on the Santa Rosa Plateau, a site in Southern California influenced by coastal climate. Seed of populations 3 and 4 were both collected in interior Northern California, at Dye Creek (Tehama County), and Parrott Ranch (Butte County), respectively. The latter two sites share the greatest climatic and geographic similarity with Davis. Columns not sharing the same letters are significantly different ($P \leq 0.05$).*

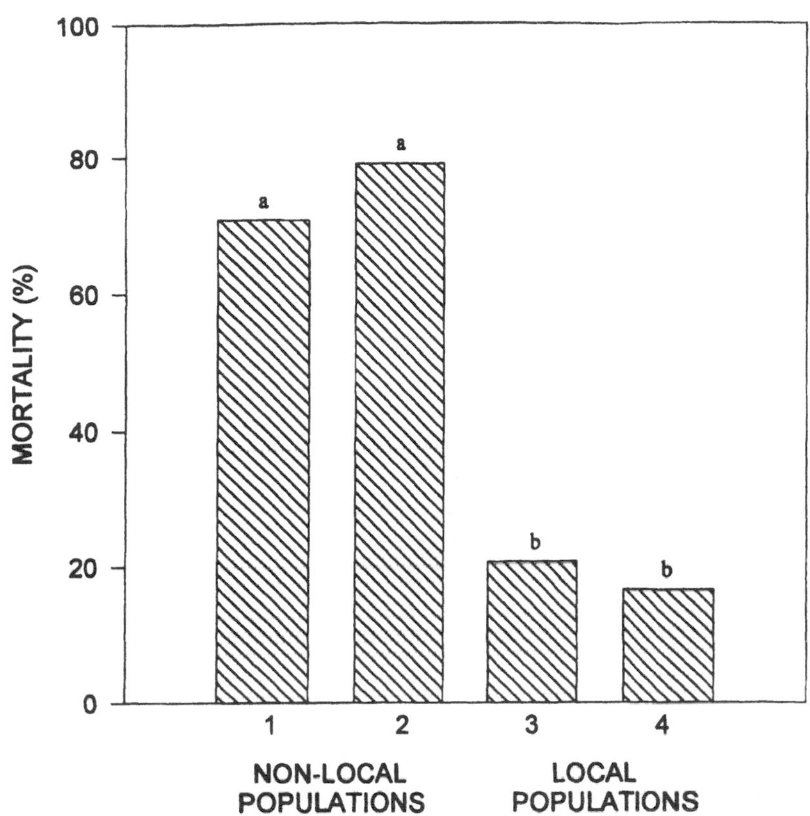

Figure 14.2
Survival of four Nassella pulchra *populations planted at Winters, California. Seed from each population was collected from different regions of Northern California and grown for three years in a grassland near Winters (25 km west of Davis), a site characterized by shallow, rocky soils of poor fertility, and competition from exotic weedy species. Seed from the Button Ranch population (1) was collected at a coastal site; seed from the Jepson Prairie population (2) was collected at a site with both coastal and interior influences plus rare edaphic (vernal pool) conditions; seed from the Parrott Ranch population (3) was collected at an interior site; and seed from the Winters population (4) was collected at a local interior site. (Young and Anderson, unpublished data). Columns not sharing the same letters are significantly different ($P \leq 0.05$).*

possible that the fitness differential between local and non-local populations (as determined by plant mortality) was "hidden" in the relatively benign environment at the Davis site.

A potential mechanism for differential mortality under competitive conditions is provided by additional data from the Davis common garden, which demonstrated that plants collected from environments similar to Davis grew nearly twice as fast as plants collected from non-local coastal environments (Figure 14.1b). In the context of the restored plant community, where conditions of intense inter-specific as well as intra-specific competition often prevail, a faster growth rate could make the difference between success or failure in a restoration planting.

These observations indicate that survival in "difficult" environments (e.g., extreme abiotic conditions and/or strong interspecific competition) might require greater specialization. A more precise or local adaptive match of plant material may therefore be necessary for restoring such harsh sites. It is also important that testing for appropriate scales of adaptation be done in environments that approximate the environment at the restoration site, because results from artificial environments (such as plantings at wide spacing in a weeded common garden) can lead to inaccurate recommendations.

Agricultural Production of Plant Materials for Restoration: Implications for Adaptation

Although restoration might seem like "agriculture using native species", there are fundamental differences between restoration and large-scale western agriculture, particularly in how the importance of adaptation and genetic variation is viewed. Both restoration and agriculture involve establishing plants over large areas and, in many cases, result in the revegetation of bare ground. However, from a genetic standpoint, the goals of restoration are generally very different than the goals of farming. In large-scale western agriculture, the ability to control the environment through leveling of the planting surface, irrigation, fertilization, cultivation, and weed control allows for the widespread production of crop varieties containing little or no genetic variation. In a sense, the agricultural environment can be manipulated to match a given crop genotype. Conversely, restorationists are faced with choosing the correct germplasm to match the existing environment(s), because long-term management of environmental variables and continued human inputs are usually not desired. In addition, natural ecosystems are rarely, if ever, composed of monotypic stands consisting of one cohort of one genotype of one species, whereas this is the rule in conventional agricultural ecosystems. Restoration at the level of the community therefore represents a major complication of the western agricultural paradigm. Ensuring that well-adapted populations are planted, and embracing diversity by using multiple species and by maintaining genetic variation within species, requires developing new strategies for the establishment and the management of restoration plantings.

Agricultural practices are often used to grow seed or container stock used in restoration (Figure 14.3). Such production of plant material is especially important for larger-scale restoration because the required quantities of propagules frequently cannot be collected sustainably from nearby existing native sources. Unfortunately, agricultural and native environments often differ in characteristics such as soil moisture, soil type, soil depth, nutrient availability, and climate, resulting in differences in selective pressures that may, after several generations, greatly change the genetic composition of the plant material. Such genetic shifts could potentially alter the adaptive potential of seed or container stock.

One example of the potential problems caused by genetic shifts is documented in a study of populations of a commercial rye (*Secale cereale*) that were grown for years in widely different environments (Hoskinson and Qualset 1967). "Balbo" rye was introduced to Tennessee from Italy in 1919 for use as both a forage and a grain crop. Planting of this variety subsequently spread as far north as Michigan and as far west as Colorado. Years later, when seed produced in these different regions were planted together in a common environment, striking differences were noted in growth habit, plant height, heading date, and time to maturity, indicating that natural selection had altered the genetic composition of the original population. Seed produced outside of Tennessee was no longer suitable for use by growers within the state, even though seed from all sources was still labeled and marketed under the same name.

Shifts in traits may be reversible, so long as genes necessary for survival of the plants in their native environment are not totally eliminated. Precautions for reducing the possibility of serious genetic shifts include managing agricultural plantings to avoid

Figure 14.3
Harvesting California melic (Melica californica) *seed using a "Flail-Vac". California melic is a native perennial grass naturally found on open hillsides, in oak woodlands, and in low-elevation coniferous forests of California. Agricultural production of native plant seed has become a common method of providing the large numbers of propagules needed for restoration. Photo courtesy of John Anderson.*

mortality, because the potential for genetic shifts is highest when selection intensities are great. If the level of mortality in a seed production planting is high, the best strategy for maintaining sufficient plant density to produce a harvestable seed crop is not by planting more seed, but by changing management practices to reduce plant mortality. This can perhaps be accomplished best through production of the seed crop in close proximity to the restoration site and by using management schemes that mimic the natural environment. Because genetic shifts are cumulative, seed from the original collection should be used to establish a seed production field. Seed harvested from a previous agricultural planting may already have undergone genetic shifts. In addition, the number of seasons that seed of perennial species is harvested from the same agricultural planting might be minimized.

Agricultural production of plant material for restoration can also result in the loss of genetic variation. The heterogeneity characteristic of many natural environments acts to maintain genetic variation (Ennos 1983), while agricultural environments typically are highly uniform which results in homogenizing selection pressures across large areas. Stabilizing selection imposed by agricultural practices may also cause loss of genetic variation. For example, mechanical harvesting (Figure 14.3) can select for uniformity in time to maturity, because seed from phenologically extreme plants (those that have already dropped their seed, or those containing immature seed) is less likely to be included. Genetic variation for plant height could also be affected, if the harvest misses seed from the tallest or most prostrate genotypes.

Loss of genetic variation is not usually a concern in agriculture because plant uniformity will generally maximize the amount of harvestable crop. Indeed, many agricultural varieties are, for this reason, bred and propagated to have little or no genetic heterogeneity. Loss of genetic variation is, however, of great concern in restoration, because restored populations may require genetic variation in order to cope with microenvironmental differences, and to evolve and adapt to new environmental challenges.

Obtaining Locally Adapted Seed: Difficulties, Compromises, and Alternative Strategies

With adequate lead time, the agricultural production of plant material from local collections can and, whenever possible, should be linked to specific restoration projects. However, emergency situations such as revegetation after a fire or other catastrophic disturbances require that sufficient stocks of locally adapted seed already exist and are immediately available. In addition, funding for mitigation-driven restoration projects often can appear (and disappear) very rapidly, making long-term planning difficult. Without adequate lead time for collecting or developing a germplasm source, restorationists may have few options other than purchasing and using whatever "native" seed or container stock is available.

From the standpoint of profit and ease, most growers of native plant seed prefer to produce large volumes from one or a few sources of each species. An uncertain future market for seed, as well as additional costs associated with storage, handling requirements, and loss of viability over time, all combine to make production of seed from many different ecotypes economically unattractive. However, from the standpoint of adaptation and, ultimately, ecosystem health, the production of seed from many different local collections would be preferable. Compromises will likely be necessary because theoretical ideals rarely coincide with the restrictions imposed by economic realities. The most conservative strategies for restoring native species will favor collection and planting over small spatial scales. It may not, however, often be feasible to develop different seed stocks for each scale at which genetic differentiation can be found.

Seed mixtures may be one method for overcoming some of the difficulties with using specific ecotypes. Mixtures of seed collected from different populations within a region could maximize the amount of available genetic variation and allow a composite population to adapt, over time, to many possible environments. Illustrating the potential of this approach is a study by Harlan and Martini (1938), in which natural selection was observed over a period of 13 years in a mixture of barley (*Hordeum vulgare*) populations. A composite population, produced by combining different varieties bred for adaptation to specific diverse environments, was grown for multiple generations in many different climates throughout the northern and western United States. Because barley is self-pollinating, competing varieties maintained their original genetic composition during the course of the experiment. Poorly adapted (non-local) germplasm was, in most cases, rapidly eliminated from the mixture by natural selection and at many of the sites the most "local" variety eventually dominated. Whether mixtures of populations can be used successfully in restoration remains to be seen.

If selection pressures are strong, evolutionary fine-tuning of mixtures may occur quite rapidly; Harlan and Martini (1938) found that the most poorly adapted genes were eliminated within the 13-year length of their study. Evolutionary response to selection has also been demonstrated to occur over short temporal scales in more natural ecosystems. Snaydon and Davies (1982) observed that populations of the perennial grass *Anthoxanthum odoratum* at the Park Grass Experiment site in Rothamsted, UK, diverged genetically for numerous traits within 6 years of environmental changes caused by artificial liming of the soil.

Unfortunately, poorly adapted genes also may persist for long periods of time. Selection is buffered by the coexistence of phenotypic plasticity and environmental heterogeneity in many natural habitats (Sultan 1987), which can slow the elimination of poorly adapted genes from the population. Reductions in population fitness caused by the presence of these genes has the potential to result in lower competitive ability, invasion of the planting with undesired exotic weeds, or soil erosion due to lack of plant cover.

In addition, hybridization between different component populations of a mixture can cause outbreeding depression (i.e., the reduction in vigor of progeny from matings between divergent populations). Outbreeding depression is thought to be due either to the lack of adaptation of the hybrid progeny to either parental environment, or due to the breakup of coadapted gene complexes (Templeton 1986; Fenster and Dudash 1994). The importance of outbreeding depression in plants and animals, and the spatial scale over which it occurs in natural ecosystems, is not well understood. Waser and Price (1994) found that outbreeding depression resulted from crosses between plants of the alpine larkspur (*Delphinium nelsonii*) growing only 30 m apart near Crested Butte, Colorado (USA). The lower fitness of progeny from crosses at this spatial scale (compared to progeny from crosses between plants growing 3 m and 10 m apart) can be attributed to a patchy and variable selective environment and limited gene flow (Waser and Price 1994). In another example, hybridization between mosquito fish (*Gambusia holbrooki*) populations collected from sites 100 km apart on the Savannah River (USA) was shown to result in outbreeding depression for rate of growth (Leberg 1993). Leberg (1993) suggested that reintroduction strategies for this species should therefore not emphasize mixtures of different stocks.

In general, however, outbreeding depression has not been detected in the majority of studies involving hybridization between divergent sources (Barrett and Kohn 1991). Additionally, heterosis (hybrid vigor) may, in many cases, outweigh any deleterious consequences of the breaking up of coadapted gene complexes (Fenster and Dudash 1994). Fenster and Dudash (1994) state that the breakup of coadapted gene complexes may actually lead to new genetic combinations that "confer even higher adaptation to the environment" (p. 47) and suggest that "the preservation of the genetic integrity of a species may be an ideal with no natural basis; therefore, it should not be used as an obstacle to the mixing of gene pools" (p. 47). Using highly diverse populations, rather than strictly local populations, might allow recovery from past episodes of directional selection and genetic drift that may currently be limiting evolutionary potential (Strauss *et al.* 1992). In addition, anthropogenic influences (e.g., introduction of exotic species, fire suppression) have, in numerous regions, altered the environment so rapidly and so completely that gene flow and natural evolutionary processes may simply be too slow and too weak for local populations to have yet adapted in response to these changes. The introduction of novel genetic combinations could facilitate this process of evolutionary adjustment. While maintaining the genetic integrity of source populations will, however, still be the goal in most cases, the use of population mixtures (containing both local and non-local genotypes) may produce superior results, particularly when restoring landscapes that have undergone recent dramatic environmental changes.

Due to the lack of data on composite mixtures in restoration, Barrett and Kohn (1991) promote experimental approaches to their use. Growing mixtures as well as populations obtained from single locations in a restoration setting could allow the relationship between genetic variation and colonization or persistence to be better tested experimentally. A compromise solution may be to create numerous different regional mixtures, within which each population is at least "coarsely adapted." By not including widely divergent populations in a mixture, the potentially negative consequences of outbreeding depression could be avoided, while still providing genetic variation for natural selection to "fine-tune" the mixture for adaptation to local microenvironmental variation.

IMPORTANCE OF GENETIC VARIATION IN RESTORATION

Genetic variation has the potential to increase the ecological amplitude of populations, enabling them to persist and thrive over a broader range of environments (Bradshaw 1984). The presence of genetic variation may therefore be especially critical for species used in restoration. Martins and Jain (1979) demonstrated that populations of rose clover (*Trifolium hirtum*) with high levels of genetic variation had greater success in colonizing new environments than populations with less genetic variation. Other evidence for the value of genetic variation comes from the many reciprocal transplant studies that have shown high levels of among population variation and fitness advantages of local populations. Such observations indicate that planting germplasm collected from a mixture of different populations might allow the growth of at least some individuals with high fitness over more possible environments (Bradshaw 1984). Studies showing that genetic uniformity (such as that often found in agricultural environments) increases the susceptibility of populations to pathogens indicates that genetic variation for disease resistance is important (Barrett 1981, Futuyma 1983). It follows that the presence of genetic variation should allow populations to survive over a greater range of environments with different disease pressures.

Despite the dearth of studies directly linking genetic variation and population performance, the desirability of genetic variation in reintroduced populations is widely agreed upon (Huenneke 1991). Maintaining genetic variation is likely to be especially critical if matching of germplasm to the environment is uncertain, or if the selective environment is complex, due to microenvironmental heterogeneity. In addition, genetic variation may buffer populations against local as well as global environmental changes by allowing populations the possibility of adaptation in response to shifting selection pressures. Since habitat fragmentation and discontinuity among sites may limit migration, adaptation may be the only means by which many restored populations can respond to environmental changes. Genetic variation may therefore be vital for the long-term persistence of restored populations.

A strategy for maintaining natural levels of genetic variation should be central to all restoration projects. Capturing genetic variation in source populations used for restoration will depend on variables such as sample size and the relative sampling intensity within and among populations. Considerations for maintaining genetic variation in reintroduced populations are covered in detail by Guerrant (1992) and in chapter 5.

Genetic Contamination of Native Gene Pools: Is It an Issue?

Introducing non-local populations into an area where native populations of the species already occur risks genetically contaminating these native populations (Figure 14.4). Concern about genetic contamination has been expressed for ethical as well as biological reasons (Millar and Libby 1989; Knapp and Rice 1994). Mixing up and potentially irreversibly altering patterns of genetic variation resulting from "natural" evolution might

Figure 14.4
Post-wildfire landscape in San Luis Obispo County, California, seeded with "Cucamonga" Brome (Bromus arizonicus). *Linear swaths of lighter colored vegetation (the planted grass) are a result of aerial application of the seed. 'Cucamonga' Brome is a Soil Conservation Service cultivar developed with seed of a collection made from a native* B. arizonicus *population near Rancho Cucamonga, California, which is over 320 km distant. The impact of large-scale plantings of non-local populations on the ecosystem and on the genetic integrity of existing native populations remains poorly understood.*

be seen as equivalent to adding modern graffiti to centuries old pictographs painted by people of native cultures. Most people would be justifiably angered by such vandalism, yet others might contend that at least some of these pictographs are essentially ancient graffiti. Whether dealing with genetic contamination or graffiti, the defensibility of current human manipulation can be based, in part, on arguments about appropriate temporal scales. Evolutionary history is full of examples of natural gene flow covering great distances, causing propagules to end up where they would not normally occur. Human transfer, whether it be from intentional planting or accidental transportation, has served to increase the magnitude of such gene flow in recent times. Unfortunately, the pace of natural evolutionary processes may, in many cases, be too slow for populations to adapt in response to the often large-scale modifications caused by modern rates of invasion by exotic genotypes.

Genetic contamination has the potential to negatively influence population and species level genetic processes. As mentioned previously in the context of composite seed mixtures, hybridization with non-local populations can result in progeny that are less fit than individuals from the local population. When the relative contribution of gametes from a poorly adapted non-local source is low, genes from non-local individuals, as well as their hybrids, might be readily removed from the population through natural selection without much consequence to population fitness. However, if large numbers of non-local organisms are introduced, the potential exists for the swamping of locally adapted genes.

Clearly, the relative size of local and non-local populations influences the direction and magnitude of gene flow between them. This concept has been demonstrated in studies of wild radish (*Raphanus sativus*), where the amount of gene flow into small populations was related more strongly to the size of the surrounding populations than to the distance between the small and the large populations (Ellstrand *et al.* 1989). The analogous situation of genetic swamping may similarly alter gene flow dynamics, which can result in the dilution of adapted genotypes from small native populations. Good examples can be found among native plant species that hybridize with crop varieties. This is due to the disparity commonly found between the potentially huge pollen pool contributed by large agricultural plantings and the often more modest pollen contribution from smaller and more diffuse stands of native species. The northern California black walnut (*Juglans hindsii*) may, in fact, be endangered due to hybridization between the small remaining populations of this species and the cultivated walnut (*J. regia*) (McGranahan *et al.* 1988). Large amounts of pollen from nearby *J. regia* orchards means that the genetic purity of nuts from *J. hindsii* is often in doubt.

In an animal example, Wayne *et al.* (1992) found evidence of high levels of hybridization with coyotes in several small and highly fragmented wolf populations, presumably as the result of genetic swamping. Recent declines in many North American wolf (*Canis lupus*) populations and expansion in numbers as well as geographic range of coyote (*C. latrans*) populations have led to an increase in the potential for genetic contamination of the native wolf populations (since the two species can interbreed). Genetic exchange between the two species is presumed to have also occurred historically, but likely with minimal effect, due to large population sizes of both species.

These examples of genetic contamination demonstrate that the dangers are highest when the local native population is isolated and surrounded by larger numbers of a non-local, or otherwise undesirable, source of genes. Such genetic swamping can also have a negative impact on local populations even if the non-local and the local sources do not hybridize. The overwhelming numbers of non-local genes will effectively lower the frequency of "native" genes in the population and genes present in low frequency are particularly vulnerable to loss through genetic drift. Thus, native genes can be lost, even if they are selectively favored. To mitigate this risk, Handel *et al.* (1994) have suggested that, when necessary, the adding of non-local sources to a native population should be done in increments over time in order to avoid the swamping of local pollen sources. The potentially deleterious impacts of genetic contamination can also be reduced by the creation of buffer zones around genetically "pristine" areas (Millar and Libby 1991).

RESTORING EVOLUTIONARY DYNAMICS TO THE COMMUNITY

Up to this point, we have devoted much of our discussion to matching germplasm to the environment and maintaining genetic variation within a population of a given species. The single-species is an attractive level at which to work because of the relative degree of simplicity. Restoration, however, is ultimately the reintroduction of one or more populations within the context of the community. Restoration should therefore ideally consider all of the species found in a community and the potential interactions between species. The importance of including all of the components of the community is particularly cogent in view of recent reports suggesting that biologically diverse communities have the highest productivity and greatest ecosystem stability (Naeem *et al.* 1994; Tilman and Downing 1994). In reality, limitations associated with the costs of restoration often result in active restoration being attempted for only a fraction of the species naturally present (Jordan *et al.* 1988).

Even if one were interested only in restoring populations of one species, interspecific interactions and the community context cannot be neglected. Handel *et al.* (1994) stress the need to consider mutualistic components, such as pollinators, seed dispersers, nitrogen-fixing bacteria, and mycorrhizae, in order to restore true native biodiversity to a community. In addition, at least some of the genetic variation that exists within populations is the result of selective pressures caused by associated organisms including prey, predators, and pathogens; without the balancing selection that these other species impose, much of this genetic variation may be lost (Futuyma 1983). Therefore, all species making up a community, including pathogens and other less "popular" organisms, warrant consideration in restoration. Failure to do so could result in the loss of genetic variation in the dominant species—genetic variation that may be vital for long-term persistence of restored populations.

Restoration of genetic structure within populations and species is especially likely to be dependent upon the existence of specific native pollinators. This possibility has been suggested in reference to vernal pool ecosystems of the arid west, a habitat type currently emphasized in restoration efforts in California. Vernal pools are seasonally flooded depressions with reduced water infiltration due to an underlying impervious soil layer, and are home to many species restricted to this unique environment. Urban and agricultural conversion has led to widespread destruction of vernal pool habitat, and the attempted off-site construction of vernal pools for purposes of mitigation has engendered considerable controversy as to whether the ecological and genetic complexity of a vernal pool ecosystem can actually be re-created. Leong (1994) and Thorp and Leong (1995) found different species of pollinators and a reduced proportion of endemic native pollinators in artificially recreated vernal pools compared to natural vernal pools. Native pollinators are often required for optimal seed set of native plant species; nonnative pollinators often have different foraging behavior and lower pollination efficiencies (Linhart 1995; Thorp and Leong 1995). Restoration of vernal pools without restoration of upland nesting habitat for native bees and flies could therefore result in the loss of native plant species within the vernal pool ecosystem. In addition, the small size and limited movement characteristic of many native vernal pool insect pollinators may have played a role in the genetic structure observed within endemic vernal pool plant species (Thorp 1976). Linhart (1974) found that *Veronica peregrina* plants on the edges of vernal pools were genetically different from plants in the center of the same pool. If native bee nesting habitat is not restored and nonnative pollinators with different flight patterns instead dominate, such within-pool genetic structure could be diluted or destroyed. In addition, the dispersal of seed (another aspect of gene flow) is often dependent on the fauna, and restoration of natural levels of gene flow will therefore require the restoration of these animal components as well.

Communities are rarely static and this reality further complicates restoration. Organisms characteristic of a community may not all be present at any one point in time or place. For example, the reintroduction of species-specific pollinators is best accomplished only after the re-introduced plant species have reached reproductive maturity. Initially established species may also directly modify environmental conditions, thereby creating an environment to which the later successional species are adapted. Evidence for such facilitation is most commonly found during the earliest stages of succession, such as what might be found when restoration starts from the substrate. For example, Gill and Marks (1991) demonstrated that some plant species grew best after other plants had first colonized bare ground, because the initial plant cover allowed these later successional species to avoid desiccation. Restoration of the entire community may require the introduction of organisms throughout the different successional stages of that community. Human intervention may be especially important in the restoration of large or isolated communities where natural dispersal of the later successional organisms from existing remnant habitat is unlikely.

CONCLUSIONS

The escalating impact of humans on the ecosystems of our planet leaves little doubt about the current and future need for restoration as a tool for reversing environmental damage and enhancing native biodiversity. A view of ecological restoration that includes a genetic component can increase our chances of successfully recreating ecosytems that function both ecologically and evolutionarily. The complexities involved are many, however. Choosing appropriate germplasm, ensuring the presence of adequate genetic variation in restored populations, and in the case of plants, avoiding genetic shifts during the production of seed, are but a few of the issues that must be considered. In this chapter, we have presented different strategies for reintroducing adapted populations. For example, the development of seed zones, based on broad ecological and geographic characteristics, may be helpful for delineating regions of adaptation outside of which propagules should not be transferred. Composite mixtures of different populations within a region might also be useful for obtaining broadly adapted populations that possess sufficient genetic variation so that selection can "fine-tune" the mixture for adaptation to different microenvironments. Mixtures may be especially useful for restoring highly disturbed sites with selective environments that may currently differ greatly from the pre-disturbance state. Care should be taken when introducing non-local genes into regions containing native local populations, but the potential negative consequences of such genetic contamination are greatest when the amount of introduced non-local genetic material is high, relative to the local source.

The temporal scale necessary for determining whether a restoration project will ultimately be successful may exceed the lifetimes of the people involved. With so much at stake in the restoration process, the extra time and effort that might be necessary to obtain the most appropriate and best-adapted genetic material will, in the end, seem inconsequential. The level of care taken will depend on the situation. For example, restoration in a national park, where native populations of the targeted species grow nearby, will undoubtedly demand a more conservative approach than would re-creating isolated habitat for waterfowl in a highly altered agricultural landscape.

Our intent, by discussing these genetic issues, is not to slow the pace of restoration, but rather to stimulate further discussion and consideration of the processes that affect the genetic structure and evolutionary dynamics of restored ecosystems. Unfortunately, much of the empirical data that tie together the successful restoration of ecosystems and underlying genetic processes remain to be gathered. The unknowns, however, should not prevent us from moving forward. The process of putting the parts of an ecological community back together may provide us with incredible opportunities for learning how ecological and evolutionary components interact in functioning ecosystems. As Bradshaw (1983) professes, "successful reconstruction of ecosystems, when it is achieved, has the reward that it is the ultimate proof of our ecological understanding" (p. 14). In this sense, knowledge gained through restoration will enable us to not just restore lost biodiversity, but to better manage existing more pristine landscapes as well.

ACKNOWLEDGMENTS

This chapter benefited greatly from discussions with many people including Subodh Jain, Steve Monsen, and Kevin Rice. Michelle Geary, Roland Knapp, Kevin Rice, and Mark Schwarz, as well as editors Peggy Fiedler and Peter Kareiva provided helpful comments on earlier versions. We thank John Anderson of Hedgerow Farms for use of the Figure 14.3 photograph. John Anderson and Bryan Young kindly allowed us to use unpublished data.

LITERATURE CITED

Antonovics, J. and A.D. Bradshaw. 1970. Evolution in closely adjacent plant populations VIII. Clinal patterns at a mine boundary. *Heredity* 25:349–362.

Barrett, J.A. 1981. The evolutionary consequences of monoculture. In *Genetic consequences of man-made change*, eds. J.A. Bishop and L.M. Cook, 209–248. London: Academic Press.

Barrett, S.C.H. and J.R. Kohn. 1991. Genetic and evolutionary consequences of small population size in plants: implications for conservation. In *Genetics and conservation of rare plants*, eds. D.A. Falk and K.E. Holsinger, 3–30. New York: Oxford University Press.

Bartley, D.M. and G.A. Gall. 1990. Genetic structure and gene flow in chinook salmon populations of California. *Transactions of the American Fisheries Society* 119:55–71.

Bradshaw, A.D. 1959. Population differentiation in *Agrostis tenuis* Sibth. I. Morphological differentiation. *New Phytologist* 58:208–227.

Bradshaw, A.D. 1983. The reconstruction of ecosystems. *Journal of Applied Ecology* 20:1–17.

Bradshaw, A.D. 1984. Ecological significance of genetic variation between populations. In *Perspectives on plant population biology*, eds. R. Dirzo and J. Sarukhan, 213–228. Sunderland, MA: Sinauer Associates.

Clary, W.P. 1975. Ecotypic adaptation in *Sitanion hystrix*. *Ecology* 56:1407–1415.

Clausen, J., D.D. Keck, and W.M. Heisey. 1940. Experimental studies on the nature of species. I. Effect of varied environments on western North American plants. *Carnegie Institute of Washington Publication 520.*

Ellstrand, N.C., B. Devlin, and D.L. Marshall. 1989. Gene flow by pollen into small populations: Data from experimental and natural stands of wild radish. *Proceedings of the National Acadamy of Sciences* 86:9044–9047.

Ennos, R.A. 1983. Maintenance of genetic variation in plant populations. *Evolutionary Biology* 16:129–155.

Fenster, C.B. and M.R. Dudash. 1994. Genetic considerations for plant population restoration and conservation, in *Restoration of endangered species*, eds. M.L. Bowles and C.J. Whelan, 34–62. Cambridge: Cambridge University Press.

Futuyma, D. J. 1983. Interspecific interactions and the maintenance of genetic diversity, in *Genetics and conservation*, eds. C.M. Schonewald-Cox, S.M. Chambers, B. MacBryde, and W.L. Thomas, 364–373. Menlo Park: Benjamin/Cummings.

Gill, D.S. and P. L. Marks. 1991. Tree and shrub seedling colonization of old fields in central New York. *Ecological Monographs* 61:183–205.

Guerrant, E.O., Jr. 1992. Genetic and demographic considerations in the sampling and reintroduction of rare plants. In *Conservation biology: The theory and practice of nature conservation, preservation, and management*, eds. P.L. Fiedler, and S.K. Jain, 321–344. New York: Chapman & Hall.

Handel, S.N., G.R. Robinson, and A.J. Beattie. 1994. Biodiversity resources for restoration ecology. *Restoration Ecology* 2:230–241.

Harlan, H.V. and M.L. Martini. 1938. The effect of natural selection in a mixture of barley varieties. *Journal of Agricultural Research* 57:189–199.

Hedrick, P.W. 1995. Gene flow and genetic restoration: The Florida panther as a case study. *Conservation Biology* 9:996–1007.

Hoskinson, P.E. and C.O. Qualset. 1967. Geographic variation in 'Balbo' rye. *Tennessee Farm and Home Science Progress Report* 62:8–9.

Huenneke, L.F. 1991. Ecological implications of genetic variation in plant populations. In *Genetics and conservation of rare plants*, eds. D.A. Falk and K.E. Holsinger, 31–44. New York: Oxford University Press.

Janzen, D.H. 1988. Tropical ecological and biocultural restoration. *Science* 239:243–244.

Jordan, W.R., R.L. Peters, and E.B. Allen. 1988. Ecological restoration as a strategy for conserving biological diversity. *Environmental Management* 12:55–72.

Kilcher, M.R. and L. Looman. 1983. Comparative performance of some native and introduced grasses in southern Saskatchewan, Canada. *Journal of Range Management* 36:654–657.

Knapp, E.E. and K.J. Rice. 1994. Starting from seed: Genetic issues in using native grasses for restoration. *Restoration and Management Notes* 12:40–45.

Knapp, E.E. and K.J. Rice. 1996. Genetic structure and gene flow in *Elymus glaucus* (blue wildrye): Implications for native grassland restoration. *Restoration Ecology* 4:1–10.

Leberg, P.L. 1993. Strategies for population reintroduction: Effects of genetic variability on population growth and size. *Conservation Biology* 7:194–199.

Ledig, F.T. 1992. Human impacts on genetic diversity in forest ecosystems. *Oikos* 63:87–108.

Leong, J.M. 1994. Pollination of a patchily-distributed plant, *Blennosperma nanum*, in natural and artificially created vernal pool habitats. Ph.D. dissertation, University of California, Davis.

Linhart, Y.B. 1974. Intra-population differentiation in annual plants. I. *Veronica peregrina* raised under non-competitive conditions. *Evolution* 28:232–243.

Linhart, Y.B. 1995. Restoration, revegetation, and the importance of genetic and evolutionary perspectives. In *Proceedings: Wildland shrub and arid land restoration symposium*, Oct. 19–21, 1993. Las Vegas, NV, eds. B.A. Roundy, E.D. McArthur, J.S. Haley, and D.K. Mann, 271–288. U.S. Dept. of Agriculture, Forest Service, Intermountain Research Station General Technical Report 315.

Loveless, M.D. and J.L. Hamrick. 1984. Ecological determinants of genetic structure in plant populations. *Annual Review of Ecology and Systematics* 15:65–95.

Maehr, D.S. and G.B. Caddick. 1995. Demographics and genetic introgression in the Florida panther. *Conservation Biology* 9:1295–1298.

Martins, P.S. and S.K. Jain. 1979. Role of genetic variation in the colonizing ability of rose clover (*Trifolium hirtum*). *American Naturalist* 114:591–595.

McGranahan, G.H., J. Hansen, and D.V. Shaw. 1988. Inter- and intraspecific variation in California black walnuts. *Journal of the American Society of Horticultural Science* 113:760–765.

McMillan, C. 1959. The role of ecotypic variation in the distribution of the central grassland of North America. *Ecological Monographs* 29:385–308.

Millar, C.I. and W.J. Libby. 1989. Disneyland or native ecosystem: Genetics and the restorationist. *Restoration and Management Notes* 7:18–24.

Millar, C.I. and W.J. Libby. 1991. Strategies for conserving clinal, ecotypic, and disjunct population diversity in widespread species. In *Genetics and conservation of rare plants*, eds. D.A. Falk and K.E. Holsinger, 149–17. New York: Oxford University Press.

Naeem, S., L.J. Thompson, S.P. Lawler, J.A. Lawton, and R.M. Woodfin. 1994. Declining biodiversity can alter the performance of ecosystems. *Nature* 368:734–737.

Noss, R.F., E.T. LaRoe III, and J.M. Scott. 1995. *Endangered ecosystems of the United States: a preliminary assessment of loss and degradation*. Biological Report 28. Washington D.C.: U.S. Dept. of the Interior, National Biological Service.

Padgett, D.J. and G.E. Crow. 1994. Foreign plant stock: Concerns for wetland mitigation. *Restoration and Management Notes* 12:168–171.

Parker, M.A. 1985. Local population differentiation for compatibility in an annual legume and its host-specific fungal pathogen. *Evolution* 39:713–723.

Rapson, G.L. and J.B. Wilson. 1988. Non-adaptation in *Agrostis capillaris* L. (Poaceae). *Functional Ecology* 2:479–490.

Rice, K.J. and R.N. Mack. 1991. Ecological genetics of *Bromus tectorum* III. The demography of reciprocally sown populations. *Oecologia* 88:91–101.

Schonewald-Cox, C.M. 1983. Guidelines to management: A beginning attempt. In *Genetics and conservation*, eds. C.M. Schonewald-Cox, S.M. Chambers, B. MacBryde, and W.L. Thomas, 14–445. Menlo Park: Benjamin/Cummings.

Snaydon, R.W. and T.M. Davies. 1982. Rapid divergence of plant populations in response to recent changes in soil conditions. *Evolution* 36:289–297.

Soil Conservation Service. 1986. 'Lassen' antelope bitterbrush. Washington, D.C.: United States Department of Agriculture.

Sorensen, F.C. 1983. Geographic variation in seedling Douglas fir (*Pseudotsuga menziesii*) from the western Siskiyou mountains of Oregon. *Ecology* 64:696–702.

Strauss, S.H., J. Bosquet, V.D. Hipkins, and Y.P. Hong. 1992. Biochemical and molecular genetic markers in biosystematic studies of forest trees. *New Forests* 6:125–158.

Sultan, S.E. 1987. Evolutionary implications of phenotypic plasticity in plants. *Evolutionary Biology* 21:127–178.

Templeton, A.R. 1986. Coadaptation and outbreeding depression. In *Conservation biology: The science of scarcity and diversity*, ed. M.E. Soulé, 105–116. Sunderland, MA: Sinauer Associates.

Thorp, R.W. 1976. Insect pollination of vernal pool flowers. In *Vernal pools, their ecology and conservation*, ed. S.K. Jain, 36–40. Davis, CA: University California, Institute of Ecology.

Thorp, R.W. and J.M. Leong. 1995. Native bee pollinators of vernal pool plants. *Fremontia* 23:3–7.

Tilman, D. and J.A. Downing. 1994. Biodiversity and stability in grasslands. *Nature* 367:363–365.

Turkington, R. and J.L. Harper. 1979. The growth, distribution, and neighbour relationships of *Trifolium repens* in a permanent pasture. *Journal of Ecology* 67:245–254.

Waser, N.M. and M.V. Price. 1994. Crossing-distance effects in *Delphinium nelsonii*: Outbreeding and inbreeding depression in progeny fitness. *Evolution* 48:842–852.

Wayne, R.K., N. Lehman, M.C. Allard, and R.L. Honeycutt. 1992. Mitochondrial DNA variability of the gray wolf: Genetic consequences of population decline and habitat fragmentation. *Conservation Biology* 6:559–569.

CHAPTER 15

Replacing Endangered Species Habitat: The Acid Test of Wetland Ecology

JOY ZEDLER

Biodiversity is of special interest in California, a state with a great number of species, a large proportion of endemics, and many taxa in jeopardy. Because major developments have occurred along the Pacific Ocean, coastal habitats have been particularly affected. It is no surprise, therefore, that ten of California's 94 endangered and threatened animal species are ones with coastal wetland affinities (Dept. of Fish and Game 1989); in southern California, these include the light-footed clapper rail (*Rallus longirostris levipes*), the California least tern (*Sterna antillarum browni*), and Belding's Savannah sparrow (*Passerculus sandwichensis beldingi*). Of the 298 coastal species considered rare by the California Native Plant Society, 17 (6%) occur in coastal wetlands.

Population statistics readily explain why so many coastal species are threatened. Over 16 million people (56% of the state's total) live in "southwest California," a narrow coastal strip that includes parts of Santa Barbara, Ventura, Orange, Riverside, and San Diego counties, but makes up just 8% of the State's area (Davis *et al.* 1995). Not only is the region densely populated, it has a long history of habitat destruction, environmental degradation, and ecosystem fragmentation. Indeed, the California State Coastal Conservancy (1989) asserts that San Diego County's wetlands may prove to be the most threatened natural resource on the California coast. Thus, there is both need and opportunity to increase the quantity and quality of wetlands in southern California (Zedler 1996a). However, efforts to restore wetland habitats lag far behind the destruction of habitat.

ACID TESTS

To be able to restore an ecosystem, an ecologist must understand its components (i.e., species and environmental conditions) and its functions (i.e., processes). An often-quoted analogy is that restoration is the "acid test of ecology" (Bradshaw 1987). Bradshaw (*op. cit.*, p. 26) elaborated that "[t]he essential element of proof of understanding by the restoration of function is that, because the object being reconstructed is complex, failure to reconstruct any part of it properly will cause it to function improperly or fail to function at all. And, generally, the more complex the object, the more stringent this test of understanding." From evaluating wetland restoration efforts in southern California, I suggest that the acid test of *wetland* ecology is the construction of habitat for target organisms with specialized and complex habitat dependencies.

Two case studies from San Diego Bay have tested our ability to satisfy highly specialized or "demanding" species. Both cases grew out of mitigation requirements. The California Department of Transportation and the U.S. Army Corps of Engineers was required to reestablish an endangered plant (salt marsh bird's-beak, *Cordylanthus maritimus maritimus* [Scrophulariaceae]) and to create habitat for an endangered bird (the Light-footed clapper rail, *Rallus longirostris levipes* [Rallidae]) to compensate for damaging their habitat during freeway widening and flood control channel construction.

The San Diego Bay site (Figure 15.1) is, in itself, a restoration challenge. The area has served as a dump for spoils dredged from San Diego Bay and as an urban landfill

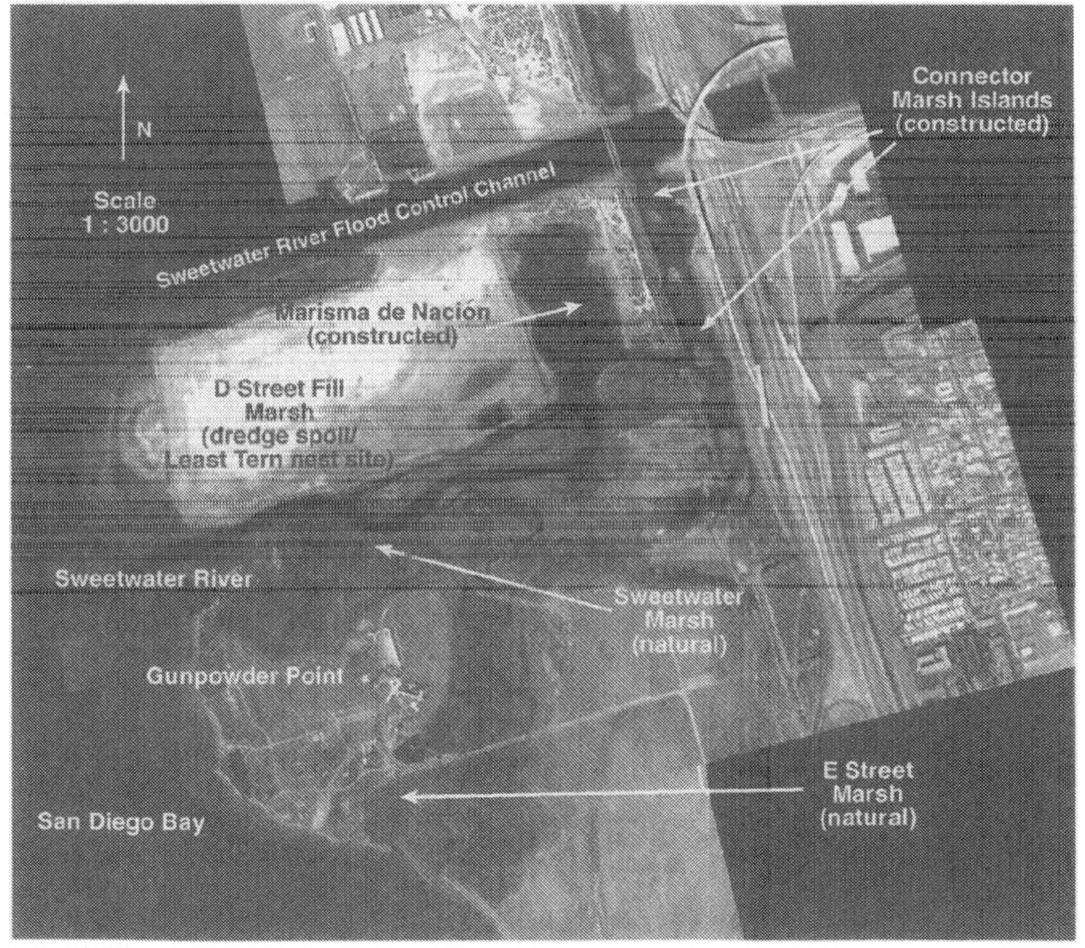

Figure 15.1
Aerial image of the central portion of Sweetwater Marsh National Wildlife Refuge, including natural and constructed wetlands. The digital multispectral image was acquired via Airbore Data Acquisition and Registration by Positive Systems, Inc. (Kalispell, Montana), provided by Dr. Doug Stow (Geography Department, San Diego State University), and annotated by Bruce Nyden (Pacific Estuarine Research Lab).

adjacent to the cities of Chula Vista and National City. The site is now part of the Sweetwater Marsh National Wildlife Refuge, a collage of natural marsh remnants, disturbed uplands, and landfills, all of which have had their hydrologic lifelines modified by an upstream dam and runoff from urban drains. Criss-crossing the refuge are an abandoned railroad, a major power line, and an interstate highway. Because so much modification has occurred in the area, it is difficult to say whether the repair work should be called habitat construction or "true" habitat restoration. In the stricter sense, it is a habitat construction effort, because wetlands were carved out of dredge spoil deposits. Yet because the project has returned tidal marsh to the same shoreline where salt marsh had been filled, it qualifies as restoration in the broader sense.

ENDANGERED SPECIES CHARACTERISTICS

Both target species have very specific requirements. The bird's-beak and the clapper rail are intertidal salt marsh residents; bird's-beak occurs along the upper marsh edge and the Clapper rail lives within the lower marsh. In addition, each relies on other habitats for critical life-history functions. Bird's-beak requires adjacent upland habitat because its pollinators live there (Parsons 1994; Parsons and Zedler in press; PERL 1995; Zedler 1996b), while the clapper rail needs upland habitat as a refuge when high waters inundate the entire marsh (Zedler 1993). As is discussed below, restoring only the habitats preferred by these species will not ensure population viability.

The resource agencies also have high standards for evaluating successful mitigation of damages to each endangered species' habitat (US FWS 1988). Standards are strict despite the fact that the highway and flood control channel projects did not cause direct mortality to either species; rather, loss of potential habitat led to a jeopardy opinion under the U.S. Endangered Species Act of 1973 and subsequent need for compensation. For bird's-beak, the regulatory requirement was to establish 5 patches of ≥20 individual plants able to sustain themselves for three years within a high-marsh community that includes at least 75% of the native high-marsh plants and has less than 20% of its area covered by weedy exotics. For the clapper rail, the U.S. Fish and Wildlife Service required 7 home ranges (each 0.8–1.6 ha, with 15% of the area vegetated with Pacific cordgrass [*Spartina foliosa*] and 15% with high marsh), each having tidal creeks, suitable forage foods, and a nesting habitat patch (≥100 m^2 of tall dense cordgrass sustainable for at least 3 years). Key features of these criteria are the detailed specifications and the inclusion of self-sustainability as a standard. This is a giant leap ahead of mitigation agreements that simply require a given level of vegetation cover or transplant survival, with replanting if standards are not met within the first year or so.

Reestablishing the Sweetwater Marsh Bird's-Beak Population

Like other species of *Cordylanthus*, salt marsh bird's-beak (Figure 15.2) is an annual, hemiparasitic herb (Chuang and Heckard 1971). Because populations must establish from seed each year, our initial approach to re-establishment was to identify constraints on seedling establishment and preferences for host plants prior to seeding the site. Bird's-beak is not host specific; neither occurrences of *C. maritimus* ssp. *maritimus* (Zedler personal observation) nor *C. maritimus* ssp. *palustris* (Kelly and Fletcher 1995) show a constancy of associated species. However, some preference is implied by results of a

Figure 15.2
Salt marsh bird's-beak, a hemiparasitic annual plant, is endangered due to substantial loss of its high intertidal marsh habitat. Photo by Dr. Paul H. Zedler, reproduced with permission.

greenhouse experiment. Bird's-beak biomass and flower production were both highest in pots with *Distichlis spicata*, and *Monanthochloe littoralis* (both grasses); less preferred hosts, in declining order, were *Salicornia virginica*, *Frankenia grandifolia* and *Arthrocnemum subterminale* (= *Salicornia subterminalis* in Hickman 1993); and *Atriplex watsonii* and bird's-beak grown by itself had the lowest biomass and flower numbers (Fink and Zedler 1990). It is understandable that grasses would be preferred hosts, as their fine root systems may be relatively easy to penetrate.

Other greenhouse experiments showed that bird's-beak grew best with intermediate shade and low soil salinity (Fink and Zedler 1990). Each of these results suggests further constraints on the regeneration niche. Hosts are beneficial, but dense host canopies can prevent establishment or limit growth. Bird's-beak is restricted to saline habitats, but lowered salinity stimulates seed germination and seedling growth. I am reminded of Goldilocks, who found little that was "just right."

Seeds of Bird's-beak are known to remain viable for multiple years in the lab (G. Noe, PERL, unpublished data) and in the field (Zedler personal observation; Kelly and Fletcher 1995). Germination is stimulated by low salinities that develop with winter rainfall, which usually occurs between December and February. Seedlings appear in February or March in small canopy gaps (Parsons 1994). Weedy annual plants also appear during winter, and in wetter years, salinities are lowered enough for several invasive species to establish and affect bird's-beak growth. Potential competitors are *Polypogon monspeliensis* (Kuhn 1995), *Parapholis incurva*, *Cotula coronopifolia*, and *Rumex crispus*, all of which are exotic species. There are no field data on competitive exclusion; however, recent mesocosm and greenhouse experiments indicate that *Polypogon monspeliensis* can reduce growth of *Salicornia virginica*, which is one of the host species used by Bird's-beak. *Salicornia virginica* biomass was 42% lower, on average, when grown with the exotic annual grass than when reared alone (Callaway and Zedler in preparation). Because the grass canopy is no taller than that of *S. virginica*, we assume that root competition (for nitrogen?) is an important aspect of the competitive interaction. Thus, the effect of weedy annuals could be indirect, operating through the host plant. I suspect that few locations have a salinity regime that allows only the bird's-beak and not the weedy annuals to germinate and establish. I do not know whether bird's-beak can compensate by parasitizing invasive winter annuals or whether annual species are unsuitable hosts because they provide resources for too short a time to benefit the bird's-beak.

As a facultative hemiparasite, Bird's-beak can germinate and grow independently, but it produces more seed capsules when grown with a host. I hypothesize that the host provides both moisture and nutrients, especially during the spring and summer months, when the surface soil is dry. The effect that bird's-beak has on its hosts, whether perennial or annual, needs study. Perhaps the parasite helps to maintain an open canopy by reducing host plant cover. If so, bird's-beak may help sustain itself by creating the small gaps needed for germination and seedling establishment.

Thus, the regeneration niche for Bird's-beak is complex. In order for seeds to germinate, a gap must appear in the canopy, but the opening must be small enough to include roots of host species and large enough to reduce shade. The appropriate-size canopy gaps are likely created by small mammals, e.g., *Thomomys bottae*, *Spermophilus beecheyi*, which often burrow in the high marsh (Cox and Zedler 1986). Populations of small mammals are, in turn, affected by their predators, which include native raptors, but also cats and dogs. If carnivory by pets and feral animals increases along with the construction of urban housing, small mammal populations may decline and the necessary canopy gaps along with them. Regrettably, the natural patterns and causes of small canopy gaps have not been quantified in this region. Deposits of wrack and urban debris can also create canopy gaps (Swift 1987), but they may not be of the proper size and distribution for use by bird's-beak.

In 1990, no naturally-occurring plants of bird's-beak remained at San Diego Bay, so seeds were imported from Tijuana Estuary (8 km to the south), the nearest naturally-occurring population. By 1992, the seeded population exceeded 5,000 plants, and we began conducting additional experiments that showed nitrogen and pollinators to be limiting to seed capsule production (Parsons and Zedler in press).

The importance of pollinators for successful seed set was postulated early in the re-establishment effort, after seeds were sown on a small island that was isolated from other marsh and upland areas by broad tidal channels. Plants established and grew to reproductive age, but few seeds were produced. We suspected a shortage of insect pollinators and shifted the reintroduction effort to a larger expanse of natural marsh with adjacent upland.

In 1991 and 1992, seeds were sown by B. Fink (Habitat Restoration, San Diego, California) over several acres of Sweetwater Marsh, and the pollinator-limitation hypothesis was tested in 1992 and 1993. Bagging of flowers (i.e., pollinators excluded) demonstrated a limited ability for self pollination, and hand-pollination revealed that seed production was greater than undisturbed flowers (Parsons 1994; Parsons and Zedler in press). Although pollinator limitation was demonstrated, the population still produced enough seed to allow the population to persist and expand (Figure 15.3).

The most important native pollinators are Halictine bees (*Dialictus* sp.), *Bombus crotchii*, *B. californicus*, and *Melissodes tepida timberlakei* (Parsons 1994). An additional pollinator, *Anthidium edwardsii*, has been described as highly effective (Lincoln 1985). Although present at Tijuana Estuary, this bee has not been observed at the reestablished bird's-beak population (Parsons and Zedler in press).

Many of the bee pollinators are believed to nest in upland habitat adjacent to the salt marsh. Any bees that nest within the intertidal zone risk mortality from high water. Parsons observed *Melissodes* individuals emerging from burrows in the high intertidal marsh near bird's-beak patches but suggested that flooding in 1993 reduced pollinator populations by inundating their burrows. Thus, adjacent upland appears to be essential for pollinator nesting. Upland habitats also provide pollinators with alternative foods when Bird's-beak flowers are rare.

Transplanting or sowing seeds from one marsh to another does not ensure that the essential food web will follow, as pollinators, predators, and other critical associates may not be able to disperse to new bird's-beak populations. Insect predators are important in controlling herbivorous insects, which can be a problem for bird's-beak. For example, the snout-nosed moth (*Liphographis fenestrella*) has been observed feeding on the seeds

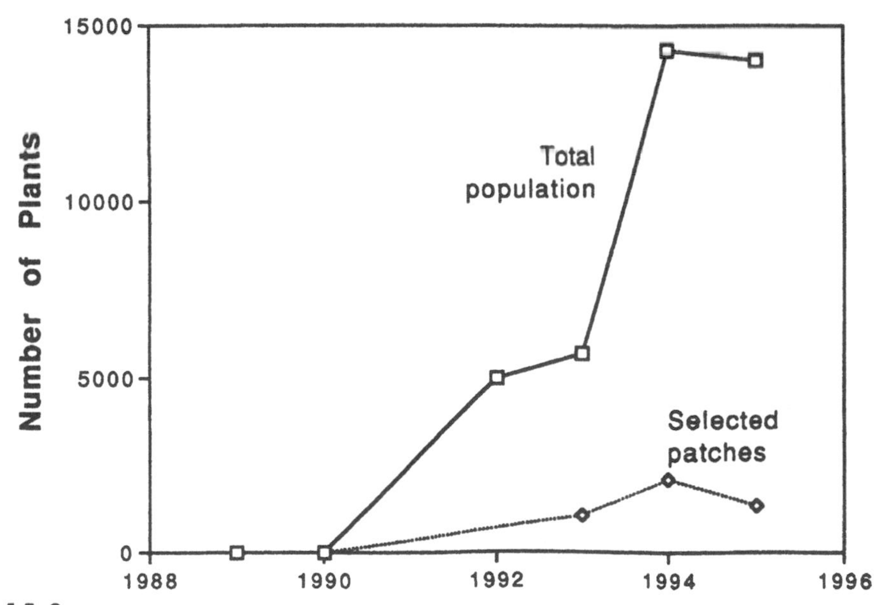

Figure 15.3
Dynamics of the bird's-beak population reestablished at San Diego Bay. Counts are provided for five patches monitored for compliance with mitigation criteria (precisely counted, dotted line) and for the entire population (includes some estimates of numbers in dense areas, solid line). Data of the Pacific Estuarine Research Lab.

of bird's-beak (Parsons 1994, B. Fink, personel communication). The moth is, in turn, fed upon by a wasp (*Euodynerus annulatus sulphureus*) which was observed at Tijuana Estuary but not at the reestablished bird's-beak population (Parsons and Zedler in press).

Our re-introduction and study of bird's-beak have made it clear that a self-sustaining population requires not only a well established, grass-dominated upper salt marsh, but also an adjacent natural upland to support pollinators and small mammals. Disturbance patches must be small and dependably frequent. In addition, winter rainfall must be sufficient to lower soil salinity, tidal inundation must restore the saline conditions in summer and reduce invasion by weedy exotics, and floods must occur on occasions to replenish marsh soils with nitrogen.

Although the mitigation criteria have been met (well over the required 100 plants have been present at San Diego Bay for the required three years; Figure 15.3), a nagging question remains: Why was the natural population at this site extirpated in the 1980's? If we knew the answer, we could better manage the reestablished population to insure its persistence. There is much to be learned about the causes of local extinction and requirements for reestablishing self-sustaining populations of endangered plants (Falk, Millar, and Olwell 1996).

Creating Habitat for the Light-footed Clapper Rail

Reproducing suitable home range and nesting habitat for the light-footed clapper rail has proven more difficult than reintroducing bird's-beak. As of fall 1995, the marshes constructed in 1984 and 1990 were not in compliance with mitigation requirements (seven home ranges, each with detailed standards that must be met for three years). There was sufficient area for seven 0.8–ha home ranges in the 12-ha site. However, not enough area was vegetated, especially in the upper intertidal zone. Salt crusts have formed and native plants have been slow to establish, even where transplanted and irrigated. In natural marshes, a bushy subshrub (*Arthrocnemum subterminale*) at the marsh-upland edge are used by clapper rails, presumably as protection from terrestrial predators during extreme high-tide events when the birds move out of the marsh. But low soil salinities are needed for high-marsh vegetation to establish (Kuhn 1995; Kuhn and Zedler, in review), and permanent salt crusts impair germination and transplant establishment in many parts of the mitigation site. The constructed marsh islands have a domed shape that may facilitate the formation and persistence of both salt crusts and hypersaline groundwater. In general, the sparse high marsh, with its disturbed urban surroundings, contrasts starkly with natural wetlands in nearby Baja California, Mexico, where extensive upland vegetation abuts broad, undisturbed uplands.

Clapper rails also require tall vegetation, especially cordgrass, in the low marsh to support their floating nests and to provide cover from aerial predators (Massey, Zembal, and Jorgensen 1984). The mitigation standard for cordgrass height (mean of 60–80 cm in the 90-m^2 nesting habitats; US FWS 1988) has nearly been met, but rails have not yet used any of the potential home ranges for nesting. The US FWS standard for height may be too lax, because rails nest primarily in cordgrass canopies where more than 30 plants/m^2 exceed 90 cm in height at the end of the growing season (Zedler 1993). The shortage of tall stems at the constructed marsh is attributed to the coarse soil, low soil organic matter content, low nitrogen supply, and low nitrogen retention (Covin and Zedler 1988; Langis, Zalejko, and Zedler 1991). In the absence of tall plants, another species, the native ladybird beetle (*Coleomegilla fuscilabris*) is also at a disadvantage, and where it is rare, a scale insect (*Haliaspis spartina*) can become a major plant pest (Boyer and Zedler 1996). When hundreds to thousands of scale insects cover cordgrass leaves, growth of short plants is impaired, and canopies senesce early. Although adding nitrogen has increased cordgrass height enough to meet mitigation criteria in experimental plots, soil nitrogen content has not accumulated, and fertilization is needed on an annual basis (Boyer and Zedler unpublished data).

Other factors are also important to clapper rails. These birds require invertebrate foods, and although most species of epibenthic invertebrates had low densities in the constructed marsh, prey animals preferred by clapper rails (crabs) were actually more abundant than in the natural marsh reference wetland (Scatolini and Zedler 1996; PERL 1994). Adjacent freeway disturbances might pose further problems for the rails, but a pair has nested among cattails (*Typha domingensis*) very near the freeway off-ramp, suggesting that a tall, dense plant canopy can ameliorate the impacts of traffic. Whether inadequate nesting canopies are the main constraint for clapper rails is not certain, but vegetation height is an obvious difference between the constructed and natural marshes.

HABITAT LINKAGES

Recreating habitat for the salt marsh bird's-beak plant and the light-footed clapper rail is difficult because these two resident species depend on connections (1) between wetland habitats; (2) between the wetland and upland; and (3) between the wetland and its watershed. While these kinds of connections were acknowledged in the mitigation design phase, the problems that developed during the assessment phase of the mitigation program earned them further attention.

Between-Habitat Linkages

Given a 12-ha mitigation site that was designed to accommodate seven home ranges, we struggled to imagine how clapper rails might divide up the space to achieve access to tidal creeks for forage, low marsh for nesting, and high marsh as a high-tide refuge. We needed to draw polygons of >0.8 ha in order to assess whether each potential home range met the standards of 15% low and 15% high marsh area. We might have subdivided each constructed marsh into oblong-shaped territories, but we guessed that birds would follow water courses while feeding and prefer not to cross open channels when moving from nest areas to high-tide refuges. In order to fit all seven home ranges into areas that had some tall cordgrass and some vegetated high marsh, we had to accept two long, narrow home ranges that may or may not be acceptable to clapper rails. It also became clear that a major gap in our knowledge about these birds is how and why they move within and between home ranges. Movements of nine light-footed clapper rails were studied by Zembal, Massey, and Fancher (1989), who used radio transmitters and showed that the birds spend most of the day in <30% of their home ranges. But, what function do their movements serve? Do they usually feed close to their nests? Do they avoid crossing broad channels? Do they prefer to travel along versus perpendicular to creeks?

Future habitat designs for clapper rails should probably be modeled more closely on natural marshes, incorporating complex tidal creek networks. And, if long, narrow home ranges prove unsuitable for home ranges, future constructed habitats should not be forced into narrow corridors, as occurred along interstate Highway 5 in the Sweetwater Marsh National Wildlife Refuge (Figure 15.1).

Wetland-Upland Linkages

At natural wetlands, the upland habitat, i.e., coastal sage or maritime scrub, probably serves as an extension of the clapper rail's high tide refuge, but we know little about how far these birds penetrate the scrub or what vegetation attributes are most useful as camouflage. Large ground birds need cover when in upland areas that are used by native predators adapted to urban landscapes: coyotes (*Canis latrans*), skunks (*Mephitis mephitis*), and possums (*Didelphis virginianus*) that hunt on the ground and northern harrier (*Circus cyaneus*), American kestrel (*Falco sparverius*), and white-tailed kite (*Elanus caeruleus*) that hunt from the air. The upland surrounding the mitigation site (Figure 15.1) may not be suitable for clapper rail use because of limited area in critical locations, such as along the highway, and because the upland is highly disturbed by people and their pets.

When intertidal wetlands are fully inundated by extreme events, clapper rails are driven into the upland. Three types of extreme events, in order of decreasing likelihood, are: (1) extreme high tides; (2) catastrophic river flooding; and (3) storm-generated sea level anomalies that coincide with high tides. Even though all these events are intermittent, a single high water level can put the Clapper rail population at risk to upland predators. Dr. A. Powell (National Biological Service, San Diego Field Station) documented a typical interaction during a January 1994 high water event (Figure 15.4). Had this dog not been chained, the birds would have been easy prey. Clearly, the clapper rail needs a

Figure 15.4
Pets pose a threat to clapper rails that move out of the intertidal marsh during extreme high water. Photo by Dr. Abby Powell, National Biological Service, San Diego Field Office, reproduced with permission.

broader buffer with fencing to exclude pets and shrub cover to provide hiding places during high water events.

The most predictable high water levels occur with the winter and summer extreme high tides. In the San Diego area, there are two unequal high tides per day, with seasonal variation in height. High tides are maximal in November, December, and January (229–238 cm above mean lower low water [MLLW]) and in May, June, and July (229–232 cm MLLW; Zedler 1993). These extreme tides are 36–50 cm above the average higher high tide. During such tidal conditions (Figure 15.5), only the tops of the high-marsh subshrubs are exposed, and birds must seek shelter beyond the salt marsh.

Catastrophic flooding and storms during high tide conditions have been unpredictably common in recent years. Major floods are infrequent on average, with only eight events in the 58-year streamflow record for Tijuana Estuary (1941, 1945, 1978, 1980, 1983, 1993, and 1995 [USGS 1936–1950, IBWC 1950–1993, pers. obs.]). However, six of the eight floods occurred in the past 17 years. Major streamflows generally persist for several days so that the marsh remains inundated during both high and low tide periods. A less likely high-water event occurs when sea storms coincide with an extreme high tide. Flick and Cayan (1984) calculated a one in 50 probability that three unusually high sea levels (21–27 cm above predicted high tide levels in November 1982, January 1983, and March 1983; San Diego tide gage) would coincide with extreme high tides; however, the actual coincidence was three times out of five. The point is that even rare and unlikely events

Figure 15.5
The light-footed clapper rail, a year-round resident of intertidal marshes in Southern California, is endangered due to substantial loss of habitat area and habitat quality. Photo by Juliet Murguia, reproduced with permission.

that threaten rare species can and do occur. The impact of high water levels on clapper rails has not been assessed; I can only speculate that frequent inundation of wetlands places considerable predation pressure on local populations.

Upland habitat is also important to the bird's-beak population as nesting areas for pollinators. We know little about the area and habitat requirements of the region's ground-nesting bees, but the highly disturbed upland surrounding Sweetwater Marsh could be a limiting factor. There is great need to improve our understanding of upland-wetland interactions and dependencies.

Wetland-Watershed Linkages

Occasional flooding brings a pulse of nitrogen to the intertidal soil while also lowering salinity stress. Studies of the 1980 flood (Zedler 1986) suggested that cordgrass growth was greatly stimulated by lowered salinity; plants increased 19% in mean height, with many plants exceeding one meter. The 1993 flood at Sweetwater Marsh coincided with a multiyear nitrogen addition experiment in a remnant natural marsh, where flood-borne nitrogen enhanced cordgrass growth. The cordgrass canopy grew so tall that we could no longer detect a fertilizer-treatment effect (PERL, unpublished data). The 1993 flood also appeared to benefit Bird's-beak. Plants with the highest biomass were located closest to the Sweetwater River, which carried a pulse of inorganic and organic nitrogen to the marsh (Parsons 1994).

Providing the essential connections among wetland habitats and between the wetland and undisturbed upland and the larger watershed will be most difficult in an urban setting. Current mitigation policy, which favors on-site over off-site compensation for damages, may need to be revised where endangered species are concerned. If the opportunities are greater in more rural settings, the urban wetlands may need to be restored for less demanding target species and communities.

MANAGEMENT AIMS FOR SOUTHERN CALIFORNIA

To sustain the biodiversity of this region's coastal wetlands, which are already greatly diminished by urban development, we need to halt losses of habitat. Next in priority is to insure that nearby and upstream land uses do not further degrade coastal wetlands. Finally, we need to replace acreage and ecological functions that have been lost.

One ambitious plan calls for restoration of approximately 200 ha of tidal wetland habitat at Tijuana Estuary by lowering the elevation of disturbed upland (Entrix *et al.* 1991; Zedler, Nordby, and Kus 1992). Much of the 1024-ha Tijuana River National Estuarine Research Reserve has become choked with sediment that entered the estuary via riverine floods and dune overwashes. But, there are two important constraints on activating plans to remove the sediments. First, funds have been identified only for the first 0.7-ha component. Second, there is no easy way to dispose of sediment spoils, because not all areas to be excavated have material sandy enough to be pumped to the ocean shore for beach replenishment, which is the most cost-effective option. Finer materials must be barged to a deep ocean fill site or trucked to inland landfills, although saline sediments are not necessarily suitable for landfills. Despite these constraints, the Tijuana Estuary Management Authority has an adaptive management program, and opportunities to expand and improve habitat for endangered and sensitive species are continually being sought.

Other actions inspire confidence in the future. A citizens' group, Save California Wetlands, is working toward a statewide wetland restoration plan. The State Coastal Conservancy, US Fish and Wildlife Service, and California Coastal Commission are collaborating on an inventory of wetland resources and potential restoration sites. Agency personnel are continually improving the regulatory process so that damages to wetlands are adequately compensated (Hymanson and Kingma-Rymek 1995).

REMAINING ISSUES

While southern California has many restoration and mitigation projects in the planning stages, there are still several region-wide issues.

Sustainability

A large unknown is whether or not the region's restored or constructed wetlands have in place the mechanisms that confer long-term sustainability of populations and ecosystem functions. The need for mitigation is usually in urban areas where development projects displace natural wetlands. On-site mitigation is usually preferred, but urban sites: (1) are usually lacking in linkages between wetlands and their watersheds; (2) have impaired hydrologic conditions (e.g., altered timing and duration of inundation), receive inflows of low-quality water; and (3) are vulnerable to human disturbances and invasions of exotic plants and animals. We know little about the width that buffer zones should be in order to protect wetlands from urban impacts, or about what qualities buffers should have. Only a few studies (e.g., White 1986) have addressed questions such as how close one can approach an endangered bird before it is disturbed. I know of none that have tested the relative effectiveness of different vegetation or fence types in screening wildlife from urban disturbance.

Habitat Configuration

Most natural tidal wetlands are broad and nearly flat, with steep-banked creeks forming dense networks. Salt marshes designed and built to replace natural wetlands often have large deep channels and mounded islands. Constructed marsh plains are likely to erode or accrete sediments. Hydrologists cannot yet predict how to construct sites with self-sustaining shape, topography, and hydrology. San Diego Bay's Marisma de Nación (Figure 15.1) was excavated to the correct elevation for low-marsh habitat, but it is eroding at the downstream end and accreting at the inland side, such that very little habitat remains suitable for the intended plant cover (J. Haltiner, Williams and Associates, Ltd., personal communication; PERL 1995). The marsh plain has an equilibrium just above mean higher high water, and *Salicornia virginica* typically dominates at this elevation. Thus, it may be impossible to construct large expanses of self-sustaining low marsh. Rather, it may be necessary to construct much larger sites with extensive tidal creek-banks, along which cordgrass might persist and clapper rails nest.

Patch Disturbance

At a smaller scale, patchy disturbances are important to the establishment and persistence of plants, especially annuals, such as salt marsh bird's-beak. If small mammals are the main natural cause of canopy openings that allow seedling recruitment, then understanding the linkages between mammals, their prey, and predators is important to predictions of long-term persistence of vegetation.

Long-term Persistence

Constraints on colonization of all salt marsh species, and especially the animals, are not well known, and long-term persistence of populations cannot be predicted without such knowledge. While the establishment of many salt marsh plant species is known to be linked to episodic rainfall and flooding, responses to chronic environmental changes, such as increasing runoff or rising sea level, are harder to document. Thus, it is not yet possible to predict how salt marsh vegetation will be able to move inland with sea level rise or other aspects of global climate change. From a technological perspective, measures that will ensure persistence of marsh vegetation are needed. For example, mitigation sites must have the appropriate wetland hydrology; fine-textured organic soil should be salvaged from the damaged wetland and used at the mitigation site; and vegetation should be salvaged (to sustain gene pools) and propagated (to reduce the need for collecting in natural wetlands) and then transplanted carefully to improve chances of long-term survival.

BROADER IMPLICATIONS

The situations described here may not translate directly to other regions, because southern California is unique. For example, California has lost 91% of its wetland area over the past 200 years, compared to an average of 53% for the lower 48 states (Dahl 1990), and the proportion of the state's area that remains in wetland (0.4%) is one of the nation's lowest (the average for all 50 states is 7.9%). Yet there are useful lessons that can be applied more broadly.

Standards for judging mitigation success should be strict. Many tidal wetlands have been created around the US, but those pronounced successful have not been judged using uniformly strict criteria, such as functional equivalency with natural wetlands or self-sustaining populations of sensitive wetland species. Goals that are difficult to achieve should have strict criteria for judging success. The San Diego Bay projects have high standards (US FWS 1988) and those for the light-footed clapper rail have not been met. One or two light-footed clapper rails have been seen in the younger mitigation marsh, but the first pair has yet to nest in transplanted cordgrass. The high standards would thus appear to be appropriate.

Restoration efforts intended to compensate for specific habitat loss need stricter standards than projects undertaken within a resource management context. The difficulties of replacing Clapper rail habitat in southern California should not discourage habitat restoration in general. While it is important to hold mitigators to tough standards, other restoration efforts need to be undertaken and are worthwhile even if the immediate result is not functionally equivalent to natural reference ecosystems. The National Research Council (1992) has set three ambitious targets for the nation, namely, the restoration of 10 million acres (~4,047,000 ha) of wetlands, 400,000 miles (~670,000 km) of river-riparian ecosystems, and 2 million acres (~810,000 ha) of lakes, all to be accomplished with 20 years. These goals deserve support whether or not resource agencies can fully restore all sites. As Wyant, Meganck, and Ham (1995) have said, restoration is a facilitating process; for terrestrial habitats, this means establishing the physical stability of the site, initiating soil development, and facilitating invasion by native vegetation. The result will probably not mimic some previous pristine condition, as we rarely know what that might have been, let alone how to achieve it quickly. An appropriate objective is thus to shorten the recovery time. When restoration takes place outside the mitigation arena, the effort may not need to pass the acid test of achieving functional equivalency with what is being destroyed.

An ounce of prevention is more prudent than a pound of cure. Regions with less habitat loss and fewer threatened species would be wise not to follow in southern California's footsteps. Where plant and animal populations have not been greatly diminished, restoration sites will be more readily colonized by the native flora and fauna. Most wetland restoration projects take place in more benign settings than along the Southern California coast (NRC 1992), that is, where landscapes are less disrupted and specific sites are more readily repaired. The southern California experience is unusual, but as such it serves as an early-warning to managers of wetlands around the nation. Preventing the plague of excessive habitat loss is preferable to the lengthy, costly, and uncertain cure of restoration.

LITERATURE CITED

Boyer, K.E. and J.B. Zedler. 1996. Cordgrass damage by scale insects in a constructed salt marsh: Effects of nitrogen additions. *Estuaries* 19:1–12.

Bradshaw, A. 1987. Restoration: An acid test for ecology. In *Restoration ecology: A synthetic approach to ecological research*, eds. W.R. Jordan III, M.E. Gilpin, and J.D. Aber, 23–29. Cambridge: Cambridge University Press.

California State Coastal Conservancy. 1989. *The coastal wetlands of San Diego County*. Oakland: State Coastal Conservancy.

Callaway, J. and J.B. Zedler. In preparation. Competitive interactions between a salt marsh native perennial (*Salicornia virginica*) and an exotic annual (*Polypogon monspeliensis*) under varied salinity and hydroperiod.

Chuang, T. and L.R. Heckard. 1971. Observations on root-parasitism in *Cordylanthus* (Scrophulariaceae). *American Journal of Botany* 58:218–228.

Covin, J.D. and J.B. Zedler. 1988. Nitrogen effects on *Spartina foliosa* and *Salicornia virginica* in the salt marsh at Tijuana Estuary, California. *Wetlands* 8:51–65.

Cox, G.W. and J.B. Zedler. 1986. The influence of mima mounds on vegetation patterns in the Tijuana Estuary salt marsh, San Diego County, California. *Bulletin of the Southern California Academy of Sciences* 85:158–172.

Dahl, T.E. 1990. *Wetlands losses in the United States 1780s to 1980s*. Washington, D.C.: U.S. Department of the Interior, Fish and Wildlife Service.

Davis, F.W., P.A. Stine, D.M. Stoms, M.I. Borchert, and A. Hollander. 1995. Gap analysis of the actual vegetation of California, 1. The southwestern region. *Madroño* 41:40–78.

Department of Fish and Game. 1989. 1988 *Annual report on the status of California's State listed threatened and endangered plants and animals*. Sacramento: California Resources Agency.

Entrix, Inc., PERL, and PWA, Ltd. 1991. *Tijuana Estuary tidal restoration program*. Draft Environmental Impact Report/Environmental Impact Statement. California Coastal Conservancy (SCC) and U.S. Fish and Wildlife Service, Lead Agencies. SCC, Oakland. Vol. I-III.

Falk, D.A., C.I. Millar, and M. Olwell, eds. 1996. *Restoring diversity: Strategies for reintroduction of endangered plants*. Washington, D.C.: Island Press.

Fink, B.H. and J.B. Zedler. 1990. Endangered plant recovery: Experimental approaches with *Cordylanthus maritimus* ssp. *maritimus*. In *Proceedings, first annual meeting of the society of ecological restoration and management*, eds. H.G. Hughes and T.M. Bonnicksen, 460–468. Madison, Wisconsin.

Flick, R.E. and D.R. Cayan. 1984. Extreme sea levels on the coast of California. In *19th coastal engineering conference proceedings*, September 3–78, 886–898. New York: American Society of Civil Engineers.

Hickman, J.C., ed. 1993. *The Jepson manual: Higher plants of California*. Berkeley: University of California Press.

Hymanson, Z.P. and H. Kingma-Rymek. 1995. *Procedural guidance for evaluating wetland mitigation projects in California's coastal zone*. San Francisco: California Coastal Commission.

International Boundary and Water Commission. 1950–1993. *Western water bulletin. Flow of the Colorado River and other western boundary streams and related data*. El Paso: IBWC.

Kelly, J.P. and G. Fletcher. 1995. Habitat correlates and distribution of *Cordylanthus maritimus* (Scrophulariaceae) on Tomales Bay, California. *Madroño* 41:316–327.

Kuhn, N.L. 1995. The effects of salinity and soil saturation on plants in the high intertidal marsh. M.S. thesis, San Diego State University, San Diego, California.

Kuhn, N.L. and J.B. Zedler. In review. Differential effects of salinity and soil saturation on native and exotic plants of a coastal salt marsh. *Estuaries*.

Langis, R., M. Zalejko, and J.B. Zedler. 1991. Nitrogen assessments in a constructed and a natural salt marsh of San Diego Bay, California. *Ecological Applications* 1:40–51.

Lincoln, P.G. 1985. *Pollinator effectiveness and ecology of seed set in* Cordylanthus maritimus *ssp.* maritimus *at Point Mugu, California*. Final report to U.S. Fish and Wildlife Service Endangered Species Office, Sacramento, California.

Massey, B.W., R.L. Zembal, and P.D. Jorgensen. 1984. Nesting habitat of the Light-footed clapper rail in Southern California. *Journal of Field Ornithology* 53:67–80.

NRC (National Research Council Committee on Restoration of Aquatic Ecosystems: Science, Technology, and Public Policy). 1992. *Restoration of aquatic ecosystems: science, technology, and public policy*. Washington, D.C.: National Academy Press.

Pacific Estuarine Research Laboratory (PERL). 1994. *Sweetwater Marsh wetland complex ecosystem assessment*. Final Report for 1994. San Diego: California Department of Transportation.

Pacific Estuarine Research Laboratory (PERL). 1995. *Sweetwater Marsh National Wildlife Refuge: ecosystem assessment for mitigation compliance*. Final Report for 1995. San Diego: California Department of Transportation.

Parsons, L.S. 1994. Re-establishment of salt marsh Bird's-beak at Sweetwater Marsh: Factors affecting reproductive success. M.S. thesis, San Diego State University, San Diego, California.

Parsons, L.S. and J.B. Zedler. In press. Factors affecting reestablishment of an endangered annual plant at a California salt marsh. *Ecological Applications*.

Scatolini, S.R. and J.B. Zedler. 1996. Epibenthic invertebrates of natural and constructed marshes of San Diego Bay. *Wetlands* 16:24–37

Swift, K.L. 1988. Salt marsh restoration: Assessing a southern California example. M.S. thesis, San Diego State University, San Diego, California.

U.S. Fish and Wildlife Service (US FWS). 1988. Biological Opinion 1-1-78-F-14-R2, The combined Sweetwater River flood control and highway project, San Diego County, California. Letter from Wally Steuke, Acting Regional Director, US FWS, Portland, Oregon, to Colonel Tadahiko Ono, District Engineer, Los Angeles District, Corps of Engineers, Mar. 30, 1988.

U.S. Geological Survey (USGS). 1971–1950 (in separate volumes). *Water resources data for California, Water Years 1971–1950*. Washington, D.C.: U.S. Geological Service.

White, A.N. 1986. Effects of habitat type and human disturbance on an endangered wetland bird: Belding's Savannah sparrow. M.S. Thesis, San Diego State University.

Wyant, J.G., R.A. Meganck, and S.H. Ham. 1995. The need for an environmental restoration decision framework. *Ecological Engineering* 5:417–420.

Zalejko, M.K. 1989. Nitrogen fixation in a natural and a constructed southern California salt marsh. M.S. thesis, San Diego State University, San Diego, California.

Zedler, J.B. 1986. Catastrophic flooding and distributional patterns of Pacific cordgrass (*Spartina foliosa* Trin.). *Bulletin of the Southern California Academy of Sciences* 85:74–86.

Zedler, J.B. 1993. Canopy architecture of natural and planted cordgrass marshes: Selecting habitat evaluation criteria. *Ecological Applications* 3:123–138.

Zedler, J.B. 1996a. Coastal mitigation in southern California: The need for a regional restoration strategy. *Ecological Applications* 6:84–93.

Zedler, J.B., principal author. 1996b. *Tidal wetland restoration: A scientific perspective and a Southern California focus.* California Sea Grant College, La Jolla. Technical Report, in press.

Zedler, J.B., C.S. Nordby, and B.E. Kus. 1992. *The ecology of Tijuana Estuary: A National Estuarine Research Reserve.* Washington, D.C.: NOAA Office of Coastal Resource Management, Sanctuaries and Reserves Division.

Zembal, R., B.W. Massey, and J.M. Fancher. 1989. Movements and activity patterns of the Light-footed clapper rail. *Journal of Wildlife Management* 53:39–42.

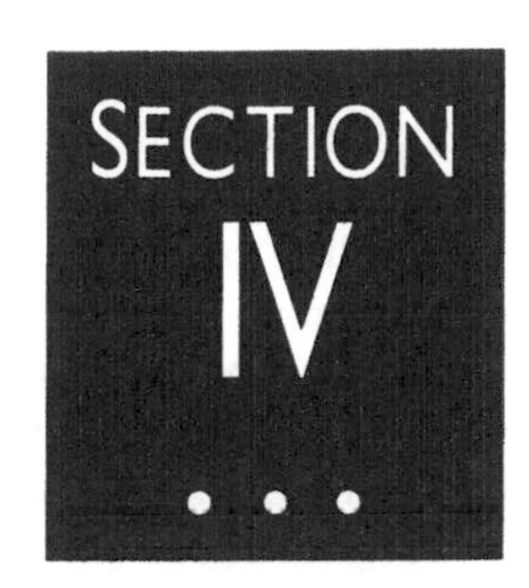

WHEN CONSERVATION MEETS THE REAL WORLD OF ECONOMICS, POLITICS, AND TRADEOFFS

• • •

Many embark on careers as conservation biologists because they want to make a difference, to do something useful. Invariably any conservation biologist who climbs down from academia's ivory towers and enters the policy or decision-making arena will encounter a stumbling block. They know what habitat is required, they know that further logging will endanger owls, and they know that scotch broom will degrade remnant prairies. But nothing can be done because of a shortage of money, because of an absence of political power, because the cost is too great in terms of jobs and development. It might seem then that the biologist can only surrender and turn the job over to politicians and activists. The contributions in this section reject such a black-and-white distinction between the "science of conservation" and the "practice of conservation." Indeed, an argument can be made that awareness of, and sensitivity to, economic and political constraints at the onset of biological research, might better direct that research toward an effective endpoint.

An inescapable, but often ignored (either out of naiveté or stubbornness) reality of conservation is that everything has economic value—development, land, species, parks, biodiversity, and so forth. Pearce and Moran point out that the concept of economic value is actually a good deal broader than the caricature many biologists have in mind. Economic value refers not just to commerce and finance, but also to human well-being. There is no question, then, that so-called "sustainable development" as embodied in the discussion of agriculture (by Carroll and colleagues) or of Patagonia's natural resources (by Pascual and colleagues) will never preserve biodiversity unless the economic value of biodiversity is appreciated. This does not mean conservation biologists have to become economists, but rather that they need to realize economic valuation is unavoidable, and should be encouraged rather than avoided.

One corollary of economic value is the concept of choices and tradeoffs—or put another way, of "costs and benefits." Sometimes conservation biologists, especially those of the more evangelical variety, lose sight of the fact that everything is not possible, and that choices must be made. Choices and tradeoffs are at the heart of all of the chapters in this section. If life on earth were simple, conservation research would be unnecessary because all of the answers are obvious—we should simply halt all logging, protect all remaining tropical rainforests, and reverse worldwide human population growth. But clearly those solutions do not reside on this planet. Instead we must figure out how to decide which parcels of forest to log, and which rainforests to protect. Biologists can factor out the consequences of these choices, but ultimately community involvement is essential. While it may be "high-powered" scientists or well-groomed bureaucrats who direct conservation policies, those policies will only succeed if the public at large is heavily involved. Thus Pascual and colleagues argue that in Argentina it is essential for conservation and resource management programs to invest heavily in public education. Similarly, Carroll and colleagues conclude that sustainable agriculture will succeed only if the farmer participants have a hand in devising their own sustainable system, as opposed to being told what to do. In general, unless the public is sold on conservation, the discounting of tomorrow's dollar versus today's dollar will always lead to biodiversity being sacrificed for immediate profit. Conservation is easily thwarted when it is unduly detached from political realities. Sometimes when reading the wish lists of noted conservation biologists, the staggering impracticality is embarrassing. Most scientists are aware of the importance of "practicality", for we have all abandoned the "perfect experiment" because it was impractical. This attention to pragmatism needs to find its way into all branches of conservation biology—a notion that is well-expressed in every contribution in the following section.

CHAPTER

16

The Economics of Biological Diversity Conservation

DAVID PEARCE
and DOMINIC MORAN

Economics is about choice. In a world where resources are finite relative to the demands that human beings make on them, choice is unavoidable. We cannot have everything. Choosing is the same as "trading off," balancing the net gains from one course of action against an alternative action. The action of conserving biological diversity is not immune from this problem of making choices, although, as we shall see, some conservation literature appears to deny the choice, while some of it argues that the choice is there but that the tradeoff in favor of diversity destruction is unacceptable—biological diversity is somehow "special." If choice is inevitable, if we cannot retain all the diversity that there is, how should such choices be made? What should be conserved and where? Few problems in economics are more complex than making choices in the context of biodiversity and no economist would argue that the problem is resolved. Nonetheless, economics offers some insights into biodiversity conservation policy and these are worth exploring.

THE THEME

The theme of this chapter can be summarised briefly. A problem is especially worth solving if it is an important problem. It therefore matters that we establish the importance of biological diversity and biological resources generally.

The first observation is that, whatever other reasons there may be for supposing biodiversity to be important, biodiversity has economic importance. 'Economic' here has a particular meaning; it does not refer to finance or commerce. It refers to human wellbeing, or, to use the economist's jargon, to "welfare" or "utility." Any contribution to human wellbeing is an economic good or benefit. Any loss of human well-being is a cost. It is important to keep these meanings in mind as our chapter proceeds.

The second observation is that policy to save biodiversity will be ineffective unless it addresses the causes of biodiversity loss. One of the most fundamental characteristics of environmental economics is that it teaches the essential interrelatedness of environment and economic activity. The workings of the economy affect the environment. Environmental change affects the economy. The fundamental causes of biodiversity loss are often therefore to be traced to the workings of the economic system or, as we shall show, to the 'misworking' or 'failure' of the economic system, its failure to account properly for all the effects on human well-being of biodiversity loss.

Third, appropriate policy measures should focus on the removal or mitigation of the causal factors generating biodiversity loss. Since those causes often lie in the economic system, so it is the correction of economic 'failure' that provides us with the most effective way of countering biodiversity loss. There is a natural sequence in these observations—importance, cause, policy—and this is the sequence we use in structuring the chapter.

THE ECONOMIC IMPORTANCE OF BIODIVERSITY

The economic importance of biodiversity relates to its direct and indirect effect on human well-being or welfare. Welfare or its equivalent measure utility, can be inferred from the money people are observed to pay in actual markets, or their stated willingness to pay—in hypothetical markets. It is this willingness to pay that provides the handle for the economic analysis of biodiversity loss.

Valuation of many biological resources is typically through established markets. Thus, even the price of the celebrated Madagascan periwinkle (*Catharanthus roseus*, Apocynaceae) as part of a prospecting arrangement for pharmaceutical trials provides one, albeit unconventional, gauge of human preferences for that plant, using money as a common measuring rod. Logically, if the periwinkle ceases to exist because of extinction, the same willingness to pay reflected in the price measures some part of the resulting loss of well-being.

Conventional markets are not always necessary to determine economic value. Indeed, the fact that many biological resources are unpriced has forced environmental economics to refine innovative methods to investigate individual preferences for states of the environment. Such methods include survey-based approaches to determine the value of resources that many people may never directly experience. Not surprisingly, the use of these approaches is the subject of considerable controversy.

For some people the whole enterprise of economic valuation of non-market goods is controversial. Equating the consequences of loss of biodiversity with a money-equivalent loss of well-being, for example, may strike some as a peculiarly anthropocentric view of the world. Economics makes no apologies for such an approach, as economic values are necessarily those held by people, and individual maximization of well-being is consistent with several ethically-based motives frequently supposed by critics to be neglected in monetary evaluation. However, it is important to appreciate the claims made for, and the reason to support, economic valuation over other approaches. Money simply translates unobserved well-being or utility into an observable quantity that is the most convenient metric for comparison of gains and losses across space and time. If money is objectionable then some other universal unit of well-being may serve the same function just as well, providing it makes such comparisons possible. Thus if it were possible to infer a value for bald eagles in Wisconsin today (and it is), the measure of value should be the same metric used to value, say, raccoons, in Florida tomorrow, (thereby facilitating the comparison of costs and benefits over respective conservation programs).

It is difficult to over-emphasize the importance for efficient resource allocation of the ability to make such tradeoffs. Unfortunately the often-cited charge of "pricing everything and valuing nothing," which arises from a confusion over the interpretation of economic value, inhibits the adoption of valuation without proposing alternative decision criteria and by even denying that tradeoffs are necessary. The use of a monetary measures does not suggest that such comparisons are always necessary, inevitable, or mutually exclusive. Nor does it claim to be completely defined to include intrinsic or spiritual values that are often invoked in the context of species preservation and our duty to future generations. Putting an economic value on biological resources is important because conservation tradeoffs of the type made above represent very real choices for allocating scarce resources. When constraints bind, spending on items perceived as providing little or zero return will be minimal. Economically informed tradeoffs are necessary for biodiversity if its services are to be compared to the very real returns from deforestation or agriculture. Other value systems may have a lot to say about the rights and wrongs of this approach, but by any objective assessment their practical role in changing the behaviour precipitating the current global biodiversity crisis has been limited. On the other hand, valuing an asset creates a rationale for conservation. The 'use it or lose it' dictum is one albeit extreme interpretation of this approach. A more sympathetic interpretation, which we hope to convey, simply seeks equal treatment for biological resources in development decisions.

TOTAL ECONOMIC VALUATION

Economic values related to biological resources can be categorized according to their underlying motives. Common components of total economic value that motivate willingness to pay are dictated by current direct and indirect use, expected future use, and existence value (Figure 16.1). Various methods can be employed to determine the magnitude of these values.

Direct Uses

Even in the most developed of societies, direct use of naturally occurring products demonstrates the fundamental contribution of biodiversity to well-being and the satisfaction of immediate needs. Direct use valuation tends to be straightforward, because many natural foods and fibers are traded in conventional markets. Even where market prices do not currently exist, as is the case for some non-timber forest products and traditional medicines, prices may be inferred from the value of close substitutes. Direct use does not have to be consumptive and therefore mutually exclusive of other values (see below). Ecotourists may, for example, view an elephant in a wildlife reserve without precluding further use. But a problem with relying on direct use as the mainstay of value is that, while many people may want to "use" elephants, the demand for use of, say, any of the 109 microorganisms in a typical gram of soil is likely to be non-existent. Similarly there has been much euphoria over the potential for direct uses of non-timber forest products as economic arguments for maintaining biodiverse environments in relatively pristine shape (Peters, Gentry, and Mendelsohn 1989). On methodological grounds such conclusions have been shown to be premature. Non-timber forest products and hunting and gathering can represent very important values in specific locations, but the findings of studies that have scrutinized the economics of such values have pointed to the need for more robust economic arguments beyond such direct uses.

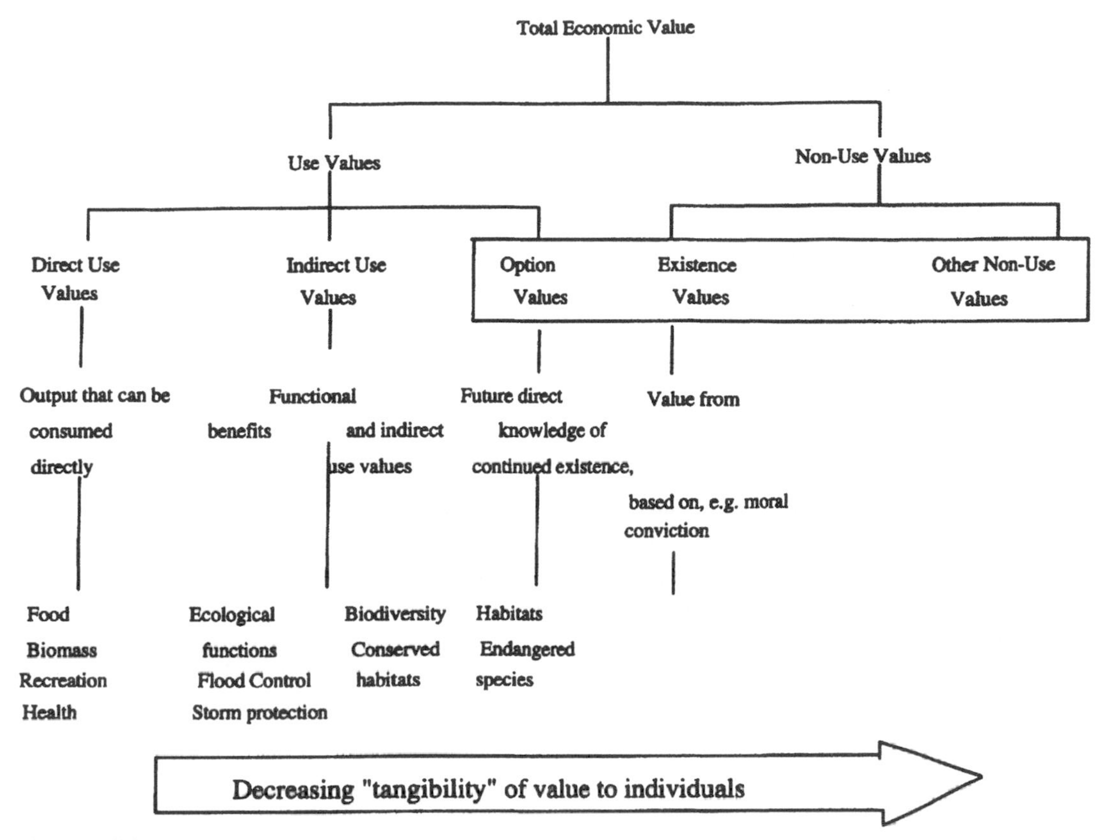

Figure 16.1
Total economic value and biological resources. Breaking down the components of total economic value for biological resources. Note that these values differ markedly with respect to how easy they are to measure (i.e., their "tangibility").

Indirect Uses

Regardless of direct uses, most people benefit indirectly from biological resources. Indirect uses are related to the functional benefits of species and ecosystems. Thus forests provide hydrological benefits, flood control, and act as oxygen sources and carbon sinks. Similarly, mangroves serve as natural storm defences and breeding grounds for aquatic species that are directly used, and natural water purification systems. Using current knowledge, the routine valuation of some of these functions may provide some of the most powerful reasons for maintaining functioning ecosystems. In fact, the realization of these values may be a sufficient conservation rationale without having to stray into the more contentious area of non-use valuation.

The measurement of indirect values is subtle, as these functions are rarely bought and sold in regular markets. Full appreciation of the extent of indirect benefits generally requires cooperation between economists and natural scientists to establish the role of functions and the extent of damage impacts resulting from their loss. Once impacts can be related to market goods or some expression of willingness to pay for resulting damages, an economic relationship is established. The degree of detective work required to attach economic values to the loss of these functions varies in complexity. Thus, the relationship between deforestation, soil erosion, sediment loading in dams, and reduced electricity generation may be straightforward relative to the assignment of a benefit for global warming damage avoided by the preservation of carbon biomass of a protected forest. In both cases, economics has devised methodologies to assign values that are now routinely included in the economic assessment of environmental projects. The inclusion of such values is proving highly significant for the economic justification of biodiversity-related projects funded by national environmental agencies and particularly the Global Environment Facility, which has a global remit.

Option Value

The concept of option value is at the heart of the issue of diversity and the maintenance of the widest portfolio of natural assets. In the face of potentially irreversible change and under present uncertainty, rational risk-averse agents may reveal an economic value for the option not to foreclose on access to the resource in question, whether this be a set of protected areas or the diversity of the genetic portfolio. This willingness to pay arises over uncertainty related to personal future use, the value associated with the option to enjoy the resource at some future point in time.

Option value closely approximates the value of diversity *per se* and is probably the value most closely associated with the popular perception of biodiversity as a source of pharmaceutical base materials. Empirical assessment of biodiversity-related option values turns out to be extremely difficult. The exact definition of diversity is uncertain and in relation to its demand, we can only suppose that, on the basis of risk aversion, non-satiated individuals want more rather than less. In order to say something about the economic consequences of reduced diversity, some common diversity attribute needs to be translated into a value-of-diversity objective (Weitzman 1992). The dilemma involving the measurement of biodiversity inevitably also arises when attempting to address priority setting for biodiversity conservation. A considerable literature covering developments in the field of systematics, particularly phylogenetic relatedness, is currently providing a bridge between biologists and economists.

Non-Use Value. Most people derive pleasure from knowing the world is a diverse place. We are psychologically diminished by the irreversible extinction of species we never see, and cherish the continued existence of the greatest variety of life on earth, irrespective of use motives. The exact motives of this "existence value" are uncertain, but they may range from pure self-interest (perhaps motivated by optional use) through to bequest motives, various forms of altruism plus rights-based and other ethical views. Equally uncertain (given the remoteness of these motives from the validation of everyday market transactions) is the question of whether such motives systematically obey any axioms of individual preferences, such as the dictates of economic theory, or whether they translate into real economic commitments. The latter issue is of particular interest, as it is possible that for many charismatic species, the aggregate global willingness to pay for existence could be enormous. The asessment of any such latent demand is obviously of considerable interest for making an economic case for conservation. If, for example, significant existence values related to the conservation of Amazon rainforests can be validated, then their realization may go some way towards easing current international funding constraints.

Careful scrutiny of existence values for reasons related to the issue of conservation has recently increased in the United States, where federal legislation admits them as compensable entities in law. In other words, damages due to oil spills, water pollution, or other impacts that diminish existence values, may be recoverable from guilty parties (see Ward and Duffield 1992; see *Exxon Valdez* discussion in chapter 6). Needless to say, the stakes involved in multi-million dollar law suits has injected considerable urgency into inquiries regarding existence values.

Whatever perspective can be taken on motives, various real economic actions such as donations or purchases of wildlife-related merchandise, seem to confirm the existence of some form of non-use value. The question then arises as to whether surrogate or hypothetical markets can be created to value the existence of unfamiliar resources or elicit unobserved option value. From recent experience in valuing non-market goods, the response would appear to be positive. Using a method known as contingent valuation (CV), economists have attempted to simulate market-like tradeoffs for non-market goods using sophisticated questionnaires and experiments (Pearce and Moran 1994). Applications of constructed market methods are growing steadily, with interesting applications over a range of species and even complex ecosystem functions. In the context of global non-use value for biological resources, the most notable application by Kramer, Mercer, and Sharma (1995) to the issue of global tropical forest conservation demonstrated a modest willingness to pay by 1,200 U.S. residents (around $30 per household). This particular study highlights many of the potential difficulties involved in valuing complex goods.

Not surprisingly, asking people what they think about biodiversity is controversial on several counts. First, do people know anything about what they are being asked to value? Second, what is to stop biased or strategic responses? Third, in the absence of market-like benchmarks for many goods now being valued using CV, can responses ever be tested, and if they can't, what is the justification of using them in resource allocation decisions?

The answers to some of these questions lie in a growing literature dedicated to reducing biases and validating stated preference methods. The resulting methodological challenges are taking CV practitioners well beyond the traditional domain of economics and into fields such as cognitive psychology. In the process, increased use of non-market methods has also fueled debate both within and outside of the economics profession over the ability to establish reliable values for environmental goods. Inevitably, criticism has centred on the displacement of an ethical stance regarding nature's value by instrumentalist consumer sovereignty. Proponents of valuation have not been slow to reclaim some of the moral high ground by pointing to the inevitable consequences for decision making when nature is inviolable and the lamentable consequences of the inaction implicit in a non-economic view of the world.

Demonstrating Economic Value

The difficulties involved in measuring the full value of biological resources points to one of the causes of the current biodiversity crisis. The values that can be demonstrated begin to put a price on the consequences of biodiversity loss. Yet it is certain that there is much that remains beyond the boundaries of economic analysis. For example, we know little about system resilience, the existence and nature of damage thresholds and system discontinuities. It is feasible that the transgression of some catastrophic threshold, for example by infraspecific genetic concentration in agriculture, sometime in the future can cause drastic food shortages at a global scale.

Following the typology of Figure 16.1, Table 16.1 offers some orders of magnitude

Table 16.1

Comparing Local and Global Conservation Values (US$/hectare, Present Values at 8%) This table is adapted from Kumari (1994), but with additional material and some changed conversions. All values are present values at 8% discount rate, but carbon values are at 3% discount rate. Uniform damage estimates at $23.3 ton of carbon have been used (Fankhauser and Pearce [1994]), so that the original carbon damage estimates in the World Bank studies have been re-estimated.

	Mexico Pearce *et al.* [1993]	Costa Rica World Bank [1992b] (carbon values adjusted)	Indonesia World Bank [1993] (carbon values adjusted)	Malaysia World Bank [1991]	Peninsular Malaysia. Kumari [1994]
Timber	—	1240	1000–2000	4075	1024
Non-timber Products	775	—	38–125	325–1238	96–487
Carbon Storage	650–3400	3046	1827–3654	1015–2709	2449
Pharmaceutical	1–90	2	—	—	1–103
Ecotourism/ Recreation	8	209	—	—	13–35
Watershed Protection	<1	—	—	—	
Option Value	80	—	—	—	
Non-use Value	15	—	—	—	

for the economic values of forest resources. Information from existing studies is patchy simply because of the paucity of good studies and the methodological difficulties encountered in applying those that have been carried out. Furthermore, Table 16.1 reports orders of magnitude for whole countries and, as such, per hectare values will vary from site to site. Several observations are noteworthy. Present values are an accounting convention to reduce an intertemporal stream of costs or benefits to a base year to facilitate comparison of investment alternatives. Some consideration of the role of time preference in biodiversity loss is provided below in the context of a simplified model. Option value as calculated in one study is based on a stylised model for pharmaceuticals that is not strictly consistent with the description of option value given previously (see Adger *et al.* 1995). The summed "sustainable" non-destructive use and non-use values are generally higher than timber values associated with destructive logging. Contrary to ethnobotanical folklore, there is little convincing evidence on the potential pharmaceutical mega-values, although the secrecy surrounding prospecting arrangements does not make such information easy to validate. On the other hand, given current scientific understanding, the relatively certain carbon storage values are clearly dominant.

THE FUNDAMENTAL CAUSES OF BIODIVERSITY LOSS

It is easy to direct attention to the proximate causes of biodiversity loss. Highly diverse ecosystems, such as tropical forests and wetlands, are the subject of various pressures. It might be logging, clearance for agriculture, fuelwood demand, overfishing. Many analyses of biodiversity loss stop at this listing of proximate factors. But it is essential to ask further questions: why should logging be inconsistent with biodiversity conservation? Why are tropical forested lands cleared for agriculture? Why does overfishing occur?

Table 16.2
World Land Conversions 1700–1980 in terms of million km² (Richards 1990).

	1700	1850	1950	1980	% Change 1700–1980
Forests and Woodland	62.1	59.7	53.9	50.5	−19%
Grasslands and pasture	68.6	68.4	67.8	67.9	−1%
Croplands	2.7	5.4	11.7	15.0	+466%

The answers to these questions depend on an analysis of the fundamental causes, the factors explaining the proximate causes.

Such an analysis suggests the following factors explaining biodiversity loss. Focusing on land-based biodiversity (the analysis is not very different when extended to other sources of biodiversity), land conversion is the major proximate factor of interest. Land uses change and they tend to change from uses where diversity is high to uses where diversity is low. Table 16.2 shows rates of land conversion in the past 300 years or so, and Table 16.3 brings the data more up to date and shows some regional trends. The main change has clearly been at the expense of forests and woodlands, generally diverse systems, and in favor of croplands, generally less diverse systems. As far as animal extinctions are concerned, it is estimated that habitat destruction accounts for about one-third of extinctions; species introduction for another third (see also chapter 11) and non-sustainable hunting for the final third (WCMC 1994). Plant losses, however, are primarily due to land conversion, as are losses of diversity among birds [perhaps accounting for three-quarters of the threats to birds (WCMC 1994)].

To find fundamental causes, however, it is necessary to ask why land conversion occurs. There are two major driving forces:

(1) At the global level, the first factor is population growth. Population growth leads to pressure on land resources, especially for food production but also for infrastructure such as roads, and urbanization; and,

(2) The second factor is the working—or rather the failure to work—of the local, national, international and even the global economy. The failing lies in the fact that the characteristics or functions of biological diversity often are not marketed, ie they are not bought and sold in the market place. There is said to be a missing market for the many outputs and values of biodiversity. The absence of a market means that the economic value of biodiversity, which exists despite the absence

Table 16.3
Land Conversions 1979/81–1991

	Million Hectares				
	Cropland	Pasture	Forest	Other	Total
Africa	+9	+8	−26	+11	+1
North and Central America	−2	+4	+2	−4	0
South America	+13	+21	−42	+11	+3
Asia	+6	+66	−26	−43	+3
Europe	−2	−3	+1	+4	0

(Note: other land includes roads, uncultivated land, wetlands, built-on land).
Source: Estimated from World Resources Institute (1994), Table 17.1.

of markets, fails to get counted when biodiversity is being "traded" against other economic goods and services, such as timber production, agricultural output, fuelwood, etc.

If the proximate cause of biodiversity loss is land conversion, then the driving force is population growth giving rise to the demands for the output of the land, combined with the fact that there is no "level playing field" between the commercial value of the converted land and the value of biological diversity on the unconverted land. The former tends to show up as a positive economic value, while the latter appears to have no economic value, or little economic value. Simple commercial forces then dictate that the highest value of the land is realized when it is converted.

Population Change

World population is expected to stabilize at around 12 billion people toward the end of the next century, but this is more than twice the number of people on earth today (Table 16.4). The fastest growth rate is in Africa, currently growing at 2.9% p.a., and heading for a population of 3 billion people towards the end of the next century, around five times the population of today. These figures suggest that sheer pressure of human beings on space will displace the habitat of other living species. Part of this displacement effect arises from the demand for extra land for food production, and part from urbanization and communications infrastructure. Must population growth always result in such effects? One apparent way of avoiding the impact of population change on the extensive use of land (converting 'unproductive' land to productive use) is through intensification of land use, i.e., the application of technology—pesticides, fertilizers, high-yielding varieties of plants, etc. Then, a given food demand can be met by limited land use, reducing the need for further land. The problem with this 'solution' is that intensification itself reduces agricultural diversity because it tends to rely on homogeneous crops (high yielding varieties, for example) and it is also likely to reduce biological diversity generally due to the impacts of pesticide contamination, nitrate pollution of water, and so forth. In short, the underlying feature, i.e., rapid population growth, remains and diversity tends to be reduced regardless of the way in which human demands are met.

Table 16.4
Population Growth

	(Billion People)			
	1960	**1990**	**2100**	**2150**
World Population (billions)	3.0	5.4	12.0	12.2
	% of the world's population in different continents			
Asia/Oceania	55.8	59.4	57.0	56.8
N & S America	13.8	13.7	11.0	10.8
Africa	9.2	11.9	23.9	24.5
Europe	21.3	15.0	8.1	7.9

(Source: World Bank, World Development Report (various issues) and World Resources Institute, World Resources 1986, Basic Books, New York, 1986.)

Economic Failure

Economic failure can be classified in three ways:

(1) Local Market Failure.

Here the local market—at community level, town, or region—ignores the economic functions served by biological diversity or by its habitat. For example, a tropical forest will regulate water supply and hence the incidence of floods. More generally, it will have watershed protection functions. These ecological functions are also economic functions. If they are not conserved, they will have to be provided in some other way, or if not provided, they will exert impacts elsewhere in the economic system. Damaging such protective functions through deforestation is said to cause an "external cost." Cost takes the form of a loss in human well-being somewhere in the economic system. The cost also is external because the tropical forest land user takes no account of these externalities in his or her 'internal' economic calculations. There are many examples of such externalities: upstream polluters pollute downstream users; emitters of air pollutants cause acidification of lakes and rivers elsewhere, and even in other countries as with transboundary pollution.

Such failures arise because of the absence of property rights in the resource in question. Downstream producers and consumers may not have rights to a clean environment that would permit them to enforce a change of behavior on an upstream polluter. If they did have such rights, they could exact compensation from the upstream polluter. Esturial fishermen do not bargain with upstream forest colonisers to reduce the effects of sedimentation on fisheries. Yet Hodgson and Dixon (1988) show that a ban on logging upstream of Bacuit Bay, Philippines, would have reduced fishery losses from sedimentation by 50%, and would have increased tourist revenues by a factor of four. In Ruitenbeek's study of the Korup rainforest in Cameroun (Ruitenbeek 1992), over 50% of the direct and indirect use values of the forest were accounted for by downstream fishery benefits. A great deal of market failure, then, arises from the failure to assign property rights to environmental quality and environmental assets, including biodiversity.

Poor or perverse land tenure is an example of the property rights problem. Poor tenure by existing occupants of forest land helps to explain reduced resistance to rival claims from those who wish to clear land. Mahar and Schneider (1994) note that only 11% of Brazilian Amazon land was titled in the early 1980s, creating an essentially "open access" resource, i.e., one that was not owned by anyone in particular and hence was accessible to all. Perverse land tenure applies when the colonizer of previously 'wild' land has an incentive to clear because clearance is evidence of tenure, as in Costa Rica, Honduras, Panama, Brazil, Ecuador. Southgate, Sierra, and Brown's (1989) study of Ecuador suggests a strong influence of tenure insecurity on deforestation.

As a renewable resource, wild species have the capability to be utilized sustainably. Of course, for this to be the case, the rate of 'harvest' must not exceed the regenerative capacity of the resource. Given the considerable uncertainties about population dynamics for most species, such a "sustainability rule" is hard to formalize accurately, and significant margins of safety are required.

Because of economic failure (wild species often reside in areas of open access where property rights are poorly defined, badly enforced, or not defined at all), such resources risk over-exploitation. Lack of information about the consequences of exploitation exacerbates the risks. Evidence of non-sustainable exploitation is widespread, from the tropical forests and coral reefs to mangrove depletion and over-hunting and harvesting. While a major part of the world demand for orchids, for example, is met from direct propagation, wild orchids are still collected and the effect has been the virtual eradication of some species. The fate of the world's fish stocks is a perennial example of failures to come to terms with open access and unregulated common property resources. Despite many international regulations, most notably the introduction of 200 mile coastal jurisdictions by nation states, a great many fishery resources are now thought to be close to their maximum catch limits, and many show signs of biological degradation (WCMC 1992). Current catch levels of around 85–90 million tons p.a. are likely to be the maximum catchable amount. Given the vital role played by fish as a food source and the context of rising world population, the gloomy prospect for sustainable fisheries management on a global scale is disturbing (see chapters 6 and 8).

(2) Global Market Failure.
Global market failure is the second form of economic failure. This is very much like local market failure, but in this case the agent responsible for the loss of biodiversity through some change of land use imposes a cost on the world in general, or on a significant group in the world. Global climate change is a good example of such a 'global failure': each and every ton of greenhouse gas emitted imposes a probable cost on all others in the world. The same is true for emitting ozone depleting substances. The general functions of the global atmospheric system and the stratospheric ozone layer are not bought and sold in markets. Hence no one individual has an incentive to take account of the global environmental effects of their actions. The global market 'fails' because those who generate benefits from biodiversity conservation cannot 'capture' or 'appropriate' those benefits through an international market. As a further example, many ecological systems have economic value to residents of other countries who value the biodiversity. Some of this value is expressed through markets—ecotourism, for example. But much of it has no apparent market, as with the widespread concern that many people feel about the fate of the tropical forests (recall that any loss of human well-being is a cost). In the absence of a market in the benefit, the user of the tropical forest fails to capture that benefit in terms meaningful to him or her, and hence his or her behavior is not changed.

(3) Intervention Failure.
Intervention failure is the third form of economic failure. It is the failure of governments to ensure that interventions in one sector of an economy do not have spillover affects that reduce biological diversity. Such failures are often also termed 'policy failure' or 'government failure.' Conspicuous examples of such failure are subsidies. Consider a subsidy to energy that keeps the price of energy below its market price. By making energy cheaper, the subsidy encourages energy use, including wasteful energy use. Energy use involves air pollution and air pollution damages ecological systems. The energy subsidy is therefore responsible indirectly for loss of biodiversity. Subsidies are widespread and are often subtle in form. For example, credit may be subsidized for land clearance. Particularly widespread are subsidies to agriculture that are often, although by no means always, inconsistent with biodiversity conservation.

Table 16.5 illustrates the extent of agricultural subsidies in OECD countries. The measure used is the Producer Subsidy Equivalent (PSE). The PSE comprises many direct and indirect support policies although not all have impact on biodiversity, and it is shown in Table 16.5 as a percentage of farm revenues. The index is, however. indicative of the pressures for land conversion and pollution, and probably correlates well with other indicators of biodiversity loss such as natural vegetation cover or endemic bird species. The table reveals the remarkably high level of subsidies in OECD countries. In Japan, for example, three-quarters of farm incomes derive from subsidies. In the United States

Table 16.5
Agricultural Support for Organization for Economic Cooperation and Development (OECD) Countries and Regions

Producer subsidy equivalents: % of farm revenue derived from agricultural support.

Country/region	1979/81	1994 (provisional)
European Community	36	50
US	15	21
Canada	20	27
Japan	60	74
Australia	8	10
New Zealand	18	3
OECD	29	43

Note: Euopean Community is 12 members post 1986.
Source: OECD, Agricultural Policies, Markets and Trade in OECD Countries: Monitoring and Outlook, OECD, Paris, 1995.

the figure is around 20%. Perhaps just as disturbing, in nearly all cases the subsidy is an increasing fraction of farm income over time. The exception is New Zealand, which embarked on an explicit policy of removing protection to its agricultural sector, with a consequent dramatic decline in the subsidy.

Table 16.6 illustrates the prevalence of energy subsidies in the world. Their scale is staggering—some $210 billion every year, roughly equal to the entire gross national product (GNP) of a country like Argentina or Belgium. Such subsidies do not just encourage the wasteful use of energy, and hence pollution, they are also a major demand on government expenditure. Reducing the subsidies would therefore release government income for other purposes, including environmental protection. Yet countries like Egypt and Poland manage to spend some 10% of their entire GNP on energy subsidies. This picture can be repeated for irrigation water (cheaper water encourages excessive use and waterlogging of soils), fertilizers (encouraging nitrate pollution), pesticides (with direct impacts on biological diversity), and so forth.

Table 16.6
World Energy Subsidies (US$ millions).

Country	Coal	Gas	Petroleum	Total
Former USSR	17,000	63,000	65,000	145,000
China	3,300		4,600	7,900
Poland	6,600	130		6,730
Brazil		50	900	950
Venezuela		1,750	3,600	5,350
Mexico	90	600	1,550	2,150
India	2,550		4,250	6,800
Egypt		350	3,000	
Iran		2,300	9,100	11,400
World	36,190	69,440	104,030	209,660

Source: B. Larsen, *World fossil fuel subsidies and global carbon emissions in a model with interfuel substitution*, World Bank, Policy Research Working Paper, No.1256, World Bank, Washington D.C., February 1994.

How Important is Economic Failure?

Measuring the extent to which economic failure explains the loss of the world's biological diversity is extremely difficult. The first thing to note is that all forms of failure can be present at the same time. Markets may not be working properly and governments may also be intervening in these imperfect markets in ways that make biodiversity loss worse. Market failure is clearly important, but the limited evidence on what the world is willing to pay to conserve biological diversity suggests that correcting for global economic failure might not be of major significance for biodiversity conservation. The extent of government intervention is startling, however. Even if all of it is not implicated in biodiversity loss, a good deal of it is. A good starting point for biodiversity conservation policy, then, is to dismantle the subsidy regimes. Many countries have begun to do this, but the growing importance of subsidies as a percentage of farm incomes (Table 16.5) shows just how hard it often is to dismantle subsidy systems.

The popular literature often cites other factors as being of importance in biodiversity loss. 'Overconsumption' refers to the per capita consumption of materials and energy in OECD countries being substantially greater than in developing countries. If this consumption was less, then few materials and less energy would be demanded, reducing pollution and resource extraction which damages biodiversity. Overconsumption is another way of looking at the lack of incentives in economic systems to lower the ratio of materials and energy use to levels of economic activity. The incentives to be 'resource and environment efficient' are in fact the same as those emerging from an analysis of economic failure. Thus environmental taxes, removing subsidies, etc., would all contribute to changing the ratio of consumption to resource use. Some people argue that residents of high income countries overconsume because they are 'too rich', i.e., that economic growth (conventionally decline as the rate of growth of GNP per capita) itself is the fundamental cause of environmental degradation. Apart from the fact that economic growth reflects the aspirations of ordinary people, the extent to which growth should be controlled in the name of biodiversity conservation is open to question, not least because the prospect of growth with conservation remains feasible. In other words, the issue reduces to (a) an empirical matter of whether the ratio of demand for environmental resources to economic activity can be lowered faster than economic activity expands, and (b) what the consequences are if economic growth truly was lowered over a long run period. Lowering economic growth in the north, for example, is likely simply to lower economic growth in the south as the north's demand for imports falls with declining income growth. Advocates of "no growth" frequently confuse growth of incomes with growth of materials and energy supplies. They are not the same.

Structural adjustment and indebtedness in the developing world are also widely cited factors in causing environmental problems, including biodiversity loss. These are complex issues. Structural adjustment is the process whereby agencies lending to indebted nations seek to attach conditions to further loans, for example the removal of economic distortions, such as subsidies. If so, structural adjustment may actually help biodiversity rather than hinder it. On the other hand, the same adjustment policies often involve reducing government expenditures and this can have a direct impact on biological resources if the budgets of environment and wildlife departments are cut accordingly. Adjustment programs may also encourage export drives that induce land conversion for export crops. In the same way, external indebtedness may encourage a similar export-oriented economic policy in an effort to secure foreign exchange to meet debt repayments. The statistical evidence on these relationships is mixed (see the essays in Brown and Pearce 1994) and it is unwise to assume that the causal connections are simple or one-way.

More recently, it has become popular to blame free trade for environmental damage and biodiversity loss. Economic theory tends to argue that free trade is generally desirable in that it maximizes human well-being by allowing countries to specialize in the production of those goods in which they have a comparative advantage. Moreover, there are many instances where protectionism—the raising of tariff barriers and other trade restrictions—can harm the environment by encouraging resource wasteful industry at home. The examples of agriculture and energy have already been given. However, there are some examples of legitimate trade restriction serving biodiversity conservation objectives, and there are cases where free trade agreements seems at odds with conservation objectives.

"Green taxes" are an example of the complex issues that the free trade arguments highlight. A green tax aims to raise the price of polluting products by taxing them. Under the General Agreement on Tariffs and Trade (GATT), if the cost of a product, say polyethylene, is increased because a government mandates pollution control on refineries, it is not 'GATT-legal' to levy an equalizing tax on polyethylene imported from countries which do not require pollution control. Many have argued that GATT is therefore sanctioning the use of lax environmental standards as a de facto pollution subsidy, i.e., it encourages countries with lax environmental standards. Such a subsidy potentially leads to heavy pollution in developing countries and reduced standards in developed countries.

Nor are the consequences of free trade agreements unambiguously environmentally friendly. The advent of the North American Free Trade Agreement (NAFTA) has already caused considerable disquiet over the issue of harmonized environmental standards in Mexico and the United States. But the underlying process of market liberalization could itself have more questionable consequences. This is particularly the case with forest management where traditional forms of resource management and tenurial rights may offer a better approximation of sustainable management than privately managed counterparts (see chapter 18). Only the presence of enabling environmental legislation can check the worse environmental excesses of competition. But states engaged in free trade are frequently reluctant to concede any relative cost advantages when environmental damages are difficult to quantify and easily disguised in official accounting procedures.

Some trade restrictions are often selective and can serve environmental objectives. Reliable eco-labelling and certification schemes may work by allowing potential importers to discern the environmental credentials of products and to redirect their purchases without recourse to damaging and potentially retaliatory trade sanctions. More obvious restrictions can have perverse effects that are not always apparent. Banning the export of whole logs is a good example. In the past, log export bans have been justified by a somewhat spurious rationale that harvesting will be reduced because of the reduction in export demand, and/or that a thriving indigenous processing sector will in the long run be more efficient and have a greater incentives to invest in stable supplies. In fact the evidence from countries as diverse as Costa Rica and Canada show how bans can approximate implicit subsidies to inefficient processors because the effect is to protect the domestic wood processing industry (Kishor and Constantino 1993). More importantly, by lowering the economic returns to forestry, they weaken the sector's competitive position relative to agriculture thereby increasing the likelihood of conversion from forest to agriculture. There is evidence that the removal of such distortions may result in considerable efficiency and environmental benefits. Such reforms generally require additional legislation to restrict exports from old growth forests rather than plantations.

Free trade might also be thought to be inconsistent with wildlife conservation where the trade is in wildlife products, such as ivory, rhinoceros horn and so forth. But there is a convincing economic rationale for legalizing trade in biological resources, contrary to the popular perception. By banning trade the economic return to the sustainable use of wildlife may easily be lowered, thereby providing an incentive to invest in more lucrative alternatives. This argument is particularly important if conservation is more land extensive and associated with unquantified benefits but high costs. Trade bans imposed on high profile species or their products, such as ivory and horn, frequently restrict harvesting and marketing by states demonstrably able to develop wildlife use as the best incentive for long term conservation. The ivory ban, for example, may simply have removed an aspect of economic value of the elephant, reducing the incentive to invest in their conservation in the way that the Southern African range states had done prior to the ban (Barbier *et al.* 1990).

The focus on economic failure and population change should also not ignore the role played by animal introductions. These may have accounted for just under 20% of animal extinctions. The introduced species often prey on endemic species, out-compete them for the available ecological niches, or cross-breed to produce hybrids (see chapter 11). Plant introductions were common with human colonization (see chapter 13). Even the forests of Mauritius are mainly introduced species. Species introductions pose less threats currently to endemic bird species in the world's islands (but see chapter 12). Nonetheless the species introduction remains an important cause of diversity loss worldwide.

Global Environmental Change. More recent threats to biodiversity come from the danger of global warming and the proven science of ozone layer depletion. The primary concern lies not with the gradual effects of projected trends in warming and increased UV radiation, but in the potential for 'collapse' among certain ecosystems as global change goes beyond the historical experience of humankind. Tropical forests, for example, might simply not survive gradual temperature increases. Significant species extinctions might be expected while other ecological communities will be disrupted and ranges of occupation changed, while indirect effects may be widespread and complex (Peters, Lovejoy, and Soulé 1992).

A SIMPLE MODEL OF ECONOMIC CAUSES OF BIODIVERSITY LOSS

The economic analysis implicit in the idea of "economic failure," the role of population change, and of other factors, can be illustrated by developing a very simple model of biodiversity loss. Focusing again on land use change as the main factor giving rose to loss of diversity, and beginning with the viewpoint of the owner or user of the land, we can say that 'development' will be preferred to "conservation" if:

$$\pi_d > \pi_c$$

where π refers to profits, **d** to development and **c** to conservation. Conservation here is taken to include sustainable uses consistent with the retention of biological diversity. The effects of various factors can now be analyzed.

(a) Growth of population will tend to increase the profitability of converting land to food production, so the first effect is that population increases π_d, making it less likely that conservation will be favored.

(b) Any subsidy to food production (or logging, or any other economic activity consistent with biodiversity reduction) will increase π_d. We might write this more formally as development being favored if:

$$[\pi_d + s] > \pi_c.$$

(c) Land clearance results in local externalities to other land users or to economic agents elsewhere who are affected by the biodiversity loss or by the land use change giving rise to biodiversity loss. If these externalities were subject to a tax, then development would be preferred by the landowner only if:

$$[\pi_d + s - t] > \pi_c$$

where t is now the externality tax.

(d) If, as argued above, we also take the subsidy away, then land use change is favored only if:

$$[\pi_d - t] > \pi_c$$

(e) The next group of problems arises from the fact that π_c will tend to be understated because so many conservation benefits (ecological functions) do not have markets. So, if π_c refers to marketed profits from, say, sustainable products, ecotourism, etc., we need to add the non-marketed benefits from conservation such as option and existence value, and the indirect value of carbon storage if it is relevant. Writing these as $\pi_{c,nm}$ (where **nm** reminds us they are non-marketed), we now have:

$$[\pi_d - t] > [\pi_c + \pi_{c,nm}]$$

as the requirement that development be preferred.

This simple model suggests that three things are very important. The first is that the competitive playing field between conservation and development is, in practice, heavily biased against conservation because of all kinds of subsidies to development. The second is that the playing field is further unbalanced because of the failure to account properly for the non-market benefits of conservation and the non-market costs of development (i.e., the externalities). These factors become important when the focus is switched from what is privately profitable for the land owner and what is socially desirable, i.e., taking into account the gains and losses to all people. The third is that population growth is likely to increase the profits from development since it adds to the pressure for more food and more land conversion.

The model is completed by the addition of some factors that are less straightforward. The first of these relates to the role played by time in the model. The simple equations above ignore time and refer to "profits." But in fact, profits occur over time, just as the benefits of conservation do. Future profits are not generally regarded as being as important as present day profits. To the landowner this "discounting of the future" can arise for several reasons. The first is sheer impatience. We all tend to prefer to have our benefits now rather than later. Second, the poorer we are, the more likely it is that we will discount the future. At the margins of poverty, the overriding concern will be with getting food and subsistence now. Tomorrow will have to take care of itself. Third, if the future in question is quite a long way off, so that the benefits and costs accrue to our children and grandchildren rather than ourselves, we may discount that future simply because we believe our descendants will be better off than we are anyway. Put another way, if future generations are going to be richer, an extra $1 to them will mean less than an extra $1 to us. We will therefore tend to 'discount' that future dollar.

How is this discussion of discounting relevant to biodiversity loss? It could be that the costs of biodiversity loss are indeed going to be felt by future generations rather than this one. If so, the phenomenon of discounting will lead us to play down those costs and to focus on the gains to this generation from converting land to "productive" use. Just as importantly, if an unsustainable land use is competing with a sustainable land use, it is likely that the sustainable use will have lower profits per hectare, but over a long period of time, compared to high returns that can be secured for only a short period of time. There is evidence to this effect in models of "nutrient mining"—i.e., the process whereby tropical forested land is burned to clear the land for crops or livestock. Initial profits may be high relative to conservation uses of the land, and they may quickly decline to zero as the nutrients released from the burned biomass into the soil are exhausted. At the point of exhaustion, or very probably before then, the landowner abandons the land and moves to the next plot (provided there is one). Despite this very short perspective, the landowner still prefers the unsustainable use because (a) the high profits now are very much more valuable than future gains, and (b) he has other land to which he can turn when the nutrient stock in the first plot of land is exhausted. Discounting alone will tend to favour the unsustainable use. Discounting and the availability of more land for conversion will make thongs worse still for conservation.

In terms of (b) above, then, we should add time so that development will be preferred by the landowner if:

$$\text{(f)}\ \mathbf{PV}[\pi_d + \mathbf{s}] > \mathrm{PV}[\pi_c]$$

and development will be preferred from a social standpoint if:

$$\text{(g)}\ \mathbf{PV}[\pi_d - \mathbf{t}] > \mathrm{PV}[\pi_c + \pi_{c,nm}].$$

where the notation **PV** refers to "present value," i.e., the value of the future stream of profits with each level of future profit being discounted in such a way that the further into the future the more discounting occurs. The *discount factor* is written as $1/(1+r)^t$, where t is time in years say, and r is the *discount rate*. So as t increases—the further into the future we look—the bigger is $(1+r)^t$ and hence the smaller is any gain or loss recorded in the future.

As long as sustainable uses promise long-lived by less profitable (in profits per annum) return than unsustainable land uses, then discounting will tend to militate against sustainable land use. Moreover, the higher the rate at which the future is discounted, the more this bias will occur. Surprisingly, we know little about the rate at which landowners and tenants in poor countries discount the future. From what evidence we do have, however, there is the suggestion that high levels of discounting take place. Thus, Cuesta, Carlson, and Lutz (1994) suggest that farmers discount the future in Costa Rica at very high rates of around 20–25% (this means that \$1 next year is regarded as being worth only \$0.8 as a 'present value' (1/1.25) and that \$1 in 5 years time is regarded as being worth only (1/1.255) now, i.e., \$0.32 now. The problem of very long-lasting effects can be illustrated. If an action now imposes a cost of \$1,000 million in 50 years time, and if the "rate of discount" is 25%, then the present value of this cost is just \$3.8 million. It is easy to see that if discount rates are this high, few people will invest even in such things as soil conservation, an activity that might be expected to be beneficial to farmers.

ECONOMIC GROWTH AND BIODIVERSITY LOSS

What little evidence we have suggests that poor people discount the future heavily. If this is correct, high discounting of the future and poverty are two sides of the same coin.

They are not separate influences: high discount rates simply reflect poverty. As it happens, the poverty—environment link is complex and illustrates again the problems of repeating popular notions about environment—economy interrelationships. We have many examples of poor communities who look after their resources that they know to be the very basis of their future livelihoods. We also have conspicuous examples of richer nations destroying their own resources. The search for some kind of 'iron rule' linking poverty and the environment is very probably misplaced, despite efforts to formulate such relationships from statistical analyses of the way in which pollution and resource degradation varies with income change. It is very likely that localized pollution worsens as economies industrialize and then gets better as environmental quality rises on the political agenda and as nations find they can afford the expenditures required to abate pollution. Poverty may also be linked to environmental degradation, and hence biodiversity loss, in indirect ways. Poor communities may 'invest' in large families in order to secure a domestic supply of labor (children work on the land, collect water and fuelwood, etc., in many poor communities), and to obtain some kind of security in old age (the children look after the parents and grandparents). These "rational" family size decisions may, however, simply degrade the environment through the resulting overpopulation.

But much depends on context. A study of Machakos District in Kenya showed that, once a community destroying its environment found that they could not extensify agriculture further, they focused on their available resources and began to augment these by tree planting and soil conservation programs, despite remaining poor (Tiffen, Mortemore, and Gichuki 1994). Some pollution is also directly correlated with income, as with carbon dioxide emissions, the main greenhouse gas, for example.

The Drive to Uniformity

The opposite of diversity is uniformity and homogenization. Are there are any underlying economic forces compelling us to be uniform in the type of economic activity we practise? It turns out that the very same forces—the need to expand production and the process of discounting the future—also explain this aspect of the loss of diversity. That is, even when land is converted from high to low diversity, so diversity may be lowered even further by the deliberate homogenization of economic activity on the converted land. The best example is the "green revolution" whereby high yielding varieties (HYVs) have taken the place of lower but more sustainable yield crops. The driving force behind this homogenization process is specialization in order to achieve higher productivity. In turn, the drive for higher productivity reflects both the underlying force of population growth and the demand for higher levels of food consumption. Species with high biological growth rates thus displace low growth species, and productivity can be further enhanced through the use of intensive agricultural techniques. Much of the drive to return to "sustainable agriculture" systems derives from the environmental and social costs of this drive to homogenization (see chapter 18). The environmental costs ensue from the use of intensive inputs—for example, groundwater contamination from pesticides and fertilizers, waterlogging from excessive use of irrigated water, water pollution from livestock waste. But there are social costs, too, in terms of health damage to farm workers. Some of the wider costs are less obvious. Homogenization is contributing to instability in food production and prices—although food production trends are rising, the variation about the trend also appears to be increasing due to the fact that any stress or shock affecting one area affects other areas as well. In any one area, lack of diversity reduces resistance and resilience to shocks such as pests, droughts and other weather events (Hazell 1989). Such loss of system resilience is aggravated by the fact that the costs of resilience loss are likely to be borne by future generations or, at least, are likely to be borne in the future rather than now.

SLOWING THE RATE OF BIODIVERSITY LOSS

The economic approach to biodiversity conservation emphasizes the need to demonstrate the economic importance of biodiversity, where, it has to be stressed again economic importance means far more than commercial or financial importance. The economic approach then focuses on the underlying causes of biodiversity loss. The simple economic model developed here suggests that economic forces intertwine with underlying population change to produce powerful incentives to destroy biodiversity. Importantly, many of those factors driving biodiversity loss are not economically justified in themselves, regardless of their effects on biodiversity. Subsidies, for example, should be removed, or at least reduced, on grounds of good fiscal practice. Their effects on biodiversity add to the economic rationale for their removal.

If this analysis is correct, and the evidence is now overwhelming that this interplay of economics and ecology is what matters, what solutions follow? Economic solutions are about devising incentive systems, systems that make people better off because biodiversity is conserved than when it is not. In turn, those incentives require the removal of economic distortions and the capture of biodiversity value. We turn briefly to these policies.

Policy Options

The discussion of economic failures is suggestive of the policy options available for arresting the erosion for biological resources. As the preceding discussion made clear, the design of such policies should be with a view to increasing (or making apparent) the economic returns to activities associated with conservation, while lowering the returns to destructive activities.

Consistent with the view of land conversion as the principle cause of habitat loss, many countries have used agricultural reform as the vehicle for new conservation programs. Most reform programs basically serve dual objectives of reducing financial support to the industry and lowering the incentives for intensive production. Such requirements are consistent with the political objectives of reducing economic distortions such as Producer Subsidy Equivalents as part of the GATT trade round agreements, and for reform of the Common Agricultural Policy of the European Union.

More ambitious schemes operating in several countries (e.g., the United States and within the European Union), offer financial incentives to meet specific conservation objectives (eg related to target species or habitats). There are dozens of such schemes, many involving direct financial payment in return for conservation actions. Their widespread adoption is encouraging even if output control remains in many cases the underlying objective. Some of the more conservation oriented schemes include:

(1) land set-aside and acreage retirement schemes;
(2) targeted conservation payments for habitats and species;
(3) land easements and covenants;
(4) habitat zoning for development restriction; and
(5) habitat mitigation plans and replacement schemes.

A fuller description of conservation incentives may be found in Clark and Downes (1995) or OECD (1996).

Many schemes typically include some element of cross-compliance to tie payments to the fulfilment of stated objectives of the program. Others may be voluntary or where necessary, backed by legal sanctions. The extension of similar schemes to biodiversity-rich developing countries is desirable, although as yet the requisite institutional structures for their operation are in many cases inadequate. In the context of the provision of global benefits however, the question arises as to the feasibility of transboundary incentive structures (Pearce 1995). As seen in the discussion on global failure, conservation in biodiversity-rich resource poor countries can be thought of as provision of a conservation service to the world at large. It was stated that the provision of this welfare-enhancing service has traditionally remained unremunerated, a result of the absence of complete markets in many environmental services. Thus what is observed is a spatial mismatch in the cost and benefits related to the conservation of global biodiversity. Those nearest the resource, perhaps a protected area, incur the greatest cost and frequently the least benefit, while those considerably removed incur little or no cost while benefitting from the knowledge that the area continues to exist intact. Even if the latter group were willing to pay for global conservation benefits the absence of clearly defined markets allowing themselves or their elected representatives to pay, presents an effective barrier to the development of global incentives schemes.

Analogous international incentive structures do exist for some global environmental services such as carbon sequestration and sulphur emission reduction. In both cases it is possible for a global objective to be met by two countries—or the private firms of two countries—jointly implementing a cost-effective emissions reduction strategy. Typically one country will pay another to reduce its own or mitigate both countries' emissions using a lower cost technology unavailable in the paying country. Preservation and restoration of forest carbon stores is a good example, and in itself may have some beneficial biodiversity side effects. There are considerably fewer similar examples of biodiversity-related incentive deals, although arguably debt-for-nature swaps and biodiversity prospecting arrangements come close to being explicit diversity transfers. There is also growing interest in ideas such as international easements, international tradeable development rights. That such structures can and do emerge in many case independently of global conventions on biodiversity and climate change, is an indication of the power of economic incentives given the appropriate legal framework. The existence of the Global Environment Facility as a facilitator of global biodiversity transfers gives further cause for optimism.

Setting Priorities

We began with the economic problem—making choices in a finite world. If the worst predictions of biodiversity loss are correct it seems clear we cannot now undertake policies that will halt that decline altogether. The "evil choice" may be that we can save some diversity but lose other diversity. In such a context, how should priorities be set? To answer this most difficult of questions it is first necessary to address a related set of issues of what exactly it is that is so vital to conserve. Beyond a kneejerk defense of keystone or charismatic species, is there a common unit of account for measuring the diversity of life on earth and can this accounting system be considered adequate for the purpose of conducting the systematic tradeoffs implied by the need to 'fit as much diversity' to a given limited budget? The short answer as yet, is no. But there is cause for optimism that a rational approach to conservation may emerge.

At this point, it is worth reiterating the broader preoccupation of much of the current chapter with the conservation of biological resources as opposed to diversity *per se*. By virtue of being more amenable to the senses, the emphasis in applied valuation work has largely been restricted to former. In other words, it has been convenient to assume that a preoccupation with forest and wilderness conservation will simultaneously address the diversity problem. However, as highlighted in the discussion of option values, the objective of a priorities exercise is more likely to be the maintenance of the widest portfolio of species for future use. In making difficult decisions to add or lose a species from a set, it is the incremental difference between the last species and its nearest neighbour in the set that we would ideally like to quantify. In economic terms, a unified theory of diversity would allow value to be associated with difference between species. The decision typically would be how to add the greatest distance at minimum cost or meet a budget constraint by losing the least distance represented in a collection of species. This type of decision is somewhat removed from the business of valuing biological resources.

Biological difference may be measured in different ways and there appears to be no *a priori* reason for favouring a numeraire, such as morphological difference over another, or such as genetic difference. It turns out that the beginnings of a difference theory can be found in the area of phylogenetics, and that the same area can also motivate an (almost) consistent economic approach to priority selection. Prior to a brief discussion of potential developments in that field however, it is instructive to consider suggested priority approaches to date.

Earliest attempts at priority setting such as species-rich "Hotspots" (Myers 1988, 1990) or "Megadiversity Areas," (Mittermeier 1988; Mittermeier and Werner 1990) were largely qualitative species count area selections. Similar approaches have been adopted using endemic indicator species (ICBP 1992). There are numerous shortcomings to these measures, not least the inability to prioritize from among priority lists themselves. Furthermore, such approaches are limited by the absence of sufficient biological or socioeconomic information necessary to determine the cost-effectiveness of selecting one area over another. If the ultimate objective is (and should be) the purchase of maximum diversity per dollar, then a priorities exercise should ideally avoid biological double-counting in successive areas selected while making some reference to the cost of successive areas.

Considerable attention has recently been focused on the development of socioeconomic approaches to priorities, not least because of the recognition that local socioeconomic conditions hold the key to the sustainability of any conservation project (Braatz *et al.* 1992; Dinerstein and Wikramanayake 1993). In suggesting an index approach to ranking priority countries for biodiversity investments, Moran, Pearce, and Wendelaar (1995) suggest a combination of elements such as the cost of an (investment) intervention, an indicator of the probability of a successful intervention, and the degree of threat posed to biodiversity in any particular country. Such an approach highlights that a priority country is more than a species rich country.

Independently of the developments noted above, a good deal of energy has simultaneously been invested into the derivation of exact biological measures of diversity to facilitate area selection (Vane-Wright, Humphries, and Williams 1991). Phylogenetic pattern has emerged as an appropriate format for measuring feature diversity (Faith 1994), and has also been related to the measurement of option value of species conservation Faith (1992) and Weitzman (1992).

As mappings of the independent evolutionary history between species, phylogenetic trees provide enough information to evaluate ancestral distance between candidate species in a conservation set. Working within a limited budget, the agony of choice involves the decision on which species to choose and which to discard from a set. The essence of such a choice this can be shown in Figure 16.2a,b where the phylogenetic model implies that the relative number of features represented by a subset of species is given by the total length of the branches spanned on the clade (tree) by the set. Weitzman (1995) likens this evolutionary process to each individual bifurcation resulting in a new entity accumulating and discarding unique beads (features) through time. Figure 16.2 can be thought to represent two areas. The first contains five species from the same family, while the second contains one species from five different families. A species richness choice cannot distinguish between these two areas. On the other hand taxonomic or phylogenetic dispersion clearly shows a preference for the widest collection of features provided in area b, and the conservation of this area can be assumed to maximize option value. Weitzman (1992, 1995) shows that under certain circumstances a decision on the inclusion or exclusion of one species from an area such as b, can be motivated by a cost-benefit decision rule. Faith (1994) discusses some difficulties with the simplifying assumptions in Weitzman's approach.

Further priority-relevant issues can be addressed in Figure 16.3 drawn from Vane-Wright, Humphries, and Williams (1991), which demonstrates another theoretical priority area analysis. Abstracting from the contentious weighting issue (see Vane-Wright, Humphries, and Williams 1991), importance weights "W" are assumed given for terminal taxa A through E. Three of the five taxa occur in three areas R1–R3, and according to the weighting system, row T gives the total aggregate scores for the occurrence of the taxa. For each of the three regions row P1 gives the percentage diversity score, indicating that R3 is the top priority region. Row P2 gives the percentage diversity scores for the remaining two regions with respect to taxa complementarity—the concept discussed earlier, which refers to the use of cost-effectiveness to select the area adding the greatest incremental species difference to an existing set. In this case having chosen R3 it is possible to see that R1 is the second priority to achieve 100% taxa coverage, rather than 89% coverage had R2 been selected.

This simplified example highlights some of the issues discussed earlier. First, that

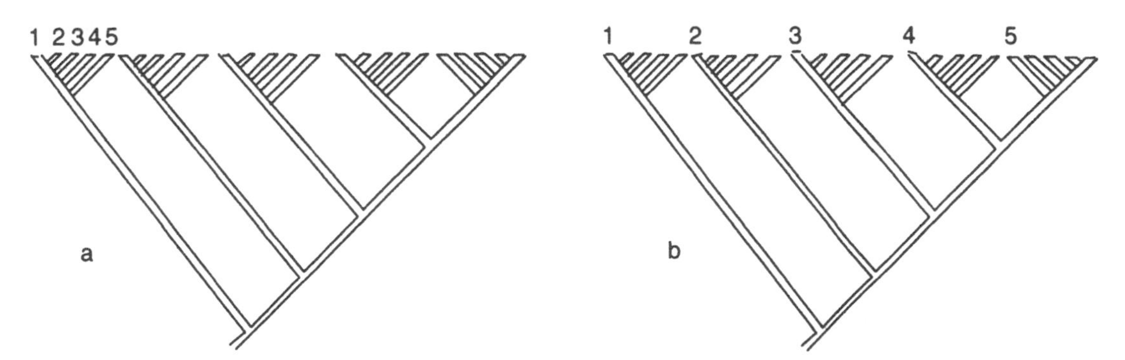

Figure 16.2
Phylogenetic pattern and area selection. The branching pattern corresponds to a phylogeny with each endpoint representing a different species. The numbers 1 through 5 are meant to represent 5 species that are found in a possible biodiversity reserve. In situation (a), the potential reserve has 5 closely related species whereas in situation (b), the potential reserve has 5 species from five different "families" or evolutionary lineages. Clearly, although a and b are equivalent in terms of number of species, they differ in their biological diversity.

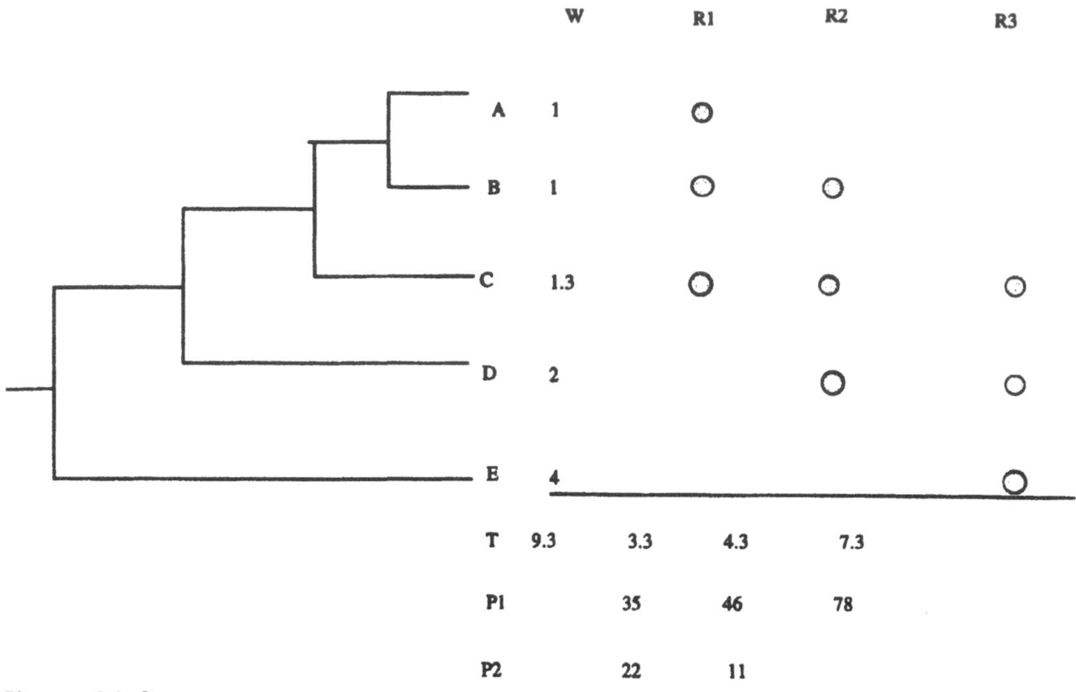

Figure 16.3
Theoretical priority area analysis (adapted from Vane-Wright et al. *[1991]). Theoretical priority analysis illustrated by a phylogeny of five species (A, B, C, D, and E) that occur as triplets in three different regions (R1, R2, and R3). In this case each species is given a different "weighting" as recorded in the W column, with the more evolutionarily distinct species (e.g. E) receiving the highest weighting. The row denoted by T gives the total score for each region (a summing of the weightings over the species found in the region, where a filled circle indicates the species occurs in a particular region). P1 gives the percentage of the total possible diversity score for the regions if only one region is picked. P2 gives the percentage diversity scores for the remaining two regions with respect to taxa complementarity. Notice that region R1 is the better "second choice" even though it comes in third in row P1.*

species richness might be a less than perfect criterion for selecting priorities. By extension the use of complementarity highlights the need to avoid double counting in character representation. Areas R1–R3 are equally species-rich, but the sequential choice matters if a budget only covers the choice of two areas. Second is the related question of cost. The simplifying assumption made here is that R1–R3 are three identical areas that may be acquired at equal cost. However it is feasible that cost differentials might make the most biologically desirable sequence in this example too expensive to implement as an area selection program. An issue that needs to be simultaneously addressed therefore, is whether cost considerations can be integrated into this form of analysis. Third, is there sufficient information available to cover the majority of taxonomic groups using such a procedure? If not, is it possible to identify likely surrogate umbrella species to use in a second best approach?

Despite high information requirements, the above form of analysis would seem to offer the beginnings of a calculus of diversity and is the basis of recent developments in algorithm-based methods for area selection (Kershaw, Williams, and Mace 1994). The area is equally a potentially fertile area for greater economic analysis. (Solow, Polasky, and Broadus 1993; Polasky, Solow, and Broadus 1993).

CONCLUSION: MORAL IMPERATIVES AND ECONOMIC CALCULUS

Economists are conscious that their trade seems morally repulsive to some non-economists. In part this has to be a misunderstanding about the nature of economics on the part of the critics. We have argued strongly that economics is not about financial balance sheets. It is about human well-being. In so far as the continued existence of biological resources contributes to that well-being then economics can help save biological resources. The viewpoint is transparently anthropocentric. This should not be taken to deny the moral view that all biota (at least) have intrinsic values, values in themselves as opposed to values for human beings. The problem is a pragmatic one, however. If it was not necessary to make choices there would be no problem. All biodiversity loss would then be a matter of ignorance, of mistakes, and all the policy we would need would consist of education, awareness raising, and moral learning. This indeed is how many people see the entire environmental problem. The problem with their perspective is that it does ignore choice. It tends to assume that people can simply switch off their choice to destroy biodiversity and switch on some choice to save it. Such a view leaves unaddressed the very incentives that give rise to the choices to destroy in the first place. Moreover, if everything has intrinsic value (it is hard to argue that some things have intrinsic value and others do not) then everything should be conserved. But everything cannot be conserved. There have to be tradeoffs; choices have to be made. In such a context, intrinsic value appears to be a largely unhelpful concept; it does not guide us at all in determining priorities. Nor does it address the issue of incentives.

The problem with "the moral view" can be put another way—it ignores cost. It is easy to see why cost appears not to matter. It is 'only money.' But money is a claim on resources and resources can be used to produce some other benefit. For every cost, then, some other moral right may be being sacrificed—rights to a livelihood in the case of the agriculturist, or a right to education or health in the case of the taxpayer. Rights have to be traded, and simply declaring that conservation is a "moral duty" becomes an escape from facing up to the reality of moral choice. That is why the economic approach to biodiversity conservation is controversial—whatever its limitations, it faces facts.

LITERATURE CITED

Adger, W.N., K. Brown, R. Cervigni, and D. Moran. 1995. Estimating the total economic value of forests in Mexico. *Ambio* 24 5:286–296.

Barbier, E., J. Burgess, T. Swanson, and D. Pearce. 1990. *Elephants, economics and ivory*. London: Earthscan.

Braatz, S., G. Davis, S. Shen, and C. Rees. 1992. *Conserving biological diversity: A strategy for protected areas in the Asia-Pacific region*. World Bank Technical Paper 193. Washington D.C.: World Bank.

Brown, K. and D.W. Pearce, eds. 1994. *The causes of tropical deforestation: The economic and statistical analysis of factors giving rise to the loss of the tropical forests*. London: University College Press and Vancouver: University of British Columbia Press.

Clark, D. and D. Downes. 1995. *What price biodiversity? Economic incentives and biodiversity conservation in the United States*. Washington D.C.: Centre for International Environmental Law.

Cuesta, M., G. Carlson, and E. Lutz. 1994. *An empirical assessment of farmers' discount rates in Costa Rica and its implication for soil conservation.* Washington D.C.: Environment Department, World Bank.

Dinerstein, E. and E.D. Wikramanayake. 1993. Beyond hotspots: How to prioritize investments to conserve biodiversity in the Indo-Pacific region. *Conservation Biology* 7:53–65.

Faith, D. 1992. Conservation evaluation and phylogenetic diversity. *Biological Conservation* 61:1–10

Faith, D. (1994), Phylogenetic pattern and the quantification of organismal biodiversity. *Philosophical Transactions of the Royal Society of London, B* 345:45–58.

Fankhauser, S. and D.W. Pearce. 1994. The social costs of greenhouse gas emissions. In *The economics of climate change*, 71–86. Paris: OECD.

Hazell, P. (1989). Changing patterns of variability in world cereal production. In *Variability in grain yields. Implications for agricultural research and policy in developing countries*, eds. J. Anderson and P. Hazell. Baltimore: John Hopkins University Press.

Hodgson, G. and J. Dixon. 1988. *Logging versus fisheries and tourism in Palawan.* Occasional Paper No.7. Honolulu, Hawaii: East-West Center.

ICBP (International Council for Bird Preservation). 1992. *Putting biodiversity on the map: Priority areas for global conservation.* Cambridge: Birdlife International.

Kershaw, M., P.H. Williams, and G.M. Mace. 1994. Conservation of Afrotropical antelopes: consequences and efficiency of using different site selection methods and diversity criteria. *Biodiversity and Conservation* 3:354–372.

Kishor, N. and L. Constantino. 1993. The costs and benefits of removing log export bans: evidence from two countries. Unpublished manuscript. Washington, D.C.: World Bank, Laten Division.

Kramer, R., E. Mercer, and N. Sharma. 1994. Valuing tropical rain forest protection using the contingent valuation method. Unpublished manuscript. Durham, N.C.: School of the Environment, Duke University.

Kumari, K. 1994. An environmental and economic assessment of forestry management options: a case study in Peninsular Malaysia. Chapter 6 of *Sustainable forest management in Malaysia.* Ph.D Thesis, University of East Anglia, United Kingdom.

Mahar, D. and R. Schneider. 1994. Incentives for Tropical Deforestation: Some examples from Latin America. In *The causes of tropical deforestation: The economic and statistical analysis of factors giving rise to the loss of the tropical forests*, eds. K. Brown and D.W. Pearce, 159–171. London: University College Press and Vancouver: University of British Columbia Press.

Mittermeier, R.A. 1988. Primate diversity and the tropical forest: case studies from Brazil and Madagascar and the importance of the megadiversity countries. In *Biodiversity*, eds. E.O. Wilson and F.M. Peter, 145–154. Washington D.C.: National Academy Press.

Mittermeier, R.A. and T.B. Werner. 1990. Wealth of plants and animals unites "megadiversity countries." *Tropicus* 4:4–5.

Moran, D., D.W. Pearce, and A. Wendelaar. 1995. Investing in Biodiversity: An Economic Perspective on Global Priorities. University College London and University of East Anglia: Centre for Social and Economic Research on the Global Environment (CSERGE).

Myers, N. 1988. Threatened biotas: "hot spots" in tropical forests. *The Environmentalist* 8:187–208.

Myers, N. 1990. The biodiversity challenge: Expanded hot-spots analysis. *The Environmentalist* 10:243–256.

OECD (Organization for Economic Cooperation and Development). 1996. *Making markets work for biological diversity: the role of economic incentive measures.* Paris: OECD.

Pearce, D.W. and D. Moran. 1994. *The economic value of biodiversity.* London: Earthscan and Island Press.

Pearce, D.W. 1995. New Ways of Financing Global Environmental Change. *Global Environmental Change* 5:1–13.

Pearce, D.W., N. Adger, K. Brown., R. Cervigni, and D. Moran. 1993. *Mexico forestry and conservation sector review: Substudy of economic valuation of forests.* Centre for Social and Economic Research on the Global Environment (CSERGE) for World Bank Latin America and Caribbean Country Department.

Pearson, P. 1994. Energy, externalities and environmental quality: Will development cure the ills it creates? *Energy Studies Review* 6:199–216.

Peters, C., A. Gentry, and R. Mendelsohn. 1989. Valuation of an Amazonian rainforest. *Nature* 339:655–666.

Peters, R., T. Lovejoy, and M. Soulé. 1992. *Global warming and biological diversity*. New Haven, CT: Yale University Press.

Polasky, S., A. Solow, and J. Broadus. 1993. Searching for uncertain benefits and the conservation of biological diversity. *Environmental and Resource Economics* 3:171–181.

Richards, J. 1990. Land transformation. In *The Earth as transformed by human action: Global and regional changes in the biosphere over the past 300 years*, eds. B. Turner, W. Clark, and W. Kates. Cambridge: Cambridge University Press.

Ruitenbeek, J. 1992. The rainforest supply price: A tool for evaluating rainforest conservation expenditures. *Ecological Economics* 1:57–78.

Schneider, R. 1994. *Government and the economy on the Amazon frontier, Latin America, and the Caribbean technical department report no. 34*. Washington, D.C.: World Bank.

Solow, A., S. Polasky, and J. Broadus. 1993. On the measurement of biological diversity. *Journal of Environmental Economics and Management* 24:60–68.

Southgate, D., R. Sierra, and L. Brown. 1989. *The causes of tropical deforestation in Ecuador: A statistical analysis*. London: London Environmental Economics Centre, International Institute for Environment and Development (IIED).

Tiffen, M., M. Mortimore, and F. Gichuki. 1994. *More people less erosion: Environmental recovery in Kenya*. London: Wylie and Sons.

Vane-Wright, R.I., C.J. Humphries, and P.H. Williams. 1991. What to protect—systematics and the agony of choice. *Biological Conservation* 55:235–254.

Ward, K.M. and J.W. Duffield. 1992. *Natural resource damages: Law and economics*. New York: John Wiley.

Weitzman, M.L. 1992. On diversity. *Quarterly Journal of Economics* 57:363–405.

Weitzman, M.L. 1995. Diversity functions. In *Biological diversity: Economic and ecological issues*, eds. C. Perrings, C. Folke, C.S. Holling, B.O. Jansson, and K.-G. Mäler. Cambridge: Cambridge University Press.

World Bank. 1991. *Malaysia: Forestry subsector study*. Report 9775-MA. Washington D.C.: World Bank.

WCMC (World Conservation Monitoring Centre). 1992. *Global biodiversity: Status of the Earth's living resources*. Cambridge: WCMC.

CHAPTER

17

The Patagonian Challenge: Melding Conservation with Development

MIGUEL A. PASCUAL
JOSÉ M. ORENSANZ
ANA M. PARMA
and SERGIO L. SABA

They come to make money, in oil, in sheep raising, or in commerce. They plan to spend a few years, then to go home, to the real homes of their hearts, and live in comfort on the well earn proceeds of the years barrenly passed in Patagonia.
George Gaylord Simpson, 1934.

Patagonia, at the southern tip of South America (Figure 17.1), is a region of over one million km^2 shared by Chile and Argentina, endowed with some of the wildest landscapes in the world. Although remote and sparsely populated, Patagonian ecosystems are not pristine—most of them have been significantly disturbed by human activities. As is the case with much of Latin America, Patagonia is experiencing a push for economic development that is imposing ever increasing strains on natural resources.

Attitudes towards the use of natural resources in Patagonia have taken two extreme forms: a crude utilitarianism, born out of the short-sighted pragmatism of the frontier—*el desierto*, and a naive conservationism, traceable in its origins to the contemplation of the bucolic scenery of southern alpine lakes and forests. Utilitarianism originated during the nineteenth century, expressed initially as a plundering by Euro-Americans of what was perceived to be a wasteland. This unsophisticated depredation gradually evolved into modern ranching, agriculture, logging, fishing, mining, and more recently tourism. Conservationism has more recent origins, having emerged in the 1920s nurtured by the desire to preserve regions of conventional scenic beauty. Vast areas of the Andean lakes and forests were designated as national parks. Analogy with the alpine landscapes of the Northern Hemisphere was eulogized, as best reflected by architecture, landscaping, and forest practices in the region. Unfortunately, the Patagonian steppe, devoid of spectacular landscapes and charismatic animals, is still largely ignored in spite of harboring areas of

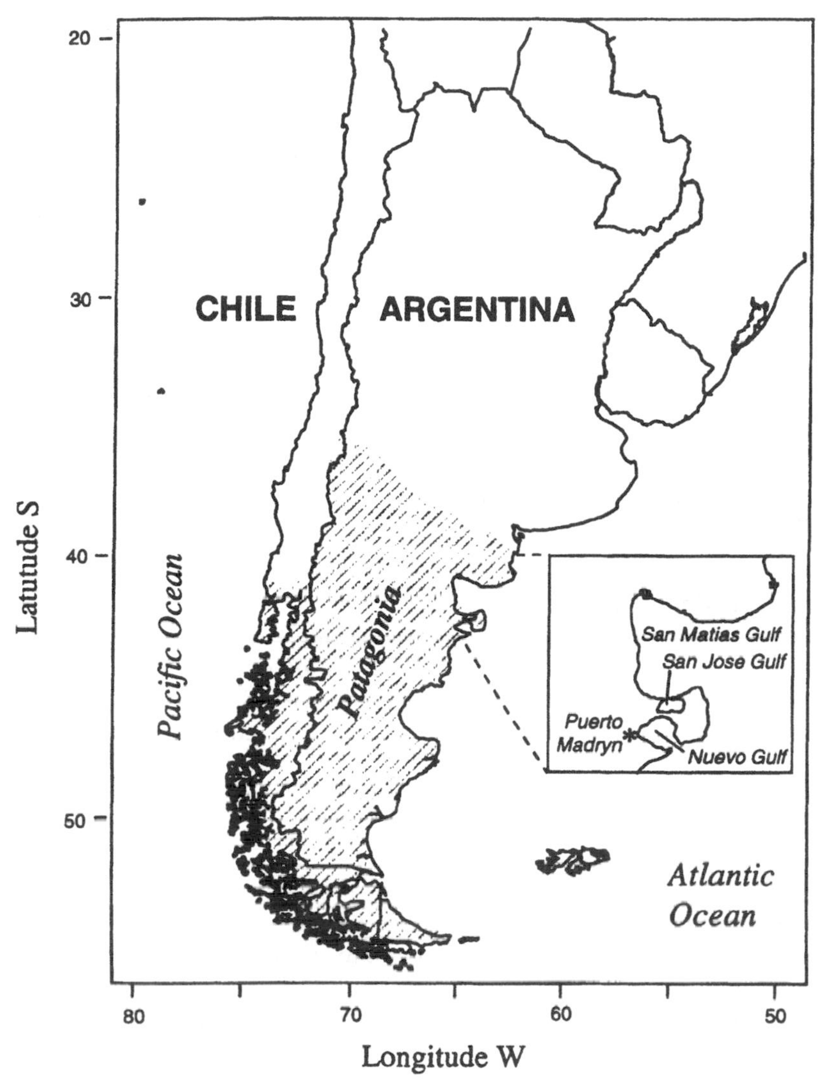

Figure 17.1
The Patagonian region. Inset shows the Valdés Peninsula complex.

unique endemic biodiversity. Gradually, however, the ethos of the environment is replacing the aesthetics of its contemplation.

Both naive conservationism and irresponsible utilitarianism are deeply flawed. Naive conservationism ignores the fact that virtually all Patagonian ecosystems already are impacted by humans. It emphasizes charismatic organisms and ecosystems, and neglects the reality that conservation and utilization are inextricable aspects of any sound management program. Irresponsible utilitarianism, on the other hand, ignores the ethical imperatives of conservation, as well as the practical societal benefits derived from sustained environmental quality. It often concentrates on introduced species, ignoring potentially valuable uses of indigenous ecosystems. It favors short-term gains over the long-term, sustainable uses of natural resources.

We argue for a perspective that transcends these extreme simplistic views, and propose that the ethos of conservation and the practicalities of exploitation ought to be inseparable components in the integration of human society with the natural world of Patagonia. Although this chapter focuses on Argentinean Patagonia, it should be obvious that Patagonia shares many of the problems of other less-developed, developing, or under-developed parts of the world—i.e., a combination of strong pressures from economic, political, or social forces for the use of its natural resources on one side, and loose or unenforced environmental legislation and weak institutions on the other. Somehow, solutions to conservation issues need to be resolved within this frail societal framework, from bottom up, if they are to be effective and lasting. In this chapter we illustrate some of the pressing environmental problems in Argentinean Patagonia. We then examine regional institutions and discuss to what extent they are capable of making development and conservation compatible. We propose that universities, by virtue of their strategic insertion within the regional society, should have a pivotal role in promoting wise resource management.

ROOTS OF PATAGONIA'S BIOTIC DIVERSITY

The unique continental biota of Patagonia reflects the dramatic history of South America during the Cenozoic (Simpson 1980; Veblen, Hill, and Read 1996). An early connection with Antarctica, New Zealand, and Australia was followed by a long period of isolation, and finally, during the Pliocene, by the establishment of the Panamanian land bridge to North America. This sequence of events molded the current tapestry of Patagonian biodiversity—*Nothofagus* forests and marsupial mammals, Neotropical endemics, and the invasion of northern forms during the upper Cenozoic (Simpson 1950, 1969; Patterson and Pascual 1968; Reig 1981).

Patagonia includes three principal terrestrial biomes (Ragonese 1968), distinguished both phytogeographically (Cabrera and Willink 1973) and zoogeographically (Cabrera and Yepes 1940; Ringuelet 1955). These include the monte desert, the Patagonian steppe, and the southern temperate rain forests. The monte, found in the northeast of Patagonia, is a warm desert supporting xerophytic vegetation. Rainfall is seasonal with summer rains averaging less than 250 mm. The Patagonian steppe lies south of the monte. It is a dry scrub desert characterized by high winds, low temperatures, and low precipitation (usually 100–270 mm) that falls mainly in winter. In stark contrast with the arid plateau, the biota of the Andean region in western Patagonia is shaped by the influence of moist Pacific winds. An uninterrupted band of dense forest dominated by southern beech (*Nothofagus* spp.), develops along the mountainous western fringe of the region, between the eastern slopes and foothills of the Andes on the Argentine side, and the Pacific coast of southern Chile. Rainfall here exceeds 5,000 mm in some localities.

There are more than 440 terrestrial vertebrate species in Patagonia (ca. 29 amphibians, 65 reptiles, 269 birds and 76 mammals), plus approximately 30 exotics, mostly mammals. Endemics are common, particularly among amphibians, reptiles, mammals, and birds (Cei 1980, 1986; Narosky and Zurieta 1993; Monjeau, Bonino, and Saba 1994). Natural insular distribution patterns are observed in all taxa, although in some cases they are the result of anthropogenic landscape fragmentation. The geographic range, abundance, and natural history of most species are little known, or are only documented for small geographic regions (Ubeda and Grigera 1995).

Unlike the terrestrial vertebrate fauna, the native freshwater fish fauna of Patagonia has a very low number of species (Ringuelet, Arámburu, and Alonso 1967). Only 18 indigenous species of fishes are found in the southern extreme of South America, representing four neotropical and three circum-antarctic families. Patagonian rivers harbor a somewhat richer assemblage of aquatic invertebrate species (Boneto 1995). In the first half of this century, lakes and rivers in the region were stocked with trout, char, and salmon, originally from the northern hemisphere. Trout populations thrived in lakes and rivers and currently sustain some of the best sport trout fisheries in the world (Leitch 1991).

The adjacent part of the southwestern Atlantic corresponds to the extensive Patagonian Shelf, part of one of the world's large marine ecosystems (Bisbal 1995). Oceanographic processes are under the influence of the Malvinas/Falkland Current, a splinter of the West Wind Drift. The marine Patagonian biota is dominated by southern components, prominently the extensive *Macrocystis* kelp forests. Towards the northern part of the Patagonian Shelf, there is an increasing number of warm-temperate West Atlantic elements. Most spectacular among the components of the marine fauna are populations of birds and mammals, including the World's largest continental colonies of penguins and elephant seals, two species of sea lions, and important calving areas of the southern right whale (Lichter 1992; Payne 1995).

THREATS TO PATAGONIA'S NATURAL RESOURCES: THREE CASES

Patagonia has a relatively short history of occupation by Europeans, a sparse human population, and majestic landscapes where the human hand is not immediately apparent. Yet it is no longer the quasi-pristine wilderness that it has often been portrayed to be; most Patagonian ecosystems have been significantly disturbed by human activities (Table 17.1). Humans first arrived in Patagonia ca. 13,000 years ago (Bryan 1986) and hunting may have played a role in the extinction of some components of the indigenous fauna of large mammals (e.g., the giant ground sloth) during the Holocene. The steppe was the

Table 17.1
Human-induced risks for Patagonia's renewable natural resources.

Environment	Source of Ecological Risk
Steppe	overgrazing and desertification
Valleys	overhunting
	species introductions (mostly grazers)
	species introduction (crop species and weeds)
	salinization
	pesticide pollution
	modification of landscape through agriculture
Forests	browsing by cattle and exotic deer
	human-induced fires
	competition of native trees with exotic species
	extensive deforestation
Lakes and rivers	species introductions (mostly salmonids)
	dams
	overfishing
	pollution
Marine coast and shelf	overfishing
	competition of fishing with marine mammals and birds
	incidental fishing mortality of non-commercial species
	oil pollution

first ecosystem to be extensively modified by the European immigration, mostly through the introduction of sheep ranching, which expanded rapidly during the last part of the 19th century. This ranching activity promoted the disorderly occupation of the hinterlands by "*estancias*" and led in the end, to overgrazing and desertification. The valleys of the Negro and Chubut Rivers were colonized by agriculturists who brought with them scores of exotic plant and animal species that spread up along the valleys.

Although the exploitation of forests and the ocean played a minor role in the early development of Patagonia, the depletion of fish stocks and forests elsewhere in the world has turned international attention to Patagonian fisheries and forests. Thus, extensive logging of the *Nothofagus* forests from Tierra del Fuego has been recently authorized by the provincial government (*Clarín*, Buenos Aires, June 30, 1996). Another example of the growing interest in non-traditional resources is provided by the fishery for toothfish (*Dissostichus eleginoides*) off southern South America (CCAMLR 1992). This fishery is rapidly becoming an unsustainable mining-like operation, showing clear signs of overcapitalization of the fleet. To highlight some of the conflicts that surround the conservation and exploitation of natural resources in Patagonia, we describe three cases in some detail.

Hunting of Indigenous Wildlife

Hunting is an important activity in continental Patagonia. Some species are targeted because of their alleged detrimental effect on cattle ranching (guanaco, *Lama guanicoe*; choique, *Pterocnemia pennata*; red fox, *Pseudalopex culpaeus*; chillá fox, *P. griseus*; puma, *Puma concolor*), others for their fur (newborn guanacos; chillá fox; pampa gray fox, *Pseudalopex gymnocercus*; skunks, *Conepatus humboldtii* and *C. chinga*; huemul *Hippocamelus bissulcus*), for sport (various Anatidae, specially the common cauquén, *Chloephaga picta*; guanaco), or for consumption (guanaco; choique; mara, *Dolichotys patagona*; copetona, *Eudromia elegans*; peludo, *Chaetophractus villosus*; pichi; various Anatidae). Provincial governments are responsible for the control and licensing of harvest and fur trading in the region. We will illustrate some of the management problems created by the conflicts between the preservation of native species and human activities by concentrating on guanaco and fox.

The guanaco is the most widely distributed of the four members of the camelid family found in South America (Franklin 1982). Its distribution extends from Northern Peru to Patagonia, which is the heartland of the guanacos distribution with 95% of the individuals. Pre-European numbers of guanaco have been estimated at 35 million (Raedeke 1979). However, by the mid-1900 most guanacos had been eliminated from the large Patagonia Pampa (Franklin and Fritz 1991). Current numbers in South America have been estimated at 600,000, but estimates vary widely (Pujalte and Reca 1985; Garrido, Mazzanti, and Garrido 1988). Reduction of the population is a result of both hunting and habitat destruction due to cattle and sheep ranching, and has led the Convention on International Trade in Endangered Species (CITES) to include it in their Appendix II (species requiring strictly regimented commerce) (Torres 1992). Despite the great reduction in numbers, guanaco is still locally abundant, however. Although most provinces forbid or limit guanaco harvest, the levels of poaching and local consumption can be quite high (Puig 1992). For example, the legal harvest of newborns for skins is a multi-million dollar industry ($3.6 million in 1979; Ojeda and Mares 1982). There is also trade in guanacos fine wool fiber, and its high quality meat is still an important source of protein in the high mountain and desert areas of Chile and Argentina (Oporto and Soto 1995). Because the guanaco represents significant economic value when harvested, there is a real risk of overharvesting. In response to this risk, it has been proposed that the best conservation strategy is to reinforce the economic incentives for its sustainable use (Franklin and Fritz 1991). Unfortunately, because we lack information on guanaco abundance and demography, harvest levels and specific regulations are based on limited data and dubious figures provided by ranch owners.

The Malvinas Islands fox, *Dusicyon australis*, is the only species of extinct terrestrial mammal in Patagonia whose demise can be attributed to harvesting by European man (Allen 1942). The three other species of foxes found in Patagonia are the large gray or pampa fox in the north of the region, the small gray or chillá fox, and the red or culpeo fox found throughout the region (Monjeau, Bonino, and Saba 1994). The International Union for the Conservation of Nature and Natural Resources (IUCN) has placed *P. griseus* in the vulnerable category and recommended further study of the populations to determine sustainable harvest levels. CITES has included both *P. griseus* and *P. culpaeus* in their Appendix II. In spite of all these restrictions, about 100,000 foxes are harvested annually in Patagonia (Rabinovich *et al.* 1987). A rural worker in Patagonia is estimated to trap an average of 10–20 foxes per season (Novaro and Funes 1993), providing a significant supplemental income to their salaries. Fox trapping is also common on Patagonian ranches purportedly to reduce predation on sheep (Bellati and Von Thungen 1987).

Fox and guanaco in Patagonia share a number of relevant characteristics that illustrate the problems of wildlife management in Patagonia:

(1) They are hunted not only for their intrinsic value, but also because they are considered harmful for sheep ranching. Harvest of these species provides a important additional income for rural workers;

(2) They have been declared vulnerable by international entities; and

(3) Provincial governments are responsible for their management, and do so based on sketchy information and without a management plan. As a result, harvest is mostly regulated by market forces, without adequate consideration of population biology.

The San José Gulf: Exploitation and Conservation

Along the east coast of Patagonia, between 42° and 44° S, the Valdés Peninsula separates two semi-enclosed coastal ecosystems: the San José Gulf to the north, the Nuevo Gulf to the south (Figure 17.1). This region harbors one of the most fascinating assemblages of natural resources and wild populations of the Patagonian coast, including a variety of shellfish and reef-fish resources, colonies of sea lions (*Otaria flavescens*), the only continental colonies of elephant seals (*Mirounga leonina*) in the World, and breeding or feeding areas of several marine birds, including the Magellanic penguin (*Spheniscus magellanicus*). The San José and Nuevo Gulfs, and Caleta Valdés, are also the most important zones of concentration for southern right whales (*Eubalaena australis*) in the southwestern Atlantic. Whales arrive between August and November to breed and calve in very shallow and sheltered inshore areas (Lichter 1992).

There is only one important urban settlement in the area, Puerto Madryn (population ca. 50,000), which besides being a major regional center for services has an aluminum plant, five fish processing plants (SAGyP 1995), and a growing tourism industry oriented towards the recreational use of the coastal zone. As a measure of the growing tourism industry, in 1992 more than 30,000 persons took part in whale sighting boat excursions and more than 80,000 visited the provincial faunistic reserves (Yorio 1993). Besides whale-watching, other tourist activities include visits to breeding colonies of birds and pinnipeds, diving, and spear- and hook-and-line fishing. Until 1970, commercial fishing activities in the region were insignificant, limited to beach seining and coastal shellfish collection for local consumption. During the late 1960s, scallop fishing boomed in adjacent San Matías Gulf in response to a marked increase in demand and the rising price of scallop meats in the U.S. market. The fishery collapsed after a few years, leaving the grounds depleted and (it has been hypothesized) ecologically degraded after intensive dredging (Orensanz, Pascual, and Fernández 1992). Fearing that the same could happen in the San José Gulf, the provincial fishery administration banned the use of dredges, and beginning in 1973, only a commercial diving fishery was allowed. Commercial diving was believed to be free from many of the problems of dredging—i.e., it can be highly size-selective, typically operates only in areas of high concentration, and does not deteriorate the sea bed or remove shell hash, a natural settlement substrate for scallop larvae (Orensanz 1986). The fishery proved sustainable with little external regulations for nearly 20 years (Ciocco 1995b), but research efforts were later discontinued and the stocks were not monitored. By 1993, harvesting intensified and became unselective following an increase in scallop price. The stocks rapidly collapsed and the fishery was closed in 1996. Fishermen (about 25 teams of artisanal commercial divers) turned to other less profitable resources (e.g., clams, mussels) and to servicing the tourism industry. In 1974 the Legislature of Chubut Province created the San José Provincial Marine Park, the only marine park in Argentina. The law, modified in 1979, allows commercial fishing but does not mention aquaculture.

Although the region as a whole is sparsely populated, the matrix of uses of the coastal zone is complex. Lines of conflict between resource conservation and resource exploitation are multiple. Following the collapse of the scallop fishery, strong pressure developed to shift the available harvesting capacity towards other shellfish resources. In the early 1990s some local entrepreneurs became interested in developing mussel aquaculture, involving the deployment of longlines of spat collectors in San José Gulf (Ciocco 1995a). Conservationists, who have long advocated the closure of the San José Gulf to all commercial operations except for the artisanal shelfishing, strongly opposed this project, claiming that it would disrupt the reproductive behavior of whales. Whale-watching and other forms of ecotourism were initially hailed by many conservationists as a pragmatic rationale for wildlife protection, an alternative to other commercial activities that could be appealing to the local community. But while there is growing interest from the industry to make the most from visitors by further developing tourism infrastructure, many in the conservation movement propose that the entire Valdés peninsula should be kept development-free. The unchecked growth of tourism (particularly whale-watching from small vessels), many now claim, is actually more disruptive than small-scale fishing or spat-collection.

While marine mammals and bird colonies have concentrated a great deal of public attention and research resources, less charismatic creatures, like the southern sand perch (*Pinguipes* spp.) and other reef fishes appear to be decimated by sport fishing. These ecological disasters can be expected to continue because there are no management plans or serious monitoring efforts aimed at these minor species.

The various sectors of the tourism industry, fishermen, and conservationists advocate their views in an increasingly intricate tapestry of priorities and influence. The tourist and fishery industries are clearly rooted in the local community, whereas conservationists have in addition, important political support at the national and international levels. In 1994 the provincial fisheries administration called together the sectors interested in the future of San José Gulf to participate in an *ad hoc* forum for discussion. The debate was remarkably open, and compromise was reached on several specific issues, particularly the zoning plan for tourism, conservation, fisheries, and aquaculture (Ciocco 1995a). This small experiment in societal dialogue illuminates two outstanding points. First, there is space for compromise between sectors with conflicting and often seemingly irreconcilable agendas, and second, there is much to be gained by having those sectors openly debate the issues and search for common ground.

The Introduction of Exotic Aquatic Species

At the turn of the century when Patagonia was still a remote and unpopulated wilderness, the federal government was already considering the introduction of fish species for sport and commercial uses. Between 1904–1910 eight major shipments of salmonid eggs were sent to Patagonia from the United States (Leitch 1991). In the next 20 years, local hatcheries complemented by private efforts from European settlers, continued to raise and plant salmonids throughout the region. As a result, landlocked salmon, rainbow, brown, and brook trout today inhabit nearly every lake and stream in Patagonia (Baigún and Quirós 1985), and sustain some of the most attractive trout fisheries in the world (Leitch 1991). Anecdotal information indicates that some of the native fish in Patagonia, such as the *peladilla* (*Haplochiton* spp.) and the pejerrey (Patagonina hatchery), were in the past very abundant in environments where they are now extremely rare (Wegrzyn and Ortubay 1991). Although introduced trout are the suspected culprits, hard data supporting this view are lacking. The Argentinean National Park Service has traditionally maintained a stand in favor of protecting the native fish and native aquatic communities within the parks, which have limited commercial value. Provincial fisheries administrations, on the other hand, adhere more closely to the values of fishermen and fishing associations, which derive most of their benefits from salmonid fishing. Accordingly, their policies tend to favor the maintenance and augmentation of salmonid populations. The lack of environmental legislation, the scarce information available on both introduced and native species, and outbreaks of long-held resentment between provincial and federal governments have made the debate less constructive that it should be (Leitch 1991). The two views, however, appear to have come closer together in recent years. The National Park Service is more accepting of the fact that trout sport fishing is a commercially significant activity in the region and provincial administrations are gradually abandoning the reckless practice of extensive and indiscriminate fish translocation that was common practice a few years back. A council for freshwater fishing, with representatives from the Patagonian provinces and the National Park Service, has been recently created with the specific purpose of establishing general fishing regulations in the region.

The import of exotic aquatic species to support the expanding aquaculture industry promises to become a major environmental issue in Patagonia. In recent years, and paralleling an explosive growth of the salmon culture industry in Chile (Achurra 1992), reports of adult Pacific salmon spawning in Patagonian rivers have become common (Grosman 1992). Due to the complexity of the potential interactions between anadromous fish and resident species (Kline *et al.* 1993; Krueger and May 1991; Willson and Halupka 1995), the effects of colonizing salmon can propagate through the ecosystem in unforeseen ways.

The Chilean success in developing the aquaculture activity in the region has raised an unprecedented interest in aquaculture. Most projects under consideration by provincial fisheries administrations in Argentina's Patagonia are based on aquaculture, and many emphasize the use of exotic species. The list of candidates is not limited to Pacific salmon, but also includes Japanese and Chilean oysters, manila clams, sturgeon, North American catfish, Atlantic halibut, and turbot. Legislation regulating the introduction of exotic species is scarce, mostly concerned with sanitary aspects of species import and with almost no contemplation for the potential ecological impact of introduced species. Only two of the five Patagonian provinces have legislation that regulates the use of exotic species and federal legislation is limited to two resolutions from the National Fisheries Secretariat.

RESOURCE MANAGEMENT IN PATAGONIA

Conflicting conservationist and utilitarian attitudes are by no means exclusive to Patagonia. Indeed, the same positions permeate the conservation debate in much of the world. As illustrated by our case studies, what is most distinctive of resource management in Patagonia is the unstructured way in which these views interplay and, ultimately, influence management decisions. Ideally, resource management should result from the interaction of four major players: the public, universities and research centers, the private sector, and the government. The government interprets the views and goals of the public and the private sector, and consequently makes specific decisions on resource use, implements management plans, and produces legislation that reflects societal values and priorities. It regulates the private sector whose interests are directly affected by the fate and use of natural resources. Universities and research centers provide technical expertise and training. In this view, two major safeguards for balanced decision-making are effective legislation and public scrutiny. Effective legislation provides a framework within which advocacy groups defend their particular views. Advisory or management councils that report to resource administrators provide a mechanism for the participation of diverse interest groups and technical experts in the analysis of management issues. The process of natural resource management in Patagonia departs from this ideal view on several counts.

Argentina in general, and Patagonian provinces in particular, lack both strong environmental legislation and a tradition for open consultation to guide management decisions. Without formal mechanisms for discussion and consultation, the interests of different advocacy groups remain more often than not irreconcilable and grounds for consensus are rarely found. In the end, the views of particular interest groups exert an inordinate influence on decision making. The Council for Freshwater Fishing, established by the provinces and the National Park Service, constitutes an auspicious attempt at regional consultation. Although the council's duties specifically focus on fishing regulations, this structure provides a good test for the viability of consulting fora to examine regional environmental problems. Another such example is the Advisory Council on Wildlife for the Patagonian Region, with participation of federal and provincial governmental agencies, Patagonian universities and research centers, conservationists organizations, and representatives of the private sector.

The role of the provincial governments in resource management and conservation in Patagonia cannot be overstated. In Argentina, 90% of the decisions about the use of forest, wildlife, fisheries, soil and water resources are made by the provinces (Tarak 1989). Since the provinces are the ultimate arbiters in environmental matters, they should be provided the best possible advice in resource management. Local universities and research centers have an obvious role to play in management decisions, as they provide a natural forum from which to generate and dispense the required technical expertise. As a result of recent economic reforms, provincial resource administrations that used to keep their own staff of biologists, rely more and more on contracted work with private consultants and universities. While this creates a novel mechanism for interaction between universities and resource administrations, the research involved typically addresses only the most immediate and short-term needs of management. The study of long-term responses of intervened systems demands more sustained and continuous research efforts. But as we discuss more extensively below, the curricula of local universities and much of the research conducted at regional centers is inadequate to respond to the long-term demands raised by resource conservation and management.

Traditionally, community organizations have not had a very active role in defining agendas for conservation and resource management in Argentina, a pattern ubiquitous throughout Latin America. Yet, in Patagonia there is a clear trend towards involvement of the public. Consider for example the case of the San José Gulf, described previously. Grassroots organizations begin to sprout everywhere. In San Antonio Oeste, a small coastal town in the Río Negro Province, as an example, the entire community was successfully led in 1995 by the local *Fundación Inalafquen* in its opposition to the construction of an oil terminal in an ecologically sensitive coastal area.

In the last few years, non-governmental organizations (NGO) have become significant players in the regional environmental arena. The Patagonian Coastal Zone Management Plan, a GEF/UNDP program, together with Fundación Patagonia Natural, an NGO with close ties to the Wildlife Conservation Society, have been successful at instilling an environmental consciousness in the region. The ecotourism industry has recently created its own NGO, *Ecovaldés*.

The Role of Universities

We believe that regional universities should play a major role in promoting sound resource use and management. This role is currently not fulfilled by the universities in Patagonia, because academic and research programs are largely detached from regional problems in resource management. Much attention has been paid to the study of natural history and basic ecology, while little consideration is given to human interventions, to the dynamic responses of intervened systems, and to the decision-making process. Basic biology programs proliferate everywhere, while there are no specialized programs with emphasis in conservation and resource management. We believe that this bias has multiple roots:

(1) Although Argentina is by its constitution a federation of provinces modeled after the United States, the system of public universities does not match this model. Public universities are all regulated by federal laws and funding is channeled through the federal administration.

(2) Following the *Reforma* of 1918 (Del Mazo 1956; Benjamin 1965), Argentinean universities have maintained a long tradition of self-government (*autonoma*). The complete lack of mechanisms for influencing research priorities implies that the provinces and local communities have virtually no input into higher education policy.

(3) In the late 1960s, a restructuring of the Argentinean research system led to the creation of research centers within the National Council for Scientific and Technical Research (best known through its Spanish acronym CONICET), divorced from the universities. In this model, copied from continental Europe, small universities

found it difficult to compete for federal research funds, which were primarily funneled to CONICET centers and major metropolitan universities.

(4) The absence of advisory councils or other forums for consultation about management decisions concerning natural resources implies that the universities and research centers have no natural means to participate in the decision process. This lack of proper mechanisms has further discouraged the development of resource-oriented research. In cases where such studies are conducted, their results often remain as matters of exclusive academic interest, going unnoticed to those making management decisions.

In order to redefine the role of universities, change should be promoted in several fronts: teaching, research, and interactions with the public, government, and private sectors. The most reasonable strategy for professional training in resource management and conservation in Patagonia is to capitalize on existing undergraduate programs and generate opportunities for specialized training at the graduate level. Graduate programs should be structured to teach students to deal with real and concrete management situations, and to generate management options through the integration of ecological knowledge and a solid training in mathematical modeling and statistics.

If science is to have an impact on management decisions, effective communication channels between the different sectors involved need to be established. One way to foster this is by channeling financial and political support through partnerships that involve scientists and provincial agencies in university-led research projects. Not only should universities generate technical expertise for resource management, they should also educate the general public, promote community involvement in resource management issues, and train future resource administrators. By creating communication channels, universities can help advance the dialogue among public, private sector, and resource administrators. Graduate programs generate further opportunities for the universities to maintain an active exchange with provincial and federal fisheries, wildlife, and tourism administrations, as well as private companies and cooperatives involved in the exploitation and conservation of natural resources.

Research and the Role of Science in Resource Management

It is widely appreciated that good science is a prerequisite for sound conservation and resource management. Yet it is not necessarily obvious what kind of science best fulfills that need (Levin 1993). Ultimately, the nature of the science taught and transferred by a university reflects the nature of the research it conducts. We believe that research directed at the optimal use and conservation of resources needs to recognize:

(1) the importance of studying the dynamics of natural systems and their response to human intervention at the appropriate scales;
(2) that decisions will have to be made in the face of great uncertainty;
(3) the experimental value of human interventions; and
(4) the need to consider possible system responses in terms of quantifiable criteria that reflect management goals and the multiple, possibly conflicting interests of the different sectors affected.

Alternative management actions need to be evaluated on the basis of a formal specification of the possible responses of the managed system to different forms of human intervention. The traditional prescription for sound resource conservation and management offered by scientists is deeper knowledge (Ludwig, Hilborn, and Walters 1993). It is optimistically believed that rigorous experimental and sampling protocols, adequate funding, and sustained study will ultimately solve major biological uncertainties, allowing scientists to predict the response of the systems of interest to various perturbations. Yet it has become increasingly apparent that ecology's ability to predict at the appropriate temporal and spatial scales is rather limited, and generally goes untested because systems cannot be experimentally perturbed at those scales. This is the case even for systems that have been extensively studied (Deriso, Hoag, and McCaughran 1986). Predictive ability is further limited when information about the dynamics of natural systems is scarce, as is usually the case in Patagonia. The quest for ever-deeper knowledge often becomes a program for inaction rather than a program for action, with the typical recommendation being to delay decision until further knowledge is acquired. The drawbacks are that (a) under strong societal pressures for development, decisions will be made anyway, and these decisions are likely to be short-sighted and lead to unsustainable use of resources, particularly in the absence of strong institutions; (b) the dynamics of natural systems cannot be understood unless the system has been manipulated to produce contrasting situations; and (c) opportunities for learning by implementing a management program that includes monitoring the system and correcting management actions in response will be missed.

When knowledge about the behavior of the intervened system is scarce, emphasis should be in the implementation of management plans that assure close monitoring of system dynamics and quick and effective management response mechanisms. The ability to react rapidly should takes precedence over the ability to predict accurately. This marks a major distinction between basic science and management, and one that is not fully appreciated by scientists brought to the managerial arena from a conventional academic background. Scientists have devoted much attention to natural phenomena, but little to the flexible procedures and mechanisms that need to be in place for policy to be adjusted to new data or insights.

In this *adaptive management* approach (Walters 1986), every intervention of a natural ecosystem is conceived as an experiment—the response of the system to the intervention is informative about its dynamics, and it must be monitored so that future management can be corrected accordingly. With rare exceptions, anthropogenic interventions in Patagonian ecosystems are poorly documented. Given the expanse of the region, it should be possible to devise management protocols that are informative over relatively short time horizons. A given region, for example, can be partitioned into a number of subregions that are subjected to contrasting intervention schedules. The opportunities for experimental management are widespread, as best exemplified by manipulating grazing in natural pastures, cuttings in forests, hunting in different areas, and fishing effort in coastal fisheries or river basins.

FROM RISK TO OPPORTUNITY

Remote Patagonia is no longer a pristine wilderness. The threats to natural resources are essentially similar to those encountered in other parts of the world—i.e., pollution, overharvesting, overgrazing, species introductions, and so on. Prospects for the sustainable use of these resources are precarious. Institutional frameworks for management are still taking shape and societal forces are still in search for mechanisms to express their agendas and influence the political process.

But we argue that conditions for progressive change in Patagonia are now ripe. First, incipient societal structures entail an opportunity—i.e., the inception of a program free of the institutional inertia of older societies. Patagonian provinces have relatively small, young, and energetic government structures. Second, the return of Argentina to democratic rule in 1983 has restored considerable leverage to the provinces regarding the establishment of policies for the conservation and management of their natural resources. Third, the public is increasingly concerned about the environment, conservation, and natural resources. The responsiveness demonstrated by the local resident community in a few test cases is encouraging. But what should be the appropriate tactic to capitalize on this opportunity?

Historically, most efforts directed towards conservation and resource management in developing countries have targeted specific resources or environmental problems with the goal of providing advice to the agencies in charge of those resources. Although potentially valuable to address short-term issues, this strategy may be insufficient to promote longer-term programs aimed at broader conservation and management agendas. In addition, the design of sound management strategies also requires a clear understanding of distinctive societal, political, and institutional realities. Ignoring the social component of conservation/management problems has often resulted in disastrous policies. Resource management needs to be based on flexible protocols that accommodate not only these social considerations, but also the process of learning from the response of natural systems to human interventions. Endowing agency personnel with technical skills is essential for advancing the goals of resource management, but concentrating training efforts in particular administrative structures is likely to have a restricted impact. In any case, training biologists and administrators addresses only part of typical resource problems. Sensible development programs should also establish mechanisms to promote and accommodate community involvement (see chapter 18).

We claim that the best strategy to advance long-term resource management and conservation in Patagonia is to establish specific programs at local universities directed at training the new scientists and managers, upgrading the background of those that are now active, and facilitating the interaction between research, government agencies, the industry, and the public. Universities have the greatest potential to influence resource conservation and management in the region. While located at the heart of communities, they are sheltered from the political sways by their statutory autonomy. In order to have an impact, universities should not only offer technical training, but also reach out to local governments and community. The interaction with other sectors of the society should be a central component of academic life, taking place in the form of workshops, seminars, extension activities, and through the consolidation of the research and advisory role of universities.

Much has been made of the lack of knowledge regarding basic biological questions pertinent to conservation. We believe that even though there is a great deal we do not know, useful knowledge and insight is readily attainable, albeit with considerable uncertainty. Much discussion has also questioned whether sustainable development is a useful ideal or a rationalization for human self-interest. We believe sustainable development is attainable—just like the required biological knowledge. Our experiences in Patagonia indicate that major limiting factors are sustained and creative input from universities, community oversight, and frameworks for interaction among the various sectors of society. If conservation and development are to be accommodated we need to tap into university scientific expertise, but to do so in a way that escapes sterile academics, and in a way that involves and serves the community at large.

ACKNOWLEDGMENTS

We thank Peter Kareiva, Claudio Chehebar, Pablo Yorio, Dee Boersma, Tony Gill, Carmen Ubeda, Joe Norman, Marcela Alvarez, and Ellen Gryj for their comments on an earlier draft. Patricia Dell'Arciprette assisted us with preparing the figure. We are specially grateful to Peter Kareiva for his enthusiasm and encouragement during our travel in Patagonia, which led to this paper.

LITERATURE CITED

Achurra, M. 1992. Gran impulso a la salmonicultura chilena. *Chile Pesquero* 71:21–25.

Allen, G.M. 1942. Extinct and vanishing mammals of the Western Hemisphere with the marine species of all oceans. *Special publication of the American commission for international wildlife protection* 11.

Baign, C.R.M. and R. Quirós. 1985. Introducción de peces exóticos en la República Argentina. Informe técnico del Departamento de Aguas Continentales no. 2, INIDEP, Mar del Plata, Argentina.

Bellati, J. and J. von Thungen. 1988. Mortalidad de corderos de hasta dos meses de edad en el oeste de la provincia de Río Negro. *Revista de agronomía y producción animal* 8:359–363.

Benjamin, H.R.W. 1965. *Higher education in the American republics.* New York: McGraw-Hill Book Co.

Bisbal, G.A. 1995. The Southeast South American shelf large marine ecosystem. *Marine Policy* 19:21–38.

Boneto, A.A. and I.R. Wais. 1995. Southern South America streams and rivers. In *River and stream ecosystems*, eds. C.E. Cushing, K.W. Cummins, and G.W. Minshall, 257–293. Amsterdam: Elsevier.

Bryan, A.L., ed. 1986. *New evidence for the Pleistocene peopling of the Americas.* Orono, Maine: Center for the Study of Early Man.

Cabrera, A.L. and A. Willink. 1973. Biogeografía de América Latina. *OEA, Monografía* 13. Serie Biología.

Cabrera, A.L. and J. Yepes. 1940. *Mamíferos Sudamericanos.* Buenos Aires.

CCAMLR (Committee for the Conservation of Antarctic Marine Living Resources). 1992. *Report of the eleventh meeting of the Scientific Committee*, Hobart, Australia, 26–30 October 1992.

Cei, J.M. 1980. Amphibians of Argentina. N.S. Monografia 2, *Monitore Zoologico Italiano*, Firenze.

Cei, J.M. 1986. Reptiles de centro, centro-oeste y sur de la Argentina. Herpetofauna de las zonas áridas y semiáridas. *Mografie IV, Museo Regionale di Science Naturali*, Torino.

Ciocco, N.F. 1995a. Primeras experiencias privadas de cultivo de bivalvos (mejillones) en los golfos San José y Nuevo (Chubut, Argentina): temporadas 93/94 y 94/95. CENPAT (Puerto Madryn). *Informes técnicos del Plan de Manejo Integrado de la Zona Costera Patagónica* (in press).

Ciocco, N.F. 1995b. La marisquería mediante buceo en el Golfo San José (Chubut, Argentina). CENPAT (Puerto Madryn). *Informes técnicos del Plan de Manejo Integrado de la Zona Costera Patagónica* (in press).

Del Mazo, G. 1956. *Estudiantes y gobierno universitario.* Buenos Aires: Librera El Ateneo.

Deriso, R.B., S.H. Hoag, and D.A. McCaughran. 1986. Two hypotheses about factors controlling production of Pacific halibut. *International North Pacific Fisheries Commission Bulletin* 47:167–173.

Franklin, W.L. 1982. Biology, ecology and relationship to man of the South American camelids. In *Mammalian biology in South America*, 457–489. Pymatuning Laboratory of Ecology and University of Pittsburg, Linesville.

Franklin, W.L. and M.A. Fritz.1991. Sustained harvesting of the Patagonia guanaco: is it possible or too late? In *Neotropical wildlife use and conservation*, eds. J.G. Robinson and K.H. Redford, 317–336. Chicago: The University of Chicago Press.

Garrido, J.L., R. Mazzanti, and D.A. Garrido. 1988, unpubl. Distribución y densidades de guanaco en la Patagonia Argentina. Informe presentado a la Dirección de Fauna de la Provincia de Río Negro.

Grosman, F. 1992. Algunos aspectos de la biología del "salmón del Pacifico" (*Oncorhynchus tshawytscha*) presente en la Provincia del Chubut. Informe técnico. Centro de Ecología Aplicada del Neuquén and Japan International Cooperation Agency, Junín de los Andes, Neuquén, Argentina.

Kline, T.C., J.J. Goering, O.A. Mathisen, P.H. Poe, and R.S. Scalan. 1993. Recycling of elements transported upstream by runs of Pacific salmon: II. δ15N and δ13C evidence in Kvichak River Watershed, Southwestern Alaska. *Canadian Journal of Fisheries and Aquatic Sciences* 50:2350–2365.

Krueger, C.C. and B. May. 1991. Ecological and genetic effects of salmonid introductions in North America. *Canadian Journal of Fisheries and Aquatic Sciences* 48 (Suppl. 1):2238–2246.

Leitch, W.C. 1991. *Argentine trout fishing: A fly fisherman's guide to Patagonia.* Portland Oregon: Frank Amato Publications.

Levin, S.A., ed. 1993. Forum: Science and sustainability. *Ecological Applications* 3:545–587.

Lichter, A.A. 1992. *Huellas en la Arena, Sombras en el Mar. Los Mamíferos Marinos de la Argentina y la Antártida.* Terra Nova Eds.

Ludwig, D., R. Hilborn, and C. Walters. 1993. Uncertainty, resource exploitation, and conservation: Lessons from history. *Science* 260:17.

Monjeau, A., N. Bonino, and S.L. Saba. 1994. Annotated checklist of the living land mammals in Patagonia. *Mastozoologa Neotropical* 1:143–156.

Narosky, T. and D. Zurieta. 1993. *Birds of Argentina and Uruguay: A field guide.* Buenos Aires: Vazquez Mazzini Editores.

Novaro, A.L. and M.C. Funes. 1993. Monitoreo anual de las poblaciones de carnívoros en la Patagonia. Informe Regional.

Ojeda R.A. and M.A. Mares. 1982. Conservation of South American Mammals: Argentina as a paradigm. In *Mammalian biology in South America*, eds. M.A. Mares and H.H. Genoways, 505–521. Pymatuning Symp. Ecol., vol. 6. Pittsburgh, PA: University of Pittsburgh.

Oporto, N. and N. Soto. 1995. Aspectos de aprovechamiento de la especie. In *Técnicas para el Manejo del Guanaco*, ed. S. Puig, Chapter 9. IUCN. Report of the Group of Specialists in South American Camelids.

Orensanz, J.M. 1986. Size, environment and density: The regulation of a scallop stock and its management implications. *Canadian Special Publications in Fisheries and Aquatic Sciences* 92:195–227.

Orensanz, J.M., M. Pascual, and M. Fernandez. 1992. Fisheries and aquaculture: Argentina. In *Scallops: biology, ecology and aquaculture*, ed. S.E. Shumway, 981–1000. Developments in Aquaculture and Fisheries 21. Amsterdam: Elsevier.

Patterson, B. and R. Pascual. The fossil mammal fauna of South America. *Quarterly Review of Biology* 43:409–451.

Payne, R. 1995. *Among whales.* New York: Scribner Publishing.

Puig, S. 1992. *South American camelids. An action plan for their conservation.* Cambridge: IUCN Publications Services Unit.

Pujalte, J.C. and A. Reca.1985. Vicuñas y guanacos, distribución y ambientes. In *Estado actual de las investigaciones sobre camélidos en la República Argentina*, eds. J.L. Cajal and J. Amaya, 25–49. Buenos Aires: Secretaría de Ciencia y Técnica.

Raedeke, K.J. 1979. Population dynamics and socioecology of the guanaco (*Lama guanicoe*) of Magellanes, Chile. Ph.D. dissertation, University of Washington, Seattle.

Ragonese, A.E. 1968. *Vegetación y ganadería en la República Argentina.* Buenos Aires: INTA.

Ravinovich, J., A. Capurro, P. Folgarait, T. Kirtzberger, A. Kramer, A. Novaro, M. Puppo, and A. Travaini. 1987. Estado del conocimiento de 12 especies de la fauna argentina de valor comercial. *Report of the second Argentinean meeting on economically valuable wildlife species*, Buenos Aires.

Reig, O.A. 1981. Teoría del origen y desarrollo de la fauna de mamíferos de América del Sur. *Monografiae Naturae*, (Museo Municipal de Ciencias Naturales Lorenzo Scaglia) 1:1–162.

Ringuelet, R.A. 1955. Vinculaciones faunísticas de la zona boscosa del Nahuel Huapi y el Dominio zoogeográfico Australcordillerano. *Notas Museo ciudad Eva Perán*, Zool. 160:81–121.

Ringuelet, R.A., R.H. Arámburu, and A. Alonso. 1967. *Los peces Argentinos de agua dulce*. Comisión de Investigaciones Científicas de la Provincia de Buenos Aires, Buenos Aires, Argentina.

SAGyP (Secretara de Agricultura, Ganadera y Pesca). 1995. Guía Pesquera Argentina. Second Edition, Buenos Aires: Masindian Cons.

Simpson, G.G. 1934. *Attending marvels: A Patagonian journal*. Chicago: The University of Chicago Press.

Simpson, G.G. 1950. History of the fauna of Latin America. *American Scientist* 38:261–389.

Simpson, G.G. 1969. South American mammals. In *Biogeography and ecology of South America*, eds. E.J. Fittkau, J. Illies, H. Klinge, G.H. Schwalbe, and H. Sioli, 879–909. The Hague: W. Junk, N.V. Publ.

Simpson, G.G. 1980. *Splendid isolation: The curious history of South American mammals*. New Haven: Yale University Press.

Tarak, A. 1989. A National Perspective. In *Conservation for the twenty-first century*, eds. D. Western and M. Pearl. Oxford: Oxford University Press.

Torres, H. 1992. *South American camelids. An action plan for their conservation*. Cambridge: IUCN Publications Services Unit.

Ubeda, C. and D. Grigera, eds. 1995. *Recalificación del estado de conservación de la fauna silvestre argentina*. Región Patagónica. Secretaría de Recursos y Ambiente Humano -Consejo Asesor Regional Patagónico de la Fauna Silvestre.

Veblen, T.T., R.S. Hill, and J. Read. 1996. *The ecology and biogeography of* Nothofagus *forests*. New Haven: Yale University Press.

Walters, C.J. 1986. *Adaptive management of renewable resources*. New York: Macmillan.

Wegrzyn, D. and S. Ortubay. 1991. *Nuestros salmónidos*. Provincia de Río Negro, Dirección de Pesca, Argentina.

Willson, M.F. and K.C. Halupka. 1995. Anadromous fish as keystone species in vertebrate communities. *Conservation Biology* 9:489–497.

Yorio, P. 1993. Plan de Manejo de la Zona Costera Patagónica- GEF/PNUD: un enfoque integral para la protección de la biodiversidad. In *On common ground: interdisciplinary approaches to biodiversity conservation and land use dynamics in the New World*, 283–291. Universidade Federal Minas Gerais and University of Florida: Conservation International.

CHAPTER

18

Tropical Agroecology and Conservation Ecology: Two Paths Toward Sustainable Development

C. RON CARROLL

ANNE M. DIX

and JAMES S. KETTLER

The amount of land that is abandoned or severely degraded is large and increasing, with especially profound implicatons for conservation in the tropics. In particular, approximately one-half of the area of tropical forest lost each year expands the base of productive agriculture, whereas the other half simply replaces agricultural land that is worn out and abandoned (Houghton 1994). Consequently, if tropical agriculture were sustainable, the "... total agricultural area could continue to grow at current rates while, at the same time, rates of deforestation could be reduced by approximately 50%" (Houghton 1994, p. 311). Clearly, for the tropics at least, sustainable agriculture is a necessary precondition for the conservation of biodiversity (see also Hoffman and Carroll 1995).

Prior to the European conquest of the Americas in the 16th and early 17th centuries, some regions of the Neotropics already were densely settled. For example, centuries before European contact great commercial cities and large rural populations lived in the inter-Andean valleys of South America, parts of lowland Mesoamerica, and in much of southern Mexico and northern Guatemala. These early populations took their toll on the natural environment. None the less, despite dense pre-Columbian settlement patterns, most of the lowland neotropical landscape was characterized by intact forest ecosystems at the time of European contact.

Upon European contact and the rise of export agriculture in the late 19th century, there was a dramatic increase in the agricultural land base at the expense of forest lands. Transformation of tropical forested landscapes to their current patterns of agricultural landuse took place in less than four human generations, and most in less than one generation. Thus, over a relatively short period of time, the social and ecological landscape was profoundly altered, with old civilizations disappearing, new civilizations spreading rapidly, international markets demanding ever greater exploitation of raw materials, and ecological degradation occurring at unprecedented levels.

Since the mid-1970s, ecologists have expressed strong concern over the erosion of biodiversity as a consequence of the few remaining tropical forests being converted to agriculture, plantations, or simply cut down. At the same time, a small but growing group of developmental economists, anthropologists, and other social scientists have raised serious questions about social equity and the long-term economic viability of existing patterns of extractive tropical landuse. We refer to these social and ecological concerns collectively as the "tropical land use crisis."

Out of the mileau of social and ecological concerns over tropical landuse, the concept of "sustainable development" has arisen to seemingly provide a general solution. However, for sustainable development to solve the tropical landuse crisis its implementation must include four essential components: (1) open access by the participants to decision-making; (2) acceptible levels of economic return; (3) environmental protection; and (4) an emphasis of economic quality over economic growth. If these essential components are not included, the call for "sustainable development" is simply a shallow "mantra" disguising "business as usual." Although we focus on the Neotropics, it should be clear that our argument applies to virtually any geographic region. We could as easily be discussing problems in achieving sustainable development among rural communities of the San Joaquin Valley in California, pueblos in New Mexico, poor counties of central Georgia (USA), villagers in central Kenya, or vegetable farmers in central France, although the details would differ. As bleak as the neotropical land use crisis might appear, there is hope, and the role for academic scientists in this process is exciting. Countries are responding by creating, and more importantly, protecting natural areas and indigenous reserves. Alternative, less chemically intensive agricultural practices are being encouraged. Governmental policies that have historically acted as disincentives to environmentally sound landuse practices are, in some places, becoming more progressive. Academics from the northern countries are forming strong collaborations with their counterparts and stewardship organizations in tropical countries.

THE ESSENTIAL COMPONENTS

Much of the literature on sustainable develoment has emphasized the important preconditions, or necessary antecedents, to achieving sustainability. We believe that the many preconditions or antecedents that are identified in this literature can be subsumed under the following four components of the development process.

First, all stakeholders (local farmers, researchers, tradespeople, public officials, etc.) in the development process should have accessable ways to participate in the decision-making process. Although participation by stakeholders in decision-making should occur in all phases and aspects of the development process, we are limiting this brief discussion to participation in the development of the reseach agenda. Note that this does not imply that every individual will have the experience, standing, or desire to contribute equally to all aspects of the research agenda. Sustainable development is an extraordinarily complex process that requires rigorous analytical information from academic scientists as well as information that derives from the local experience of farmers. Local beliefs may, of course, be incorrect; but, neither is academic information always correct. The important point is that a research agenda in which local stakeholders have input has the best chance of leading to a sound policy for sustainable development.

There is another reason why stakeholders should have access to the decision-making process. People are more likely to accept restrictions to their personal license if they have the opportunity to contribute in a meaningful way to the decision-making process. In the two case studies, we emphasize the value of wedding local or indigenous knowledge with analytical-mechanistic science.

Second, economic returns from the development process must meet local needs and goals. However, economic development that is environmentally sustainable will generally not maximize short-term rates of economic return. Thus, one of the major social barriers to sustainable development is the understandable difficulty of impoverished peoples to take long-term economic perspectives (see chapter 16). The history of economic development in the tropics is full of examples where high rates of economic return in the short-term have eroded the environmental basis for sustainability. On the other hand, environmentally sustainable projects often offer meager rates of economic return. Furthermore, sustainable development is made more vulnerable when the market demand is large, relative to the pool of rural producers, and when markets are distant and insulated from the needs of the producers. One reason this occurs is because large market demand often leads to economic specialization in the rural economy (e.g., a landscape of sugarcane fields) and less diversity. Rural producers may then become overly dependent upon urban and foreign markets that keep their supply costs low by forcing producers to compete among themselves. Such highly specialized markets can be especially vulnerable to cycles in supply and demand. For example, when periodic frosts damage the large southern Brazilian coffee or citrus harvests, world market prices rise and coffee and citrus producers in Central America benefit. Conversely, Central American producers suffer when Brazilian harvests are abundant.

In our two case studies, we emphasize how modifying traditional practices to achieve better environmental sustainability can be accomplished in the context of meeting local economic and social needs. We also show that achieving sustainable development through the production of a few export crops is problematic.

Third, the environment must not be degraded by the development process. The most apparent examples of failure to protect the environment are found among the many case studies of hydroelectric development projects. There are few cases in Latin America where a project to build a dam for hydroelectrical power generation has been tied adequately to protection of watersheds. Often when such projects are initiated, however, speculators move into the area and establish extensive cattle ranches, usually with accompanying deforestation. The environmental consequences are that erosion in the watershed increases, sedimentation of the reservoir accelerates, and the economic lifetime of the project often dramatically decreases. On the positive side, the need to protect watersheds to extend the life of hydroelectric facilities is now better understood.

Fourth, the "sustainability" of the development process must be achieved by improving the *quality* of development as opposed to accelerating the *rate* of economic development. Indeed, a long-term goal for a nation's sustainable development agenda ought to be to increase per-capita quality of life while not encouraging unlimited growth of per capita gross national product. This implies a zero population growth rate *at some point* because per-capita reductions in consumption will inevitably otherwise be offset by aggregate consumption from an expanding population.

Generally, sustainable development is more likely to be achieved when local economies are diverse, adaptive, and not overly dependent on single external markets. However, a farm family that wishes to diversify its agriculture must deal with three pervasive constraining forces: (a) poorly developed markets for small farm production, (b) limited information about environmentally sound production practices, and (c) restricted access to credit. Ecologists can help ease the latter two of these constraints and thereby help progressive farmers move towards sustainability. Environmentally sound farming substitutes management of agroecosystems for excessive reliance on commercial fertilizers, pesticides, and tillage (for example, see Carroll, Vandemeer, and Rossett 1990). Agricultural credit lenders survive when loan principals and especially their accumulated interest, are paid back by farmers. Because farming is inherently a risky operation with small profit margins, agricultural credit lenders tend to be conservative in their lending practices, preferring to deal with large farms that use well-established production practices. Credit lenders are therefore seldom agents of agricultural innovation. By focusing research goals towards finding environmentally sound production practices that provide good and stable economic return, ecologists can address the risk-adverse behavior of credit lenders.

We illustrate our four essential components of sustainable development through two case studies. The first focuses on the development of pest management strategies in a traditional Guatemalan region that is beginning to produce a non-traditional export crop. The second case study describes a modified traditional slash-mulch production system in Costa Rica. In our two case studies we show how farming risk may be reduced, access to employment and new revenue improved, and exposure to pesticides reduced. We do not mean to imply that these projects are perfect examples of sustainable development, but they contain elements of the essential components and provide a useful backdrop for discussion.

CASE STUDIES

Case Study One. *Broccoli: A Traditional Community Deals with a Non-Traditional Export Crop*

Over the past decade, tropical countries have placed great emphasis on the promise of non-traditional export agriculture, including mariculture, to generate new sources of income. A non-traditional agricultural export is simply any product that has not been previously exploited for export. Some examples for Latin America include broccoli, snow peas, berries, melons, cut flowers, and mariculture shrimp production. Foreign assistance programs have encouraged this emphasis on new export products. For example, in Honduras, US AID's Honduras Export Development and Services Project has promoted melons and mariculture shrimp for export since the mid-1980s.

Non-traditional export crops often require blemish-free products, a practice which strains production practices. For example, farmers use pesticides to achieve blemish-free harvests, yet pesticide residues found during random tests of shipments will trigger rejection of the entire shipment. For large commercial growers there is a meticulous paper trail from the shipment to the grower, hence the reasons for rejecting a shipment during a border check can usually be traced back to the practices of a particular large producer. The high level of accountability for their producton practices, as well as their access to information and financial capital, have encouraged large commercial producers to find successful production practices that minimize the use of pesticides. For small producers, no such paper trail exists, hence, they have low accountability for high pesticide residues in their produce. Furthermore, knowledge about the risks associated with the use of pesticides is minimal. Their main concern is to grow crops that are attractive to local buyers. Hence, small growers have a financial incentive to apply pesticides and lack the knowledge and financial capital to find ways to minimize the use of pesticides.

In Guatemala one of the more significant non-traditional crops is broccoli, sold mostly to the U.S. market. The introduction of this new crop throughout the highlands has led to the adoption of intensive agricultural practices, where traditional agriculture was formerly the norm. In the past ten years the total area in Guatemala cultivated for broccoli has more than quadrupled (GEXPRONT 1994).

Broccoli production has given many Guatemalan farmers access to large markets and higher incomes. With greater income, adequate food is now available to families throughout the year and people have less need to migrate seasonally for work. This is the case with farmers in our project area in Chilasco, Baja Verapaz, a community that is in the buffer zone of the Sierra de las Minas Biosphere Reserve, Guatemala. The Sierra de las Minas Biosphere Reserve is used by the community as a source of bamboo vines for woven baskets and tree ferns whose trunks are used to make decorative flower pots. These minor cottage industry products are sold to wholesale buyers. In Chilasco, broccoli is still largely viewed as the miracle crop that has allowed them to buy shoes for their children and provide health care for their families. However, change has involved tradeoffs in terms of family health and environmental quality. In some cases children's education may be affected because broccoli production requires continual field maintenance and children often leave school to tend the fields. People are exposed to far more agricultural chemicals, especially pesticides, than ever before, and the exposure risk is especially great among women and children. Men use backpack sprayers to apply pesticides and microbial pathogens of crop pests. Some safety training is available through governmental programs to the men engaged in pesticide spraying to control foliar pests. However, while personal safety of the men may be somewhat provided, there are still environmental concerns. For example, backpack sprayers are typically cleaned in streams.

For the control of pests that damage root systems of broccoli, the situation is quite different. When the root systems are damaged at low levels, the weight of the broccoli heads are reduced, while at higher damage levels the entire plant dies. Farmers respond to this risk of root damage by dipping the roots of seedling broccoli plants in highly toxic pesticides (e.g., organophosphate terbufos) prior to transplanting. The dipping is typically done by women and children without protective gloves and without access to water in the field for washing.

> Although in the short-term economic levels have risen, the longer-term future is more problematic. Consider this perception by a broccoli farmer in Patzun, Chimaltenango, where broccoli has been produced since 1978. "We now live in worse conditions than we did before. We are now dependent on external inputs, land prices have increased, labor costs have increased, we now work more and the real prices paid for our products are less. We are so preocuppied with caring for our product, that we sacrifice even Sundays. We have no time to get together and worry about our neighbors. We have become individualistic and don't worry about others. Five per cent of the farmers became richer and ninety-five per cent of the farmers became poorer after the introduction of broccoli" (Mucia 1994).

New agricultural practices have also led to major secondary pest problems on broccoli, principally root-damaging beetle grubs (*Phylophaga*, *Anomala*, and *Cyclocephala* spp., Scarabaeidae) that are commonly found on the roots of wild plants, especially grasses as well as on many crop plants. Because broccoli is an important crop, and is grown at considerable risk, farmers are particularly interested in finding a solution to the grub problem. Farmers estimate that they spend 75% of their income from the harvest on inputs, and can make a 25% profit if the harvest goes well. However, when things go poorly, as often happens with infestation by beetle grubs, farmers lose up to 25% of their harvest before it is ready to sell. Coupled with the cosmetic problems that come with the presence of other foliar pests such as caterpillars and aphids, farmers have a strong incentive to apply the pesticides in a preventative manner.

In addition, the use of large amounts of chicken manure as a fertilizer has resulted in major housefly infestations, which the farmers claim were not a problem before the introduction of broccoli. Because there is a residual fertilizer effect from broccoli production, farmers plant four-month maize, a newly introduced variety, between broccoli cycles and the traditional nine-month maize is now planted less frequently and is relegated to marginal fields. As a result, stored maize is present in large amounts year round. Farmers claim that they lose larger amounts of stored maize to moth larvae and rats than previously when maize was only seasonally abundant. Consequently farmers also apply highly toxic pesticides in their houses to control flies and maize pests. Thus, the highland farmers of Guatemala have a change in agricultural practices that is coupled with a surge of new problems (Dix 1995; Dix and Carroll 1995).

Farmers are faced with the conflict between what outsiders recommend they do and practices they feel are necessary due to their experiences with the problem. For example, in the case of white grubs, farmers have observed that the pest seems to be most abundant where chicken manure has been used as a fertilizer and where there is buried maize stalks and roots (stover). We used this farmer-based information as a starting point for a research agenda, which could be later used to develop new technologies, such as using areas of high organic matter as traps for pests in the fields.

In the first investigation, we sampled random sites in fields that had just been planted to broccoli. Our primary question was, "What is the relationship between the distribution of beetle larvae and the distribution of buried maize stover?" Because the buried stover was very patchily distributed, it was possible to make comparisons between sites with and without buried stover. We found that the larvae were highly associated with stover and that this association was especially strong for first instar larvae. Three broccoli fields were sampled. Out of 41 soil samples (each of 31.8 liters) with stover and 41 soil samples without stover, we found a total of 360 first instar larvae, 94% were associated with stover (Dix *et al.* 1995). This finding suggested that the ovipositing females might seek clumps of stover in which to lay their eggs.

The next questions were, "Do the larvae actively seek maize stover?" and "What order of preference do they show among stover, maize roots, broccoli, chicken manure, and composted chicken manure?" Larvae were placed in the center of dirt-filled boxes and given a choice of four directions to move. Two sides were not planted and served as controls. The other two sides were planted in combinations of the above five buried materials. As an additional control, some boxes contained only soil and no other additional buried material. Experimental results confirmed what we found in the field, i.e., larvae were strongly attracted to maize stover and to chicken manure and less attracted to the other materials. We are continuing to investigate the ecological mechanisms that influence the patchy distribution of larvae observed in the field, but we now have the elements for a future multiple-objective experiment in biological control. Instead of leaving maize stover buried in clumps, farmers will bury the stover in rows perpendicular to the slope of the field. This will act as a check against soil erosion and will position the attractive zones for the beetles. Some of the stover rows will be inoculated with *Beauvaria*, a fungal pathogen of beetle larvae that can be cultured on maize stover; others will have no additional treatment; and still others will be treated with either a mild pesticide or with another microbial pesticide as yet to be determined. This future experiment will build upon farmers' knowledge and address several of their needs: lowering pesticide use, providing the farmers with their own means for controlling these pests, and providing a check against soil erosion on steep slopes.

Participatory research with farmer input has helped validate the farmers' hypotheses and has led to the development of a mechanistic model to explain particular pest problems. Armed with this knowledge, the farmers can now make sound decisions in the face of conflicting information, and have learned some tools they can use in solving future problems. By coupling scientists with local people, the research not only becomes much more responsive to local needs, but also gives both parties access to information that might not otherwise be taken into account.

Importantly, farmers were involved in the research project during its inception and information exchange was invaluable in order to develop technologies that were appropriate and tto leave within the community the capacity to do further problem-solving at the end of the project. Information about pest life cycles helped farmers make the connection between the adult beetles and the grubs that were infesting their fields. Farmers observations that there was a relationship between grubs and buried maize stover and between the incidence of grubs and the use of chicken manure as fertilizer proved to be crucial for arriving at an understanding of the mechanism to explain why some fields were more infested than others. Scientists were able to help farmers prove what they already believed to be true, but more importantly, to take these observations a step further in search of a solution for the pest problem.

Case Study Two. *Modifying a Traditional Slash-Mulch System: Sustainable Land Use for Fragile Lands*

Slash-mulch agriculture in the Americas predates European contact (Patiño 1965) and continues to be widespread throughout the tropical parts of North, Meso, and South America (Thurston 1994). The purpose of the three-year research reported here was to analyze the production potential of a traditional Mesoamerican slash-mulch system known as "*frijol tapado*" (literally, covered bean) and to investigate modifications that might build upon the positive attributes of the traditional system. The research site was located in the southern highlands of Costa Rica near the town of San Vito in the canton of *Coto Brus*.

The potential production problems faced by resource-limited farmers that use fragile lands are the focus of this case study (for details see Kettler 1995a, 1995b, 1996). The *frijol tapado* form of slash-mulch consists of farmers selecting the appropriate field site, using vegetation, slope, and aspect as the selection criteria. Passageways are chopped through the second-growth vegetation to create access to the land for sowing of the seed. Bean seeds, usually an indeterminate bush variety of *Phaseolus vulgaris*, are broadcast into the standing vegetation at densities of 180,000 to 400,000 seeds per hectare. Second growth vegetation is then chopped to form a more or less compacted mulch layer through which the bean seedlings emerge. The fields are left untouched until harvest (about 12 weeks later) when the bean plants are pulled, dried in the field, and thrashed. The land is left fallow for approximately eight months and the agricultural cycle repeated for another two to three years.

The advantages of slash-mulch agriculture include the large input of organic matter to the soil, reduced erosion, facilitation of nutrient cycling, absence of inorganic fertilizers, weed suppression, and in dry periods, maintenance of soil moisture (Bunch 1994). Disadvantages include susceptibility to certain pests (primarily slugs), loss of production during the fallow periods, and the need to allow for a much longer fallow period of two to three years after three to four years of production. Recently, increasing pressure for land has obligated many bean farmers to attempt shorter rotations or even continuous production, thus leading to general soil degradation (Bellows 1992).

Fragile lands, here defined as lands that are highly susceptible to large rates of soil erosion, are characteristic of most sloping lands in the humid tropics. In Costa Rica over 17% of the country's soils are severely eroded and 24% are moderately eroded (Tropical Science Center 1991). Agricultural policies in many tropical countries are often part of the complex socio-economic web driving the cycle of degradation on fragile lands (Bojo 1991; Carriere 1991). These policies often emphasize export agriculture for economic development and to service external debts (Duncan 1991). The outcome is that highly capitalized export-focused farms expand and displace small farmers on to less productive, fragile lands that are far more susceptible to erosion following deforestation. In addition, small farmers have the important economic function of being the primary food producers and their decline has negatively affected domestic food production. To compensate, many tropical countries have been increasing food imports, primarily from the United States, and subsequently increasing their already large external debt that perpetuates the cycle of degradation (Garst and Barry 1990).

Traditional forms of crop production are often said to be inferior to modern, more high-input, productions systems. While this may often be true, at least for short-term comparisons, it does not appear to be the case with *frijol tapado*. When *frijol tapado* systems are compared with modern high-input systems of bean production on equivalent land, there are no significant differences in yield (Table 18.1). The lower values reported in the literature for *frijol tapado* are a consequence of traditional farmers being displaced to more marginal lands or of farmers choosing to use marginal lands for bean production while keeping more productive lands for higher income generating crops such as coffee. Thus, country-wide production values compare yields of high-input systems on high quality land with the yields of *frijol tapado* systems on generally lower quality land. However, when yields on a particular field are averaged over a long time period, e.g., ten years, yields of the *frijol tapado* system, with its need for several years of fallow following several production years will be much lower than yields of high-input systems.

The principal limiting factor of production for *frijol tapado* systems and for other slash-mulch systems is the need to leave fields fallow. Thus, one critical research question involves exploration of ways to enhance the nutrient recovery processes during the fallow period, with the goal of reducing the length of the fallow period and thereby lowering the rate at which forested lands are cleared for production. To address this need, we studied particular combinations of nitrogen-fixing tree species as an addition to the

Table 18.1
A comparison between modern high-input and traditional low-input bean production in Costa Rica (Source: Kettler 1995).

	Modern*	Traditional
Time/Unit Area (min/m^2)	8.56	1.60
Cost/Unit Area (colones/m^2)	41.89	7.59
Time/Unit Yield (min/kg dm beans)	60.57	11.62
Cost/Unit Yield (colones/kg dm beans)	296.28	55.08
Average Yield (kg/ha)	1444	1378

*Fertilized at 32.5 kg N, 97.5 kg P, 32.5 kg K per ha.

traditional production system. Our idea was to use trees that produced material for slash and mulch that would decompose at fast and slow rates, thereby providing nutrient release to the growing bean plants over a longer period of their growth. We found that the combination of fast-growing leguminous trees *Inga edulis* and *Calliandra calothrysus* provided large amounts of available nitrogen and phosphorus, nearly reaching the amounts used elsewhere in the application of synthetic fertilizers in commercial production. Furthermore, the labor and time demands of the improved systems, i.e., yield per unit effort, were superior to that of the modern high-input system (Table 18.1). Therefore, with relatively minor modifications, the traditional slash/mulch system with its low demand for pesticides and greater protection of the soil against erosion is equivalent to the modern high-input system that requires large amounts of pesticides and which provides little protection against erosion.

The traditional *frijol tapado* production system has low maintenance once the field is prepared. This allows the farmer to seek additional wage labor in the commercial coffee plantations in the Costa Rican highlands. In contrast modern high-input bean production requires continuous maintenance throughout the year, thereby denying the farmer the opportunity to work in the coffee plantations. Farmers who elect to use high-input modern bean production practices must rely on their farm for the majority of their income. Consequently, these farmers use frequent applications of pesticide in order to reduce the risk of crop failure. The modern Costa Rican bean farmer gambles on a high-risk, high-gain production strategy with some environmental contamination by pesticides. The traditional bean farmer has lower risk of crop failure and access to off-farm income but suffers income loss during fallow years.

CONCLUSIONS

Contributions to conservation ecology, as a science and as a knowledge base for the protection of biodiversity, are not restricted to studies of natural biota in protected areas. The great demand that agriculture places on the land base indicates that efforts directed towards finding more sustainable ways to do agriculture will also contribute to sustaining a larger natural land base to maintain biodiversity. We believe that it is important to work with the people who have to make a living in the agroecosystem, to collaborate with them to build the appropriate research agenda, and to leave the stakeholders with the tools to solve their own problems.

Participatory research is not easy because a wide cultural gap can separate the western analytical scientist from the tropical small farmer who makes a living off the land. Often, the farmer believes that the scientist is the font of all important knowledge and retreats into an embarassed silence if his/her observations and beliefs are questioned. On the other hand, scientists have sometimes overly romanticized farmers' beliefs. Resource-poor farmers are unlikely to understand the microbial processes in the soil that influence maize yields, but, with few resources available to them they are still good maize farmers. That is, the scientist knows why maize grows; the farmer knows how to grow maize. The interaction between the theoretical/analytical and the local/practical worldview can both advance science and contribute towards achieving sustainable land-use. If conservation ecology embraces agroecology and farmer-back-to-farmer participatory approaches to research, we believe that both the quality of the scientific effort and conservation goals will be strengthened.

LITERATURE CITED

Bellows, B. 1992. Sustainability of steep land bean (*Phaseolus vulgaris* L.) farming in Costa Rica: an agronomic and socio-economic assessment. Ph.D. Dissertation, University of Florida, Gainesville.

Bentley, J.W. and K.L. Andrews 1991. Pests, peasants and publications: anthropological and entomological views of an integrated pest management program for small-scale Honduran farmers. *Human Organization* 50:113–123.

Bojo, J.P. 1991. Economics and land degradation. *Ambio* 10:75–79.

Bunch, R. 1994. The potential of slash/mulch for relieving poverty and environmental degradation. In *Slash/Mulch: How Farmers Use It and What Researchers Know About It*, eds. H.D. Thurston, M. Smith, G. Abawi, and S. Kearl. Ithaca, NY: Cornell International Institute for Food, Agriculture and Development, Cornell University.

Carriere, J. 1991. The crisis in Costa Rica: An ecological perspective. In *Environment and Development in Latin America: The politics of sustainability*, eds. D. Goodman and M. Redclift. New York: Manchester University Press.

Carroll, C.R., J.H. Vandermeer, and P.M. Rossett, eds. 1990. *Agroecology*. New York: McGraw-Hill.

den Biggelaar, C. 1991. Farming Systems Development: Synthesizing Indigenous and Scientific Knowledge Systems. *Agriculture and Human Values* 8:25–36.

Dix, A.M. 1995. IPM in non-traditional export agriculture in Guatemala: the case of Chilasco, Baja Verapaz, Guatemala. XIX International Congress, Latin American Studies Association, Washington D.C., September 28–30 1995 (proceedings in press).

Dix, A.M. and C.R. Carroll. 1995. IPM in non-traditional export agriculture in Guatemala: the case of Chilasco, Baja Verapaz, Guatemala. IPM CRSP Working Paper 95-4, IPM CRSP, Office of International Research and Development, Virginia Tech University, Blacksburg, VA 24061-0334.

Dix, A.M., C.R. Carroll, M.W. Dix, and G. Dal Bosco. 1995. Corn stalks influence patchy distribution of white grubs in broccoli fields. IPM CRSP Working Paper 95-5, IPM CRSP, Office of International Research and Development, Virginia Tech University, Blacksburg, VA 24061-0334.

Duncan, C. 1991. Agriculture, export diversification, and the environment in Central America. In *Modernization and stagnation: Latin American agriculture into the 1990's*, eds. M.J. Twomey and A. Helwedge. New York: Greenwood Press.,

Fujisaka, S. 1992. Farmer knowledge and sustainability in rice-farming systems: Blending science and indigenous innovation. In *Diversity, Farmer Knowledge, and Sustainability*, eds. J.L. Moock and R.E. Rhoades. Ithaca: Cornell University Press.

Garst, R. and R. Barry. 1990. *Feeding the crisis: U.S. food aid and farm policy in Central America*. Lincoln: University of Nebraska Press.

GEXPRONT 1994. *Bases para la elaboracion de una politica de desarrollo economico atraves de las exportaciones*. Guatemala, Mayo de 1994 (Authors' note: GEXPRONT is the growers association that promotes non-traditional export crops and offers technical help on production practices.)

Hoffman, C.A. and C.R. Carroll. 1995. Can we sustain the biological basis of agriculture? *Annual Review of Ecology and Systematics* 26:69–92.

Houghton R.A. 1994. The worldwide extent of land-use change. *BioScience* 44:305–13.

Kettler, J.S. 1995a. A sustainable strategy for fragile lands: Fallow enrichment of a traditional slash-mulch system in Coto Brus, Costa Rica. Ph.D. Dissertation, University of Georgia, Athens, Georgia.

Kettler, J.S. 1995b. A new insect pest for *Erythrina peoppigiana* in Costa Rica. *Nitrogen Fixing Tree Association Research Reports* 13:51–53.

Kettler, J.S. 1996. Weeds in the traditional slash/mulch practice of *frijol tapado*: Indigenous characterization and ecological implications. *Weed Research* (in press).

Molnar, J.J., P.A. Duffy, K.A. Cummins, and E. Van Santan. 1992. Agricultural science and agricultural counterculture: Paradigms in search of a future. *Rural Sociology* 57: 83–91.

Mucia, M. 1994. *Sostenibilidad Social: La Experiencia de los Productores de Patzun, Chimaltenango.* In *Sostenibilidad de la produccion agricola no-tradicional de exportacion por pequenos productores en Guatemala*, 6–9. Seminario-Taller en Antigua Guatemala. 20–22 de Septiembre 1993.

O'Brien, W.E. and C.B. Flora. 1992. Selling appropriate development vs. selling-out rural communities: empowerment and control in indigenous knowledge discourse. *Agriculture and Human Values* 9:95–102.

Patiño, V.M. 1965. *Historia de la Actividad Agropecuaria en America Equinoccial.* Cali, Colombia: Imprenta Departmental.

Thurston, H.D. 1994. Slash/mulch systems: Neglected sustainable tropical agroecosystems. In: *Slash/mulch: How farmers use it and what researchers know about it*, eds. H.D. Thurston, M. Smith, G. Abawi, and S. Kearl. Cornell University, Ithaca, NY: Cornell International Institute for Food, Agriculture and Development.

Tropical Science Center. 1991. Accounts Overdue: Costa Rica Natural Resource Accounting Study. Tropical Science Center. San José, Costa Rica.

NEW TECHNOLOGIES AND NOVEL PERSPECTIVES FOR THE NEXT GENERATION OF CONSERVATION BIOLOGY

Conservation biology will always be about real estate, demography, restoration, economic value, biodiversity, and ecosystem functioning, although the jargon may change. But, as is the case with any science, there is also the possibility of fresh insights as a result of new techniques and innovative connections to other disciplines. The challenge is in distinguishing trendy hyperbole from real potential. We invited four contributions to do just that—comment on some of those "headline advances" that currently are being sold as the most important new angles or approaches to conservation: molecular biology applied to the genetics of threatened species, stress endocrinology as an indicator of population well-being, and computerized geographic information systems applied to landscape ecology. The fourth contribution in this section quite literally addresses a "hot topic"—global climate change and species interactions as the Achilles heel of our best-planned reserves. The measured skepticism of the four chapters in this section is probably not surprising. After all, scientists are a conservative group and always chafe at the selling of any discovery or "breakthrough" (particularly one that is not their own). What is a little surprising is how the contributions share some of the same deep dissatisfactions with our ability to anticipate and monitor threats to species and ecosystems.

Earlier in the book, we saw that viability analyses typically were starved for data, that they could be misleading if the limitations of short-term studies were not appreciated, and that they sometimes were essentially impossible. Molecular genetics, endocrinology, and landscape ecology all promise to complement standard viability studies and endeavor to fill huge gaps associated with short term, localized, demographic analyses. The key notion motivating studies of genetic structure is that quantitative assessments of genetic structure today can tell us about future threats and past demographies. But as Steinberg and Jordan show us—during the disequilibrium stage when threatened species play out their fates—this is unlikely to be possible without a great deal of additional ecological knowledge and a custom-made ecological/genetic model, as opposed to generic "textbook" statistics. The possibilities for "conservation endocrinology", which promises to provide early warning signals for populations in trouble (before they have plummeted because hormone titers reflect reproductive vigor, and reproductive vigor determines future population growth), are more optimistic than analyses of genetic structure. However this field is still so new it is not clear to what extent assays of steroid levels, for example, will totally fail to anticipate a population's risk, or alternatively, how often hormone assessments will yield false alarms. Schoech and Lipar do make it clear, however, that the disruption of hormone systems as is hypothesized to be associated with environmental pollutants, could potentially produce severe threats to species survival. This increasing involvement of endocrinologists in conservation efforts may help unite conservation biology and more "chemically" oriented environmental biology, two fields that are ironically quite disparate. Similarly, the recent rapid development of the field of landscape ecology is a response to the inadequacy of simple PVA's—species exist in large, complex landscapes and no PVA can be realistic and thus useful unless the landscape is incorporated into the analysis. Of course, if we lack demographic data for single local populations, one has to wonder whether we could ever build convincing landscape models—an issue tackled by Meir and Kareiva. They argue that detailed spatial models may find their best application as guides to monitoring strategies and not as predictive tools in and of themselves. Finally, Gryj's essay regarding implications of global climate change to species interactions raises the specter of uncertainty that this debate about detailed demographic modeling might be a moot point anyway—i.e., the environment could be changing so rapidly that all of our measurements of today's populations will be outdated within a few decades.

The worst thing that can happen to a conservation biologist is to be caught off guard about threats to species or ecosystems. After all, even when we know the threats to biodiversity, it is hard to produce effective countermeasures. Thus while the "new" approaches or perspectives in this section may come across as being somewhat negative, they all offer some sort of "early warning system." For example, evaluations of patterns of genetic variation may tell us that a species is vulnerable because genetic exchange is not occuring among populations. Titers of reproductive hormones can warn us that a nonlethal environmental disruption is placing a population at risk. Monitoring patches for presence or absence may indicate whether a metapopulation is at risk. And subtle alterations in phenology or development rates can foreshadow major shifts in pestilence and disease.

The last theme that emerges from these discussions of "trendy topics" is that the allure of shortcuts and an easy way out is fantasy. New ideas are often appealing because they promise (sometimes almost subliminally) answers with a minimum of work. But genetic structure can tell us very little of practical value to conservation without supplemental ecological information and specific models imbued with detailed biology. Endless samples of steroid levels in threatened populations will be uninformative unless proper control populations are also studied. Gloriously colored computer maps with GIS layer after GIS layer will be totally useless without hard-won natural history data on the habitat preferences and dispersal behavior of animals. Studies of photosynthesis rates as a function of CO_2 and temperature are unreliable without scaled up versions that attempt to predict the response of suites of interacting species. New ideas and new technologies do offer us new insights, but they will not do so without a great deal of complementary work. Frustration with the limitations of current conservation studies and the desire for simple solutions are a major part of the appeal of the new approaches and perspectives in this section.

Regardless of the hyperbole, nothing discussed in this section is a passing fad. Future generations of conservation biologists will undoubtedly be knowledgeable about molecular population genetics, endocrinology, GIS and landscape models, and climate change scenarios. The contributions that follow provide assessments of the potential for these technologies and viewpoints. The bottom line is that the reader is urged to be critical when deciding how best to incorporate new ideas and tools into one's own conservation endeavors.

CHAPTER

19

Using Molecular Genetics to Learn About the Ecology of Threatened Species: The Allure and The Illusion of Measuring Genetic Structure In Natural Populations

ELEANOR K. STEINBERG
and CHRISTOPHER E. JORDAN

Conservation of threatened and endangered species depends upon understanding the contribution of migration and local demography to population change. Unfortunately, studies of species at risk tend to be plagued by logistic problems, including limited access to populations, small sample sizes, and restrictions prohibiting manipulative experimentation. Thus, even the most basic demographic data (e.g., birth and death rates)—and certainly data regarding migration—can be difficult to acquire. Technological advances such as radio telemetry and geographic positioning systems have improved somewhat our ability to pursue field demography (i.e., McKelvey *et al.* 1993; Lahaye *et al.* 1994), but in general studies that employ such technology remain extremely expensive and logistically difficult. In contrast, recent technological advances in molecular population genetics have greatly reduced the cost and simultaneously increased the ease of field genetic studies. For example, the recent development of the polymerase chain reaction (PCR) allows amplification of DNA from tiny skin biopsies, individual hairs, or even scat. These non-intrusive sampling methods mean we can obtain genetic data on highly endangered species without sacrificing a single individual. In addition, easy-to-use computer packages are readily available to translate genetic data from individuals into assessments of population genetic structure (i.e., BIOSYS by Swofford and Selander 1981; GDA by Lewis and Zaykin 1996). Most importantly, the analysis of genetic structure does not require tracking the fate of individuals, or even capturing individuals more than once. Thus, it is not surprising that we know more about the genetic structure of many endangered species than we do about their fundamental demographic processes.

In conservation biology, a central tenet involving population ecology is that as our landscapes become increasingly fragmented by habitat destruction, understanding movement of individuals between remaining isolated habitat patches will be critical for management programs (i.e., Harrison *et al.* 1993; Fahrig and Merriam 1994; Dunning *et al.* 1995). Thus, based upon the fact that migration is the force that moves genes between populations, many conservation biologists have sought insight into migration by quantifying patterns of genetic variation in space (e.g., Bowen *et al.* 1993; Fleischer *et al.* 1995; Paetkau *et al.* 1995). Patterns of genetic structure also have been used as an indicator of historical influences on populations, such as the likelihood of past population bottlenecks (e.g., Bonnell and Selander 1974; Fleischer *et al.* 1995).

In this chapter we evaluate opportunities and insights potentially gained by analyses of population genetic structure. More specifically, we focus our critical discussion on three issues that we believe should be of particular concern to conservation practitioners. These are:

(1) Estimation:
Because large errors can be associated with subsampling populations and because of the inherent variability of random processes (e.g., genetic drift), simple measures of genetic structure can not be taken at "face value." While this point may be well-known to evolutionary biologists, it is occasionally glossed over or ignored in the conservation literature.

(2) Interpretation:
There are many alternative possible causes for any estimated genetic structure (i.e., limited dispersal, historic population bottlenecks, high variation in population size, etc.). Therefore, using statistics of population genetic structure (particularly statistics that assume equilibrium conditions) to determine specific ecological details about species or populations can be problematical.

(3) Application:
Even if we have good estimates of genetic structure, it is not immediately obvious how they can be practically applied.

Before tackling these three issues, we introduce the primary statistics used to summarize population genetic structure and describe their origins in evolutionary theory, because it helps to understand the scientific roots of these concepts as a foundation for understanding modern applications. We pay particular attention to the theory that links F-statistics to population size and migration. Given this background in the fundamentals, we review a sampling of case studies in conservation genetics. Finally, we present an original model that details both genetics and demography and can be used to explore scenarios of change in genetic structure as a result of habitat fragmentation or population catastrophes. We use this model to illuminate limitations in using F-statistics as a generic tool, particularly for management applications, and to suggest promising future directions by which measures of genetic structure might offer better insight to conservation efforts.

GENETIC STRUCTURE AS THE KEY TO EVOLUTION

Genetic measures of population structure played a central role in the development of the evolutionary synthesis that took place during the first half of the twentieth century. We discuss two measures here: effective population size and F_{ST}. Effective population size, typically symbolized N_e, roughly estimates the number of breeders contributing to a population's pool of gametes each generation. F_{ST} describes the degree to which genetic variation is partitioned among different populations within a species. Although N_e and F_{ST} originated as concepts in population genetics theory, they receive considerable attention from conservation biologists.

Effective population size (N_e) is important because it determines the relative influence of genetic drift as a process promoting change in gene frequencies (Wright 1931). If N_e is small, chance events control patterns of neutral genetic variation, resulting in the loss of some alleles while others become fixed (present as homozygotes in every individual). In addition, when N_e is small, inbreeding (i.e., mating between close relatives) is more likely and the erosion of heterozygosity will be hastened. More plainly, a small effective population size favors the prevalence of random drift and the loss of genetic variation, often removing alleles favored by weak selection. Given this "bottom line" interpretation of N_e, it is not surprising that discussions of "genetic viability" for threatened species often mention low effective population size as a liability to species persistence (see Lande and Barrowclough 1987). Unfortunately, while the theoretical importance of N_e is undisputed, actually quantifying N_e for populations in the wild can be quite challenging.

The term "effective population size" can be easily misinterpreted because it sounds much like "population size." However, the ecologist's measure of population size is very different from what a population geneticist or evolutionary biologist means when referring to N_e. While generally pertaining to the "effective number of successful breeders" in a population, N_e is actually a more complicated genetic concept representing the expected importance of random genetic drift in producing gene frequency changes and ultimately loss of genetic diversity. Small effective population sizes imply that severe loss of genetic variation will occur through drift, whereas large N_e's imply the opposite. N_e is thus commonly expressed in terms of expected decline in heterozygosity (H). In particular, for an "ideal" population (i.e., panmictic, equal sex ratios, constant population sizes, etc., see Hartl and Clark 1989):

$$H_t = H_0 [1-1/(2N)]^t \qquad (1)$$

where t is time in generations, H_i is the estimated average proportion of genes in an individual that are heterozygous in generation i, and N is the number of interbreeding individuals. Deviations from the idealized assumptions cause H to decline at a different rate than that predicted by equation (1). Given these deviations, N_e is estimated by the N that would have to be substituted into equation (1) to generate the expected change in heterozygosity associated with a particular "real-world" population. In other words, effective population size is a calibration number, representing the size an "ideal" (see above) population would have to be to recreate the same genetic properties as are found in the non-ideal real-world population.

The importance of calculating F_{ST} to quantify genetic differentiation among populations was championed by Sewall Wright in the 1940s (Wright 1943). Wright proposed a theory of evolution in which the partitioning of genetic variation among subpopulations was central. In particular, he believed that the most favorable condition for evolution was the partial isolation of many subpopulations, each differing genetically from one another by a modest but significant amount (Wright 1951). To summarize the pattern of genetic variation among subpopulations, Wright advocated the use of F-statistics. The general idea behind this approach is that random genetic drift will act independently in isolated subsets of populations, resulting in differentiation (in terms of local gene frequencies) among these subpopulations. Applying F-statistics, one can explore how genetic variation is partitioned at various levels of organization. In particular, "F_{ST}" describes how variation is partitioned at the subpopulation ("s") level, relative to the total population ("T") level (i.e., T = all subpopulations lumped together as though they actually are one panmictic population). If genetic drift strongly influences subpopulations independently, F_{ST} will be higher than if the subpopulations are instead interconnected by a high rate of gene flow. At the extremes, $F_{ST} = 1$ indicates (in a two subpopulation system) that one allele is fixed in one population and a different allele is fixed in the other population, whereas $F_{ST} = 0$ indicates that the frequency of alleles is identical in the two populations. When F_{ST}'s are reported for species or populations they are typically averaged over many different loci—but they always describe the partitioning of genetic variation. For more detailed explanations of F-statistics, see Weir and Cockerham 1984 and chapter six of Hartl and Clark 1989.

ESTIMATING AND APPLYING GENETIC STATISTICS

Two primary approaches are used to estimate effective population size, and considerable debate exists regarding which is the best approach (see Husband and Barrett 1992; Nunney 1995). The "genetic" procedure uses variance in neutral markers across generations to estimate N_e (see Waples 1989). This approach requires population samples from multiple generations; therefore, many studies lack the necessary data. An alternative "ecological" approach does not require any genetic data. Instead the ecological procedure uses equations that incorporate information on fluctuations in population size, variance in family size, reproductive success, sex ratios, etc. (Lande and Barrowclough 1987; Harris and Allendorf 1989; Nunney and Elam 1994). All deviations from a constant idealized population cause N_e to be lower than a simple count of the number of breeders. The implication of varying population size for N_e is especially important to conservation biology, because so many threatened species have experienced dramatic population changes as a result of human disturbance. Such directional change, or even simply fluctuations about a mean, dramatically reduced N_e. In particular, N_e for a varying population is given by the equation:

$$N_e = 1/2[1 - \{1 - 1/2N_e(i))\}^{1/t}] \qquad (2)$$

(Lande and Barrowclough 1987), where t is the total number of generations over which N_e is to be calculated, and $N_e(i)$ is the effective population size in the ith generation. A key feature of equation 2 is that generations with the smallest population sizes have a disproportionate influence on the long-term value of N_e. For example, if we imagine data representing four successive generations with the N_e's of 10, 100, 10, 100, then equation 2 yields an N_e of 18.01, a value substantially lower than 55, which is the arithmetic mean of 10, 10, 100, and 100. In the conservation literature, equation 2 is often approximated by the harmonic mean. This approximation is appropriate for fluctuations occurring on a short time scale, with gene frequencies remaining relatively constant, but should not be used as a long-term measure of N_e (Lande and Barrowclough 1987). Excellent general discussions of estimation procedures for N_e that account for a wide range of ecological complications are found in Lande and Barrowclough (1987) and Harris and Allendorf (1989).

F_{ST}, which is the degree to which genetic variation is partitioned among subsets of populations, is typically estimated in one of two ways: either (i) by the deficiency of heterozygotes in subpopulations compared to expected heterozygosity for one panmictic population, or (ii) by the ratio in variance of allele frequencies among subpopulations relative to total variation. Regardless of how it is measured, the key relationship between F_{ST} and N_e that appears in most population genetics textbooks (e.g., Hartl and Clark 1989) is:

$$F_{ST} = 1/(4N_em + 4N_eu + 1) \quad (3)$$

with m and u respectively representing average migration and mutation rates. Because mutation is often assumed to occur at a very low rate, equation (3) is often reduced to the more familiar form in the absence of mutation:

$$F_{ST} = (1/(4N_em + 1) \quad (4)$$

Notice that equations (3) and (4) explicitly connect something that is readily estimated, F_{ST}, to processes that are extraordinarily difficult to observe (i.e., migration and mutation). Thus, it is not surprising that the F_{ST} equations are commonly used to make educated guesses about N or m.

The interpretation of F_{ST} in terms of N_e or m is, unfortunately, not as simple as equations (3) and (4) might seem to indicate. The main problem with interpretation is that *these equations hold only if several key assumption are met.* Specifically, derivations of (3) and (4) assume that there is a *large number of subpopulations* and an *infinite number of possible alleles at each locus,* or alternatively, if mutation is not included (because it is assumed to be insignificant), that there is an *infinite number of subpopulations.* These are the sorts of mathematical assumptions that sound worse than they are, because a modestly "large number" may suffice in lieu of infinite.

More problematic is the assumption that *migration is equal among all subpopulations*, or to put it another way, that migrants enter one common "bath" of dispersers, which then rains down upon the subpopulations equally. This treatment of dispersal can be very far off the mark, especially if dispersal is highly localized. However, the most damning assumption in conservation applications is that of genetic equilibrium—i.e., a balance in the addition and loss of alleles throughout the population. *Only after genetic equilibrium has been reached will equations 3 or 4 hold.* Simulations have shown that it can take an inordinately long time to reach equilibrium(Slatkin 1985; Varvio *et al.* 1986; Slatkin and Barton 1989; Figure 19.1). It is particularly hard to imagine threatened and endangered species reaching genetic equilibrium because the primary reasons they are at risk are usually due to recent changes. Indeed, population dynamics and ecological features may themselves be consistently changing so rapidly that the genetic parameters for populations are always playing "catch-up" with the moving target of some idealized equilibrium. In summary, for species having experienced major recent changes or species with highly localized dispersal, one should be very careful about using F-statistics to infer ecological parameters. It is worth noting, however, that although equations (3) and (4) dominate the conservation literature, there are several theoretical alternatives that permit more realistic life histories and models of migration (e.g., Maruyama 1971; Malecot 1975; Crow and Aoki 1984; Slatkin 1985 and 1989; Lande and Barrowclough 1987). Importantly, *all of the alternative approaches still require that a genetic equilibrium has been reached before any simple relationships between* F_{ST} and N_e or m can be established.

Similar to the case with N_e, relationships involving F_{ST} can also be misleading because of the different connotations of the words "population size" and "migration rate" in ecology as opposed to population genetics. Whereas an ecologist uses population size to simply quantify the total number of individuals in some specified region, the genetic notion of effective population size is not something that can be directly measured. Instead,

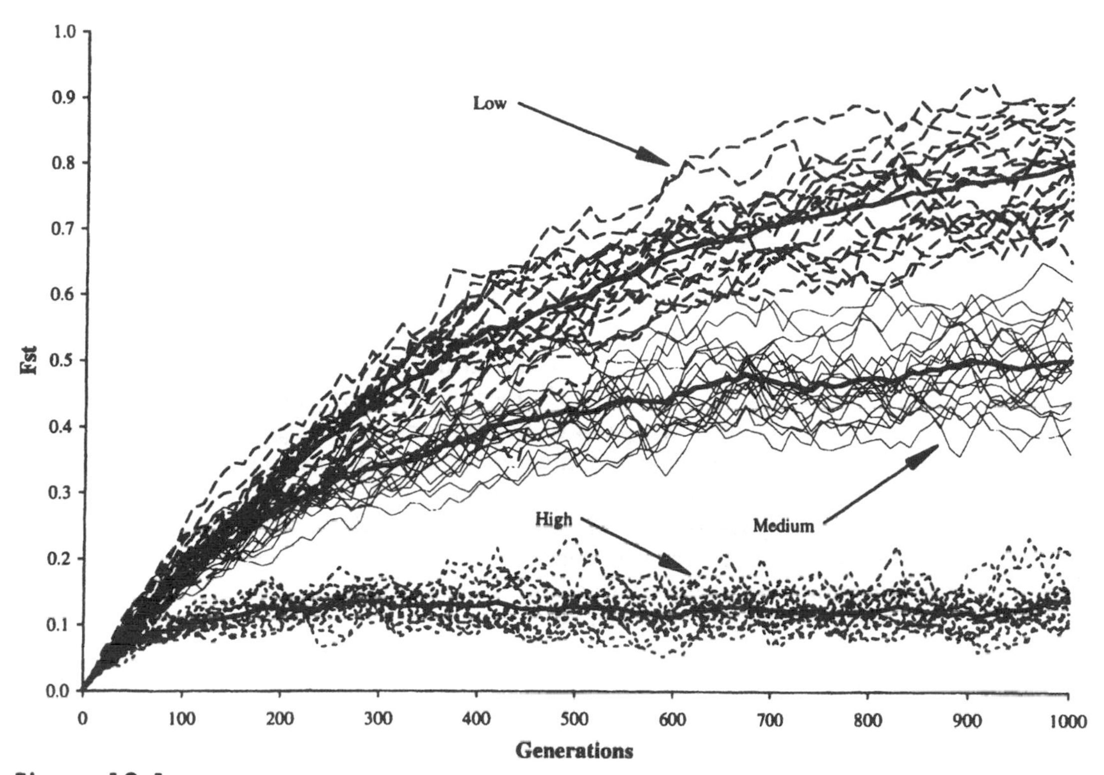

Figure 19.1
Development of genetic substructure as a function of migration rate. F_{ST} calculated for five diploid loci over all individuals (~800) in the four sectors of the simulated population. The figure shows 20 replicate simulations (light lines) and their means (heavy lines) at three different migration rates over 1,000 generations. We used the following migration success rates: low = 0.1% of the potential interpopulational dispersers (ID's) succeed per generation, medium = 1.0% of potential ID's succeed per generation, high = 10% of potential ID's succeed per generation. For the runs reported here, the potential ID's represent one quarter of the dispersing individuals that encounter a sector edge (the rest of the dispersing individuals that encounter an edge either die before attempting to migrate or return to the sector without migrating). *For example, the medium migration rate translates to roughly two individuals successfully migrating per sector per generation. The high and low migration rates are 10 times higher and lower, respectively. F_{ST} increases from an initial condition of 0 (no structure) both as a function of time, and inversely as a function of dispersal rate.*

as we stated previously, it is a calibration number, representing the requisite size an "ideal" population must be to recreate the same genetic properties as are found in the real-world population. Similarly, "m" is not simply the ecologists' notion of the proportion of a population that is immigrants—instead it is the "effective" proportion of the population that is immigrants. This requires not only immigration, but also a contribution of "new blood" (i.e., novel alleles) to the breeding pool. *In general, both N and m in equation (4) will be much smaller than standard ecological usage implies.*

We focus on simple F-statistics and estimates of effective population size because they represent the most prevalent approaches found in the conservation biology literature.

These measures of genetic structure date back more than half a century (e.g., Wright 1931) and have a long history of application. Another reason for their widespread use is undoubtedly because they can be easily computed using computer programs and simple equations, such as those mentioned previously. It has been suggested that alternative analysis methods may provide additional insight for conservation applications (see Milligan *et al.* 1994). Without going into great detail, we touch upon three of the newer approaches for genetic structure analysis (rare alleles analysis, cladistic analysis, and regression models for isolation by distance). The rare alleles approach (Slatkin 1985) uses the distribution of "private" or rare alleles (i.e., alleles found in some but not all locations) to estimate gene flow. Cladistic methods (Slatkin and Madison 1990) generally involve construction of phylogenetic trees using DNA sequence data and then "mapping" the geographic locations of samples onto those phylogenies to determine the most parsimonious migration scenario that could have produced the phylogeny. Finally, isolation by distance approaches (e.g., Slatkin 1993; Hellberg 1994) regress estimates of genetic differentiation, such as F_{ST}, between pairs of populations against geographical distances between those populations. The slope and intercept of these regressions offer some insight into the extent of dispersal and effective population size.

It is important to emphasize that although all three of these newer methods take advantage of more detail in the data or apply more sophisticated analyses, they all also incur problems in estimation and interpretation. As an example, the private alleles method is particularly sensitive to sampling errors and to errors in coding common alleles as novel (hence private) alleles. However, most importantly, even if the estimation challenges are overcome, *all of these methods still represent indirect assessments of dispersal, with the unavoidable difficulty of interpreting which of several scenarios might explain patterns uncovered. Also problematic is the fact that genetic structure estimates may represent hidden pasts rather than current situations.*

One final but very important point to remember when applying theoretical population genetics models to real-world scenarios is that it is theory based on probabilities and estimators. In the derivation of equations for estimators such as F_{ST}, population geneticists consider probabilities of genes drawn from random populations and from random replicate realizations of the genetic process. *Particular individual realizations of the identical process can be very different due to the randomness of genetic drift and mutation.* Thus, N_e and F_{ST} describe "expected courses of events and not the fluctuations we should expect around those averages" (Felsenstein 1995).

USING MEASURES OF GENETIC STRUCTURE TO SUGGEST CONSERVATION PRACTICE: SOME CASE STUDIES

A recent trend in conservation studies is the estimation of genetic structure to assess impacts of habitat disturbances or local rarity on threatened taxa (e.g., Lesica *et al.* 1988; Stiven and Bruce 1988; Ashley *et al.* 1990; Dinerstein and Mccracken 1990; Leberg 1991; Soltis *et al.* 1992; Stangel *et al.* 1992; Wayne *et al.* 1992; Sarre 1995; Hall *et al.* 1996; Young and Brown 1996). Molecular markers used in these studies include allozymes, DNA sequences, restriction fragment length polymorphisms (RFLPs), minisatellites, and microsatellites. Several excellent references are available on the details of the various molecular methods (Avise 1994; Hillis *et al.* 1996); therefore, in discussing examples we will not describe the techniques. Instead, we will focus on the challenges of estimation, interpretation, and finding practical applications associated with conservation-related studies of genetic structure.

An interesting early study that used estimates of genetic structure to explore consequences of land management involved the salamander *Desmognatus quadramaculatus* in the southern Appalachian forests of eastern north America (Stiven and Bruce 1988). Here, the researchers were able to garner allozyme data from populations in watersheds with known historical differences inland management. Salamanders were sampled in three pairs of control and logged watersheds in two different forests (two pairs in Coweeta Forest and one pair in the Great Smoky Mountains National Park). The disturbance involved different mixes of logging, road building and erosion. Based on 14 allozyme loci, they found significantly higher levels of heterozygosity in the disturbed (primarily logged) watersheds than in the paired control watersheds. This might lead one to conclude that "disturbance" increased genetic variability. However, when the data were pooled and contrasted among forests, populations from the more disturbed Coweeta Forest had lower heterozygosity than populations from the less disturbed Great Smoky Mountains National Park forest, allowing the authors to speculate about the effect of disturbance on genetic polymorphism at one spatial scale. This left the dilemma of why disturbance did not reduce heterozygosity at the scale of watersheds, but did supposedly reduce heterozygosity at the scale of entire forests.

The authors turned to estimates of F-statistics as one approach to gain further insight, but in doing so they encountered some fairly typical limitations of estimation and interpretation with respect to analyses of genetic structure. Specifically, following the general approach of equation (4), F_{ST} values were calculated and used to estimate the "effective number of migrants exchanged among populations" (Stiven and Bruce 1988, p. 203), which the authors determined were consistent with data from other salamander species. However, comparisons of estimates have little value if confidence intervals are not considered. If more modern techniques for bootstrapping standard errors had been used (see Weir 1996), it might have completely undermined any confidence in the salamander estimates of genetic structure. Further, even if one accepts a particular F_{ST} value as being accurate, its interpretation in this salamander system is confounded by the recent disturbance history of the system, with some watersheds having been massively logged less than five salamander generations ago. This genetic system was thus most likely quite far from a state of genetic equilibrium. To their credit, Stiven and Bruce avoided making any practical management suggestions on the basis of their data, perhaps because they recognized the weak links implicit insuch a leap. Instead they used their results to urge further work connecting perceived patterns of genetic structure to disturbance regimes.

A much more recent study that similarly adopted a control/disturbed contrast and estimated genetic structure to investigate the effect of fragmentation on geckos in Australia using RFLP analysis of mitochondrial DNA (Sarre 1995). The pertinent contrast was between several remnant and fragmented populations occupying small islands of suitable vegetation and thriving populations in large nature reserves. There was an additional complexity in this study in that the different populations were assigned to three distinct geographical regions. First, Sarre concluded that geckos in the nature reserves were more genetically diverse than geckos occupying remnant habitats. This conclusion was based on calculations of haplotype diversity for three populations in nature reserves and for nine remnant populations not in nature reserves. While it was certainly true that the average within-population haplotype diversity was higher for nature reserve populations than remnant populations, there were fragmented populations with haplotype diversities as high as any of the nature reserve populations. Indeed the second highest haplotype diversity over all populations including nature reserve geckos was derived from an isolated remnant population of geckos. Moreover, if the goal is to figure out how to protect overall diversity, then the appropriate comparison might have been haplotype diversity of all fragmented subpopulations in a region lumped together versus haplotype diversity of all nature reserve populations lumped together—a comparison not presented. In addition to examining local amounts of genetic variability, Sarre calculated F-statistics within each region using the haplotype data. By comparing observed F_{ST} estimates to those calculated by randomly subsampling haplotype data using a bootstrap procedure, Sarre found evidence of greater than expected (by chance) genetic structure within each of the three regions. These results led to an apparent contradiction between high levels of estimated gene flow (based on back-calculating Nm from F_{ST}) and what was presumed to be much lower gene flow "in reality" due to fragmentation. This contradiction noted by Sarre is a perfect example of how hard it can be to infer migration from F-statistics. In this case, according to Sarre, the apparent paradox arose because the appropriate genetic equilibrium had not yet been attained and hence the gene flow estimates possibly represented historical conditions as opposed to the current fragmented situation—a point we illustrate in the next section, using a simulation model (Figure 19.2).

In many cases, conservation biologists simply want to compare patterns of genetic structure among species as a way of identifying potentially interesting patterns. For example, with the goal of contrasting patterns of genetic variation in rare species to common species, Hickey *et al.* (1991) used allozyme data to estimate F-statistics in four species of clover (*Trifolium*) from Ohio. Their system included one endangered native species, one rare native species, and two common introduced species. They found a great deal of variability in the amount and distribution of genetic diversity among the species, with the rare native species being the least variable (it was uniform across all loci in the one population sampled) and the two introduced species being the most variable. In terms of genetic structure, the F_{ST} estimate for the endangered species (0.206) was greater than that of the two introduced species sampled (0.036 and 0.021, respectively), although the statistical significance of these comparisons is not clear because standard errors were not presented. This study further illustrates the problems one can encounter interpreting F_{ST}'s in that, as the authors pointed out, the lack of genetic structure in the introduced species could have been due to a highly diverse original gene pool, to multiple introductions from the original gene pool, or to both. The weakness of this study in terms of meeting its objectives derives from selecting introduced species as points of comparison to endangered species, because introduction itself involves many special ecological circumstances that are likely to produce peculiar genetic effects. Without additional knowledge of the species' ecology and demography, similarities or differences between threatened and introduced species tell us very little. A better comparison would involve native endangered and native nonthreatened species, a comparison pursued by our next two examples.

Taylor, Sherwin, and Wagner (1994) used 16 microsatellite DNA loci to contrast genetic variability in the last remaining population of the northern hairy-nosed wombat (*Lasiorhinus krefftii*), with genetic variability in its closely related congener, the abundant southern hairy-nosed wombat (*L. latifrons*). They also analyzed genetic material from museum specimens from a now extinct population of northern hairy-nosed wombat. As might be expected, the southern hairy-nosed wombat populations had higher heterozygosities and greater allelic diversity then did populations of the endangered northern hairy-nosed wombat. In addition, Taylor and colleagues estimated heterozygosity in the southern hairy-nosed wombat as a baseline from which to interpret heterozygosity in the northern

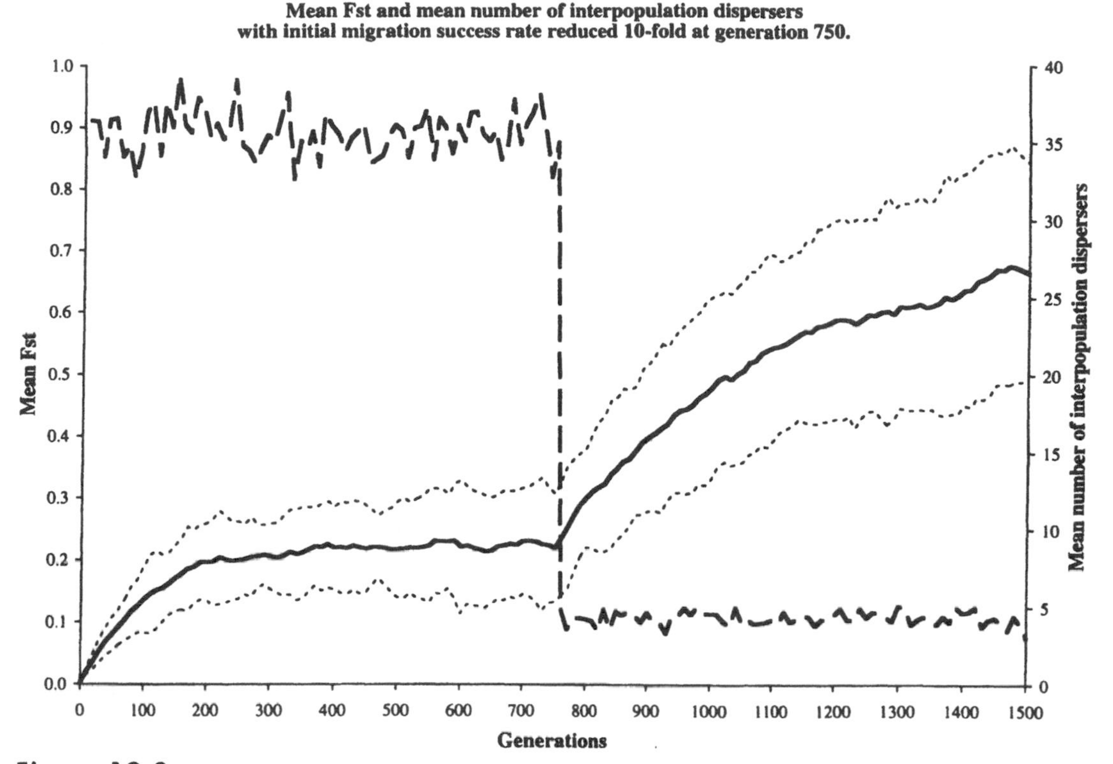

Figure 19.2
The effect of dispersal barriers on the development of genetic substructure. A simulated population with a migration success rate of 5% (see legend to Figure 19.1) per generation was perturbed after reaching genetic equilibrium by reducing the migration success rate ten-fold to 0.5% per generation at t = 750. The mean F_{ST} (+/− 2 s.d.) for 20 replicate simulations is shown (left hand scale: heavy line; 95% confidence interval, light dashed lines), as well as the mean number of dispersers (right hand scale: heavy dashed line). F_{ST} was calculated for five diploid loci over all individuals (~800) in the four sectors of the population.

hairy-nosed wombat. Based on expected levels of heterozygosity in model populations (see eq. 1), they treated the southern hairy-nosed variation as the likely "initial condition" for the northern hairy-nosed wombat (i.e., H_0), and by assuming that the northern hairy-nosed wombat had been declining over about 120 years, calculated the effective population size of the northern hairy-nosed wombat during the decline. In this case, a seemingly well-chosen reference population enhanced the information value of genetic data from a threatened species. Moreover, in this paper, the authors *did* calculate and present error estimates, and these errors indicated that their effective population size estimates were robust. Of course, also implicit is the assumption that no historical events left hidden imprints on southern hairy-nosed wombat genetic data.

Sherwin *et al.* (1991) attempted a similar approach using allozyme loci to understand the genetic structure of an endangered population of eastern barred bandicoot (*Perameles gunnii*), a marsupial endemic to Victoria, Australia. They used a conspecific, widespread, dense population in Tasmania as a reference and compared patterns of genetic variation between the populations. Unfortunately, no between-population divergence or within population variation was detected in any of the loci, which precluded further analysis.

One reason for the lack of variation in this study was that the markers they used were invariant. Interestingly, when this analysis was repeated with variable number of tandem repeat (VNTR) markers, which tend to be more variable than allozymes, Robinson *et al.* (1993) found *more* genetic variability among endangered mainland populations than among non-endangered Tasmanian populations. They thus did not use the Tasmanian population to extrapolate baseline data for the endangered population, but instead discussed possible scenarios that could explain the observed pattern. While uncovering unexpected patterns can be very useful in terms of generating new hypothesis about populations of conservation concern, this study highlights the point that without knowing historical factors, it is difficult to know if one has selected reasonable reference populations.

An interesting subtlety associated with studies of genetic structure is the extent to which results can depend upon which populations are sampled, how many populations are sampled, and the spatial scale of sampling. This is well illustrated in a comparative study of genetic structure in a rare woodland shrub of New South Wales, *Daviesia suaveolens*, and its common sympatric congenor, *D. mimosoides* (Young and Brown 1996). In examining the total range of the rare species versus the total range of the common species, the authors determined that populations of the two species exhibited similar average amounts of genetic variation and levels of genetic structure as estimated by F-statistics (F_{ST} = .12 for *D. suaveolens* versus F_{ST} = .11 for *D. mimosoides*). Based on these data (for which they included error estimates), one might conclude that the rare species was no more subjected to limited gene flow than the common species. However, when F_{ST} was estimated for populations of the common species at the equivalent scale of the restricted rare species, its resulting F_{ST} was only .04—one third of the F_{ST} for the rare species! These results illustrate several interesting points. First and foremost, estimates of genetic structure only tell us how genetic variation is partitioned *among the populations sampled,* and *not* for a species as a whole. Many studies (e.g., Stiven and Bruce 1988; Leberg 1991; Fleischer *et al.* 1995) refer to F_{ST} estimates from other species, as though the estimates should be comparable in some general way. The *Davesia* study highlights the fact that without information about the scale at which species have been sampled, it is not necessarily obvious how genetic estimates should be interpreted. For management considerations in particular, the results of this study make it clear that sampling design is critical in studies of genetic structure. Based on the design of their study, Young and Brown (1996) were able to suggest that because the amount of genetic variation detected in populations of the rare species was similar to the amount of variation in populations of the common species, that special attention need not be focused on preserving the genetic variability of particular populations of the rare species. Instead, because the rare species was significantly more genetically structured on the same spatial scale as the common species, management efforts would be more wisely directed towards conserving populations from the extremes of the range of the species, thus maximizing maintenance of overall genetic variation.

The Importance of Natural History

Some of the problems in interpreting estimates of genetic structure derive from gaps in our natural history understanding. For example, if we know absolutely nothing about dispersal, it is impossible to know whether dispersal could explain a particular set of F-statistics. However, genetic statistics can be much more informative if combined with insightful natural history. An excellent example of this is found in a series of papers regarding the conservation genetics of desert fishes (e.g., Meffe and Vrijenhoek 1988; Vrijenhoek 1996). Here, by contrasting two different models for fish populations, very clear patterns of genetic structure are delineated. One model was for a group of fish species occupying a series of ponds that are now totally isolated, but were once part of a much larger interbreeding population. An alternative model had fish occupying a hierarchy of streams with varying degrees of isolation caused mainly by dams. For the completely isolated fish populations, Meffe and Vrijenhoek predict low diversity of neutral genes, high between-population variation, (i.e. higher F_{ST}'s) and opportunity for adaptive divergence. For the stream hierarchy model, Meffe and Vrijenhoek predicted a hierarchical partitioning of genetic variation, with the greatest differences expected between geographic regions. This simple verbal model yields testable hypotheses about genetic structure, which ultimately proved to be of practical use. Indeed, after conducting a hierarchical partitioning of genetic variation using allozyme data for the Sonoran topminnow (*Poeciliopsis occidentalis*), it was clear that the majority (53%) of genetic diversity was represented by differences among three geographic regions (Meffe and Vrijenhoek 1988). The practical implication of this finding is that topminnows should probably *not* be moved between regions, even though this may sometimes seem reasonable from a demographic standpoint. In fact, moving fish between populations is a common management practice. Clearly, this model-driven desert fish analysis is more sophisticated than standard "cookbook" F-statistic analyses, and illustrates that more readily interpretable and useful information can be gathered from partitioning genetic variation when the process is connected to biologically specified models.

Estimation and interpretation are certainly major challenges in the use of genetic structure data. But the biggest challenge is determining how to use estimates of genetic statistics to solve practical conservation questions. Indeed, it is rarely clear *when* estimates of genetic structure should be applied to management issues, much less *how* they should be applied. A good example of this dilemma is provided by the clapper rail (*Rallus longirostris*) in southern California. Because clapper rails are listed as an endangered species under the U.S. Endangered Species Act of 1973, and occupy a habitat under intense development pressure (i.e., coastal wetlands and marshes), this species is often the focus of the mitigation or restoration projects (see chapter 15). Occasionally possibilities such as translocation of individuals are considered as management options. A recent genetic study by Fleischer, Fuller, and Ledig (1995) served as a springboard for comments regarding the advisability of translocation. In particular, minisatellite (DNA fingerprinting) data were used to examine patterns of variation in endangered clapper rail populations in southern California. Fleischer and colleagues assessed the variation expressed in DNA fingerprinting (minisatellite) data from two endangered subspecies of the clapper rail; a highly fragmented coastal subspecies for which four subpopulations were sampled, and a more continuously distributed subspecies in California's interior for which one subpopulation was sampled.

At first glance, the results of this study were simple and striking. The subpopulations of the fragmented coastal rails revealed almost no genetic diversity as quantified by "band-sharing" for DNA fingerprints (i.e., all the birds tended to share similar banding patterns). The inland subspecies, however, exhibited much less band-sharing among individuals. Using an approach pioneered by Lynch (1991), Fleischer, Fuller, and Ledig converted band-sharing into "similarity indices", which can in turn provide estimates of effective population size assuming a mutation-drift equilibrium and a known mutation rate. The estimated effective population sizes were 65–420 for the coastal subspecies and

507–1140 for the inland subspecies. Given these estimates of effective population size, migration frequency was estimated by substituting N_e into equation 4 (after calculating F_{ST}'s following Lynch (1991).

While the intricacy of this approach may seem bewildering, there is a central logical error that is starkly clear. Without auxiliary information, genetic structure data do not allow independent estimation of effective population size and migration rate. Specifically, one cannot assume zero migration to calculate effective population size, and then plug that effectively "migrationless" estimate of effective population size into an equation that is used to estimate a non-zero migration rate. The questionable estimation of migration rate is not a trivial problem because it produced a supposed contradiction between the amount of differentiation found among rail subpopulations and estimated migration. In particular, because estimated migration was low yet there was minimal differentiation among coastal rail populations, Fleischer and colleagues suggested that coastal rail populations may have gone through a recent bottleneck or had a special metapopulation structure precluding differentiation among populations.

If all of this were simply hypothesizing in an academic arena, then there would be little cause for concern. However, in this case the genetic analyses were used to actually suggest that translocations be conducted with wild populations to increase genetic variability in coastal rails. Indeed, the researchers indicated that translocations would be a way of increasing effective population size without cost, because with so little haplotype divergence there would be little chance that maladapted genotypes would be introduced. Of course, assessments of genetic structure say nothing about adaptation to local environments, and the extent to which translocation increases effective population size depends on the likelihood that the transplanted individuals carry and pass on novel alleles into the recipient populations. These assessments also do not say anything about other important factors to consider, such as the risk of spreading parasites or diseases by moving individuals between populations. Unfortunately, these various critical aspects of translocations were not apparent from the available data. Clearly Fleischer and colleagues were in a fairly typical conservation situation—i.e., having to suggest management options without sufficient supporting ecological data and with an abundance of genetic data whose interpretation was ambiguous. However, given all of the potential pitfalls of genetic structure analyses, it is important that the temptation to over-interpret data be resisted.

These case studies may seem to indicate that there is not a great deal of utility in genetic structure analyses of threatened populations. However, even if we cannot easily infer *specific* ecological parameters from population genetic studies, we do not believe that this is the case. First, simply identifying the scale at which genetic differentiation is manifested tells us the appropriate scale to direct our management efforts if our goal is to maintain genetic diversity. More subtly, the very fact that F_{ST}'s *do* reflect history can be an asset. For example, imagine a fragmented landscape, dotted with small isolated subpopulations. If such a complex of subpopulations shows a lack of genetic structure (i.e. a F_{ST} near zero), then we have learned something important. Specifically, we can be sure that historically, fragmentation with low dispersal was most likely *not* the situation. In addition, if we find small isolated populations with unusually high amounts of genetic variation relative to the rest of the populations studied, this could indicate that there are undetected populations in the vicinity (and thus more surveys might be in order), or that something ecologically interesting is taking place (and thus those particular populations might warrant additional study). However, we also believe that there is room for improvement in studies of genetic structure. Meffe and Vrijenhoek's (1988) desert fish example illustrated out how partitioning genetic variation following a biologically meaningful model based on practical management units can help direct conservation strategies. In the next section we explore ways that individual-based, spatially-explicit simulation models can be employed to further enhance the practical utility of genetic structure approaches.

A DETAILED SIMULATION MODEL EXPLORING THE CONNECTION BETWEEN ECOLOGICAL PROCESSES AND MEASURES OF GENETIC STRUCTURE

The three forces involved in structuring neutral genetic variation are drift, migration, and mutation. While we can predict how differences in each of these factors should influence genetic structure, it is difficult to tease apart the effects of any one of them from any other simply on the basis of estimates of genetic structure. Both migration and mutation introduce variation into populations slowing down differentiation, while maintenance of large population sizes over time reduces the influence of genetic drift and thus slows down spatial differentiation as well as loss of genetic variation. At any one time point the same genetic structure can be obtained from small population size and high migration or from large population size and low migration. To make matters more complicated, because changes in genetic structure do not instantaneously track changes in population size or connectivity, a currently small or isolated population could exhibit the genetic structure of the large or interconnected population it was long ago.

One way to see both the limitations and possibilities of genetic structure measures is to explore a model in which one knows both the genetics and the ecology (e.g., Rohlf and Schnell 1985; Harris and Allendorf 1989; Lesica and Allendorf 1992; Leberg, 1992; de Jong *et al.* 1994; Richards and Leberg 1996). To explore the problems of estimation and inference using estimates of genetic structure, we have developed an individually-based model for diploid organisms in which migration, population collapse, and habitat fragmentation can be explicitly modeled, while tracking matings and changes in gene frequencies. Instead of producing a general abstract model, we kept in mind a particular animal (the pocket gopher, a fossorial rodent in the family Geomyidae) for which there are existing data (e.g., Patton and Feder 1981; Daly and Patton 1990) and for which we have been collecting additional data of the type we have been discussing (Steinberg 1995). Although our choice of pocket gophers as model organisms may seem idiosyncratic, these organisms have previously proven to be good model subjects for the study of genetic structure (e.g., Slatkin 1993). More importantly, we have practical applications of this model in mind, because our genetic and ecological data are from threatened populations of pocket gophers (*Thomomys mazama*) from Washington state that occur in rapidly disappearing remnant prairies. In addition, it is helpful to have in mind the biology of a specific organism if the goal is to assess the link between changes in ecology and changes in genetic structure.

The behavior that we give our model individuals is that of an asocial territorial animal. Each individual is assigned a genotype for between five and 15 independently segregating loci. We vary this number because the number of loci sampled influences the strength of our inferences about genetic structure and diversity. For each locus we allow between five and 20 alleles (the lower number might represent allozyme data, whereas the higher numbers represent typical microsatellite data). The runs reported here are simulated based on microsatellite variants, where the variation is in terms of length polymorphisms. We assume a Gaussian initial frequency distribution of variants with a specified mean and variance, such that the genotype of each locus in each individual begins as a diploid pair of random draws from this distribution. This initializes the model with genetic variation, but no genetic structure (i.e., $F_{ST} = 0$). The length variation is grouped into integer values so that a mean of 5 and a standard deviation of 2 would commonly produce alleles of length 5, less so of lengths 4 and 6, even less so of lengths 3 and 7, and so on. The mutation process is viewed as slippage during DNA replication (Levinson and Gutman 1987; Schlotterer and Tautz 1992) and mutation is thus depicted in our model as length changes of one unit (e.g., from 4 to 3) at a rate of 10^{-4} per allele per gamete—a commonly reported mutation rate for microsatellite data (see Levinson and Gutman 1987).

Space is represented in our model as a lattice. Our lattice comprises four separate sectors, each sector is 20 by 20 cells or nodes (400 potential "territories" in each sector). Dispersal within each sector, which is meant to correspond to "within subpopulation," occurs at the same rate within all four sectors. However, dispersal between the four sectors is much reduced compared to within-sector movement, and occurs at a rate that we manipulate in different simulation runs to examine the consequences of habitat fragmentation. Our organisms are territorial and cannot live through the winter unless they hold an exclusive territory. The requirement of a territory to survive the winter introduces a strongly stabilizing form of density-regulation.

Although the genetics is based upon an annual breeding cycle, the model actually runs on a much finer time-scale so that movement and survival of individuals can be tracked in a manner that meshes with the type of data field biologists are likely to gather (e.g., Daly and Patton 1990). Our organisms live for up to four years; each year broken into winter, spring, and summer; and individual behavioral options depending upon the season as well as the sex and age of the individual. During their first year, juveniles disperse from their natal site in search of open territories, while all older individuals hold a territory, otherwise they would have died the preceding winter. Dispersal is by a simple random walk, with the disperser checking the status of the cell in which it lands at each step to see whether it is "open." All dispersing individuals are subjected to elevated mortality compared to territory-holders. When a disperser encounters the edge of its habitat sector, it either turns around, attempts to migrate to a new sector, or dies. We can manipulate the "successful" migration probability to simulate dispersal barriers between sectors by specifying the fraction of the dispersers that find any one of the three other habitat sectors.

The other key feature of our model is mating. All females one year and older are able to reproduce, while only males two years and older can reproduce (following Daly and Patton 1986). To obtain mates, the females search outwards from their home territory in an ever widening circular path up to ten cells or nodes in diameter until they find an eligible male. All mating encounters are successful, producing a litter of four to six juveniles with a primary sex ratio of 1:1 and randomly assorted allelic values for each juvenile. Juveniles occupy their mothers' territories for one month before leaving in search of their own territories.

All simulations reported here started with a total of 800 individuals spread among our four subpopulations (200 per sector, which represents near-equilibrium population size for this model). Initial locations for individuals were randomly assigned. Because the simulations involve several stochastic processes (random walk movement, random mortality, random mating, mutation, etc.), we ran twenty replicate simulations for each combination of parameters. The output of each simulation run included the number and position of individuals in each subpopulation, gene frequencies for each subpopulation, and F-statistics every one to ten years estimated from samples of varying sizes. We also recorded the same output for the entire system, as though every individual had been sampled.

First, perhaps the most mundane, but nonetheless important point is that when neutral allelic variation is being examined, the outcome of any particular population trajectory can be quite different from the mean expected value over many replicate trajectories. Thus, *by chance alone, it is possible to generate F_{ST}'s that depart substantially from the expected F_{ST} for some specified ecology and demography.* This indicates that calculating backwards from an observed F_{ST} to thenumber of migrants per generation or effective population size can be precarious business. For example, in Figure 19.1 we display twenty different realizations of the same ecological process (the conditions of the model described above), as summarized by F_{ST}. It is not uncommon to see different random trajectories varying by a factor of two, even though Figure 19.1 illustrates a "best case" scenario for our simulation, because every virtual gopher was sampled genetically. The variance will be even greater in typical cases when only a small fraction of a population is sampled. In the real world, variability between replicate evolutionary outcomes depends on a number of different factors, including dispersal, population size (the populations we modeled were typical of threatened species, but undoubtedly small compared to populations of many common species), and the process by which populations are founded. Figure 19.1 thus serves as a demonstration of a particular case; however, it is certainly possible to use this approach to incorporate the effects of chance into any empirical analysis, given various different models of potential historic factors and processes (Steinberg and Jordan in preparation).

More relevant to conservation concerns, however, is the tendency to interpret F_{ST}'s and other genetic indices in the context of current ecologies. *Genetic structure does not respond immediately to change in population dynamics*—in fact, given changing environmental conditions, genetic structure may never "catch up" to population structure. To illustrate this point we simulated a "fragmentation event" that decreased the fraction of between-sector migrants that succeed by 90%, a process roughly equivalent to cutting the rate of dispersal success between subpopulations from .5% per year to .05% per year. We found that F_{ST}'s eventually tripled as a result of this impedance of dispersal, but that the effect on the time scale of 10 to 20 generations following the "fragmentation" is negligible (Figure 19.2). Thus while an ecological parameter (in this case the mean number of between-subpopulation dispersers) shows a precipitous and dramatic response, the genetic response lags far behind (Figure 19.2).

Finally, our discussion thus far has emphasized simulation results based on sampling every single individual in all populations. In the real world, field studies of threatened species tend to be plagued with small sample sizes. If one runs our simulation and samples only 20 individuals from each subpopulation of 200, it becomes even more difficult to detect statistically significant changes in F_{ST} (Figure 19.3). Here, we randomly sampled individuals from a population and then analyzed our samples using Lewis and Zaykin's (1996) computer program Genetic Data Analysis (GDA), which allowed us to both calculate estimated F_{ST}'s and to provide confidence intervals for those sample estimates (see Figure 19.3). Actually, Figure 19.3 is somewhat optimistic, since in reality, researchers often don't have access to loci with the amount of variability we modeled (on average 10 alleles/locus). Indeed, careful consideration of sampling design and the constraints of

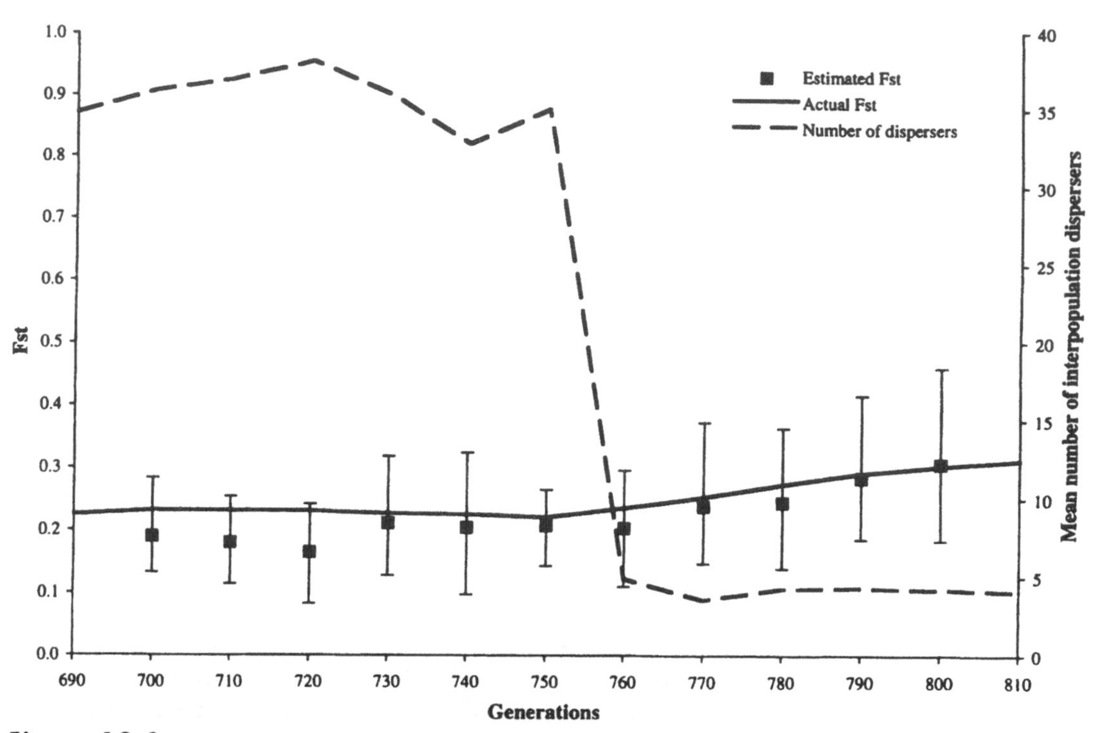

Figure 19.3
Detail of dispersal perturbation event. Mean F_{ST} and mean ID's 50 generations before and after reduction of migration rates from 5% per generation (see legend to Figure 19.2 for interpretation) to 0.5% per generation at generation 750. The solid and dashed lines are as in Figure 19.2. In addition we calculate F_{ST} with five diploid loci for a subsample of the total population (20 individuals per sector, N = 80) every ten generations (solid squares, GDA v1.0 (Lewis and Zaykin 1996)). The error bars represent 95% confidence intervals on the subsampled F_{ST} calculated by numerically resampling the data set (1000 bootstrap replicates).

sparse sampling is generally lacking in the conservation genetics literature. On the positive side, the modeling approach we use can be easily applied to identify optimal sampling strategies for particular systems (Steinberg and Jordan in preparation).

CONCLUSIONS

Our ability to quantify genetic variation at the molecular level has spawned an industry of papers reporting the genetic structure of natural populations. These reports have merit because genetic structure represents valuable baseline data against which changes through time might be detected, and because assessments of genetic structure can provide a platform for the exploration of assorted evolutionary and ecological hypotheses. However, there is a tendency in much of the literature to take too literally abstract models that predict genetic structure as a function of ecological and population parameters. The models that inspire such thinking are insightful, but they are intended to work in one direction—i.e., given some assumptions about population biology, what can we predict about genetics? The models do not work in the reverse. In short, there is not just one way to get a particular pattern of genetic structure, and thus we can not easily assign ecological parameters to data on genetic structure. However, all is not without hope. *More profound inference may be possible if instead of general population genetic models, we explore particular genetic models coupling ecology and genetics.* Given such specific models, the range of possible causes for a realized genetic structure are constrained by particular biological parameters. If a species is clearly characterized by high dispersal and yet a high F_{ST} is estimated from genetic data, then one can rule out low dispersal for causing the pattern, and the explanation for the high F_{ST} can be sought elsewhere. Of course, this use of models and data on genetic structure does not obviate the inherent problem of demographic parameters being difficult to obtain for most threatened species. It may well be that genetics cannot teach us what we want to know about demography; although specific models, and more sophisticated approaches might at least identify which demographic data are most crucial to obtain.

A second, more subtle problem is associated with the seemingly straightforward "comparative" use of genetic structure indices. For example, one might find small F_{ST}'s among large subpopulations that are connected by corridors, and large F_{ST}'s among smaller, more fragmented populations. It is natural to attribute these differences to the effect of fragmentation. As long as one engages some form of numerical subsampling (e.g., bootstrap analysis) to make sure the two F_{ST}'s are significantly different, then one can conclude the two groups of populations are indeed significantly different in terms of genetic structure. Too often, however, after doing rigorous statistics to test for significant differences among F_{ST}'s, researchers then ignore the fact that a particular F_{ST} is but one realization of a stochastic process and report Nm's as a single number with no confidence intervals. This data presentation can be very misleading. To make matters much worse, there is the hazard of observing genetics that reflect historical circumstances, having nothing to do with the current situation. It is important that these two issues be taken into consideration when data are being interpreted to suggest management practices.

It is an alluring idea that data describing the distribution of genetic variation within a species can be analyzed in a simple way to determine otherwise elusive demographic parameters. However, we think this idea is generally misguided—especially within a conservation framework where populations are unstable at best. Given the vast number of conditions that can lead to particular patterns of genetic structure, and the general violation of critical assumptions of equilibrium conditions necessary to apply F_{ST}-based models, we suggest an alternative approach. Specifically, we advocate constructing models that match the natural history of the organism of interest as closely as possible, and then exploring parameters of those models in the context of observed genetic structure data—asking which specific explanations might be plausible explanations and which are not. Never will data on genetic structure cure deficiencies in ecological data, nor will answers ever be found in the simple equations drawn from basic population genetics textbooks. But the combination of genetic data and ecological data, in concert with specific models, could provide important insight into likely consequences of problems that so often worry conservation biologists.

ACKNOWLEDGMENTS

During the preparation of this chapter E. Steinberg was supported by a training grant from NSF in mathematical biology, a research grant from The Nature Conservancy, and a research grant to P. Kareiva by NSF. We thank P. Kareiva, J.L. Patton, and four anonymous reviewers for helpful comments on the manuscript and P. Lewis for making GDA available for our use.

LITERATURE CITED

Allendorf, F.W. 1986. Genetic drift and loss of alleles versus heterozygosity. *Zoological Biology* 5:181–190.

Amos, B., C. Schlotterer, and D. Tautz. 1993. Social structure of pilot whales revealed by analytical DNA profiling. *Science* 260:670–672.

Ashley, M.V., D.J. Melnick, and D. Western. Conservation genetics of the black rhinoceros (*Diceros bicornis*), I: Evidence from the mitochondrial DNA of three populations. *Conservation Biology* 4: 71–77.

Avise, J.C. 1994. *Molecular markers, natural history and evolution*. Chapman & Hall, New York, NY.

Bonnell, M.T. and R.K. Selander. 1974. Elephant seals: Genetic variation and near extinction. *Science* 184:908–909.

Bowen, B., J.C. Avise, J.I. Richardson, A.B. Meylan, D. Margaritoulis, S.R. Hopkins-Murphy. 1993. Population structure of loggerhead turtles (*Caretta caretta*) in northwestern Atlantic Ocean and Mediterranean sea. *Conservation Biology* 7:834–844.

Crow, J.F. and K. Aoki. 1984. Group selection for a polygenic behavioral trait: Estimating the degree of population subdivision. *Proceedings of the National Academy of Sciences* 81:6073–6077.

Daly, J.C. and J.L. Patton. 1986. Growth, reproduction, and sexual dimorphism in *Thomomys bottae* pocket gophers. *Journal of Mammalogy* 67:256–265.

Daly, J.C. and J.L. Patton. 1990. Dispersal, gene flow, and allelic diversity between local populations of *Thomomys bottae* pocket gophers in the coastal ranges of California. *Evolution* 44:1283–1294.

Dinerstein, E. and G.F. McCracken. 1990. Endangered greater one-horned rhinoceros carry high levels of genetic variation. *Conservation Biology* 4:417–422.

Dunning, J., D.J. Stewart, B.J. Danielson, B.R. Noon, T.L. Root, R.H. Lamberson, and E.E. Stevens. 1995. Spatially explicit population models: current forms and future uses. *Ecological Applications* 5:3–11.

Fahrig, L. and G. Merriam. 1994. Conservation of fragmented populations. *Conservation Biology* 8:50–59.

Felsenstein, J. 1995. *Theoretical evolutionary genetics*. Seattle: ASUW Publishing.

Fleischer, R.C., G. Fuller, and D.B. Ledig. 1995. Genetic structure of endangered Clapper Rail (*Rallus longirostris*) populations in southern California. *Conservation Biology* 9:1234–1243.

Hall, P., S. Walker, and K. Bawa. 1996. Effect of forest fragmentation on genetic diversity and mating system in a tropical tree, *Pithecellobium elegans*. *Conservation Biology* 10:757–768.

Harris, R.B. and F.W. Allendorf. 1989. Genetically effective population size of large mammals: an assessment of estimators. *Conservation Biology* 3:181–191.

Harrison, S., A. Stahl, and D. Doak. 1993. Spatial models and spotted owls: exploring some biological issues behind recent events. *Conservation Biology* 7:950–953.

Hartl, D.L. and A.G. Clark. 1989. *Principles of population genetics*. Sunderland, MA: Sinauer Associates.

Hellberg, M.E. 1994. Relationships between inferred levels of gene flow and geographic distance in a philopatric coral, *Balanophyllia elegans*. *Evolution* 48:1829–1854.

Hickey, R.J., M.A. Vincent, and S.I. Guttman. 1991. Genetic variation in Running Buffalo Clover (*Trifolium stoloniferum*, Fabaceae). *Conservation Biology* 5:309–316.

Hillis, D.M., C. Moritz, and B.K. Mable, eds. 1996. *Molecular systematics*, 2nd Edition, Sunderland, MA: Sinauer Associates.

Holt, R.D., S.W. Pacala, T.W. Smith, and J. Liu. 1995. Linking contemporary vegetation models with spatially explicit animal population models. *Ecological Applications* 5:20–27.

Husband, B. and S.C.H. Barrett. 1992. Effective population size an genetic drift in tristylous *Eichhornia paniculata* (Pontederiaceae). *Evolution* 46:1875–1890.

Jong, G. de, J.R. de Ruiter, and R. Haring. 1994. Genetic structure of a population with social structure and migration. In *Conservation genetics,* eds. V. Loeschcke, J. Tomiuk, and S.K. Jain, 147–164. Basel, Switzerland: Birkhauser Verlag.

Karl, S.A., B.W. Bowen, and J.C. Avise. 1992. Global population genetic structure and male-mediated gene flow in the green turtle (*Chelonia mydas*): RFLP analysis of anonymous nuclear loci. *Genetics* 131:163–173.

Lande, R. and G.F. Barrowclough. 1987. Effective population size, genetic variation, and their use in population management. In *Viable populations for conservation,* ed. M. Soulé, 87–123. Cambridge: Cambridge University Press.

Lahaye, W., R. Gutierrez, and H. Akcakaya. 1994. Spotted owl metapopulation dynamics in southern California. *Journal of Animal Ecology* 63:775–785.

Leberg, P.L. 1991. Influence of fragmentation and bottlenecks on genetic divergence of wild turkeys. *Conservation Biology* 5:522–530.

Lesica, P. and F.W. Allendorf. 1992. Are small populations of plants worth preserving? *Conservation Biology* 6:135–139

Levinson, G. and G.A. Gutman. 1987. Slipped-strand mispairing: A major mechanism for DNA sequence evolution. *Molecular Biology and Evolution* 4:203–221.

Lewis, P.O. and D. Zaykin. 1996. Genetic data analysis II: Software for the analysis of discrete genetic data. Computer program distributed by the authors.

Lynch, M. 1991. Analysis of population genetic structure by DNA fingerprinting. In *DNA fingerprinting: approaches and applications*, eds. T. Burke, G. Dolf, A.J. Jeffreys, and R. Wolff, 113–126. Basel, Switzerland: Birkhauser Verlag.

Malecot, G. 1975. Heterozygosity and relationship in a regularly subdivided population. *Theoretical Population Biology* 8:212–241.

Maruyama, T. 1971. Rate of decrease of genetic variability in a two-dimensional continuous population of finite size. *Genetics* 70:639–651.

McKelvey, K., B.R. Noon, and R.H. Lamberson. 1993. Conservation planning for species occupying fragmented landscapes: the case of the northern spotted owl. In *Biotic Interactions and Global Change*, eds. P.M. Kareiva, J.G. Kingsolver, and R.B. Huey, 424–450. Sunderland, MA: Sinauer Associates.

Meffe, G.K. and R.C. Vrijenhoek. 1988. Conservation genetics in the management of desert fishes. *Conservation Biology* 2:157–169.

Milligan, B.G., J. Leebens-Mack, and A.E. Strand. Conservation genetics: beyond the maintenance of marker diversity. *Molecular Ecology* 3:423–435.

Nunney, L. 1995. Measuring the ratio of effective population size to adult numbers using genetic and ecological data. *Evolution* 49:389–392.

Nunney, L. and D.R. Elam. 1994. Estimating the effective population size of conserved populations. *Conservation Biology* 8:175–184.

Paetkau, D., W. Calvert, I. Stirling, and C. Strobeck. 1995. Microsatellite analysis of population structure in Canadian polar bears. 1993. *Molecular Ecology* 4:347–354.

Patton, J.L. and J.H. Feder. 1981. Microspatial genetic heterogeneity in pocket gophers: Non-random breeding and drift. *Evolution* 35:912–920.

Richards, C. and P.L. Leberg. 1996. Temporal changes in allele frequencies and a population's history of severe bottlenecks. *Conservation Biology* 10:832–839.

Robinson, N.A., N.D. Murray, and W.B. Sherwin. 1993. VNTR loci reveal differentiation between and structure within populations of the eastern barred bandicoot *Perameles gunnii. Molecular Ecology* 2:195–207.

Rohlf, F.J. and G.D. Schnell. 1985. An investigation of the isolation by distance model. *American Naturalist* 105:295–324.

Sangel, P.W., M.R. Lennartz, and M.H. Smith. 1992. Genetic variation and population structure of Red-cockaded woodpeckers. *Conservation Biology* 6:283–292.

Sarre, S. 1995. Mitochondrial DNA variation among populations of *Oedura reticulata* (Gekkonidae) in remnant vegetation: Implications for metapopulation structure and population decline. *Molecular Ecology* 4:395–405.

Schlotterer, C. and D. Tautz. 1992. Slippage synthesis of simple sequence DNA. *Nucleic Acids Research* 20:211–215.

Sherwin W.B., N.D. Murray, J.A. Marshall Graves, and P.R. Brown. 1991. Measurement of genetic variation in endangered populations: Bandicoots (Marsupialia: Peramelidae) as an example. *Conservation Biology* 5:103–108.

Slatkin, M. 1985a. Rare alleles as indicators of gene flow. *Evolution* 39:53–65.

Slatkin, M. 1985b. Gene flow in natural populations. *Annual Review of Ecology and Systematics* 16:393–430.

Slatkin, M. and N.H. Barton. 1989. A comparison of three indirect methods for estimating average levels of gene flow. *Evolution* 43:1349–1368.

Slatkin, M. and W.P. Maddison. 1990. A cladistic measure of gene flow inferred from the phylogenies of alleles. *Genetics* 123:603–613.

Slatkin, M. 1993. Isolation by distance in equilibrium and non-equilibrium populations. *Evolution* 47:264–279.

Soltis, P.S., D.E. Soltis, T.L. Tucker, and F.A. Lang. 1992. Allozyme variability is absent in the narrow endemic *Bensoniella oregona* (Saxifragaceae). *Conservation Biology* 6:131–134.

Steinberg, E.K. 1995. *A study of genetic differentiation and variation in the Mazama pocket gopher (*Thomomys mazama*) with an emphasis on Fort Lewis populations*. Final Report on Contract #WAFO 100193 submitted to The Nature Conservancy.

Steinberg, E.K. and C.E. Jordan. Using individual-based simulation models to design sampling methods and generate explicit testable hypotheses in conservation genetics. In prep.

Stone, G.N. and P. Sunnucks. 1993. Genetic consequences of an invasion through a patchy environment—the cynipid gallwasp *Andricus quercuscalicis* (Hymenoptera: Cynipidae). *Molecular Ecology* 2:251–268.

Stiven, A. E. and R.C. Bruce. 1988. Ecological genetics of the salamander *Desmognathus quadramaculatus* from disturbed watersheds in the southern Appalachian biosphere reserve cluster. *Conservation Biology* 2:194–205.

Swofford, D.L. and R.B. Selander. 1981. BIOSYS-1: A FORTRAN program for the comprehensive analysis of electrophoretic data in population genetics and systematics. *Journal of Heredity* 72:281–283.

Taylor, A.C., W.B. Sherwin, and R.K. Wayne. 1994. Genetic variation of microsatellite loci in a bottlenecked species: The northern hairy-nosed wombat *Lasiorhinus krefftii*. *Molecular Ecology* 3:277–290.

Varvio, S., R. Chakraborty, and M. Nei. 1986. Genetic variation in subdivided populations and conservation genetics. *Heredity* 57:189–198.

Waples, R.S. 1989. A generalized approach for estimating effective population size from temporal changes in allele frequency. *Genetics* 121:379–391.

Wayne, R.K., N. Lehman, M.W. Allard, and R.L. Honeycutt. 1992. Mitochondrial DNA variability of the gray wolf: Genetic consequences of population decline and habitat fragmentation. *Conservation Biology* 6:559–569.

Weir, B.S. 1996. *Genetic Data Analysis II*. Sinauer, Sunderland, MA.

Weir, B.S. and C.C. Cockerham. 1984. Estimating F-statistics for the analysis of population structure. *Evolution* 38:1358–1370.

Wetton, J.H., R.E. Carter, D.T. Parkin, and D. Walters. 1987. Demographic study of a wild house sparrow population by DNA fingerprinting. *Nature* 327:147–149.

Wright, S. 1931. Evolution in Mendelian populations. *Genetics* 16:97–159.

Wright, S. 1943. Isolation by distance. *Genetics* 28:114–138.

Wright, S. 1951. The genetic structure of populations. *Annals of Eugenics* 15:323–354.

Wright, S. 1969. *Evolution and the genetics of populations. Vol. 2. The theory of gene frequencies.* Chicago: University of Chicago Press.

Wright, S. 1978. *Evolution and the genetics of populations. Vol. 4. Evolution and the genetics of populations.* Chicago: University of Chicago Press.

Young, A.G. and H.D. Brown. 1996. Comparative population genetic structure of the rare woodland shrub *Daviesia suaveolens* and its common congener *D. mimosoids*. *Conservation Biology* 10:1220–1228.

CHAPTER

20

Conservation Endocrinology: Field Endocrinology Meets Conservation Biology

STEPHAN J. SCHOECH
and JOSEPH L. LIPAR

Conservation biology as a scientific discipline is dominated by natural history and population biology. However, there are several challenges and questions in conservation biology that can be elegantly addressed with new techniques from endocrinology. The major change in endocrinology that makes this possible is the development of field techniques that allow us to probe an animal's hormonal status while the animal ranges free in the wild (Wingfield and Farner 1976). Clearly hormones are as crucial an attribute to an animal as are its body size, general health, and reproductive rates. In fact, from one perspective, hormones might be the most fundamental measure of an animal's likely success. After all, it is hormones that largely control reproduction and coordinate the physiological responses necessary for survival in a stressful environment. Thus, endocrinology offers us a window to better understand the factors impairing a species' demographic vitality, and it may even offer us early-warning signals of a risk before survivorship or reproductive rates plummet. In this chapter we sketch the recent advances in endocrinology that have the greatest potential as tools in the service of conservation biology. Before turning to case studies that document the role that endocrinology can play in conservation, we briefly review some pertinent aspects of vertebrate endocrine systems.

GENERAL ASPECTS OF VERTEBRATE ENDOCRINOLOGY

The hypothalamo-pituitary-gonadal (HPG) axis consists of the brain region known as the hypothalamus, the pituitary, and the gonads of an organism. In response to endogenous or environmental cues, 'higher brain centers' stimulate the hypothalamus to release gonadotropin-releasing hormone (GnRH). GnRH travels to the anterior pituitary where it stimulates the release of the gonadotropins, luteinizing and follicle-stimulating hormone (LH and FSH, respectively). During puberty or during the annual reactivation of the reproductive tract in a seasonal breeder, these blood-borne hormones stimulate gonadal maturation.

In an animal whose gonads are mature, LH and FSH maintain gonadal function. Additionally, the female's estrous or menstrual cycles in many mammalian species are temporally synchronized through differential release of LH, FSH and sex steroid hormones.

Mature testes and ovaries, in addition to producing the respective male and female gametes, also produce and secrete sex steroid hormones. The primary sex steroids, testosterone and estradiol in males and females, respectively, travel via the blood where they affect (often in synergy with other hormones) the development of secondary sex characteristics, e.g., facial hair and breasts in humans and 'nuptial' plumage of male birds. Additionally, there are binding sites for sex steroid hormones in several areas of the brain where they typically induce behaviors that are associated with the reproductive requirements of the individual (see reviews in Follett 1984; Ball 1993; Wingfield and Farner 1993).

In addition to the "activational" functions of reproductive hormones, sex steroid hormones also have profound "organizational" roles in sex determination. In general, developmental exposure to the "correct" sex steroid hormones is essential for the normal development (i.e., organization) of a fully functional reproductive system. Exposure to the 'wrong' exogenous hormones or, alternately, hormone mimics, during the organization of the reproductive system of an individual can result in masculinized females or feminized males at the level of the brain or gonad (see *American Alligators* Case Study below).

Another relevant endocrine axis is the hypothalamo-pituitary-adrenal axis (HPA axis). In much the same fashion as the HPG axis, an endocrine cascade is initiated by 'higher brain centers' in response to endogenous or environmental cues. In response to the cue, the hypothalamus secretes corticotropin releasing hormone (CRH) which then travels to the pituitary. The pituitary responds by releasing adrenocorticotropin (ACTH) into the peripheral blood stream where it travels to, and subsequently acts upon, the adrenal cortex. ACTH stimulates adrenal release of glucocorticoids, primarily corticosterone in birds, reptiles, and amphibians, and cortisol in mammals and fishes. Interestingly, some organisms can have both, and within classes there are exceptions, e.g., most rodents have corticosterone (see Gorbman *et al.* 1983). These adrenal steroid hormones are secreted in response to a wide variety of stressful stimuli and are generally thought to be an essential component of an individual's acclimatization or response to stress (Harvey *et al.* 1974; Siegel 1980; Greenberg and Wingfield 1987; Wingfield 1988; Moore, Thompson, and Marler 1991). Gluconeogenesis, one of the primary actions of glucocorticoids, serves to increase blood glucose levels to provide the animal with ready energy to fuel a response to the stressor. It is generally believed that the adrenocortical response facilitates survival of the individual by allowing the animal to respond to environmental challenges. However, prolonged or repeated stimulation of the adrenocortical system can be debilitating and, in extreme cases, can result in the death of the animal (Selye 1971; Sapolsky 1987; Wingfield *et al.* in press).

In addition to effecting the catabolic machinery of an animal, glucocorticoids have been shown to inhibit reproductive behavior and physiology in a wide variety of taxa (Wilson and Follett 1975; Siegel 1980; Moore and Miller 1984; Moore and Zoeller 1985; Wingfield 1988). In birds (and other animals), elevated corticosterone can cause low levels of sex steroid and luteinizing hormones, as well as incomplete gonadal development (see review in Wingfield and Farner 1993).

In short, it is easy to see that knowledge of the endocrine state of an animal might be important to those interested in conserving a species. If the objective is to assure the survival of a species or population, we must guarantee that conditions are correct for individuals of the species to reproduce. Although not a panacea, field endocrinology can be added to the arsenal of tools employed by conservation biologists. Endocrine methods can be used to assess the reproductive status of individuals within a population to gauge the overall health of a population. Additionally, by monitoring glucocorticoid levels, we might learn when a population is stressed and this, in turn, may help us make informed management decisions to protect the species in question.

FIELD ENDOCRINE CASE STUDIES

The following case studies offer examples of how field endocrinology can be used in conservation biology. This is not a comprehensive review of conservation endocrinology, but rather a series of examples that provide a glimpse of what can be learned through the application of field endocrine methodologies. Although not all of the studies featured are directly related to conservation biology, the techniques used and information learned are (or may be) directly applicable to conservation biology.

Florida Scrub-Jays

The Florida scrub-jay (*Aphelocoma coerulescens*) is a cooperatively breeding species that is limited to xeric scrub oak habitat in Florida. Prior to its recent designation as a species (American Ornithologists' Union 1995), the Florida scrub-jay was federally and state listed as a threatened subspecies. The overall numbers of this species are in steady decline in large part because of habitat loss to agriculture and development for housing and recreation, as well as fire suppression practices that allow their preferred scrub habitat to be overgrown and, hence, unusable. Recent studies of the reproductive physiology of this jay provide crucial insights for conservation (Schoech, Mumme, and Moore 1991; Schoech 1996; Schoech, Mumme, and Wingfield 1996a, 1996b, 1997).

Florida scrub-jays are characterized by nonbreeding individuals (helpers) that assist breeding pairs in rearing young (Woolfenden and Fitzpatrick 1984, 1990). Although nonbreeders may remain reproductively inactive for several years, the majority attain breeder status in their second year. A key question is whether nonbreeders, especially one-year-olds, are physiologically capable of reproducing. Recent field endocrine research reveals temporal patterns of testosterone secretion in nonbreeders that suggest fully functional testes (Figure 20.1; Schoech, Mumme, and Wingfield 1996a). Similarly, estradiol levels among nonbreeders and breeders are equivalent (Figure 20.1). Finally, although nonbreeders have smaller testes and ovarian follicles than breeders, the gonads of one-year-olds are significantly more developed than those found in birds during the winter months when all jays are reproductively quiescent. In summary, the endocrine and gonadal morphology findings suggest that one-year-old jays are fully capable of breeding.

This has interesting implications for reintroduction programs, which is one tool considered for helping this taxon recover from endangerment. Interestingly, however, so far a reintroduction program with one-year-old birds has had limited success (Mumme and Below 1995). Following translocation to Rookery Bay near Naples, Florida—a location at which scrub-jays had been extirpated in the 1920s—five of nine older nonbreeders paired and established territories. However, none of the four first-year jays that were translocated either paired or established territories.

There is considerable evidence that naturally occurring fluctuations in food abundance can affect when or whether an individual breeds in a given year (Daan *et al.* 1986; Wingfield and Kenagy 1991). In poor years individuals typically delay their reproductive effort and some might fail to breed altogether. Conversely, during years with abundant

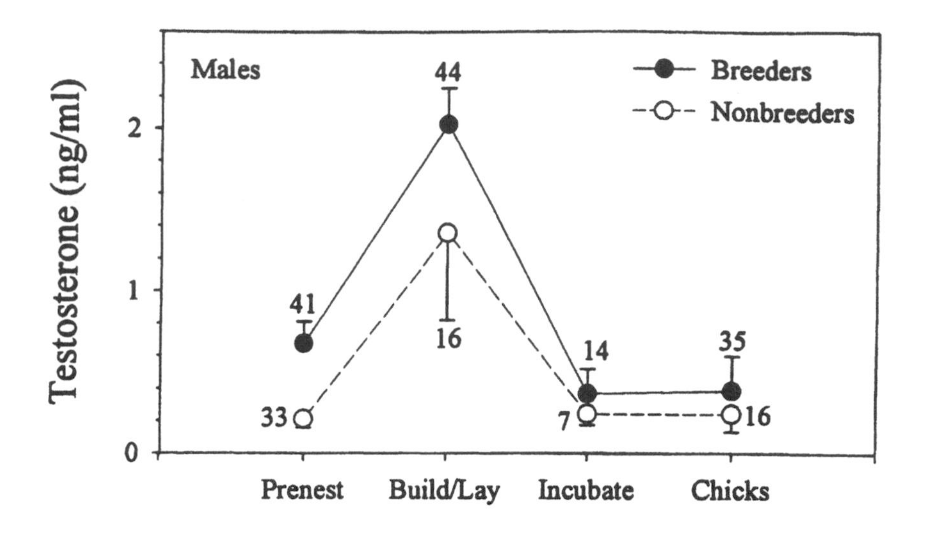

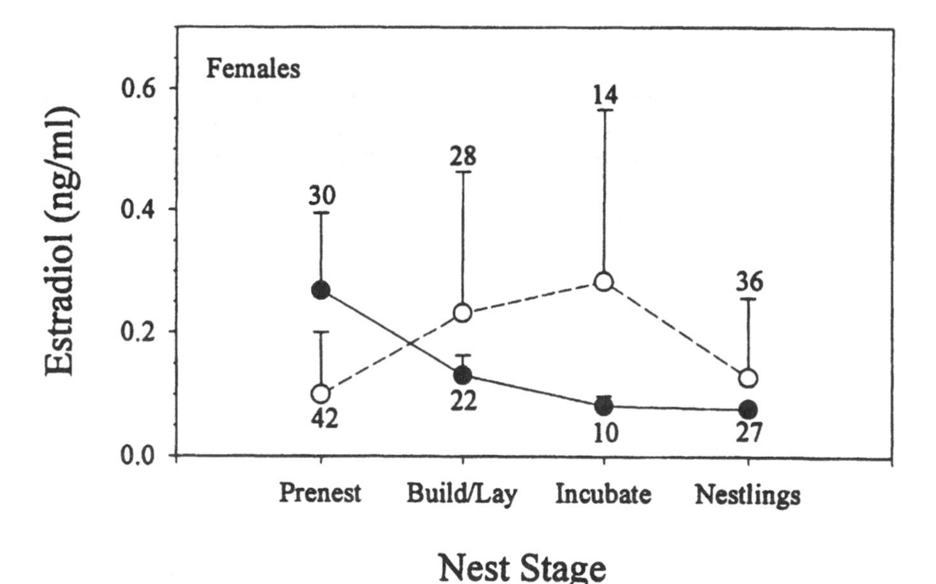

Figure 20.1
Florida scrub-jay sex steroid hormone levels. Upper Panel: Seasonal plasma testosterone levels (means ± SE) of male breeders and nonbreeders when data from 1992–1994 are combined. Although breeders' T levels are higher than nonbreeders, the profile and absolute levels of nonbreeders demonstrates that their testis are producing steroids. Lower panel: Female breeders and nonbreeders had plasma estradiol levels (means ± SE) that did not statistically differ. The estradiol titres seen in nonbreeders show that their ovaries are functional. Adapted from Schoech, Mumme, and Wingfield 1996a and reprinted with permission of Springer-Verlag.

food individuals will generally breed earlier, and most within a population obtain adequate nutrition to initiate reproduction. It might be that the extraordinarily high number of one-year-old Florida scrub-jays that bred in 1993 and 1994, two years that were also notable for exceptionally early clutch initiation, can be attributed to increased food resources (Schoech 1996). Additional evidence of the importance of food in the timing of the reproductive effort can be found in numerous food supplementation experiments, including one with Florida scrub-jays (see Schoech 1996 and citations within). Further support for the importance of food and body condition in Florida scrub-jays comes from data demonstrating that nonbreeders have lower body mass and lower percentages of body lipids than breeders (Schoech 1996). Unfortunately, neither body mass nor condition were measured in either the one-year-old breeders in 1993 and 1994 or in the translocated first year jays. When considered together, the above findings suggest that although first-year birds are capable of breeding, this may occur only in exceptionally good years or with exceptionally healthy individuals. For the purpose of reintroduction efforts, there would be a clear advantage to researchers to be able to use relatively common one-year-old Florida scrub-jays rather than being constrained to using the less common older nonbreeders or valuable established breeders. Clearly, the first management priority should be to protect existing populations against the habitat conversion. Additionally, with an enlightened management reintroduction program that 1) reclaims overgrown scrub and maintains the reclaimed habitat with controlled burning (see Fitzpatrick and Woolfenden 1986; Woolfenden and Fitzpatrick 1990), and 2) helps translocated birds make the transition to their new locales by providing supplemental food, this species might remain off the "critical" list. The key to making this strategy work might well be using one-year-olds with supplemental food—a possibility identified by the application of field endocrinology.

Western Fence Lizards

Western fence lizards (*Sceloporus occidentalis*) are a common species throughout most of their extensive range in the western United States. A series of papers by Dunlap and colleagues demonstrate techniques and pose questions that may benefit conservation efforts with less commonly occurring species (see Dunlap 1995a, 1995b; Dunlap and Schall 1995; Dunlap and Wingfield 1995; Dunlap and Church 1996; Wingfield *et al.* in press).

In particular, Dunlap and Schall's (1995) study of a blood parasite's (*Plasmodium mexicanum*) impact upon the endocrine physiology of lizards could teach us some general lessons of value to conservation biology. Many species of birds, mammals, and reptiles are susceptible to malarial parasites of the genus *Plasmodium*, and western fence lizards provide an excellent model with which to study the behavioral and physiological consequences of infection with this relatively common blood parasite. Schall (1990a, 1990b) noted numerous effects of infection, including decreased fecundity due to lesser fat stores in females, and in males, reduced territorial and courtship behaviors, sexual coloration (thus reducing their attractiveness to females), and testes size. Given these types of effects, it is easy to imagine the potential damage that this blood parasite could inflict upon a population of an endangered species. For example, what might the recent introduction of avian malaria to the Hawaiian Islands do to the beleaguered and highly sensitive avifauna of this archipelago (see Chapter 12).

To better understand the mechanisms underlying the disease symptoms noted by Schall (1990a, 1990b), Dunlap and Schall (1995) compared plasma levels of corticosterone (both basal and capture stress levels, see below) and testosterone in malaria-infected and control lizards. Infected lizards demonstrated lower circulating testosterone titres as well as more pronounced stress responses as indicated by their more elevated corticosterone levels following one hour of confinement (Figure 20.2).

Whereas basal glucocorticoid levels reveal the current (or immediate past) status of adrenocortical activity, sampling an individual following confinement (i.e., capture stress) can provide additional information about adrenocortical responsiveness. This in turn can reveal how an individual might be affected by other stressful stimuli, such as inclement weather, aggressive social interactions, disease, starvation, or human perturbations (see Wingfield 1994). It should be noted that almost without exception, animals in good physiological condition (as determined by body mass or fat depots) have less acute responses to capture stress than individuals in poor condition (Wingfield *et al.* 1994a, 1994b; Dunlap 1995; Schoech, Mumme, and Wingfield 1997). Dunlap (1995) also noted that fence lizards sampled during drought conditions in the Mojave Desert exhibited corticosterone patterns similar to those seen in lizards that were infected with *Plasmodium*, i.e., basal levels that did not differ from controls but an increased stress response.This is further evidence that many different types of stressors induce a similar endocrine response.

To determine whether the low testosterone levels and smaller testes of infected animals were due to direct actions of the parasite upon the reproductive axis, Dunlap and Schall (1995) implanted uninfected individuals with exogenous corticosterone. Captive animals held in enclosures and treated with corticosterone had lower levels of plasma testosterone, smaller testes, and reduced fat stores when compared to controls. The findings of the same characteristics in corticosterone-treated and malaria-infected lizards suggest that impairment of reproductive function is mediated through the adrenocortical response. Although not conclusive, the captive lizard study supports Dunlap and Schall's (1995) hypothesis that the parasite infection caused heightened adrenal response to acute stressors

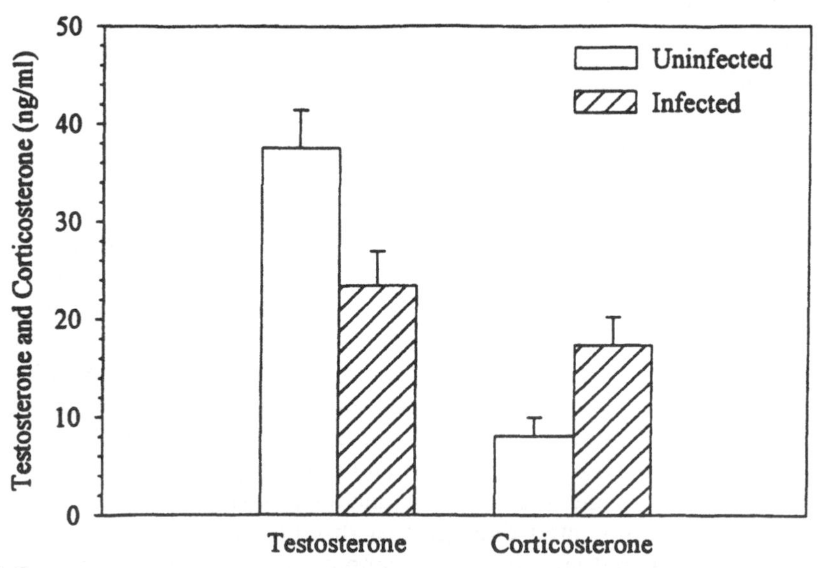

Figure 20.2
Western fence lizards that were infected with the blood parasite, P. mexicanum, *had lower basal testosterone and higher stress corticosterone levels than uninfected individuals (means ± SE). For testosterone 24 infected and 28 uninfected lizards were sampled and for corticosterone 21 infected and 25 uninfected lizards were sampled. Adapted from Dunlap and Schall (1995) and reprinted with permission of University of Chicago Press.*

and this, in turn, resulted in the observed suppression of testosterone and the subsequent inhibition of reproductive behaviors.

An associated study compared hormone levels of populations that were on the periphery of the species range with those of more centrally located populations. Although widespread, the distribution of western fence lizards (indeed, virtually all species) is not continuous but rather, consists of disjunct populations that are isolated, to varying degrees, due to the irregularity of their required habitat. As is often the case, animals living on the periphery of a species' range might occupy habitat that is suboptimal and, therefore, they might be regularly exposed to harsher environmental conditions than those that are more centrally located. However, because individuals that comprise a peripheral subpopulation may have adapted behavioral or physiological means to deal with extreme conditions, they may not necessarily be stressed by their marginal environments. Dunlap (1995) and Dunlap and Wingfield (1995) measured several physiologic factors that can indicate whether an lizard is experiencing stressful conditions. Individual lizards sampled at the edge of their distribution in the Mojave Desert of California were compared with more centrally located populations in California and Oregon. Western fence lizards from the Mojave population showed significant seasonal changes in several measures of blood composition, e.g., osmolality, plasma protein, and hematocrit, whereas lizards in the central part of the range did not.

Additionally, the Mojave lizards demonstrated significantly greater adrenocortical responses even after the body condition of the sampled individuals was controlled (i.e., analyses adjusted for size differences by using a snout-vent length/body mass measure that considered both size and condition). Interestingly, these endocrine differences also persisted when lizards from the central and peripheral populations were brought into the controlled environment of the laboratory. The fact that peripheral populations of lizards are characterized by increased responsiveness of the adrenocortical system to stressful stimuli may have practical applications for managers of endangered species. Although we do not yet know whether this is a widespread phenomenon in vertebrates, it may be common to many populations at the periphery of the species' range, in sub-optimal habitat, or both. Even if the adrenal responsiveness of the species in question has not been determined, it may be advisable to attempt to minimize their exposure to unnecessary stressors; especially if the population of concern is currently being challenged by their environment. For example, a manager of a species in decline might restrict access or close a refuge to human visitation during drought or other potentially stressing environmental condition. This might be of greatest benefit during an animals' breeding season when the impact of elevated glucocorticoid secretion upon the reproductive axis might result in abandonment of the reproductive effort.

Maned Wolves

Maned wolves (*Chrysocyon brachyurus*) are endangered canids found primarily in the grasslands of Brazil, but whose geographic range also extends into parts of Peru, Uruguay, Bolivia, Paraguay, and Argentina. As with many critically endangered species that are faced with extensive and rapid habitat loss throughout their natural ranges, the survival of the maned wolf may hinge on the cooperative efforts of captive breeding programs at zoos and wildlife parks. Maned wolves, however, do not readily breed in captivity, making a reliable way of monitoring the reproductive cycle of females in a population essential. The capture and handling necessary to measure plasma levels of reproductive hormones to accurately characterize the female's cycle may cause further stress in an animal that may already be severely stressed due to its captive status. Such a task can be further complicated because animal keepers or field researchers, even when equipped with the knowledge of the length of the gestation period may be known for a species, may not always be able to observe copulation, and even when observed, copulation may not result in fertilization. Therefore, it is clear that a tool that allows one to track the reproductive cycle of a species would be of value. In mammals, the estrous and menstrual cycles, as well as the gestation period can often be accurately tracked by monitoring relative amounts and temporal surges of circulating estrogens and progestins.

Motivated by a desire to develop a less invasive method to assess hormone levels, Wasser and colleagues have developed techniques for measuring steroid hormones using fecal or urine samples which might be of particular benefit to managers of species that are sensitive to human disturbance, difficult (or dangerous) to capture, or difficult to observe in their natural habitat (Wasser 1995; Wasser, Monfort, and Wildt 1991; Wasser, de Lemos Velloso, and Rodden 1995; Wasser *et al.* 1996). Interestingly, Wasser, de Lemos Velloso, and Rodden (1995) noted that several female maned wolves that had given birth unnoticed by zoo staff subsequently ate their pups. If zoo personnel had known that these females were on the verge of delivering a litter of valuable young, they could have taken measures to minimize the probability of infanticide by the mothers.

Estrogen and progestin concentrations were monitored in the feces of eight mature (> 2 years of age) and one sexually immature (1.5 years) female maned wolves. Coincidental with estrus, which is evidenced by increased playfulness and scent marking as well as mating, the four mature females that conceived and whelped pups exhibited surges of estrogens that were 4 to 10 fold higher than pre-estrus levels. Concurrent with the estrogen spikes, progestins increased approximately an order of magnitude. However, whereas estrogen levels returned to roughly pre-estrus levels, progestins remained elevated throughout gestation. In contrast, three of the four mature females that failed to conceive also failed to exhibit pronounced estrogen surges, and in all four the increases in their progestin levels were significantly less than that of the females that were impregnated. The immature female was acyclic with low levels of both estrogens and progestins throughout the monitoring period of approximately four months.

The aforementioned problems with captive breeding in this and other species argues persuasively for a non-invasive technique for assessing the reproductive status of individual females. The fecal steroid differences noted between females that had conceived and those that had not could allow managers to better use their time to monitor only impregnated females (Wasser, de Lemos Velloso, and Rodden 1995). Additionally, understanding the nature of the reproductive failure in females that do not bear young would allow researchers to designate them for hormone therapy or, alternately, infertile females could be housed in facilities that are not devoted to captive breeding.

Wasser and his colleagues (1995) also note that fecal steroid analysis can provide considerable information about free-living species that are difficult to observe. Observing maned wolves in the wild is exceptionally difficult but, because monogamous pairs occupy discrete territories and repeatedly use designated "latrines", it is possible to reliably collect feces for analysis. Even in the absence of extensive behavioral observations researchers may be able to monitor the reproductive and stress physiology of members of a pair. The physiologic data may then be correlated to ecological conditions. This could provide managers of this critically endangered species with measures of how environmental conditions, particularly human disturbances, affect maned wolves. These techniques are currently being applied to maned wolves in the wild.

Magellanic Penguins

Colonial breeding oceanic species like the Magellanic penguin (*Spheniscus magellanicus*) are particularly vulnerable to localized environmental perturbations. Because great numbers of individuals congregate within a relatively small area during the breeding season, there is potential for an environmental disaster to have catastrophic consequences. For example, some colonial or semi-colonial alcids (e.g., Xantus' murrelet, *Synthliboramphus brevirostris*, whiskered auklet, *Aethia pygmaea*, and Craveri's murrelet *Endomychura craveri*) breed on only a few islands and, therefore, a disaster near a breeding colony could decimate an entire species.

As a species that spends the major portion of their life at the ocean surface, Magellanic penguins are especially vulnerable to the effects of increasingly common oil spills. Oil fouling of marine mammals and birds can have dramatic effects on an animal at both the physical and physiological level (see chapter 6). At high latitudes where many vulnerable colonial breeding species are concentrated, perhaps the most acute problem that results from oil fouling is loss of thermoregulatory capabilities due to physical alteration of the feather or fur "coats" of affected individuals. If the combination of cold water and loss of thermal insulation does not kill a seriously fouled animal, it is likely to ingest fatal amounts of oil as it attempts to clean itself. However, not all animals exposed to oil spills are fouled to this extreme degree. As part of a long-term study on the reproductive endocrinology of Magellanic penguins at Punta Tombo, Argentina, Fowler and colleagues measured the effects of a relatively light oil covering upon circulating hormone levels in this species (Fowler 1993; Fowler *et al.* 1994; Fowler, Wingfield, and Boersma 1995).

The Patagonian coast where the Punta Tombo colony is found supports a heavy tanker traffic and spills of varying severity occur regularly where oiled penguins can be found in most years (Boersma 1987; Gandini *et al.* 1994; see chapter 17). During the 1991 breeding season a relatively large spill near the colony affected approximately 17,000 Magellanic penguins (Fowler, Wingfield, and Boersma 1995). Although substantial numbers of heavily oiled penguins were too impaired to attempt to breed and remained moribund on the beach, 'lightly' fouled birds (i.e., median percent of body covered with oil was 20%) entered the breeding colony and attempted to reproduce.

Oiled birds (especially females) when compared with birds without visibly oiled feathers differed in virtually all of the variables that were measured. For example, oiled female penguins weighed, on average, 40% less than non-oiled females (3 kg vs. 5 kg; Figure 20.3); oiled males were approximately 20% lighter than their non-oiled colony-mates (4 kg vs. 5 kg). Because individuals rely on fat and protein stores during the extended incubation bouts that are common in this and other penguin species, it is clear that even if a clutch were successfully laid, incubation by oil-fouled birds might be compromised. None of the oiled females sampled by Fowler, Wingfield, and Boersma (1995) had eggs in nests by mid-October, whereas most of the non-oiled controls had produced eggs by this time.

The pronounced negative effects of oil were also reflected in the hormone titres of the Punta Tombo penguins (Figure 20.3). Females and males that were lightly oiled had plasma estradiol and testosterone levels, respectively, that were three-fold lower than those of birds with no visible oil on their plumage. Also, luteinizing hormone (LH) levels of oiled females were significantly lower than in control females, although LH levels did not differ in males. Fowler and his colleagues (1995) also measured basal plasma levels of corticosterone and although oiled females had significantly elevated levels, oiled and

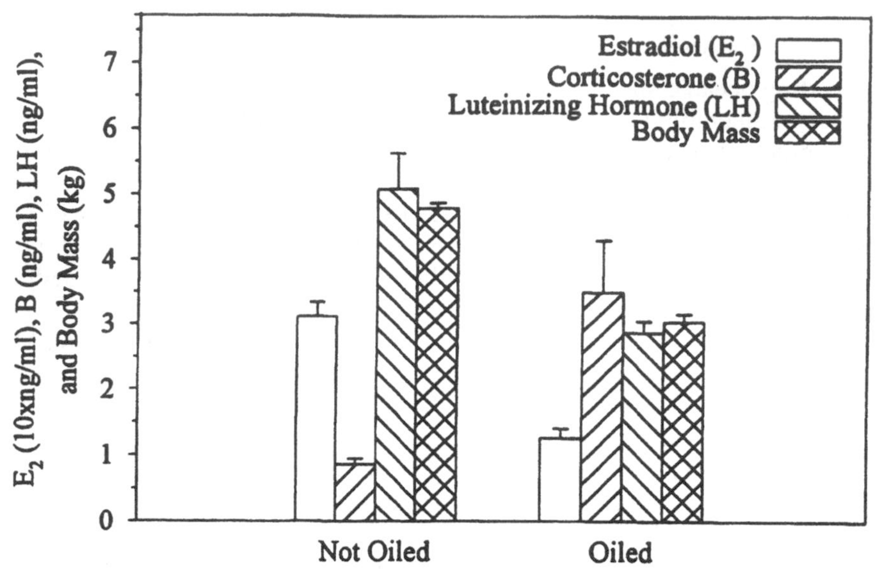

Figure 20.3
Lightly oiled female Magellanic penguins had higher circulating levels of corticosterone but had lower levels of luteinizing hormone (LH) and estradiol and weighed less than females that were free of oil (means ± SE). Note: estradiol levels have been multiplied by ten. Sample sizes are 11 in all cases. Adapted from et al. *(1995) and reprinted with permission of the American Ornithologists' Union.*

oil-free males had equivalent levels. All of these data concur with and help explain the above noted failure of oiled females to produce clutches.

Although attempts to save animals that have been fouled by oil may be necessary, two lines of evidence suggest that rehabilitation efforts may not achieve the desired results and that the methods used to clean oiled birds need reevaluation. First, Fowler and colleagues collected blood samples from oiled penguins that were being treated at a rehabilitation facility at Punta Tomba. Oiled penguins that were being held for cleaning and those that had been cleaned but were being held prior to release had very high levels of corticosterone—approximately five to seven times higher than controls. Birds that had been washed had significantly higher corticosterone levels, thus reflecting that they were more stressed than birds prior to washing. It should be noted that all of these birds had been held captive for three to four weeks. Clearly, animals that are already extremely stressed due to oiling should be processed more rapidly, thus minimizing the deleterious effects of chronically elevated corticosterone titres. Second, additional evidence that rehabilitation following oiling may not achieve the desired goals comes from a recent study that examined survival following cleaning in 13 species of marine birds in North America. Sharp (1996) found that although post-treatment release rates of cleaned birds are generally used as a measure of the success of cleaning, analyses based on recapture data show a very low post-release survival rate for cleaned and released birds.

The massive numbers of dead birds following major spills provide compelling evidence of the deadly effects of heavy oiling in numerous species of seabirds. Fowler's work on the hormone and body mass data from Magellanic penguins demonstrate that even birds with minimal exposure to oil may suffer reproductive failure, at least in the short-term. We agree that "... conservation efforts should stress prevention of oil pollution at all levels, from seepage to major spills."

American Alligators

Wetlands historically have been perceived as wastelands and have often been used as dumping grounds. Because of this mistreatment and because of their natural filtering properties, many wetlands have been the recipients of large amounts of chemical pollutants. Lake Apopka, one of Florida's largest and most polluted lakes, has extremely high levels of nutrient and pesticide contaminants. Immediately adjacent to a United States Environmental Protection Agency Superfund site, it has high levels of the pesticides dicofol, DDT (and metabolites), and sulfuric acid. Other contaminants in the lake originate from a nearby sewage treatment facility and fromagricultural runoff. Due to years of heavy pollution from these sources, levels of organochlorine pesticides in the lake are among the highest in Florida.

Recent studies have shown that exposure to environmental contaminants can affect endocrine and reproductive function at both the organizational and activational levels (see Colborn, Dumanoski, and Meyers 1996). Many organic pesticides, including organochlorines such as dicofol, DDT, and DDT metabolites, can act as xenobiotics (i.e., bioreactive exogenous compounds) and disrupt endocrine function by binding primarily to estrogen receptors. In doing so, they may cause the hormonal milieu of the organism to become estrogenic, thus impeding normal hormone function (Rories and Spelsberg 1989; McLachlan *et al.* 1992). Because the interactions between steroid hormones and receptors mediate the development of many tissues, endocrine disruptors have the potential to cause developmental abnormalities that may be irreversible and not expressed until an individual is fully developed. For example, when American alligator embryos (*Alligator mississippiensis*) are exposed to estrogens, an abnormally high ratio of females results, even at temperatures where males would normally develop (Bull, Gutzke, and Crews 1988). Additionally, if estrogen concentrations are below a certain threshold value, many intersex individuals, i.e., individuals that exhibit a phenotype with both male and female characteristics, occur. These findings show that developmental exposure to estrogens (and likely estrogen mimics) can result in "feminized" males.

As a result of bioaccumulation, top-level predators have an especially increased risk of chemical contamination. As a top-level carnivore that is found in Lake Apopka and other lakes in Florida, the American alligator is a good candidate for studies of the effects of pollutants on reproduction and survivorship. An early study of alligators in Lake Apopka documented a lower hatching rate, a higher incidence of deformed embryos, and a lower rate of overall reproduction than alligator populations from relatively unpolluted lakes (Jennings, Percival, and Woodward 1988). During the years that Jennings, Percival, and Woodward (1988) studied the alligators of Lake Apopka, the population declined by 90% and it remains low to date. One likely cause of the decline in reproduction is *p,p'-DDE*, a DDT metabolite that is one of the most prevalent organochlorines in Lake Apopka. This compound was also found in high concentrations in the eggs of the lake's alligators (Heinz, Percival, and Jennings 1991). Although these findings suggest that environmental contaminants were the cause of the reproductive failure in Lake Apopka's alligators, the mechanisms underlying this link remained to be demonstrated.

Over the last decade, Guillette and his associates have investigated the decline of Lake Apopka's alligators (Guillette *et al.* 1994, 1995a, 1995b, 1996). Their hypothesis that estrogenic contaminants are responsible for the decline in reproductive success led them to measure plasma steroid hormone levels and several morphological characters of the reproductive tracts of young alligators of both sexes. Guillette et al. (1994) found that juvenile females had plasma estradiol concentrations that were nearly twice as high as those in similar aged females from a control lake with minimal contamination (Figure 20.4). Additionally, the morphology of the females' ovaries was abnormal, with numerous polyovular follicles and polynuclear oocytes. Juvenile males in Lake Apopka had plasma

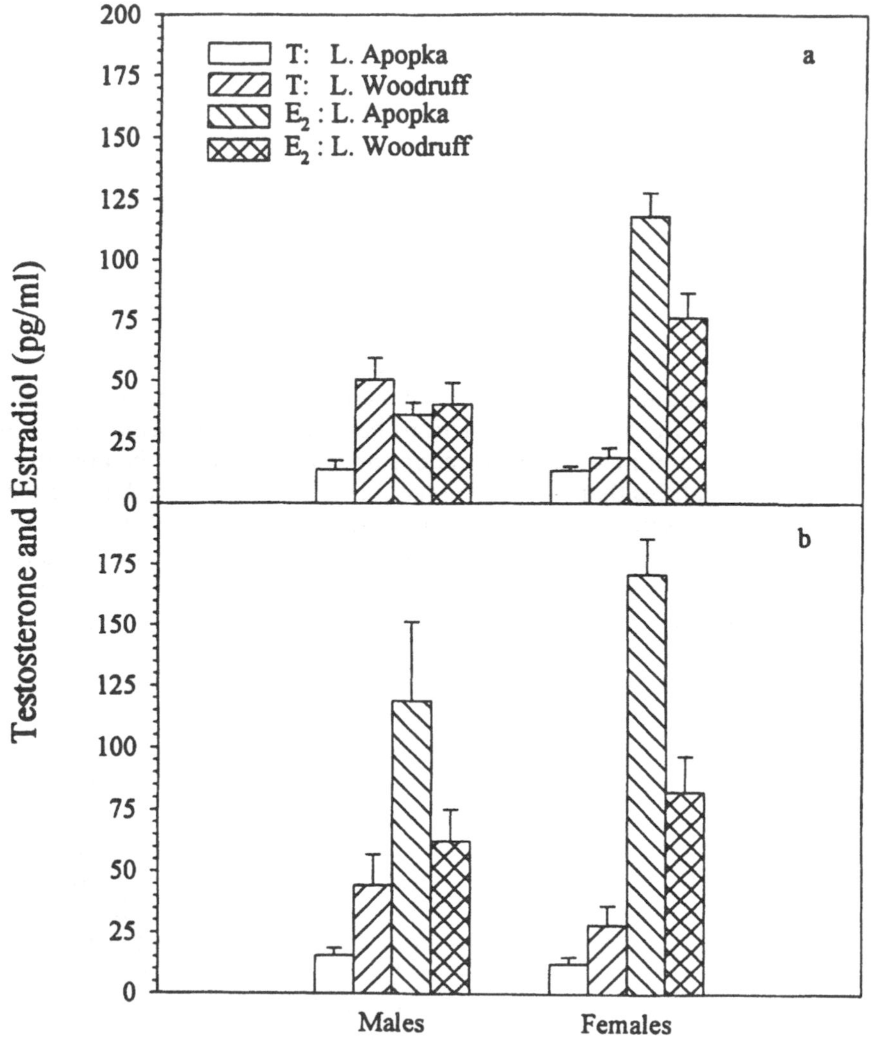

Figure 20.4

Plasma levels of testosterone (T) and estradiol (E_2) in six-month old male and female alligators from highly polluted Lake Apopka and the control Lake Woodruff (means ± SE). The upper panel shows plasma hormone levels before a luteinizing hormone (LH) challenge,whereas the lower panel shows levels following the LH challenge. Five males and eleven females were sample from Lake Apopka and nine males and 13 females were sampled from Lake Woodruff. Adapted from data in Guillette et al. *(1994) and reprinted with permission of Environmental Health Perspectives.*

testosterone levels that were three times lower than control males and equivalent to testosterone levels measured in control females (Figure 20.4). These male alligators also had poorly developed testes and penises that were on average 24% smaller than those of males from control lakes (Guillette *et al.* 1996).

To test whether the gonads of the Lake Apopka alligators had been permanently modified during development, Guillette *et al.* (1994) injected alligators with the gonadotropin, luteinizing hormone (LH). The LH challenge evoked a marked increase in estradiol levels in both males and females from Lake Apopka, thus suggesting that the gonads had been altered as a result of developmental exposure to contaminants (Figure 20.4). To further examine gonadal function, Guillette *et al.* (1995a) measured the *in vitro* production of steroid hormones by testes and ovaries from juvenile alligators. In direct contrast to the LH challenge data, *in vitro* testes of males from Lake Apopka produced testosterone that was equal to control males. Also, ovaries from Lake Apopka females synthesized estradiol at a lower rate than those of control females. Interestingly, both of these studies were performed on the same individuals.

Despite the paradoxical differences noted in the *in vivo* and *in vitro* studies, it is clear that something is seriously amiss with the endocrine and reproductive systems of Lake Apopka's alligators. That this is not just a phenomenon of alligators is demonstrated by the findings that virtually all of the red-eared turtles (*Chrysemis nelsoni*) that were sampled in Lake Apopka were either normal females or intersex; there were very few males and none of these showed normal androgen synthesis (Gross and Guillette, unpublished data, as cited in Guillette *et al.* 1994). Additionally, an increasing body of evidence suggests that the effects of environmental pollutants may be responsible to some degree for alarming trends of decreased fertility in many species worldwide (Guillette *et al.* 1995b; for review see Colborn, Dumanski, and Meyers 1996). Another example from Florida of decreased fertility that may be due to endocrine disruptors can be found in the critically endangered Florida panther (*Felis concolor coryi*). Facemire, Gross, and Guilette (1995) speculate that the relatively high plasma estradiol levels and subsequent feminization of males of this subspecies is the result of exposure to estrogenic endocrine disrupters. Although American alligators are no longer considered critically endangered, the information gathered and methods used in the Lake Apopka studies should be helpful in formulating conservation strategies for the Florida panther and other endangered species, as well as in providing critical information about the overall health of ecosystems under continued release of man-made chemicals into the environment.

CONCLUDING REMARKS

The case studies presented in this chapter were selected to give an overview of how field endocrine methods can be used to further conservation goals. Studies of Florida scrub-jays, western fence lizards, and Magellanic penguins, for example, employed well-established blood sampling protocols to quantify reproductive and stress hormones. The fecal steroid technique used in the maned wolf case study provides an alternate, less invasive method of gathering potentially critical information about the reproductive state of animals. This technique has recently been and continues to be applied to conservation issues in critical species. The American alligator case study presents evidence of the damage upon living systems that can be wrought by man-made chemical substances released into the environment. The effects of these contaminants upon all life forms sharing the biosphere should be of concern to all that are interested in conserving species and our environment.

In closing, Wingfield *et al.* (in press) suggested five ways that endocrine physiology may be applied to issues in conservation biology. These are:

(1) the endocrine characteristics of a species' breeding cycle can be identified;
(2) the effect of environmental pollutants upon the reproductive maturation of individuals within a population can be determined;
(3) the effects of environmental disturbances from either natural or human sources can be learned;
(4) hormone therapy can be used to treat endocrine disorders; and
(5) reproductive hormone titres can be altered to enhance or decrease the reproductive output of a species.

The case studies presented within deal primarily with the first three issues, but include allusions to how numbers (4) and (5) might be implemented (see citations within Wingfield *et al.* in press). The examples selected from birds, mammals, and reptiles illustrate how field endocrine methods can be applied to conservation biology. Although we have frequently speculated about the applicability of some of these techniques to conservation biology, our speculations are firmly rooted in the belief that the basic science of today becomes the applied science of tomorrow.

ACKNOWLEDGMENTS

We thank Peter Kareiva, Martha Groom, and Ellie Steinberg for inviting us to contribute to this volume. Sarah Kistler's and Peter Kareiva's editorial comments on, and proof reading of the manuscript were also of great help. The following people provided helpful comments and reprints on the respective case studies: Ron Mumme (Florida scrub-jays), Kent Dunlap (Western Fence Lizards), Gene Fowler (Magellanic Penguins), and Sam Wasser (Maned Wolves). Lou Guillette (American Alligator) kindly provided re- and pre-prints that were of great help. Thanks to all!

LITERATURE CITED

American Ornithologists' Union. 1995. Fortieth supplement to the American Ornithologists' Union Check-list of North American Birds. *Auk* 112:819–830.

Ball, G.F. 1993. The neural integration of environmental information by seasonally breeding birds. *American Zoologist* 33:185–199.

Boersma, P.D. 1987. Penguins oiled in Argentina. *Science* 236:135.

Bull, J.J., W.H.N. Gutzke, and D. Crews. 1988. Sex reversal by estradiol in three reptilian orders. *General and Comparative Endocrinology* 70:425–428.

Colborn, T., D. Dumanoski, and J.P. Meyers. 1996. *Our stolen future: Are we threatening our fertility, intelligence, and survival?: A scientific detective story.* New York: Penguin Books.

Daan, S., C. Dijkstra, R. Drent, and T. Meijer. 1986. Food supply and the annual timing of avian reproduction. In *Acta XIX International Ornithological Congress*, ed. H. Ouellet, 392–407. Ontario, Canada: University of Ottawa Press.

Dunlap, K.D. 1995a. External and internal influences on indices of physiological stress: II. Seasonal and size-related variations in blood composition in free-living lizards, *Sceloporus occidentalis*. *Journal of Experimental Zoology* 272:85–94.

Dunlap, K.D. 1995b. Hormonal and behavioral responses to food and water deprivation in a lizard (*Sceloporus occidentalis*): implications for assessing stress in a natural population. *Journal of Herpetology* 29:345–354.

Dunlap, K.D. and D.R. Church. 1996. Interleukin-1β reduces daily activity level in male lizards, *Sceloporus occidentalis. Brain, Behavior, and Immunity* 10:68–73.

Dunlap, K.D. and J.J. Schall. 1995. Hormonal alterations and reproductive inhibition in male fence lizards (*Sceloporus occidentalis*) infected with the malarial parasite *Plasmodium mexicanum. Physiological Zoology* 68:608–621.

Dunlap, K.D. and J.C. Wingfield. 1995. External and internal influences on indices of physiological stress: I. Seasonal and population variation in adrenocortical secretion of free-living lizards, *Sceloporus occidentalis. Journal of Experimental Zoology* 271:36–46.

Facemire, C.F., Gross, T.S., and L.J. Guillette, Jr. 1995. Reproductive impairment in the Florida panther: Nature or nurture. *Environmental Health Perspectives* 103:79–86.

Fitzpatrick, J.W. and G.E. Woolfenden. 1986. Demographic routes to cooperative breeding in some New World jays. In *Evolution of animal behavior*, eds. M. Nitecki and J. Kitchell, 137–160. Chicago: University of Chicago Press.

Follett, B.K. 1984. Birds. In *Marshall's physiology of reproduction I. Reproductive cycles of vertebrates*, ed. G.E. Lamming, 283–350. Edinburgh: Churchill-Livingston Press.

Fowler, G.S. 1993. Ecological and endocrinological aspects of long-term pair bonds in the Magellanic Penguin (*Spheniscus magellanicus*). Ph.D. Dissertation, University of Washington, Seattle.

Fowler, G.S., J.C. Wingfield, P.D. Boersma, and R.A. Sousa. 1994. Reproductive endocrinology and weight change in relation to reproductive success in the Magellanic Penguin (*Spheniscus magellanicus*). *General and Comparative Endocrinology* 94:305–315.

Fowler, G.S., J.C. Wingfield, and P.D. Boersma. 1995. Hormonal and reproductive effects of low levels of petroleum fouling in Magellanic penguins (*Spheniscus magellanicus*). *Auk* 112:382–389.

Gandini, P., P.D. Boersma, E. Frere, M. Gandini, T. Holik, and V. Lichtschein. 1994. Magellanic penguins (*Spheniscus magellanicus*) affected by chronic oil pollution along coast of Chubut, Argentina. *Auk* 111:20–27.

Gorbman, A., W.W. Dickhoff, S.R. Vigna, N.B. Clark, and C.L. Ralph. 1983. *Comparative endocrinology*. New York: John Wiley and Sons.

Greenberg, N. and J.C. Wingfield. 1987. Stress and reproduction: Reciprocal relationships. In *Hormones and reproduction in fishes, amphibians and reptiles*, eds. D.O. Norris and R.E. Jones, 461–505. New York: Plenum Press.

Guillette, L.J., Jr., T.S. Gross, G.R. Masson, J.M. Matter, H.F. Percival, and A.R. Woodward. 1994. Developmental abnormalities of the gonad and abnormal sex hormone concentrations in juvenile alligators from contaminated and control lakes in Florida. *Environmental Health Perspectives* 102:680–688.

Guillette, L.J., Jr., T.S. Gross, D.A. Gross, A.A. Rooney, and H.F. Percival. 1995a. Gonadol steroidogenesis *in vitro* from juvenile alligators obtained from contaminated or control lakes. *Environmental Health Perspectives* 103 (Suppl 3):31–36.

Guillette, L.J., Jr., D.A. Crain, A.A. Rooney, and D.B. Pickford. 1995b. Organization versus activation: the role of endocrine-disrupting contaminants (EDCs) during embryonic development in wildlife. *Environmental Health Perspectives* 103 (Suppl 3):157–164.

Guillette, L.J., Jr., D.B. Pickford, D.A. Crain, A.A. Rooney, and H.F. Percival.1996. Reduction in penis size and plasma testosterone concentrations in juvenile alligators living in a contaminated environment. *General and Comparative Endocrinology* 101:32–42.

Harvey, S., J.G. Phillips, A. Rees, and T.R. Hall. 1974. Stress and adrenal function. *Journal of Experimental Zoology* 232:633–645.

Heinz, G.H., H.F. Percival, and M.L. Jennings. 1991. Contaminants in American alligator eggs from Lakes Apopka, Griffin, and Okeechobee, Florida. *Environmental Monitor and Assessment* 16:277–285.

Jennings, M.L., H.F. Percival, and A.R. Woodward. 1988. Evaluation of alligator hatchling and egg removal from three Florida lakes. *Southeast Association of Fish and Wildlife Agencies* 42:283–294.

McLachlan, J.A., R.R. Newbold, C.T. Teng, and K.S. Korach. 1992. Environmental estrogens: Orphan receptors and genetic imprinting. In *Chemically-induced alterations in sexual and functional development: the wildlife/human connection*, Vol. 21., eds. T. Colborn and C. Clement, 107–122. Princeton, New Jersey: Princeton Publications.

Moore, F.L. and L.J. Miller. 1984. Stress-induced inhibition of sexual behavior: Corticosterone inhibits courtship behaviors of male amphibians. *Hormones and Behavior* 18:400–410.

Moore, F.L. and R.T. Zoeller. 1985. Stress-induced inhibition of reproduction: Evidence of suppressed secretion of LH-RH in an amphibian. *General and Comparative Endocrinology* 60:252–258.

Moore, M.C., C.W. Thompson, and C.A. Marler. 1991. Reciprocal changes in corticosterone and testosterone levels following acute and chronic handling stress in the Tree Lizard, *Urosaurus ornatus*. *General and Comparative Endocrinology* 81:217–226.

Mumme, R.L. and T.H. Below. 1995. *Relocation as a management technique for the threatened Florida scrub jay*. Florida game and freshwater fish commission nongame wildlife program project report. Tallahassee, Florida, USA.

Rories, C. and T.C. Spelsberg. 1989. Ovarian steroid action on gene expression: Mechanisms and models. *Annual Review of Physiology* 51:653–681.

Sapolsky, R.M. 1987. Stress, social status, and reproductive physiology in free-living baboons. In *Psychobiology of reproductive behavior: An evolutionary perspective*, ed. D. Crews, 291–321. New Jersey: Prentice-Hall.

Schall, J.J. 1990a. The ecology of lizard malaria. *Parasitology Today* 6:264–269.

Schall, J.J. 1990b. Virulence of lizard malaria: The evolutionary ecology of an ancient parasite-host association. *Parasitology* 100:s35-s52.

Schoech, S.J., R.L. Mumme, and M.C. Moore. 1991. Reproductive endocrinology and mechanisms of breeding inhibition in cooperatively breeding Florida scrub jays (*Aphelocoma c. coerulescens*). *Condor* 93:354–364.

Schoech, S.J. 1996. The effect of supplemental food on body condition and the timing of reproduction in a cooperative breeder, the Florida scrub-jay (*Aphelocoma coerulescens*). *Condor* 98:234–244.

Schoech, S.J., R.L. Mumme, and J.C. Wingfield. 1996a. Delayed breeding in the cooperatively breeding Florida scrub-jay (*Aphelocoma coerulescens*): inhibition or the absence of stimulation. *Behavioral Ecology and Sociobiology* 39:77–90.

Schoech, S.J., R.L. Mumme, and J.C. Wingfield. 1996b. Prolactin and helping behaviour in the cooperatively breeding Florida scrub-jay (*Aphelocoma coerulescens*). *Animal Behaviour* 52 (in press).

Schoech, S.J., R.L. Mumme, and J.C. Wingfield. 1997. Breeding status, corticosterone, and body mass in the cooperatively breeding Florida scrub-jay (*Aphelocoma coerulescens*). *Physiological Zoology* 70 (in press).

Selye, H. 1971. *Hormones and resistance*. Berlin: Springer-Verlag.

Sharp, B.E. 1996. Post-release survival of oiled, cleaned seabirds in North America. *Ibis* 138:222–228.

Siegel, H.S. 1980. Physiological stress in birds. *Bioscience* 30:529–534.

Wasser, S.K. 1995. Costs of conception in baboons. *Nature* 376:219–220.

Wasser, S.K., A. de Lemos Velloso, and M.D. Rodden. 1995. Using fecal steroids to evaluate reproductive function in female maned wolves. *Journal of Wildlife Management* 59:889–894.

Wasser, S.K., S.L. Monfort, and D.E. Wildt. 1991. Rapid extraction of faecal steroids for measuring reproductive cyclicity and early pregnancy in free-ranging yellow baboons (*Papio cynocephalus cynocephalus*). *Journal of Reproduction and Fertility* 92:415–423.

Wasser, S.K., S.L. Monfort, J. Southers, and D.E. Wildt. 1994. Excretion rates and metabolites of oestradiol and progesterone in baboon (*Papio cynocephalus*) faeces. *Journal of Reproduction and Fertility* 101:213–220.

Wasser, S.K., S. Papageorge, C. Foley, and J.L. Brown. 1996. Excretory fate of estradiol and progesterone in the Africa elephant (*Loxodonta africana*) and patterns of fecal steroid concentrations throughout the estrous cycle. *General and Comparative Endocrinology* 102:255–262.

Wilson, F.E. and B.K. Follett. 1975. Corticosterone-induced gonadosuppression in photostimulated tree sparrows. *Life Science* 17:1451–1456.

Wingfield, J.C. 1988. Changes in reproductive function of free-living birds in direct response to environmental perturbations. In *Processing of environmental information in vertebrates*, ed. M.H. Stetson, 121–148. Berlin: Springer-Verlag.

Wingfield, J.C. 1994. Modulation of the adrenocortical response to stress in birds. In *Perspectives in comparative endocrinology*, 520–528. National Research Council of Canada.

Wingfield, J.C. and D.S. Farner. 1976. Avian endocrinology—field investigations and methods. *Condor* 78:570–573.

Wingfield, J.C. and G.J. Kenagy. 1991. Natural regulation of reproductive cycles. In *Vertebrate endocrinology: fundamentals and biomedical implications*, eds. M.P. Schreibman and R.E. Jones, 181–241. New York: Academic Press.

Wingfield, J.C. and D.S. Farner. 1993. Endocrinology of reproduction in wild species. In *Avian biology, Volume IX*, eds. D.S. Farner, J.R. King, and K.C. Parks, 63–327. New York: Academic Press.

Wingfield, J.C., P. Deviche, S. Sharbaugh, L.B. Astheimer, R. Holberton, R. Suydam, and K. Hunt. 1994a. Seasonal changes of the adrenocortical responses to stress in redpolls, *Acanthis flammea*, in Alaska. *Journal of Experimental Zoology* 270:372–380.

Wingfield, J.C., R. Suydam, and K. Hunt. 1994b. The adrenocortical responses to stress in snow buntings (*Plectrophenax nivalis*) and Lapland longspurs (*Calcarius lapponicus*) at Barrow, Alaska. *Journal of Comparative Biochemistry and Physiology, C* 108:299–306.

Wingfield, J.C., K. Hunt, C. Breuner, K. Dunlap, G.S. Fowler, L. Freed, and J. Lepson. In press. Environmental stress, field endocrinology and conservation biology. In *Behavioral approaches to conservation in the wild*, eds. R. Bucholz and J. Clemmons. Cambridge: Cambridge University Press.

Woolfenden, G.E. and J.W. Fitzpatrick. 1984. *The Florida scrub jay: Demography of a cooperative-breeding bird.* Princeton University Press, New Jersey, USA.

Woolfenden, G.E. and J.W. Fitzpatrick. 1990. Florida scrub jays: A synopsis after 18 years of study, p 241–266. In *Cooperative breeding in birds: Long-term studies of ecology and behavior*, eds. P.B. Stacey and W.D. Koenig, 241–266. Cambridge: Cambridge University Press.

CHAPTER

21

Global Climate Change and Species Interactions

ELLEN GRYJ

A major environmental concern of this century is the impact of human activity on the Earth's climate, with prospects of major biological upheaval. The combined effects of deforestation and release of chemicals to the environment are likely to dramatically modify global temperature, and wind and rain patterns within the next century (Schneider 1989; Wetherald, 1991). However, most conservation biology proceeds without reference to climate shifts. This may be because the major threats to biodiversity (e.g., habitat destruction) occur quite independently of climate. Unfortunately, however, many of our solutions to biodiversity threats, such as the creation of national parks and reserves, are designed on the assumption that the climate will stay as it currently is. In this chapter I review the commonly accepted scenarios for climate change, sketch some research regarding impact on different organisms, and focus on two issues: how might climate change threaten the persistence of species and what surprises might await us as a result of climate change.

Since the industrial revolution, gas emissions linked to human activities have resulted in a 25% increase in CO_2 in the atmosphere (Root and Schneider 1993). This elevated CO_2 and other gases "trap heat" in the atmosphere, and an increase in average temperature is expected as "greenhouse gases" continue to accumulate in the atmosphere. The main sources of CO_2 are the combustion of fossil fuels and the destruction of tropical forests, which release 5–6 and 1–3 billion tons of carbon per year, respectively (Tregarthen 1988; Legget 1990; Rubin *et al.* 1992; Skole and Tucker 1993). Most models predict that if the gas release continues at the same rate, there will be an average increase of temperature on Earth between 2.5° and 4.5°C by the middle of next century (Schneider 1989; Davis 1990; Legget 1990; Wetherald 1991). Already in the 20th century there apparently has been a warming trend of 0.5 to 2°C (Root and Schneider 1993). During the Ice Ages in the Pleistocene the average temperature change was 5°C over 10,000 years, and this caused major changes in the distribution and abundance of organisms during that period (Woodward 1987; Schneider 1989). A similar temperature change occurring ten to one hundred times more rapidly could have devastating effects on many species, and will almost certainly change community and ecosystem dynamics.

Of course, climate change will not affect all organisms equally. Some may be able to disperse northward to escape the "heat"; others may possess physiological tolerance to the altered climate; and some may have a greater ability to evolve in response to the changing environment. Because species will differ in their responses to climate change, there is certain to be a reshuffling of species associations even if there are not massive extinctions. One then has to wonder whether goals such as preserving selected natural communities are feasible. Perhaps communities and ecosystems as we know them must inevitably cease to exist as a result of climate change.

DIRECT EFFECTS OF GLOBAL CHANGE ON ORGANISMS

Abiotic variables, such as temperature and humidity, largely determine which organisms succeed in a given area. Thus a change in the physical environment is expected to cause changes in survival, reproduction and physiological responses, as well as shifts in the geographic distribution of those organisms. Measuring the response of individual species to changes in abiotic factors should stimulate hypotheses about the effect of global change on different organisms.

Direct Effects of Global Change on Plants

One way plants can respond to climate change is to shift their distributions via dispersal and colonization of new regions. Unfortunately, the capacity of organisms to disperse among habitat fragments may be so poor that this response is not a "real option" (Groom and Schumaker 1993). In fact, limited dispersal abilities apparently caused massive extinction of plant species during previous climate change events (Gates 1993; Knoll 1984). For example, the paleontological data suggest that the taxonomic composition of the temperate forests of Eastern North America and Northwestern Europe were very similar during the Pliocene. However, the present woody flora of Northwestern Europe is very species-poor relative to Eastern North America. This is apparently due to the difference in the topography of the two regions: in America, the Appalachian Mountains run in north-south direction, whereas the Alps run from west to east. During the Ice Ages in the Pleistocene, the Alps constituted a barrier for dispersal to warmer climates, whereas seed dispersal was much more successful in America without such west-east mountain barriers (Knoll 1984).

It is important to realize that the expected climate change in the next 60 years would be at least two orders of magnitude faster than in the Pleistocene. Thus, to avoid massive extinctions, seed dispersal and establishment rates of new populations would have to be much higher than anything known from the past. Assuming that (1) a 1°C temperature increase approximately corresponds to 100 km in latitude (Pacala and Hurtt 1993), (2) temperatures will increase approximately 2.5°C over the next century, and (3) plants disperse at historical rates, it is obvious that the pace of warming will be too fast for natural dispersal of plants to keep up with predicted changes in temperature (see Table 21.1) (Schneider 1989; Gates 1993; Morse *et al.* 1993).

Given the likely inadequacy of dispersal, it is therefore interesting to ask to what extent current distributions of species overlap with predicted future regions of suitable climate. The Nature Conservancy (TNC) has developed a "climate envelope" analysis to address this question. The analysis includes all the native vascular plants of North America north of Mexico: a total of 15,148 species (Morse *et al.* 1993). The "climate envelope" of each species is defined by the range of climates in which the species is currently found. This envelope is then compared to the geographic area that would be within this

Table 21.1
Calculated dispersal rates during the Holocene Epoch for a woody taxa. The dispersal rates are given in kilometers/century.

Taxon	Rate of Range Extension	Source
Jack/red pine	40 km/century	Gates 1993
White pine	30–35 km/century	Gates 1993
Scotch pine	37.5–80 km/century	Morse *et al.* 1993
Oak	35 km/century	Gates 1993
Spruce	25 km/century	Gates 1993
Larch	25 km/century	Gates 1993
Elm	25 km/century	Gates 1993
Hemlock	20–25 km/century	Gates 1993
Hickory	20–25 km/century	Gates 1993
Balsam fir	20 km/century	Gates 1993
Maple	20 km/century	Gates 1993
Beech	20 km/century	Gates 1993
Chestnut	10 km/century	Gates 1993

temperature range under a uniform 3°C increase. Under the most conservative scenarios, approximately 7% of the species would be completely out of their present climate envelope. Endemics and many threatened species are disproportionately represented within this group of species that will be outside their climate envelope (Morse *et al.* 1993). It should be emphasized that the TNC analysis very likely underestimates the probability of extinction of some groups of species, because the analysis is based on a relatively small number of variables. For instance, according to the analysis, wetland species are not as vulnerable as upland species. However, this may be a reflection of the wide distribution of many wetland species along coastlines and watersheds. Because wetlands are discontinuous, dispersal among patches is difficult, and thus the probability of extinction of wetland species found is probably much higher than what a simple analysis suggests (Morse *et al.* 1993). It is certainly clear that a geographic shift of climate envelopes could reduce the effectiveness of reserve systems, since the new climate envelopes may not overlap with the location of the reserves that currently protect particular species (Morse *et al.* 1993).

Climate change will also induce physiological changes in plants. For example, it has been shown that high levels of CO_2 increase the growth rates and yield of many species. However, contrary to what has sometimes been suggested (Nordhaus 1990; Schelling 1992), this "fertilization" could have detrimental effects. First, because not all species show an increase in growth and photosynthesis with CO_2, the species composition of communities will change. In particular, those species not able to take advantage of the CO_2 fertilization could disappear due to changes in competition regimes (Bazzaz and Fajer 1992). An increase in atmospheric CO_2 could also alter the competitive balance through its effects on stomatal regulation and water use (Bazzaz and McConnaughay 1992). Second, in systems where CO_2 is not the most limiting resource for plant growth there may be no increase in productivity (Bazzaz 1990).

Changes in CO_2 also modify the chemical composition of leaves and other plant structures. In many species, the nitrogen content of leaves decreases as atmospheric CO_2 increases (Lincoln, Couvet, and Sionit 1986; Fajer, Bowers, and Bazzaz 1989; Fajer 1989; Couteaux *et al.* 1991; Ayres 1993; Williams, Lincoln, and Thomas 1994) (see Table 21.2). In a later section I discuss how nitrogen content of leaves is likely to affect herbivory. A more direct consequence, however, is reflected in the litter decomposition

Table 21.2
Effect of increase in levels of atmospheric CO_2 on several attributes of different species of plants. The following symbols are used: − = decrease; + = increase; 0 = no observed effect in the attribute; f = faster; s = slower. The references by species are Glycine max *(Lincoln, Couvet and Sionit, 1986);* Pinus taeda *(Williams, Lincoln and Thomas, 1994);* Plantago lanceolata *(Fajer, 1989; Fajer, Bowers, and Bazzaz, 1990); and Bazzaz 1990 for the rest of the species shown in the table.*

Attribute	Plant Species	Effect of CO_2 Increase
Biomass		
	Acer rubrum (seedling)	0
	Acer saccarum (seedling)	+
	Amaranthus sp.	+
	Betula papyrifera	0
	Bouteloa gracilis	+
	Bromus tectorum	+
	Fagus grandifolia (seedling)	+
	Pinus strobus (seedling)	+
	Prunus serotina (seedling)	+
	Scirpus olneyi (marsh plant)	+
	Tsuga canadensis (seedling)	+
Growth Rates		
	Gallardia pulchela	0
	Gaura brachycarpa	0
	Plantago lanceolata	+
Nitrogen Content of Leaf		
	Glycine max (soybean)	−
	Pinus taeda	−
	Plantago lanceolata	−
	Scirpus olneyi	− (% but not content)
Water concentration		
	Glycine max	+
	Plantago lanceolata	−
Mortality		
	Plantago lanceolata	+
Time to begin flowering		
	Gallardia pulchela	f
	Gaura brachycarpa	f
	Lupinus texensis	s
	Oenothera laniculata	0
Number of flowers		
	Abutilon theophrastis	−
	Phlox drumondii	+
Number of fruits		
	Abutilon theophrastis	−
Number of seeds		
	Abutilon theophrastis	−
Weight of weeds		
	Abutilon theophrastis	+

rate—i.e., when nitrogen decreases relative to carbon, decomposition slows down (Woodward 1993). This could potentially decrease soil fertility due to nutrients being trapped in the litter for longer times, thereby being inaccessible to plants (Bazzaz and Fajer 1992).

Direct Effects of Climate Change on Vertebrates and Insects

Temperature also has a very important role determining the distribution of animals, and influencing their growth and reproduction. Responses by animals to climate change will differ from plants in many aspects, and within animals we will probably see many differences between vertebrates and invertebrates.

Insects in particular are very sensitive to temperature changes. The developmental rates of many insects increase with temperature, and survival and reproductive success are also affected (see Table 21.3). At a first glance, it would seem that an increase in temperature would be favorable for insects, by allowing them to develop and grow faster. However, early development could make insects out of synchrony with the appearance and availability of food and other resources. For example, an increase in environmental temperature can have conflicting effects in the threatened Bay checkerspot butterfly (*Euphydryas editha bayensis*). These butterflies are restricted to patches of their annual host plants (*Plantago erecta*, *Castilleja purpurascens* and *C. densiflora*) in central California (Weiss *et al.* 1987; Murphy, Freas, and Weiss 1990). In this region there is a cool rainy season (October to April) and a warm summer drought (May to September) that governs the phenology of the plants and the butterflies (Weiss *et al.* 1987; Murphy, Freas, and Weiss 1990). Adult butterflies lay their eggs between February and May, and the newly hatched larvae feed on the plants, transition through four larval instars, and enter diapause at the beginning of the dry season. During this period, individual larvae move and feed on several different individual plants. In years of drought and high temperatures, the host plants senesce earlier in the season, before most larvae can complete their prediapausal development. Plants on cooler slopes survive for longer periods, and it is in these areas where most larvae survive in especially hot years (Murphy, Freas, and Weiss 1990). Global warming would probably exaggerate this effect and place more butterflies at risk of having the plants senesce too rapidly. Diapause is broken with the onset of the autumn rains, which also trigger the germination of the host plants. The post-diapausal larvae resume feeding, pupate, and emerge as adults. Post-diapausal development speed increases with temperature, and this translates into earlier adult emergence and reproduction. Early reproduction is advantageous in this rapidly drying environment as it gives a headstart to the offspring that begin feeding earlier in the season (Weiss *et al.* 1987; Murphy, Freas, and Weiss 1990). From this perspective, global warming could be seen as advantageous to butterflies. Thus, global change is likely to modify microenvironmental conditions and affect species like the Bay checkerspot in ways that are very hard to predict. One might expect that terrestrial vertebrates would be comparatively insensitive to global warming because they can regulate their body temperature and often are capable of extensive long-distance dispersal. However, the story is not that simple. For example, many vertebrates restrict their movement to regions of undisturbed vegetation or die if they venture to areas of major disturbances (Brown and Gibson 1983; Janzen 1986). Also, the range of many vertebrates is limited by minimum temperatures. For instance, at least 51 species of songbirds have a northern range limited by January minimum temperatures (Root and Schneider 1993) and their distribution could shift under climate change.

Climate change is also likely to have a high impact in areas where species distributions show elevation zonation, such as montane regions. Isolated mountains and mountain

Table 21.3

Effect of changes in ambient temperature on some plant species and the insects that interact with them.

Plant sp	Animal sp	Temperature Increase—Effects on Plant	Temperature Increase—Effects on Interaction	Temperature Increase—Effects on Insect	Other Interactions	Source
Ficus citrifolia	*Pegoscapus assuetus*			Faster development		Bronstein and Patel 1992
Ficus aurea	*Pegoscapus jimenezi*			Faster development		Bronstein and Patel 1992
Populus tremuloides	*Malacosoma disstria* (forest tent caterpillar)			Faster development	Higher probability of escape from insect pathogens (due to less exposure time)	Roland 1993
Plantago erecta (also *Orthocarpus densifloras* and *C. purpurescens)*	*Euphydryas editha bayensis* (Bay checkerspot butterfly)	Combined with drought, causes early senescence	Synchrony of plant and insect is important: Extreme conditions decrease survival of butterfly	Faster larval and pupal development; earlier adult emergence and reproduction		Murphy and Weiss 1992; Weiss *et al*. 1993

ranges are basically islands of cold weather where the warmer lowlands act as barriers for small terrestrial mammals, amphibians and other vertebrates (Brown and Gibson 1983). In the tropics mountaintops often harbor high numbers of endemic species (Brown and Gibson 1983; Soulé 1991). If temperature increased, the habitat of species restricted to mountaintops would be reduced. Moreover, species that are now limited only by temperature would tend to expand their range and could displace the species that are now restricted to high elevations (McDonald and Brown 1992). Table 21.4 shows some examples of vertebrates that are likely to be affected by climate change.

Temperature changes can influence animals through impacts on vegetation as well as via the more obvious direct effects. For example, the Kirtland's warbler is restricted to a narrow area in northern Michigan where jack pine trees grow in sandy soils (Root and Schneider 1993). These birds construct their nests on the ground under young pine trees. Since the pine trees are very sensitive to temperature changes, even small changes are likely to cause their local extinction and possibly a range shift to the North. In the more northerly areas however, the soil is not sandy enough to allow drainage for successful fledging of the birds. As a consequence, even a small increase in local temperature would cause the extinction of this warbler (Root and Schneider 1993). In general, birds with specific nesting requirements, such as the Kirtland's and Blackburnian warblers, and the red-cockaded woodpecker, will probably go extinct if the trees they depend on are displaced by other trees adapted to warmer climates (Rodenhouse 1992; Root and Schneider 1993).

Effects of Climate Change on Competitive Interactions

Because species can react idiosyncratically to climate change, it is hard to predict how communities or even pairs of competing species will be influenced by phenomena such as warmer temperatures or elevated CO_2. The best we can do is propose plausible scenarios. For example, photosynthetic systems offer a variable that could help us predict community changes. In particular, C3 plants tend to increase growth as CO_2 increases, whereas C4 plants are much less responsive (Bazzaz 1990; Woodward, Thompson, and McKee 1991; Bazzaz and McConnaughay 1992). At the community level this could be manifested in at least two different ways: (1) an increase in productivity accompanied by a relative increase of C3 plants; and (2) no increase in productivity but nonetheless an increase in C3 plants. Most experiments suggest that there will be a change in species composition with increases in CO_2. In contrast, the effects of CO_2 on productivity are ambiguous, as experiments sometimes reveal an increase in productivity and other times do not (Bazzaz 1990; Bazzaz and McConnaughay 1992). Furthermore, the fertilization effect of CO_2 could be lost over time due to acclimation and negative feedback mechanisms of the plants (Bazzaz 1990; Bazzaz and Fajer 1992).

The increased photosynthetic rates of plants may have effects other than simply increasing plant growth. For instance, increases in CO_2 also affect nitrogen accumulation and stomatal behavior (Bazzaz 1990; Woodward, Thompson, and McKee 1991; Bazzaz and McConnaughay 1992). We can expect very complex interactions between all these factors. For example, an increase in CO_2 could decrease the instantaneous water loss by decreasing stomatal transpiration rates. However, increased growth would also increase shading on plants or leaves in the understory, which could also influence stomatal behavior (Woodward, Thompson, and McKee 1991; Bazzaz and McConnaughay 1992). There will probably be changes in water use at a community level, but their direction is difficult to predict. Additionally, a reduction in the water loss rate may affect temperature regulation

Table 21.4
Examples of vertebrates whose distribution is likely to change with global change.

Vertebrate Species	Limiting Factors or range	Predicted Effect	Notes	Source
Sayornis phoebe (Eastern Phoebe)	Limiting Temperature: 4°C	Could expand its range	Can not overwinter in areas colder than 4°C	Root and Schneider 1993
Dendroica kirtlandi (Kirtland's Warbler)	Jack Pine Trees + Sandy Soils	Extinction	When trees expand North, they will not be on areas with sandy soil, and the Warbler will not be able to nest	Root and Schneider 1993
Dendroica fusca (Blackburnian warbler)	*Picea rubens* (Red Spruce)	Extinction	*Picea rubens* Replaced by Hardwoods with 2°C	Rodenhouse 1992
Dendroica caerulescens (Blackthroated Blue Warbler)	Breeding grounds: hardwood forests	Global warming may favor bird reproduction by enhancing food production	Warmer temperatures will increase the number of caterpillars in the breeding site of birds	Rodenhouse 1992
Ochotona princeps (Pika)	Mountaintops of Great Basin	Extinction	Habitat in Mountain tops will be reduced	McDonald and Brown 1992
Zapus princeps (Western jumping mouse)	Mountaintops of Great Basin	Extinction	Habitat in Mountain tops will be reduced	McDonald and Brown 1992
Spermophilus beldingi (Belding ground squirrel)	Mountaintops of Great Basin	Extinction	Habitat in Mountain tops will be reduced	McDonald and Brown 1992
Lepus townsendii (White-tailed hare)	Mountaintops of Great Basin	Extinction	Habitat in Mountain tops will be reduced	McDonald and Brown 1992

in some plants, and this could increase mortality rates in some species (Bazzaz and McConnaughay 1992). The effect of carbon dioxide also varies depending on the life stage of the plant. In some species, CO_2 affects seed germination (Woodward 1991), whereas in others it alters mainly seedling growth (Bazzaz 1990; Bazzaz and McConnaughay 1992). This could have profound implications in some systems, such as tropical rainforests, where growth rates of seedlings and saplings are crucial in determining establishment in forest gaps (Denslow 1987).

Physiological studies of plant responses to climate change document clearly that different species respond in dramatically different directions and degrees (Tables 21.2 and 21.5). As mentioned before, this means that some species will be favored and others will be placed at a disadvantage. At a community level, it is reasonable to expect changes in competition hierarchies due to differences in growth rates (Bazzaz 1990), modification of canopy architecture (Bazzaz and McConnaughay 1992), and changes in phenological patterns (Bazzaz 1990; Murphy and Weiss 1992; Gates 1993; Weiss *et al.* 1993). These changes highlight the unpredictability of the effects of climate change in species composition and productivity.

Effects of Climate Change on Plant-Insect Interactions

Not only will climate shifts alter competitive interactions among plants, they will also alter between trophic level interactions. The best studied of these possibilities involve plant-insect interactions. As discussed earlier, the chemical composition of plant structures is modified by increases in atmospheric CO_2. One of the most common changes is the decrease in nitrogen content of leaves (Lincoln, Couvet, and Sionit 1986; Fajer 1989; Fajer, Bowers, and Bazzaz 1989; Coteaux *et al.* 1991; Ayres 1993; Williams, Lincoln, and Thomas 1994). The importance of nitrogen content regulating the performance of herbivorous insects is well documented (Mattson 1980; Benedict and Hatfield 1988). Nitrogen is an essential element for insect larvae, and its effects on development and growth rates often translate in adult survival and reproductive success (Lincoln, Couvet, and Sionit 1986; Fajer 1989; Fajer, Bowers, and Bazzaz 1989). In general, foliage consumption increases as nitrogen content decreases (Mattson 1980; Fajer 1989), but the developmental, growth and survival rates of insect larvae decrease when fed on leaves with low nitrogen (Lincoln, Couvet, and Sionit 1986; Fajer 1989; Fajer, Bowers, and Bazzaz 1989; Ayres 1993; Williams, Lincoln, and Thomas 1994) (see Table 21.6). For example, the leaf consumption of larvae of the butterfly *Junonia coenia* increases by 40% when fed on low nitrogen foliage. However, these larvae gain less weight and have longer developmental times than larvae grown on high nitrogen foliage (Fajer 1989). The changes in foliage production and leaf composition will probably affect herbivory patterns and insect survival in natural settings.

Biological control is another arena where the changes in consumption patterns of herbivores could have drastic consequences. *Salvinia molesta*, an aquatic fern that became a pest in Australia, serves to illustrate this point. Two herbivores were introduced in Australia in an attempt to control *S. molesta*: the weevil *Crytobagus salviniae* and the moth *Samea multiplicalis* (Room 1990). When the insects were released, they were effective at controlling plants with high nitrogen, but had little success when plants had low nitrogen content. The weevil populations declined in plant populations that had low nitrogen content and the moths discriminated between oviposition sites based on the nitrogen content of the hosts (Room 1990). Thus, in the absence of herbivores, high nitrogen content promoted invasiveness by increasing vegetative growth rates of the plant.

Table 21.5
Temperature limitations for some plants reported in the literature.

Species	Minimum temperature tolerance	Maximum temperature tolerance	Required Growing degree days*	Other temperature limitations	Consequences of temperature Increase	Source
Picea rubens (Red Spruce)					Replaced by Hardwoods with a 2°C increase	Rodenhouse 1992
Most tropical plants (chilling sensitive)	10°C				Potential increase in range	Gates 1993
Tropical rain forest trees			10,000		Could cause changes in phenology	Gates 1993
Koenigia islandica (tundra plant)			700		Could cause changes in phenology	Gates 1993
Tilia cordata (Lime Tree)		No germination success above 35°C	2,000	Pollen will not germinate below 15°C (optimum = 19°C); optimum seed germination temperature 17–22°C	Potential increase in range	Gates 1993
Plantago erecta					Combined with drought, produces early senescence	Murphy and Weiss 1992; Weiss *et al.*, 1993

*Growing degree-days (GDD) = unit of total growing days needed by a given species to complete its vegetative and reproductive life cycle. The mean temperature of a GDD has to be above 0°C (Gates 1993)

Table 21.6
Effect of an enriched CO_2 atmosphere on some plants species and the insects that interact with them.

Plant	Insect	CO_2 Increase—Effects on plant	CO_2 Increase—Effects on the interaction	CO_2 Increase—Effects on Insect	Source
Plantago lanceolata (Plantaginaceae)	*Junonia coenia* (Lepidoptera, Nymphalidae)	Decrease of nitrogen in leaves	Increase in foliage consumed by larvae	Decrease in larval growth, increase in larval mortality and developmental time	Fajer 1989; Fajer, Bowers, and Bazzaz 1989
Pinus taeda	*Neodiprion lecontei* (sawfly)	Decrease of nitrogen in needles	Increase in consumption rate of larvae	No change in larval growth rates	Williams, Lincoln, and Thomas 1994
Glycine max	*Pseudoplusia includens* (Lepidoptera: Noctuidae)	Decrease of nitrogen	Increase in consumption rate of larvae	Less growth, lower food efficiency conversion	Lincoln, Couvet, and Sionit 1986

However in the presence of herbivores, the most efficient control was achieved when plants were rich in nitrogen, since the damage the insects produced was able to overwhelm any growth on the part of the plant (Room 1990).

Tropical Diseases and Climate Change

One of the most important consequences of climate change is likely to be a dramatic expansion of the geographic distribution of tropical diseases. It is important to realize that many of the world's most important parasitic diseases are currently restricted to the tropical and subtropical regions of the world (Hinz 1988; Markell, Voge and John 1992). Figure 21.1 illustrates the combined distribution of leishmaniasis, trypanosomiasis, malaria, schitosomiasis, hookworms, filariasis, onchocerciasis, and loiasis in the world (based on Markell, Voge, and John 1992). The transmission of all these diseases depends either on relatively specific insect vectors or on the ability of the pathogen to survive outside of the host (Table 21.7). As a consequence, the range of these diseases is determined by the range of the vectors and free-living stages of the pathogens. Remarkably, temperature is probably the single most important factor limiting the distribution of the vectors and free-living stages of these pathogens (Olsen 1974; Hinz 1988; Dobson and Carper 1992). In particular, the distribution of many of these diseases coincides with specific minimum high temperatures. For example, the distribution of leishmania in the Mediterranean does not go beyond the 10°C isotherm, which is the temperature that limits the distribution of the phlebotomes (sandflies) that transmit the disease (Hinz 1988). The transmission of *Leishmania donovani* in India is limited to times when temperatures range between 27–29°C because this is when the vector, *Phlebotomus argentipes*, can lay more than

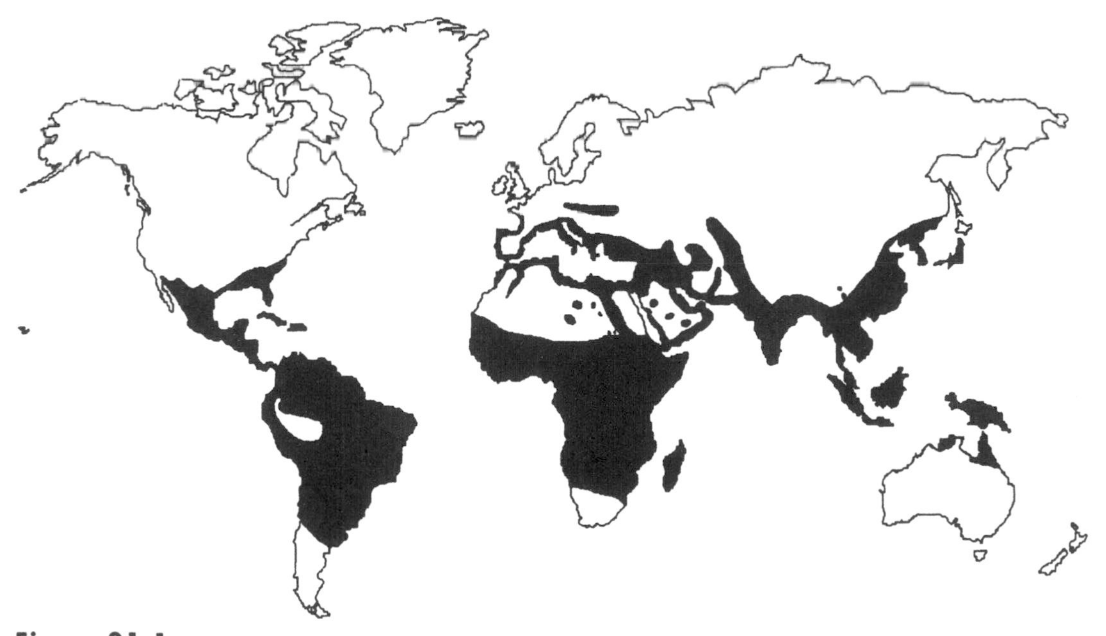

Figure 21.1
Combined distribution of some important parasitic diseases in the world: leishmaniasis, trypanosomiasis, malaria, schistosomiasis, hookworms, filariasis, onchocerciasis, and loiasis (based on Markell, Voge, and John 1992).

Table 21.7

Some important parasitic diseases that are presently restricted to the tropics and subtropics.

Disease	Parasite	Vector	Original Distribution	Present Distribution	Likelihood of expanding range	Temperature limitation	Notes	Source
African trypanosomiasis	*Trypanosoma brucei*	*Glossina morsitans* (tsetse fly)	Tropical Africa	Tropical Africa	Likely			Rogers and Packer 1993; Mehlhorn and Waldorf 1988
Chagas disease	*Trypanosoma cruzi*	Reduviid bugs	Tropical America	Tropical America	Unknown			Mehlron and Waldorf 1988; Markell, Voge, and John 1992
Dengue	arbovirus	*Aedes albopictus* and *A. aegypti*	Asia	Asia, Brazil and U.S. (just mosquito in U.S.), Australia	Unknown	*A. aegypti*: minimum of 22°C to hatch; minimum thresholds for larval development: 13.4°C (*A. aegypti*) and 11°C (*A. albopictus*)		Rogers and Packer 1993; Focks *et al.* 1993; Dobson and Carper 1993; Markell, Voge, and John 1992
Hookworms	*Ancylostoma duodenale* and *Necator americanus*		Probably Africa	Tropics and Subtropics	Unknown			Jeffrey and Leach 1991; Markell, Voge, and John 1992
Leishmaniasis	*Leishmana tropica*	*Phlebotomus sergenti* (sandfly)	Asia and Southern Europe	Asia and Southern Europe	Unknown	optimum transmission: 30°C		Rogers and Packer 1993; Mehlhorn and Waldorf 1988; Hinz 1988
Leishmaniasis	*Leishmania* sp.	*Lutzomyia* spp. and *Phlebotomus* spp. (sandflies)	Africa and South America	Africa and South America	Unknown	Distribution limits in Mediterranean: 10°C isotherm		Mehlhorn and Waldorf 1988; Rogers and Packer 1993
Leishmaniasis	*Leishmania donovani*	*Phlebotomus argentipes*	Probably India	Tropics	Unknown	27°C when laying more than one clutch		Hinz 1988; Markell, Voge, and John 1992
Leishmaniasis	*Leishmania mexicana*	*Lutzomyia* spp.	Tropical America	Tropical America	Unknown	optimum transmission: 25°C		Hinz 1988; Markell, Voge, and John 1992
Lymphatic filariasis	*Wucheria bancrofti*, *Brugia malayi* and *B. timori* (nematodes)	*Culex* spp., *Aedes* spp., *Mansonia* spp., *Anopheles* spp.	Tropics, probably Asia	Tropics and Subtropics	Likely			Busvine 1975; Mehlhorn and Waldorf 1988; Jeffrey and Leach 1991; Markell, Voge, and John 1992; Rogers and Packer 1993

Malaria	*Plasmodium falciparum* and *P. vivax*	*Anopheles* spp. (mosquitoes)	Tropics and Subtropics	Tropics and Subtropics	highly likely	Optimum temperature for development of *P. falciparum*: 26°C	Mosquitoes in England could act as vectors	Rogers and Packer 1993; Mehlhorn and Waldorf 1988; Pearce 1992
Onchocerciasis	*Onchocerca volvolus* (nematode)	*Simulium* spp. (blackflies)	Africa and Latin America	Africa and Latin America	likely			Rogers and Packer 1993; Busvine 1975
Schistosomiasis	Flatworms	Tropical snails	Tropics and Subtropics	Tropics and Subtropics	very likely			Rogers and Packer 1993; Pearce 1992
Yellowfever	Virus	*Aedes albopictus* and *A. aegypti*	Probably Asia or Africa	Africa, and Neotropics (the vectors are present in the U.S.)	Unknown	(see under Dengue)		Rogers and Packer 1993; Dobson and Carper 1993; Busvine 1993

one clutch of eggs (Hinz 1988). The mosquitoes that transmit dengue and yellow fever are limited by temperatures between 11°C and 13.4°C (Rogers and Packer 1993; Focks *et al.* 1993; Dobson and Carper 1993). Temperature also limits the distribution of reduviid bugs and tsetse flies, vectors of Chagas disease and sleeping sickness respectively (Hinz 1988) (see also Table 21.7).

Additionally, higher temperatures can increase the developmental and survival rates of vectors (Busvine 1975; Dobson and Carper 1993). For example, larval development of mosquitoes slows down with cooler temperatures, and in this case, the larvae are restricted to large permanent bodies of water. In contrast, when conditions are warmer, larvae develop much faster and can complete their development in puddles and water-filled containers (Dobson and Carper 1993). In general, if the projections of global change models are correct, we can expect that within the next thirty years the range of many tropical diseases will expand to now parasite-free areas.

Transmission rates of diseases may also be influenced by temperature and CO_2 concentration. Biting rates of adult mosquitoes increase with temperature, but because mortality also increases, the net effect of temperature on transmission rate depends on the relative change of each parameter (Rogers and Packer 1993). In addition to the effects of temperature, pulses of carbon dioxide induce flight activity in mosquitoes and tsetse flies (Haas and Voigt 1988). An increase in atmospheric CO_2 could promote higher activity rates in the insects, but it is not clear whether this would affect transmission rates in any way.

SURPRISES AND THE DIFFICULTY IN ANTICIPATING THOSE SURPRISES

Several lines of evidence suggest that global climate change will have a major impact on natural systems. This is true in spite of the fact that different global change models have major discrepancies regarding the effect that an increase in carbon dioxide is predicted to have on temperature, rainfall patterns, wind, and ocean circulation (Schneider 1989; Legget 1990). The important thing to remember is that all models predict that the increase in greenhouse gases will cause major environmental changes. In translating these climatic shifts to the natural world, much effort has been invested in measuring the responses of individual organisms to predicted environmental changes. There have also been attempts to measure the responses of relatively simple communities. So far it appears that studies based on single species have very little predictive power concerning the modifications that communities will undergo under climate change (Bazzaz 1990; Bazzaz and McConnaughay 1992).

Nonetheless, even with this degree of uncertainty, we can make some general predictions about how climate change will affect organisms and communities. For example, species with narrow ranges are especially likely to be prone to extinction as a result of climate change because they may not be able to reach and establish in new sites at a fast enough rate, or their habitat may completely disappear. Organisms that depend closely on other species are also likely to have high risks of extinction because the species upon which they depend may have different dispersal rates, or may have responses that could set the two groups of species out of synchrony.

On the other hand, the disappearance of some organisms could facilitate the explosive increase of others. Organisms that are closely associated with human establishments would fall in this category. The range and abundance of these organisms will probably increase, especially if temperature is the main factor limiting their present distribution.

Table 21.8
Attributes that may favor organisms to cope with global change

Attribute	Rationale
Short generation times	More chances to produce mutations in a short time, allowing them to adapt to new conditions
Populations with high genetic variability	Higher chances of having individuals that can survive in the new conditions
High phenotypic plasticity	Organisms can change to cope with new conditions
High dispersal ability	Enables them to get to new suitable areas
Populations can persist with few individuals	Habitat reduction may produce small isolated populations
Prey that escape due to high rates of dispersal relative to their predators	This will hold if the predators are the factor limiting the prey population size
No obligate dependence on other organisms	Organisms with which they interact are likely to have different generation times and/or dispersal abilities
Widespread organisms	Higher chances of preserving part of their habitat
Organisms normally transported by humans or associated with human activities or settlements	Area occupied or used by humans is likely to keep increasing

Agricultural pests and many pathogens are prime examples of organisms that have a high chance of being favored by climate change (see Table 21.8).

We can expect climate change to disrupt natural communities, alter species distributions, favor some species, threaten other species, and produce many surprises for which we cannot possibly plan. Perhaps the best way to think of climate change is simply as a massive unpredictable environmental variation. Conservation biologists can mitigate the impact of this perturbation in the same way they mitigate the impact of all environmental variation, by promoting:

(1) Large population sizes;

(2) Ample genetic variation; and

(3) Networks of reserves with opportunities of dispersal.

The more areas and larger areas we set apart as reserves, the higher the chances we have of achieving our conservation goals. In addition, conservation efforts should not only focus on species that are already declining, but should also pay attention to groups of species that are prone to be negatively affected by fragmentation and climate change. The information on individual species should be used to look at general trends, to locate areas of high endemism or high biodiversity. The bottom line is that climate change should prompt us simply to be even more conservative in our viability analyses, in our reserve designs, and in general, in our conservation efforts.

ACKNOWLEDGMENTS

The discussions and editorial comments provided by Peter Kareiva greatly improved the manuscript. Lynn Kutner kindly provided a copy of the TNC document on Climate

Envelopes Analysis. I have been partially supported by CONACyT, Mexico (fellowship no. 56181).

LITERATURE CITED

Ayres, M.P. 1993. Plant defense, herbivory, and climate change. In *Biotic interactions and global change*, eds. P.M. Kareiva, J.G. Kingsolver, and R.B. Huey, 75–94. Sunderland, MA: Sinauer Associates.

Bazzaz, F.A. 1990. The response of natural ecosystems to the rising global CO_2 levels. *Annual Review of Ecology and Systematics* 21:167–96.

Bazzaz, F.A. and E.D. Fajer. 1992. Plant life in a CO_2-rich world. *Scientific American* January 1992:68–74.

Bazzaz, F.A. and K.D.M. McConnaughay. 1992. Plant-plant interactions in elevated CO_2 environments. *Australian Journal of Botany* 40:547–63.

Benedict, J.H. and J.L. Hatfield. 1988. Influence of temperature-induced stress on host plants suitability to insects. In *Plant stress-insect interactions*, ed. E.A. Heinrich, 139–66. New York: John Wiley and Sons.

Bronstein, J.L. and A. Patel. 1992. Temperature-sensitive development: Consequences for local persistence of two subtropical fig wasp species. *American Midland Naturalist* 128:397–403.

Brown, J.H. and A.C. Gibson. 1983. *Biogeography*. St. Louis: The C.V. Mosby Company.

Busvine, J.R. 1975. *Arthropod vectors of disease*. New York: Crane, Russak, and Company, Inc.

Busvine, J.R. 1993. *Disease transmission by insects: Its discovery and 90 years of effort to prevent it*. New York: Springer-Verlag.

Couteaux, M.M., M. Mousseau, M.L. Célérier, and P. Bottner. 1991. Increased atmospheric CO_2 and litter quality: Decomposition of sweet chestnut leaf litter with animal food webs of different complexities. *Oikos* 61:54–64.

Davis, M.B. 1990. Climatic change and the survival of forest species. In *The Earth in transition: Patterns and processes of biotic impoverishment*, ed. G.M. Woodwell. New York: Cambridge University Press.

Denslow, J.S. 1987. Tropical rainforest gaps and tree species diversity. *Annual Review of Ecology and Systematics* 18:431–51.

Dobson, A. and R. Carper. 1992. Global warming and potential changes in host-parasite and disease-vector relationships. In *Global warming and biodiversity*, eds. R. Peters and T. Lovejoy, 201–217. New Haven: Yale University Press.

Dobson, A. and R. Carper. 1993. Health and climate change: Biodiversity. *The Lancet* 342: 1096–1099.

Fajer, E.D. 1989. The effects of enriched CO_2 atmospheres on plant-insect herbivore interactions: growth responses of larvae of the specialist butterfly, *Junonia coenia* (Lepidoptera: Nymphalidae). *Oecologia* 81:514–20.

Fajer, E.D., M.D. Bowers, and F.A. Bazzaz. 1989. The effects of enriched carbon dioxide atmospheres on plant-insect herbivore interactions. *Science* 243:1198–200.

Focks, D.A., D.G. Haile, E. Daniels, and G.A. Mount. 1993. Dynamic life table model for *Aedes aegypti* (Diptera: Culicidae): Analysis of the literature and model development. *Journal of Medical Entomology* 30: 1003–17.

Gates, D.M. 1993. *Climate change and its biological consequences*. Sunderland, MA: Sinauer Associates.

Groom, M. and N. Schumaker. 1993. Patterns of worldwide deforestation and local fragmentation. In *Biotic interactions and global change*, eds. P.M. Kareiva, J.G. Kingsolver, and R.B. Huey, 24–44. Sunderland, MA: Sinauer Associates.

Janzen, D. 1986. The eternal external threat. In *Conservation biology: The science of scarcity and diversity*, ed. M.E. Soulé, 286–303. Sunderland, MA: Sinauer Associates.

Jeffrey, H.C. and R.M. Leach. 1991. *Atlas of medical helminthology and protozoology*. New York: Churchill Livingstone.

Haas, W. and W.P. Voigt. 1988. Host finding—a physiological effect. In *Parasitology in focus: Facts and trends*, ed. H. Mehlhorn, 454–464. Berlin: Springer-Verlag.

Hinz, E. 1988. Geomedical aspects of parasitology. In *Parasitology in focus: Facts and trends*, ed. H. Mehlhorn, 607–618. Berlin: Springer-Verlag.

Knoll, A.H. 1984. Patterns of extinction in the fossil record of vascular plants. In *Extinctions*, ed. M.H. Nitecki, 21–68. Chicago: The University of Chicago Press.

Legget, J. 1990. The nature of the greenhouse threat. In *Global warming: The Greenpeace report*, ed. J. Legget, 14–43. New York: Oxford University Press.

Lincoln, D.E., D. Couvet, and N. Sionit. 1986. Response of an insect herbivore to host plants grown in carbon dioxide enriched atmospheres. *Oecologia* 69:556–60.

Markell, E.K., M. Voge, and D.T. John. 1992. *Medical parasitology*. Philadelphia: W.B. Saunders Company.

Mattson, W.J., Jr. 1980. Herbivory in relation to plant nitrogen content. *Annual Review of Ecology and Systematics* 11:119–161.

McDonald, K.A. and J.H. Brown. 1992. Using montane mammals to model extinctions due to global change. *Conservation Biology* 6:409–15

Mehlhorn, H. and V. Walldorf. 1988. Life cycles. In *Parasitology in focus: Facts and trends*, ed. H. Mehlhorn, 1–148. Berlin: Springer-Verlag.

Murphy, D.D., K.E. Freas, and S.B. Weiss. 1990. An environment-metapopulation approach to population viability analysis for a threatened invertebrate. *Conservation Biology* 4:41–51.

Murphy, D.D. and S.B. Weiss. 1992. Effects of climate change on biological diversity in Western North America: Species losses and mechanisms. In *Global warming and biological diversity*, eds. R.L. Peters and T.E. Lovejoy, 355–368. New Haven: Yale University Press.

Nordhaus, W.D. 1990. Greenhouse economics: count before you leap. *The Economist* July: 21–3.

Olsen, O.W. 1974. *Animal parasites: Their life cycles and ecology*. Baltimore: University Park Press.

Pacala, S.W. and G.C. Hurtt. 1993. Terrestrial vegetation and climate change: Integrating models and experiments. In *Biotic interactions and global change*, eds. P.M. Kareiva, J.G. Kingsolver and R.B. Huey, 57–74. Sunderland, MA: Sinauer Associates.

Pearce, F. 1992. A plague on global warming. *New Scientist* December:12–3.

Rodenhouse, N.L. 1992. Potential effects of climatic change on a neotropical migrant landbird. *Conservation Biology* 6:263–72.

Rogers, D.J. and M.J. Packer. 1993. Vector-borne diseases, models and global change. *The Lancet* 342:1282–5.

Roland, J. 1993. Large-scale forest fragmentation increases the duration of tent caterpillar outbreak. *Oecologia* 93:25–30.

Room, P.M. 1990. Ecology of a simple plant-herbivore system: Biological control of *Salvinia*. *Trends in Ecology and Evolution* 5:74–9

Root, T.L. and S.H. Schneider. 1993. Can large-scale climatic models be linked with multiscale ecological studies? *Conservation Biology* 7:256–70.

Rubin, E.S., R.N. Cooper, R.A. Frosch, T.H. Lee, G.Marland, A.H. Rosenfeld, and D.D. Stine. 1992. Realistic mitigation option for global warming. *Science* 257:148–149, 261–266.

Schelling, T.C. 1992. Some economics of global warming. *The American Economic Review* 82:1–14.

Schneider, S.H. 1989. *Global warming: Are we entering the greenhouse century?* New York: Vintage Books.

Skole, D. and C. Tucker. 1993. Tropical deforestation and habitat fragmentation in the Amazon: Satellite data from 1978 to 1988. *Science* 260:1905–10.

Soulé, M.E. 1991. Conservation: tactics for a constant crisis. *Science* 253:744–50.

Morse, L.E., L.S. Kutner, G.D. Maddox, L.L. Honey, C.M. Thurman, J.T. Kartesz, and S.J. Chaplin. 1993. *The potential effects of climate change on the native vascular flora of North America: A preliminary climate envelopes analysis*. Arlington, Virginia: The Nature Conservancy. Research Project 3041–03.

Tregarthen, T. 1988. The warming of the earth. *The Margin* Nov/Dec:23–25.

Weiss, S.B., R.B. White, D.D. Murphy, and P. Ehrlich. 1987. Growth and dispersal of the checkerspot butterfly *Euphydryas editha*. *Oikos* 50:161–6.

Weiss, S.B., D.D. Murphy, P.R. Ehrlich, and C.F. Metzler. 1993. Adult emergence phenology in checkerspot butterflies: The effects of macroclimate, topoclimate, and population history. *Oecologia* 96:261–270.

Wetherald, R.T. 1991. Changes of temperature and hydrology caused by an increase of atmospheric carbon dioxide as predicted by general circulation models. In *Global climate change and life on earth*, ed. R.L. Wyman, 1–17. New York: Chapman & Hall.

Williams, R.S., D.E. Lincoln, and R.B. Thomas. 1994. Loblolly pine grown under elevated CO_2 affects early instar pine sawfly performance. *Oecologia* 98:64–71.

Woodward, F.I. 1987. *Climate and plant distribution*. Cambridge: Cambridge University Press.

Woodward, F.I. 1993. How many species are required for a functional ecosystem? In *Biodiversity and ecosystem function*, eds. E.D. Schulze and H.A. Mooney, 271–291. Berlin: Springer-Verlag.

Woodward, F.I., G.B. Thompson, and I.F. McKee. 1991. The effects of elevated concentrations of carbon dioxide on individual plants, populations, communities and ecosystems. *Annals of Botany* 67 (Supplement 1):23–38.

CHAPTER 22

Contributions of Spatially Explicit Landscape Models To Conservation Biology

ELI MEIR
and PETER M. KAREIVA

The practice of conservation is often a form of land management. One of the most powerful approaches for connecting the needs of a particular species with land usage is the linking of biologically-detailed models of that species dispersal and demography with geographic information systems (GIS). For example, juvenile spotted owls must depart their birthplace in search of unoccupied expanses of old growth forest. Maps that detail the scarcity, fragmentation, and location of remnant old growth stands dramatize how difficult a search these juvenile owls may face in heavily logged portions of the Pacific Northwest. By connecting these spatially detailed maps with a model of how owls disperse and reproduce, managers can construct logging plans that make the best of what little old growth might remain. We call such approaches spatially explicit population models (or SEPMs) because they assign habitats and owls to particular locations in space, and depending upon the number and placement of individuals, they predict population change as a result of dispersal, mortality, and reproduction. The emergence of user-friendly GIS software, the maturing of ecological theory pertaining to population dynamics in fragmented habitats, and the increased popularity of individual behavior simulation models have combined to produce a tremendous enthusiasm for SEPM's (see *Ecological Applications*, issue #1, volume 7, 1995).

There is no question that spatially explicit models and landscape ecology have much to offer to conservation. There is some question, however, regarding the practicality of detailed landscape models as a management tool. Kareiva and colleagues (Kareiva, Skelly, and Ruckelshaus 1997; Ruckelshaus, Hartway, and Kareiva 1997) have argued that we lack (and forever will lack) the quality of dispersal data needed to support predictive landscape models, or even models that can identify some optimal pattern of land usage. Specifically, modest errors in estimating mortality while dispersing result in large errors in predicting the fraction of animals that find a habitat before dying (Figure 22.1), and hence in connecting demography to landscape patterning. This pessimism does not apply to landscapes with a third or more of their area represented by suitable habitat. But that caveat offers little reassurance, because those situations in which we turn to spatially-

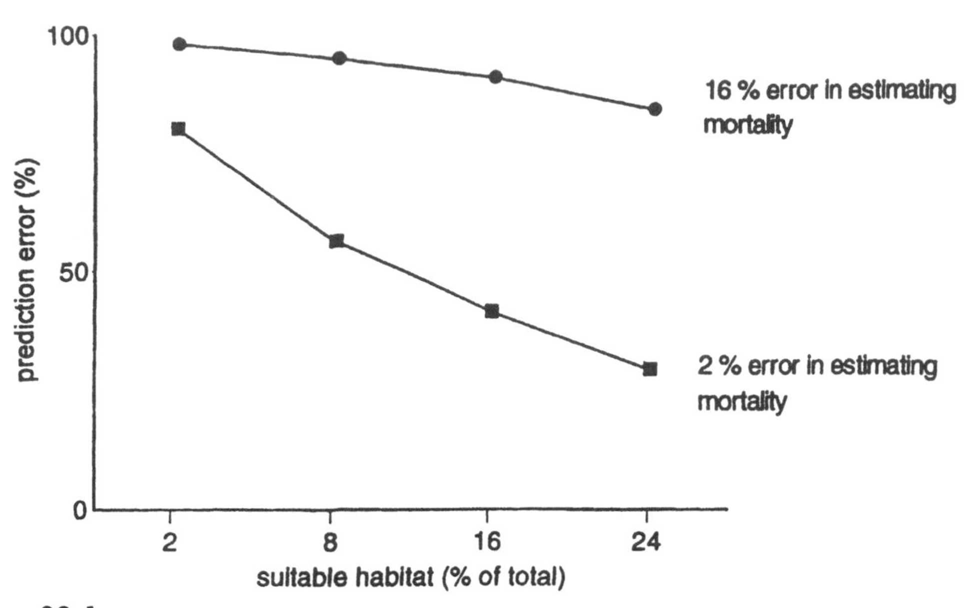

Figure 22.1
Propagation of prediction errors for a spatially explicit model of an animal searching in a fragmented landscape. The two lines correspond to different percentages by which we overestimate the mortality rate suffered by an animal while dispersing, because animals that leave a study area are often mistakenly assumed to have died, or because transmitters on animals can cause unnatural deaths. The prediction error is simply the percent by which the animal's success is consequently underestimated, compared to what it would be if the model had the correct data as its input. The horizontal axis is the percentage of the total landscape that represents suitable habitat. Propagation of errors due to mistaken parameter estimates declines with increasing habitat availability. (Simulation results are taken from Ruckelshaus, Hartway, and Kareiva [1997]).

explicit models for threatened species typically correspond to landscapes with small fractions of remaining "good" habitat.

In this chapter we avoid questions of predicting or "optimizing" population dynamics models. Instead we explore an alternative use of landscape models as possible tools for interpreting distributional data and for designing monitoring schemes. This application of modeling represents a particularly promising avenue for SEPMs because it is tied very tightly to data and has as its goal getting the most out of spatially structured information. We begin by using a model to illustrate the problems of detecting change in spatially structured environments. Only with a model for which we "know" what change is being simulated can we clearly see the perils of different monitoring schemes. In the second half of our chapter we review some pioneering studies of real-world spatial systems in which ingenious quantitative approaches are applied to spatially-structured data sets as a way of investigating hypothesized conservation hazards.

USING A SPATIALLY-EXPLICIT MODEL TO EVALUATE THE POWER OF MONITORING PROGRAMS

In conservation biology, it is a truism that "more habitat" is better than "less habitat." As habitat disappears, however, details about how fragments of remnant habitat are distributed could dramatically alter a species' success. One way of minimizing the adverse effects of habitat loss is to manage land such that remaining habitat for threatened species is arranged in a way that optimizes the success with which searching organisms find parcels of suitable habitat (Dunning *et al.* 1995; Liu, Dunning, and Pulliam 1995). The intuitive appeal of this line of reasoning is strong, especially if it is presented using computerized maps showing different geometries of habitat overlain with hypothesized movements of animals. Detailed SEPMs that link habitat maps with a species' demography also draw inspiration from more abstract metapopulation theory. This theory has at its core a key idea that is germane to all landscape models—i.e, the idea that there is some "threshold" of habitat availability below which a population is doomed to extinction because colonization rates cannot balance local extinction events. The idea of a threshold for habitat availability extends to any landscape model in which organisms must disperse to find habitats (for breeding territories or simply to find adequate resources), as long as there is some mortality or energetic cost to the dispersal process.

Indeed, this threshold phenomenon is quite general. A similar threshold is seen, for example, in epidemiology, where a shortage of susceptible individuals (i.e., habitats for disease) can cause a disease to disappear completely from a population. The threshold phenomenon implicit in almost all of these models suggests a major practical problem that extends beyond the theoretical question of landscape design. Suppose over time we are destroying habitat to the point that the threshold for extinction is crossed, thereby pushing a species towards inevitable doom, although not immediately and perhaps even agonizingly slowly. Are we likely to be able to document this imminent collapse from readily available census data? Or will we typically only discover that species are marching towards extinction long after the threshold of habitat availability has been crossed, when it is too late to do anything about it?

To evaluate our ability to detect a species' collapse following excessive loss of habitat, we analyzed a spatially explicit individual behavior model of animals in a patchy habitat. Details of the model are available from a web site (see legend to Figure 22.2), but a few key features are worth discussing. For the runs reported here, the model represents the world as 4,900 cells (i.e., 70 × 70) designated as suitable and unsuitable habitat. The suitable habitat is initially arranged in 100 patches, which as an ensemble consistently sustains a relatively stable population for the demographic parameters underlying the results reported in Figures 22.2 and 22.3 (though chance local extinction and recolonization is a regular event within particular patches). Our simulated animals can reproduce only within suitable habitat, but can move between patches (with some mortality cost). All movement is random, with the exception that there is a tendency to remain within suitable habitat rather than to venture into the less desirable habitat, where mortality is higher. In addition to being influenced by habitat, mortality is also influenced by crowding, which provides a density-dependent feedback that tends to stabilize population dynamics. The simulation tracks the fate and movement of each individual animal in the landscape, with birth, death, and movement occurring as stochastic processes.

After identifying the threshold of habitat destruction, above which our 'virtual animal'

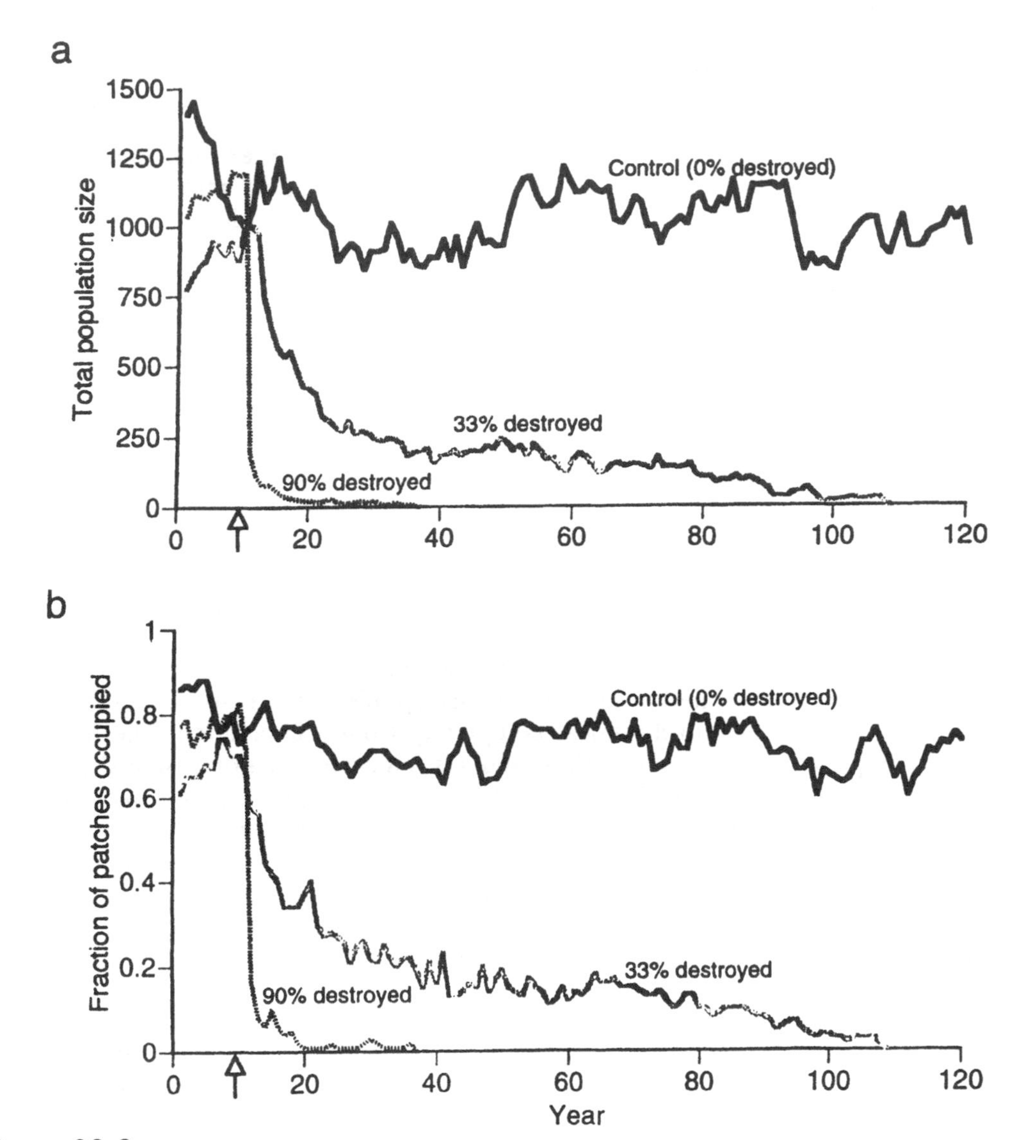

Figure 22.2
The collapse of a population facing either 33% or 90% habitat destruction compared to a "control" landscape lacking any habitat destruction. In (a) the status of the metapopulation is summarized by the total number of individuals in the entire landscape, whereas in panel (b) the metapopulation is characterized by the fraction of habitat patches containing at least one individual. The model landscape is originally comprised of 100 equal-sized square habitat patches that are reduced to 66 or 10 habitat patches following habitat destruction. Timing of the habitat destruction is indicated by the arrow. Each individual has death rates of 0.08 and 0.15 per month in good and bad habitat, respectively, and gives birth to 4 new individuals once a year if residing in good habitat. The death rate increases by 0.2 per month if the individual is too close to another. Additional details of the model can downloaded from the website http://www.webcom.com/sinauer/ecobeaker.html.

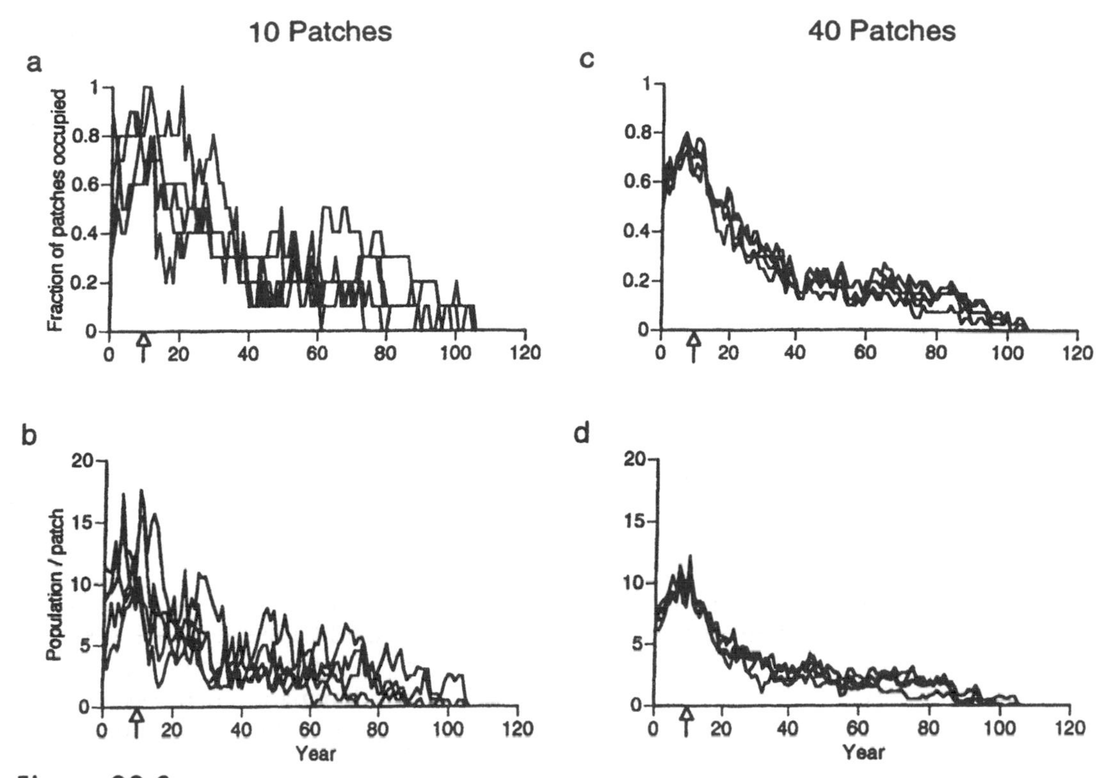

Figure 22.3

The effectiveness with which different monitoring programs reveal a metapopulation's ultimate collapse. Panel (a) depicts monitoring of presence or absence in ten different habitat patches (i.e., 15% of the available habitat); (b) depicts monitoring of population size in ten different patches (15% of the available habitat); (c) depicts monitoring of presence or absence in forty different habitat patches (60% of the available habitat); and (d) depicts monitoring of population size in forty different habitat patches (60% of the available habitat). Each line is one stochastic realization of the model for the parameters described in Figure 22.2, and one realization of a monitoring program, with the patches selected for monitoring being randomly picked at the beginning of the simulation.

could not persist, we removed just enough habitat to exceed this threshold (i.e., removing 33%, given a threshold of approximately 28% habitat loss), and also removed habitat to far above this threshold (90% removed). We sampled both the soon-to-be-extinct populations and a control population in which no habitat was removed, and we recorded two measures of population performance or status: (1) numbers per habitat patch and (2) the fraction of patches occupied by the species. Figure 22.2 illustrates the population collapse given complete data from every parcel of habitat in the landscape and assuming no sampling or observation error, and we see that both the total number of individuals and the fraction of patches occupied declines rapidly following habitat destruction. This immediate decline occurs because when patches are destroyed in our simulation, the organisms left stranded in poor habitat suddenly have higher mortality rates and cannot reproduce unless they manage to find their way to an undisturbed habitat patch.

The question is whether we could discern this decline and ultimate collapse via a pragmatic sampling program that garnered information from only a few of the habitat patches. The most obvious monitoring approach would be to count the number of organisms in habitat patches. Unfortunately, such population estimates can be very costly and time-consuming, and a less expensive and more convenient approach might be to take advantage of the fact that the fraction of remaining patches occupied also declines as an indication of the population's collapse (Figure 22.2b). Thus, instead of censusing populations in order to estimate population size, it might suffice to simply record presence or absence and tabulate the fraction of habitat patches that are occupied. After destroying 33% of the habitat in our model, we recorded both population sizes and presence/absence of populations in each patch. We then randomly sampled either 10 or 40 habitat patches, corresponding to sampling 15% or 60% of the remaining habitat. Both monitoring approaches clearly pick up the population decline when viewed as an average of many different monitoring attempts, even with as little as 15% (ten patches) of the habitat sampled. However, for any particular realization of the stochastic process, the variability is large, and sampling occasionally produces runs of data that fail to indicate a decline until ten to twenty years have elapsed (Figure 22.3a and 22.3b). The most interesting result of this analysis is that "fraction of occupied" patches could be a legitimate monitoring approach, because such a monitoring program would be much more practical than censusing populations across a large landscape.

The key to evaluating the merits of presence/absence monitoring is a formal analysis of the relative costs of alternative sampling programs and the likelihood of being wrong by chance. For instance, it is possible to record a species as present based on the siting of a transient individual, even though the patch lacks a breeding pair, much less a viable population. The key question is: Given the choice between censusing animal populations in five patches versus recording presence or absence in 40 patches, what approach will provide the best early warning signal of a population in decline due to habitat loss?

Although monitoring may be viewed as a boring topic, monitoring programs are essential to conservation, and raise more challenging theoretical questions than is usually recognized. In particular, effective monitoring is not easy to design, and there is always the risk that the data gathered will not provide the desired signal, or that money is wasted on inefficient sampling efforts when much simpler approaches could have yielded even better information. Monitoring is especially important, because when special interest groups such as "developers" and "preservationists" clash, compromise solutions are often implemented, with the caveat that the "situation will be monitored" to insure, for example, that too much land has not been surrendered or too much degradation has not been tolerated. Much can be gained by using landscape models to evaluate monitoring programs, and this application of models does not require accurately knowing dispersal rates, or that the models be predictors of the future. However, understanding the ramifications of errors and biases in monitoring programs does require considerable knowledge of a species' natural history.

OTHER DATA-PRACTICAL APPLICATIONS OF SPATIALLY EXPLICIT MODELS

Several researchers have recently proposed techniques for using spatially explicit models to evaluate spatially-structured data. The best developed ideas can be found in the work of Ilkka Hanski and colleagues regarding metapopulation models applied to butterfly populations (Hanski 1996; Hanski 1997). Hanski assumes that a system, or what he calls a "network", of habitat islands has reached a stochastic equilibrium in the sense that the fraction of patches occupied by butterflies remains roughly constant, although populations may fluctuate and extinction may occur in any single island of butterfly habitat. Hanski (1997) categorizes each habitat island by its area and isolation from other islands. Both extinction rates and colonization rates are likely to be influenced in a regular fashion by these island attributes. Once a regression model is developed relating extinction and colonization to patch attributes, it is possible to use the incidence of a species occurrence in a large ensemble of habitat patches to predict the consequences of changes in habitat availability. Although there are several assumptions underlying Hanski's approach and several decisions that must be made about the shape of functions relating isolation to colonization (as just one example), the connection to data is always transparent. If a particular functional form is flawed, little of the collected data will be explained and there is not much chance of proceeding down a misguided path. In addition, because Hanski's approach provides estimates of patch specific extinction and colonization rates on an annual basis, his approach can be clearly validated (or rejected) by having two successive samples of a species incidence in a large ensemble. The major limitation of Hanski's approach is that it requires a large number of habitat islands with presence or absence data in order to build the model in the first place; subsequently any test of the model requires fairly high turnover as a check against predicted incidence records. It is no accident that Hanski developed his approach for butterflies whose habitat is a well-defined foodplant with many discrete patches harboring breeding populations of butterflies.

A second elegant application of landscape analyses involves source-sink models and the impact of degraded habitats on populations of animals in nearby preferred habitats (Doak 1996). A decade ago Pulliam (1988) suggested that marginal habitats could have a large impact on the population dynamics of organisms in their 'core' or preferred habitats, and that a failure to appreciate the connection between source and sink habitats could yield severe misunderstandings of a species' population biology. Pulliam's original work was very general and abstract, although its pertinence to real-world situations was transparent. More recently, several researchers have built on Pulliam's insight in special cases of conservation concern. One of the best examples is Doak's (1996) study of source-sink models and grizzly bears in Yellowstone Park. If bears move frequently between good and degraded habitat patches, then gradual habitat degradation can prompt a precipitous collapse in bear populations, which would be extremely difficult to anticipate by existing programs for censusing bear populations. Unfortunately, we do not know at what rates bears do disperse, but at least the modeling indicates an order of magnitude rates of dispersal that would be a cause for concern, thereby directing research efforts. In addition, Doak's source-sink model could be used to develop monitoring programs that targeted the type of population collapse associated with slow habitat degradation.

A final unique landscape analysis involves the rule-based modeling of Skelly and Meir (Skelly and Meir 1997). As mentioned previously, the presence or absence of species from particular habitat patches is easier to collect than density estimates, but it remains unclear whether such data give enough information to help manage threatened species. In addition to simply detecting dangerous population declines, we would also like to be able to identify plausible causes of the decline. Taking advantage of a unique data set on the distribution of fourteen amphibian species from thirty-two ponds in Michigan, with population censuses spanning a 25-year period, Skelly and Meir (1997) combined a spatial analysis with habitat categorization to explore the importance of "pond isolation." The amphibians exhibited high rates of turnover between the two major census periods, with a total of 36 extinctions and 40 colonizations. The null hypothesis is that the changes in distribution were simply due to random events. A second possibility is that the spatial arrangement of ponds was responsible for the changing amphibian distributions, perhaps because more isolated ponds were difficult to colonize or suffered more extinctions with less frequent rescues provided by immigrants. A third possibility is that successional changes in the vegetation surrounding selected ponds drove most of the extinctions and colonizations. Skelly and Meir explored this question by constructing stochastic rules for extinction and recolonization using the turnover data and asking how well these rules explained observed data compared to a random process.

The specific approach adopted by Skelly and Meir involved a combination of isolation measurements and habitat categorizations. They constructed a model based on the 1974 amphibian distribution and the turnover events between censuses, with the model parameterized in accord with specific hypotheses. For example, to formalize the hypothesis that pond isolation is a key determinant of amphibian distributions, Skelly and Meir formulated rules for colonization and extinction that categorized ponds into three isolation classes (i.e., <80 m, 80–420m, and >420m). They then ran this parameterized stochastic model 10,000 times, generating a frequency distribution for "mistakes" made in terms of predicted amphibian distributions, a mistake being wrongly predicting species' occurrences or absences in ponds that actually contained or lacked those species in 1992. The performance of a biological rule-based model (i.e., isolation or habitat succession) was compared to a null model parameterized from the same data set. The difference in number of mistakes between the null model and rule-based models is a measure of the importance of the biological rules being investigated. Using this approach it was possible to identify species whose changing distribution was influenced by successional changes, by isolation, or for which no biological rules produced better explanatory power than a null random model (Figure 22.4). If we were concerned with landscape-level analysis for these Michigan amphibians, such an analysis would indicate that isolation is a potential factor in the extinction/colonization process for *Ambystoma maculatum*, but not for *Pseudacris crucifer* or *Rana pipiens* (at least at the scale of the 540-ha network of ponds). Thus, an explicit model of turnover allows us to identify which amphibian species are most sensitive to the spatial arrangement of ponds. A key point is that spatial models are not only useful when we believe isolation and habitat distribution are important. The models also can be used to demonstrate statistically that space does not matter.

HOW MUCH SPACE IS NECESSARY?

The visual impact of maps and striking aerial photos of fragmented forests tend to generate a great deal of enthusiasm for spatially explicit models in conservation, especially for

Figure 22.4 ▶
The number of mistakes made by contrasting rule-based models for distributional change applied to three different amphibian species. The filled triangles refer to a succession based model; the hollow squares refer to a spatial isolation model; the hollow circles represent a null model in which ponds were randomly divided into three different "classes," and the turnover rates for those randomly designated classes were estimated and used for prediction (see Skelly and Meir 1997). The horizontal axis is the number of mistakes made in a single run of the model, and each model was run 10,000 times to obtain the complete frequency distribution. Peaks closer to the origin represent models that were more effective at explaining distributional changes in the species.

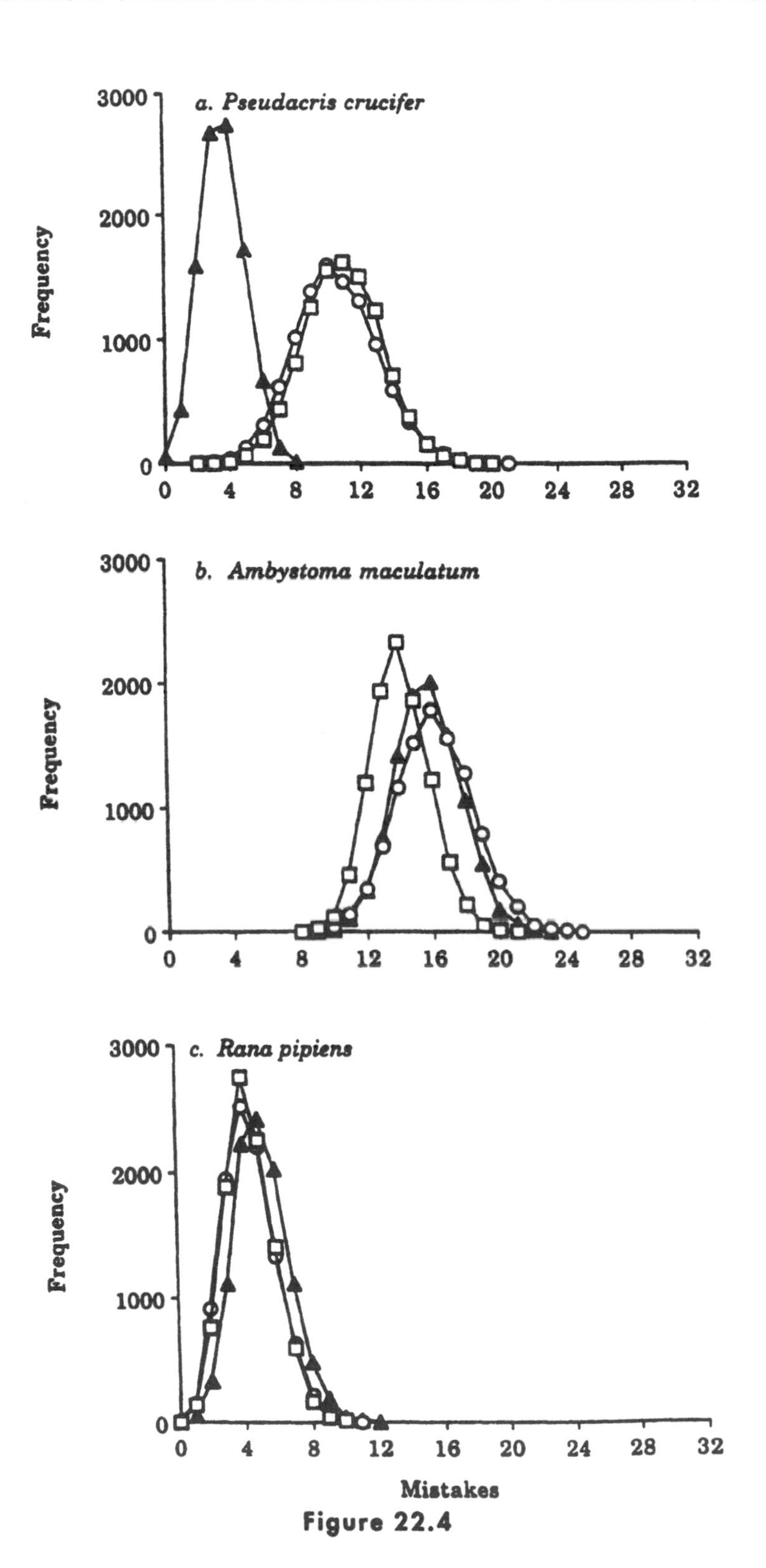
a. Pseudacris crucifer
b. Ambystoma maculatum
c. Rana pipiens
Frequency
3000
2000
1000
0
0
4
8
12
16
20
24
28
32
Mistakes

Figure 22.4

the connection of those models to GIS data bases on land usage (Liu, Dunning, and Pulliam 1995). Certainly, habitat destruction is a major threat to biodiversity and to the persistence of numerous species. But this does not mean we necessarily need elaborate spatial models or high-tech GIS systems. Something as straightforward as the total area of old growth may tell the whole story for spotted owls, regardless of all the connectivity indices, fragmentation measures or fractal dimensions that one can calculate from a GIS data base (Groom and Schumaker 1993). The power of landscape models and SEPM's is widely appreciated; however, this fashion needs to be tempered by the realization that it is very difficult to connect spatially explicit models with real data in anything but a trivial way (see Steinberg and Kareiva 1997), and that spatial models are not always necessary. The tremendous power and expense of spatially explicit models in conservation should be reserved for problems where practical answers are truly forthcoming, and not just campaign promises. We believe that monitoring and interpretation of data represent such problems, much more so than do questions about optimal landscape design.

ACKNOWLEDGMENTS

Peter Kareiva was supported by an NSF grant and Eli Meir was supported as a Howard Hughes Predoctoral Fellow. E.K. Steinberg and M. Ruckelshaus offered editorial comments and technical assistance.

LITERATURE CITED

Doak, D. 1996. Source-sink models and the problem of habitat degradation: General models and applications to the Yellowstone grizzly. *Conservation Biology* 9:1370–1379.

Dunning, J.B., D.J. Stewart, B.J. Danielson, B.R. Noon, T.L. Root, R.H. Lamberson, and E.E. Stevens. 1995. Spatially explicit population models: Current forms and future uses. *Ecological Applications* 5:3–11.

Groom, M. and N. Schumaker. 1993. Evaluating landscape change: Patterns of worldwide deforestation and local fragmentation. In *Biotic interactions and global change*, eds. P. Kareiva, J. Kingsolver, and R. Huey, 25–40, Sunderland, MA: Sinauer Associates.

Hanski, I., A. Moilanen, T. Pakkala, and M. Kusari. 1996. The quantitative incidence function model and persistence of an endangered butterfly metapopulation. *Conservation Biology* 10: 578–590.

Hanski, I. 1997. Predictive and practical metapopulation models: the incidence function approach. In *Spatial ecology: the role of space in population dynamics and interspecific interactions*, eds. D. Tilman and P. Kareiva. Princeton, NJ: Princeton University Press.

Kareiva, P., D. Skelly, and M. Ruckelshaus. 1997. Reevaluating the use of models to predict the consequences of habitat loss and fragmentation. In *Enhancing the ecological basis of conservation: heterogenity, ecosystem function, and biodiversity*, eds. S.T.A. Pickett, R.S. Ostfeld, H. Schchak, and G.E. Likens. New York: Chapman and Hall.

Lamberson, R., R. McKelvey, B. R. Noon, C. Voss. 1992. A dynamic analysis of Northern Spotted owl viability in a fragmented forest landscape. *Conservation Biology* 6:505–512.

Liu, J., B. Dunning, and H. Pulliam. 1995. Potential effects of a forest management plan on Bachman's sparrows: linking a spatially explicit model with GIS. *Conservation Biology* 9:62–75.

McKelvey, K., B. Noon, and R. Lamberson. 1993. Conservation planning for species occupying fragmented landscapes: the case of the Northern Spotted Owl. In *Biotic interactions and global change*, eds. P. Kareiva, J. Kingsolver and R. Huey, 424–450. Sunderland, MA: Sinauer Associates.

Pulliam, R. 1988. Sources, sinks, and population regulation. *American Naturalist* 132:652–661.

Ruckelshaus, M., C. Hartway, and P. Kareiva 1997. Assessing the data requirements of spatially explicit dispersal models. *Conservation Biology*, in press.

Skelly, D. and E. Meir. 1997. Rule-based models for evaluating mechanisms of distributional change. *Conservation Biology*, in press.

Steinberg, E. and P. Kareiva. 1997. Challenges and opportunities for empirical evaluation of "spatial theory." In *Spatial ecology: The role of space in population dynamics and interspecific interactions*, eds. D. Tilman and P. Kareiva. Princeton: Princeton University Press.

Wahlberg, N., A. Moilanen, and I. Hanski. 1996. Predicting the occurrence of endangered species in fragmented landscapes. *Science* 273:1536–1538.

Wootton, J. and D. Bell. 1992. A metapopulation model of the peregrine falcon in California: viability and management strategies. *Ecological Applications* 2:307–321.

EPILOGUE

• • •

A Retrospective "Gap Analysis"

Every reader will find some favorite topic in conservation biology missing from this book. Where is the discussion of biodiversity and its function, for example? Why isn't there a chapter on ecosystem management, because we all know that single-species conservation is passé? Why is there no explicit discussion of mapping biodiversity? or of remote sensing in conservation? Those are all real gaps, and we want to use these final remarks to suggest where else the reader might turn to build a firm and complete understanding of conservation biology in the frontlines of both practice and research these days.

First, we have entirely neglected discussions of biodiversity function. For many, this is one of the most interesting areas of ecological research, and we are discovering that biodiversity does make a difference to processes such as nutrient retention, productivity, and recovery from catastrophe (Silver, Brown, and Lugo 1996; Tilman 1966; Tilman, Wedin, and Knops 1996). In addition, disruption of ecological linkages among trophic levels (e.g, the loss of pollinations of economically important plants; see Buckmann and Nabhan 1996) may have devastating consequences in both the short- and long-term. These works set the stage for *why* we should want to preserve biodiversity.

Second, the book is lacking in explicit discussions of humans as part of the ecosystems in which they live, outside of their overall negative influences on ecosystem processes. Especially notable omissions include (1) the role of humans in maintaining seminatural landscapes; (2) ecotourism as a tool for conservation, education, and profit; (3) ethnobiology and it impact on *in situ* conservation; (4) environmental legislation; (5) zoo biology; and (6) the ethics of conservation.

The importance of seminatural landscapes to biodiversity is thoroughly covered inBerglund (1991), with several expecially intriguing discussions on landscape, landuse, and vegetation by A. Gunilla Olsson. More general landscape ecology perspectives that address humans in the landscape include Zonneveld and Formann (1990), Hansen and di Castri (1991), Hudson (1991), or the review by Turner (1989) for advances in landscape ecology. So-called "ecosystem management" also touches on the human-landscape interface (e.g., Ecological Society of America's *Ecological Applications* volume 6, number 3, 1996; Meffe and Carroll 1997).

Ecotourism often falls into the larger discussions of sustainable development (see Sandlund, Hindar, and Brown 1992), but a more narrowly defined view of ecotourism (i.e., wildlife as "sport") can be found in the wildlife management literature, which was recently detailed in the volumes by Knight and Gutzwiller (1995) and Caughley and Sinclair (1994). For the high drama of illicit trafficking in wildlife, we recommend Reisner (1991).

Ethnobotany, or more broadly, ethnobiology, has a long and distinctive intellectual history, which recently has become especially alluring not only to pharmaceutical companies, but to graduate students as well. Recent and compelling literature that blends the human element into the botanical realm is proferred by Davis (1985, 1988, 1996) or

Plotkin (1993). The works of Schultes (e.g., Schultes and Hofman 1980, Schultes and Raffauf 1990), Plotkin (e.g., Plotkin and Famolare 1992), and Heywood and Synge (1991) are critical references for medicinal plants, their ethnobiology, and tropical origins. For a highly readable introduction to the broad field of ethnobotany in general, we recommend Balick and Cox (1996).

Environmental legislation is also a common topic in beginning courses in conservation biology, and much has been written to provide accessible analyses of the environmental laws, their accompanying regulations, and their implementation. For a thorough summary of U.S. legislation, Snape and Houck (1996) have written an excellent volume. Of particular interest for many, however, is the U.S. Endangered Species Act of 1973. For a solid foundation in the federal legislation, consult Rohlf (1989); for commentary on its impact, see the controversial book entitled *Noah's Choice* by Mann and Plummer (1995).

The biology and conservation of captive populations is another topic that links directly human activities and the fate of animal species. While much of the literature on this broad topic relates to enhancement of the immediate environment of captive animals, a recent article (Snyder *et al.* 1995) on the captive breeding programs calls into question the efficacy of such efforts in endangered species recovery. The newly published volume by Norton *et al.* (1995) provokes reflection regarding our motives for establishing and maintaining captive populations.

Ethics, of course, is a burgeoning topic within the broader field of conservation. Standard texts on environmental ethics include Norton (1986) and Callicott (1989), but a significant number of more recent texts address a wide range of ethical considerations. For contemplating the value of life, see Kellert (1996); for the interrationship between economics and ethics, consult the works of Hardin (1993) or Borman and Kellert (1991); for grappling with the philosophical concept of wilderness, Oelschlager (1991) should be of interest; and finally, for the more recent discussions of our postmodern society and nature, see Oelschlager (1995) and Soulé and Lease (1994).

For all our claims about practicality, it is perhaps embarrassing that we have also neglected any discussion of the concrete tools any conservation biologists should have—i.e., sampling theory, for example, or how to establish and interpret monitoring programs, mark-recapture methods, and so forth. These are essential. It would be nice if there were a book entitled, *Tools for Conservation*, but there is not. We can, however, point the reader to some excellent references. For sampling vegetation, guidance can be found in the "industry standards," *Aims and Methods of Vegetation Ecology* (Mueller-Dombois and Ellenberg 1974), *Quantitative Plant Ecology* (Grieg-Smith 1983), and *Methods in Plant Ecology* (Moore and Chapman 1986). For sampling mobile animals, excellent discussions are in Begon, Harper, and Townsend (1990), Krebs (1989), and White *et al.* (1982), among others. Mark-recapture methodologies and application are superbly reviewed in Burnham, Anderson, and White (1996) and Lebreton *et al.* (1992). Monitoring programs for conservation and ecology are aptly explained by a series of experts in Spellerberg, Goldsmith, and Morris (1990), Goldsmith (1991), and Heyer *et al.* (1994). Selected examples of excellent monitoring programs can be found in a variety of recent symposia volumes, including those edited by Bowles and Whelan (1994) and Falk, Millar, and Olwell (1995), among others.

Everything up until now has been very academic. We have been able to identify important references and key researchers. All of this citing and discussing neglects probably the biggest limiting factor in solid conservation, and something that does not lend itself easily to learning about from "reading the literature"—that is, natural history. Often, the key to useful practical advice in conservation is knowing about an orchid's germination requirements, understanding a gopher's habitat preferences, or knowing how far juvenile spotted owls can move in search of old growth. Taxonomic- and organismal-based courses are in decline at many major North American universities. Field stations have to fight shrinking budgets to stay open. But without good natural history, conservation biology will go nowhere. Thus, if the budding conservationist is not extremely knowledgeable about some taxon, or some ecosystem and its natural history, then the budding conservationist better be extremely adept at computers and modeling if she or he hopes to make a contribution. All the eigenvalues and bootstrapping and asymptotic analyses in the world will not go anywhere without ample grounding in natural history.

We cannot emphasize this point enough. Certain traditions even we hold dear. Thus for us, it seems self-evident that if conservation biology is to succeed in the coming decade, then detailed knowledge about a taxonomic group or habitat must also provide the grounding for each and every conservation biologist of this next generation.

LITERATURE CITED

Agee, J.K. and D.R. Johnson, eds. 1988. *Ecosystem management for parks and wilderness.* Seattle: University of Washington Press.

Balik, M.J. and H.S. Beck. 1990. *Useful palms of the world: A synoptic bibliography.* New York: Columbia University Press.

Balik, M.J. and P.A. Cox. 1996. *Plants, people, and culture: The science of ethnobotany.* New York: W.H. Freeman and Company.

Begon, M., J.L. Harper, and C.R. Townsend. 1990. *Ecology: Individuals, populations, and communities.* 2nd edition. New York: Blackwell Scientific Publications.

Bergman, B.E., ed. *The cultural landscape during 6000 years in southern Sweden.* Ecological Bulletins 41, Munksgaard International Booksellers, Copenhagen.

Bowles, M.L. and C.J. Whelan, eds. 1994. *Restoration of endangered species. Conceptual issues, planning, and implementation.* Cambridge: Cambridge University Press.

Bormann, F.H. and S.R. Kellert, eds. 1991. *Ecology, economics, ethics: The broken circle.* New Haven: Yale University Press.

Brush, S.B. and D. Stabinsky. 1996. *Valuing local knowledge. Indigenous people and intellectual property rights.* Washington, D.C.: Island Press

Buckmann, S.L. and G.P. Nabhan. 1996. *The forgotten pollinators.* Washington, D.C.: Island Press/ Shearwater Books.

Burnham, K., D. Anderson, and G. White. 1996. Meta-analysis of vital rates of the Northern spotted owl. *Studies in Avian Biology* 17:92–101.

Callicott, J.B. 1989. *In defense of the land ethic.* Albany: State University of New York.

Caughley, G. and A.R.E. Sinclair. 1994. *Wildlife ecology and management.* Cambridge, MA: Blackwell Science.

Davis, W. 1985. *The serpent and the rainbow.* New York: Simon and Schuster.

Davis, W. 1988. *Passage of Darkness. The ethnobiology of the Haitian zombie.* Chapel Hill: The University of North Carolina Press.

Davis, W. 1996. *One river.* New York: Simon and Schuster.

Ecological Society of America. 1996. Special edition on ecosystem management. *Ecological Applications* 6(3).

Ehrenfeld, D.W. 1981. *The arrogance of humanism.* New York: Oxford University Press.

Falk, D.A., C.I. Millar, and M. Olwell, eds. 1996. *Restoring diversity. Strategies for reintroduction of endangered plants.* Washington, D.C.: Island Press.

Goldsmith, F.B., ed. 1991. *Monitoring for conservation and ecology.* London: Chapman & Hall.

Grieg-Smith, P. 1983. *Quantitative plant ecology.* Oxford: Blackwell.

Hansen, A.J. and F. di Castri, eds. 1991. *Landscape boundaries: Consequences for biodiversity and ecological flows*. New York: Springer-Verlag.

Hardin, G. 1993. *Living Within Limits: Ecology, economics, and population taboos*. New York: Oxford University Press.

Heyer, W.R., M.A. Donnelly, R.W. McDiarmid, L-A.C. Hayek, and M.S. Foster, eds. 1994. *Measuring and monitoring biological diversity. Standard methods for amphibians*. Washington, D.C.: Smithsonian Institution Press.

Heywood, V. and H. Synge, eds. 1991. *The conservation of medicinal plants*. Cambridge: Cambridge University Press.

Hudson, W., ed. 1991. *Landscape linkages and biodiversity*. Washington, D.C.: Island Press.

Kellert, S.R. 1996. *The value of life. Biological diversity and human society*. Washington, D.C.: Island Press/Shearwater Books.

Knight, R.L. and K.J. Gutzwiller, eds. 1995. *Wildlife and recreationists*. Washington, D.C.: Island Press.

Krebs, C.J. 1989. *Ecological methodology*. New York: Harper Collins.

Lebreton, J., K. Burnhan, J. Clobert, and D. Anderson. 1992. Modeling survival and testing biological hypotheses using marked animals: a unified approach with case studies. *Ecological Monographs* 62:67–118.

Mann, C.C. and M.L. Plummer. 1995. *Noah's choice. The future of endangered species*. New York: Alfred A. Knopf.

McNeely, J.A. 1995. *Expanding partnerships in conservation*. Washington, D.C.: Island Press.

Meffee, G. and C.R. Carroll. 1997. *Principles of conservation biology*. 2nd ed. Sunderland, MA: Sinauer Associates.

Moore, P.D. and S.B. Chapman, eds. 1986. *Methods in plant ecology*. 2nd edition. Oxford: Blackwell Scientific Publications.

Mueller-Dombois, D. and H. Ellenberg. 1974. *Aims and Methods of Vegetation Ecology*. New York: John Wiley and Sons.

Norton, B.G., ed. 1986. *The preservation of species: The value of biological diversity*. Princeton: Princeton University Press.

Norton, B.G., M. Hutchins, E.F. Stevens, and T.L. Maples, eds. 1995. *Ethics on the ark: Zoos, animal welfare, and wildlife corridors*. Washington, D.C.: Smithsonian Institution Press.

Oelschlaeger, M. 1991. *The idea of wilderness: From prehistory to the age of ecology*. New Haven: Yale University Press.

Oelschlaeger, M., ed. 1995. *Postmodern environmental ethics*. Albany: State University of New York Press.

Plotkin, M.J. 1993. *Tales of a shaman's apprentice*. New York: Penguin Books.

Plotkin, M.J. and L. Famolare, eds. 1992. *Sustainable harvest and marketing of rain forest products*. Washington, D.C.: Island Press.

Reisner, M. 1991. *Game wars*. New York: Viking Penguin Books.

Rohlf, D.J. 1989. *The Endangered Species Act. A guide to its protections and implementation*. Stanford: Stanford Environmental Law Society.

Rolston, H., III. 1988. *Environmental ethics. Duties to and values in the natural world*. Philadelphia: Temple University Press.

Sandlund, O.T., K. Hindar, and A.H.D. Brown, eds. 1992. *Conservation of biodiversity for sustainable development*. Oslo: Scandanavian University Press.

Schultes, R.E. and A. Hofman. 1980. *The botany and chemistry of hallucinogens*. Springfield, Illinois: Charles C. Thomas.

Schultes, R.E. and R. Raffauf. 1990. *The healing forest: Medicinal and toxic plants of the Northwest Amazon*. Portland: Dioscorides Press.

Silver, W.L., S. Brown, and A.E. Lugo. 1996. Effects of changes in biodiversity on ecosystem function in tropical forests. *Conservation Biology* 10:17–24.

Snape, W.J., III and O.A. Houck. 1996. *Biodiversity and the law*. Washington, D.C.: Island Press.

Soulé, M.E. and G. Lease, eds. 1994. *Reinventing nature? Responses to postmodern destruction.* Washington, D.C.: Island Press.

Spellerberg, I., F.B. Goldsmith, and M.G. Morris, eds. 1990. *Scientific management of temperate communities for conservation.* British Ecological Society Symposium, No. 31. Oxford: Blackwell.

Synder, N.F.R., S.R. Derrickson, S.R. Beissinger, J.W. Wiley, T.B. Smith, W.D. Toone, and B. Miller. 1996. Limits of captive breeding in endangered species recovery. *Conservation Biology* 10:338–348.

Tilman, D. 1966. Biodiversity: population versus ecosystem stability. *Ecology* 77:350–363.

Tilman, D., D. Wedin, and J. Knops. 1996. Productivity and sustainability influenced by biodiversity in grassland ecosystems. *Nature* 379:718–720.

Turner, M.G. 1989. Landscape ecology: The effect of pattern on process. *Annual Review of Ecology and Systematics* 20:171–197.

White, G., D. Anderson, K. Burnham, and D. Otis. 1982. *Capture-recapture and removal methods for sampling closed populations.* Los Alamos National Laboratory, LC–8787-NERP, Los Alamos, New Mexico, USA.

Zonneveld, I.S. and R.T.T. Forman, eds. 1990. *Changing landscapes: An ecological perspective.* New York: Springer-Verlag.

INDEX

INDEX

Lightning Source UK Ltd.
Milton Keynes UK
UKOW07n2313290615

254305UK00001B/22/P